METHODS IN MOLECULAR BIOLOGY™

Series Editor
**John M. Walker
School of Life Sciences
University of Hertfordshire
Hatfield, Hertfordshire, AL10 9AB, UK**

For other titles published in this series, go to
www.springer.com/series/7651

In Vivo NMR Imaging

Methods and Protocols

Edited by

Leif Schröder

ERC Project Biosensor Imaging, Leibniz-Institut für Molekulare Pharmakologie, Berlin, Germany

Cornelius Faber

Westfälische Wilhelms-Universität Münster, Münster, Germany

Editors
Leif Schröder
ERC Project Biosensor Imaging
Leibniz-Institut für Molekulare Pharmakologie
Berlin, 13125, Germany

Cornelius Faber
AG Experimentelle Magnetische
 Kernresonanz
Institut für Klinische Radiologie
Westfälische Wilhelms-Universität
 Münster
Münster, 48149, Germany

ISSN 1064-3745   e-ISSN 1940-6029
ISBN 978-1-61779-218-2   e-ISBN 978-1-61779-219-9
DOI 10.1007/978-1-61779-219-9
Springer New York Dordrecht Heidelberg London

Library of Congress Control Number: 2011934742

© Springer Science+Business Media, LLC 2011
All rights reserved. This work may not be translated or copied in whole or in part without the written permission of the publisher (Humana Press, c/o Springer Science+Business Media, LLC, 233 Spring Street, New York, NY 10013, USA), except for brief excerpts in connection with reviews or scholarly analysis. Use in connection with any form of information storage and retrieval, electronic adaptation, computer software, or by similar or dissimilar methodology now known or hereafter developed is forbidden.
The use in this publication of trade names, trademarks, service marks, and similar terms, even if they are not identified as such, is not to be taken as an expression of opinion as to whether or not they are subject to proprietary rights.
While the advice and information in this book are believed to be true and accurate at the date of going to press, neither the authors nor the editors nor the publisher can accept any legal responsibility for any errors or omissions that may be made. The publisher makes no warranty, express or implied, with respect to the material contained herein.

Printed on acid-free paper

Humana Press is part of Springer Science+Business Media (www.springer.com)

Preface

The history of nuclear magnetic resonance (NMR) is a unique success story of translating basic research from the fields of Physics and Chemistry into applied life sciences. The 1952 Nobel laureate Edward Purcell somehow anticipated the incomparable potential of NMR when stating "I am sure we have only begun to explore the domain of very weak interactions – the 'audio spectrum' of molecules, if I may call it that." Several Nobel Prizes followed for related research and the latest one in 2003 acknowledges a technique that is today indispensable in everyday biomedical diagnostics: nuclear magnetic resonance imaging, also known as MRI in clinical context to avoid misleading connotations to nuclear science.

From the perspective of traditional (optical) imaging it may appear unusual that electromagnetic radiation with a wavelength in the range of one meter can be used to reveal structures on the micrometer scale. But thanks to the numerous ways to manipulate and detect an NMR signal, it is possible to obtain a variety of information with excellent spatial and temporal resolution. All this comes with the advantage of harmless, non-ionizing radiation and includes revealing of processes even down to the molecular level. Today's MRI techniques go far beyond the illustration of pure anatomical structures. Contributions from scientists of a very diverse background helped to increase the versatility of the method tremendously and allow drawing a rather detailed picture of what is going on in living tissue. Consequently, this modality attracts great attention from many researchers not originally trained in NMR/MRI method development, but seeking to use this powerful tool to address their biomedical questions. The number of small animal imaging centers relying on MRI as a key method for preclinical research to understand diseases and to test for novel treatments is growing rapidly. This emphasizes the ever expanding community of MRI users in both academic environments and research departments of biotech companies.

In Vivo NMR Imaging is written as an experimental laboratory text to provide a descriptive approach of the various applications of magnetic resonance imaging and its underlying principles. In order to provide the reader with a descriptive compendium of modern in vivo NMR imaging, the book is structured in three parts:

1. Starting with Section I as a compact introduction of basic NMR physics and image-encoding techniques, the underlying principles of hardware setup and contrast generation are explained. Information about practical aspects of designing experimental studies that follow the special conditions for micro-imaging setups are also provided.

2. In the second part (Section II), advanced concepts of generating contrast in MR images will be introduced and corresponding protocols will be provided. These include some more recent developments in contrast generation based on special preparation of the magnetization that carries image information.

3. In the applications part (Section III-X), the authors cover an interdisciplinary range of problems to be addressed by this non-invasive technique, including study protocols for addressing morphological, physiological, functional, and biochemical aspects of various tissues in living organisms. Recent developments will be addressed with an additional focus on novel techniques for molecular imaging and new protocols for imaging metabolism and molecular markers.

Modern NMR imaging covers so many aspects that it is almost impossible to have detailed experimental experience in all its variety. To give a broad overview to the NMR novice was therefore only one aspect of designing this book. Furthermore, it is hoped that even the versed MRI scientist will find some techniques useful that he has not yet implemented in his research. The successful protocol style of this book series will surely facilitate experimental design for both types of audiences.

We are grateful that so many leading experts in their particular fields agreed to participate in this project. Their diverse experimental experience allowed gathering protocols for a broad audience and we would like to thank them for sharing this priceless knowledge.

Cornelius Faber, Leif Schröder

Contents

Preface . . . . *v*
Contributors . . . . *xi*

SECTION I: BASICS OF MRI

1. Physical Basics of NMR 3
   Rolf Pohmann

2. Spatial Encoding – Basic Imaging Sequences 23
   Rolf Pohmann

3. Basic Contrast Mechanisms 45
   Leif Schröder and Cornelius Faber

4. Scanner Components 69
   Volker C. Behr

5. Small Animal Preparation and Handling in MRI 89
   Patrick McConville

SECTION II: SPECIAL TECHNIQUES

6. Cerebral Perfusion MRI in Mice 117
   Frank Kober, Guillaume Duhamel, and Virginie Callot

7. High Field Diffusion Tensor Imaging in Small Animals and Excised Tissue 139
   Bernadette Erokwu, Chris Flask, and Vikas Gulani

8. The BOLD Effect 153
   Joan M. Greve

9. Screening of CEST MR Contrast Agents 171
   Xiaolei Song, Kannie W.Y. Chan, and Michael T. McMahon

10. Hyperpolarized Noble Gases as Contrast Agents 189
    Xin Zhou

11. Hyperpolarized Molecules in Solution 205
    **Jan Henrik Ardenkjaer-Larsen, Haukur Jóhannesson,
    J. Stefan Petersson, and Jan Wolber**

12. MR Oximetry 227
    Jeff F. Dunn

13. MRI Using Intermolecular Multiple-Quantum Coherences 241
    Rosa Tamara Branca

SECTION III: BRAIN APPLICATIONS

14. Experimental Stroke Research: The Contributions of In Vivo MRI 255
    Therése Kallur and Mathias Hoehn

15. Volumetry and Other Quantitative Measurements to Assess the Rodent Brain . . . 277
    Alize Scheenstra, Jouke Dijkstra, and Louise van der Weerd

16. Models of Neurodegenerative Disease – Alzheimer's Anatomical
    and Amyloid Plaque Imaging. 293
    Alexandra Petiet, Benoit Delatour, and Marc Dhenain

17. MRI in Animal Models of Psychiatric Disorders 309
    Dana S. Poole, Melly S. Oitzl, and Louise van der Weerd

18. Spectroscopic Imaging of the Mouse Brain 337
    Dennis W.J. Klomp and W. Klaas Jan Renema

SECTION IV: SPINE APPLICATIONS

19. Spinal Cord – MR of Rodent Models 355
    Virginie Callot, Guillaume Duhamel, and Frank Kober

SECTION V: CARDIOVASCULAR APPLICATIONS

20. Assessment of Global Cardiac Function 387
    Jürgen E. Schneider

21. Plaque Imaging in Murine Models of Cardiovascular Disease 407
    Gert Klug, Volker Herold, and Karl-Heinz Hiller

22. Interventional MRI in the Cardiovascular System 421
    Harald H. Quick

23. MR for the Investigation of Murine Vasculature 439
    Christoph Jacoby and Ulrich Flögel

SECTION VI: LUNG APPLICATIONS

24. MRI of the Lung: Non-invasive Protocols and Applications to Small
    Animal Models of Lung Disease 459
    Magdalena Zurek and Yannick Crémillieux

SECTION VII: CANCER MODELS

25. Characterization of Tumor Vasculature in Mouse Brain by USPIO
    Contrast-Enhanced MRI 477
    Giulio Gambarota and William Leenders

26. Cancer Models—Multiparametric Applications of Clinical MRI in Rodent
    Hepatic Tumor Model 489
    Feng Chen, Frederik De Keyzer, and Yicheng Ni

Section VIII: Functional MRI

27. BOLD MRI Applied to a Murine Model of Peripheral Artery Disease 511
    Joan M. Greve

28. Manganese-Enhanced Magnetic Resonance Imaging 531
    Susann Boretius and Jens Frahm

29. Spin Echo BOLD fMRI on Songbirds 569
    Colline Poirier and Anne-Marie Van der Linden

Section IX: Phenotyping

30. MRI to Study Embryonic Development 579
    Bianca Hogers

31. Mouse Phenotyping with MRI 595
    X. Josette Chen and Brian J. Nieman

32. Analysis of Freshly Fixed and Museum Invertebrate Specimens Using High-Resolution, High-Throughput MRI 633
    Alexander Ziegler and Susanne Mueller

Section X: Metabolic and Targeted Imaging

33. Applications of Hyperpolarized Agents in Solutions 655
    Jan Henrik Ardenkjaer-Larsen, Haukur Jóhannesson, J. Stefan Petersson, and Jan Wolber

34. Target-Specific Paramagnetic and Superparamagnetic Micelles for Molecular MR Imaging 691
    Roel Straathof, Gustav J. Strijkers, and Klaas Nicolay

35. Tracking Transplanted Cells by MRI – Methods and Protocols 717
    Michel Modo

36. MRI of CEST-Based Reporter Gene 733
    Guanshu Liu and Assaf A. Gilad

Index 747

Contributors

JAN HENRIK ARDENKJAER-LARSEN • *GE Healthcare, Frederiksberg C, Denmark*

VOLKER C. BEHR • *Experimentelle Physik V (Biophysik), Universität Würzburg, Würzburg, Germany*

SUSANN BORETIUS • *Biomedizinische NMR Forschungs GmbH am Max-Planck-Institut für biophysikalische Chemie, Göttingen, Germany*

ROSA TAMARA BRANCA • *Department of Chemistry, Duke University, Durham, NC, USA*

VIRGINIE CALLOT • *Centre de Résonance Magnétique Biologique et Médicale (CRMBM), UMR 6612, CNRS, Université de la Méditerranée, Marseille Cedex 05, France*

KANNIE W.Y. CHAN • *Division of MR Research, The Russell H. Morgan Department of Radiology and Radiological Sciences, The Johns Hopkins School of Medicine, Baltimore, MD, USA*

FENG CHEN • *Section of Radiology, Department of Medical Diagnostic Sciences, University of Leuven, Leuven, Belgium; Department of Radiology, Zhong Da Hospital, Southeast University, Nanjing, China*

X. JOSETTE CHEN • *Mouse Imaging Centre, Hospital for Sick Children, Toronto, ON, Canada*

YANNICK CRÉMILLIEUX • *Centre de Recherche Cardio-Thoracique, Université Bordeaux 2, Bordeaux, France*

BENOIT DELATOUR • *Laboratoire de Neuropathologie Raymond Escourolle, Paris, France*

FREDERIK DE KEYZER • *Section of Radiology, Department of Medical Diagnostic Sciences, University of Leuven, Leuven, Belgium*

MARC DHENAIN • *MIRCen, CEA/CNRS URA 2210, Fontenay-aux-Roses, France*

JOUKE DIJKSTRA • *Division of Image Processing, Department of Radiology, Leiden University Medical Center, Leiden, The Netherlands*

GUILLAUME DUHAMEL • *Centre de Résonance Magnétique Biologique et Médicale (CRMBM), UMR 6612, CNRS, Université de la Méditerranée, Marseille Cedex 05, France*

JEFF F. DUNN • *Department of Radiology, Physiology, and Biophysics, Faculty of Medicine, University of Calgary, Calgary, Canada*

BERNADETTE EROKWU • *Department of Radiology, Case Western Reserve University, Cleveland, OH, USA*

CORNELIUS FABER • *AG Experimentelle Magnetische Kernresonanz, Institut für Klinische Radiologie, Westfälische Wilhelms-Universität Münster, Münster, Germany*

CHRIS FLASK • *Case Western Reserve University, Cleveland, OH, USA*

ULRICH FLÖGEL • *Institut für Herz- und Kreislaufphysiologie, Heinrich-Heine-Universität Düsseldorf, Düsseldorf, Germany*

JENS FRAHM • *Biomedizinische NMR Forschungs GmbH am Max-Planck-Institut für biophysikalische Chemie, Göttingen, Germany*

GIULIO GAMBAROTA • *Department of Radiology, Radboud University Nijmegen Medical Center, Nijmegen, The Netherlands*

Contributors

ASSAF A. GILAD • *Institute for Cell Engineering, F.M. Kirby Research Center for Functional Brain Imaging, Johns Hopkins University School of Medicine, Baltimore, MD, USA*

JOAN M. GREVE • *Biomedical Imaging, Genentech, Inc., South San Francisco, CA, USA*

VIKAS GULANI • *Department of Radiology, University Hospitals, Cleveland, OH, USA*

VOLKER HEROLD • *Experimentelle Physik V (Biophysik), Universität Würzburg, Würzburg, Germany*

KARL-HEINZ HILLER • *Research Center Magnetic Resonance Bavaria, Würzburg, Germany*

MATHIAS HOEHN • *In-vivo-NMR-Laboratory, Max Planck Institute for Neurological Research, Köln, Germany*

BIANCA HOGERS • *Department of Anatomy, Leiden University Medical Center, Leiden, The Netherlands*

CHRISTOPH JACOBY • *Institut für Herz- und Kreislaufphysiologie, Heinrich-Heine-Universität, Düsseldorf, Germany*

HAUKUR JÓHANNESSON • *REAC Fuel AB, Lund, Sweden*

THERÉSE KALLUR • *In-vivo-NMR-Laboratory, Max Planck Institute for Neurological Research, Köln, Germany*

DENNIS W.J. KLOMP • *Department of Radiology, University Medical Center, Utrecht, The Netherlands*

GERT KLUG • *Clinical Division of Cardiology, Medical University, Innsbruck, Austria*

FRANK KOBER • *Centre de Résonance Magnétique Biologique et Médicale (CRMBM), UMB 6612, CNRS, Université de la Méditerranée, Marseille Cedex 05, France*

WILLIAM LEENDERS • *Department of Pathology, Radboud University Nijmegen Medical Center, Nijmegen, The Netherlands*

GUANSHU LIU • *Department of Radiology, Institute for Cell Engineering, Johns Hopkins University School of Medicine, Baltimore, MD, USA*

PATRICK McCONVILLE • *Molecular Imaging Research, Inc., 800 Technology Drive, Ann Arbor, MI, USA*

MICHAEL T. McMAHON • *Division of MR Research, The Russell H. Morgan Department of Radiology and Radiological Sciences, Johns Hopkins University School of Medicine, Baltimore, MD, USA; F.M. Kirby Research Center for Functional Brain Imaging, Kennedy Krieger Institute, Baltimore, MD, USA*

MICHEL MODO • *King's College London, Institute of Psychiatry, Centre for the Cellular Basis of Behaviour, London, UK*

SUSANNE MUELLER • *Centrum für Schlaganfallforschung Berlin, Charité Universitätsmedizin Berlin, Humboldt-Universität zu Berlin, Berlin, Germany*

YICHENG NI • *Section of Radiology, Department of Medical Diagnostic Sciences, University of Leuven, Leuven, Belgium*

KLAAS NICOLAY • *Biomedical NMR, Department of Biomedical Engineering, Eindhoven University of Technology, Eindhoven, The Netherlands*

BRIAN J. NIEMAN • *Mouse Imaging Centre, Hospital for Sick Children, Toronto, ON, Canada*

MELLY S. OITZL • *Division of Medical Pharmacology, University of Leiden, Leiden, The Netherlands*

J. STEFAN PETERSSON • *GE Healthcare, Stockholm, Sweden*

ALEXANDRA PETIET • *CNRS, URA 2210, Fontenay aux Roses, France*

Rolf Pohmann • *Max Planck Institute for Biological Cybernetics, Magnetic Resonance Center, Tübingen, Germany*
Colline Poirier • *Bio-Imaging Lab, University of Antwerp, Antwerp, Belgium*
Dana S. Poole • *Department of Radiology, Leiden University Medical Center, Leiden, The Netherlands*
Harald H. Quick • *Institute of Medical Physics, Friedrich-Alexander-University Erlangen-Nürnberg, Erlangen, Germany*
W. Klaas Jan Renema • *Department of Radiology, Radboud University Nijmegen Medical Center, Nijmegen, The Netherlands*
Alize Scheenstra • *Division of Image Processing, Department of Radiology, Leiden University Medical Center, Leiden, The Netherlands*
Jürgen E. Schneider • *British Heart Foundation Experimental MR Unit (BMRU), Department of Cardiovascular Medicine, University of Oxford, Headington, Oxford, UK*
Leif Schröder • *ERC Project Biosensor Imaging, Leibniz-Institut für Molekulare Pharmakologie, Berlin, Germany*
Xiaolei Song • *Division of MR Research, The Russell H. Morgan Department of Radiology and Radiological Sciences, The Johns Hopkins School of Medicine, Baltimore, MD, USA*
Roel Straathof • *Biomedical NMR, Department of Biomedical Engineering, Eindhoven University of Technology, Eindhoven, The Netherlands*
Gustav J. Strijkers • *Biomedical NMR, Department of Biomedical Engineering, Eindhoven University of Technology, Eindhoven, The Netherlands*
Anne-Marie van der Linden • *Bio-Imaging Lab, Department of Biomedical Sciences, University of Antwerp, Antwerp, Belgium*
Louise van der Weerd • *Department of Radiology, Anatomy and Embryology, Leiden University Medical Center, Leiden, The Netherlands*
Jan Wolber • *GE Healthcare, Amersham, UK*
Xin Zhou • *Wuhan Institute of Physics and Mathematics, Chinese Academy of Sciences, Wuhan, Hubei Province, China*
Alexander Ziegler • *Museum of Comparative Zoology, Harvard University, Cambridge, MA, USA*
Magdalena Zurek • *Université Lyon 1, Creatis – LRMN, Lyon, France*

Section I

Basics of MRI

Chapter 1

Physical Basics of NMR

Rolf Pohmann

Abstract

This chapter gives a short introduction to the physical and technical basics of nuclear magnetic resonance. It describes the formation of the NMR signal from the generation of the magnetization to the detection in the spectrometer. The behaviour of nuclear spins in a magnetic field is shown based on classical dynamics. The formation of the free induction decay and the spin echoes, as well as the concepts of longitudinal and transverse relaxation are explained. The basics of signal acquisition and reconstruction are presented. The concept of chemical shift is introduced with its application in NMR spectroscopy.

Key words: Nuclear magnetic resonance, NMR, nuclear spin, Zeeman effect, relaxation, chemical shift.

1. Introduction

The phenomenon of nuclear magnetic resonance (NMR) was first observed by Rabi (1, 2) when he shot lithium chloride molecules through two inhomogeneous magnetic fields surrounding a third, homogeneous field, while at the same time applying a radiofrequency radiation. He observed that for a certain relationship between magnetic field strength and radiation frequency, the magnetic properties of the nuclei change, which was explained by flips between the quantum states of the nuclear magnetic moment that are induced by the radiation. In solids, nuclear magnetic resonance was first detected by Purcell (3), who measured absorption of photons with a certain frequency by paraffin in a strong magnetic field. At the same time, Bloch (4) used a similar experiment to observe the effect in fluids by measuring the re-emission of the absorbed photons in water. The rapid rise of NMR in laboratories was first launched by the discovery of the chemical shift (5, 6), which allows to assign footprints to different molecules and thus identify the substances that are present in a given sample. Further milestones, like the discovery of the spin echo by Hahn (7), the development of techniques based on the Fourier transformation (8) and the introduction of two-dimensional spectroscopy (9), made NMR an invaluable tool in chemical and biological research.

A whole additional field of applications opened up when Lauterbur (10) and Mansfield (11) discovered the possibility of generating images of samples and even living objects by superimposing the constant and homogeneous magnetic field by additional fields that vary in space. This new technique, by now usually called MRI (*magnetic resonance imaging*, the word *nuclear* having been dropped for the sake of anxious patients), became even more interesting with Damadian's discovery that NMR parameters of pathologies differ from those of healthy tissue (12), making MRI a valuable tool also in medical diagnosis and therapy. But even beyond that, the great versatility of magnetic resonance techniques opens widespread possibilities to measure parameters like perfusion, diffusion, brain activation, blood flow velocity, metabolite concentrations and much more, thus giving rise to the current surge in MR scanners at increasing magnetic fields.

This chapter will give a short and simplified introduction into the physical background of NMR experiments. More elaborate explanations can be found in the introductory textbooks on NMR (13–15), or, for a detailed and accurate physical treatment, in one of the classical monographs (16–18).

2. Spins and Magnets

Some of the basic aspects of magnetic resonance can be visualized by the motion of a spinning top: when placed on a level surface, the top, not standing exactly upright, will not fall down, but start to rotate around a vertical axis through its foot. This movement is called *precession* and it is due to the angular momentum of the rapidly rotating toy (**Fig. 1a**). The torque applied by the gravitational force operates perpendicular to both the rotational axis of the top and gravity itself, causing a motion in a tangential direction. So instead of falling down, it precesses. If we now imagine that the top was carved out of a bar magnet and rotates in a magnetic field, it will behave similarly, even when not subjected to gravity. Now it is the magnetic field that imposes a torque on the top, trying to make it align to its own direction. This magnetic top then again reacts by precessing around the direction of the field (**Fig. 1b**). Many nuclei behave similarly

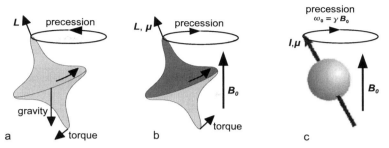

Fig. 1. Precession of (**a**) a spinning top with angular momentum L under the influence of gravity, (**b**) a magnetic top with angular momentum L and magnetic moment μ inside a magnetic field and (**c**) a nucleus with spin I and an associated magnetic moment μ inside a magnetic field B_0.

when placed into a magnetic field. Like the spinning top, they have an angular momentum and, connected to that, a magnetic moment. These properties make them behave like tiny, rotating magnets. The external field imposes a torque on these magnets and thus induces precession in a similar way as seen with the top. The frequency of this rotation is constant and depends only on the species of the nucleus and the magnetic field (**Fig. 1c**):

$$\omega_0 = \gamma B_0, \quad \text{or} \quad \nu_0 = \frac{\gamma}{2\pi} B_0, \quad [1]$$

where B_0 is the strength of the external magnetic field (measured in Tesla), ω_0 and ν_0 are the angular velocity and frequency of the precession, respectively, and γ is the gyromagnetic ratio, depending only on the nuclear species. ν_0 is called the Larmor frequency for the given nucleus at field B_0. **Table 1** contains the gyromagnetic ratios for a few of the most used nuclei in NMR.

In contrast to the spinning top, the nucleus is subject to the laws of quantum mechanics, which change its behaviour in many ways. The analogy of the magnetic top, however, can be (and effectively is often) used to qualitatively describe many of the phenomena associated with NMR.

2.1. Zeeman Effect

For a better understanding of the principles of nuclear magnetism, a short look into the quantum mechanical properties of atomic nuclei is necessary: Many nuclei possess an intrinsic angular momentum, also called *spin*, and an associated magnetic moment μ. It is described by a spin quantum number I, which can only assume integer and half integer values and is a constant for the species of the nucleus. In the classical picture, this would mean that each nucleus of a certain species always rotates with the same, constant frequency. The ^{1}H-nucleus, for example, has spin 1/2, while ^{23}Na has spin 3/2 and ^{17}O, 5/2. Other nuclei, like

Table 1
Gyromagnetic ratio, spin quantum number and natural abundance of some of the most used nuclei in in vivo NMR

Nucleus	γ [10^6 rad/T]	$\gamma/2\pi$ [MHz/T]	Spin	Natural abundance [%]
^{1}H	267.522	42.576	1/2	100
^{2}H	41.066	6.536	1	0.015
^{13}C	67.283	10.705	1/2	1.1
^{19}F	251.815	40.078	1/2	100
^{31}P	108.394	17.235	1/2	100
^{23}Na	70.808	11.262	3/2	100
^{17}O	−36.281	−5.7716	5/2	0.04
^{3}He	−203.789	−32.434	1/2	100
^{129}Xe	−74.521	−11.777	1/2	24.4

^{12}C or ^{16}O, however, have a spin of zero, meaning that they have no angular momentum and do not precess, which makes them invisible to NMR.

Due to their magnetic moment, the nuclei with a spin larger than zero behave like tiny bar magnets. However, in the absence of an external magnetic field, this is not discernible from the outside, since their axes are arbitrarily aligned in space, all orientations having the same energy. Macroscopically, these randomly oriented magnetic moments will cancel and no effect is observed. When placed inside a magnetic field, however, the behaviour of these nuclei will change: The spins now may only assume a limited set of spin states, the number of possible states depending on the spin quantum number: A nucleus with spin I can assume $(2I+1)$ separate states, described by numbers from $-I$ to $+I$, with integer steps between the states. The ^{1}H-nucleus can thus assume two levels, labelled $-1/2$ and $+1/2$, while ^{23}Na can have four states between $-3/2$ and $+3/2$. Each of these states corresponds to a separate orientation of the axes of angular momentum and magnetic moment in space, described by the angle to the magnetic field direction: For ^{1}H, the rotational axis either deflects by an angle of 54.7° from the direction of the magnetic field, or by the same angle from the opposite direction. These two different states differ in energy, an effect that is called *Zeeman splitting* (**Fig. 2**): The spins with axes oriented in direction of the field have a slightly lower energy than those pointed in the opposite direction. Although this energy difference is very small, it is the basis for all NMR experiments. If all states were populated by the same number of spins, the net effect would cancel, yielding no macroscopic magnetization. However, the small energy gap causes a tiny difference in the population numbers of the states. According to

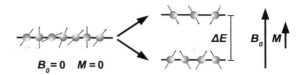

Fig. 2. The Zeeman effect: While in absence of a magnetic field all spins have the same energy (*left*) and are oriented randomly, in a magnetic field the energy levels split up (*right*). The energy gap causes a population difference which generates a macroscopic magnetization.

Boltzmann's law, the population difference between two energy levels is given by:

$$\frac{n_1}{n_2} = e^{\frac{\Delta E}{k \cdot T}} = e^{7.048 \cdot 10^{-6} \cdot B_0}, \qquad [2]$$

where n_1 and n_2 are the population numbers of the two states separated by the energy difference ΔE, k is Boltzmann's constant and T is the temperature. While the left-hand part of Eq. [2] is valid for all nuclei, the number in the exponential on the right was calculated specifically for protons at room temperature.

Thus, the energy gap between the spin states causes slightly more spins to be oriented in the direction of the field than in the opposite one, leaving a small net effect that is used in magnetic resonance techniques. This difference is very small. For example, of one million hydrogen nuclei in a magnetic field of 3 T, only about 11 contribute to the NMR signal. This explains the limited sensitivity of NMR, which is its main drawback and poses strong constraints on the spatial resolution that can be reached in MRI experiments or on the minimum concentrations of metabolites detectable in MR spectroscopy studies.

While the magnetic moments of the individual spins thus have different orientations and behave according to the laws of quantum mechanics, the macroscopic magnetization that arises from the superposition of all the single magnetic moments is, in equilibrium, oriented parallel to the external magnetic field and behaves, under certain conditions, similar to the magnetic spinning top described in the last section. This analogy is thus often used to describe NMR experiments. However, when using this picture, keep in mind that

- it describes the macroscopic magnetization and not the individual spins, which are governed by quantum mechanics,
- this magnetization is much smaller than the sum of the magnetic moments of all nuclei, since it is formed only by the small surplus of nuclei in the lower energy level,
- it is only valid for spins that can be assumed not to interact with any other spins. Many phenomena that play an important role in NMR experiments (e.g. spin–spin coupling or the Nuclear Overhauser effect) cannot be explained in

this image. Even a thorough examination of the relaxation (which will be discussed briefly later in this section) requires quantum mechanics.

Mathematically, the motion of the magnetization in the classical picture is described by the Bloch equations, which are often used to compute the behaviour of the spins in an experiment.

2.2. Excitation

As shown above, in equilibrium, the macroscopic magnetization M_0 that is generated by the combined magnetic moments of the spinning nuclei is *longitudinal*, which means that it is oriented in the z-direction, parallel to the magnetic field. This corresponds to a spinning top that is standing exactly upright. It will not experience a torque and thus not precess at all. Thus, it can not be detected in an NMR experiment. To obtain a signal, the magnetization has to be forced to precess, which is, just as for the spinning top, achieved by tilting the rotational axis with respect to the external magnetic field B_0. For this, again, precession is used: Applying another external magnetic field, usually labelled B_1, perpendicular to B_0, induces an additional rotation about its direction. The magnetization will deflect from M_0 and immediately start precessing with the Larmor frequency around the direction of B_0. To obtain a steady motion away from the z-axis, the direction of this additional field has to rotate with ω_0 around the z-axis, thus following the precession of the magnetization. Only if the frequency of the second field is exactly equal to the Larmor frequency will there be an effect, which is the reason why the phenomenon is termed *resonance*. The necessary frequency for protons is between 63.9 MHz for 1.5 T and 1 GHz (for 23.5 T) and is thus in the radiofrequency (rf) range.

Subjected to an rf field with the correct frequency, the precession around B_0 is superimposed by a tilt towards the transverse plane (**Fig. 3a**). This motion is also governed by Eq. [**1**], though here the B_1-field determines the precession frequency:

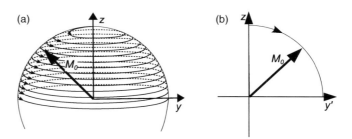

Fig. 3. A 90° pulse in the laboratory frame (**a**) and in the rotating coordinate system (**b**). In the laboratory frame, the magnetization moves in spirals, which result as a superposition of the precession around B_0 with frequency ω_0 and that around the rotating B_1 with ω_1. In the rotating coordinate system, the B_0-field and the corresponding precession disappear, leaving just the simple rotation around the B_1-direction.

$$\nu_1 = \frac{\gamma}{2\pi} B_1. \qquad [3]$$

Usually, B_1 is much smaller than B_0 and ν_1 is in the order of some hundred Hertz, leading to pulse durations in the millisecond range.

If the B_1-field is turned off after time τ, the magnetization will be tilted against the z-direction by an angle $\alpha = \gamma B_1 \tau$. In an NMR experiment, only the transverse component of the magnetization can be detected, which again is proportional to $\sin \alpha$. Maximum signal thus is obtained after a pulse with a flip angle of 90° (therefore called a 90° or a $\pi/2$ pulse) or 270°, while a 180° or 360° pulse will yield no signal at all, the magnetization being oriented towards the negative or positive z-axis, respectively.

For easier examination of the different precession components, it is practical to observe the magnetization from a rotating coordinate system. Described by coordinate axes usually labelled x', y' and z, it rotates around the z-axis with the nucleus' resonance frequency ω_0. Now, the effect of the B_0-field is just compensated by the rotation of the coordinate system, thus eliminating the precession around the z-axis. An on-resonance rf irradiation with the Larmor frequency then corresponds to a static field perpendicular to the z-axis and thus induces a rotation around its direction with the constant frequency ν_1 (*see* **Fig. 3b**).

In the quantum mechanical interpretation of the excitation, the B_1-field induces transitions between the different spin quantum states: Electromagnetic radiation with the frequency ν carries an energy given by $E = h\nu$, h being Planck's constant. If ν is equal to the Larmor frequency, this energy is just equal to the energy difference between the spin states. The field will thus induce transitions between the levels and cause changes in their populations.

2.3. The Free Induction Decay

The simplest NMR experiment consists of a single excitation pulse, which flips the magnetization by an angle of, e.g. 90°. It is then oriented in the transverse direction, precessing around the z-axis. The detected signal $S(t)$ will also oscillate with ω_0 and fall off exponentially:

$$S(t) = S_0 \cdot \cos(\omega_0 t) \cdot e^{-t/T_2^*}, \qquad [4]$$

with S_0 being the initial signal amplitude and T_2^* the usual term for the decay rate, the *apparent transverse relaxation time* (**Fig. 4**). This signal is called *Free Induction Decay* (FID).

The decay is caused by dephasing of the spins: Due to variations in the magnetic fields experienced by the individual spins, their precession frequencies differ, causing them to fan out. The superposition of all spins forming the macroscopic magnetization will therefore decay and disappear.

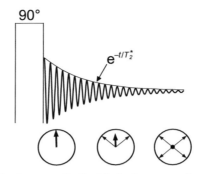

Fig. 4. A 90° pulse flips the magnetization into the transverse plane, where it precesses. The single spins contributing to the signal have slightly different frequencies, so that their phases diverge, causing the total magnetization to decay.

2.4. Longitudinal Relaxation

After the excitation with a 90° pulse, all magnetization is oriented in the transverse plane. Another 90° pulse will thus not yield any signal, because there is no longitudinal magnetization left to excite. With time, however, the longitudinal magnetization, which is zero after a 90° and $-M_0$ after a 180° pulse, will re-grow until it reaches again its equilibrium value M_0. The speed, with which the magnetization recovers, is described by the longitudinal relaxation time T_1. The behaviour of the z-component of the magnetization after a pulse with arbitrary pulse angle is given by (**Fig. 5**):

$$M_z(t) = M_z(t=0) \cdot e^{-t/T_1} + M_0 \cdot \left(1 - e^{-t/T_1}\right), \quad [5]$$

where $M_z(t=0)$ is the initial longitudinal magnetization.

During this relaxation process, those spins that were transferred to the higher energy level during excitation will fall back

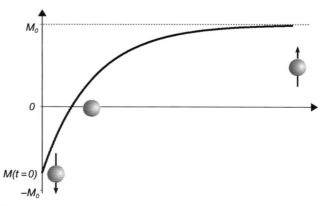

Fig. 5. T_1-relaxation: After a pulse with a flip angle of here almost 180°, the magnetization returns exponentially to the equilibrium value.

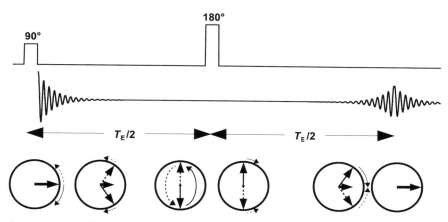

Fig. 6. A spin echo appears after a pair of pulses, one of 90° and the other of 180°. After the excitation, the FID decays because of locally varying magnetic fields: some spins move slightly faster than others (*broken arrow*), causing dephasing and thus a decay of the magnetization. The 180° pulse inverts the phases; due to the unchanged field distribution, the spins rephase after T_E.

to reach the equilibrium distribution. The excess energy is transferred to the molecular surroundings, the lattice. The longitudinal relaxation thus is also called *spin–lattice relaxation*. In in vivo experiments, T_1 is in most tissues in the order of seconds.

2.5. Spin Echo

In addition to the FID, another type of signal can be observed in NMR experiments: the *spin echo*. To generate a basic spin echo, an excitation pulse with a flip angle of 90° or less is followed by a second pulse with a flip angle of 180° (**Fig. 6**). The excitation pulse generates an FID that disappears with T_2^*. The second pulse does not create any immediate signal, but flips the magnetization around the transverse plane. After this, a signal will be observed that builds up exponentially, reaching its maximum after twice the pulse spacing, and then decaying again like an FID. The time between the excitation pulse and the maximum of the spin echo is called the *echo time* T_E; the maximum amplitude of the echo is lower than that of the FID, falling exponentially with increasing echo time with a time constant called the *transverse relaxation time* T_2:

$$A(T_E) = A_{\text{FID}} \cdot e^{-T_E/T_2}, \qquad [6]$$

where $A(T_E)$ is the amplitude of the spin echo and A_{FID} is the amplitude of the initial FID. T_2 is also called *spin–spin relaxation time*, since it is caused by interaction between the spins. In living tissue, it is often in the order of a few hundred milliseconds.

The spin echo appears because the 180° pulse reverses some of the processes that are responsible for the spin dephasing that causes the T_2^*-decay. These are, on the one hand, temporal fluctuations that are generated by microscopic motion inside the sample. This effect is irreversible and only depends on its intrinsic properties. On the other hand, spatial variations of the field also cause differences in the precession frequencies and thus dephase the spins. They can be due to inhomogeneities of the external B_0-field, but also to internal structures that modify the local field. These effects do not vary with time, so that spins experience consistently a somewhat different field and thus precess with a frequency that is slightly higher or lower than ν_0.

The phase of a faster spin will, after some time, be higher than that of the slower one. The 180° pulse, also called the *refocusing pulse*, now flips the spins over the transverse plane, which effectively causes an inversion of the phases. The fast spin now has the smaller phase, while the slow spin has the higher phase. The refocusing pulse, however, did not change the position or the precession frequency of the spins: The formerly fast one is still faster and thus it will now catch up on the slower ones. The phase difference between the spins will decrease again, until, just after the same time that was used for dephasing, all spins will be in phase again. At that time, the spin echo will reach its maximum. Since only the local variations are cancelled, while the temporal field fluctuations cannot be refocused, the spin echo will always be smaller than the original FID. While T_2 is a tissue-specific constant, T_2^* does not only depend on the sample but also on the homogeneity of the external B_0-field, the vicinity of the observed volume or even the spatial resolution of the experiment.

A different type of echo is the *stimulated echo*. In contrast to the spin echo, it takes at least three pulses to generate a stimulated echo; maximum amplitude is reached when all three have a flip angle of 90° (**Fig. 7**). After the first pulse, the spins dephase with T_2^*. The second 90° pulse flips the dephased spins towards the z-axis. The state of the spins is now stored in the longitudinal direction. During this time, transverse relaxation is interrupted; the magnetization relaxes with T_1. The third pulse moves the magnetization back to the transverse plane. An echo is generated,

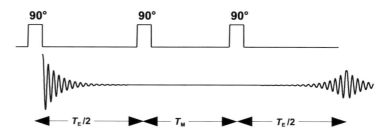

Fig. 7. A stimulated echo is generated by three 90° pulses. The time from the third pulse to the echo maximum is equal to that from the first to the second pulse, independent of the mixing time T_M. The amplitude of the echo is only half that of a spin echo with equal T_E.

the maximum of which appears just as long after the third pulse as the first and second pulse are apart, independent of the spacing between the second and third pulse, which is usually called *mixing time* T_M. The sum of the durations between the first two pulses and between the third pulse and the echo maximum is again called the echo time T_E. The echo amplitude depends on T_E and T_M, but at maximum is half the FID amplitude.

Although spin echoes reach their maximum amplitude for pulse angles of 90° and 180°, every pulse that is not exactly a 180° pulse and is followed by a second one with a flip angle of not exactly 90°, will create a spin echo, and if it is followed by two more pulses of not exactly 180°, it will generate a stimulated echo. Since rf pulses never are exactly 180° or 90° pulses everywhere in the sample, every sequence of two or more pulses will generate spin and stimulated echoes. A sequence consisting of a larger number of pulses thus generates a huge number of signals, all of which have to be kept in mind when designing or applying a multi-pulse sequence.

3. Chemical Shift and Spectroscopy

The simple equation for the Larmor precession, Eq. [1], does not at first glance point out the manifold applications of this technique. If a nucleus, described by its gyromagnetic factor γ is observed in a constant magnetic field B_0, what could be observed except for a precession with a single frequency? In fact, two basically different applications have been developed taking advantage of differences in the magnetic field that a nucleus is effectively exposed to. While imaging techniques use an artificially generated spatial inhomogeneity of the external field, the older technique of NMR spectroscopy measures microscopic internal differences in the magnetic field at the position of the nuclei due to the different chemical environment. The electron shell of the nucleus influences the field it is actually subjected to. The external magnetic field slightly modifies the electron's orbit which again has the effect of generating a secondary magnetic field that, according to Lenz's rule, is opposite to the original one. The nucleus will therefore experience a slightly lower field. How strong this shielding is depends on the electron density around the nucleus, which again is determined by its chemical surroundings: If other nuclei in the molecule have a high electronegativity, they will attract the electrons, leaving the observed nucleus relatively unshielded and thus exposed to a slightly higher field (and therefore precessing with a slightly higher frequency) than another nucleus of the same species but with a less electronegative chemical environment. Nuclei in different compounds will therefore have somewhat different resonances, allowing their identification by NMR spectroscopy. This difference in frequency is called *chemical shift* and is very small compared to the external B_0. It is described by the shielding constant σ, which modifies the magnetic field at the position of the nucleus as

$$B = (1 - \sigma) \cdot B_0. \quad [7]$$

The chemical shift δ is usually defined as the difference between the resonance frequency of the observed compound and that of a reference molecule:

$$\delta = \frac{\omega - \omega_{\text{reference}}}{\omega_0} \approx \sigma_{\text{reference}} - \sigma \quad [8]$$

and is specified in units of ppm (parts per million), since this is independent of the magnetic field strength. In principle, any arbitrary resonance can be used as reference; however, certain substances have by convention become the usual references. These resonances have to be clearly defined single lines that do not shift with variations in the environment. In proton spectroscopy, water, as the prominent peak in most applications, would appear the obvious choice. However, the water peak changes its frequency, e.g. with varying temperature, making this peak unsuitable as reference. Instead, it is common to choose the resonance of tetramethylsilane (TMS) as reference. The water line then appears in the spectrum at a chemical shift of around 4.7 ppm.

A simple spectroscopic experiment consists of an excitation over a wide range of frequencies. The resulting signal is recorded, being a superposition of the signals from all molecules containing nuclei of the examined species, and thus a mixture of frequencies.

A simple example, the NMR spectrum of ethanol, is shown in **Fig. 8**. Each of the three groups has protons with equal chemical environment: The proton in the hydroxyl group is bound to the oxygen nucleus, which has a high electronegativity and thus leaves the proton relatively unshielded. The signal of this proton therefore has a high frequency and appears in the spectrum on the left side – for historical reasons, NMR spectra are plotted with the high frequencies on the left and the low frequencies on the right. The two protons in the methylene group experience stronger shielding due to the higher electron density in their surroundings. Their signal therefore appears at lower frequencies (more to the right of the spectrum) and is twice as high as the signal of the hydroxyl proton, since signals from two hydrogen nuclei contribute. The signal from the three protons of the methyl group is yet higher and shifted to lower frequencies due to the less electronegative environment.

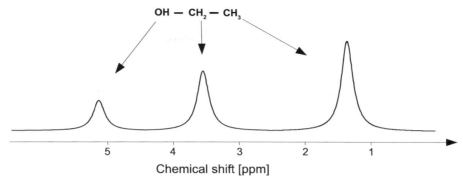

Fig. 8. A low-resolution spectrum of ethanol. High resonance frequencies are on the *left*. The single proton of the OH-group is least shielded and thus has the highest precession frequency. The area under the lines increases with the number of equivalent protons.

A highly resolved in vivo spectrum obtained from a rat brain is displayed in **Fig. 9**. More than 19 metabolites can be quantified at high magnetic fields. The frequency dispersion in proton spectra is quite small, especially at lower magnetic field strength, making the identification of the individual peaks difficult. In addition, due to the inherent limited sensitivity of magnetic resonance techniques, only metabolites with a fairly high concentration ($\approx$mM) can be detected in vivo.

3.1. Signal Generation, Detection and Processing

While the previous sections have focussed on the physical aspects of magnetic resonance, describing the basics of the dynamics of the spins under the influence of the static and the rotating magnetic fields, we will now briefly sketch the more technical aspects

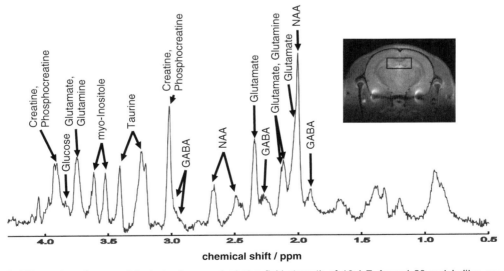

Fig. 9. MR spectrum from a rat brain in vivo, acquired at a field strength of 16.4 T. Around 20 metabolites can be quantified at high field. Spectrum courtesy of S.-T. Hong.

of how the signal is generated, detected and processed in an NMR scanner.

3.1.1. Excitation

Magnetic fields are usually generated by solenoid coils. This is in principle not different for the B_1 field that is required for excitation. However, since this field has to be oriented perpendicular to B_0, the solenoid would have to be oriented at a right angle to the magnet bore, which would make it impossible to place a subject inside the coil. Thus, it is often reduced to one or two windings placed close to or around the examined volume (*surface* or *Helmholtz* coil). In addition, alternative coil configurations were developed, which generate a field in the right orientation, but still allow positioning a long sample inside the magnet. Since the B_1-field has to rotate with the Larmor frequency in order to excite the spins, coil setups are used that generate circularly polarized and thus rotating fields (*quadrature coil*). However, by just applying an alternating current with ν_0 on a simple, single loop produces a field with a rotating component that can induce excitation.

3.1.2. Detection

After excitation with a 90° pulse, the magnetization precesses within the transverse plane, generating a varying magnetic field in its proximity. This is equivalent to the situation encountered in one of the classic experiments in physics: If a coil is positioned in a changing magnetic field, an electric current is induced. This phenomenon is described by Faraday's law of induction. A conductor loop parallel to the B_0-field experiences a sinusoidal magnetic flux Φ, causing the generation of a voltage in the loop (**Fig. 10**):

$$U_{\mathrm{ind}} = -\frac{\mathrm{d}\phi}{\mathrm{d}t} = -\frac{\mathrm{d}B}{\mathrm{d}t} \cdot A \cdot \cos(\vec{B},\vec{A}) \propto B \cdot \omega_0, \qquad [9]$$

where A is the area of the coil and the cosine is applied to the angle between the normal of the coil and the oscillating magnetic field B generated by the spin. This illustrates the fact that only the transverse component of the magnetization contributes to the signal.

Thus, in a coil placed close to the sample, the precessing magnetization induces an alternating current which constitutes

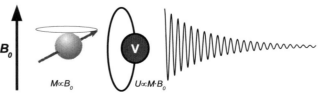

Fig. 10. The precessing magnetization after an excitation pulse generates an oscillating magnetic field. In a coil positioned close to the sample, this field induces a voltage that alternates with ω_0.

3.1.3. Phase-Sensitive Detection

The signal that is recorded by the receive coil is an oscillation with frequencies close to ω_0, which is in the MHz-range. Since signals with such high frequencies are difficult to process and interpret, while the frequency range of an NMR experiment is very small in comparison, one of the first processing steps performed by the spectrometer is to reduce the frequency. This is done by a process called *mixing*, which effectively results in a subtraction of ω_0 from the sampled signal frequencies. It corresponds to a transformation into the rotating coordinate frame, where the direct influence of the B_0-field disappears. This, however, poses the danger of losing information: If a cosine wave is shifted down by the Larmor frequency, it is no longer possible to differentiate between frequency components below and above ω_0, which now corresponds to negative and positive frequencies. This is avoided by describing the rotation by two terms, corresponding to the coordinates in a two-dimensional coordinate system (**Fig. 11**). To simplify further processing, the signal is then represented as a complex number. A signal with original frequency ω is thus output as $s(t) = \cos((\omega - \omega_0)t) + i\sin((\omega - \omega_0)t)$, or, equivalently, $s(t) = \exp(i(\omega - \omega_0)t)$. This type of signal processing is termed *phase-sensitive detection* or, alternatively, *quadrature detection* (which must not be confused with detection by a quadrature coil). The signal that is generated by the scanner thus is complex and consists of frequencies that range around zero.

3.1.4. The Fourier Transformation

The acquired signal is a mixture of oscillations with different frequencies. To analyse it for its frequency components, the time domain signal $s(t)$ is subjected to a mathematical procedure called

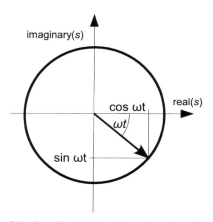

Fig. 11. A rotation is fully described by the sine and cosine of the rotation angle. By setting the former as the real, the latter as the imaginary part of a complex number, a single complex function is sufficient.

Fourier transformation, which is defined as:

$$S(\nu) = \int_{-\infty}^{\infty} e^{-2i\pi\nu t} S(t)\, dt. \qquad [10]$$

If the signal is an FID with frequency ν_0, described by $s(t) = A \cdot e^{2i\pi \nu_0 t - t/T_2^*}$, the result of the Fourier-transformation in the frequency domain is

$$S(\nu) = \int_{-\infty}^{\infty} e^{-2i\pi\nu t} A \cdot e^{2i\pi \nu_0 t - t/T_2^*}\, dt = \frac{AT_S^*}{1/T_2^* + 2i\pi(\nu - \nu_0)}. \qquad [11]$$

A plot of this function is shown in **Fig. 12**. Its real part has a Lorentzian lineshape with a maximum at ν_0 and a linewidth that increases with decreasing T_2^*. Correspondingly, a mixture of two or more frequencies results in a spectrum with several lines (**Fig. 13**), the positions, amplitudes and linewidths of which contain the information needed for analysis.

3.1.5. Digitalization and Nyquist

NMR signals are acquired and analysed by computers and are therefore not sampled continuously, but in discrete steps. According to the Shannon sampling theorem, this digitalization does not degrade the signal, as long as the frequency range

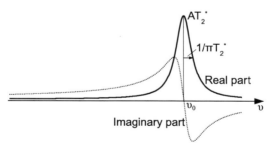

Fig. 12. Real and imaginary parts of the Fourier transform of an FID with a single frequency υ_0 and an apparent transverse relaxation time T_2^*. The position of the peak depends on υ_0, its linewidth on T_2^*.

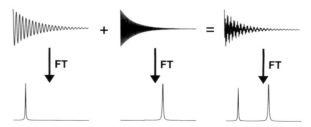

Fig. 13. Fourier transformation analyses the signals for the frequencies it contains. The line position depends on the frequency, the linewidth on the decay rate.

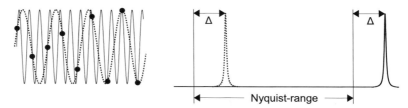

Fig. 14. The Nyquist condition limits the frequency range that can be reconstructed correctly: The *continuous lines* show the signal (*left*) and the corresponding correct spectrum (*right*). If this FID is sampled with too large a dwell time (*dots*), the signal is ambiguous and will be reconstructed with a lower frequency (*dotted line, right*). The spectral line thus is projected into the Nyquist range.

(the bandwidth F) of the signals is smaller than the inverse of the time steps the signal is sampled with (the dwell time Δt):

$$F \leq \frac{1}{\Delta t}. \qquad [12]$$

The highest allowed frequency is also called the Nyquist frequency. All signals outside this range are projected into it (*aliasing*, **Fig. 14**). The Nyquist condition has extensive consequences in NMR experiments:

- The dwell time has to be small enough such that all frequencies appearing in the signal are covered by the bandwidth. Frequencies outside this range may falsely appear somewhere in the spectrum or image.
- Even if there are no signals from the sample outside the selected bandwidth, it has to be made sure that these frequencies are suppressed by bandpass filters. The reason for this is that even if there is no signal, there will be noise, which is projected into the observed frequency range and thus will corrupt the experiment.

3.2. Field Dependence

During the entire history of NMR, the strengths of the magnetic fields that are used have been increasing, currently moving towards a maximum of 11.7 T for human research and exceeding 20 T for small animal imaging. The main reason for this trend towards high fields is the increased signal intensity, which depends on the population difference described by Eq. [**2**] and which grows linearly with increasing magnetic field strength. In addition, the law of induction, Eq. [**9**], also is proportional to ω_0, causing another linear signal increase with field strength. In total, the signal thus grows with the square of the magnetic field. This signal growth is the fundamental reason for the present rush towards higher fields.

The contribution of noise, on the other hand, also gets higher as the field strength increases (19, 20). Noise is generated by two factors: First, the coil itself produces noise due to its finite temperature, causing electron fluctuations in the detection circuit. The level of this *Johnson noise* or simply *coil noise* depends on the square roots of the temperature and resistance of the coil and the bandwidth of the acquisition, which again is proportional to the magnetic field.

When a conducting sample, like a human subject or an animal is examined, additional noise is generated by fluctuations in the sample. The noise contributions from this source are proportional to B_0.

How the signal-to-noise ratio (SNR) varies with field strength depends on which of these two noise contributions is dominant: for the examination of non-conductive, but also of small conducting samples at low fields, the coil noise prevails, causing a total SNR dependence on B_0 of SNR $\propto B_0^{7/4}$. For in vivo experiments on small animals, especially at lower field strength, SNR often is determined by this relationship. In that case, decreasing the coil resistance by using cooled coils can help to improve the sensitivity.

Examinations on large samples at high field are sample dominated, which means that the noise contribution of the sample is higher than that of the coil, causing a total linear dependence of SNR on the field strength. This applies in almost all cases to experiments on humans. In those measurements, the signal-to-noise ratio grows linearly with the field: SNR $\propto B_0$.

References

1. Rabi, I. I. (1937) Space quantization in a gyrating magnetic field. *Phys Rev* 51, 0652–0654.
2. Rabi, I. I., Zacharias, J. R., Millman, S., Kusch, P. (1938) A new method of measuring nuclear magnetic moment. *Phys Rev* 53, 318–318.
3. Purcell, E. M., Torrey, H. C., Pound, R. V. (1946) Resonance absorption by nuclear magnetic moments in a solid. *Phys Rev* 69, 37–38.
4. Bloch, F., Hansen, W. W., Packard, M. (1946) Nuclear Induction. *Phys Rev* 69, 127.
5. Proctor, W. G., Yu, F. C. (1950) The dependence of a nuclear magnetic resonance frequency upon chemical compound. *Phys Rev* 77, 717.
6. Dickinson, W. C. (1950) Dependence of the F-19 nuclear resonance position on chemical compound. *Phys Rev* 77, 736–737.
7. Hahn, E. L. (1950) Spin echoes. *Phys Rev* 80, 580–594.
8. Ernst, R. R., Anderson, W. A. (1966) Application of Fourier transform spectroscopy to magnetic resonance. *Rev Sci Instr* 37, 93–102.
9. Aue, W. P., Bartholdi, E., Ernst, R. R. (1976) 2-dimensional spectroscopy – application to nuclear magnetic resonance. *J Chem Phys* 64, 2229–2246.
10. Lauterbur, P. C. (1973) Image formation by induced local interactions – examples employing nuclear magnetic resonance. *Nature* 242, 190–191.
11. Mansfield, P., Grannell, P. K. (1973) NMR Diffraction in Solids. *J Phys C Solid State Phys* 6, L422–L426.
12. Damadian, R. (1971) Tumor detection by nuclear magnetic resonance. *Science* 171, 1151–1153.
13. Hashemi, R. H., Bradley, W. G., Lisanti, C. J. (2004) MRI: the basics. Philadelphia: Lippincott Williams & Wilkins.
14. Weishaupt, D., Köchli V. D., Marincek, B. (2006) How does MRI work? An introduction to the physics and function of magnetic resonance imaging. Berlin, New York: Springer.

15. Levitt, M. H. (2001) Spin dynamics: Basics of nuclear magnetic resonance. Chichester, New York: Wiley.
16. Ernst, R. R., Bodenhausen, G., Wokaun, A. (1987) Principles of nuclear magnetic resonance in one and two dimensions. Oxford: Oxford University Press.
17. Abragam, A. (1961) The prinicples of nuclear magnetism. Oxford: Clarendon Press.
18. Slichter, C. P. (1990) Principles of magnetic resonance. Berlin, New York: Springer.
19. Edelstein, W. A., Glover, G. H., Hardy, C. J., Redington, R. W. (1986) The intrinsic signal-to-noise ratio in NMR imaging. *Magn Reson Med* **3**, 604–618.
20. Hoult, D. I., Lauterbur, P. C. (1979) Sensitivity of the zeugmatographic experiment involving human samples. *J Magn Reson* **34**, 425–433.

Chapter 2

Spatial Encoding – Basic Imaging Sequences

Rolf Pohmann

Abstract

This chapter presents the basic techniques for generating images with magnetic resonance. First, the usage of gradients for slice selection, frequency and phase encoding is explained. The concept of k-space is introduced and imperfections of the encoding methods are demonstrated by means of the point spread function. The most prominent imaging techniques based on gradient and spin echo signals are presented. Finally, alternative encoding procedures are described briefly.

Key words: Magnetic resonance imaging, MRI, slice selection, frequency encoding, phase encoding, imaging techniques.

1. Introduction

The goal of imaging is to obtain information about the spatial distribution of spins. However, the Larmor equation, $\omega_0 = \gamma B_0$, does not contain any position-dependent terms. It is thus necessary to introduce some spatially varying factor that allows distinguishing between spins at different positions in space. This is achieved by abandoning the strict spatial homogeneity of the main magnetic field by superimposing an additional, spatially dependent field. These fields, called gradients, are generated by coils installed inside the magnet bore, which produce magnetic fields that vary linearly in space. An MRI scanner is usually equipped with three gradient coils, producing fields that vary in x-, y- and z-directions. In contrast to the main magnetic field B_0, these additional fields are not constantly active, but are turned on and off in short, well-controlled gradient pulses. Since fast switching of these coils inside the B_0 field generates strong forces on the coils, they start to vibrate, thus generating the loud noise that accompanies MRI examinations.

In the presence of a gradient field G_x that varies linearily in the x-direction, the total magnetic field has the shape $B(x) = B_0 + G_x(t) \cdot x$, where t in the parentheses emphasizes that the gradient varies with time. The spins thus precess with a spatially dependent frequency

$$\nu(x) = \frac{\gamma}{2\pi} B(x) = \nu_0 + \frac{\gamma}{2\pi} \cdot G(t) \cdot x. \qquad [1]$$

This is used in three different ways for spatial encoding, which can be combined to resolve all three spatial dimensions in one experiment: *Slice selection* is used to selectively excite the spins in a well-defined plane, while *frequency* and *phase encoding* produce an image in two or three dimensions.

2. Spatial Encoding

2.1. Slice Selection

The space dependence of the Larmor frequency in the presence of a gradient can be used to selectively excite the spins within a slice perpendicular to the gradient direction. Since only the excited spins generate a signal, this is used to restrict the imaged volume to a well-defined region. Position and thickness of this slice can be adjusted freely.

The principle of slice selection is simple: If the Larmor frequency changes with position, a radio frequency (rf) pulse with a single frequency only selects those spins that are at the appropriate position (**Fig. 1**) where the Larmor frequency is equal to that of the pulse. Changing the frequency of the pulse modifies

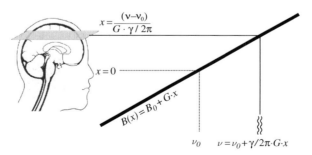

Fig. 1. Slice selection: A gradient G imposes a space dependence on the main magnetic field. An rf pulse with the single frequency ν then will excite only those spins for which the Larmor condition is fulfilled. The slice position thus depends on the strength of the gradient and the pulse frequency.

Fig. 2. Slice profile: The Fourier transform, and thus the frequency distribution, of a block pulse is described by a sinc function. When this pulse is used for slice selection, the spatial distribution of the excited spins looks similar.

the position of the excited slice, while its thickness is determined by the *bandwidth*, the frequency range, of the pulse.

While this principle seems simple, a few things have to be paid attention to in order to ensure good slice selection.

The excitation profile is never the perfect rectangle that would be optimal for slice selection. Pulses in NMR are usually short (in the order of milliseconds or less) and thus always contain a mixture of many frequencies. To analyse the frequency composition of an rf pulse is again a job for the Fourier transformation: **Figure 2** shows the simple case of a short pulse with constant amplitude over its entire duration (*block pulse*). The Fourier transform and thus the frequency distribution of this pulse is not at all uniform, but follows a sinc function, defined as $\sin(x)/x$. When we use this pulse for excitation, the slice profile will look similar, which means that the signal is not restricted to the narrow plane in the centre, but also comes from regions of the sample that are distant from the intended slice. Actually, the Fourier transform of the pulse accurately describes the slice profile for low excitation angles (low enough so that $\sin(x) \approx x$). For higher flip angles, the slice profile will often be significantly different (*see* **Fig. 3**).

In addition to obtaining a good excitation profile, rf pulses have to meet additional requirements. In most cases, they should be short and not require too much power. Additional requirements may come up, e.g. when surface coils with an inhomogeneous B_1 field are used, or when the application requires special pulse characteristics. To satisfy these demands, the modulation of the pulse amplitude or, less frequently, its frequency is adapted to modify the behaviour of the pulse. In addition, its properties depend on whether it is used for excitation, with flip angles below 90°, refocusing, saturation or inversion. Many pulse shapes have been developed for different applications. Among the most often used pulses are the following (*see* **Fig. 3**):

The block pulse: As described above, this pulse has constant amplitude over its entire duration and a bad slice profile. Its advantage is the low power or short duration required to reach a certain pulse angle. It is usually used for three-dimensional imaging, where the entire sample is excited simultaneously in the absence of a gradient.

The sinc pulse: For small flip angles, the slice profile is the Fourier transform of the pulse envelope. For a perfect, rectangular

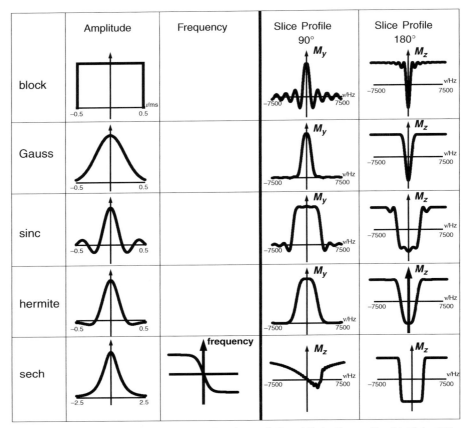

Fig. 3. Pulse shapes: Some of the most prominent pulse shapes (*left*) and their slice profiles (*right*) for 90° and 180° pulses. The bandwidths in the pulse profiles are calculated for pulse durations of 1 ms (sech: 5 ms); increasing the duration decreases the bandwidth. The sech pulse is an adiabatic pulse and the only one that also modulates the frequency. The pulse shown is the full-passage 180° pulse; the half-passage pulse is obtained by cutting off the pulse in the middle.

slice profile, this would be the sinc function. Correspondingly, its slice profile for low angles is quite good, although not perfect due to the finite duration of the pulse which requires an early cut-off of the sinc function. The sinc pulse requires a relatively high power.

The hermite pulse: is a numerically optimized pulse shape for 90° and 180° pulses. The refocusing profile especially is considerably improved in comparison to the sinc pulse. In addition, the required pulse power is reduced.

The Gauss pulse: Its envelope is a Gaussian and so is its Fourier transform. This pulse thus has a definitely non-rectangular excitation shape, but is quite well localized and requires only little pulse power.

The sech pulse: This pulse has the shape of a hyperbolic secant function and is the most prominent example of the group of adiabatic pulses. In contrast to the pulses presented before, not only the amplitude, but also the rf is shaped during the pulse. This gives adiabatic pulses the special characteristics of being able to exactly reach a certain pulse angle, usually 90° or 180°, over a large volume, even if the B_1 field strength varies considerably over this region. This can be guaranteed due to the property of the adiabatic pulses to never exceed the intended flip angle. Sech pulses can be designed for 90° or 180° angles; however, due to their phase properties, they are of only limited use for excitation and refocusing and are usually used for either saturation or inversion.

Many other pulse shapes and modifications of the presented ones have been designed for improving the slice profile or for special purposes.

When using any of these pulses for excitation, an additional gradient has to be switched to adjust the phase of the magnetization: right after the pulse (in the presence of a gradient), the slice profile is superimposed by a linear phase dispersion in the direction of the gradient, which is due to the varying precession frequency caused by the gradient during the excitation process. To recover the full signal, this has to be reversed by inverting the gradient after the end of the pulse for half the pulse duration. This rewinds the phase dispersion and the signal corresponds to the desired slice profile (**Fig. 4**). The usual excitation module in any imaging sequence thus consists of a slice selection gradient, during which the pulse is played, followed by a rephasing gradient, generated by inverting the first one until phase coherence is reached.

2.2. Frequency Encoding

To obtain an image, the excitation (slice-selective or not) has to be coupled with a procedure that achieves spatial resolution in two or three dimensions. One technique that is often used for one direction is turning on a gradient during signal acquisition.

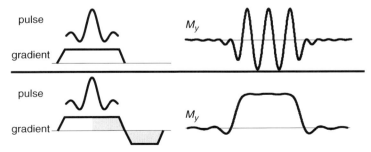

Fig. 4. Slice selection module: Directly after the excitation pulse, the phase dispersion within the slice causes a loss in signal (*top*). The phase is corrected by inverting the gradient after the end of the pulse (*bottom*). The integral of this rephasing gradient is equal to that of the slice selection gradient starting at the centre of the pulse.

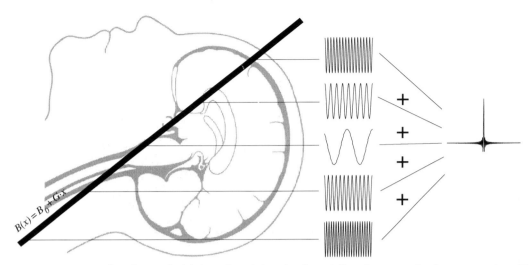

Fig. 5. Frequency encoding: If a gradient is turned on during signal acquisition, the precession frequency varies with position. The signal contributions with different frequencies add up to give the acquired signal (the gradient echo). The Fourier transformation is used to recover the frequency and thus the spatial distribution.

The effect of this is to introduce a spatially dependent frequency (**Fig. 5**). Similar to spectroscopy, a Fourier transformation then recovers the frequency and, accordingly, the spatial information.

The signal in presence of the gradient is written as

$$s(t) = \int \rho(x) \cdot e^{i\gamma Gtx} dx, \qquad [2]$$

where $\rho(x)$ is the spin density, describing the spatial distribution of the spins in the sample, and the gradient is assumed to be constant and oriented in the x-direction. Again, the signal is not sampled continuously, but in discrete steps separated by the dwell time Δt. Analogous to spectroscopy, the Nyquist criterion now determines the value of Δt that is necessary to image a certain volume, the field of view (FOV):

$$\text{FOV} = \frac{2\pi}{\gamma G \cdot \Delta t}. \qquad [3]$$

This relation follows directly from the definition of the Nyquist frequency, when the bandwidth is replaced by the gradient term. Thus, a small FOV requires a large gradient or a large dwell time (which also leads to a long acquisition).

In practice, it is often advantageous to first dephase the spins with one gradient, invert the gradient and thus observe the spins first rephase and then dephase again (**Fig. 6**). The signal then grows until the effects of the two gradients compensate and is

Spatial Encoding – Basic Imaging Sequences

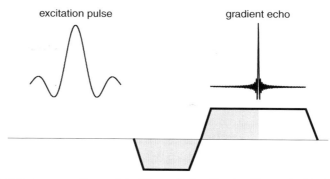

Fig. 6. Frequency-encoding module: After excitation, the signal is acquired in the presence of a read gradient. First, another gradient in opposite direction dephases the magnetization to allow for sampling of the entire, symmetric echo.

dephased again under the influence of the second gradient. Since this signal has the shape of an echo, it is called *gradient echo*.

2.3. Phase Encoding

The frequency-encoded signal results in the projection of the sample profile perpendicular to the gradient direction. Since it only works in one dimension at a time, the second and third spatial dimensions have to be resolved differently. The most common technique for this is *phase encoding*. Here, a gradient in the direction to be resolved is turned on for a short time between excitation and acquisition. This gradient will introduce a spatially varying phase

$$s = \int \rho(y) \cdot e^{i\gamma \cdot y \int G(t)dt} dy \qquad [4]$$

on the entire acquired signal. Here, the gradient is placed under integration to emphasize that it is not constant. The phase of the signal now depends on position. This experiment is repeated with varying phase by changing the strength of the gradient $G(t)$ and thus the magnitude of the exponential in Eq. [4] (**Fig. 7**). Since one phase-encoding gradient can only modulate one signal,

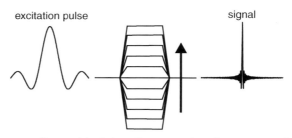

Fig. 7. Phase-encoding module: Schematic presentation of a phase-encoding gradient in a sequence diagram. The gradient is plotted with different sizes to illustrate that its strength changes with repetition. The *arrow* indicates the order of the phase-encoding steps and is used only if this is important for the sequence.

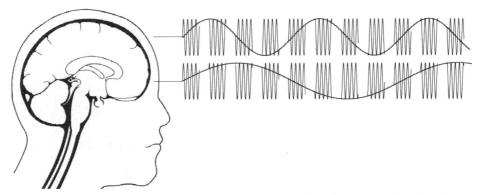

Fig. 8. Phase encoding: Phase encoding requires several scans, each one with a different strength of the phase-encoding gradient, thus imposing spatially dependent phase shifts between the signals. For two positions, the phase of one point of the signal is plotted for a series of phase-encoding steps. The frequency of the modulations is higher in positions further away from the centre.

N separate and different phase-encoded signals are required to obtain N points (voxels) in the image.

Figure 8 demonstrates the signal modulation during a phase-encoded measurement by following the phase of the first point of the signal through several phase-encoding steps and for two different positions. The phase modulation gets faster with increasing distance from the centre. This is similar to the frequency distribution within a single signal in frequency encoding.

Again, the difference in the gradient strength between the phase encoding steps is determined by the Nyquist limit as

$$\text{FOV} = \frac{2\pi}{\gamma \int G(t)\,dt} = \frac{2\pi}{\gamma G \tau_G}, \qquad [5]$$

where the last term considers the special case of a phase gradient with constant strength during its entire duration τ_G.

3. The k-Space

Equations [2] and [4] show clear similarities which indicate that frequency and phase encoding really are complementary ways for accomplishing the same effect. It is therefore practical to treat both approaches with a common formalism. This is done with the concept of k-space, defined as

$$\mathbf{k}(t) = \frac{\gamma}{2\pi} \int_0^t \mathbf{G}(t')\,dt'. \qquad [6]$$

Both frequency and phase encoding now are described by

$$s(t) = \int \rho(x) \cdot e^{2\pi i k x} dx, \qquad [7]$$

which shows the analogy to the spectral signal (*see* Eq. [11] in **Chapter 1**). Analogously, the Nyquist criterion now determines the distance Δk between two acquired points in k-space as

$$\text{FOV} = \frac{1}{\Delta k}, \qquad [8]$$

which applies both to frequency encoding, where it describes the dwell time and thus the bandwidth, and to phase encoding, where it determines the differences in the strengths of the phase-encoding gradients.

Alternatively, this relation can be based on the spatial resolution:

$$\Delta x = \frac{1}{K}, \qquad [9]$$

where Δx is the size of one voxel and K is the width of the k-space region that is sampled in the entire experiment.

To acquire a two-dimensional image with $N_x \times N_y$ voxels with the described techniques, the experiment has to sample a region in k-space, the size of which is determined by Eq. [9], while the step size is given by Eq. [8]. This can be visualized by the k-space diagram (**Fig. 9**). k-Space is sampled in lines that are acquired by frequency encoding and columns that are covered by

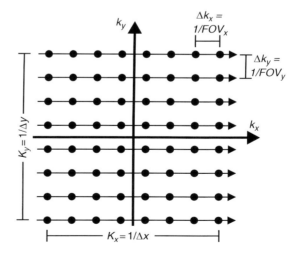

Fig. 9. *k*-Space diagram: This kind of plot is used to illustrate the *k*-space coverage of an imaging sequence. The sampled *k*-space points are plotted as dots; the *arrows* show the trajectories during one frequency-encoding scan, corresponding to one phase-encoding step.

phase encoding and thus require a separate signal for each line. This kind of diagram, although not very exciting in this simple case, is often used to describe alternative encoding schemes and experimental procedures.

3.1. The Point Spread Function

The signal in Eq. [7] is Fourier-transformed to yield the spatial spin distribution. However, the result will only be equal to $\rho(x)$, if the integration is performed from $-\infty$ to ∞, which requires acquisition of the entire k-space. Since this is obviously not possible within a finite experimental time, the Fourier transform will be convoluted with a weighting function that describes the deviation from the ideal, infinite k-space signal. In the simplest case, this weighting function is just a window function that is 1 over the sampled region of k-space and 0 everywhere else. The signal of an imaginary point source at position x_0 is then:

$$s(t) = \int_{-K/2}^{K/2} e^{2\pi i k x_0} dx, \qquad [10]$$

and its Fourier transform:

$$s(x) = \int_{-K/2}^{K/2} e^{2\pi i k x_0} e^{-2\pi i k x} dk = K \cdot \frac{\sin(\pi K(x - x_0))}{\pi K(x - x_0)}. \qquad [11]$$

The calculated signal of a point source is called the *point spread function* (PSF) and is an important parameter that describes the behaviour of a sequence or encoding scheme. In the simple case of an unweighted but cut-off acquisition, it is a sinc function (**Fig. 10**). The PSF is easily calculated by Fourier-transforming the weighting function; in this case, a simple window function.

This shape of the PSF has effects on the appearance and interpretation of an imaging experiment:

- In addition to the central peak, that generates an image at the correct position, wiggles of the PSF extend far into the neighbouring voxels. Due to the shape of the curve – in every voxel there are positive and negative contributions – this is not always visible in the images. In case of a sharply outlined structure, however, these wiggles can be seen as shifted ghosts, often referred to as Gibbs ringing.
- The central peak of the PSF is considerably wider than the nominal voxel size (which is just FOV/number of voxels). However, the spatial resolution (defined as the distance that two objects have to be apart to be distinguishable in the image), indeed is equal to the voxel size for this sinc-shaped

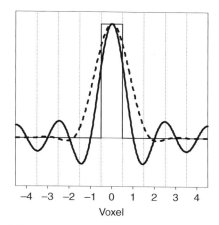

Fig. 10. Point spread function of an MR imaging experiment: The ideal PSF would be 1 inside the voxel and 0 everywhere else (*thin line*). The PSF of an actual, unweighted measurement is the sinc function, shown as *thick line*. The wiggles outside the targeted voxel can be suppressed by applying a filter, which, however, increases the linewidth and thus affects the spatial resolution (*broken line*).

PSF. However, in techniques where the PSF is changed by additional weighting, it may get wider, thus degrading the spatial resolution.

The shape of the PSF can be manipulated even after the experiment: Multiplying the k-space signal with a filter function modifies the PSF and can be used to suppress the wiggles. While this is used to efficiently reduce the Gibbs-ringing artifact, it also widens the central peak of the PSF and thus affects the spatial resolution (**Fig. 10**).

4. Basic Imaging Techniques

To generate an image, the spatial encoding techniques explained above have to be combined to obtain spatial resolution in all dimensions. Usually, either two-dimensional images are acquired by slice selection in the first, phase encoding in the second and frequency encoding in the third dimension, or three-dimensional images are generated by phase encoding in two and frequency encoding in the third direction. In the first case, the number of (differently encoded) signals is equal to the number of desired image points (voxels) in the phase direction, in the 3D case it is the product of the voxels in the two phase-encoding directions.

There are a few different possibilities of how to generate these signals, based mainly on gradient or spin echoes. Some of

4.1. Gradient Echo Techniques

the most often used techniques will be presented briefly in the following sections.

The simplest way to generate N gradient echoes is through a sequence of N simple excitation pulses, each followed by a phase-encoding gradient and a frequency-encoded readout. A sequence diagram for this type of experiment is plotted in **Fig. 11**. However, as simple as this experiment appears, there are a few things to consider.

Figure 12 shows a few pulses of a gradient echo experiment: After the first pulse, we obtain an FID. The second pulse also generates an FID, but also refocuses part of the magnetization of the first signal and thus generates an echo that just appears around the third pulse and thus interferes with the third FID. The spatial encoding of this echo, however, will be different, since it has experienced the phase gradients of the first and second repetitions. The third echo will again generate spin echoes from all earlier signals, and, in addition, a stimulated echo. Again, these signals will have wrong phase encoding, thus causing artifact in the image.

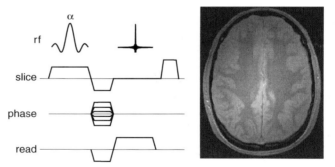

Fig. 11. Gradient echo (FLASH) sequence: Slice selection, frequency and phase encoding are combined to form a complete two-dimensional image. The plotted sequence section is repeated for all different values of the phase-encoding gradient. On the *right* is a typical image of the human brain as acquired with FLASH.

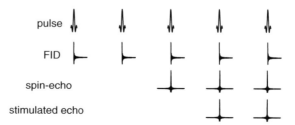

Fig. 12. Gradient echo signals: A gradient echo sequence generates FIDs (*top row*), spin echoes (*middle*) and stimulated echoes (*bottom*) that overlap. If phase encoded differently, the overlapping signals generate artifact.

Two different ways of dealing with these signals are used: Either the echoes are suppressed or they are used to enhance the signal.

In the first type of sequences, known as FLASH (fast low angle shot (1)), all the echoes caused by refocusing of earlier signals have to be destroyed; only the FID of each pulse is left and sampled as gradient echo. Two approaches to suppress the spurious signals are possible.

In the first one, a gradient pulse is inserted before each excitation. This gradient causes dephasing of all the remaining transverse magnetization and thus suppresses the unwanted echoes. These gradients are called spoiler gradients, since their aim is to destroy magnetization. This procedure is shown in the sequence depicted in **Fig. 11**. An alternative technique is known under the name of rf spoiling: Here the phase of each excitation (the angle between the *x*-axis and the transverse component of the magnetization) is varied in each scan in an optimized way which prevents all the echoes from turning up (2).

In a FLASH experiment, signal intensity depends on the pulse angle that is used for excitation. The maximum signal is obtained by using a 90° pulse. However, after the pulse, there is no magnetization left in the longitudinal direction which might generate signal in the next phase-encoding step. Using 90° pulses, we thus would have to apply a repetition time (TR) of around T_1 before the next phase encoding step can be performed, which would considerably prolong measurement time. On the other hand, using a low flip angle will generate less signal but leave more magnetization for the next excitation.

The optimum between the two extremes and thus the flip angle that produces the maximum signal for a given T_1 and repetition time T_R, is called the Ernst angle and is defined as

$$\cos \alpha_{\text{Ernst}} = e^{-T_R/T_1}, \qquad [12]$$

and thus decreases with increasing T_1 and decreasing T_R. Experiments that use short repetition times with low flip angles will assume, after a few repetitions, a steady state where every excitation yields the same signal amplitude. Thus, these first excitations are usually not used for acquiring signal (*dummy scans*).

FLASH images have the big advantage of ease of use and a relatively high degree of fail-safeness. For this reason, this sequence is always used for localizer scans. The contrast in FLASH images can be based on T_1, especially when high flip angles above the Ernst angle are used. Alternatively, T_2^* contrast is possible by using long echo times.

The second group of gradient echo sequences takes advantage of all the echoes appearing in an excitation pulse train to obtain different contrast or to increase the signal. These techniques are called SSFP (steady-state free precession) and acquire either the FID (FISP: fast imaging with steady state precession (3)), the echo (PSIF) or both (TrueFISP (4)). The last variant is of special interest because of its ability to obtain high SNR within short experimental durations. Instead of suppressing all the spurious signals with crusher gradients or rf spoiling, all the echoes and the FID are added. To avoid artifacts, it is necessary to make sure that all signals that appear simultaneously have experienced the same phase-encoding gradient. One way to achieve this is by rewinding the phase after each scan. After the acquisition and before the next excitation, a gradient in the phase direction is switched, having the same duration and magnitude, but different sign, as the previous phase-encoding gradient (**Fig. 13**). Thus, at the time of the excitation, the phase of all remaining transverse magnetization components is reset and the net phase is determined only by the gradient of the current encoding step.

Since echoes and FIDs overlap, the contrast depends on both T_1 and T_2. Assuming short repetition times ($T_R \ll T_2$), the signal becomes a function of T_1/T_2 (5):

$$S = \frac{M_0 \sin \alpha}{T_1/T_2 \cdot (1 - \cos \alpha) + (1 + \cos \alpha)} e^{-T_E/T_2}. \quad [13]$$

A disadvantage of TrueFISP in some applications is its sensitivity to frequency differences. The signal as a function of the frequency is not homogeneous, but shows dark bands. While this does not affect the quality of the images in a homogeneous magnetic field, it causes dark stripes in the image in regions where the field strength varies, e.g. because of susceptibility differences caused by tissue/air boundaries, especially at high field.

An even faster possibility to obtain a gradient echo image is the EPI (echo planar imaging (6)) sequence (**Fig. 14**). Here only one excitation is needed for an entire, two-dimensional image. The magnetization is de- and rephased by a gradient pair to

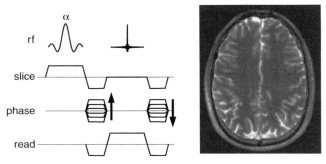

Fig. 13. TrueFISP sequence. All gradients are perfectly symmetric; TR between the phase-encoding steps is fixed and very short.

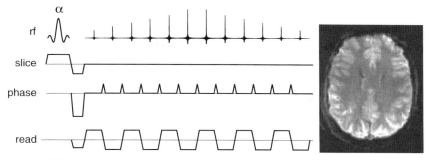

Fig. 14. EPI sequence. The echoes are generated by a repeatedly inverted read gradient; phase encoding is achieved by small gradient blips each of which moves the k-space trajectories one line further.

generate a gradient echo. Then, the dephased magnetization is once more rephased by inverting the gradient again. In this way, the required number of signals is generated by repeatedly inverting the read gradient. Each time, an echo is created, the amplitude of which decays with T_2^*. To finish the entire experiment within the short time available before the signal disappears requires gradients that can be switched very fast. However, the fast decay of the signal still imposes strong constraints on the spatial resolution. EPI experiments often are restricted to resolutions of 64^2 and seldom exceed 128^2 voxels. The echoes are phase encoded by gradient blips. Before the first echo, a strong phase gradient dephases the magnetization to reach the farthest necessary position in k-space. After each echo, a short, small gradient moves the magnetization one step further in k-space. Since only a small phase gradient is needed for these blips, they can be very short and do not slow down the sequence. The centre of k-space thus is reached with the central echo. EPI produces strongly T_2^*-weighted images. Due to the long acquisition per excitation, it is extremely susceptible to B_0 inhomogeneities. Variations in the field strength as caused by susceptibility differences or bad shim cause distortion or even cancellation of the signal. For these reasons, EPI images are only used in experiments where either extremely short acquisition duration is required, or where a strong T_2^* weighting is desired. The latter is the case in functional MRI (fMRI) experiments, where the BOLD effect causes a change in T_2^* which is to be imaged. Thus, EPI is used in most fMRI applications.

It should be noted that due to the T_2^* decay during the EPI echo train, the point spread function becomes a convolution of the sinc function with a Lorentzian, causing further broadening of its central peak. The spatial resolution thus is lower than the nominal voxel size.

To increase the speed of the EPI scan, the acquisition is often started before the gradient has reached its plateau. The data points acquired on the gradient ramp then have to be regridded onto the regularly spaced grid points of k-space during reconstruction to avoid artifacts. It is even possible to use a completely different gradient waveform and acquire the data points, e.g. during a sinusoidal gradient shape.

This flexibility in the acquisition is used for even faster scanning in *spiral imaging* techniques by completely giving up the separation between read and phase encoding. Here, after excitation, both gradients run in sinusoidal waveforms, thus traveling on a spiral path through k-space. In contrast to standard EPI acquisition, this has the advantage of starting at the centre of k-space, thereby reaching very short echo times and thus low T_2^*-weighting. Alternatively, it is possible to move to the corner of k-space first and run on the spiral path from outside to the centre, thereby getting a very high echo time. For reconstruction, it is necessary to first measure the exact gradient trajectories during the acquisition.

4.2. Spin Echo Techniques

In contrast to gradient echoes, generating a spin echo requires at least two pulses, often with flip angles of 90° and 180°. Images acquired with spin echo methods can have a T_2-based contrast.

The simplest spin-echo experiment consists of a sequence of excitation / refocusing pulse pairs, each phase encoded differently (**Fig. 15**). Due to the 180° pulse, the use of small flip angles (like the Ernst angle in gradient echo imaging) is not possible. This sequence thus requires relatively long delays in between two scans to allow for sufficient T_1-relaxation, leading to often unacceptable measuring times.

To accelerate the acquisition of T_2-weighted images, a train of spin echoes can be acquired by applying a series of equidistant refocusing pulses after a single excitation, each pulse refocusing the preceding echo (**Fig. 16**). The spin echoes that appear between the refocusing pulses are phase encoded individually to form the image. After each acquisition, the phase-encoding gradient is rewound to set the phase back to zero. As in TrueFISP, this avoids the superposition of differently encoded signals. The

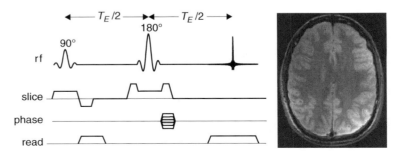

Fig. 15. Spin echo sequence. This technique requires long repetition times for recovery of the magnetization.

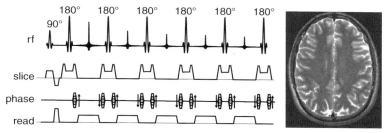

Fig. 16. RARE sequence. The signals are generated by a series of 180° refocusing pulses. Additional gradients in phase direction are switched prior to each pulse rephasing the spins to avoid overlap of differently encoded echoes.

number of echoes following a single excitation pulse can range from two up to the total number of phase encodes required for the image, thus enabling a single-shot acquisition. This type of experiment was first published as RARE (rapid acquisition with relaxation enhancement (7)) and is also known as turbo spin echo (TSE) or fast spin echo (FSE). While it is often used to acquire T_2-weighted images, other contrasts are also possible, especially if short echo times are used.

The main problem of RARE sequences for use in experiments on humans is the high number of 180° pulses that are required and cause a large deposition of energy in the subject, which can lead to unacceptable heating of the examined tissue. Especially for multi-slice experiments with a high number of refocusing pulses per excitation, this can be a limiting factor in human applications. A possible solution is using lower flip angles for refocusing (U-FLARE, ultrafast low angle RARE (8)). For angles that can go down to less than 90°, an equilibrium with the required stable echo amplitudes is obtained after a few pulses. The amplitudes of these echoes are smaller than those obtained with 180° pulses, but the lower refocusing flip angles cause a slower decay and thus a longer echo train.

Even more signal can be recovered by varying the flip angle throughout the echo train: Starting with high flip angles which get smaller during the pulse train will lead to significantly higher echo amplitudes (9).

4.3. Chemical Shift Imaging

Chemical shift imaging (CSI) is the combination of spectroscopy with imaging (10). The goal is to obtain spatially resolved spectra, which means that for a grid of voxels we simultaneously obtain an entire spectrum for each voxel.

In spectroscopy, the spectral information is recovered from an FID acquired without gradients. Thus, the simplest way to acquire spatially resolved spectra is to do without the read gradient and encode all two or three spatial dimensions with phase encoding (**Fig. 17**). The disadvantage of this approach is the long experimental time. For each dimension, the number of

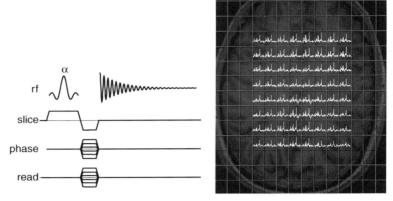

Fig. 17. Chemical shift imaging. In the sequence shown here, FIDs are acquired. Alternatively, a spin echo can be generated by a 90°–180° sequence. Due to the pure phase encoding, a large number of repetitions is required. This technique obtains a spectrum for each voxel (*right*).

scans required is equal to the number of voxels in the respective direction. The entire number of scans for a chemical shift imaging experiment is thus the product of the number of voxels per dimension and thus equal to the total number of voxels. A 2D experiment with a resolution of only 32 × 32 voxels thus requires 1024 scans, which, with a repetition time of 1 s, takes a total of 17 min. However, for a spectroscopic experiment, scan times in the order of 15 min or more are common, since the low concentration of the examined metabolites requires adding a large number of signals to obtain sufficient amplitude. Thus, CSI often does not take longer than single-voxel spectroscopy, but acquires much more information in the same time. Both duration and signal-to-noise ratio cause restrictions on the voxel size; experiments with as few as 16 × 16 voxels or less are common in CSI.

To decrease the duration of CSI measurements, the technique can be combined with SSFP to obtain increased signal with shorter repetition times (*11*). In addition, especially for experiments with higher spatial resolutions, fast techniques have been developed, that combine spatial encoding by read gradients with alternative ways to obtain the spectroscopic information (for an overview, *see* (*12*)).

4.4. Alternative Encoding Strategies

While the encoding techniques described in the previous sections are applied in the vast majority of MRI examinations, other methods have been developed that replace or enhance the conventional ones, with the goal of reducing the time needed to acquire an image or of optimizing the image characteristics for a certain application. A few prominent examples of alternative techniques will be briefly presented here:

Projection Reconstruction (Radial Imaging): This was the reconstruction technique used in the very first MR images (13). It currently experiences renewed interest because of its ability to generate gradient echo images with very short echo times. The echo is generated and acquired in presence of a read gradient. The direction of the read gradient is rotated in a number of subsequent repetitions, thus yielding a number of projections through the sample in different directions, from which the image can be reconstructed with an algorithm called *filtered backprojection*. An advantage of this technique is, in addition to the shorter echo times that can be reached, a reduced sensitivity to motion. On the other hand, projection reconstruction images can show blurring artifacts due to B_0 inhomogeneities and gradient nonlinearities. In addition, the number of repetitions is usually higher than for Fourier techniques.

Gridded Reconstruction: We have already encountered gridding when discussing spiral imaging. It is, however, always used whenever the sequence applied samples the k-space in non-equidistant points. To allow using the Fourier transformation for reconstruction of the images, the k-space points on the rectangular grid are recovered by a mathematical algorithm (14). This is used, e.g. in EPI-experiments, when the acquisition is sped up by starting to acquire before the gradients have reached their plateau, as an alternative reconstruction option for radial imaging or for other techniques which require uneven sampling of k-space.

Partial Fourier Acquisition: In theory, k-space shows strong Hermitian symmetry (which means that the real part is symmetric and the imaginary part is antisymmetric) if a real object is imaged. This would mean that only half of k-space is sufficient to reconstruct the entire image. In practice, phase fluctuations due to susceptibility differences as well as motion, eddy currents or other hardware imperfections cause the image to be not entirely real. While this is not a problem in full k-space imaging, it weakens the k-space symmetry and makes image reconstruction from just half of the k-space impossible. Thus, instead of scanning just half of k-space, some lines in the other half (called *overscan data*) are also acquired and serve for determining and correcting those errors. Actual partial Fourier techniques thus sample usually between 5/8 and 7/8 of k-space.

Parallel Imaging: For many applications, the use of phased array coils has become standard in MR imaging. The signals from several small coils are acquired and combined to form a single image. While the main reason for this is the signal gain that can be achieved compared to acquisition with a single, larger coil, this kind of acquisition can also be used to speed up the experiment. The signals from such coils have different spatial distributions, which can be used to recover additional, not measured k-space lines. Thus, after acquiring only a fraction of the k-space steps, interjacent points can be reconstructed. In practice, two approaches to parallel imaging are currently used: SENSE (sensitivity encoding (15)) operates in image space, using the coil sensitivities to remove aliasing from undersampled data. GRAPPA (generalized autocalibrating partially parallel acquisitions (16)) works in k-space by recovering the missing k-space lines before Fourier transformation. The factor by which the acquisition is sped up is called the acceleration factor. The acceleration factor that can be reached in an experiment depends for both techniques on the coil geometry and the number of coils used and can be considerably more than two. The SNR of images acquired with parallel imaging is usually lower than that of conventionally generated images, which is mainly due to the reduced scan time. However, an additional SNR loss due to the parallel imaging reconstruction is added, which depends on the coil geometry and the acceleration factor and is described by the geometry factor g.

Compressed Sensing: Similar to parallel imaging, compressed sensing serves to reconstruct the entire image from incomplete, undersampled k-space data. In contrast to parallel imaging, the missing points are not recovered, but fundamental image properties are used to correct for the missing data by using specialized reconstruction algorithms. They take advantage of the fact that all natural images have certain statistical characteristics (17). The images are reconstructed by selecting from all possible solutions the one that most closely complies with these assumed properties. The characteristic mostly used is *sparsity*. The image or, more often, one of its transforms (e.g. the wavelet transform or simply the next-neighbour difference) contains more often values close to zero than a random distribution would do. This works best for highly sparse datasets, like angiography where large fractions of the image are black. But even in normal images, a time reduction of a factor of 2 and more is possible. In contrast to parallel imaging, the acquisition should not cover k-space in regular intervals, but should sample the k-space lines in irregular steps for optimum results.

References

1. Frahm, J., Haase, A., Matthaei, D. (1986) Rapid NMR imaging of dynamic processes using the FLASH technique. *Magn Reson Med* **3**, 321–327.
2. Crawley, A. P., Wood, M. L., Henkelman, R. M. (1988) Elimination of transverse coherences in FLASH MRI. *Magn Reson Med* **8**, 248–260.
3. Oppelt, A., Graumann, R., Barfuss, H., Fischer, H., Hartl, W., Shajor, W. (1986) FISP: a new fast MRI sequence. *Electromedica* **54**, 15–18.
4. Duerk, J. L., Lewin, J. S., Wendt, M., Petersilge, C. (1998) Remember true FISP? A high SNR, near 1-second imaging method for T2-like contrast in interventional MRI at .2 T. *J Magn Reson Imaging* **8**, 203–208.
5. Scheffler, K., Hennig, J. (2003) Is TrueFISP a gradient-echo or a spin-echo sequence? *Magn Reson Med* **49**, 395–397.
6. Mansfield, P. (1977) Multi-planar image formation using NMR spin echoes. *J Phys C: Solid State Phys* **10**, L55–58.
7. Hennig, J., Nauerth, A., Friedburg, H. (1986) RARE imaging: A fast imaging method for clinical MR. *Magn Reson Med* **3**, 823–833.
8. Norris, D. G. (1991) Ultrafast low-angle RARE: U-FLARE. *Magn Reson Med* **17**, 539–542.
9. Hennig, J., Weigel, M., Scheffler, K. (2003) Multiecho sequences with variable refocusing flip angles: optimization of signal behavior using smooth transitions between pseudo steady states (TRAPS). *Magn Reson Med* **49**, 527–535.
10. Brown, T. R., Kincaid, B. M., Ugurbil, K. (1982) NMR chemical shift imaging in three dimensions. *Proc Natl Acad Sci USA* **79**, 3523–3526.
11. Schuster, C., Dreher, W., Geppert, C., Leibfritz, D. (2007) Fast 3D 1H spectroscopic imaging at 3 Tesla using spectroscopic missing-pulse SSFP with 3D spatial preselection. *Magn Reson Med* **57**, 82–89.
12. Pohmann, R., von Kienlin, M., Haase, A. (1997) Theoretical evaluation and comparison of fast chemical shift imaging methods. *J Magn Reson* **129**, 145–160.
13. Lauterbur, P. C. (1973) Image formation by induced local interactions: examples employing nuclear magnetic resonance. *Nature* **242**, 190–191.
14. Matej, S., Bajla, I. (1990) A high-speed reconstruction from projections using direct Fourier method with optimized parameters – an experimental analysis. *IEEE Trans Med Imaging* **9**, 421–429.
15. Pruessmann, K. P., Weiger, M., Scheidegger, M. B., Boesiger, P. (1999) SENSE: sensitivity encoding for fast MRI. *Magn Reson Med* **42**, 952–962.
16. Griswold, M. A., Jakob, P. M., Heidemann, R. M., Nittka, M., Jellus, V., Wang, J., Kiefer, B., Haase, A. (2002) Generalized autocalibrating partially parallel acquisitions (GRAPPA). *Magn Reson Med* **47**, 1202–1210.
17. Lustig, M., Donoho, D., Pauly, J. M. (2007) Sparse MRI: the application of compressed sensing for rapid MR imaging. *Magn Reson Med* **58**, 1182–1195.

Chapter 3

Basic Contrast Mechanisms

Leif Schröder and Cornelius Faber

Abstract

This chapter provides an overview of how contrast in MR images can be achieved. The physical origin of the most basic contrast mechanisms is briefly explained and experiments to exploit these are discussed. Furthermore, the concept of using exogenous contrast agents is introduced.

Key words: Proton density, T_1, T_2, magnetization transfer contrast, contrast agent.

1. Relaxation Mechanisms for Contrast Generation

1.1. Definition of Contrast and Its Parameter Space

In order to discriminate various types of tissue in MR scans, their signal intensity has to be different, i.e. the detected magnetizations must be distinguishable. There are several parameters that influence this quantity and the choice of the imaging pulse sequence will eventually generate a contrast that is predominantly based on one of them. To quantify the contrast C between two adjacent tissue structures, it is useful to define the following ratio based on their signal contributions S_1 and S_2

$$C = \left| \frac{S_1 - S_2}{S_1 + S_2} \right|$$

The normalization takes into account that small signal differences can easily be masked when the overall signal intensity is high. It should be kept in mind that the ability to delineate two tissue structures also depends on the signal-to-noise ratio (SNR) of an image. The SNR is defined by the mean signal in a region of interest (ROI) normalized by the standard deviation (SD) of the background noise (**Fig. 1**).

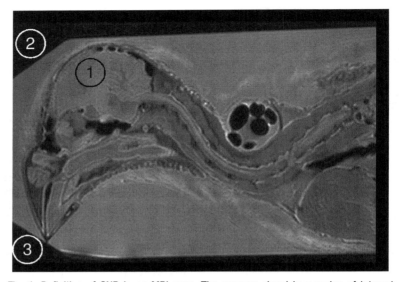

Fig. 1. Definition of SNR in an MRI scan. The average signal in a region of interest (ROI 1), here in the brain of a fixed zebra finch specimen, is divided by the standard deviation in an image region devoid of spins, but inside the FOV (ROI 2 and ROI 3). To minimize systematic noise sources, noise is taken from several ROIs.

$$\text{SNR} = \frac{\text{mean ROI signal}}{\text{SD of noise}}$$

A high SNR value is no guarantee to delineate two different tissues when the signal intensities in the two ROIs are practically the same. On the other hand, a measurable signal difference that is on the same order as the noise of the data makes it also difficult to identify different structures – the contrast between two pixels might be reasonable but it does not mean anything if it does not stand out from the dynamic range of the noise. The introduction of the so-called contrast-to-noise ratio (CNR) therefore becomes increasingly popular (1). However, the usefulness of this quantity in terms of an objective measure is not generally accepted since the ability to discriminate details from the noise also depends on the individual visual reception by the observer.

The great advantage of MR imaging is the variety of parameters to generate contrast. Compared to X-ray imaging, where the contrast is solely based on the electron density of the material that causes an attenuation of the radiation that propagates through the tissue, MR signals are not exclusively determined by the density distribution of nuclear magnetic moments. This density depends on the concentration of the molecules that comprise the MR-sensitive nuclei. Since most MR images are based on the detection of hydrogen nuclei (i.e. protons), the proton density according to the water and fat concentration of the tissue is just one intrinsic contrast parameter (other molecules are practically not directly detectable; *see* **Note 1**). Beyond this unchangeable parameter, the pulse sequences in MR imaging can be used to actively manipulate the detectable magnetization. This precise manipulation is achieved during the following steps that characterize every MR experiment:

- radiofrequency (rf) *excitation of the system* to generate detectable magnetization
- *evolution of the magnetization* that drives the system back towards its thermodynamic equilibrium
- *encoding of the spatial information and readout* of the manipulated magnetization

These three steps influence each other, especially since many MRI techniques are based on a repetitive acquisition of data, thus making the detected magnetization depend on the previous cycle of the experiment. We have to distinguish between two types of magnetization in order to understand the generation of contrast.

- *Longitudinal magnetization*, which arises when the sample reaches its thermodynamic equilibrium in the external magnetic field and which is aligned with the latter. Longitudinal magnetization does not induce a detectable signal but can be conceived as the reservoir that is available for each rf excitation to generate detectable magnetization. The course of the MR experiment will influence the amount of this reservoir since each rf excitation can use up more or less of the available magnetization (*see* **Note 2**).

- *Transverse magnetization*, which is generated by manipulating longitudinal magnetization with a short rf pulse. The so-called flip angle indicates how much the longitudinal magnetization is tipped off the z-axis, thereby producing a transverse component in the xy-plane (**Fig. 2**). This one precesses around the external field, thus inducing a detectable AC voltage signal in the detection antenna. It is this magnetization that is controlled during the spatial encoding by means of additional rf and magnetic field pulses and that decays to a certain extent before and during detection.

Each rf pulse represents a perturbation of the system which is followed by relaxation processes of the magnetization back to its thermal equilibrium. These processes are governed by internuclear and inter-molecular mechanisms which depend on the physico-chemical environment of the spins (2). It is therefore possible to differentiate various types of tissue and visualize abnormal areas because of the underlying differences in relaxation. The reader should realize that these processes occur on a scale that is much smaller than the eventual image resolution.

In addition to the initial equilibrium magnetization (determined by the proton density, PD), all the three steps mentioned

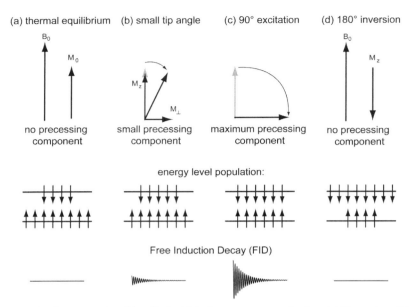

Fig. 2. Various scenarios for conversion of longitudinal into transverse magnetization. (**a**) Thermal equilibrium with no detectable transverse component. (**b**) Small tip angle excitation that reduces the population difference between the two energy levels and generates a small FID signal. (**c**) A 90° excitation that causes equal population (so-called saturation) and the maximum available FID signal. (**d**) Inversion experiment with no detectable FID signal.

above have an influence on the two types of magnetization because they determine the initial disturbance of the system and the subsequent schedule of relaxation events. Hence, the interplay of sequence timing and relaxation mechanisms contributes to image contrast.

1.2. General Considerations on Image Quality

It should be mentioned at this point that the multi-dimensional parameter space for MRI contrast generation is in general accompanied by a trade-off between image quality in terms of SNR, spatial resolution and acquisition speed, i.e. temporal resolution. Several acquisition parameters influence the final result (*see* **Note 3**). Besides the sequence timing that controls influence of relaxation mechanisms, the following parameters are also important:

- Field of view
- Matrix size (i.e. the number of pixel elements)
- Slice thickness
- Number of averages
- Receiver bandwidth

The first three are geometric parameters that simply determine how many spins contribute to the signal in each subunit of the image. Better spatial resolution therefore cuts down the received induction signal per pixel. This can be compensated for by signal averaging since the real signal accumulates linearly whereas contributions to the noise partially cancel each other and increase only with the square root of the number of acquisitions. Hence, the number of acquisitions has to be quadrupled to maintain a certain SNR when the spatial resolution improves by a factor of 2 (e.g. halving the slice thickness or changing from 100 to 50 μm in one dimension). The receiver bandwidth must be large enough to allow for recording of all frequency components from the entire field of view (*see* **Note 4**), but the acquisition of noise also increases with the square root of this parameter.

Although strong gradients are a prerequisite to improve spatial resolution for resolving small structures, they are, however, useless if they are not paired with rf-receiving hardware that ensures the necessary sensitivity to detect a decreasing number of spins. Hence, the detectable SNR at a certain spatial resolution is an important parameter to judge the quality of an MRI system. The achievable resolution depends on the object that is imaged and on the optimized detectors. The state of the art is reflected in the examples given in the following chapters of this book (3). Current progress is also based on a new type of coils that reduces noise by cooling the rf circuits with helium or liquid nitrogen. These so-called cryo-probes gain a factor of 2–3 in sensitivity that can be used to improve spatial or temporal resolution (4).

1.3. T_1 Relaxation

1.3.1. Physical Background

T_1 relaxation is the process that is responsible for recovery of longitudinal magnetization after excitation by an rf pulse. Since longitudinal magnetization is defined by the difference of spins in parallel and antiparellel orientation, it is related to the energy of the spin system and the corresponding relaxation is the recovery from an excited state back to thermal equilibrium through loss of energy. This is an exponential process (*see* **Fig. 3**) that is characterized by the time constant T_1 (*see* **Note 5**). The full range of relaxation occurs after an inversion pulse with a zero passage at $t = T_1 \cdot \ln 2$ (inversion-recovery experiment). Almost complete

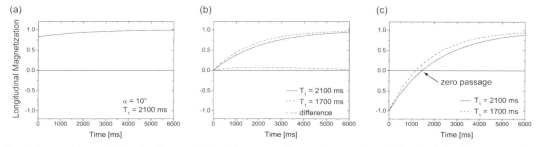

Fig. 3. Different longitudinal relaxation scenarios. (**a**) recovery after small tip angle excitation. (**b**) Saturation recovery for two different relaxation times T_1. (**c**) Inversion recovery for a fast and a slow component, illustrating the expected signal difference when the slow component has its zero-passage.

recovery after maximum signal generation by a 90° pulse that leaves no longitudinal magnetization (the so-called saturation-recovery experiment) is achieved after ca. $3T_1$.

Exchange of energy occurs with the neighbouring nuclei and molecules that are sometimes summarized in the generalized term "lattice" (*see* **Note 6**). T_1 relaxation is therefore also called spin–lattice relaxation. The source of the relaxation process is the fluctuating fields caused by the sum of the fields attributed to surrounding electrons and nuclei. The detected nuclei experience an oscillating field due to molecular tumbling. If the spectrum of the fluctuations caused by the surrounding molecules contains components at the NMR resonance frequency ω (ca. 50–500 MHz) or 2ω it can absorb energy from the detected spin system and therefore drive the system back into its thermodynamic equilibrium. Components contributing at 2ω are actually weighted four times as much as components at ω (2). T_1 relaxation increases as the molecular mobility decreases but becomes very slow again for extreme conditions such as those in solids. Therefore, areas with less restriction in molecular tumbling (such as the cerebrospinal fluid) are characterized by long T_1 values whereas tissue rich in microstructure (e.g. muscle) exhibit fast T_1 relaxation. Some examples are given in **Table 1**.

1.3.2. Repetitive Nature of Data Acquisition: How to Use T_1 Relaxation

Most standard MR-imaging sequences require a repetitive excitation of the magnetization because the spatial encoding is not done in a single shot (5). The raw data space from which the image is finally computed is filled up in a line-by-line scheme, thus introducing a dependence of each line's signal intensity on the previous acquisition step. Sufficient spatial resolution often requires the acquisition of tens to hundreds of lines in raw data space and the desired acquisition time ranges from the sub-second regime to the order of several minutes or even hours. This combination yields repetition times (TR) that are on the order of T_1 relaxation and therefore automatically allows using this effect for generation of contrast by adjusting TR properly.

Table 1.
Relaxation times for some mouse tissues at 9.4 T according to Kuo (1)

Tissue	T_1 [ms]	T_2 [ms]
Grey matter	1900	40
White matter	1700	32
Muscle	2100	20
Fat	850	30

A fast repetition of the pulse sequence can "saturate" the spin system (*see* **Note** 7) and therefore impair the subsequent generation of transverse magnetization to encode the image information. In contrast, very slow repetition will eliminate the relaxation influence and the differences in T_1 between various tissues fade out, thus yielding a method to eliminate this contribution to emphasize other types of contrast.

1.3.3. Manipulating T_1 Relaxation: Contrast Agents and External Field

The fact that relaxation depends on the fluctuating magnetic fields experienced by the nuclear spins introduces the opportunity to enhance natural T_1 relaxation by introducing additional magnetic moments that influence the detected protons in water (and fat) molecules. Such magnetic moments are provided by unpaired electrons, e.g. in Fe^{3+}, Mn^{2+} or Gd^{3+} ions (6). The paramagnetic effect from electron-rich systems like lanthanide complexes is especially efficient since the magnetic moment of each electron is ca. 600 times larger than that of protons. Gadolinium, e.g. has seven unpaired electrons that alter the relaxation of adjacent protons very efficiently. The element itself is toxic and has to be made harmless by complex formation, e.g. with hydrophilic poly(aminocarboxylate) ligands. One or two coordination sites are occupied by water molecules (7) that are temporarily bound to the Gd ion and therefore experience efficient dipolar interactions with the electron spins. Such interactions fall off rapidly with increasing distance but water molecules in the inner hydration shell of these complexes are also affected to some extent by relaxation effects.

Such complexes influence both T_1 and T_2 relaxation but the dominant effect is shortening the slower longitudinal relaxation. Gd chelates therefore cause bright signal contributions in T_1-weighted MR images (so-called positive contrast). This can be used to increase contrast between two types of tissue or allow retaining contrast while speeding up the acquisition by using shorter repetition times. The effect can be amplified even more by "adjusting" the correlation time of such molecules: macromolecular contrast agents have a higher relaxivity due to their reduced tumbling rate and can therefore be detected at lower concentrations (*see* **Note** 8). In any case, this type of contrast agents only affects areas where it is directly accessible since the water needs to get close to the paramagnetic centre.

Since there is no way to reduce the number of fluctuating magnetic moments in tissue, a reduction of T_1 relaxation can only be achieved by reducing the matching frequency components in the above-mentioned range of fluctuations. By increasing the field strength B_0 of the scanner system, the resonance frequency of the detected water molecules is shifted against the fixed spectrum of field fluctuations due to molecular motion. Resonant contributions are less pronounced and T_1 relaxation times increase when switching from standard clinical field strengths of 1.5–3 T to modern high-field scanners of 7 T or more (8).

1.3.4. Practical Considerations: Acquisition Time

Several experimental setups require a certain temporal resolution for consecutive imaging or are subject to other constraints that determine an upper limit for the total acquisition time. Since longitudinal relaxation is a property of the tissue that can be manipulated by contrast agents only to some extent, choice of the suitable imaging sequence can depend on the speed of the encoding scheme. The total acquisition time is given by the product of the repetition time (TR), the number of phase encoding steps (N_{ph}) and the number of acquisitions (N_{aq}; see **Note 9**):

$$T = \text{TR} \cdot N_{ph} \cdot N_{aq}$$

For some applications, it is therefore necessary to use sequences that achieve the desired contrast even for relatively short TR, i.e. acquisition schemes that minimize the disturbance of the longitudinal magnetization.

It should be mentioned at this point that T_1 relaxation of protons is relatively favourable for most in vivo applications (see **Table 1**). Other nuclei such as ^{31}P have longer relaxation times (9) (see **Note 10**). Given the fact that the relative sensitivity for such nuclei is much smaller (see **Note 11**), small flip angle excitations often yield insufficient signal and the required 90° excitation is then related to very long TR or substantial saturation effects. This aggravates the problem for many nuclei other than ^{1}H to collect MRI datasets in a reasonable amount of time.

1.4. T_2 Relaxation

1.4.1. Physical Background

The signal strength induced by the precessing transverse magnetization depends on the phase coherence among the different spin packages. Loss of this coherence is called transverse relaxation. The amplitude of the vector sum of all magnetizations is decreasing exponentially with a time constant T_2 (**Fig. 4a, b**).

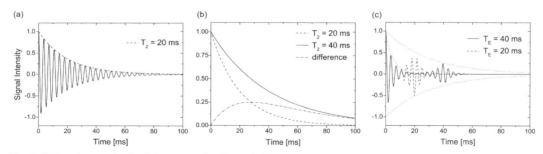

Fig. 4. Different scenarios for transverse relaxation. (**a**) FID signal showing T_2 relaxation during signal acquisition. (**b**) Two T_2 decays and corresponding signal difference illustrating the optimum timing for maximum contrast. (**c**) FID and echo formation with two different echo times TE.

The process is also called spin–spin relaxation and is not related to loss of energy (like T_1 relaxation). Compared to longitudinal relaxation, this process occurs more rapidly because additional mechanisms contribute. Microscopic processes such as interactions with the local magnetic fields of neighbouring nuclei are governed by the molecular tumbling rate and the spectral density of different tumbling modes. Transverse relaxation is influenced by spectral components at ω, 2ω (less weighted than the previous one) and low-frequency components (intermediate weight) (2). Rapid motion causes averaging of spin–spin interactions over time and makes such processes less efficient, thereby slowing down T_2 relaxation. Relaxation times for free water molecules are therefore relatively long (ca. 3 s and are mostly independent of the magnetic field strength) whereas T_2 values for macromolecules bound to cell membranes are on the order of a few microseconds only. In the latter case, relaxation is more efficient because of the longer "contact time" between molecules. Skeletal muscle tissue is an example for tissue with relative short T_2 values (*see* **Table 1**).

Very rapid T_1 relaxation usually does not impose a problem (in fact, it is sometimes desired and generated with suitable contrast agents); however, extensive T_2 relaxation can impede acquisition of sufficient signal intensity even for short echo times. The scanner hardware is limited in terms of arbitrarily fast pulse sequences due to minimum times for switching the rf and gradient pulses on and off. Hence, some nuclei with very short T_2 (e.g. sodium or even protons in case they are in a macromolecular environment) are hard to detect and give only small signal contributions.

1.4.2. T_2^* Relaxation

Loss of phase coherence between the different magnetizations within a sample happens not only as an intrinsic process in terms of spin–spin interactions, it is also a consequence of local field inhomogeneities ΔB that cause small differences in precession frequencies of different spin packages. This process therefore enhances transverse relaxation. It is called T_2^* relaxation and happens on a timescale faster than pure T_2 processes. T_2 and T_2^* relaxation rates, i.e. the inverse of the relaxation times, are linked through the following equation:

$$\frac{1}{T_2^*} = \frac{1}{T_2} + \frac{\gamma \Delta B}{2}$$

Since we deal with the sum of the rates, the differences between T_2 and T_2^* are less pronounced for fast-relaxing tissue for a given field inhomogeneity. In general, imperfections of the magnetic field increase with the distance from the centre of the magnet (the so-called iso-centre or sweet spot; *see* **Note 12**).

Inhomogeneities can also be caused by exogenous para- or ferromagnetic material (see **Note 13**).

Changes in the local concentration of endogenous paramagnetic substances such as oxy-haemoglobin can be used to generate contrast (see corresponding **Chapters 8**, **27**, and **29** on the BOLD effect). An intrinsic source of local field distortions are the so-called susceptibility effects: even an otherwise homogenous magnetic field results in local inhomogeneities at the interfaces of anatomical structures when those tissues differ in their magnetic susceptibility (the ability of material to become magnetized by an external field). Air-tissue transitions especially can cause signal loss due to enhanced T_2^* relaxation in some sequences. We will see later how different MRI schemes have different sensitivities to the difference in T_2 and T_2^* relaxations.

1.4.3. Echo Formation: How to Use T_2/T_2^* Relaxation

Similar to the method of using longitudinal relaxation for generation of contrast by adjusting the repetition time of the MRI experiment, there is also a way to exploit transverse relaxation. The acquired signal is usually a so-called echo signal that arises from transient restoration of the phase coherence between the precessing transverse magnetization components. This process of refocusing and subsequent dephasing can be achieved either by a 180° rf pulse (so-called spin echo, SE) or a bipolar gradient (gradient echo, GE). In any case, the sequence timing allows generating the echo by enabling the rf pulse or gradient sooner or later after the transverse magnetization is generated by the initial rf pulse. The time between this initial pulse and the maximum of the echo is called echo time, TE (see **Fig. 4c**). Since the transverse magnetization decreases continuously in magnitude while precessing, a short TE conserves more detectable magnetization for echo formation than a long TE.

1.4.4. Manipulating T_2/T_2^* Relaxation: Contrast Agents

The fact that local field inhomogeneities enhance transverse relaxation can be exploited to enhance contrast by magnetic nanoparticles (also called negative contrast agents) that locally induce strong field gradients. Such nanoparticles are much larger than Gd ion chelates and their field gradients can extend into adjacent tissue even if the agent itself has no access into such areas (e.g. into the extravascular space while the magnetic particles reside inside a blood vessel). Water molecules experience accelerated T_2/T_2^* relaxation when diffusing through those gradients and the corresponding areas appear hypo-intense on T_2-weighted MR images. Hence, extravascular transverse relaxation can be influenced in neuroimaging, whereas extravascular T_1 relaxation remains unaffected as long as the blood–brain barrier is intact. Such substances usually possess superparagmagnetic properties (see **Note 14**) and can be loaded into cells for cell tracking (see below) or used as

1.4.5. Revealing "MR-Invisible" Spins

It was mentioned previously that molecular mobility influences T_2 relaxation. Protons in macromolecules with relatively restricted mobility or those water molecules that are themselves temporarily bound to macromolecules are characterized by very short relaxation times (hundreds of microseconds) that make them practically undetectable with conventional excitation-detection schemes. The fact that the pool of such immobile protons is linked to the pool of mobile, detectable water through chemical exchange and spin–spin interactions enables indirect detection of the immobilized protons. Since their resonance line spreads over several tens of kilohertz, it is possible to disturb their magnetization selectively with a frequency-selective rf pulse that does not directly affect the water resonance. The interaction between the two pools eventually causes a change in the acquired water signal, thereby making macromolecular contributions detectable (*see* **Fig. 5**). This method is called magnetization transfer (*see* below).

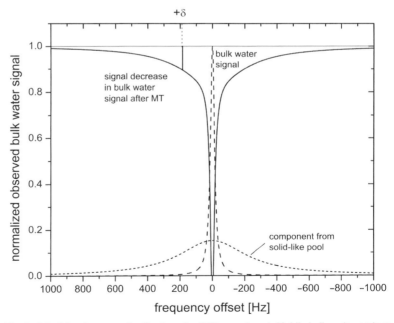

Fig. 5. Principle of a magnetization transfer (MT) experiment. Mobile bulk water protons give a narrow, directly detectable signal that is used for reference here at 0 Hz offset. Immobile protons form a resonance that cannot be detected directly due to fast T_2 relaxation. Application of a narrow bandwidth presaturation pulse affects the magnetization of immobile protons and causes a signal decrease of the bulk water signal (e.g. 10% in this case when a frequency offset is set to δ).

2. Choice of Encoding Scheme and Contrast Agent

2.1. Acquisition Schemes Exploiting Endogenous Contrast

All available modern MR scanners and spectrometers with imaging capability have a wide collection of pulse sequences that emphasize a certain contrast depending on the biomedical question to be addressed. As described above, T_1 contrast is adjusted by means of the repetition time whereas T_2 contrast can be influenced through the echo time. Hence, the applied pulse sequences have a certain range in which these parameters can be adjusted. Scanners usually come with software that also contains an easy to use graphical interface for adjusting TE and TR before executing the experiment. In its most basic version, a scanner should be equipped with the following sequences (the different acquisition schemes are discussed in more detailed in **Chapter 2** in this book and in (5, 10)):

2.1.1. Spin Echo (SE) Sequence

This is a general tool to generate images with either T_1, T_2 or proton density contrast. Although the classic spin echo sequence is too slow for some acquisitions, it has, however, some advantages that are still quite valuable in certain experiments: it is relatively insensitive to field inhomogeneities and other imperfections of the scanner system. Each excitation uses the entire magnetization reservoir, thus causing a need for relatively slow repetition to allow for sufficient T_1 relaxation. Formation of the NMR signal is achieved by refocusing the transverse magnetization with a 180° pulse which also eliminates contribution from local field inhomogeneities.

2.1.2. Gradient Echo (GE) Sequence

FLASH is a common protocol for this purpose. This type of sequence uses only small flip angles, thereby conserving the major part of the longitudinal magnetization reservoir and allowing shorter repetition times and fast MR imaging. The available magnetization reaches a steady state after a few excitations and the combination of low flip angles with fast repetition makes this scheme relatively insensitive to T_1 contrast even for TR $< T_1$. In order to avoid further disturbance of the longitudinal magnetization, the detected echo cannot be formed using a 180° pulse. It is therefore generated through a bipolar gradient that induces dephasing and refocusing of the magnetization. Because of the accelerated repetition, any remaining transverse magnetization is destroyed by so-called spoiler gradients at the end of the signal acquisition. Due to the lack of the 180° pulse, relaxation due to local field inhomogeneities is not eliminated and gradient echo sequences are therefore T_2^* sensitive.

2.1.3. Inversion Recovery Sequence

This scheme contains an additional 180° pulse for inversion of the longitudinal magnetization prior to the excitation pulse of the imaging sequence. This is followed by an evolution time, also called inversion time TI, to generate a certain constellation of longitudinal magnetizations in different tissues before the signal is spatially encoded either by a spin echo or gradient echo sequence. Since the additional inversion pulse is only linked to further T_1 effects, this sequence is mainly used for T_1-weighted imaging. Due to the inversion, the dynamic range of the longitudinal magnetization is in principle twice as big compared to the use of an excitation pulse only (*see* **Note 15**). One major feature of this type of sequence is the elimination of a specific tissue contribution when the corresponding magnetization has its zero-passage (*see* **Fig. 2c**).

2.1.4. Magnetization Transfer (MT)

This sequence must be capable of enabling a so-called saturation pulse several kHz off-resonant from the detected water resonance. Since the latter has a linewidth of usually less than 100 Hz in vivo, it remains unaffected by such a pulse, whereas the broad underlying signal of immobilized protons can absorb this rf irradiation and approaches a saturated state, i.e. a vanishing longitudinal magnetization. The corresponding signal decrease is eventually transferred onto the narrow water resonance. This technique requires the acquisition of a reference dataset with signal intensity S_0 and an MT dataset with reduced intensity S_1 due to the saturation pulse that is turned on for a few seconds. Quantification of the MT effect is done with the so-called magnetization transfer ratio (MTR):

$$\mathrm{MTR} = \frac{S_0 - S_1}{S_0}$$

This parameter is determined by the relative size of the two pools and the exchange rate between them, two quantities that can be altered by tissue pathology.

2.2. Contrast Agents

In order to amplify contrast that goes beyond the intrinsic relaxation behaviour of different tissues, contrast agents (CA) can be used (*see* **Note 16**). Such substances can improve both sensitivity and specificity. Unlike in X-ray imaging, the contrast agents are not detected directly. It is rather their influence on surrounding spins (i.e., water protons in most applications) and the related change in signal intensity that reveals areas that accumulate CAs over tissue that is not doped with such substances.

To address various problems in a flexible setup, the following types of contrast agents should be considered.

2.2.1. Positive Contrast Agents: Gadolinium Chelates

Depending on the desired application, different positive contrast applications are available. Since the relaxation effect is only transferred onto protons that have relatively close access to the paramagnetic centre, the distribution of the CA plays an important role. Two major groups can be distinguished:
- Extracellular unspecific agents
- Blood pool agents

The first group comprises low molecular weight agents that can exist as ionic and non-ionic molecules. The currently commercially available substances (provided, e.g. by Bayer Schering Pharma AG, Germany; Bracco Imaging, Italy; GE Healthcare, UK; Laboratoires Guerbet, France) are characterized by T_1 relaxivities of ca. 3–5 $s^{-1}mM^{-1}$ at body temperature and 1.5 T (11). They are of great importance for neuroimaging: because the integrity of the blood–brain barrier is compromised in lesions, these agents extravasate and accumulate in such areas whereas they remain in the bloodstream when passing through healthy tissue. The typical dosage ranges between 0.1 and 0.3 mmol/kg body weight (12). Helpful lists with additional properties are given in (11, 12). These agents are subject to rapid excretion through the kidneys (half life of ca. 45 min in humans; for rodents clearance may be much faster and be nearly completed after 30–60 min).

The second group differs from the first one by showing only very limited or no transition from the intravascular space into the extracellular fluid space. They are well suited for angiography and perfusion studies and are related to increase SNR due to the lack of extravasation. Agents of this type also often come with an increased relaxivity (up to 50 $s^{-1}mM^{-1}$) due to reversal binding to plasma proteins such as albumin (13).

2.2.2. Negative Contrast Agents: Superparamagnetic Particles

Various agents have been developed that consist of iron oxides and therefore possess superparamagnetic properties. In contrast to paramagnetic, positive agents, this class does not require close proximity or even complex formation between the water molecules and the agent. It is rather a long-distance effect that includes all water molecules that diffuse through the local field gradient. The iron oxide core is coated with polymers (e.g. dextran) or small molecules (e.g. silane, citrate) to achieve stability in biological materials/tissue (12, 14). Different vendors (e.g. Ferropharm GmbH, Germany; Miltenyi Biotec, Germany; Chemicell GmbH, Germany) provide various types that also differ in size. Particles in the micrometre range are classified as MPIO (micrometre-sized iron oxide), particles > 50 nm are classified as SPIO (superparamagnetic iron oxide), those below 50 nm as USPIO (ultra-small particles of iron oxide), and those around 10 nm as VSOPs (very small iron oxide particles). Large particles are phagocytosed rapidly by the reticuloendothelial system (RES) in the liver and the spleen. The smaller type of particles is mainly phagocytosed in lymph nodes and remains longer in the bloodstream than SPIOs (half-life of several hours in rats). Hence, the latter type can be used as a blood pool agent, whereas the former one is a more specific RES agent.

A list of the most common types of negative contrast agents is given in (12, 14). Their T_2 relaxivity ranges between 30 and 190 $s^{-1}mM^{-1}$ at 1.5 T and 30 and 205 $s^{-1}mM^{-1}$ at 7 T for water (both at 37°C); the typical dosage is 10–70 μmol Fe/kg. Some of these nanoparticles exist as a functionalized version (e.g. with an affinity for the transferrin receptor (15)). However, such special applications are beyond the scope of this chapter.

3. Contrast Selection with Basic Pulse Sequences

3.1. Essentials for Contrast Adjustment

In order to provide a quick overview with general concepts of contrast selection, **Fig. 6** summarizes the influence of TR and TE on signals from different tissues with different examples, independent of the sequence-specific details that will be given in the following sections.

3.2. Spin Echo Sequence Characteristics

By choosing the appropriate values for TR and TE, this sequence can be used to generate both T_1 and T_2 contrasts and proton density (PD, ρ) contrast. The signal depends as follows on these three parameters:

$$S = \varrho \left(1 - e^{-\frac{TR}{T_1}}\right) e^{-\frac{TE}{T_2}}$$

3.2.1. Achieving T_1 Contrast

Assuming comparable proton density, signal differences attributed to the different T_1 relaxations of two structures grow initially with TR and finally vanish again when both

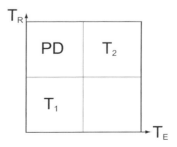

Fig. 6. Quick orientation for adjusting contrast by varying TR and TE.

magnetizations reach their thermal equilibrium. To achieve best T_1 contrast, choose a TR that is between the two T_1 values of the different tissues to be considered (e.g. *see* **Table 1** and **Fig. 4b**). To avoid T_2 contrast contributions, choose short echo times TE $<< T_2$.

3.2.2. Achieving T_2 Contrast

Similar to T_1 contrast, T_2 contrast is best when TE is adjusted to a value that is between the two T_2 values of the tissue (*see* **Table 1**). To avoid T_1 contrast contributions, choose a TR that is long enough ($\geq 3 \cdot T_1$) to allow for sufficient longitudinal relaxation.

3.2.3. Displaying PD

The influence of the proton density on the signal intensity cannot be changed. Hence, only the relaxation contributions can be eliminated to get contrast that is predominantly based on ρ. To do so, choose a long TR ($> 3 \cdot T_1$) and a short TE $<< T_2$.

3.3. Gradient Echo Sequence Characteristics

The gradient echo technique with low flip angles was designed for applications where images are acquired with TR $< T_1$ to achieve, e.g. better temporal resolution while still having the ability to work with T_1, T_2^*, and PD contrast. It has therefore one more parameter that influences the detectable magnetization compared to the spin echo sequence: the flip angle α, or more precisely, the combination of α and TR. The larger the flip angle, the more we depend on the TR to achieve sufficient recovery of the magnetization for the next acquisition. Hence, the T_1 contribution for the signal of the GE sequence becomes a function of α:

$$S = \varrho \frac{\left(1 - e^{-\frac{TR}{T_1}}\right) \sin \alpha}{1 - e^{-\frac{TR}{T_1}} \cos \alpha} e^{-\frac{TE}{T_2^*}}$$

3.3.1. Achieving T_1 Contrast

For low flip angles ($\alpha \approx 10°$), T_1 contrast can only be achieved for very short TR (ca. 20 ms). An increase in α will shift the maximum of T_1 contrast towards longer TR. The optimum flip angle for a given ratio TR/T_1 can be calculated and is called Ernst angle:

$$\alpha = \cos^{-1}\left(e^{-\frac{TR}{T_1}}\right)$$

In general, T_1 contrast improves with larger flip angles. This sets, however, some limitations to the minimum TR in order to regain reasonable magnetization for the next acquisition. Flip angles of $\alpha \approx 60°$ are therefore appropriate when combined with TR $\approx 20 - 300$ ms. Similar to T_1-weighted spin echo acquisition, the echo time must be short, TE $< T_2^*$, to eliminate differences in transverse magnetization.

3.3.2. Achieving T_2^* Contrast

In order to minimize T_1 contrast, an appropriate combination of α and TR has to be chosen. As we saw above, a small α allows for relatively fast repetitions. However, this will result in low SNR. A value of $\alpha \approx 15°$ combined with TR $= 50 - 500$ ms is sufficient to reduce T_1 contrast and emphasize T_2^* pronounced signal differences by using TE $= 15 - 40$ ms. It was mentioned above, that local field distortions can cause T_2^* relaxation that results in unwanted signal loss in some areas. Shortening TE helps to minimize such effects (*see* **Note 17**) in case they are unwanted. Otherwise, GE sequences can be explicitly used to visualize magnetic susceptibility effects as with superparamagnetic nanoparticles because this acquisition type is more sensitive to such effects than the SE sequence.

3.3.3. Displaying PD

This can be achieved by following similar criteria to those described above for the SE sequence. However, elimination of T_2^* contrast requires even shorter TE since T_2^* relaxation happens faster than T_2 relaxation. TR can be reduced to 50–500 ms when using flip angles $\alpha \approx 15°$.

3.4. Special Magnetization Preparation

The additional inversion pulse followed by the inversion time TI of this sequence reflects in an additional term for the generated signal intensity when compared to the SE sequence:

$$S = \varrho \left(1 - 2e^{-\frac{TI}{T_1}} + e^{-\frac{TR}{T_1}}\right) e^{-\frac{TE}{T_2}}$$

3.4.1. Inversion Recovery

It provides therefore three parameters (TR, TE, and TI) to manipulate contrast.

(a) *Achieving T_1 contrast*: Starting with negative magnetization, this sequence offers one special condition of T_1 contrast, i.e. the situation when one type of magnetization has its zero-passage. Many applications work with TR $\gg T_1$ and adjust T_1 contrast by means of TI. To eliminate the signal contribution of one type of tissue (e.g. otherwise intense contributions from subcutaneous fat), TI has to be adjusted depending on the T_1 of that very tissue according to

$$TI = T_1 \ln 2$$

Contrast between two non-vanishing signal contributions depends on the sequence type, i.e. if it is possible to identify the sign of the longitudinal magnetization. A so-called phase-sensitive sequence conserves the sign of the magnetization when it is converted into detectable transverse components. Two exponential recoveries starting with same amplitude do not intersect in this case (*see* **Fig. 3**). When evaluating only the absolute values, however, the two exponential curves initially decrease to zero before approaching the equilibrium value with the consequence of intersecting each other at one point. This mode is called magnitude reconstruction and it features a condition for destructive T_1 contrast, namely, the TI that corresponds to the intersection of the two curves. Good contrast is in general achieved for a TI that is in between the two T_1 values of the tissues to be distinguished. However, to avoid cancellation of T_1 contrast, TI should be chosen sufficiently long to pass the zero-passage of the slower relaxing component, T_{1s}, which is given by

$$\text{TI} = T_{1s} \ln \frac{2}{1 + e^{-\frac{TR}{T_{1s}}}}$$

The echo time is again chosen according to TE $<<$ T_2 to minimize T_2 contrast contributions.

(b) *Achieving T_2 contrast*: In contrast to the SE sequence, evolution during TE can start with different magnitudes of transverse magnetizations due to different relaxation during TI (*see* **Fig. 3**). In order to eliminate this effect, TI $>>$ T_1 has to be chosen (since we always have TR $>$ TI, the repetition time has no further relevant influence). The other way to eliminate T_1 contrast is to choose a short inversion time, TI $<<$ T_1, combined with a long TR $>>$ T_1. The desired T_2 contrast contribution is achieved by adjusting TE in between the T_2 values of the tissues of interest.

Displaying PD

This can be achieved by following similar criteria as described in **Section 3.2.3** for the SE sequence.

3.4.2. Ultra Short Echo Time (UTE) MRI

Echo times below 100µs can be achieved by combining short rf pulses with radial read out schemes. These so-called UTE sequences (16) allow for visualization of short T_2 components in tissue such as cartilage, tendons, or lung parenchyma. Although UTE techniques do not apply a special preparation of magnetization, they render spins detectable that would be invisible otherwise. The application of UTE has recently been suggested to be useful in combination with iron oxide–based contrast agents. Owing to the short TE, spins with shortened T_2 in the vicinity of the paramagnetic particles become detectable. At the same time, T_1-shortening of these spins enables very fast repetition. The resulting images show signal in the vicinity of the particles while tissue with normal relaxation times is strongly saturated and appears dark.

3.4.3. Magnetization Transfer

For this scheme, a relatively strong selective saturation pulse of usually a few microteslas in amplitude is turned on for a few seconds before each rf excitation of the following imaging sequence. The carrier frequency of the transmitter must be explicitly set off-resonant from the water proton frequency by a few kilohertz in order to avoid direct saturation. For the reference dataset, the saturation pulse is turned off while otherwise keeping the timing of the pulse sequence unchanged in order to have comparable relaxation conditions. The MTR is then calculated from the two datasets either for an entire ROI or pixel by pixel in terms of a parameter map. A high MTR indicates a large fraction of immobilized protons/macromolecular structures and/or a fast exchange with the mobile water protons and damaged tissue microstructure therefore causes a reduced MTR. The reader is referred to the corresponding application **Chapters 9** and **36** for more detailed information on MT experiments.

3.5. Parameter Maps

For the sake of completeness, it should be mentioned that tissue contrast can also be visualized by so-called parameter maps. These are datasets that consist of a whole series of T_1 or T_2-weighted images by sweeping through various settings for TR or TE, respectively. The change in signal intensity for each pixel is then fitted to the above-mentioned exponential functions to extract T_1 or T_2 for each element. The experimental details are beyond the scope of this introductory paper but are discussed, e.g. in **Chapter 25** of this book.

3.6. Relaxation Enhancement Contrast Agents

3.6.1. Gadolinium-Based T_1 Agents

This type of CA is usually given by i.v. or i.p. injection. Depending on the type of the agent, i.e. its characteristics of extravasation, it can be either used to highlight areas with increased vascular permeability or to perform angiographic studies. The former type of experiment is usually performed by acquiring an image when the CA concentration has reached its maximum level in the tissue of interest. Angiographic studies, however, can be enriched by so-called dynamic studies that monitor the bolus passage by taking several acquisitions, starting before administration of the CA and lasting until the washout of the paramagnetic substance. This allows revealing haemodynamic parameters, e.g. concerning the integrity of the blood–brain barrier (the reader is referred to the corresponding application chapters).

3.6.2. T_2 Agents: Iron Oxide Nanoparticles

Paramagnetic nanoparticles can be used for various purposes. Current applications include liver and gastrointestinal imaging, perfusion and blood volume studies, cell tracking, as well as macrophage labelling for displaying inflammatory lesions. Since physicochemical properties like size and coating influence the uptake by phagocytic cells, intracellular trafficking and metabolism, such aspects have to be taken into account for addressing different questions where i.v. injection is the method of administration. In vitro labelling of cells can be achieved through different routes of internalization (phagocytosis or pinocytosis, depending on the size of the particles). The blood half-life is dose dependent for all iron oxide nanoparticles because of saturation of macrophage uptake and it is generally significantly shorter in mice and rat (on the order of an hour) compared to humans (14). Applications like liver and lymph node imaging are based on reduced nanoparticle uptake in tumour tissue due to the lack of macrophages.

4. Notes

1. In spectroscopic imaging (SI), various metabolites can be detected by exploiting the different resonance frequencies of protons bound in different molecules. However, signal suppression for water and lipid resonances is necessary since the concentrations of the metabolites are ca. 5 orders of magnitudes lower. Spatial resolution is significantly lower than conventional, water-based MRI.

2. The extent to what this magnetization reservoir is reduced is quantified by the so-called flip angle which is the angle by which the longitudinal magnetization is tipped into the transverse plane. Maximum conversion is achieved by a 90° pulse that leaves no longitudinal magnetization. Further increase of the flip angle causes an increased amount of negative longitudinal magnetization (i.e. anti-parallel to the external field), thus a 180° pulse completely inverts the initial magnetization.

3. A good introduction into the complex parameter space of MRI in general, including various sample images, is also given on the following websites:
   http://www.cis.rit.edu/htbooks/mri "The Basics of MRI" by J. P. Hornak

   http://www.spincore.com/nmrinfo "NMR Information Server" by SpinCore Technologies, Inc.

   http://www.med.harvard.edu/AANLIB/home.html "The Whole Brain Atlas" by K. A. Johnson and J. A. Becker

4. As explained earlier in this book, the spatial encoding in MRI is done by applying additional magnetic field gradients that spread the resonance frequency of the water protons over a certain range and allow assigning each pixel in the image to a distinct resonance frequency. Hence, the

field of view together with the applied gradient strength gives the required receiver bandwidth.

5. A pure monoexponential recovery is in principle only observable for the relaxation between identical spins (2). However, most practical applications are not affected by this approximation.

6. This expression is still used for historic reasons, though it might be confusing in this context. It is not a real solid state component that has to be imagined for the relaxation process but rather a generalized thermodynamic reservoir that can exchange energy with the spins.

7. Saturation is achieved when an equal number of nuclei are in the two energy levels caused by the external magnetic field. The so-called population difference vanishes and so does the detectable magnetization. Repetitive rf pulses shuffle nuclei from the usually slightly over-populated lower energy level to the upper one. This process competes with relaxation and can significantly reduce the longitudinal magnetization that is needed for generation of detectable transverse magnetization.

8. The relaxivity of a substance is defined as the change in relaxation rate (given in units of s^{-1}) per concentration of that substance (given in millimolars). It is usually given in $s^{-1} mM^{-1}$.

9. For conditions of low SNR, repetitive acquisitions for signal averaging improve the image quality. The SNR increases with the number of acquisitions according to $SNR \sim \sqrt{N_{aq}}$.

10. ^{31}P MRI is often implemented in terms of spectroscopic imaging to visualize high-energy phosphorous metabolism or measure pH in vivo. The T_1 of ^{31}P nuclei in, e.g. phosphocreatine is ca. 3 s.

11. Nuclei other than ^1H have a smaller gyromagnetic ratio. This has several consequences, including a smaller detectable magnetization and a lower resonance frequency which is more difficult to detect as an oscillating signal caused by Faraday induction. Also, their concentration is much smaller than the ca. 90 M of water protons in vivo.

12. Depending on the construction of the magnet, the transition from the homogeneous field to areas with noticeable inhomogeneities can happen relatively close to the magnet centre. Vertical magnets that were specially designed initially for NMR spectroscopy and that are upgraded with micro-imaging accessories show a relatively fast drop-off of the magnetic field which might limit the suitable field of view to ca. 50 mm along the z-axis. Positioning slices over

13. a larger area of an object could sometimes require repositioning to minimize distortions.

13. Ensure that no para- or ferromagnetic materials that are needed to perform animal studies (e.g. injection needles) are brought close to the imaging area or even into the magnet bore. Replace the materials with non-magnetic items or keep them outside the safe 5 Gauss area of the magnet, even if you feel only a weak attractive force.

14. Superparamagnetism is a form of magnetism, which appears in ferromagnetic or ferrimagnetic nanoparticles. In such particles, magnetization can randomly flip direction under the influence of temperature. In the absence of an external magnetic field, their magnetization appears to be on average zero. In this state, an external magnetic field is able to magnetize the nanoparticles, similarly to a paramagnet. However, their magnetic susceptibility is much larger than that of paramagnets.

15. To distinguish eventually between parallel and anti-parallel magnetization in the image, the sequence needs to be phase sensitive, as described in **Section 3.4.1(a)**.

16. Because of its excellent intrinsic soft tissue contrast, MR was long believed to be a method that can abstain from exogenous contrast agents. However, in the 1980s, the potential of substances like gadolinium chelates was discovered and made available to increase contrast or reduce acquisition time.

17. This is one method to reduce the dephasing of spins per voxel which otherwise causes loss of signal. An alternative is to reduce the voxel size, e.g. by cutting down the slice thickness – provided the overall reduction of SNR does not impair image quality too much.

References

1. Kuo, Y.-T., Herlihy, A. H. (2008) Optimization of MRI contrast for pre-clinical studies at high magnetic field. In Webb, G. A. (ed.), *Modern Magnetic Resonance*, Vol. I, pp. 759–768. Springer, Dordrecht.
2. de Graaf, R. A. (2007) In vivo NMR spectroscopy – dynamic aspects, Chapter 3. In de Graaf, R. A. (ed.), *In Vivo NMR Spectroscopy, Principles and Techniques*. Wiley, West Sussex.
3. Neuberger, T., Webb, A. (2009) Radiofrequency coils for magnetic resonance microscopy. *NMR Biomed.* **22**, 975–981.
4. Baltes, C., Radzwill, N., Bosshard, S., Marek, D., Rudin, M. (2009) Micro MRI of the mouse brain using a novel 400 MHz cryogenic quadrature RF probe. *NMR Biomed.* **22**, 834–842.
5. Callaghan, P. T. (1991) *Principles of Nuclear Magnetic Resonance Microscopy*. Clarendon Press, Oxford.
6. Krause, W. (ed.). (2002) *Contrast Agents I – Magnetic Resonance Imaging*. Springer, Berlin.
7. Caravan, P. (2006) Strategies for increasing the sensitivity of gadolinium based contrast agents. *Chem. Soc. Rev.* **35**, 512–523.

8. Ugurbil, K., Adriany, K., Akgün, C., Andersen, P., Chen, W., Garwood, M., Gruetter, R., Henry, P.-G., Marjanska, M., Moeller, S., Van de Moortele, P.-F., Prüssmann, K., Tkac, I., Vaughan, J. T., Wiesinger, F., Yacoub, E., Zhu, X.-H. (2006) High magnetic fields for imaging cerebral morphology, function, and biochemistry. In Robitaille, P.-M., Berliner, L. (eds.), *Ultra High Field Magnetic Resonance Imaging*, pp. 285–343. Springer, New York, NY.
9. Lei, H., Zhu, X.-H., Zhang, X.-L., Ugurbil, K., Chen, W. (2003) In vivo ^{31}P magnetic resonance spectroscopy of human brain at 7 T: an initial experience *Magn. Reson. Med.* **49**, 199.
10. Reiser, M. F., Semmler, W., Hricak, H. (eds.). (2008) *Magnetic Resonance Tomography*. Springer, Berlin.
11. Huppertz, A., Zech, C. J. (2008) Contrast agents. In Reiser, M. F., Semmler, W., Hricak, H. (eds.), *Magnetic Resonance Tomography*, pp. 92–113. Springer, Berlin.
12. Burtea, C., Laurent, S., Van der Elst, L., Muller, R. N. (2008) Contrast agents: Magnetic resonance. In Semmler, W., Schwaiger, M. (eds.), *Molecular Imaging I*, Handbook of Experimental Pharmacology 185/I, pp. 135–165. Springer, Berlin.
13. Tóth, E., Helm, L., Merbach, A. E. (2002) Relaxivity of MRI contrast agents. In Krause, W. (ed.), *Contrast Agents I – Magnetic Resonance Imaging*, pp. 61–101. Springer, Berlin.
14. Corot, C., Port, M., Guilbert, I., Robert, P., Raynal, I., Robic, C., Raynaud, J.-S., Prigent, P., Dencausse, A., Idée, J.-M. (2007) Superparamagnetic contrast agents. In Modo, M. M. J., Bulte, J. W. M. (eds.), *Molecular and Cellular MR Imaging*, pp. 59–83. CRC Press, Taylor & Francis Group, Boca Raton.
15. Weissleder, R., Moore, R., Mahmood, U., Bhorade, R., Benveniste, H., Chiocca, E. A., Basilion, J. P. (2000) In vivo magnetic resonance imaging of transgene expression. *Nat. Med.* **6**, 351.
16. Bergin, C. J., Pauly, J. M., Macovski, A. (1991) Lung parenchyma: projection, reconstruction MR imaging. *Radiology* **179**, 777–781.

Chapter 4

Scanner Components

Volker C. Behr

Abstract

New developments in magnetic resonance imaging (MRI) are being achieved in two fields: methodological and technological innovations. This chapter will focus on the technological aspects of scanners, explain concepts, and give hints on how to deal with hardware-related issues. First, magnets used in MRI and gradient units will be introduced. Second, the radio frequency (rf) hardware will be described and explained. It has an often underestimated impact on imaging quality and can be improved by custom-built devices if one knows how to do it.

Key words: Magnets, shim coils, gradient coils, gradient amplifiers, drift compensation, pre-emphasis, rf coils, surface coils, volume resonators, phased arrays, B_1-mapping, field focusing, dielectric resonance, transmitters, receivers, preamplifiers.

1. Magnets

Almost any NMR experiment requires a magnet that offers a highly homogeneous field over the sample volume. For many experiments, homogeneities up to one part in 10^9 are required and—for whole body scanners—volumes of more than 100 l have to be covered.

Additional information on magnets can be found in (1).

1.1. Magnet Types

There are three basic kinds of magnets that are employed in NMR: permanent magnets, resistive magnets, and superconducting magnets.

Permanent magnets are often used in small and mobile systems, but also in low-field scanners for larger samples. They offer moderate field strength (typically on the order of 0.5 T) and

L. Schröder, C. Faber (eds.), *In Vivo NMR Imaging*, Methods in Molecular Biology 771,
DOI 10.1007/978-1-61779-219-9_4, © Springer Science+Business Media, LLC 2011

good stability, and require no special equipment. For long-term experiments, the ambient temperature has to be stable since the field strength drifts with temperature. Many permanent magnets today are put together from small very strong elements (typically made from some alloys containing neodymium, iron, or cobalt), in order to shape their field optimally. This allows assembling, e.g., single-sided permanent magnets (2) like the Halbach design (3), which can in advanced mechanical setups create magnets that allow changing their field strength continuously by moving parts with respect to each other (4).

The second group of magnets are resistive electromagnets. While being the workhorse of NMR up to the early 1980s, they are in most cases replaced by superconducting magnets today. Resistive magnets come with or without an iron core, depending on whether higher field strength or larger sample volume is required. Since—especially for iron core magnets—the field strength depends on the permeability of the iron inside the magnet which in turn varies with temperature, most of those systems come with feedback circuits that stabilize the field. Since resistive electromagnets rely on current flowing in its wires, a lot of heat is dissipated, which can reach several tens of kilowatts. To cool these systems, often additional water cooling has to be installed. The one application these magnets can still be found are experiments that want to ramp up and down their fields during the experiment. Sometimes, electromagnets are used in conjunction with other kinds of magnets for that purpose.

Finally, superconducting magnets are found in the majority of today's imaging and spectroscopic scanners. The superconducting wires are made of alloys of niobium, titanium, and tin embedded in a copper matrix. Since field strengths of more than 20 T require currents of several hundreds of amperes, until now, no high-temperature superconductors could be used. Therefore the magnet coil has to be kept at liquid helium temperature (approximately 4.2 K) or even below that. In the latter case, the vapor above the helium is actively pumped in order to reduce its pressure and further cool down the helium. To avoid high thermal losses, the helium chamber is surrounded by an evacuated space. Nevertheless, loss due to thermal radiation, which is proportional to temperature to the fourth power, would still cause too much helium boil-off. To counteract this, the inner vacuum chamber is placed inside a tank of liquid nitrogen which in turn again is surrounded by vacuum as depicted in **Fig. 1**. This greatly reduces the boil-off of liquid helium at the cost of comparably very inexpensive liquid nitrogen. To bring the magnet to field, it is hooked up to a power supply that connects over a small patch of the coil that is made non-superconducting. Once the final current is reached, the small patch is cooled down to superconducting temperature again, effectively shorting the power supply which

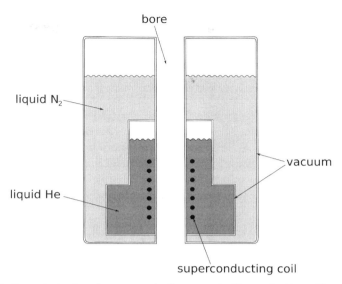

Fig. 1. Conceptual setup of a superconducting magnet with a vertical bore. The superconducting coils that cylindrically surround the bore are placed in a chamber filled with liquid helium. To avoid heat transfer to the surroundings, the chamber is enclosed in vacuum and to reduce radiative losses an outer compartment containing liquid nitrogen with another vacuum shielding is installed.

can be removed. Liquid nitrogen and liquid helium have to be refilled at regular intervals to keep the coils cooled down. Newer systems come with their own closed helium circuit that brings back boiled-off helium into the magnet and have maintenance cycles that are quite long. Should ever the liquid helium boil off and fail to cool the conductor to superconductivity, a quench would result from the heat dissipated in the now-resistive conductor. This poses a danger to people nearby due to the risk of suffocation in the escaping helium gas and to the system itself, since the superconducting wires can melt and be destroyed in the process, rendering the magnet useless. Also, ramping the field up and down is a lengthy procedure which is usually not being done for individual experiments. Nevertheless, superconducting magnets offer high field homogeneity paired with good stability and the highest available field strengths. This is the reason why they are the predominant kind of magnet in NMR spectrometers.

Due to the very high fields achievable with superconducting magnets, stray fields also increase and for several years now, manufacturers have been producing shielded systems, that essentially create on the outside a counter-field to compensate the stray field. This sacrifices a small amount of maximum field strength but allows integrating those magnets in environments that cannot offer sufficient room to place these magnets far away from all other equipment and thus increases safety by reducing the risk of

1.2. Shim Coils

accidental entry into the magnet's stray field with ferromagnetic objects.

Since most magnets (of either kind) do not offer sufficient homogeneity for NMR studies, systems are equipped with shim coils. Originally, the word *shim* referred to small pieces of metal that were placed close to the magnet to optimize its field. This is still done in some magnets as a first step of homogenizing the field. For superconducting magnets, the most important shims are a set of superconducting shim coils—cryo-shims—that are adjusted when installing the magnet to make the field as homogeneous as possible. A set of room-temperature shim coils is common to all kinds of magnets. These coils are used to fine-tune the system once a sample is in place. For whole body systems, this process usually is fully automated and uses shim directions of the first or at most second order. Especially in high-field systems that are employed for spectroscopy and imaging, one can adjust those shims and even higher order shims by hand. Shimming the systems well will lead to narrow linewidths, which is also important in spectroscopic and microscopic imaging. Two more tasks that are performed by the room-temperature shim coils are drift compensation and frequency locking. A drift compensation adds some additional field strength parallel to the magnets own field to compensate for the unavoidable losses that occur in superconducting systems, thereby keeping the Larmor frequency stable over longer periods of time. Experiments that are very sensitive to off-resonances will require even better stabilization during the experiment, which can be provided by a frequency lock. This depends on a specially prepared sample containing ^{2}H, whose Larmor frequency is monitored independently of the experiment and any frequency changes are counteracted by a feedback circuit powering up the room-temperature shims accordingly.

For some systems, e.g., small mobile systems, there are no separate shim coils or only a reduced set of separate shim coils and shimming is done by the gradient coils (*see* **Section 2**).

2. Gradient Coils

For all imaging applications, magnetic field gradients are required to encode the spatial information. Usually, gradient coils consist of three sets of coils that can create gradients in three orthogonal directions. By superimposing those fields one can achieve any desired gradient direction. It is important to keep in mind that all gradients always create only additional field along the main field axis. It is that field's amplitude that varies with the gradient's

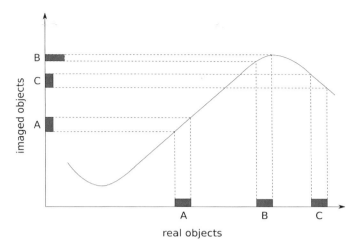

Fig. 2. Spatial encoding inside and outside the linear regime of a gradient (gradient values are depicted by the solid curve). Object A is encoded in the linear regime of the gradient and mapped to its image properly. Object B resides close to the maximum of the gradient at a smaller slope than in the linear regime. In the image, it appears squeezed and with very high intensity. Finally object C is located even beyond the maximum of the gradient; though its intensity appears almost correctly, it is displaced and flipped.

direction being zero in the center of the system and maximum at its borders. Of course, gradient coils cannot maintain a linear field ramp over infinite distance. At least along the bore of the magnet, one might experience artifacts due to the non-linearity of gradients if the imaged sample extends too far outside. Then, the spatial encoding will fail and objects will appear distorted or even in entirely wrong positions as illustrated in **Fig. 2**.

Gradient coils are characterized by the maximum achievable field difference over the sample, which they can create. For most whole body systems, gradient strengths are on the order of several tens of milliteslas per meter while for high-field systems for microscopic imaging, even a few teslas per meter can be reached.

An ideal gradient should be the shape of a boxcar function, that is, it should switch on and off in zero time and be stable at its set value. In reality, gradients can be switched on a scale of a fraction of a millisecond, but they will not immediately, once they reach their maximum value, be flat but overshoot by some amount and oscillate for a short period in time. This is due to coupling in other elements close to the gradient coils (e.g., the shim coils) and also due to limitations of the gradient's power amplifiers. To counteract these unwanted distortions, systems allow setting a pre-emphasis, which is essentially a kind of pre-distortion that is overlaid over the desired gradient currents to compensate for the aforementioned effects. This pre-emphasis has to be adjusted for each gradient unit and system individually to obtain maximum efficiency. In addition, gradient coils today are built as shielded coils to avoid unwanted coupling into other parts of the systems and further reduce distortions.

More considerations on gradient systems can be found in (1).

3. RF System

The rf system in MRI has to fulfill very special requirements: on the one hand, during transmission, it has to handle voltages of up to several hundreds of volts and on the other hand, during reception, it has to be sensitive to smallest signals down in the range of micro- or even nano-volts (5). The first two sections will comment on the transmit and receive chains in general. After that, in far greater detail, probe bases and rf coils will be discussed including hands-on approaches on how to manufacture and improve these components. Finally a method on how to characterize rf coils in the scanner will be presented.

Suggested reading on rf systems include (1, 6–8).

3.1. The Transmit Chain

3.1.1. Hardware

The transmit chain of a modern MRI system consists of a frequency source and a power amplifier. Usually, highly stable synthesizers are employed as frequency source, which provide the reference frequency and phase for the measurement. A very high phase stability is essential for being able to correctly time pulse sequences. The synthesizer feeds through a phase shifter and an rf gate into the power amplifiers of the MRI system. These amplifiers handle transmit powers up to several kilowatts for whole body scanners. Most systems will come with at least one power amplifier for ^{1}H measurements. This amplifier can—due to the gyromagnetic ratios of ^{1}H and ^{19}F which differ less than 10%—usually also handle transmission for ^{19}F. For all other X-nuclei, different transmitters can be built into the system. These will not reach the high frequencies of ^{1}H and ^{19}F but will offer highly linear amplification for other nuclei.

While transmit power is for most systems not a critical resource, one should keep in mind that for coaxial cable losses are non-negligible. For RG 58, a commonly used cable type in MRI, with a wave impedance of 50 Ω, the attenuation at 300 MHz is approximately 0.3 dB/m, which means that at a distance of only 10 m, half the power will dissipate in the cable. Attenuation will get worse with higher frequencies. So in order to characterize a transmit chain one should measure the power at its end where it connects to the probe. How to perform such measurements is described in **Section 3.1.2**.

One more issue to address in the transmit line is noise from the transmitter during reception. Transmitters—even when

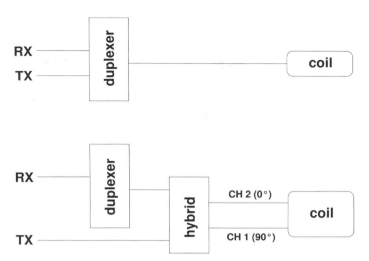

Fig. 3. *Top*: single coil connected via duplexer to transmitter and receiver. *Bottom*: quadrature coil connected to a hybrid. The transmitter is directly connected to the hybrid and therefore decoupling circuits within the duplexer will not block any transmitter noise during reception.

blanked—will still yield low noise on their outputs at the Larmor frequency. Commercial systems usually offer decoupling switches in their hardware that will separate the transmitter from the receive line during acquisition but the use of power splitters in order to drive several transmit coils with a single transmitter or quadrature hybrids (*see* **Section 3.2**) often bypasses those switches as shown in **Fig. 3**. In that case, one has to add a decoupling mechanism to the transmit chain as described in **Section 3.1.3** to avoid additional noise from the transmitter.

3.1.2. Characterization of the Transmit Chain

To perform such measurements in one's own lab, one will need a directional coupler, an oscilloscope, which does not need to be able to resolve the actual transmit frequency, a couple of attenuators, and 50-Ω loads. Connect the transmit cable as shown in **Fig. 4** to one input of the directional coupler, the through path goes into an attenuator which can handle the full transmit power—e.g., a liquid-cooled attenuator or an actual rf coil. In any case, make sure to terminate this line with a 50-Ω load. The fork with only a small fraction of the power goes into the oscilloscope. Make sure the split part is sufficiently low not to ruin the oscilloscope; one may need to inset an additional pass-through attenuator. Finally the fork for the returned power is terminated with a 50-Ω load. Since there should be a good 50 Ω at the end of the line, there should be only very little reflected power. Now one can measure, for different transmit settings the power at the end of the transmit chain and depict it in a graph to characterize the power output.

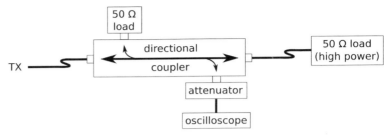

Fig. 4. Setup for measuring the output power at the end of the transmit chain. Via the bidirectional coupler a well-defined fraction of the forward power is redirected from the line into the oscilloscope. The remaining forward power is dissipated in the 50 Ω load at the end of the setup. The 50 Ω load that would receive a fraction of the reflected power is in place to properly terminate the setup.

3.1.3. Decoupling of Transmit Lines During Reception

The transmitter can be decoupled from the reception line by a passive element which is simply plugged into the transmit line. Its basic layout is drawn in **Fig. 5**.

During transmission, due to the high transmit power, all diodes will be switched through. In that case, the resonant circuit formed by the inductor, the capacitor, and the diodes will be off-resonant and the rf can pass unhindered. The diodes at the end of the quarter-wavelength cable will form a short which in turn is transformed by the cable to an open end, rendering this part of the setup "invisible" during transmission. During reception, the signal power will be too low to switch through the diodes. Now those diodes in combination with the capacitor and the inductor form an on-resonant circuit thereby effectively blocking any rf noise from the transmitter. From the receiver's point of view,

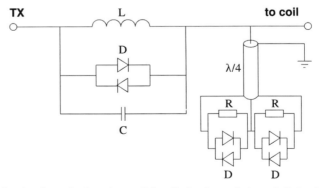

Fig. 5. Passive decoupler for a transmit line. During transmission, all diodes D will be switched through due to the high power level. The resonant circuit formed of the inductance L, capacitance C, and the diodes D will be off-resonant and the short at the end of the λ/4 line will be transformed to an open end. This allows full transmission of the pulses. During reception, the resonant circuit will be on-resonant, thereby blocking any noise from the transmitter with high impedance. The two resistors R at the end of the λ/4 line yield a proper 50 Ω termination for the coil's side of the system.

there will be an open end. The diodes at the end of the quarter-wavelength cable will also be blocking and the additional resistors load the cable at its end with 50 Ω. Since it is a 50-Ω wave impedance cable, these 50 Ω will be also visible at the other end of the cable, thereby properly terminating the system from the receiver's side.

3.2. The Receive Chain

3.2.1. Hardware

The very first element of the receive chain is a low-noise preamplifier. This amplifier is—especially in phased arrays (*see* **Section 3.3**)—sometimes included in the coil but may also be found as a integral part of the receive chain itself. This preamplifier is the one which determines the noise of the entire system, since it will get the very low signal from the detector and amplify it adding its own noise that may be comparable in magnitude. Any later stage of preamplifiers will already see a higher input signal that outweighs their own noise by far. To determine the intrinsic noise that comes from the receive chain, a simple experiment as described in **Section 3.2.2** can be performed. Since this first preamplifier is built to handle very low input voltages on the order of micro- or even nano-volts, it has to be protected during transmission to avoid high voltages from burning the preamplifier. This is done by a so-called duplexer, which either with active or passive circuitry, separates transmit from receive chain. A more advanced version of such a duplexer is a quadrature hybrid that can be used with quadrature coils (*see* **Section 3.3**). Some considerations on these hybrids and their layout will be discussed in **Section 3.2.3**.

Since acquired data are processed by a computer, they need to be converted from analog signals to digital ones. This is done by an analog-digital-converter (ADC), which can typically handle frequencies up to several hundreds of kilohertz or a few megahertz. Therefore, the original signal at the Larmor frequency of several tens or even hundreds of megahertz has to be modified in order to fit into this regime. Downsampling is done by mixing the high-frequency signal with a reference signal from the synthesizer mentioned already in **Section 3.1**. A schematic setup of the receive chain is displayed in **Fig. 6**. As shown, the signal is not only mixed with the reference frequency but also with a 90° phase-shifted reference. Mixing the signal with a reference will yield a new signal at the frequency difference between the original Larmor frequency and the reference that is on the order of kilohertz and thus well within the limits of an ADC.

$$\omega_{\text{mixed}} = \omega_0 - \omega_{\text{ref}}$$

Nevertheless, having just one mixed signal will lead to ambiguities in case there are frequencies present in the spectrum that are slightly below and others that are slightly above the reference.

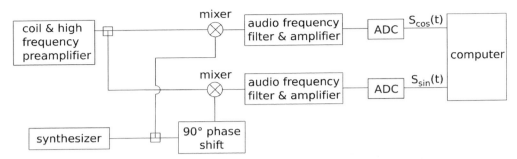

Fig. 6. Schematic drawing of the receive chain for quadrature detection. The signal from the coil is split up and mixed with two reference signals that are 90° out of phase. After mixing, the signal is sampled down to the audio range and can be further filtered and amplified. After digitizing the signal in the ADC, two signals that constitute the real and imaginary part of the full signal are passed to the computer for further processing (e.g., Fourier transformation).

These will not be distinguishable by fitting the data to the following expression for the signal strength in a free induction decay (FID):

$$S_{\cos}(t) = A \cdot \cos(\omega_{\text{mixed}} t) \cdot e^{-\lambda t}$$

Using the second signal mixed with the phase-shifted reference, one will obtain additional information on the real signal:

$$S_{\sin}(t) = A \cdot \sin(\omega_{\text{mixed}} t) \cdot e^{-\lambda t}$$

Now, frequencies below and above the reference are distinguishable and, in addition, one gains a factor of $\sqrt{2}$ in SNR. This can be easily seen when interpreting the two above expressions as the real ($S_{\cos}$) and the imaginary ($S_{\sin}$) part of a complex signal:

$$S(t) = S_{\cos}(t) + i S_{\sin}(t) = A \cdot e^{i \omega_{\text{mixed}} t - \lambda t}$$

Now one gets information from the full signal including its phase compared to having only one linear component, which corresponds effectively to two averages yielding the above mentioned $\sqrt{2}$ in SNR. For this reason, this kind of detector is called a phase-sensitive detector (PSD) or quadrature detector (do not confuse with quadrature coils!) and today is standard equipment in all NMR spectrometers.

If the two channels of a quadrature detector are imperfectly balanced, i.e., they vary in amplitude or are not at exactly in 90° phase of each other, one will experience quadrature artifacts in the image or spectrum by additional components of the counter-rotating field in the laboratory frame. These artifacts will appear at minus the frequency of the real signal. To avoid these, one has to properly adjust the two detector channels or use phase-cycling techniques in the experiments.

3.2.2. Determining the Noise Figure of a Receive Chain

It is important to know the noise figure of the system's receive chain to be able to judge what the minimal signal detectable will be and validate the receiver chain operates at full performance in case one has to track down a source of noise in experiments. Therefore, this experiment should be performed as soon as possible after setting up a new system, since once it is damaged, one cannot know what should have been the correct noise figure unless measured beforehand.

To measure the noise figure, a "hot–cold" measurement with a noise source is performed. As noise source, a 50-Ω terminator connected to a coaxial cable can be employed. Make sure it is not a carbon resistor since those will have significant temperature dependence in their resistance—metal film resistors are suited well for this experiment. Furthermore, a small dewar with liquid nitrogen will be needed to cool down the terminator.

To obtain the noise figure, two measurements are necessary: for the first one, the terminator at room temperature is connected to the receive chain instead of an NMR coil. Now a spectrum at the frequency of interest with a sweep width of 10 kHz or more is acquired at a receiver gain setting that fills the receiver at least 50% with noise. Then the measurement has to be repeated with identical settings and with the terminator cooled down to the temperature of liquid nitrogen. Next, the spectra are Fourier transformed without applying any line broadening and the root mean square (RMS) of the noise is computed. Any scaling or normalization constants during processing have to be identical in both cases.

Each noise source—in this case the terminator—has a so-called excess noise ratio (ENR) that is defined as

$$\text{ENR} = \frac{P_H - P_C}{P_H}$$

where P_H is the noise power generated at high temperature (which is room temperature in this case) and P_C the noise power at low temperature (liquid nitrogen). Taking into account that $P = kT\Delta f$ where k denotes the Boltzmann constant, T the temperature, and Δf the bandwidth, all powers in the above equation can be replaced by the corresponding temperatures. To calculate the noise figure, which represents the reduction in the excess noise by the device under test, one has to divide the ENR by ratio of the noise powers of the device:

$$\text{NF} = \frac{\text{ENR}}{\frac{\text{RMS}_H^2 - \text{RMS}_C^2}{\text{RMS}_H^2}}$$

Typically, the noise figure is given in decibels rather than linear units, which leads to the following equation:

$$\mathrm{NF_{dB}} = 10 \cdot \log\left(1 - \frac{T_\mathrm{C}}{T_\mathrm{H}}\right) - 10 \cdot \log\left(1 - \frac{\mathrm{RMS}_\mathrm{C}^2}{\mathrm{RMS}_\mathrm{H}^2}\right)$$

For an intact spectrometer, this value should be in the range of 1 dB up to a maximum of 3 dB. If the noise figure cannot be reproduced or is too high, a closer check of the receive chain is highly advised.

The most critical parts are those closest to the end of the chain: if the preamplifier is not working correctly it will degrade the noise figure of the entire chain. The same holds true for parts of the duplexer or any filters in front of the preamplifier. All those components should have noise figures significantly below 1 dB. To check filter, duplexer, or residual transmitter noise (*see* **Section 3.1.3**), one can simply remove these parts and redo the measurement. If the noise figure improves substantially this narrows down the places to look for the issue. To check the preamplifier, one can either use a different channel on multi-channel systems or test with a different nucleus that has its own preamplifier. An improvement in noise figure hints at issues with the preamplifier.

3.2.3. Quadrature Hybrids

Most systems are shipped with more than one receiver channel. Thus, array receive coils (see **Section 3.3**) can be directly attached to the individual receivers. Nevertheless, this means each coil will generate its own dataset, all of which have to be combined afterwards in the postprocessing. If SNR is low, one might desire to improve the gain of single-channel coils, too. Additionally, only a few systems come with multiple transmit channels and so multi-element transmit–receive coils cannot be used directly.

One approach from the time when only single-channel systems were available still can greatly improve SNR: not only using one linearly polarized coil but two orthogonal ones with a 90° phase shift. This will create a circularly polarized field and yield a factor of $\sqrt{2}$ in SNR. To use these, one will need a splitter and combiner for the signal, first in transmission and then back in reception. Furthermore, this hybrid also has to act as duplexer, meaning, it has to separate the transmit and receive chains.

Such a setup is called quadrature hybrid (in combination with a quadrature coil, *see* **Section 3.3**). This must not be confused with quadrature detection as described in **Section 3.2.1**, which can also yield a SNR increase of a factor of $\sqrt{2}$ independently.

There are several realizations of quadrature hybrids consisting of discrete elements (capacitors, inductors) but building it out of cable is better suited to explain its function. The layout

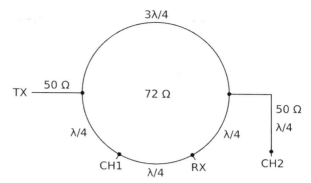

Fig. 7. Layout of a quadrature hybrid. Signal coming from the 50-Ω transmit line (TX) enters a ring consisting of 72-Ω cable. During transmission, constructive interference occurs at channel 1 (CH 1) and channel 2 (CH 2) at phases of 90° and 270°, respectively. The additional 50-Ω delay line at CH 2 brings the signal there to 0°, yielding two channels with 90° phase offset. The receiver is protected by destructive interference. During transmission, in reverse, constructive interference occurs only at the output to the receiver (RX), recombining the two signals with proper phases.

of the hybrid is shown in **Fig. 7**. To understand the hybrid, one has to look at the signal phases for transmission and reception. During transmission, signal from the transmit port (TX) can (and will) travel two ways to reach the other ports. Calculating the phases relative to the reference 0° at the TX port, the phases at CH1 are 90° and 450° (=90°), and those at CH2 are 270° each, which will interfere constructively. The additional λ/4 delay line at CH2 will bring the signal to a phase of 360° (=0°) and thereby the two channels give an output with a 90° phase shift as required for a quadrature coil. The receive port (RX) will during transmission see phases of 180° and 360°, which will lead to destructive interference and thereby decouple the transmitter from the receiver. Switching to reception, to CH1 the coil will deliver a signal with a phase of 90°, to the delay line at CH2 with a phase of 0°. Doing the same calculations as before, one will see that now there is constructive interference with two times 180° only at RX; while at TX, CH1, and CH2 interference will be destructive. All detected signal will go to the RX port as desired. To test the quadrature hybrid, one can do measurements with a network analyzer. While transmitting into TX, there should be very low signal at RX (significantly below −60 dB to protect the receiver) and signal with half its amplitude (i.e., −3 dB) at CH1 and CH2 with phases of 90° and 0°, respectively. Transmitting into RX (to simulate reception) or into one of the channels can test the other signal pathways. Alternatively, one can do a quick test at the scanner: the hybrid will work with quadrature coils that create a circularly polarized magnetic field. If one connects the two channels of the coil "the wrong way", the generated field will be counter-rotating to the

Larmor precession and thereby be off-resonance by twice the Larmor frequency. Doing an image with this setup should yield (ideally) noise only. Connecting only one channel should yield a factor of $\sqrt{2}$ more current required for the same flip angle as compared to proper quadrature operation. Since many spectrometers only show the power used to form a pulse, keep in mind that a factor of $\sqrt{2}$ in current corresponds to 2 times the power.

3.3. Coils and Resonators

In any NMR experiment, a key role is played by the coil or resonator. It creates the B_1 field during excitation and receives the signal during acquisition. Depending on the experiment, one will find either separate transmit and receive coils as is the case in most whole body scanners or one coil that does the full job as typically found in high-field setups for microscopy. The advantage of having different coils during transmission and reception is the ability to optimize each coil for its purpose. A transmit coil is expected to create as uniform as possible a field over the sample; often the entire sample, e.g., in case experiments are to be done that require inversion pulses applied to the entire sample as preparation. On the other hand, the receive coil has only to be sensitive to the region of interest, which in most cases is only a small part of the sample. Having a smaller coil for reception will increase the SNR in two ways: first, it can be located closer to the region of interest. Second, it will be insensitive to noise coming from other regions of the sample. In systems with small bore sizes, as typically found in high-field applications or mobile systems, often due to spatial constraints, only one single coil can be placed that has to meet both requirements mentioned above as closely as possible.

3.3.1. Coil Designs

Coils are designed to meet the criteria of a specific kind of experiments or even one single experiment. Building general-purpose coils can be feasible at low frequencies and large bore size-to-sample volume ratio but almost impossible at very high frequencies and small bore size-to-sample volume ratio. In any case, it is a good idea to use optimized coils for experiments. This will lead to a good filling factor, i.e., the sample will occupy a high fraction of the coil's sensitive area thereby reducing noise from outside the sample. Additionally, the sensitivity of the coil can be optimized for the region of interest.

In general, there are two kinds of coils: surface coils and volume coils, also called volume resonators.

A surface coil can be as simple as a single planar resonant loop. This will offer very good sensitivity on the surface of the sample at the cost of lower sensitivity deep within the sample and at the cost of field homogeneity. A surface coil creates a very inhomogeneous field that strongly depends on the distance from the loop. The most common application of such coils is in combination with a larger transmit coil as local detector for small regions of interest. To accommodate for sample geometry, those coils can also be non-planar or for specific applications, advanced designs like microstrips (9) or hole-slotted coils (10) can be used.

The second kind of coils is comprised of volume resonators. The basic requirement for a volume coil is to generate a homogeneous field over the sample for excitation and also, in setups as mentioned before, for reception. The intuitive approach to create a uniform and strong field in a given volume would be to enclose this volume with a solenoid coil. Indeed, this setup is used in many open and portable systems. Nevertheless, superconducting magnets with a field along the magnet's bore require rf coils that create fields orthogonal to the bore. Using solenoids here would severely hamper loading and unloading the coil in the magnet and put strong restraints on the maximum length of the sample. The theoretical solution to the question how to create a homogeneous rf field perpendicular to the bore of a cylinder is an infinite conducting cylindrical surface with a sinusoidal current distribution on its surface. This idea is realized by reducing the cylinder to a discrete set of rungs connected by two common end-rings (11, 12). For its likeness, the setup is called a birdcage and, depending on the location of the capacitors, either called high-pass (capacitors in the end-rings only), low-pass (capacitors in the rungs only), or band-pass (capacitors in both end-rings and rungs). Looking at the eigenmodes of this resonant structure, one can identify two modes that fulfill the condition of sinusoidal current distribution: one shaped like a sine, the other orthogonal like a cosine (**Fig. 8**). These two modes create perpendicular fields with high homogeneity within the resonator. The high-pass design offers the additional benefit of having no capacitors close to the center of the resonator where usually the sample is

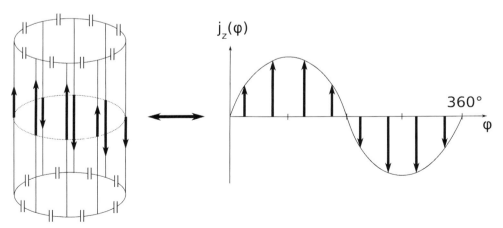

Fig. 8. Schematic drawing of an eight-rung high-pass birdcage resonator (*left*) and its current distribution (*right*). The resonant mode forms a sinusoidal current distribution over the rungs thereby creating a homogeneous field inside the resonator. An independent orthogonal mode can form at a 90° phase offset following the form of a cosine.

located; thereby the coupling of the electric fields close to the capacitors into the sample is reduced—reducing noise via that pathway. The two modes of a birdcage can be used to build a quadrature coil that uses two orthogonal linearly polarized fields to create a circularly polarized field (see **Section 3.2.3**). Another option is to use these two modes for double-tuning the birdcage to two frequencies in X-nuclei experiments (13). Similar in design are resonators as the saddle coil with only 4 "rungs" or the predecessor of the birdcage, the Alderman-Grant resonator (14). A more recent design, the Litz coil or Litzcage, uses the mutual capacitance of overlapping conductors in order to design optimally adjustable coils for very high frequencies (15).

As a result of trying to combine high local sensitivity of surface coils with good sample coverage of volume resonators the NMR phased array was developed (16). Originally designed as receive-only coils by now, especially in very high field whole body scanners that do not have a body coil of their own, phased arrays are used in transmission as well. A phased array basically consists of a number of individual surface coils that cover a larger area of the sample or even encompass it fully. One can now get information of each of the elements that correspond to the information obtained by a surface coil in the respective spot. Combining all the data accounting for the phase differences between the elements yields images of larger portions of the sample. Important for all phased arrays is a good decoupling of the individual elements to get uncorrelated noise information and thereby be able to improve SNR. Since knowledge of the individual element's sensitivity can give an additional source of information, these arrays can also be used to speed up imaging by acquiring only parts of the k-space and reconstructing the missing parts with the help of the additional coil information (partial parallel acquisition (PPA)) (17, 18).

3.3.2. Coil Characterization

There are different ways to characterize coils. On the workbench, one can determine a resonant setup's Q-factor, a measure of the energy stored in the resonator in relation to the dissipated energy. This factor can be written as

$$Q = \frac{X}{R} = \frac{\omega_0 L}{R} = \frac{\omega_0}{\Delta \omega}$$

with X being the reactance, R the resistance, and L the inductance of the setup. ω_0 denotes the frequency of resonance and $\Delta \omega$ the full width half maximum at ω_0 as measured in a transmission experiment. For such an experiment for a transmit coil, a network analyzer (or an oscilloscope and a frequency generator) are required. The coil is connected to the rf output and via a "pick-up" loop (19), the transmission from the coil is detected.

A high absolute value of Q relates to a very effective coil, generating high fields—and in turn being very sensitive (20)—but at the cost of long ring-down times. So for some applications that need the transmit coil to quickly ring down after excitation, it might be desirable to reduce the Q on purpose, e.g., by adding a resistor to the circuit. Anyhow, the more important figure is the drop in Q once the coil is loaded, since this figure relates to the noise generated by the coil and by the sample. A Q-drop of 50% indicates equal contribution of noise from the coil and from the sample. For coils that are large compared to the sample, it is hard to achieve good Q-drops but as a rule of thumb, reducing the noise generated by the coil becomes negligible once a Q-drop of a factor of 4 or 5 is reached.

An alternative way to characterize coils is via mapping their B_1 fields. A method for doing this is presented in the following section.

3.3.3. B_1-Mapping

B_1-mapping, also known as rotating field imaging is a technique to characterize the field homogeneity of an rf coil. Instead of information on the sample, information on the rf distribution across the sample is displayed. The technique is called rotating field imaging, since the precession of the magnetization around the direction of B_1 is measured, which can be easiest depicted in the rotating frame (21, 22).

While the maps can be created using the sample that is to be examined in a study employing one specific coil, to get an idea about the general homogeneity of the field distribution of a coil, one should use a homogeneous sample containing a high spin density in order to get ample signal. The relaxation properties of the phantom should be chosen according to the imaging method used (see below). As far as the scanner itself or the coil is concerned, no special requirements beyond a regular setup for an NMR experiment are required.

3.3.3.1. Procedure of B_1-Mapping

A rotating frame experiment consists of a preparation module followed by an imaging module. During the preparation, a nonselective pulse of fixed power but variable duration is applied, followed by strong spoilers in all three directions to eliminate any FIDs resulting from this pulse. After this preparation, an image is acquired. For this image, one can either use a fast gradient echo sequence (FLASH) (23), to get the image in a single shot. For this to work, T_1 of the phantom has to be long compared to the acquisition time of the image to avoid relaxation between preparation and acquisition. In case this cannot be fulfilled for the chosen phantom, the alternative is to use a spin echo sequence (24). With an individual preparation for each individual line in k-space (or at least for a group of lines in k-space that again fulfill the above mentioned condition), this sequence is more robust than the FLASH sequence at the cost of increased experimental time. Another advantage of the spin echo technique is that effects of inhomogeneities in B_0 will be eliminated. In case of even more rapid relaxation in the phantom, single-point imaging techniques can be used at the cost of highly increased measurement time. Prior to each repeated preparation, a delay has to be applied, which is sufficiently long for the phantom to be fully relaxed (approximately $5T_1$). This experiment is repeated several times with different flip angles in the preparation pulse—the flip angles should cover a range from almost 0° up to 720° or even more. Since the coil will not generate a completely uniform field distribution, in each of the images one will see variations in signal due to different flip angle caused by the preparation pulse.

In the full set of images, the intensity of individual pixels will vary due to the different flip angles the pixel has seen during preparation. These flip angles are proportional to the duration of the preparation pulse. Fitting the data of each pixel in all images with the function

$$I(t) = A \cdot |\sin(\omega_1 t + \Phi)|$$

with $I(t)$ being the intensity of a pixel, t the duration of the preparation pulse, and A, ω_1, and Φ the fitting parameters for maximal signal amplitude, precession frequency of the magnetization around B_1, and initial phase, respectively. The map obtained by this will show the distribution of field strength with high field strength corresponding to high precession frequency. To get to absolute values of the field, the Larmor equation has to be applied:

$$B_1 = \frac{\omega_1}{\gamma}.$$

3.3.3.2. Notes on B_1-Mapping

1. To obtain a stable fit, it is recommended to acquire 2–4 half-periods of the absolute sine function with at least 10 data points per arch.
2. Especially at very high field strengths (10 T and above) special care has to be taken not to form standing waves in a phantom, i.e., the wavelength inside the medium should be large compared to the dimensions of the phantom. Therefore, phantoms should be chosen to have low values of relative permittivity. Water with an ε_r of 81 is not a good choice. The effects of standing waves on field homogeneity maps can be seen in **Fig. 9**. Vegetable oils make good phantoms with low permittivity and still sufficiently long T_1.

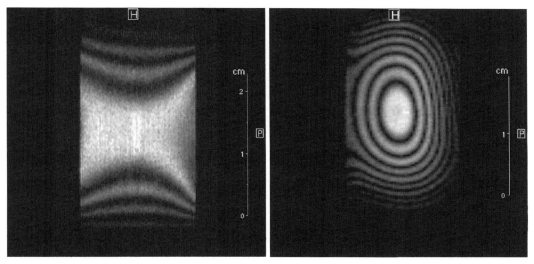

Fig. 9. (a) Relative B_1 field map acquired inside a 27-mm birdcage resonator with a 20-mm phantom filled with vegetable oil at 17.6 T using a 900° preparation pulse. From bright to bright region, B_1 varies by 20%. (b) Map for the identical setup and experiment with water ($\varepsilon_r = 81$) instead of oil. Due to strong dielectric resonance it is no longer possible to discern the original B_1 field distribution.

3. In case the phantom shows more than one line and therefore visible chemical shift artifacts, one can counteract as follows:
   a. Large spectral width will cause the artifacts to converge and not disturb the mapping sequence at the cost of increased noise.
   b. If small spectral widths allow to completely separate the overlapping images, this can be used but at the cost of long acquisition times.
   c. chemical shift imaging can be used to overcome the issue at the cost of greatly increased experiment time.
4. If a coil has to be strongly detuned or matched in order to accommodate a phantom this might break the homogeneity of the coil, though it would have been fine with the sample.
5. A good coil should vary no more than about 10% in field strength over its main field of view. Close to the conductors of the coil, the field variation increases dramatically.
6. A high power during transmission relates to a high sensitivity during reception as stated by the principle of reciprocity (20).

Acknowledgements

The author would like to thank Mirjam Falge for her assistance in design of the figures.

References

1. Fukushima, E., and Roeder, S. B. W. (1981) *Experimental Pulse NMR – A Nuts and Bolts Approach*. Perseus Books, Reading, MA.
2. Mallinson, J. C. (1973) One-sided fluxes – a magnetic curiosity? *IEEE Trans. Magn.* **9**, 678–682.
3. Halbach, K. (1980) Design of permanent multipole magnets with oriented rare earth cobalt material. *Nucl. Inst. Meth.* **169**, 1–10.
4. Bauer, C., Raich, H., Jeschke, G., and Blümler, P. (2009) Design of a permanent magnet with a mechanical sweep suitable for variable-temperature continuous-wave and pulsed EPR spectroscopy. *J. Magn. Reson.* **198**(2), 222–227.
5. Hoult, D. I. (1978) The NMR receiver: A description and analysis of design. *Prog. NMR Spect.* **12**(1), 41–77.
6. Levitt, M. H. (2005) *Spin Dynamics – Basics of nuclear magnetic resonance*. Wiley, Chichester, West Sussex, England.
7. Chen, C. N., and Hoult, D. I. (1989) *Biomedical Magnetic Resonance Technology*. Adam Hilger, Bristol, UK, New York, NY.
8. Silver, H. W., and Wilson, M. (2009) *The ARRL Handbook for Radio Communications*. American Radio Relay League, Newington, CT, USA.
9. Zhang, X., Ugurbil, K., and Chen, W. (2001) Microstrip RF surface coil design for extremely high-field MRI and spectroscopy. *Magn. Reson. Med.* **46**(3), 443–450.
10. Rodriguez, A. O. (2006) Magnetron surface coil for brain MR imaging. *Arch. Med. Res.* **37**(6), 804–807.
11. Hayes, C. E., Edelstein, W. A., Schenck, J. F., Mueller, O. M., and Eash, M. (1985) An efficient, highly homogeneous radiofrequency coil for whole-body NMR imaging at 1.5 T. *J. Magn. Reson.* **63**(3), 622–628.
12. Tropp, J. (1989) The theory of the bird-cage resonator. *J. Magn. Reson.* **82**(1), 51–62.
13. Lanz, T., Ruff, J., Weisser, A., and Haase, A. (2001) Double tuned ^{23}Na and ^{1}H nuclear magnetic resonance birdcage for application on mice in vivo. *Rev. Sci. Instrum.* **72**(5), 2508–2510.
14. Alderman, D. W., and Grant, D. M. (1979) An efficient decoupler coil design which reduces heating in conductive samples in superconducting spectrometers. *J. Magn. Reson.* **36**(3), 447–451.
15. Doty, F. D., Entzminger, G. Jr., and Hauck, C. D. (1999) Error-tolerant RF litz coils for NMR/MRI. *J. Magn. Reson.* **140**(1), 17–31.
16. Roemer, P. B., Edelstein, W. A., Heyes C. E., Souza, S. P., and Mueller, O. M. (1990) The NMR phased array. *Magn. Reson. Med.* **16**(2), 192–225.
17. Pruessmann, K. P., Weigner, M., Scheidegger, M. B., and Boesiger, P. (1999) SENSE: Sensitivity encoding for fast MRI. *Magn. Reson. Med.* **42**(5), 952–962.
18. Griswold, M. A., Jakob, P. M., Heidemann, R. M., Nittka, M., Jellus, V., Wang, J., Kiefer, B., and Haase, A. (2002) Generalized autocalibrating partially parallel acquisitions (GRAPPA). *Magn. Reson. Med.* **47**(6), 1202–1210.
19. Burgess, R. E. (1939) The screened loop aerial. *Wireless Engineer* October, 492–499.
20. Hoult, D. I., and Richards, R. E. (1976) The signal-to-noise ratio of the nuclear magnetic resonance experiment. *J. Magn. Reson.* **24**(1), 71–85.
21. Hoult, D. I. (1979) Rotating frame zeugmatography. *J. Magn. Reson.* **33**(1), 183–197.
22. Murphy-Boesch, J., So, G. J., and James, T. L. (1987) Precision mapping of the B1 field using the rotating-frame experiment. *J. Magn. Reson.* **73**(2), 293–303.
23. Haase, A., Frahm, J., Matthaei, D., Hänicke, W., and Merbold, K. D. (1986) FLASH imaging: Rapid NMR pulse imaging using low tip angle pulses. *J. Magn Reson.* **67**(2), 258–266.
24. Hahn, E. L. (1950) Spin echoes. *Phys. Rev.* **80**(4), 580–594.

Chapter 5

Small Animal Preparation and Handling in MRI

Patrick McConville

Abstract

Animal handling and preparation is one of the most critical aspects of in vivo NMR imaging in small animals, and involves a broad spectrum of challenges, any of which could affect data quality and reproducibility. This chapter will outline the most critical considerations in animal handling for in vivo MRI experimentation in rodent models. Highly accurate and reproducible positioning is one of the most important aspects, since sensitivity, motion and susceptibility artifacts, animal imaging throughput, and ease of data quantification are all dependent on it. A variety of devices exist today that assist in several aspects of animal handling and positioning, each with its own advantages and limitations. This chapter will detail many of the devices that are commercially available and how they have dealt with integration of RF coil technology, restraint, anesthesia, fiducial markers, warming, and physiological monitoring. The chapter will additionally detail various aspects of animal anesthesia, maintenance of core body temperature, physiological monitoring, intubation and ventilation, and systemic contrast agent administration. An increasingly important factor in running a small animal MRI laboratory, facility biosecurity, will also be reviewed.

Key words: Small animal MRI, in vivo, mouse, rat, physiological monitoring, animal preparation, animal handling, high throughput, imaging.

1. Introduction

Animal handling and preparation is one of the most important aspects of successful in vivo NMR imaging, as it has the potential not only to affect the image quality in a variety of ways, but also to affect the animal physiology, and therefore the intended end point for the experiment. Small animals account for the vast majority of in vivo MR-based research applications, with mouse models providing the broadest and most flexible platform for medical research. The smaller the subject, the greater the challenges for sensitivity and resolution, and the greater the role animal positioning and handling plays in determining data quality. While an imaging facility may spend significant resources in optimizing hardware and software for imaging protocol development, the benefits of this optimization may not be realized if sufficient attention is not given to animal preparation and handling.

A successful in vivo MRI laboratory will have necessarily carefully paid attention to the key aspects of animal handling, including anesthesia, animal core temperature maintenance, and physiological monitoring. Not all laboratories will, or should take the same approaches, but the implemented solutions should carefully address the specific needs based on the research streams, model types, and imaging protocols that will be used at the facility. Currently available animal handling technology, if carefully chosen and implemented, can enhance data quality, reproducibility, and animal imaging throughput simultaneously.

This chapter will outline the key aspects of animal handling and positioning in MRI and associated current techniques and technologies for controlling in vivo image quality and animal physiology through NMR-compatible methodology. Other major considerations relating to the use of animals in an MRI laboratory will also be outlined and discussed.

2. Animal Positioning Challenges

Animal positioning is one of the most critical, yet often not well considered and poorly understood aspects of in vivo MRI. Position inherently affects the RF coil loading and Q factor (*see* **Note 1**). RF tuning and matching and pulse powers can therefore be affected by the positioning of the animal in the three-dimensional space within the coil and to what degree the animal body is included in the active coil volume. Normal RF coil heterogeneity makes the precise location of the animal in the coil even more critical, as there could be severe sensitivity losses if the anatomy of interest is too far outside the RF coil homogeneous window or "sweet spot".

The symmetry and ultimate geometrical shape of the animal, which also depends on positioning, will also affect the three-dimensional nature of air/tissue interfaces, and therefore directly affect local susceptibility gradients and magnetic field inhomogeneities. The degree to which shimming can offset these effects is therefore dependent on the animal position which influences signal linewidth and ultimately, sensitivity.

Animal positioning will therefore inherently affect sensitivity of the experiment as well as susceptibility artifacts, which can have severe detrimental effects at the relatively high fields generally used for animal research (4.7–9.4 T) (*see* **Note 2**). If sufficient attention is not paid to animal positioning, it can severely affect animal-to-animal variability and quality of an image data set or could even render an image as being non-quantifiable.

Most animal imaging is necessarily performed under anesthesia. While this provides important control of animal physiology and is intended to allow inherent control over absolute position, muscular relaxation presents manual challenges related to relative position of limbs and tissues that affect sensitivity and accuracy of the experiment. The most important basic aspects of positioning for small animals are centralization of the animal's body in all three dimensions and relative position of the head, torso, and limbs. By accurate control and reproducibility of these parameters, the quality and reproducibility of the subsequent image data can be optimized.

3. Animal Positioning Technology

Optimal animal positioning for MRI generally requires the use of a "bed" of some kind, analogous to the way in which patients are positioned and inserted into a magnet on a bed during a

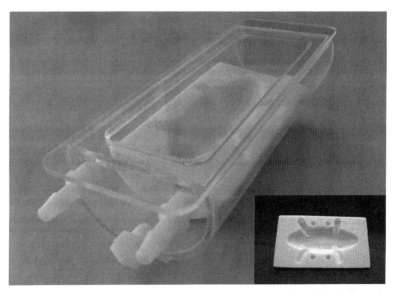

Fig. 1. Example of an MRI-compatible animal imaging chamber, with features for accurate body centralization and limb positioning. The sealed chamber ensures stable anesthetic delivery. Slots can be included for fiducial markers (*see inset*), either by insertion of small plastic tubes that contain a signal standard or by insertion of standard materials. The foam insert is disposable, enabling inherent control of biosecurity and prevention of animal cross-contamination. Images courtesy of Numira Biosciences.

clinical MRI procedure. The bed may be as simple as a small rigid platform that the animal rests on with placement of a surface coil adjacent to the anatomy of interest, or insertion of the platform into a volume coil. On the other hand, more elaborate animal beds are now commonly in use that provide simple mechanisms for precise location of the three-dimensional position of the body (1). Animal management devices (*see* **Note 3**) can also include features that control the relative positions of the head, torso, and limbs, such as body molds (2) (**Figs. 1** and **2**). The degree to which animal body and limb position should be controlled will depend on the application. For example, for precise brain imaging, stereotactic devices may be necessary (3).

Consistency in animal to animal positioning in order to reduce inter-subject variability and random error in the measurement is as important as accurate and well thought out positioning

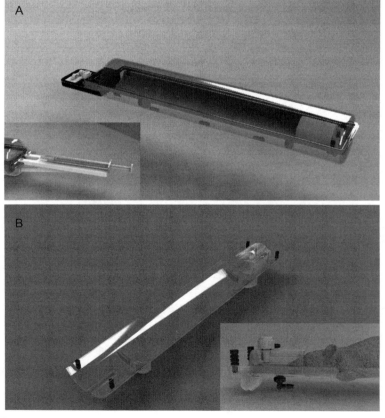

Fig. 2. (**a**) An animal management system that incorporates a modular approach that facilitates parallel animal preparation and multiple modality translation. Functionality for an attachable syringe holder is shown (*inset*). The management system includes a modular "sled" (**b**) that is inserted into the device. The sled incorporates head restraint features as shown in a close up (*inset*). Images courtesy of m2m Imaging.

of the animal in the RF coil. Animal management devices that enable easy and accurate control of the above aspects of positioning also inherently ensure reproducibility. Not only does this increase data quality, it can lead to dramatic decreases in the time spent in preparing the animal physically and preparing the MR scan through software, including the time required for shimming, pulse calibration, and slice positioning. Accurate positioning of both the coil in the magnet and animal with respect to the coil to within ~1 mm can make slice packet and read offset adjustment redundant as well as ensure that little or no adjustment of tuning, matching, shimming, and pulse powers are necessary.

The critical features of an MRI animal bed are outlined in **Table 1**. New animal management technologies are becoming available that include integrated packaging of many of the features described in **Table 1** (**Figs. 1–4**). An imaging facility focused on throughput will generally desire to have more than one animal bed, so that while an animal is being imaged, the next animal can be anesthetized and positioned. An attractive method for this is use of a modular "sled" or smaller bed which is inserted into and connected with the larger, more elaborate management system (**Fig. 2**). This can simplify the process of animal preparation as well as enable animals for future imaging to be prepared in advance. It can also facilitate translation of the bed for use in other imaging systems, even while the animal remains under anesthesia.

4. Integration of RF Coils with Animal Management Technology

Integration of the RF coil and the bed is a key challenge, due to the wide variety of coil designs from a number of different manufacturers. While coil integration may not be possible across all coil platforms, the bed should at least address the need for using different coils for different applications, including both volume and surface coils (*see* **Note 4**). One approach is the use of mounting hardware that allows the independent placement of both the coil and the animal bed (**Fig. 3**). An alternative approach is use of an animal bed platform that can accommodate interchangeable RF coils (**Fig. 4**).

While RF coil integration can provide an optimal method for addressing the unique demands of MRI as it relates to animal position and ease of animal preparation, it can also lead to necessary compromise in sensitivity as the coil may need to be larger than ideal, due to having to accommodate hardware for animal positioning. Another disadvantage of this approach is that it may restrict compatibility with imaging systems across other modalities, unless a modular approach is taken, where the bed itself can

Table 1
Desirable features for animal positioning and handling management technology in MRI (*see* Note 3)

Feature	Purpose	Advantages for
Mechanism for locating coil in magnet	Ensure RF coil is precisely centered in magnet bore	Sensitivity
Mechanism for 3D location of subject in coil	Ensure animal is centered in all three dimensions, facilitate fine adjustment for multiple exams	Sensitivity and homogeneity
Nose cone and bite bar	Center animal head and ensure reproducibility between subjects	Image preparation time, robustness of anesthesia
Head restraint or ear bars	Precise horizontal location of head and motion restriction for brain imaging	Image preparation time, motion artifact prevention
Mechanism for limb and body localization	Ensure reproducible shape of body and relative positions of all limbs	Sensitivity, homogeneity, image preparation time
Fiducial marker incorporation	Enable robust knowledge of relative position across different imaging sessions and subjects	Co-registration of images over time, images in different animals or between different modalities
Anesthesia integration	Remove need for external tubing to carry anesthetic	Anesthesia reliability, animal preparation time
Warming capability	Maintain animal core temperature	Animal well-being, physiology, data quality
Physiological sensor integration	Inbuilt sensors for respiration, ECG and temperature	Monitoring reliability, animal preparation time, inherent physiological characterization
Physiological monitoring electronics integration	Ensure robust monitoring and remove need for external cables for monitoring signals	Monitoring reliability, animal preparation time, inherent physiological characterization
Intravenous drug injection functionality	Allow guides for localization of a drug infusion line	Robustness of intravenous drug injection or infusion
RF coil integration	Functionality for changing RF coils efficiently while ensuring bed integration	Flexibility for use across multiple applications
Multiple modality compatibility	Enable use in other imaging modalities, e.g., PET, SPECT, CT, optical	Multi-modal imaging and co-registration
Inclusion of modular animal "sled"	Enables rapid animal positioning and use of multiple sleds for positioning of other animals while imaging is ongoing	Efficiency and translation of sled use across other imaging modalities

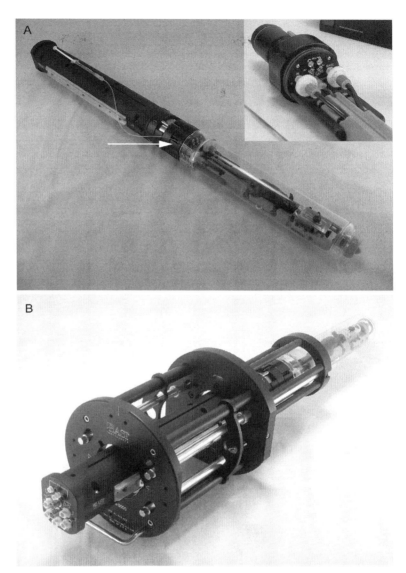

Fig. 3. Example of an advanced animal management system that incorporates modular technology for (**a**) the animal bed and (**b**) a mounting device that holds an RF volume coil and localizes it in the axial direction. The animal bed with sealed chamber is then conveniently inserted using the guiding rungs of the mounting device, which is semi-permanently fixed to the magnet bore. The animal bed shown (a) incorporates multiple channels for physiological monitoring and use of multi-channel surface coils (*see inset*). A feature for drug infusion is also included (*see arrow*). A head fixation device and body mold with inclusion of basic limb restraint ensures central and reproducible animal positioning and motion restraint. Images courtesy of ASI Instruments.

be separated from the coil. A core facility that operates in multiple disease areas and in a variety of models may prefer manageable compromises on sensitivity to enable maximum flexibility for use with different coils and across modalities. On the other hand, a

Fig. 4. Example of an MRI animal bed platform with integrated anesthesia and functionality for animal axial positioning and head immobilization. The bed is designed to allow straightforward removal and exchange of multiple phased array RF coils for specific applications including (**a**) head imaging and (**b**) cardiac imaging. Images courtesy of m2m Imaging.

smaller facility focused on more specific research topics may prefer customized animal management for each RF coil in use.

4.1. Co-registration and Fiducial Markers

Multi-modality capabilities are rapidly becoming the standard for in vivo imaging facilities. As the preclinical imaging field has matured, the power of multi-modality imaging has been demonstrated, and today PET, SPECT, optical, ultrasound, and CT imaging modalities are commonly used in tandem with MRI. Additionally, MRI itself is commonly used in a longitudinal fashion in the same animals over time. The application of multiple scans or MRI protocols in a given imaging subject is also becoming the standard, rather than the exception. All of these trends highlight the importance of image co-registration. While software based methods for co-registration using non-rigid body approaches such as mutual information (4) are now commonly available (*see* **Note 5**), the use of fixed fiducial markers has become prevalent and can aid simple, robust image co-registration between MRI scans and between modalities. This functionality can be conveniently incorporated into the animal bed (**Fig. 1**) through the use of fixed standard materials or fillable volumes where liquids can be used to provide standard localized signal for later co-registraton.

5. Anesthesia

Today, the vast majority of imaging applications rely on anesthesia (5–8). The breathable halogenated ether isoflurane (6, 7) is the most commonly used anesthetic due to rapid effect, rapid recovery time, excellent tolerance by mammalian species, and low cost. Sevoflurane (7) is another halogenated ether used for this purpose. Both have little or no known toxic side effects as they are not metabolized. They are also relatively insoluble in the blood, which promotes rapid recovery when the gas delivery is turned off.

Isoflurane can be delivered in either oxygen or air, with dosage level accurately controlled in real time by a vaporizer that mixes vapor-phase isoflurane with the breathable gas of choice. This provides advantages over injectable methods of anesthetic, as the dosage level can be easily adjusted based on the specifics of the species, model, and treatment being used. The dose can also be adjusted over time after an initially high induction dose of up to 4–5% concentration, allowing fine control of respiratory rate and health of a research animal over several hours if necessary. Modern imaging labs are therefore normally fitted with centrally supplied compressed air and/or oxygen, multiple isoflurane vaporizers and gas piping that integrates the anesthetic with the imaging system(s). It is then a simple matter to open the appropriate valve to deliver the anesthetic to the induction chamber, animal bed or imaging bore.

The predominant alternative to breathable anesthesia is injectable anesthetic and most commonly the dissociative anesthetic ketamine is used in combination with the sedative/relaxant, xylazine. Use of ketamine/xylazine anesthesia can be preferred in cases where cardiac physiology is critical, as isoflurane is known to affect cardiac function (9). Ketamine/xyalzine also results in more relaxed respiration, compared with isoflurane, which can cause erratic respiration. In MRI, this can lead to undesirable motion artifacts, and therefore isoflurane anesthesia may not be preferred in thoracic imaging applications. The potential physiological side effects of short- or long-duration anesthesia and use in oxygen versus air should be well understood within the context of the end points being quantified (8). Animal dehydration during anesthesia is a primary concern and can be addressed by the use of eye lubrication and post-anesthetic intraperitoneal saline injection for rehydration and recovery.

6. Anesthesia System Design

A basic anesthesia system comprises just a few simple components, including these:
- Air or oxygen supply: by a centralized compressor delivered through piping or by individual gas tanks. Centralized compressed air removes the need for frequent tank exchange and the risk of a tank emptying during an imaging run.
- Vaporizer: accurately mixes anesthetic with input air or oxygen with output of anesthetic gas at known concentration and flow rate.
- Induction chamber: takes gas from the vaporizer for induction of anesthesia
- Tubing from vaporizer to animal bed: takes anesthetic gas to the animal bed/RF coil/magnet

However, design and implementation of a more elaborate system, while more costly, can result in large increases in animal anesthesia reliability, efficiency, and flexibility, and also in enhanced occupational health and safety of laboratory workers. Thought should be given to capacity for induction of multiple animals simultaneously, capacity for lab bench animal procedures under anesthesia (e.g., catheterization), and delivery of anesthetic via ceiling drops at the magnet bore opening and other animal work areas. New products are now available that provide management of multiple anesthetic lines via a small consolidated unit with easy adjustment and accurate control of flow rates, and switching between multiple lines that control supply for induction, procedure, and imaging (**Fig. 5**). The following components are features of an optimally designed MRI anesthesia system:
- Robust delivery: The breathable anesthetic gas to the animal must be delivered efficiently to the animal's snout. Commonly, this is achieved using a well-designed nose cone (**Figs. 2** and **3**) or by using a sealed imaging chamber (**Fig. 1**).
- Integrated animal warming: Warming should be incorporated in all induction chambers, procedure stations, and animal beds by warm water, warm air or heat lamp.
- Excess anesthetic gas removal: Non-breathed waste gas must be removed either passively or actively using a suction pump and can be absorbed by charcoal canisters or vented to the

outside air. This includes the animal bed nose cone, which should be designed as a closed circuit, such that any non-breathed anesthetic gas is channeled for absorption or to the outside world. Centralized suction can greatly facilitate excess gas removal, including waste gas removal from the magnet bore.

- Devices for mounting charcoal canisters and tubing management can greatly aid organization and make optimal use of space, which is commonly limiting.
- Gas flow regulation: Use of devices that regulate gas flow increase the robustness of the anesthesia, ensuring adequate volume delivery while also not allowing excess volume delivery that could result in unnecessarily large waste gas volumes and potential environmental exposure. Commonly, the effect of tubing bore size and length in determining resistance and flow rate is neglected leading to potentially inadequate or excessive flow rates. Technology that relies on a pressure-driven approach that is independent of traditional flow meters can ensure accurate flow regardless of tube size, length, and load (**Fig. 5**).

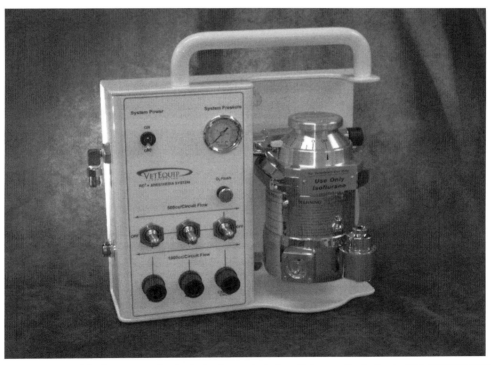

Fig. 5. A multi-channel isoflurane anesthesia device for control of multiple anesthesia lines using constant pressure technology that ensures accurate simultaneous flow of anesthetic gas to induction chambers, procedural workstations and imaging systems. This device was designed to control simultaneous anesthesia delivery to multiple imaging systems. Image courtesy of VetEquip.

- Filters: Filters for particulates, water, and oil should be incorporated into all incoming gas lines to ensure animal health and also to reduce fire hazard where oxygen is being used.

While cheap and commonly available materials can provide functional anesthesia delivery temporarily, it pays to use tested materials that will resist wear and rupture, preventing compromised anesthesia delivery and potential exposure hazards. While isoflurane is considered a non-toxic chemical, studies have demonstrated potential safety concerns and minimizing environmental exposure of technicians during usage should be an area of focus for an MRI lab. For example, induction boxes that utilize active suction to minimize exposure during chamber opening are now available (*see* **Note 6**). Monitoring actual exposure through sensors that are worn on external clothing can be a reliable way to demonstrate acceptable exposure and track improved exposure in problem areas (*see* **Note 7**).

7. Motion Restriction and Imaging of Conscious Animals

While in vivo MRI is predominantly achieved using anesthetized animals, the use of MRI in conscious animals is increasing in parallel with recent advancements in MRI-compatible motion restriction devices (*see* also **Chapter 17**). Imaging in conscious animals is a particularly important consideration in functional MRI applications where brain activation in response to stimulus is key to the experiment. As functional MRI protocols are extended increasingly to small animal models, imaging in awake animals will also increase (10–12). Devices for this purpose are usually based on traditional stereotactic technology (**Fig. 6**) such as bite bars, nose cones, and ear bars (3, 13).

Fig. 6. Example of an RF volume coil with integrated stereotactic functionality (*see arrows*) that can be used for awake animal brain imaging. This strategy can also be used with anesthetized animals for precise brain imaging. Images courtesy of EKAM Imaging.

Of course, these methods have more general applications in motion restriction for MRI, even in anesthetized animals. Respiratory motion is a large concern in MRI, where even small motion during the scan can translate to significant phase artifacts that increase data variability or may even render a slice or entire image as being non-quantifiable. Respiratory motion commonly extends throughout the anesthetized animal's body and can even create severe motion artifacts in brain imaging. In small animals, even the use of tape and immobilization splints can localize full body respiratory motion to the thorax, immobilizing the head or lower abdomen. Fortunately, the incorporation of simple mechanisms for motion restriction such as bite bars and nose cones are now commonly incorporated into MRI animal bed and/or RF coil design negating the need for an enormous inventory of surgical tapes in today's imaging lab! These features not only help to reduce animal respiratory motion, but also inherently increase the reproducibility of the animal position. For example, stereotactic devices ensure the animal's head is fixed in space both during a procedure and *between* procedures, providing inherent registration of images and minimizing animal preparation and setup time.

8. Maintenance of Core Body Temperature

Maintenance of animal body temperature is an important consideration during imaging procedures, especially during the use of anesthesia which generally causes a decrease in core body temperature. There are therefore a variety of MR-compatible techniques that are used for warming of the animal while in the RF coil and magnet. These include the following:

– Warm water: This relies on a warm water bath and pump. The warm water is pumped and recirculated to a water-filled pad that makes contact with the animal. The temperature of the water in the bath may need to be adjusted to be above body temperature to account for heat loss from the bath to the animal pad. A simple water-heating pad can be constructed using standard thin plastic tubing or there are a number of commercially available water-filled pads available today (*see* **Note 8**) that are thin to minimize space usage and flexible to enable use within imaging bores and coils. It is important to ensure good contact between the pad and the animal and prevent any blockage of the water flow, which will reduce the effective temperature of the water in the pad.

– Warm air: This method involves blowing of warm air over the animal and is most commonly achieved by introducing warm air into the magnet bore, effectively warming the entire bore of the magnet.

- Electronically induced heating: A variety of devices are available that result in electric current-induced warming of a material that makes contact with the animal (*see* **Note 9**). Materials have been developed for this approach that are thin, flexible and can be heated to very high temperatures with small DC currents. Use of non-metallic materials, wires, and contacts, and prevention of electric current-induced artifacts are challenges that this methodology presents (*see* **Note 10**), but are challenges that are being overcome through current developments.
- Chemically induced heating: There are a number of materials that produce body temperature heat for several hours or more when exposed to air (*see* **Note 11**). Commercially available products based on this methodology are commonly used in human body warming applications in freezing conditions. Some of these products are non-metallic and can be successfully applied for animal warming in MRI.

9. Physiological Monitoring and Gating

An essential aspect of animal imaging is physiological monitoring, both to ensure animal health and to characterize and control physiology, ultimately optimizing data quality. A challenging aspect of animal monitoring for MRI is maintaining MR-compatibility through the use of non-metallic materials where possible and/or strategies for precluding inducement of noise or artifacts if metals are used. The following subsections will discuss the major categories of animal monitoring for MRI and considerations in their undertaking.

9.1. Temperature Monitoring

As discussed above, control of animal core temperature is a necessity in animal imaging, especially during lengthy imaging sessions, and especially when physiological end points are being generated. The optimal solution for animal temperature maintenance should integrate temperature monitoring, both to enable explicit characterization of temperature throughout an imaging session and to provide a method for automated control of temperature through an electronic feedback loop. Real-time monitoring of animal temperature can be used to control fine adjustment of the degree of heating, especially when warm air or electronic methods are used for heating.

The most common method for small animal temperature monitoring is the rectal probe, due to robustness of contact and accuracy. Traditional rectal probes are based on metallic thermocouples, but MR-compatible versions (e.g., fiber optic) are commercially available (*see* **Note 12**). MR-compatible rectal probes are typically fragile and expensive. Alternative strategies that can avoid the need for an external probe include thermocouples that measure abdominal temperature. The advantage of this method is that the probe can be integrated into the animal bed.

9.2. Respiratory Monitoring

Monitoring of respiration during MRI is probably the simplest way to manage overall animal health. The major general risk to animal health during an MRI procedure is the anesthesia, and respiratory rate is the most reliable method for monitoring the potentially deleterious effects of anesthesia. If done through robust means, respiratory monitoring can also inherently enable respiratory gated image acquisition. This is achieved through simple external electronics that convert analog respiratory signals into digital pulses that are sent to the MR scanner to trigger excitations and subsequent MR signal acquisitions and synchronize acquisition with the respiratory cycle. Commonly, the acquisition is triggered a short delay after the end of expiration, which is the most motion-free segment of the respiratory cycle.

The most common strategy for respiratory monitoring is the use of an air-filled "pillow" that is connected to a sensitive pressure transducer via well-sealed, air-filled tubing. Neonatal pressure sensors can work well for this purpose, providing good sensitivity in mice, but must be well positioned (*see* **Note 13**) for maximal sensitivity. Alternative methods include the use of mechanical means to induce electric current in response to respiratory motion (14) and use of optical methods (15). These methods for respiratory monitoring can be sensitive enough to also detect cardiac motion, providing a potential alternative to traditional ECG monitoring of the cardiac cycle (14).

9.3. ECG and Blood Oxygenation Monitoring

Monitoring of the cardiac cycle may be necessary for applications in which heart rate is a critical parameter. However, the most common need for cardiac monitoring in MRI is for cardiac gating in cardiovascular imaging applications. The traditional method for detecting the cardiac cycle is the use of ECG leads that make electrical contact on either side of the body. While metallic leads are generally problematic due to electrical noise and resulting image artifacts, carbon leads provide a robust alternative with the necessary conduction properties. One of the most common sources of failure in cardiac monitoring is loss of electrical contact. Needle electrodes have been used for this purpose, as have carbon fiber pads, but both are prone to contact failure. Recently, highly adhesive conducting pads (*see* **Note 14**) have enabled improvement in robustness of MR-compatible ECG probes. These pads can be used in a disposable manner.

An alternative method for cardiac monitoring is the use of pulse-oximetry technology commonly in use in veterinary and clinical medicine. This technique measures infrared and red emitted light absorption through a translucent region of the body, typically in extremities such as paws or ears. Since absorption properties differ for oxyhemoglobin and deoxyhemoglobin at these wavelengths, blood oxygen levels can be calculated in real time. The technology is also sensitive to expansion and contraction of vessels that occurs with the cardiac cycle and can therefore be used to monitor the cardiac cycle. Recently, pulse-oximetry technology has been developed to be sensitive enough for robust use in mice. The probes are typically attached via a spring-loaded clip and are therefore conveniently placed and removed.

Analogous to methodology for respiratory gating, simple external electronics can be used to convert analog cardiac pulses to digital pulses that can be sent to the MR scanner to trigger excitation and subsequent data acquisition, synchronizing image acquisition with the cardiac cycle to enable precise imaging of the thoracic region, including the myocardium and major vessels at specific time points in the cardiac cycle.

9.4. Physiological Monitoring Software

Monitoring of physiological parameters requires capability for analog-to-digital conversion of the biological signals, and real-time display and data time course recording for multiple signals simultaneously. There are a variety of packages now available (*see* **Note 15**) that bundle sensors, electronics, and software. Standard features include automatic and manual gain adjustment, signal polarity control, noise filtering, and automatic baseline adjustment. Most software packages will also include gating functionality, including ability to set signal thresholds and timing for gating events and output of TTL digital pulses that can be sent directly to the standard gating input port on the MRI spectrometer.

9.5. Intubation and Ventilator Control of Respiration

An alternative procedure that can be used to prevent respiratory artifacts and enable inherent ability to image the body during stillness is endotracheal intubation and use of a ventilator. This procedure involves introduction of a tube via the trachea by a skilled technician. This allows control over lung volume and pressure using a ventilator. Common ventilators allow flexibility in controlling the duration and frequency of respiration and can normally be operated in modes that allow an inspiratory pause as well as fine control over lung volume at the end of inspiration. This control can be leveraged during MRI procedures to enable imaging during lung stillness. The ventilator also inherently provides an opportunity for gating by output of digital pulses to synchronize the respiratory cycle with MR data acquisition. The main drawback of intubation and ventilator respiration control is the complexity and invasiveness of the procedure, placing limits on imaging throughput and also risking the health and well-being of the animal if training and knowledge of the operators is not adequate.

10. Systemic Administration of Contrast Agents and Therapeutics

A common need in in vivo MRI is systemic administration of MRI contrast agents. This is most commonly achieved via an intravenous injection. Since the experiment normally relies on quantitative comparison before and after injection, ideally, the injection occurs without displacing the animal from the RF coil and magnet. This enables scanning to occur before, during and after the course of the injection, as is necessary during widely used dynamic contrast-enhancement (DCE) protocols. In DCE MRI protocols, a highly time-resolved tracer concentration time course can be used to calculate tissue vascular parameters that measure permeability, blood flow, and vascular surface area (*see* **Chapters 19** and **26**).

Intravenous agent administration in rodents generally occurs via either of the two tail veins. To enable an injection while the animal is in the magnet, the most common approach is to use a tail vein catheter, introduced by one of several available means. Fine-gauge pediatric catheters (24–26 GA) serve this purpose well in mice and rats. These commercially available Teflon catheters are introduced using needles. Introduction can be achieved reliably and robustly in both mice and rats by skilled technicians using a similar technique to that used for syringe-based tail vein injections. An increasing number of products that facilitate easy and robust introduction of catheters through more complex mechanisms are now commercially available but are generally more expensive than standard pediatric catheters.

During insertion, tail vein dilation through warming can help and careful methods for fixation of the catheter (e.g., using tape) must be incorporated; otherwise, the catheter will be prone to dislodging from the vein during subsequent animal preparation before imaging. After insertion and fixation of the catheter, the catheter hub must be closed. This can be accomplished using screw-on infusion plugs. Testing of the catheter at this point by injection of a small amount of saline can help avoid wasted time in cases where there is a problem. A small amount of heparin can be used to help prevent clotting near the catheter, but potential side effects of heparin that could influence the experiment outcome, including anti-inflammatory and angiogenic effects (16, 17) should be considered and well understood. Before the imaging session and injection can begin, the catheter must first be connected to a sufficiently long infusion line. Microbore tubing can be used for this and it minimizes the total volume needed for priming the line, a potentially important consideration for expensive or limited-supply contrast agents. At all stages of the animal catheter preparation, prevention of air bubble entry or removal of air bubbles where necessary is critical for animal health. Contrast agent leakage before commencement of imaging must also be prevented, for example, by clamping the infusion line prior to the start of the injection.

Due to the small total volumes generally used in mice and rats, the accuracy of the volume and dose delivered is another important consideration. To this end, dead space in the catheter hub must also be accounted for. In DCE MRI, a rapid bolus is normally desirable and therefore the volume cannot be too high, such that the injection cannot be achieved within the required period (typically 5–30 s). Syringe pumps are best used for the injection in order to maintain accuracy and reproducibility of the arterial input function. The potential for error in the arterial input function due to tail vein localization of the contrast agent after the injection should also be considered, and if necessary, a subsequent saline flush used to clear residual contrast from the tail vein.

11. Multiple Animal Imaging

There have been several recent implementations of strategies to enable MR imaging of multiple animals simultaneously (18–22) (*see* also **Chapter 31**). The most prominent development of such technology has been by Henkelman and co-workers who designed and built an RF coil array that enables imaging of up to 16 mice in a standard 7-T MRI system (18–21). This technology has produced significant increases in throughput without sacrificing data quality, particularly in high-resolution three-dimensional isotropic imaging (23). Other approaches including the use of clinical magnets and single, large RF coils for multiple animal imaging have been demonstrated (24, 25). These approaches highlight a future trend that will require parallel innovation in animal handling, positioning, anesthesia, and monitoring methodology. If multiple animal handling efficiency is not sufficiently addressed, the maximum potential of a multiple mouse imaging strategy may not be realized. To this end, Henkelman and co-workers have designed hardware to facilitate multiple mouse preparation that incorporates a single, large anesthesia-induction chamber and modular animal "sleds" that incorporate an inlet for anesthesia as well as integrated temperature, respiratory, and cardiac sensors (26).

12. Facility Biosecurity

In the past decade, as animal research has expanded across the biomedical field, facility biosecurity has seen rapidly increasing focus. This is especially true in facilities where immune compromised mouse strains (the mainstay of oncology research) require specific pathogen free (SPF) environments. Frequent examples where entire facilities have been closed for many months due to the introduction of a colony infection have highlighted the need for adequate methodology to maximize biosecurity. Inadequate biosecurity can not only compromise image data quality, but can lead to the imaging laboratory being a weak link in the operation of an in vivo research unit.

The necessity for animals from multiple studies and sources to enter the imaging lab and make contact with imaging beds and RF coils makes it difficult to maintain imaging lab biosecurity at the same level that can be achieved in animal housing spaces. However, there are many ways in which this potential weakness can be compensated for:

- HEPA filtering: Use of air recirculating and HEPA filtering (*see* **Note 16**) equipment can ensure many complete HEPA-filtered air changes in a standard laboratory per hour. This will actively remove common airborne pathogens and contagions.

- Gowning procedures: Imaging technicians and staff should be required to wear a disposable gown, head cover, face mask, gloves, and shoe covers. This helps prevent introduction of a pathogen from the external environment.

- Entry and exit procedures: Strict protocols should be in place for entry and exit, including access by service technicians and visitors. Incoming gas and cryogens tanks should be appropriately decontaminated using anti-microbial, anti-bacterial solutions before entry.

- Use of decontamination procedures for common equipment: Where possible, imaging equipment should be disinfected daily using anti-microbial, anti-bacterial solutions with adequate contact times. Since it is not possible to do this for imaging equipment that might be sensitive to liquids or corrosion, alternative decontamination strategies include the use of UV lamps and hydrogen peroxide vapor gas.

- Use of disposable supplies: Use of an animal bed that incorporates a disposable surface where the animal contacts the bed (**Fig. 1**) helps to prevent animal-to-animal contact and potential cross contamination.

13. Future Trends and Conclusion

As preclinical in vivo imaging has matured through technological improvements in imaging hardware and software, in parallel, the needs for efficient, reliable, and well-designed methodology for animal handling have been highlighted. In the last ten years, advances have occurred in our understanding of how animal handling affects data quality and throughput. This has driven access to a wide array of technologies that support more effective and reproducible animal imaging by addressing the entire process of an animal entering the imaging lab through to it leaving the lab after the imaging procedure.

Current trends will continue to drive MR applications from being anatomically focused to being more enabling of functional, physiological, and molecular end points. This trend will require even more attention to animal handling, physiology, and well-being. The success in applying MRI-based protocols for quantitative assessment of drug pharmacology is now pointing toward similar opportunities in drug safety and toxicology assessments. This may eventually lead to the introduction of regulated imaging laboratories that operate under GLP (good laboratory practice) protocols. This will further drive the need for certification, quality control, and quality assurance in all aspects of imaging animal handling.

14. Notes

1. The Q factor or "quality factor" is a property of a resonant circuit that defines the fraction of energy dissipated in one oscillatory cycle and for an RF coil, affects the coil sensitivity. For a parallel tuned circuit RF coil, $Q = \omega L/R$, where $\omega =$ the angular frequency, $L =$ coil inductance, and $R =$ coil resistance. The coil signal-to-noise ratio (SNR) is proportional to $\sqrt{Q}$.

2. Susceptibility is a magnetic property of materials and tissue that describes the magnetization induced in the material by a magnetic field, with a higher magnetic field resulting in greater magnetization. Differential susceptibility (e.g., at a tissue/air interface) can give rise to susceptibility artifacts due to local distortion of the static magnetic field. The distortions and resulting artifacts are greater at higher static magnetic fields.

3. A number of manufacturers provide commercial access to animal handling, positioning and management technologies for MRI and other imaging modalities. These devices address one or more of the features described in **Table 1**. Examples of such technology providers include ASI Instruments (http://www.asi-instruments.com/), m2m Imaging (http://www.m2mimaging.com/), Agilent (http://www.chem.agilent.com/Library/datasheets/Public/SI-0374.pdf), Bruker (http://www.bruker-biospin.com/mri_accessories.html), Rapid Biomedical (http://www.rapidbiomed.com/pages/english/animal-coils/equipment/air-heated-holder.php), Minerve (http://www.bioscan.com/molecular-imaging/minerve), and Numira Biosciences (http://www.numirabio.com/site/services/imaging-chamber).

4. MRI RF coils can be broadly categorized into two major types: surface coils and volume coils. A surface coil involves a wire loop design, where the subject is positioned adjacent to the loop plane, with the loop magnetic field induced into the subject. A volume coil, commonly of cylindrical design, involves placement of the subject inside the coil, with the coil magnetic field induced within the coil volume (and the subject). While surface coils generally provide the highest sensitivity (close to the coil loop), they also result in rapidly diminishing magnetic field with depth into the subject and greater coil field inhomogeneity, compared with volume coils. Surface coils are suitable for localized imaging applications (where the tissue of interest is close to the surface of the subject), whereas volume coils are more suitable for larger volume imaging applications, such as whole body imaging.

5. Traditional image co-registration approaches involve "rigid" three-dimensional translations and is commonly achieved using fiducial markers (external objects used as landmarks) or anatomical landmarks. However, there is no deformation of the image volumes during the registration process. Recently, non-rigid or "elastic" approaches have shown utility (e.g., mutual information (4) approaches). These involve locally warping the target image to account for normal tissue deformations or geometric differences that occur from image to image in the same subject or even between different subjects where co-registration is needed.

6. An induction box is used to induce gas anesthesia in the animal. Typically, the box is a transparent plastic container with a hinge-based or sliding mechanism for opening and closing the lid. The procedure of opening the box creates

a risk of anesthetic gas exposure for the operator. Devices have therefore been developed to circumvent this occupational hazard. One example is to use a valve that creates a pressure differential to manually flush the air/anesthetic mixture from the box just prior to opening. A more practical method (e.g., VetEquip, Inc: http://www.vetequip.com) utilizes an active suction placed just above the opening to the induction box. Upon opening the induction box lid, only the anesthetic that escapes the box is safely removed. This has the advantage of preventing full removal of the gas, so that the anesthetic environment in the box is maintained for use with the next animal.

7. Monitoring worker isoflurane exposure is an important aspect of health and safety in an imaging lab that utilizes isoflurane anesthesia. This can be accomplished via a badge that can be worn on the chest and that absorbs halogenated hydrocarbons such as isoflurane and halothane. The badges can be worn by workers during anesthetic procedures and shipped for laboratory testing for exposure levels.

8. Water-filled tubing pads provide a convenient method for animal warming (e.g., Gaymar: https://www.gaymar.com). These rely on circulation of warm water (typically from a laboratory heating bath) through tubing that can be configured in a flexible pad and draped underneath or around the animal.

9. Thin materials that can be heated to sufficiently high temperatures by a DC current (e.g., battery) provide a convenient means for animal warming, especially if controlled by a feedback loop in conjunction with an animal temperature probe. These materials can be flexible enough to drape around an animal and are thin enough to position over or underneath the animal. MRI-compatible examples include devices based on etched-foil resistive elements laminated between layers of flexible insulation (e.g., Minco: http://www.minco.com). Aerospace technology has also provided heating elements based on thin, conductive textile consisting of nickel-coated carbon fibers formed into a nonwoven fabric (e.g., Thermion: http://www.jjgalleher.com/thermion.html). These can provide significant uniform heat across a thin flexible material when connected to a DC current.

10. Metal objects cause artifacts in MRI due to large susceptibility artifacts (*see* **Note 2**) and also the induction of eddy currents during gradient switching which can cause image artifacts as well as substantial electrical noise in the wires. For this reason, metallic wires can be problematic in MRI even if contained outside the field of view. Effective

blanking can provide a method for precluding the effect of the gradient noise.

11. Chemical heating can also provide a method for small animal warming. Such devices are commonly used for glove and shoe warming, and can be found in camping or outdoor recreation stores in disposable packs. Some of these devices are based on exothermic oxidation of iron after exposure to air and can output heat at body temperature for extended periods of 8 h or more. However, being ferromagnetic, iron can create MRI artifacts and therefore some of these products are not suitable for all MRI applications. Similar products that have been used in medical applications create heat based on exothermic crystallization (e.g., Hood Thermo-Pad Canada Ltd: http://thermo-pad.com/) and unlike the iron-based heating devices, can be reusable.

12. A convenient MRI-compatible, non-electrical method for temperature monitoring that does not involve any metallic parts is based on a fiber optic sensor for detection of temperature-based changes in the length of a Fabry–Perot cavity, and detection of these fine changes using a silicon diaphragm (e.g., FISO Technologies Inc: http://www.fiso.com/). These sensors can be used in MRI applications and include FDA-approved devices for patient temperature monitoring.

13. Detection of breathing motion can be accomplished in small animals by use of an air-filled pressure sensor placed in contact with the abdomen or thorax. Sensitive pressure sensors of this type have been developed for neonatal respiration monitoring (e.g., as manufactured by Graseby) and can be used for mouse and rat respiratory monitoring when connected to a pressure transducer and analog-to-digital converter.

14. Monitoring of the cardiac cycle traditionally involves the use of electrodes placed on opposing sides of the body to detect cardiac electrical signals in the skin (e.g., paws). It is difficult to use these devices in small animals due to difficulty in ensuring and maintaining adequate electrical coupling between the skin and the electrode. However, self-adhesive conductive pads (e.g., 3M: http://solutions.3m.com) provide a convenient product for this purpose and are also cheap and disposable.

15. A number of manufacturers (e.g., Small Animal Instruments Inc: http://www.i4sa.com, m2m Imaging: http://www.m2mimaging.com/, Biopac Systems Inc: http://www.biopac.com and Rapid Biomedical: http://www.rapidbiomed.de), now provide integrated multi-faceted

physiological monitoring systems that are tailored to small animals. These normally include sensors for respiration, ECG and temperature (commonly all MRI-compatible), signal transduction and analog-to-digital conversion electronics, and software for real-time monitoring of the physiological systems. Most of these systems also include one or more general analog inputs to allow simultaneous use and monitoring of signals from a broad variety of devices. Commonly, these systems also include animal warming devices that can be controlled via a feedback loop with the temperature sensor, and software and hardware for generating digital pulses that can be sent to MR spectrometers for gating.

16. HEPA is an acronym for high-efficiency particulate air. HEPA filtering results in >95% removal of airborne particles of size >=0.3 μm. In an animal vivarium or animal research environment, this can greatly help prevent the spread of pathogens between animals which commonly occurs via airborne particles and dander. This is particularly important in a facility using immune-compromised animals, which encompass the standard strains used for today's animal tumor models, as these animals are highly susceptible to infection. Well-designed imaging facilities dealing with immune-compromised animals should therefore be designed to be specific pathogen free (SPF) and one important component of this is effective HEPA filtering of the laboratory air.

References

1. Zhang, M., Huang, M., Le, C., Zanzonico, P. B., Claus, F., Kolbert, K. S., Martin, K., Ling, C. C., Koutcher, J. A., and Humm, J. L. (2008) Accuracy and reproducibility of tumor positioning during prolonged and multi-modality animal imaging studies. *Physics in Medicine and Biology* **53**, 5867–5882.
2. Zanzonico, P., Campa, J., Polycarpe-Holman, D., Forster, G., Finn, R., Larson, S., Humm, J., and Ling, C. (2006) Animal-specific positioning molds for registration of repeat imaging studies: comparative microPET imaging of F18-labeled fluoro-deoxyglucose and fluoro-misonidazole in rodent tumors. *Nuclear Medicine and Biology* **33**, 65–70.
3. Fricke, S. T., Vink, R., Chiodo, C., Cernak, I., Ileva, L., and Faden, A. I. (2004) Consistent and reproducible slice selection in rodent brain using a novel stereotaxic device for MRI. *Journal of Neuroscience Methods* **136**, 99–102.
4. Pluim, J. P., Maintz, J. B., and Viergever, M. A. (2003) Mutual-information-based registration of medical images: a survey *IEEE Transactions on Medical Imaging* **22**, 986–1004.
5. Davis, J. A. (2008) Mouse and rat anesthesia and analgesia. In *Current Protocols in Neuroscience/Editorial Board, Jacqueline N. Crawley et al.*, Vol. 42, J. N. Crawley, Editor, Wiley, A. 4B.1–A.4B.21, http://onlinelibrary.wiley.com/doi/10.1002/0471142301.nsa04bs42/abstract.
6. Eger, E. I., 2nd (1981) Isoflurane: a review. *Anesthesiology* **55**, 559–576.
7. Terrell, R. C. (2008) The invention and development of enflurane, isoflurane, sevoflurane, and desflurane. *Anesthesiology* **108**, 531–533.

8. Hildebrandt, I. J., Su, H., and Weber, W. A. (2008) Anesthesia and other considerations for in vivo imaging of small animals. *ILAR Journal/National Research Council, Institute of Laboratory Animal Resources* **49**, 17–26.
9. Gare, M., Schwabe, D. A., Hettrick, D. A., Kersten, J. R., Warltier, D. C., and Pagel, P. S. (2001) Desflurane, sevoflurane, and isoflurane affect left atrial active and passive mechanical properties and impair left atrial-left ventricular coupling in vivo: analysis using pressure-volume relations. *Anesthesiology* **95**, 689–698.
10. Lahti, K. M., Ferris, C. F., Li, F., Sotak, C. H., and King, J. A. (1999) Comparison of evoked cortical activity in conscious and propofol-anesthetized rats using functional MRI. *Magnetic Resonance in Medicine* **41**, 412–416.
11. Sicard, K., Shen, Q., Brevard, M. E., Sullivan, R., Ferris, C. F., King, J. A., and Duong, T. Q. (2003) Regional cerebral blood flow and BOLD responses in conscious and anesthetized rats under basal and hypercapnic conditions: implications for functional MRI studies. *Journal of Cerebral Blood Flow & Metabolism* **23**, 472–481.
12. Tenney, J. R., Duong, T. Q., King, J. A., Ludwig, R., and Ferris, C. F. (2003) Corticothalamic modulation during absence seizures in rats: a functional MRI assessment. *Epilepsia* **44**, 1133–1140.
13. Ludwig, R., Bodgdanov, G., King, J., Allard, A., and Ferris, C. F. (2004) A dual RF resonator system for high-field functional magnetic resonance imaging of small animals. *Journal of Neuroscience Methods* **132**, 125–135.
14. Fishbein, K. W., McConville, P., and Spencer, R. G. (2001) The lever-coil: A simple, inexpensive sensor for respiratory and cardiac motion in MRI experiments. *Magnetic Resonance Imaging* **19**, 881–889.
15. Burdett, N. G., Carpenter, T. A., and Hall, L. D. (1993) A simple device for respiratory gating for the MRI of laboratory animals. *Magnetic Resonance Imaging* **11**, 897–901.
16. Li, J. P. and Vlodavsky, I. (2009) Heparin, heparan sulfate and heparanase in inflammatory reactions. *Thrombosis and Haemostasis* **102**, 823–828.
17. Vlodavsky, I., Ilan, N., Nadir, Y., Brenner, B., Katz, B. Z., Naggi, A., Torri, G., Casu, B., and Sasisekharan, R. (2007) Heparanase, heparin and the coagulation system in cancer progression. *Thrombosis Research* **120**(Suppl 2), S112–120.
18. Bock, N. A., Konyer, N. B., and Henkelman, R. M. (2003) Multiple-mouse MRI. *Magnetic Resonance in Medicine* **49**, 158–167.
19. Bock, N. A., Zadeh, G., Davidson, L. M., Qian, B., Sled, J. G., Guha, A., and Henkelman, R. M. (2003) High-resolution longitudinal screening with magnetic resonance imaging in a murine brain cancer model. *Neoplasia (New York, NY)* **5**, 546–554.
20. Nieman, B. J., Bishop, J., Dazai, J., Bock, N. A., Lerch, J. P., Feintuch, A., Chen, X. J., Sled, J. G., and Henkelman, R. M. (2007) MR technology for biological studies in mice. *NMR in Biomedicine* **20**, 291–303.
21. Bock, N. A., Nieman, B. J., Bishop, J. B., and Mark Henkelman, R. (2005) In vivo multiple-mouse MRI at 7 Tesla. *Magnetic Resonance in Medicine* **54**, 1311–1316.
22. Ramirez, M. S., Ragan, D. K., Kundra, V., and Bankson, J. A. (2007) Feasibility of multiple-mouse dynamic contrast-enhanced MRI. *Magnetic Resonance in Medicine* **58**, 610–615.
23. Nieman, B. J., Flenniken, A. M., Adamson, S. L., Henkelman, R. M., and Sled, J. G. (2006) Anatomical phenotyping in the brain and skull of a mutant mouse by magnetic resonance imaging and computed tomography. *Physiological Genomics* **24**, 154–162.
24. Xu, S., Gade, T. P., Matei, C., Zakian, K., Alfieri, A. A., Hu, X., Holland, E. C., Soghomonian, S., Tjuvajev, J., Ballon, D., and Koutcher, J. A. (2003) In vivo multiple-mouse imaging at 1.5 T. *Magnetic Resonance in Medicine* **49**, 551–557.
25. Beuf, O., Jaillon, F., and Saint-Jalmes, H. (2006) Small-animal MRI: signal-to-noise ratio comparison at 7 and 1.5 T with multiple-animal acquisition strategies. *Magma (New York, NY)* **19**, 202–208.
26. Dazai, J., Bock, N. A., Nieman, B. J., Davidson, L. M., Henkelman, R. M., and Chen, X. J. (2004) Multiple mouse biological loading and monitoring system for MRI. *Magnetic Resonance in Medicine* **52**, 709–715.

Section II

Special Techniques

Chapter 6

Cerebral Perfusion MRI in Mice

Frank Kober, Guillaume Duhamel, and Virginie Callot

Abstract

Perfusion MRI is a tool to assess the spatial distribution of microvascular blood flow. Arterial spin labeling (ASL) is shown here to be advantageous for quantification of cerebral microvascular blood flow (CBF) in rodents. This technique is today ready for assessment of a variety of murine models of human pathology including those associated with diffuse microvascular dysfunction. This chapter provides an introduction to the principles of CBF measurements by MRI along with a short overview over applications in which these measurements were found useful. The basics of commonly employed specific arterial spin-labeling techniques are described and theory is outlined in order to give the reader the ability to set up adequate post-processing tools. Three typical MR protocols for pulsed ASL on two different MRI systems are described in detail along with all necessary sequence parameters and technical requirements. The importance of the different parameters entering theory is discussed. Particular steps for animal preparation and maintenance during the experiment are given, since CBF regulation is sensitive to a number of experimental physiological parameters and influenced mainly by anesthesia and body temperature.

Key words: Cerebral blood flow, arterial spin labeling, perfusion, anesthesia, flow-alternating inversion-recovery.

1. Introduction

Unlike magnetic resonance angiography that deals with direct visualization of macroscopic vessels, perfusion MRI attempts to gather information about the status and function of capillary microcirculation within the brain tissue. Microvascular alterations play a key role in numerous pathologies and the assessment of perfusion can significantly contribute to a better characterization of human disease in animal models. In stroke, territorial microvascular changes are a direct consequence of macrovascular obstruction. Beyond the study of pathology, dynamic mapping of microvascular changes can also be used as a measure of brain activity during stimuli (functional MRI) or during pharmacological stress.

Several measures are useful for microvascular characterization, the most known being capillary blood flow (in the brain, one uses the term cerebral blood flow (CBF)), which represents the volume of blood exchanged per unit time in a given mass of tissue (mL/g/min). Besides CBF, cerebral blood volume (CBV) is another important parameter of microcirculation. CBV is expressed as the fraction of capillary blood volume per tissue mass (mL/g). Finally, capillary vascular resistance (CVR) can be assessed as a test of microvascular function by submitting the organism to an external pharmacological stress, which in the brain can be obtained, for instance, by an increased CO_2 concentration in the breathing air.

Assessment of tissue perfusion in the brain has been in the focus of many clinical and research studies and a variety of MRI techniques for measuring capillary blood flow have been used in the past for both human and animal studies.

Dynamic susceptibility contrast (DSC) MRI uses a bolus injection of a contrast agent, whose concentration dynamics are observed through sequential imaging with fast repetition. Tissue perfusion can be assessed by evaluation of the image signal dynamics during the first passage of the contrast agent at the location of interest. The bolus-tracking technique has entered clinical routine as a strong tool for accessing relative capillary blood flow and capillary blood volume in human brain. While DSC is a technique sensitive enough to provide reliable relative blood flow and blood volume measurements even in low-flow conditions in humans, absolute quantification of CBF is submitted to a number of assumptions that are still under intensive investigation. Since capillary blood flow in small rodents is roughly five times higher than in humans, a dynamic first-pass measurement of the contrast agent through the organ requires extremely short bolus injections and imaging times.

Arterial spin labeling (ASL) MRI uses magnetically labeled blood as an endogenous tracer. The labeling consists of a local magnetization inversion of arterial blood that is achieved prior to image acquisition either by using dedicated rf pulses (pulsed ASL, PASL) or through a continuous wave irradiation at a location upstream of the imaging slice(s) (continuous ASL, CASL). Similar to DSC imaging, ASL techniques have also been widely applied in clinical research studies, but they were originally set up in small animal studies (1, 2), where they appear particularly well adapted. In ASL, the signal is proportional to capillary blood flow, which is comparatively high in rodents. In addition to pure MR signal considerations, the lifetime of the magnetic labeling is determined by the longitudinal relaxation time of blood, which increases with field strength. Both the high blood flow observed in rodents and the high field strength commonly used for their exploration are therefore an advantage for the ASL technique. Compared with DSC, ASL permits relatively easy and reliable quantitative CBF measurements that can be repeated during the same experiment. The primary intention of using ASL is CBF assessment, although theoretically, changes in CBV can also be measured through a combination of ASL with blood oxygenation level-dependent (BOLD) contrast techniques. For a quantitative assessment of CBV, however, the use of contrast agent techniques is recommended, the two alternatives being dynamic first-pass assessment using a standard contrast agent or T_1 mapping before and after injection of an intravascular agent (3, 4).

1.1. A Short Overview of Applications

A variety of pathophysiological studies have used perfusion MRI, often along with other MR modalities. The most prominent application is imaging of stroke models by occlusion of the middle cerebral artery (MCAO). Adapted from clinical research, the correlation and mismatch between T_2-weighted MRI and diffusion and perfusion MRI of the affected ischemic territory yield important information about the time course of brain injury following stroke (5–7) and efficacy of drug treatments. Quantitative CBF measurements have provided information on microvascular alterations in mouse models of cerebral malaria (8) and Alzheimer's disease (9, 10). Perfusion mapping can also be used for studying brain activity. Functional MRI (11, 12) studies have been mainly carried out on rats and although some examples can be found in the literature, fMRI studies in mice are scarce so far and appear to be more difficult than in rats. A few studies on tumor models have been published to date (13, 14).

2. Materials

2.1. Autoradiography: A Standard of Reference

Autoradiographic measurements are still recognized as the gold standard for CBF quantification, against which non-invasive techniques are compared. **Table 1** shows regional CBF values found by Frietsch et al. (15) using autoradiography. These values were obtained in conscious state after temporary isoflurane anesthesia during surgery. MRI CBF values obtained with isoflurane anesthesia are therefore generally higher due to its vasodilating properties (*see* **Note 1**).

We note that fluorescent microsphere microembolization has also been recognized as a standard of reference and can be carried out in the rodent brain (16).

Table 1
Regional CBF given in mL/100 g/min and measured with autoradiography in conscious healthy mice. Adapted from [15] with permission

Mean CBF	121 ± 14	**Telencephalon**	
Regional CBF		Hippocampus CA1	114 ± 15
Cerebellum		Hippocampus CA2	98 ± 17
Cerebellar cortex	145 ± 17	Hippocampus CA3	119 ± 21
Dentate nuclei	245 ± 45	Hippocampus CA4	120 ± 21
		Dentate gyrus	104 ± 15
Medulla-pons		Amygdaloid complex	82 ± 15
Vestibular nucleus	256 ± 54	Globus pallidus	111 ± 17
Cochlear nucleus	271 ± 89	Caudate nucleus	128 ± 15
Superior olive	239 ± 59	Nucleus accumbens	122 ± 10
Pontine gray	136 ± 19	Visual cortex	124 ± 28
Lateral lemniscus	217 ± 60	Auditory cortex	157 ± 33
		Parietal cortex	166 ± 44
Mesencephalon		Sensory motor cortex	171 ± 35
Inferior colliculus	265 ± 79	Frontal cortex	165 ± 39
Superior colliculus	155 ± 18	Cingulate cortex	198 ± 53
Substantia nigra c.p.	172 ± 24	Piriform cortex	121 ± 13
Substantia nigra r.p.	115 ± 17	Lateral septal nuclei	111 ± 16
Diencephalon		**Myelinated fiber tracts**	
Medial geniculate body	186 ± 26	Internal capsule	58 ± 10
Lateral geniculate body	178 ± 35	Medial habenulae	183 ± 41
Mammillary body	199 ± 62	Lateral habenulae	224 ± 44
Hypothalamus	108 ± 19	Corpus callosum	79 ± 18
Ventral thalamus	188 ± 38	Genu of corpus callosum	71 ± 10
Lateral thalamus	203 ± 47	Cerebellar white matter	65 ± 8

2.2. MR Magnet, Gradient System, and rf Coils

1. *MR system* with high-field magnet (vertical or horizontal, **Note 2**)
2. *Gradient coils* >200 mT/m, ramp times <200 μs
3. *Pre-emphasis correction* compatible with small animal EPI studies
4. *rf coils:*
   - Transmit: resonator, homogeneous length must be greater than 3 cm
   - Receive: a decoupled dedicated brain surface coil or small decoupled resonator is advantageous

Table 2
Two dedicated small animal MR systems used for the example protocols presented in this chapter

MR System	Bruker 4.7 T/30 horizontal Biospec	Bruker 11.75 T/89 mm vertical Avance WB
Gradients	Shielded 116 mm bore 200 mT/m	Shielded 50 mm bore 1 T/m
EPI ramp time	110 μs	80 μs
Transmit rf coil	6 cm diameter birdcage	TX/RX 3 cm diameter, 5 cm length, birdcage
Receive rf coil	Decoupled pre-amplified 2 cm diameter surface coil	TX/RX 3 cm diameter 5 cm length, birdcage
Respiratory monitoring	Biopac acquisition system, pressure balloon on abdomen	SA Instruments acquisition system, pressure balloon on abdomen
Anesthesia	Isoflurane, standard vaporizer	Isoflurane, pump-injector vaporizer/Univentor

The protocol described in the **Section 3** is tailored to the two dedicated small animal MR systems, gradients, and rf coils shown here in **Table 2**.

2.3. Software

1. *Acquisition:* Paravision 3.1 (Bruker, Ettlingen, Germany)
2. *Post-processing:* Home-built using Interactive Data Language (IDL, ITT Visual Solutions, Boulder, CO, USA). The software should be capable of performing pixel-by-pixel fitting of the functions described in the theory section to the serial image data acquired with the MR system. It should be capable of processing multiple parameter maps (ΔM and T_1^{app} maps) obtained by the pixel-by-pixel analysis done beforehand. Analysis of regions of interest on the resulting parameter maps should be provided.

2.4. Other Equipment Necessary for ASL of the Rodent Brain

1. *Isoflurane vaporizer* (classical airstream vaporizer or Univentor pump injector vaporizer) if isoflurane anesthesia has been chosen.
2. *Respiration monitoring:* pressure sensor and dedicated acquisition system.
3. *Warming pad:* thermostatic heating blanket, either electrical (Pelletier) heating or circulating hot water.
4. *Temperature monitoring:* a rectal probe connected to an electrically isolated temperature amplifier should be used: body temperature maintenance at 37°C is recommended.

3. Methods

The two major labeling schemes used in arterial spin labeling are depicted in **Fig. 1**.

3.1. Continuous, Pseudo-continuous, and Dynamic ASL

With these three continuous ASL (CASL) techniques, magnetic labeling is achieved distal from the slice(s) of observation. This can either be accomplished by a dedicated labeling coil (two-coil CASL) or using the imaging coil itself (single-coil CASL). All the CASL techniques were shown in theory and experimentally to provide significantly better sensitivity than their pulsed ASL counterparts, although a quantitative assessment of blood flow appears less straight in absolute terms. Two-coil continuous techniques (17, 18) require dedicated labeling coils and drivers that are not commonly installed in small animal imaging setups. Also, in mice, the distance from the labeling coil to the imaging coil becomes small, making interference between both coils more difficult to avoid. Therefore, Williams et al. (1) have proposed to achieve the label with the imaging coil itself (single-coil CASL). In this case, the labeling induces off-resonance saturation of the magnetization outside the labeling slice including the image slice location, which influences the detected on-resonance signal by magnetization transfer (MT). This effect has to be compensated through an adequate control experiment. The control experiment can be accomplished with a labeling location on the opposite side of the imaging slice or at the same labeling location

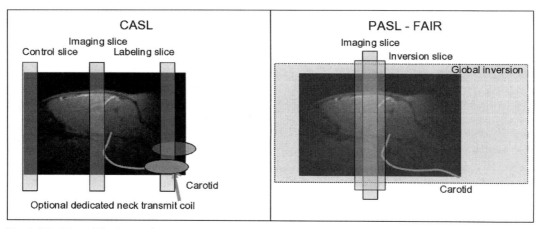

Fig. 1. Principles of the two major arterial spin-labeling techniques for CBF measurement. *Left scheme*: CASL uses either a labeling slice placed at neck level and a control slice at the opposite side of the imaging slice, or, as an option, a dedicated coil produces the rf transmission for labeling in which case the control experiment employs no labeling. *Right scheme*: FAIR pulsed ASL compares the magnetization after either global (control) or slice-selective inversion produced symmetrically around the imaging slice.

using dedicated amplitude-modulated rf emissions. The first solution is compromised in multi-slice studies for reasons of asymmetry and the second solution affects the labeling efficiency of the experiment.

For advanced studies, dynamic ASL (DASL) has been proposed (19) as a modification of CASL. It employs specifically adapted dynamic labeling functions so that beyond capillary blood flow, the involved relaxation times of blood and tissue can be extracted from the measurement data simultaneously.

More recently, pseudo-continuous (pCASL) techniques (20, 21) were proposed mimicking a continuous label with pulsed rf and gradient fields. The pseudo-continuous technique promises to become the most interesting technique since it has the potential to combine the advantages of PASL and CASL providing a good balance between tagging efficiency and sensitivity. It overcomes the problem of off-resonance saturation intrinsically and therefore has multi-slice capability without loss of efficiency.

3.2. Pulsed ASL—Flow-Sensitive Alternating Inversion-Recovery (FAIR)

In pulsed ASL (PASL), the magnetization of arterial blood is labeled using a short selective or global inversion pulse. Although a number of preparation modules and readout schemes have been proposed for pulsed ASL in clinical work, flow-sensitive alternating inversion recovery (22) is the most commonly used labeling scheme for pulsed ASL in animal studies. The CBF-dependent magnetization difference is produced by alternating acquisitions after global spin inversion of the largest possible volume of the animal and slice-selective inversion of the magnetization only in the imaging slice. In the latter case, blood entering the imaging slice during the inversion time TI was not inverted and therefore contributes to an accelerated observed relaxation in the slice. After global inversion, blood entering the imaging slice has been inverted as well and contributes to slow down the apparent tissue relaxation at field strengths at which blood T_1 is longer than tissue T_1. At the inversion time TI, a difference in longitudinal magnetization will be present between the two preparations. This difference is proportional to CBF, as shown in detail below.

PASL is less efficient than CASL due to long recovery periods, during which no signal is acquired, and due to a magnetic "arterial input function" that decays by longitudinal relaxation of blood (*see also* **Note 3**). Unlike in CASL techniques, the labeling zone for the FAIR technique is, however, placed symmetrically and close to the observation slice, which leads to better robustness with respect to arterial transit times (**Note 4**) and to an easier modeling (*see also* **Note 5**). Look–Locker inversion-recovery techniques were first proposed for rodent cardiac applications (23, 24), but also had their entry into brain studies. Despite higher sensitivity to bulk flow in larger vessels, they were shown to provide slightly better sensitivity than classical FAIR techniques

(25, 26) and to open new ways to the exploration of the magnetization dynamics after labeling (27), including assessment of the shape of the magnetic "arterial input function." Look–Locker techniques also provide an inherent and simultaneous measurement of T_1, which enters CBF quantification, and which might be submitted to changes during the protocol.

3.3. Technical Requirements for Arterial Spin Labeling MRI of the Mouse Brain

Arterial spin labeling can be performed with any MRI setup provided that a homogeneous excitation radiofrequency coil is available. The necessary sensitive length of the excitation coil depends on the particular technique employed. Pulsed ASL techniques require coverage from brain to abdomen such that a sufficient amount of blood can be inverted. The reason for this necessity is that blood turnover in the rodent body is fast compared with humans so that non-inverted blood would enter the imaging slice before image acquisition, should the spatial coverage of the inversion be insufficient. This situation is shown in **Fig. 2**.

For continuous ASL techniques, in which labeling is provided during a longer time period, the rf excitation coverage has to include at least the animal's neck for efficient labeling, which takes place in a thin slice usually placed at carotid level.

As outlined in the introduction of this chapter, high field strengths are advantageous since they provide higher MR signal and longer T_1 values. Very high static fields, however, also impose more constraints on the robustness of rapid imaging techniques

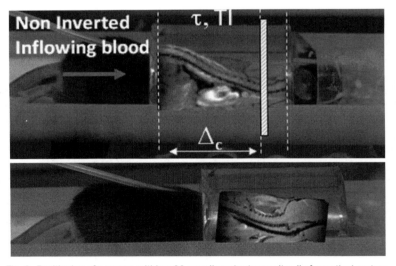

Fig. 2. Positioning of a mouse within a 30 mm diameter transmit coil of a vertical system. *Top*: If the heart is included in the sensitive rf volume of the coil, global spin inversions are effective on a sufficient amount of blood, but if the rf coil is too short to produce a sufficiently large inversion volume, the image location of interest would be outside the sensitive volume. *Bottom*: In this example, it is not possible to place the animal such that both location of interest (brain) and heart are included in the rf volume, which would preclude efficient pulsed ASL.

like EPI or spiral acquisition making the MR measurement more difficult in practice. Also, T_2 relaxation is faster at very high field, which partly counterbalances the achieved signal gain, and which in addition leads to underestimation of perfusion due to the presence of blood in the imaging signal if not accounted for in the model used for data analysis (28). Efficient perfusion imaging using arterial spin labeling can be carried out in mice at field strengths as low as 4.7 T.

3.4. Theory and Post-Processing

The theory of arterial spin labeling is based on the modified Bloch equations [2] which describe the tissue magnetization $M_b(t)$ in the presence of inflowing spins.

$$\frac{dM_b(t)}{dt} = \frac{M_b^0 - M_b}{T_{1b}} + fM_a - fM_v \quad [1]$$

where f is the blood flow, T_{1b} is the T_1 of brain tissue, M_b^0 is the equilibrium brain tissue magnetization. M_a and M_v are the arterial and venous blood magnetization, respectively.

Quantification of blood flow with ASL relies on subtracting two images acquired with different states for arterial blood magnetization $\left(\Delta M_b(t) = M_b^{lab}(t) - M_b^{ctrl}(t)\right)$, so that equation [1] can be rewritten to

$$\frac{d\Delta M_b}{dt} + \frac{\Delta M_b}{T_{1b}} = f\left(M_a^{lab} - M_a^{ctrl}\right) - f\left(M_v^{lab} - M_v^{ctrl}\right) \quad [2]$$

A standard kinetic model (29, 30) can be used to consider the magnetization difference ΔM_b as a tracer delivered by arterial blood flow (delivery function $c(t)$), which exchanges between blood and tissue compartment and which is cleared by a combination of T_1 decay (relaxation function $A(t)$) and venous outflow (residue function $r(t)$). The signal difference ΔM_b can then be expressed as a convolution of the three previously defined functions,

$$\Delta M_b = c(t) \otimes (r(t) \cdot A(t)) = \int_0^t c(t')\, r(t-t')\, A(t-t')\, dt' \quad [3]$$

Three main assumptions (*see also* **Note 5**) are made in the classical ASL model to calculate ΔM_b.
1. The arrival of the labeled blood in a brain tissue voxel is considered to be uniform. Because there is a gap between the labeling region and the imaging plane, a transit delay, δ, is necessary for the labeled blood before it arrives in a brain tissue voxel. This leads to the situation where labeled

blood enters the voxel between $t = \delta$ and $t = \tau + \delta$; τ being the duration of the labeling. Outside this time interval, the blood entering the voxel is unlabeled. For PASL techniques the gap between labeling and imaging regions, applied to avoid effects of imperfect slice profile, is usually very small leading to short δ (**Note 4**). In CASL, the labeling plane can be distant from the imaging region (e.g., tagging in the neck for imaging in the brain) leading to higher δ values. The duration of the labeling, τ, is set by the extent of the inversion area for PASL techniques, whereas it is equal to the duration of the rf pulse for CASL. Consequent to the assumption with regard to the transit delay, the arterial delivery function can be defined for both PASL techniques and CASL as

$$c(t) = \begin{cases} 0 & 0 < t < \delta \\ 2M_a^0 f \cdot \alpha \cdot e^{\frac{-t}{T_{1a}}} \quad \text{(PASL)} & \delta < t < \tau + \delta \\ 2M_a^0 f \cdot \beta \cdot e^{\frac{-\delta}{T_{1a}}} \quad \text{(CASL)} & \\ 0 & \tau + \delta < t \end{cases} \quad [4]$$

where T_{1a} is the blood relaxation time. α and β represent the inversion efficiencies for PASL techniques and CASL respectively (α and β are equal to 1 for a complete inversion). Equation [4] shows that the amount of labeled blood obtained after the inversion, ($2 M_a^0 \alpha$ for PASL and $2 M_a^0 \beta$ for CASL) is delivered to brain tissue with blood flow f. The exponential terms account for blood relaxation time, T_{1a}, after the labeling. For PASL techniques, since the labeling is performed instantaneously at time $t = 0$, relaxation starts at $t = 0$. For CASL, since the blood is continuously labeled at the inversion plane, relaxation only starts at $t = \delta$.

2. The second assumption considers water as a freely diffusible tracer, meaning that the water exchange between tissue and blood can be described by a single-compartment model. In particular, it is assumed that the tracer concentration ratio of tissue compartment to venous compartment is a constant equal to the tissue/blood partition coefficient λ (in volume of water per mass of tissue divided by the volume of water per volume of blood). With this assumption, the residue function simply reduces to

$$r(t) = e^{-\frac{f}{\lambda}t} \quad [5]$$

whereas the equilibrium blood magnetization M_a^0 is equal to M_b^0/λ the equilibrium brain tissue magnetization divided by the partition coefficient.

3. The third assumption considers that once the labeled blood has reached the tissue voxel, there is a complete and instantaneous extraction of water from the vascular space (fast exchange), so that relaxation is driven by tissue relaxation time, T_{1b}. With this assumption, the relaxation function is

$$A(t) = e^{-\frac{t}{T_{1b}}} \quad [6]$$

Including Eqs. [4], [5] and [6] in Eq. [3] leads to the following values for ΔM_b:

For PASL,

$$\Delta M_b = \begin{cases} 0 & 0 < t < \delta \\ 2\frac{M_b^0}{\lambda} f \cdot \alpha \cdot \frac{e^{-\frac{\delta}{T_{1a}}}}{\frac{1}{T_{1app}} - \frac{1}{T_{1a}}} \cdot \left(e^{-\frac{t-\delta}{T_{1a}}} - e^{-\frac{t-\delta}{T_{1app}}}\right) & \delta < t < \tau + \delta \\ 2\frac{M_b^0}{\lambda} f \cdot \alpha \cdot \frac{e^{-\frac{t}{T_{1app}}}}{\frac{1}{T_{1app}} - \frac{1}{T_{1a}}} \cdot e^{-\delta\left(\frac{1}{T_{1a}} - \frac{1}{T_{1app}}\right)} \left(e^{-\tau\left(\frac{1}{T_{1a}} - \frac{1}{T_{1app}}\right)} - 1\right) & \tau + \delta < t \end{cases}$$

[7]

with

$$\frac{1}{T_{1app}} = \frac{1}{T_{1b}} + \frac{f}{\lambda}. \quad [8]$$

For CASL, the magnetization difference becomes

$$\Delta M_b = \begin{cases} 0 & 0 < t < \delta \\ 2\frac{M_b^0}{\lambda} f \cdot \beta \cdot T_{1app} \cdot e^{-\frac{\delta}{T_{1a}}} \cdot \left(1 - e^{-\frac{t-\delta}{T_{1app}}}\right) & \delta < t < \tau + \delta \\ 2\frac{M_b^0}{\lambda} f \cdot \beta \cdot T_{1app} \cdot e^{-\frac{\delta}{T_{1a}}} \cdot e^{-\frac{t-\tau-\delta}{T_{1app}}} \cdot \left(1 - e^{-\frac{\tau}{T_{1app}}}\right) & \tau + \delta < t \end{cases}$$

[9]

For both techniques PASL and CASL, absolute quantification of blood flow will in addition require the determination of ΔM_b, T_{1app}, and of the inversion efficiency (α or β). This is usually achieved by the acquisition of an extra scan dedicated to the measurement of these parameters.

3.5. Some Sensitivity Considerations

Equations [8] and [9] show that in theory, CASL gives higher signal than PASL. This is illustrated in **Fig. 3a**, which shows simulated ΔM_b curves obtained for PASL and CASL with

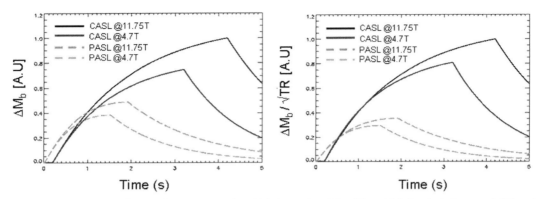

Fig. 3. (a) Magnetization difference between control and label experiment for different labeling techniques and different field strengths as a function of post-labeling time. At longer post-labeling times, the CASL signal still benefits from inflowing blood whereas the PASL signal diminishes due to relaxation. (b) $\Delta M_b/\sqrt{TR}$ sensitivity curves for TR=7 s/10 s, which are typical repetition times used at 4.7 T and 11.75 T to be in full relaxation conditions. These plots clearly highlight the advantage of CASL over PASL in terms of sensitivity.

characteristic parameters obtained at 4.7 T/11.75 T (26, 31): $T_{1app} = 1.3$ s/1.7 s, $T_{1a} = 1.7$ s/2.1 s. τ was chosen to be 1.5 s/1.9 s for PASL and 3 s/4 s for CASL. δ was estimated to be 0.02 s for PASL and 0.2 s for CASL. Finally, perfect inversion ($\alpha = \beta = 1$) was assumed. Although dependent on the values of f and T_{1app}, the quantity $\Delta M_b/M_b^0$ is usually of the order of a few percent making several signal averages necessary to obtain a sufficient signal-to-noise ratio (SNR).

Since SNR is proportional to the square root of the number of averages, each of which taking one repetition time (TR) for acquisition, the value can be considered as the time-independent sensitivity of a method, with TR being the minimum time between successive averages. For CASL, since blood magnetization is only locally affected at the inversion plane and when the labeling rf is on, TR should in principle be simply equal to the sum of the labeling time τ and the delay before imaging $t - \tau$. For PASL, however, a large volume of blood is affected by the inversion rf pulse due to its large frequency profile. At time t after the image acquisition, the blood might not have relaxed entirely so that an extra recovery time is usually necessary between successive averages. This recovery time increases with the field as shown in **Fig. 3b**.

There is a strong interest in optimizing PASL sequences to increase their sensitivity $\Delta M_b/\sqrt{TR}$ and several strategies have already been developed, among which are presaturation pulses allowing faster repetition in FAIR techniques (31, 32) or Look–Locker approaches (25, 26). It is important to note that the strategies employed to enhance the efficiency can considerably modify the quantification models due to the nonfull relaxation conditions. Some aspects of acquiring data in multiple slices are discussed in **Note 6**.

3.6. Experimental Procedures

3.6.1. Animal Preparation

1. isoflurane induction at 3% isoflurane concentration in an induction chamber
2. after visual detection of stable breathing motion at ~60 bpm, the mouse can be transferred to the MR setup
3. either room air or pure oxygen gas stream of at least 300 mL/min with maintenance of approximately 1.5% isoflurane concentration administered through a mouse face mask (**Note 1**)
4. respiration monitor balloon placed on the abdomen with light pressure
5. adjust isoflurane concentration to obtain breath rates around 120 bpm (**Note 7**)
6. additional fixation by rods pressed against the skull is helpful to avoid breath motion, the face mask itself is in general not sufficient to avoid all motion artifacts with brain imaging
7. warming pad: thermostatic heating blanket, either electrical (Pelletier) heating or circulating hot water
8. temperature monitoring: a rectal probe connected to an electrically isolated temperature amplifier should be used: body temperature maintenance at 37°C is recommended (**Note 8**)
9. Throughout the experiment, stable physiology is necessary (consider **Note 9**)

3.6.2. MR Scans

The following two tables describe the MR protocol divided in two parts: preparation scans (**Table 3**) and the actual perfusion MRI acquisition (**Table 4**).

3.6.3. Post-Processing

1. Accomplish all steps necessary to obtain satisfactory raw images from the acquired data. For EPI, this may include ghost corrections and gridding for data points acquired during ramp time. It is important that the raw images have at least 16 bit resolution.
2. Apply a mask to all raw images based on pixel intensity to avoid post-processing of noise.
3. Use pixel-by-pixel exponential fitting of the T_1^{app} measurement data. This data may also provide a map of the inversion efficiency α.
4. Obtain a value for the T_1 of arterial blood T_1^a at your field strength (**Note 10**).
5. Compute ΔM difference maps between label and control scans.

Table 3
Preparative scan protocol for MR imaging of CBF in the mouse. The protocol is shown for vertical systems with small homogeneous TX/RX resonators and for horizontal systems with decoupled RX coils and larger TX resonators

4.7 T horizontal Surface Rx coil		11.75 T vertical Homogeneous Tx/Rx coil
FAIR-EPI-1	**FAIR-LL**	**Presat-FAIR**
Preparation/Localization **Shim**: EPI readout modules require a relatively good shim quality over the imaged volume. The use of higher order shims might, however, be problematic due to their effects on the volume outside the imaging slice and on the global inversion efficiency. The authors of this chapter only used manual first-order shimming for their studies.		
Scout image (three-plane gradient echo)		
FLASH image (Saggital) for longitudinal localization		
FOV 25 × 25 mm², matrix 128 × 128 *slice thickness 2.0 mm*		*FOV 30 × 30 mm², matrix 256 × 128* *slice thickness 1.0 mm*
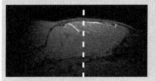 (image is cropped)		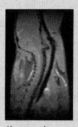 (image is cropped)
Positioning of the axial perfusion MRI location at the longitudinal level required by the study		
RARE T2-weighted image (axial) as anatomical reference		
FOV 25 × 25 mm², matrix 256 × 256 slice thickness 1.0 mm TR/TE = 2000/40 ms		FOV 20 × 20 mm², matrix 256 × 256 slice thickness 0.65 mm, TR/TE = 5000/9 ms
 (image is cropped)		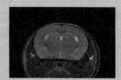 (image is cropped)

6. In case of multiple-TI FAIR, use pixel-by-pixel fitting of the ΔM data with the appropriate model and the previously obtained T_1^{app} and α maps to obtain maps of the blood flow f. In case of single-TI FAIR, obtain maps of the blood flow f by inverting the appropriate model equation and inserting the previously obtained T_1^{app} and α maps.

Table 4
Perfusion MR protocol shown with example parameters for two measurement techniques and the two MR systems described in Table 3

Perfusion MRI		
EPI adjustments are done using the same image location as the T_2-weighted image. At this time, the shim can be adjusted again to optimize the EPI image quality.		
FAIREPI-1TI (4.7 T)	**LLFAIREPI (4.7 T)**	**Presat-FAIR-1TI (11.75 T)**
Readout: SE-EPI / 2 segments Matrix: 128 × 64 FOV: 25 × 25 mm^2 BW 200 kHz TE: 18 ms ST: 1.0 mm Exc; Gauss 0.5 ms/90°	*Readout:* GE-EPI / 2 segments Matrix: 128 × 64 FOV: 25 × 25 mm^2 BW: 200 kHz TE: 9 ms ST: 1.0 mm Exc; Gauss 0.5 ms/~10°	*Readout:* SE-EPI / 2 segments Matrix: 128 × 128 FOV: 25 × 25 mm^2 BW: 400 kHz TE: 14.9 ms ST: 1.0 mm Exc; Gauss 0.5 ms/90°
ASL Inv: Sech 4 ms Inv/Im ST ratio: 3:2 TI: 1.3 s TR: 7.5 s Temp. resolution: 35 s Acq. time : ~10 min	*ASL* Inv: Sech 4 ms Inv/Im ST ratio: 3:2 Ni x ΔTI: 50 × 150 ms TR: 3 s Temp. resolution: 42 s Acq. time : ~10 min	*ASL* Inv: Sech 4 ms Inv/Im ST ratio: 3:2 TI: 1.7 s τ: 3.4 s Temp. resolution: 21 s Acq. time: 10 min
Calibration scans: The following complementary information is necessary in order to quantify CBF		
T_{1app}, α, M_0 *Slice selective IR-EPI* TI = [100, 400, 800, 1000, 1300, 1500, 3000, 7500] ms TR = 7.5 s	flip angle θ, sat. fact: ξ *No extra scan* But flip angle calibration required, using a reference phantom of known T1 within FOV.	T_{1app}, α, M_0 *Slice selective IR-EPI* TI = [130, 530, 900, 1300, 1700, 2000, 4000, 10000] ms TR = 10 s
Processing Points acquired during the 220 µs EPI gradient switch are discarded before reconstruction		
Analytical solution $\Delta M = 2\frac{M_b^0}{\lambda} f \cdot$ $\alpha \cdot \left(e^{-TI/T_{1a}} - e^{-TI/T_{1app}}\right) \cdot$ $\left(1/T_{1app} - 1/T_{1a}\right)^{-1}$	Least-square fitting $\Delta M = \frac{M_b^{0,eff}}{\lambda} f \cdot \left(2\alpha^{eff} - 1 + \frac{1}{\xi}\right) \cdot$ $\left(e^{-TI/T_{1a}} - e^{-TI/T_{1app}^{eff}}\right) \cdot$ $\left(1/T_{1app}^{eff} - 1/T_{1a}\right)^{-1}$	Analytical solution $\Delta M = 2\frac{M_b^0}{\lambda} f \cdot \alpha \cdot$ $\left(e^{-TI/T_{1a}} - e^{-TI/T_{1app}}\right) \cdot$ $\left(1/T_{1app} - 1/T_{1a}\right)^{-1} \cdot$ $\left(1 - e^{-\frac{\tau}{T_{1a}}}\right)$

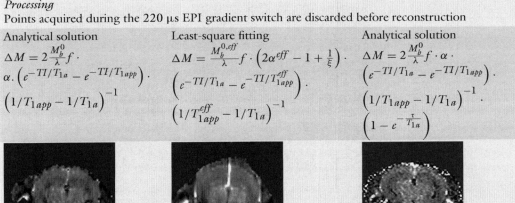

CBF maps obtained with each of the methods

4. Notes

1. *Anesthesia*: Almost all in vivo magnetic resonance studies require immobilization of the animal during image acquisition. Anesthetics are known to modify a number of physiological parameters in different ways. The most commonly used anesthetic in MR studies is isoflurane, which is a potent vasodilator if applied in too high concentrations, but also to a minor degree in normal concentrations with respect to the conscious state (33, 34). On the other hand, mixtures of ketamine and xylazine injected intraperitoneally result in reduced CBF values (35). This is illustrated in **Fig. 4** showing typical CBF maps using the same ASL technique (Look–Locker FAIR gradient echo) on mice under isoflurane and ketamine/xylazine anesthesia. Propofol, which is preferable for fMRI investigation also seems to contribute to diminished CBF as shown in rats using microspheres (36).

   The use of N_2O as an addition to the oxygen or airstream containing the anesthetic has also been shown to significantly increase CBF by up to 100% (37) in rats when no other anesthetic was used in parallel.

   Although there is no recommendation on the use of anesthetics for CBF measurements that can be given in general terms, isoflurane, sevoflurane, and propofol best preserve hemodynamics in terms of cardiac function and

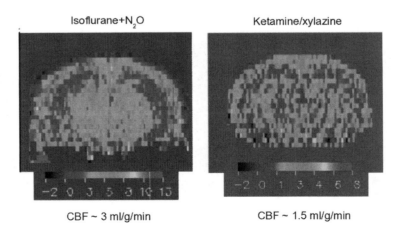

Fig. 4. Mouse cerebral blood flow maps measured using a Look–Locker FAIR Gradient Echo MR technique in mice anesthetized with isoflurane (*left*) and N_2O or ketamine/xylazine (*right*). Note the different color scales. While both isoflurane and N_2O are acting as potent vasodilators, ketamine/xylazine reduces CBF and general hemodynamics (Courtesy of C. Laigle, MF. Penet, A. Viola, Centre de Résonance Magnétique Biologique et Médicale, Marseille, France).

blood pressure. Isoflurane has an advantage over all gaseous anesthetics in that it is easily administered and controlled during the experiment, whereas the liquid propofol is in general administered through a catheter.

When interpreting results, the potential influence of the anesthetic must be taken into account.

2. *Vertical versus horizontal magnets*: A recent study has shown dependency of CBF on the animal's position. CBF reductions of up to 40% were reported in vertically positioned animals compared with the horizontal position (38). While results obtained in vertical systems are still meaningful, the reduction of perfusion has to be taken into account in the interpretation of the results.

3. *Labeling efficiency*: The labeling efficiency α for PASL should be of the order of 0.95 using adiabatic inversion pulses, such as the hyperbolic secant pulse proposed in the experimental description. Neglecting imperfect inversion in cases of PASL does therefore not lead to substantial errors in quantification. For CASL, the inversion is flow driven, and its efficiency therefore depends on the B_0 and B_1 field strengths, the blood flow velocity in the labeling zone, and the available gradient strength. As a general rule, efficient arterial blood inversion can be reached if $1/T_1 < 1/T_2 << G_{ave}v/B_{1ave} << \gamma B_{1ave}$. At very high field, blood T_2 is short whereas flow velocities in mice are comparable to human arterial flow velocities. High gradient strengths and B_1 fields are therefore necessary to make this inequality hold in very high field animal systems.

4. *Arterial transit time*: This term has to be clearly distinguished from the mean transit time (MTT), which is often measured in clinical studies, and which describes the mean time a spin needs to pass from the entry of the arterial to the venous compartment. The arterial transit time describes the time of travel from the labeling location to the entry of the capillary compartment at the location of interest. Both PASL and CASL are sensitive to this transit time, which has to be estimated to avoid errors in CBF quantification. The use of several inversion times (TI) in PASL techniques can provide a measurement of the transit time along with CBF (32). However, this approach has the clear disadvantage of requiring long acquisition times. Other approaches have been suggested to make the experiments less sensitive to transit time variations. For CASL, a delay w, introduced after the labeling has shown to reduce the transit-time sensitivity of the CBF measurement (39). If w is greater than δ, the third regime of equation [9] applies, and therefore CBF

would depend on δ times the difference between apparent tissue and blood relaxation rates. Since these rates are relatively close to each other ($T_{1app}/T_{1a} \sim 1.5/1.7$ s at 4.7 T, $T_{1app}/T_{1a} \sim 1.8/2.1$ s at 11.75 T) the dependence of CBF quantification on transit time is reduced. For PASL, a modification called QUIPSS II (quantitative imaging of perfusion using a single subtraction—second version) has been proposed (40). QUIPSS II uses a saturation of the tagged region after an initial delay (TI_1) and allows an additional delay ΔTI before the image acquisition. This leads to the effective creation of a finite bolus of tagged blood. The transit time sensitivity and the intravascular tagged blood effect are therefore reduced if TI_1 and ΔTI are optimized. In mice, the transit time can be considered as short given that in larger arteries, blood velocities are comparable with the situation in humans, whereas the distances to travel are an order of magnitude below the human situation.

5. *Quantification model*: The main ASL experiments and perfusion quantification attempts rely on a model that describes the exchange of a freely diffusible tracer (water) between capillaries and the tissue (Kety model). This model is, however, still questioned and its validity challenged by experiments, theory, or different approaches (27, 28, 41). None of the proposed alternatives has, however, reached a clear conclusion and each bears limitations in the process of quantification.

6. *Multi-slice readouts*: CASL performed with a separate labeling coil does not face any special challenges other than the technical requirement for multi-slice perfusion imaging. In particular, there is no off-resonance saturation of the tissue of interest by the long rf pulse. For a single-coil system, this saturation cannot be corrected accurately across multiple slices, leading to quantitation errors. Modifications of the CASL scheme for the control experiment have been proposed to overcome this limitation (e.g., amplitude modulation of the labeling rf for the control (42), but usually at the cost of a significant loss in the labeling efficiency (43–45). Again, because by nature the pCASL experiment compensates the problem of off-resonance saturation intrinsically without a loss of efficiency, it should be a method of choice for multi-slice continuous ASL experiments. For PASL, any labeling module (FAIR, PICORE, EPISTAR) can be used with mutli-slice readouts, provided that it is modified to a QUIPSSII version. The main difficulty stands in the choice of appropriate values for TI_1 and TI_2 for small animals (29, 40, 46).

7. *Breathing patterns*: As a general rule, isoflurane should be applied in minimal concentration by monitoring the breath rate and breathing pattern. Breath rates below 70 min^{-1} indicate an unnecessarily high isoflurane concentration, while breath rates above 120 min^{-1} are an indication of insufficient concentration. The breathing pattern during stable anesthesia under isoflurane is characterized by comparatively long expiration states and short inspiration/expiration phases. A more sinusoidal pattern along with breath rates above 120 min^{-1} appear at too low concentrations.

8. *Body temperature*: Under isoflurane, the body temperature of rodents becomes sensitive to the room temperature and has to be maintained carefully, since it has a significant influence on CBF. The animal therefore has to be warmed during the experiment under permanent control of the body temperature. The latter is best achieved using a rectal probe connected to an electrically isolated amplifier.

9. *Influence of physiology on absolute CBF*: The vascular system is controlled and auto-controlled in a wide dynamic range. It is permanently adapting itself to the external conditions. CBF quantification errors made by the different measurement techniques are therefore in most cases overtaken by true CBF variations as a consequence of differences in the physiological conditions. Comparisons with literature data can only be made if the physiological conditions match those in which the reference values were obtained. An appropriate discussion with respect to animal physiology should therefore be provided with all CBF studies.

10. *T_1 of arterial blood*: The equations in the methods section show that blood T_1 (T_1^a) enters blood flow quantification. Direct measurement of T_1^a in arteries is difficult in mice due to the small diameter of the vessels. In general, literature values are used for this parameter, but we state here that Thomas et al. (47) have proposed a repeated modified FAIR sequence allowing measurement of T_1^a indirectly. Direct measurement of T_1^a can otherwise be accomplished in the left ventricular chamber using a cardiac-gated inversion-recovery experiment.

References

1. Williams, D. S., Detre, J. A., Leigh, J. S., Koretsky, A. P. (1992) Magnetic resonance imaging of perfusion using spin inversion of arterial water. *Proc. Natl. Acad. Sci. USA.* **89**, 212–216.

2. Detre, J. A., Leigh, J. S., Williams, D. S., Koretsky, A. P. (1992) Perfusion imaging. *Magn. Reson. Med.* **23**, 37–45.

3. Perles-Barbacaru, A. T., Lahrech, H. (2007) A new magnetic resonance imaging

method for mapping the cerebral blood volume fraction: the rapid steady-state T1 method. *J. Cereb. Blood Flow Metab.* **27**, 618–631.

4. Schwarzbauer, C., Syha, J., Haase, A. (1993) Quantification of regional blood volumes by rapid T1 mapping. *Magn. Reson. Med.* **29**, 709–712.

5. Leithner, C., Gertz, K., Schrock, H., Priller, J., Prass, K., Steinbrink, J., Villringer, A., Endres, M., Lindauer, U., Dirnagl, U., Royl, G. (2008) A flow sensitive alternating inversion recovery (FAIR)-MRI protocol to measure hemispheric cerebral blood flow in a mouse stroke model. *Exp. Neurol.* **210**, 118–127.

6. Prass, K., Royl, G., Lindauer, U., Freyer, D., Megow, D., Dirnagl, U., Stockler-Ipsiroglu, G., Wallimann, T., Priller, J. (2007) Improved reperfusion and neuroprotection by creatine in a mouse model of stroke. *J. Cereb. Blood Flow Metab.* **27**, 452–459.

7. van Dorsten, F. A., Olah, L., Schwindt, W., Grune, M., Uhlenkuken, U., Pillekamp, F., Hossmann, K. A., Hoehn, M. (2002) Dynamic changes of ADC, perfusion, and NMR relaxation parameters in transient focal ischemia of rat brain. *Magn. Reson. Med.* **47**, 97–104.

8. Penet, M. F., Viola, A., Confort-Gouny, S., Le Fur, Y., Duhamel, G., Kober, F., Ibarrola, D., Izquierdo, M., Coltel, N., Gharib, B., Grau, G. E., Cozzone, P. J. (2005) Imaging experimental cerebral malaria in vivo: significant role of ischemic brain edema. *J. Neurosci.* **25**, 7352–7358.

9. Weidensteiner, C., Metzger, F., Bruns, A., Bohrmann, B., Kuennecke, B., von Kienlin, M. (2009) Cortical hypoperfusion in the B6.PS2APP mouse model for Alzheimer's disease: comprehensive phenotyping of vascular and tissular parameters by MRI. *Magn. Reson. Med.* **62**, 35–45.

10. Faure, A., Verret, L., Bozon, B., El Tannir El Tayara, N., Ly, M., Kober, F., Dhenain, M., Rampon, C., Delatour, B. (2009) Impaired neurogenesis, neuronal loss, and brain functional deficits in the APPxPS1-Ki mouse model of Alzheimer's disease. *Neurobiol. Ageing.* doi:10.1016/j.neurobiolaging.2009.03.009.

11. Belliveau, J. W., Kennedy, D. N., Jr., McKinstry, R. C., Buchbinder, B. R., Weisskoff, R. M., Cohen, M. S., Vevea, J. M., Brady, T. J., Rosen, B. R. (1991) Functional mapping of the human visual cortex by magnetic resonance imaging. *Science* **254**, 716–719.

12. Ogawa, S., Lee, T. M., Kay, A. R., Tank, D. W. (1990) Brain magnetic resonance imaging with contrast dependent on blood oxygenation. *Proc. Natl. Acad. Sci. USA.* **87**, 9868–9872.

13. Sun, Y., Schmidt, N. O., Schmidt, K., Doshi, S., Rubin, J. B., Mulkern, R. V., Carroll, R., Ziu, M., Erkmen, K., Poussaint, T. Y., Black, P., Albert, M., Burstein, D., Kieran, M. W. (2004) Perfusion MRI of U87 brain tumors in a mouse model. *Magn. Reson. Med.* **51**, 893–899.

14. Silva, A. C., Kim, S. G., Garwood, M. (2000) Imaging blood flow in brain tumors using arterial spin labelling. *Magn. Reson. Med.* **44**, 169–173.

15. Frietsch, T., Maurer, M. H., Vogel, J., Gassmann, M., Kuschinsky, W., Waschke, K. F. (2007) Reduced cerebral blood flow but elevated cerebral glucose metabolic rate in erythropoietin overexpressing transgenic mice with excessive erythrocytosis. *J. Cereb. Blood Flow Metab.* **27**, 469–476.

16. De Visscher, G., Haseldonckx, M., Flameng, W., Borgers, M., Reneman, R. S., van Rossem, K. (2003) Development of a novel fluorescent microsphere technique to combine serial cerebral blood flow measurements with histology in the rat. *J. Neurosci. Methods* **122**, 149–156.

17. Silva, A. C., Zhang, W., Williams, D. S., Koretsky, A. P. (1995) Multi-slice MRI of rat brain perfusion during amphetamine stimulation using arterial spin labelling. *Magn. Reson. Med.* **33**, 209–214.

18. Dixon, W. T., Du, L. N., Faul, D. D., Gado, M., Rossnick, S. (1986) Projection angiograms of blood labeled by adiabatic fast passage. *Magn. Reson. Med.* **3**, 454–462.

19. Barbier, E. L., Silva, A. C., Kim, H. J., Williams, D. S., Koretsky, A. P. (1999) Perfusion analysis using dynamic arterial spin labeling (DASL). *Magn. Reson. Med.* **41**, 299–308.

20. Dai, W., Garcia, D., de Bazelaire, C., Alsop, D. C. (2008) Continuous flow-driven inversion for arterial spin labeling using pulsed radio frequency and gradient fields. *Magn. Reson. Med.* **60**, 1488–1497.

21. Wu, W. C., Fernandez-Seara, M., Detre, J. A., Wehrli, F. W., Wang, J. (2007) A theoretical and experimental investigation of the tagging efficiency of pseudocontinuous arterial spin labelling. *Magn. Reson. Med.* **58**, 1020–1027.

22. Kim, S. G. (1995) Quantification of relative cerebral blood flow change by flow-sensitive alternating inversion recovery

(FAIR) technique: application to functional mapping. *Magn. Reson. Med.* **34**, 293–301.

23. Belle, V., Kahler, E., Waller, C., Rommel, E., Voll, S., Hiller, K., Bauer, W., Haase, A. (1998) In vivo quantitative mapping of cardiac perfusion in rats using a noninvasive MR spin-labeling method. *J. Magn. Reson. Imaging* **8**, 1240–1245.

24. Kober, F., Iltis, I., Izquierdo, M., Desrois, M., Ibarrola, D., Cozzone, P. J., Bernard, M. (2004) High-resolution myocardial perfusion mapping in small animals in vivo by spin-labeling gradient-echo imaging. *Magn. Reson. Med.* **51**, 62–67.

25. Gunther, M., Bock, M., Schad, L. R. (2001) Arterial spin labeling in combination with a Look–Locker sampling strategy: inflow turbo-sampling EPI-FAIR (ITS-FAIR). *Magn. Reson. Med.* **46**, 974–984.

26. Kober, F., Duhamel, G., Cozzone, P. J. (2008) Experimental comparison of four FAIR arterial spin labeling techniques for quantification of mouse cerebral blood flow at 4.7 T. *NMR Biomed.* **21**, 781–792.

27. Petersen, E. T., Lim, T., Golay, X. (2006) Model-free arterial spin labeling quantification approach for perfusion MRI. *Magn. Reson. Med.* **55**, 219–232.

28. St Lawrence, K. S., Wang, J. (2005) Effects of the apparent transverse relaxation time on cerebral blood flow measurements obtained by arterial spin labeling. *Magn. Reson. Med.* **53**, 425–433.

29. Wong, E. C., Buxton, R. B., Frank, L. R. (1998) A theoretical and experimental comparison of continuous and pulsed arterial spin labeling techniques for quantitative perfusion imaging. *Magn. Reson. Med.* **40**, 348–355.

30. Buxton, R. B., Frank, L. R., Wong, E. C., Siewert, B., Warach, S., Edelman, R. R. (1998) A general kinetic model for quantitative perfusion imaging with arterial spin labeling. *Magn. Reson. Med.* **40**, 383–396.

31. Duhamel, G., Callot, V., Cozzone, P. J., Kober, F. (2008) Spinal cord blood flow measurement by arterial spin labelling. *Magn. Reson. Med.* **59**, 846–854.

32. Pell, G. S., Thomas, D. L., Lythgoe, M. F., Calamante, F., Howseman, A. M., Gadian, D. G., Ordidge, R. J. (1999) Implementation of quantitative FAIR perfusion imaging with a short repetition time in time-course studies. *Magn. Reson. Med.* **41**, 829–840.

33. Okamoto, H., Meng, W., Ma, J., Ayata, C., Roman, R. J., Bosnjak, Z. J., Kampine, J. P., Huang, P. L., Moskowitz, M. A., Hudetz, A. G. (1997) Isoflurane-induced cerebral hyperemia in neuronal nitric oxide synthase gene deficient mice. *Anesthesiology* **86**, 875–884.

34. Hendrich, K. S., Kochanek, P. M., Melick, J. A., Schiding, J. K., Statler, K. D., Williams, D. S., Marion, D. W., Ho, C. (2001) Cerebral perfusion during anesthesia with fentanyl, isoflurane, or pentobarbital in normal rats studied by arterial spin-labeled MRI. *Magn. Reson. Med.* **46**, 202–206.

35. Lei, H., Grinberg, O., Nwaigwe, C. I., Hou, H. G., Williams, H., Swartz, H. M., Dunn, J. F. (2001) The effects of ketamine-xylazine anesthesia on cerebral blood flow and oxygenation observed using nuclear magnetic resonance perfusion imaging and electron paramagnetic resonance oximetry. *Brain Res.* **913**, 174–179.

36. Werner, C., Hoffman, W. E., Kochs, E., Schulte am Esch, J., Albrecht, R. F. (1993) The effects of propofol on cerebral and spinal cord blood flow in rats. *Anesth. Analg.* **76**, 971–975.

37. Baughman, V. L., Hoffman, W. E., Miletich, D. J., Albrecht, R. F. (1990) Cerebrovascular and cerebral metabolic effects of N_2O in unrestrained rats. *Anesthesiology* **73**, 269–272.

38. Foley, L. M., Hitchens, T. K., Kochanek, P. M., Melick, J. A., Jackson, E. K., Ho, C. (2005) Murine orthostatic response during prolonged vertical studies: effect on cerebral blood flow measured by arterial spin-labeled MRI. *Magn. Reson. Med.* **54**, 798–806.

39. Alsop, D. C., Detre, J. A. (1996) Reduced transit-time sensitivity in noninvasive magnetic resonance imaging of human cerebral blood flow. *J. Cereb. Blood Flow Metab.* **16**, 1236–1249.

40. Wong, E. C., Buxton, R. B., Frank, L. R. (1998) Quantitative imaging of perfusion using a single subtraction (QUIPSS and QUIPSS II). *Magn. Reson. Med.* **39**, 702–708.

41. St. Lawrence, K. S., Frank, J. A., McLaughlin, A. C. (2000) Effect of restricted water exchange on cerebral blood flow values calculated with arterial spin tagging: a theoretical investigation. *Magn. Reson. Med.* **44**, 440–449.

42. Alsop, D. C., Detre, J. A. (1998) Multisection cerebral blood flow MR imaging with continuous arterial spin labelling. *Radiology* **208**, 410–416.

43. O'Gorman, R. L., Summers, P. E., Zelaya, F. O., Williams, S. C., Alsop, D. C., Lythgoe, D. J. (2006) In vivo estimation of the flow-driven adiabatic inversion efficiency

for continuous arterial spin labeling: a method using phase contrast magnetic resonance angiography. *Magn. Reson. Med.* **55**, 1291–1297.
44. Wang, J., Zhang, Y., Wolf, R. L., Roc, A. C., Alsop, D. C., Detre, J. A. (2005) Amplitude-modulated continuous arterial spin-labeling 3.0-T perfusion MR imaging with a single coil: feasibility study. *Radiology* **235**, 218–228.
45. Werner, R., Norris, D. G., Alfke, K., Mehdorn, H. M., Jansen, O. (2005) Improving the amplitude-modulated control experiment for multislice continuous arterial spin labeling. *Magn. Reson. Med.* **53**, 1096–1102.
46. Wong, E. C., Buxton, R. B., Frank, L. R. (1997) Implementation of quantitative perfusion imaging techniques for functional brain mapping using pulsed arterial spin labelling. *NMR Biomed.* **10**, 237–249.
47. Thomas, D. L., Lythgoe, M. F., Gadian, D. G., Ordidge, R. J. (2006) In vivo measurement of the longitudinal relaxation time of arterial blood (T1a) in the mouse using a pulsed arterial spin labeling approach. *Magn. Reson. Med.* **55**, 943–947.

Chapter 7

High Field Diffusion Tensor Imaging in Small Animals and Excised Tissue

Bernadette Erokwu, Chris Flask, and Vikas Gulani

Abstract

Molecular diffusion plays an important role in many biological phenomena. Magnetic Resonance (MR) imaging is inherently sensitive to diffusion and can be used to help understand diffusion processes. Diffusion MR imaging is most widely used for imaging the ischemic brain. Diffusion imaging and diffusion tensor imaging (DTI) have also found clinical application in areas such as tumor characterization throughout the body, imaging of demyelinating disorders, and fiber tract mapping. DTI is also now widely used in small animal imaging—both in vivo and in characterizing excised tissue. DTI studies in these settings can be accomplished with high resolution and can offer exquisite contrast, but the technical and practical challenges can sometimes be different than those seen on clinical MRI scanners. Here, a stepwise methodology is presented for using small-bore, high field strength scanners (>3 T) for DTI. This chapter is aimed at addressing readers with no prior knowledge of DTI and we present both a basic explanation of underlying principles and a practical approach to the experiment.

Key words: Diffusion tensor imaging, small animal imaging, trace, tensor, fractional anisotropy (FA), mean diffusivity.

1. Introduction

Diffusion is a process by which molecules, such as water, are transported due to random Brownian motion in a given medium (1). Two major, complementary descriptions of diffusion have been presented by Fick and Einstein (2, 3). Fick, working from an analogy to transport of heat, described diffusion as the transport of material from an area of high concentration to an area of low concentration. This is known as Fick's First Law of Diffusion. In his second law, Fick described the time dependence of the diffusion phenomenon. This description works beautifully in understanding the concept of diffusive mass transfer, but for NMR/MRI, where the most common form of diffusion being studied is the self-diffusion of water, this description is difficult to apply. An alternative formulation for studying diffusion involves a statistical approach toward studying the phenomenon. Specifically, the probability of a molecule traveling a distance r in time t can be calculated (4–6). For a liquid such as water in an unrestricted environment, this probability distribution $p(r,t)$ is a Gaussian with a mean of 0 and is given by the following equation in the three-dimensional case, where D is the diffusivity (3–5):

$$p(r,t) = \frac{e^{-r^2/4Dt}}{4\pi Dt^{\frac{3}{2}}} \quad [1]$$

The mean squared displacement can then be calculated from the second moment of this distribution:

$$<r^2(t)> = \left(\iiint\right)_{-\infty}^{\infty} \left[(x^2 + y^2 + z^2)\right] p(r,t)\, dxdydz = 6Dt \quad [2]$$

This is the well-known Einstein equation of diffusion and this model is used to derive the NMR signal attenuation due to diffusion of molecules (4, 7).

Most modern diffusion imaging techniques employ variations of bipolar pulsed field gradient methods (8) to obtain diffusion sensitivity. In these sequences, two magnetic field gradients are applied sequentially to dephase and then rephase spins or protons. Any spins that have moved in the interval between the two gradients do not experience similar magnetic field as during the first gradient and thus they do not get rephased, resulting in a net loss of spin coherence and a quantifiable loss in the MRI signal (9) which can be related to the apparent diffusion coefficient or ADC. In the simplest case, with spin-echo MRI and considering for the moment only a single diffusion coefficient per voxel, it can be shown (starting from the Einstein formulation of diffusion) that the signal and the ADC are related as follows (10):

$$S = S_0 e^{-bD} \quad [3]$$

where S is the measured signal, S_0 the signal in absence of the gradients, and D is the ADC. The b-factor in this equation is a function of the gradient strength G, duration δ, and time separation Δ between the gradient pulses. The b-factor must be calculated for each sequence, which can be non-trivial when the diffusion gradients interact with the imaging gradients in a complicated fashion. For the simple case where imaging gradients and the resultant cross terms in the b-factor calculations are ignored, $b = \gamma^2 G^2 \delta^2 (\Delta - \delta/3)$, (10) where γ is a physical constant, the gyromagnetic ratio. D can be determined from the above equation if at least two images are acquired with different diffusion weightings (and thus different b-factors). It also becomes evident from this equation that any movement of spins during the diffusion time Δ will result in MRI signal loss. Thus diffusion MRI is sensitive to all motion of molecules during the sequence, and not just purely diffusive motion. Moreover, the observed diffusion coefficients in tissues reflect diffusion in several compartments—extracellular and intracellular, intranuclear, mitochondrial, etc. — which may all have different diffusivities if one could resolve them. It is for these reasons that the diffusion coefficient and tensor as measured with MRI are called the apparent diffusion coefficient (ADC) and apparent diffusion tensor (ADT), respectively.

The diffusivity of a substance is not always identical in all directions. For example, diffusion coefficients can be relatively large in one direction and small in another. Diffusion is more generally characterized by a second-order tensor, or a matrix of nine coefficients:

$$\boldsymbol{D} = \begin{pmatrix} D_{xx} & D_{xy} & D_{xz} \\ D_{yx} & D_{yy} & D_{yz} \\ D_{zx} & D_{zy} & D_{zz} \end{pmatrix} \qquad [4]$$

This implies that nine diffusivities would have to be determined to measure the tensor. However, the diffusion tensor is symmetric, meaning that $D_{xy} = D_{yx}$, $D_{xz} = D_{zx}$, and $D_{yz} = D_{zy}$. Thus, only six diffusivities need to be determined. When the axes of the measurement system coincide exactly with the axes of the fiber or object being studied, the off-diagonal elements of the tensor (i.e., D_{xy}, D_{xz}, and D_{yz}) become zero, and the measured diagonal diffusivites D_{xx}, D_{yy}, and D_{zz} are termed the principal diffusivities, often annotated D_{11}, D_{22}, and D_{33} or λ_1, λ_2, and λ_3. Of course, in the general case, the measurement and principal axes of diffusion do not coincide for most fibers. In such a case, these principal diffusivities, which are also called the eigenvalues of the diffusion tensor, along with the eigenvectors of the system (defining the directions of the principal diffusivities), can be calculated from the measured tensor by a simple mathematical transformation (*see* **Note 1**).

In an environment where the barriers to diffusion are identical in all directions, diffusion is considered isotropic. This is the case, for example, in a cup of water, where even if molecules are allowed to diffuse for relatively long times, they are unlikely to encounter a barrier to diffusion so that movement in one direction is not favored over any other. In such situations, the diffusion coefficients in all directions are equivalent (i.e., $D_{xx} = D_{yy} = D_{zz}$, etc.).

For isotropic cases, measurement of a single scalar coefficient suffices to completely characterize the system. This condition is often not met in tissue measurements where cell membranes, packing, and other ordering in tissue structures promote directed water diffusion. For medical imaging applications, this is the case for example in white matter, muscle, cartilage, kidneys, tumors, and in the lens of the eye (11–19). In such settings, diffusion is said to be anisotropic, reflecting the non-equal diffusivities in various directions.

The coefficients along the x, y, and z-directions (D_{xx}, D_{yy}, and D_{zz}) are simply the diagonal terms of the tensor and are the principal diffusivities of the system only when the laboratory coordinate frame coincides with that of the sample orientation. In the general case where the laboratory axes are different from the orientation of the sample, the principal diffusivities are found by determining the eigenvalues of the measured diffusion matrix (1, 15, 16). These principal diffusivities are independent of the relation between the coordinate system of the laboratory and the coordinate system of the sample (1, 16).

The tensor nature of diffusion must be taken into account in the mathematical framework for describing diffusion. Torrey derived the Bloch equations with diffusion terms and included the rudiments of the tensor analysis needed to consider anisotropic diffusion (20). In 1965, Stejskal extended this analysis to give a formal description of the effect of anisotropic diffusion on the NMR signal (7). Basser et al. further extended this treatment and measured the ADT by applying gradients in various directions (15, 16). Starting with the mathematical expressions for the description of magnetization in the presence of spin diffusion (20) they derived the expressions necessary to calculate the various terms of **D** (or ADT) from an MRI experiment. They showed that

$$S = S_0 \, e^{-\sum_{i=1}^{3}\sum_{j=1}^{3} b_{ij} D_{ij}} \qquad [5]$$

where

$$b_{ij} = \gamma \iint G_i dt' \int G_j dt' \, dt \qquad [6]$$

Here, the indices i and j represent the measurement axes (1, 2, and 3 being x, y, and z, respectively). G_i and G_j thus represent the applied diffusion gradients in various directions. This equation bears obvious resemblance to the case for measuring a single diffusion coefficient. However, all six unique diffusivities must be accounted for, and thus this implies that at least seven images must be acquired to completely determine the seven variables in this equation (S_0 and the six unique diffusivities). Obviously, the calculation of the b-matrix as per Eq. [6] is considerably more complex than that of *b*-factors for a uni-directional experiment and care must be taken to account for the interaction between the various diffusion and imaging gradients.

2. Materials

2.1. In Vivo Imaging

The following MR-system-compatible materials are essential for running a DTI experiment in small rodents. The procedure for setting up the experiment on excised tissue (excision, fixation, insertion, etc.) is not described, but the mechanics of the DTI experiment would be similar other than the obviation of the need for anesthesia equipment.

1. A small-bore high field strength whole body animal MRI scanner—typically, scanners between 4.7 T and 17.5 T and having maximum gradient capabilities between 400 and 1000 mT/m are used for rodent diffusion tensor imaging.

2. Calibrated small rodent inhalation anesthesia equipment such as (E-Z Anesthesia, Euthanex Corporation, Palmer, PA) located outside the MR room, is used for both induction and maintenance of anesthesia.

3. A small animal monitoring and gating system (Model 1025 by SA Instruments, Inc., Stony Brook, NY) consisting of a fan that drives a set percentage of heat to the animal in the scanner is typically employed. This continuously monitors several of the animal's physiological parameters such as respiration, body temperature, heart rate, and ECG during the experiment. The system should be equipped with a gating module which can be set to selectively acquire data during a desired phase of respiration if needed.

4. A rat or mouse bed furnished with nonmagnetic ear and tooth bars (Bruker Biospin MRI, Inc., Billerica, MA).

5. A rectal temperature probe (SA Instruments, Inc., Stony Brook, NY).

6. A respiration pillow sensor pad (SA Instruments, Inc., Stony Brook, NY).

7. Masking tape (Scotch 234) used to tape down the respiration pad on to the chest of the animal and to restrict excessive motion from other regions of the animal.

8. A heating module fan serving as a direct source of heat to the animal (SA Instruments, Inc., Stony Brook, NY).

9. Acquisition and data processing software, e.g., MATLAB (The Mathworks, Natick, MA) or ParaVision 4.0 software implemented on the MRI scanner.

2.2. Transcardial Perfusion for Ex Vivo Sample Preparation

1. Ringer's solution: Prepare 7.2 g NaCL, 0.17 g $CaCl_2$, 0.37 g KCl and dissolve in reagent-grade H_2O, and bring the final volume to 1 L. Adjust the pH to 7.3–7.4. Once thoroughly dissolved, filter through a 0.22-μm filter, aliquot into single-use volumes (25–50 mL), and autoclave.
2. BD 30 G 1 inch PrecisionGlide needle from Fisher Scientific.
3. Surgical instruments for transcardial perfusion such as thumb forceps, 120 mm; tissue forceps, 115 mm; and eye scissors, 115 mm (Harvard Apparatus).
4. Mini-pump variable flow (Fisher Scientific).
5. Phosphate-buffered saline (PBS): Prepare 0.13 M NaCl (Sigma), 0.0027 M KCl (Sigma), 0.01 M Na_2HPO_4 (Sigma), 0.0018 M KH_2PO_4 (Sigma) in 1 L dd H_2O. Adjust pH to 7.4 with HCl and NaOH. Autoclave before storage at room temperature.
6. Four percent paraformaldehyde (Sigma): Prepare a 4% (w/v) solution in PBS. The solution has to be carefully heated and not allowed to boil using a stirring hot plate in a fume hood) to dissolve. The solution is filtered and then cooled to room temperature for use.

3. Methods

3.1. Pre-scan Check

1. MRI scanners are large, powerful magnets and their magnetic strength should never be underestimated. Any and all loose metals from both the operator and the animal (such as ear tags) must be removed prior to entry into the scanner room, with or without the animal.
2. Apart from strict daily quality checks for proper running and maintenance of the scanner hardware, steps should be taken to check the system for smooth running of experiments (*see* **Note 2**).
3. It is important to ensure that there is enough anesthetic (i.e., isoflurane) for the duration of the experiment and that all physiological monitors are working properly before an animal is anesthetized.
4. The animal should be positioned in either rat or mouse coil and at the specified distance from the isocenter of the magnet. A phantom can be used in addition to boost signal-to-noise ratio. Improper measurement or calculation (*see* **Note 3**) will result in inability to acquire ideal data and is unnecessary waste of time. This is because the animal has to

be pulled out during the course of the experiment and the repositioning process restarted.

3.2. Animals

1. An animal's identification is correctly entered into the software and desired protocol selected before the start of experiment.

2. Inhalation anesthetic is generally preferred to an injectible one because it allows for safer, faster induction and quick recovery. The animal is placed in an induction chamber and anesthesia is rapidly induced using 3% isoflurane in oxygen at a flow rate of 2 L/min (typical). It is very important to watch the animal very closely at this time because of varied strain and individual differences in response to anesthesia. It should be ensured that the animal reaches the desired depth of anesthesia, which can be tested by the pedal withdrawal reflex. A slight pressure is placed on the animal's hind limb and if this causes the foot to be withdrawn, the animal is lightly anesthetized. However, this reflex is absent in deeply anesthetized animals.

3. After induction, the animal is immediately placed in either the prone or the supine position on a bed connected to another oxygen and isoflurane inlet and anesthesia is maintained at 1 to 1.5% at the same flow rate as at induction.

4. A rectal thermometer is inserted and its temperature should be registered on the monitor. The body temperature is maintained at 37°C or at the initial starting temperature. At this point, the source of heat should be directed toward the animal (but not too close) and temperature carefully and constantly monitored and controlled to avoid induction of hypothermia or hyperthermia which affects the outcome of MRI data.

5. The respiration pad is placed and adhered to the chest area with the aid of a masking tape to sense and record the animal's respiration rate. Typically, respiration frequencies of about 80 and 60 breaths/min are maintained in mice and rats, respectively. Much higher respiration rates may result in movement of the animal in the scanner and consequent change of initial positioning and geometry. This, of course, delays and prolongs the experiment in that the animal will either have to be anesthetized deeper or moved out of the scanner and repositioned, before resumption of data acquisition. On the other hand, lower respiration rates can lead to death of the animal.

6. It is critical at this point to ensure that the head or region of interest is stabilized either by using non-magnetic ear bars equipped on the bed or by placing extra pieces of scotch tape around the area.

7. The animal's physiological parameters should be cautiously maintained during duration of scan.

3.3. Fixed Tissue Imaging

1. This section is included in case fixed tissue imaging, rather than in vivo imaging, is required. Fixing tissue cross-links proteins and alters the local microenvironment and there is work being done to determine the effects of fixation on the ADC and ADT. Although these effects mean that the tissue being studied is obviously not in its native environment, fixation offers a powerful tool for microscopy as it offers the ability to image for significantly longer, allowing more signal averaging, better resolution, better SNR, and no physiological motion.

2. Typically, tissue is fixed, starting with mammalian Ringer's solution. This is accomplished by placing a needle into the left ventricle of the in situ beating heart. The right atrium is then cut to allow the blood to drain out. The Ringer's solution is then allowed to flow in.

3. Fixation can be accomplished by using a pump, or gravity (1.5 m above the heart). Some feel that the latter is a more gentle method for fixing tissue.

4. After the Ringer's solution has flowed through the tissue and the fluid return into the heart is clear, the solution is replaced with 4% paraformaldehyde in phosphate-buffered saline (PBS).

5. Finally, the tissue of interest is extracted and kept in the paraformaldehyde solution for storage. A few hours before imaging, the tissue is often transferred to simple PBS to minimize T_2/T_2^* shortening effects.

3.4. Automatic Adjustments

At this time, complete automatic adjustments are enabled to boost signal-to-noise ratio during data acquisition.

1. Global shimming is done from a saved file prior to placement of the animal.

2. Tuning a probe adjusts its current resonance frequency to that of the subject.

3. Matching a probe ensures similar impedance with the subject in order to diffuse reflected power. If a probe is not properly matched, an additional radiofrequency is emitted to the subject or sample, thereby increasing its temperature.

3.5. Data Acquisition with DTI Pulse Sequence

The most common diffusion sequence used today is the bipolar pulse field gradient (PFG) diffusion sequence described by Stejskal and Tanner (8). This sequence was applied to MR imaging in the 1980s (21–24). It is essentially a spin echo imaging sequence with a pair of stepped, identical gradients of duration and magnitude added on either side of a 180° RF pulse. These gradients allow an image to be made "diffusion-weighted" in a given direction. By applying gradients in different combinations in various directions, weighting can be achieved at different vector orientations. At least seven different directions are needed to encode the tensor and typically 9–12 directions are used. However, tens to hundreds of encoding directions can potentially be necessary, for example, in applications such as high angular resolution diffusion imaging (HARDI) (25, 26).

3.6. Sample DTI Parameters for Both Rats and Mice

Parameters	Rat/Mouse
Measuring method	DTI
Sequence	Diffusion weighted spin echo, gradient echo, or spin echo/gradient echo–echo planar imaging
Slice thickness	0.1–1 mm (typical)
Diffusion gradient duration	2–7 ms (typical)
Diffusion gradient separation	5–14 ms (typical)
Number of diffusion directions	9–12 (typical)
b values (s/mm^2)	0–2000 (typical)

3.7. DTI Acquisition

2D or 3D (27) data should be acquired using optimized DTI parameters to yield a high signal-to-noise ratio. DTI protocols in small animal imaging often call for long acquisition times of 30 min or more, high SNR images with fewer artifacts to aid in post-processing, and accurate calculation of diffusivities and secondarily calculated parameters such as scalar invariants (*see* Step 2 in **Section 3.8**). Echo planar imaging (EPI) offers much faster diffusion acquisition. EPI as a gradient echo technique has its own challenges, including eddy current-induced distortions and magnetic susceptibility artifacts (*see* **Section 1**). However, these artifacts can be partially alleviated by using distortion correction, parallel imaging or multi-shot EPI techniques (28, 29). Artifacts seen inherently in typical DTI acquisitions are caused by eddy currents generated from switching of strong diffusion gradient pulses. These currents can be reduced by using diffusion gradients with optimized pre-emphasis adjustment and twice-refocused spin echo sequences for data acquisition (30). Other possible artifacts include motion (respiratory and cardiac), background radio frequency noise, miscalibrated gradients, background gradients, partial volume contamination, magnetic susceptibility variations, complications from B_1 inhomogeneities (19) and gradient nonlinearity. These must be recognized and corrected for, but a detailed individual discussion of each is beyond the scope of this chapter (31–33). Gating and fat suppression protocols can be employed for significant reduction of artifacts but for a much longer acquisition time. Experiments should be repeated in order to assess reproducibility and stability of the technique.

3.8. Post-processing and Data Analysis

1. Typically, the acquired images are read into an image analysis software suite, such as MATLAB (The Mathworks, Natick, MA). Images are Fourier-transformed and the various diffusivities fit from the acquired images and the b-matrix as determined from the acquisition parameters and the sequence employed (using Eqs. [5] and [6]). The eigenvalues of the diffusion matrix are determined on a pixelwise basis. Diffusion tensor eigenvalues and eigenvectors are then readily calculated as needed for analysis.

2. Scalar invariants are calculated from the measured diffusivities obtained from a DTI experiment and are not dependent on gradient directions and their relationship to the subject. Several such invariants, which are easy to conceptualize and display, have been used in DTI. These quantities are useful in that they provide structural information about the diffusion system; and along with the eigenvalues and eigenvectors of the system, they are independent of the measurement axes across data sets. The diffusion trace ($Tr(\boldsymbol{D})$) is calculated as follows:

$$Tr(\boldsymbol{D}) = \lambda_1 + \lambda_2 + \lambda_3 \quad [7]$$

The trace, or mean diffusivity ($<\lambda> = Tr/3$), gives information about the overall mobility of water molecules in the environment being studied. Several additional scalar invariants have been suggested to give information about diffusion anisotropy. The most commonly used of these quantities are fractional anisotropy (FA) and relative anisotropy (RA), which are calculated as follows:

$$FA = \sqrt{\frac{\frac{3}{2}[(\lambda_1 - <\lambda>) + (\lambda_2 - <\lambda>) + (\lambda_3 - <\lambda>)]}{\lambda_1^2 + \lambda_2^2 + \lambda_3^2}} \quad [8]$$

$$RA = \sqrt{\frac{[(\lambda_1 - <\lambda>)^2 + (\lambda_2 - <\lambda>)^2 + (\lambda_3 - <\lambda>)^2]}{3\lambda}}$$

FA and RA are similar in that both quantitate the degree of diffusion anisotropy in the tensor. FA ranges from 0 (completely isotropic) to 1 (completely anisotropic diffusion), while RA ranges from 0 to $\sqrt{2}$. FA and RA maps are often depicted for DTI data, as they, along with the

<λ>, offer computationally and conceptually the simplest yet quantitative insight into diffusion anisotropy in the system. Numerous other scalar invariants have been described, but a discussion of these is beyond the scope of this chapter. Of these, the most frequently encountered is A_σ, not shown here, which is closely related to RA.

3. Depending on the application at hand, further analysis such as percent changes in the calculated parameters can be performed, and statistical testing applied (t-tests, ANOVA, etc.).

4. Not discussed here are techniques such as HARDI or q-space imaging, multi-exponential analysis (34–39) etc., which may require the acquisition of hundreds of images and/or more complicated post-processing. The reader is referred to the rich developing literature on these subjects, well beyond what can be discussed in an introductory chapter.

3.9. DTI/FA Map Images

As examples of DTI and secondarily calculated maps, two sets of images from published DTI literature are presented here—from the cervical enlargement of an excised, fixed rat spinal cord at 17.6 T (*see* **Fig. 1**) (19) and from a neonatal rat brain at 9.4 T (*see* **Fig. 2**) (27).

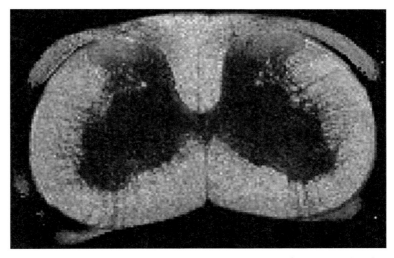

Fig. 1. FA map of an excised and fixed (4% paraformaldehyde followed by phosphate buffered saline) cervical enlargement of a rat spinal cord. Imaging parameters include: Bruker Biospin 750 MHz spectrometer, 0.7 T/m maximum applied gradient strength, home built Alterman–Grant resonator, b of 0–1995 s/mm^2, resolution 23 × 23 × 500 μm, reconstructed from 39 spin and gradient echo images. Reproduced from (19) with permission from Elsevier Science.)

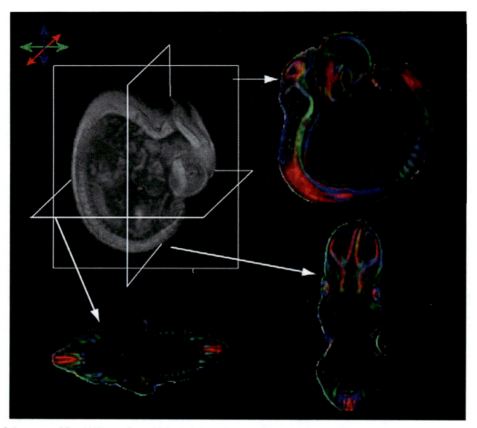

Fig. 2. Color maps of fixed (4% paraformaldehyde followed by phosphate-buffered saline and fomblin) E12 mouse embryo from a 3D DTI acquisition. Imaging parameters include: GE Omega 399 MHz spectrometer, custom-made solenoid volume coil, 3D multiple echo sequence with navigator-echo phase correction, 8 h acquisition, $37 \times 43 \times 42$ μm resolution after zero-filling. Color maps are reconstructed by assigning *red*, *green* or *blue* (*see colored arrows*) to each pixel according to the orientation of the primary eigenvector, and the signal intensity according to FA. Three representative slices extracted from the 3D data are shown in this image. (Reproduced with permission from Ref. (27) from Elsevier Science).

4. Notes

1. The eigenvalues of a 3×3 matrix

$$A = \begin{bmatrix} a & b & c \\ d & e & f \\ g & h & i \end{bmatrix}$$

can be determined by the characteristic polynomial given by

$$\det \begin{bmatrix} a-\lambda & b & c \\ d & e-\lambda & f \\ g & h & i-\lambda \end{bmatrix} = -\lambda^3 + \lambda^2(a+e+i)$$
$$+\lambda\left(db+gc+fh-ae-ai-ei\right)$$
$$+\left(aei-afh-dbi+dch+gbf-gce\right).$$

The eigenvalues are then the roots of this polynomial, which can be found using the method for solving cubic equations.

2. A typical example of a setback in continuous data acquisition occurs when there is not enough disk space in the computer for data storage.

3. A malpositioning of the animal/coil or miscalculation from a specified distance such as between a region of interest/coil and edge of the magnet, will result in an inability to properly tune and match the coil.

References

1. Crank, J. (1975) *The mathematics of diffusion*. Oxford, England: Oxford University Press.
2. Fick, A. (1855) Uber diffusion. *Poggendorffs Annalen der Physik und Chemie* **94**, 59–86.
3. Einstein, A. (1956) *Investigations on the theory of Brownian movement* R. Furth TbADC, editor. New York, NY: Dover.
4. Le Bihan, D. (1991) Molecular diffusion nuclear magnetic resonance imaging. *Magn. Res. Quar.* **7**, 1–30.
5. Kimmich, R. (1997) *NMR tomography, diffusometry, relaxometry*. Berlin, Germany: Springer.
6. Le Bihan, D. (1992) *Diffusion imaging in magnetic resonance imaging*. D. D. Stark and W. G. Bradley J, editors. St. Louis, MO: Mosby Year-Book.
7. Stejskal, E. O. (1965) Use of spin echoes in a pulsed magnetic-field gradient to study anisotropic, restricted diffusion and flow. *J. Chem. Phys.* **43**, 3597–3603.
8. Stejskal, E. O. and Tanner, J. E. (1965) Spin diffusion measurements-spin echoes in presence of a time-dependent field gradient. *J. Chem. Phys.* **42**, 288–292.
9. Chen, F., De Keyzer, F., Wang, H., Vandecaveye, V., Landuyt, W., Bosmans, H., et al. (2007) Diffusion weighted imaging in small rodents using clinical MRI scanners. *Methods* **43**(1), 12–20.
10. Gulani, V. and Sundgren, P. C. (2006) Diffusion tensor magnetic resonance imaging. *J. Neuroophthalmol.* **26**(1), 51–60.
11. Filidoro, L., Dietrich, O., Weber, J., Rauch, E., Oerther, T., Wick, M., et al. (2005) High-resolution diffusion tensor imaging of human patellar cartilage: feasibility and preliminary findings. *Magn. Reson. Med.* **53**(5), 993–998.
12. Moffat, B. A. and Pope, J. M. (2002) Anisotropic water transport in the human eye lens studied by diffusion tensor NMR microimaging. *Exp. Eye Res.* **74**(6), 677–687.
13. Beaulieu, C. and Allen, P. S. (1994) Determinants of anisotropic water diffusion in nerves. *Magn. Reson. Med.* **31**(4), 394–400.
14. Le Bihan, D., Mangin, J. F., Poupon, C., Clark, C. A., Pappata, S., Molko, N., and Chabriat, H. (2001) Diffusion tensor imaging: concepts and applications. *J. Magn. Reson. Imaging* **13**(4), 534–546.
15. Basser, P. J., Mattiello, J., and LeBihan, D. (1994) Estimation of the effective self-diffusion tensor from the NMR spin-echo. *J. Magn. Reson. Ser. B* **103**(3), 247–254.
16. Basser, P. J., Mattiello, J., and LeBihan, D. (1994) MR diffusion tenser spectroscopy and imaging. *Biophys. J.* **66**(1), 259–267.
17. Cleveland, G. G., Chang, D. C., Hazlewood, C. F., and Rorschach, H. E. (1976) Nuclear magnetic resonance measurement of skeletal muscle: anisotrophy of the

diffusion coefficient of the intracellular water. *Biophys. J.* **16**(9), 1043–1053.
18. Damon, B. M., Ding, Z., Anderson, A. W., Freyer, A. S., and Gore, J. C. (2002) Validation of diffusion tensor MRI-based muscle fiber tracking. *Magn. Reson. Med.* **48**(1), 97–104.
19. Gulani, V., Weber, T., Neuberger, T., and Webb, A. G. (2005) Improved time efficiency and accuracy in diffusion tensor microimaging with multiple-echo acquisition. *J. Magn. Reson.* **177**(2), 329–335.
20. Torrey, H. C. (1956) Bloch equations with diffusion terms. *Phys. Rev.* **104**(3), 563–565.
21. Le Bihan, D., Breton, E., Lallemand, D., Grenier, P., Cabanis, E., and Laval-Jeantet, M. (1986) MR imaging of intravoxel incoherent motions: application to diffusion and perfusion in neurologic disorders. *Radiology* **161**(2), 401–407.
22. Merboldt, K. D., Hanicke, W. and Frahm, J. (1985) Self-diffusion NMR imaging using stimulated echoes. *J. Magn. Reson.* **64**, 479–486.
23. Taylor, D. G. and Bushell, M. C. (1985) The spatial mapping of translational diffusion coefficients by the NMR imaging technique. *Phys. Med. Biol.* **30**(4), 345–349.
24. Le Bihan, D. (1985) Imagerie de diffusion in vivo par resonance magnetique nucleaire. *Cr Acad Sci (Paris)* **301**, 1109–1112.
25. Frank, L. R. (2002) Characterization of anisotropy in high angular resolution diffusion-weighted MRI. *Magn. Reson. Med.* **47**(6), 1083–1099.
26. Tuch, D. S., Reese, T. G, Wiegell, M. R., Makris, N., Belliveau, J. W., and Wedeen, V. J. (2002) High angular resolution diffusion imaging reveals intravoxel white matter fiber heterogeneity. *Magn. Reson. Med.* **48**(4), 577–582.
27. Zhang, J., Richards, L. J., Yarowsky, P., Huang, H., van Zijl, P. C. M., and Mori, S. (2003) Three-dimensional anatomical characterization of the developing mouse brain by diffusion tensor microimaging. *NeuroImage* **20**(3), 1639–1648.
28. Wang, F. N., Huang, T. Y., Lin, F. H., Chuang, T. C., Chen, N. K., Chung, H. W., et al. (2005) Propeller EPI: an MRI taechnique suitable for diffusion tensor imaging at high field strength with reduced geometric distortions. *Magn. Reson. Med.* **54**(5), 1232–1240.
29. Liu, X., Zhu, T., Gu, T., and Zhong, J. (2009) A practical approach to in vivo high-resolution diffusion tensor imaging of rhesus monkeys on a 3-T human scanner. *Magn. Res. Imaging* **27**(3), 335–346.
30. Reese, T. G., Heidi, O., Weiskoff, R. M., and Weeden, V. J. (2003) Reduction of eddy current-induced distortion in diffusion MRI using a twice-refocused spin echo. *Magn. Reson. Med.* **49**(1), 177–182.
31. Boretius, S., Natt, O., Watanabe, T., Tammer, R., Ehrenreich, L., Frahm, J., and Michaelis, T. (2004) In vivo diffusion tensor mapping of the brain of squirrel monkey, rat, and mouse using single-shot STEAM MRI. *MAGMA* **17**(3–6), 339–347.
32. Kennedy, S. D. and Zhong, J. (2004) Diffusion measurements free of motion artifacts using intermolecular dipole-dipole interactions. *Magn. Reson. Med.* **52**(1), 1–6.
33. Boretius, S., Wurfel, J., Zipp, F., Frahm, J., and Michaelis, T. (2007) High-field diffusion tensor imaging of mouse brain in vivo using single-shot STEAM MRI. *J. Neurosci. Methods* **161**(1), 112–117.
34. King, M. D., Houseman, J., Gadian, D. G., and Connelly, A. (1997) Localized q-space of the mouse brain. *Magn. Reson. Med.* **38**(6), 930–937.
35. Bilton, I. E., Duncan, I. D., and Cohen, Y. (2006) High b-value q-space diffusion in myelin-deficient rat spinal cords. *Magn. Reson. Imaging* **24**(2), 161–166.
36. Nossin-Manor, R., Duvdevani, R., and Cohen, Y. (2007) Spatial and temporal damage evolution after hemi-crush injury in rata spinal cord obtained by high value q-space diffusion magnetic resonance imaging. *J. Neurotrauma* **24**(3), 481–491.
37. Cohen, Y. and Assaf, Y. (2002) High b-value q-space analyzed diffusion-weighted MRS and MRI in neuronal tissues – a technical review. *NMR Biomed.* **15**(7–8), 516–542.
38. Ong, H. H., Wright, A. C., Wehril, S. L., Souza, A., Schwartz, E. D., et al. (2008) Indirect measurement of regional axon diameter in excised mouse spinal cord with q-space imaging: simulation and experimental studies. *Neuroimage* **40**(4), 1619–1632.
39. Hikishima, K., Yagi, K., Numano, T., Homma, K., Nitta, N., et al. (2008) Volumetric q-space imaging by 3D diffusion-weighted MRI. *Magn. Reson. Imaging* **26**(4), 437–445.

Chapter 8

The BOLD Effect

Joan M. Greve

Abstract

The purpose of this chapter is to introduce the novice NMR imager to blood oxygen level dependent (BOLD) contrast as well as remind the seasoned veteran of its beauty. Introduction to many of the factors that influence the BOLD signal is given higher priority than pursuing any subset in exquisite detail. Instead, references are given for readers seeking intense investigations into a given aspect. The hope is that this overview inspires the reader with the elegant simplicity of BOLD contrast while not, at first, intimidating too much with the underlying complexity. As one's knowledge of NMR matures so too will one's understanding, appreciation, and application of BOLD MRI. BOLD contrast derives from variations in the magnetic susceptibility of blood due to variations in the concentration of deoxyhemoglobin. These magnetic susceptibility effects produce local magnetic fields around blood vessels that can result in phase dispersion of nearby spins and, therefore, changes in signal intensity in NMR images. After providing brief historical context for BOLD, this chapter will follow the trail of magnetic susceptibility through definition, its source and location in vivo, and how the source and location in vivo interact with anatomical (e.g., blood vessel size) and imaging considerations (e.g., pulse sequence) to influence the BOLD signal. We will conclude by briefly highlighting clinical and preclinical applications using BOLD contrast.

Key words: BOLD, magnetic susceptibility, deoxyhemoglobin.

1. Introduction

1.1. Historical Context for BOLD

Pauling and Coryell are the authors of what might be one of the oldest, yet still easily accessible, publications on the magnetic properties of various states of hemoglobin (1). However, they note in their very first sentence that even Faraday, one of the forefathers of electromagnetism, had done some investigation into this arena as far back as 1845. It took another 50 years from Pauling and Coryell's landmark 1936 paper, along with harbingers such as Thulborn et al. (2), to inaugurate the term BOLD contrast in NMR imaging, ironically, in the same journal in which Pauling and Coryell's original work was published (3). Subsequently, throughout the 1990s, work was published on BOLD studies performed in the brain, the field that was to become the most prolific in vivo application of BOLD contrast (4–6). Shortly thereafter, the complexity of understanding what contributes to BOLD contrast came to the forefront. Investigation and comprehension of the factors that influence the BOLD signal have required in vitro, in vivo, and in silico methods.

1.2. What is Magnetic Susceptibility?

An intuitive sense of magnetic susceptibility (χ) is likely to develop shortly after one's introduction to NMR imaging. In physical terms, it is easy to understand magnetic susceptibility by taking an everyday steel wrench near a magnet and feeling significant forces pulling and/or torquing the tool (DO NOT try this at home!). In imaging terms, a novice NMR imager rapidly becomes aware of magnetic susceptibility as the signal loss near the sinuses in human imaging and in the abdomen in murine imaging. As a result, a novice imager is likely to consider differences in magnetic susceptibility an annoyance. However, differences in magnetic susceptibility between tissues on a microscopic scale is the very property upon which BOLD contrast is based (7). In this section, the goal is to elucidate the various forms of magnetism and their contribution to magnetic susceptibility. As noted by Schenck (7), there is some inconsistency in the units and definitions of magnetic parameters, including susceptibility. This chapter does not attempt to clarify those (refer to the Schenck reference for an intense read), but instead seeks to provide an intuitive understanding of susceptibility that will enable cogent discussion related to the BOLD effect.

Perhaps the most intuitive definition of magnetic susceptibility is that it is a "measure of the degree to which an object can perturb [ΔB] an applied field [B_o]" (7). Schenck goes on to highlight that only a perturbation parallel (ΔB_z) to the applied field is relevant in NMR and that the perturbation is strongest within or at the surface of the object leading to $-\chi B_o < \Delta B_z < \chi B_o$ and an estimation of $\chi \simeq \Delta B_{max}/B_o$. From this definition, it is easily surmised that a material with magnetic susceptibility 10 times that of another material can perturb the applied field, and hence resonant frequency of nearby spins, ten times as much.

During imaging, every proton inside the body is exposed to a magnetic field which, at any given time, could be composed of the static main field, the RF excitation field, gradient fields to achieve spatial localization, and a localized internal field based on that proton's location relative to other endogenous and exogenous tissues and materials. The localized internal field is dominated by the neighboring atomic electrons (8) and associated magnetic moments. Following Lenz's law, all of the electrons exposed to a magnetic field experience an alteration in their orbital motion which results in the induced magnetic field opposing the applied field. This mechanism of an opposing magnetization is called diamagnetism and would result in $\Delta B_{max} < 0$, i.e., $\chi < 0$. Diamagnetism is ubiquitous in all materials and therefore is aptly described by Schenck (7) as the "default mechanism" that determines a material's magnetic susceptibility, with the caveat, "unless it is overridden by some more powerful mechanism," e.g., paramagnetism.

Paramagnetism arises when materials are present that have unpaired electrons. Returning to one's first chemistry class, electrons in an atom are organized such that the spin moments of each pair cancel. Non-zero magnetic moments are therefore only present in materials with unpaired electrons. In the absence of an externally applied field, these moments are randomly distributed. However, upon interaction with an externally applied field, the moments tend to align with the applied field, thereby augmenting the local magnetic field, i.e., $\Delta B_{max} > 0$ and $\chi > 0$. The paramagnetic response of unpaired electrons (spin alignment) can swamp the diamagnetic response of paired electrons (orbital motion) in surrounding material/tissue. Although a single paramagnetic ion could cancel the diamagnetism of thousands of surrounding water molecules, more often than not, such ions are found in trace amounts in tissue and therefore largely do not supersede the bulk diamagnetism. As a result, the magnetic susceptibility of most tissues falls within the range of: $-7 \times 10^{-6} < \chi < -11 \times 10^{-6}$ (7).

An example of an exogenous paramagnetic material commonly used in NMR imaging to achieve certain contrast in the image is gadolinium which has unpaired electrons in the f-shell. An example of an endogenous paramagnetic material commonly used in NMR imaging to achieve certain contrast in the image is deoxyhemoglobin which has unpaired electrons present in the heme complex.

The third magnetic susceptibility class of materials, ferromagnetic materials, are not discussed since no such materials are naturally found in vivo. Ferromagnetic materials such as nickel and iron have very high susceptibilities of ~600 and 200,000, respectively, due to unpaired electrons in ordered structures (7).

2. Magnetic Susceptibility In Vivo

2.1. What Is the Source of Magnetic Susceptibility In Vivo?

The proteins of the globin superfamily may not share an evolutionarily conserved sequence but they certainly share structural and functional characteristics (9). Perhaps the most well-known function of globins is to reversibly bind oxygen for transport throughout the cardiovascular system. The term "globin" derives from a three-dimensional folding confirmation of the polypeptide that is constructed mainly of alpha helices. The globin-folded protein is non-covalently bound to a heme moiety consisting of a porphyrin (macrocyclic molecule) coordinated to a ferrous iron (Fe^{2+}) in the center. The heme moiety is a chemically active group and needs to be surrounded by a hydrophobic environment, provided by the 3D globin fold, to retain its activity.

Hemoglobin (Hb) is a remarkable molecule. It is probably the most well-known globin and was one of the first protein structures determined by X-ray crystallography (10). The most common form of adult hemoglobin contains two polypeptides folded in the α form and two folded in the β form accompanied by a heme moiety to which oxygen (O_2) attaches via the ferrous iron. Each hemoglobin molecule can therefore carry four molecules of O_2, Hb_4O_8. The presence of hemoglobin in the blood increases its O_2-carrying capacity approximately 70-fold (11). When O_2 is not present, the globin units are tightly bound reducing the affinity for O_2. Upon binding of an oxygen molecule, there is a change in configuration which exposes the other oxygen binding sites, resulting in an increasing affinity of each heme for each subsequent O_2 molecule. This effect is represented by the oxygen–hemoglobin dissociation curve having a sigmoidal shape. To underscore the exquisite balance of structure and function, consider the fact that a single amino acid mutation in hemoglobin results in sickle-cell anemia.

What is of most relevance to BOLD contrast in NMR imaging is the amount of deoxyhemoglobin in the blood vessels of a given volume of tissue because it is paramagnetic thanks to the Fe^{2+} and lack of oxygen while oxyhemoglobin has a susceptibility approximately equal to the surrounding tissue. As the amount of deoxyhemoglobin increases in the blood vessels of a given volume of tissue, so too does the localized internal field which is experienced by surrounding spins. The ultimate result of this is a loss of signal intensity in that region of the NMR image due to loss of phase coherence amongst spins (**Fig. 1**).

2.2. Where Is the Source of Magnetic Susceptibility Found In Vivo?

Iron is the most abundant paramagnetic ion in the body with approximately 70% of it residing within hemoglobin (7) compartmentalized in red blood cells (RBCs; a.k.a. erythrocytes), which themselves are contained inside blood vessels along with plasma. The proportion of blood volume occupied by RBCs is defined by the metric hematocrit (Hct) which averages 40–50% in humans (11). RBCs are approximately 6–8 μm in diameter, 2 μm thick, and can contain approximately 280 million hemoglobin molecules (7). RBCs are more homogeneously distributed during bulk flow through large vessels but take on a single-file alignment when traversing capillaries and venules which have diameters similar to the RBCs themselves (12, 13). Deoxyhemoglobin

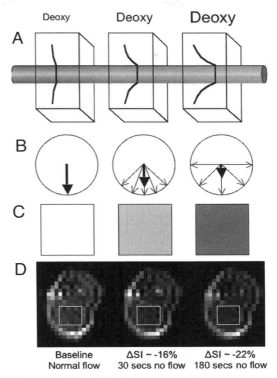

Fig. 1. Graphical representation of the basic concept of BOLD contrast. (**a**) Increasing the amount of deoxyhemoglobin (deoxy) produces larger local magnetic field gradients around the blood vessel. (**b**) The larger the magnetic field gradients, the more the phase dispersion of nearby spins. (**c**) Greater phase dispersion results in greater loss in signal intensity in the NMR images. (**d**) Empirical example derived from quantifying the signal intensity in a region of interest centered in the gastrocnemius muscle of a mouse (*rectangle*) during normal blood flow and 30 s and 180 s after complete cessation of flow to the hindlimb by occluding the aorta and vena cava while the animal was in the magnet (*see* **Chapter 27** or Greve et al. (38)).

is also spatially distributed along the length of the vascular network; hemoglobin in arterial blood is approximately 97% saturated with oxygen, while that in venous blood is about 75% saturated (11). The difference in magnetic susceptibility between tissue and blood that is approximately 60% saturated with oxygen has been estimated at 3×10^{-8} (14).

What is of most relevance to BOLD contrast in NMR imaging is the fact that the source of magnetic susceptibility in vivo, deoxyhemoglobin, has a very specific spatial distribution. It is compartmentalized intravascularly within the red blood cell and its basal level concentration depends on location along the vascular network. The water molecules residing within the RBCs, the plasma, and extravascularly all experience the paramagnetism of deoxyhemoglobin. To be explored in the next section are situations that alter the "what and where" of magnetic susceptibility (i.e., deoxyhemoglobin) in vivo, e.g., increases in blood flow that change the concentration of deoxyhemoglobin and the size of the vessel where the deoxyhemoglobin resides, respectively.

3. BOLD Contrast Generation

The aim of this section is to explain how alterations in the "what and where" of magnetic susceptibility in vivo alter the contrast in NMR images.

3.1. Things That Change the Amount of Deoxyhemoglobin

The factors that change the amount of deoxyhemoglobin are oxygen utilization, blood volume, blood flow. Similar to Norris (15), we will start by taking each parameter individually while holding the other two constant.

- Increased oxygen utilization will increase the concentration of deoxyhemoglobin, increasing the amount of paramagnetic material present and local magnetic field inhomogeneity, thereby leading to a reduction in signal intensity in the NMR images.

- Without changing the concentration of deoxyhemoglobin, increased blood volume will increase the total amount of paramagnetic material present and local magnetic field inhomogeneity, thereby leading to a reduction in signal intensity in the NMR images.

- Increased blood flow will increase the washout of deoxyhemoglobin, decreasing the amount of paramagnetic material present and local magnetic field inhomogeneity, thereby leading to an increase in signal intensity in the NMR images.

Taken individually, things are simple enough. However, these parameters usually change concurrently. In fMRI experiments, for example, neuronal activation tends to increase all three. The local magnetic field increase resulting from O_2 utilization and increased blood volume is overcome by the local magnetic field decrease resulting from increased blood flow, resulting in an overall increase in signal intensity (16). Norris reflects the sentiment of the fMRI field when he describes the situation as a "near paradox" (15). Even Ogawa et al. originally hypothesized that "the active region could show darker lines in the image because of the increased level of deoxyhemoglobin resulting from higher oxygen consumption" (3). Although it is possible they were envisioning the "fast response" of the BOLD signal changes (beyond the scope of this chapter; for an excellent brief introduction see Buxton's commentary (17)), it seems more likely they initially did not realize that increased blood flow would far surpass volume and

O_2 utilization effects. This learning curve, even for the pioneers of the BOLD effect, is respectfully highlighted to emphasize the complicated nature of the physiology that underlies the contrast mechanism and buoy the confidence of novice imagers that time can provide clarity.

For the physiologists entering the NMR imaging realm, there is likely one more consideration in mind: changes in blood volume due to recruitment or vasodilation. Boxerman et al. (14) addressed BOLD signal changes due to recruitment as defined by increased number of perfused vessels or increased blood flow velocity through vessels. The focus of the latter was changes in RBC velocity through the capillaries and how that might affect the BOLD signal changes. Changes in R_2^* and R_2 linearly increased with physiologically relevant increases in blood volume. There was very little difference between whether the volume was increased due to recruitment (increased number of perfused vessels) or vasodilation. For recruitment as defined by increased blood flow/RBC velocity, changes in R_2^* and R_2 remained at the same level as those seen for typical velocities through the capillaries. It should also be noted that changes in R_2^* and R_2 calculated for typical RBC velocities through the capillaries are slightly smaller compared to models run with the non-physiologic situation of stationary RBCs. Macrovascular velocity effects were not studied due to more homogeneous distribution of RBCs during bulk flow through large vessels as mentioned above.

3.2. Things That Relate to the Localized Nature of Deoxyhemoglobin

Although deoxyhemoglobin is contained intravascularly within the RBCs, its field perturbations extend into the extravascular space, giving BOLD contrast an intravascular as well as an extravascular component.

3.2.1. Intravascular Relaxation Mechanisms

To begin the discussion of these two contributing components, it is worthwhile to remember that the contrast in MRI derives from water molecules and how their frequencies and phases are altered based on the magnetic field they experience locally. Therefore, we should begin by noting where the water molecules can reside:

(1) Within the vessel in the plasma.

(2) Within the vessel in the RBCs.

(3) Extravascularly in the tissue.

There is a difference in magnetic susceptibility between the RBC and plasma itself. Therefore, one can imagine that spins contained within the RBC experience a different field compared to spins contained in the plasma. This simple idea becomes more complicated when one considers the fact that spins are not stationary. There are two options put forth to help conceptualize this last fact:

(1) One can think of the signal loss resulting from the water molecules in the plasma diffusing through fields produced

by the paramagnetic deoxyhemoglobin contained within the RBC.

(2) Alternatively, one can think of the signal loss resulting from two-site exchange. Spins could be contained within an RBC or the plasma (experiencing the respective magnetic susceptibility) and then could exchange to the alternative compartment to experience a completely different magnetic susceptibility.

Modeling to explore each of these concepts has been performed (18).

A second intravascular component comes into focus with a slightly bigger picture view. A blood vessel containing some concentration of deoxyhemoglobin can be thought of as a cylinder made of a paramagnetic material. The frequency shift inside the cylinder depends on both the size of the cylinder and its orientation relative to the main field (4). This effect is important for larger vessels, radii > 10 μm, and is reduced when diffusion/two-site exchange is included in the model (4) (*see* **Section 3.2.4**). This frequency shift is a static mechanism, will be refocused during a spin echo acquisition, and therefore will not contribute to changes in R_2.

3.2.2. Extravascular Relaxation Mechanisms

With regard to water exchange between either intravascular compartment (RBC or plasma) and the extravascular space, Boxerman et al. (14) showed virtually no effect on changes in R_2^* or R_2 whether the vessel was modeled as impermeable or with a level of permeability that is physiologically relevant. Although the model was not specific to the brain, per se, the argument of blood vessel impermeability may be due to the blood–brain barrier. Weisskoff (19) also notes this and follows with empirical evidence of the fact that there is no T_1 enhancement in healthy brain tissue upon injection of non-blood-pool gadolinium agents. In organs where vascular permeability to water is higher (heart, kidney, muscle), it may contribute to the BOLD effect.

Beyond vessel permeability, what is important when considering extravascular relaxation mechanisms are the field gradients that extend into the extravascular space and diffusion of water molecules through these gradients. This will be discussed in more detail in the next section as it leads to the dependence of BOLD contrast on vessel size and pulse sequence.

3.2.3. Intra- vs. Extravascular Contributions to BOLD Contrast

Boxerman et al. (16) combined Monte Carlo simulations with an fMRI simulation emulating visual stimulation to assess the intravascular and extravascular contributions to signal changes on gradient echo images acquired at 1.5 T. They concluded from this work that approximately 2/3 of the change in signal intensity can be attributed to intravascular spins. They elegantly went on to emphasize this point empirically by acquiring fMRI data with and without diffusion gradients present that were set at values suitable to null signal from blood, thereby removing the intravascular contribution to signal changes that occur upon visual stimulation. In a similar manner, they observed that changes in signal intensity upon visual stimulation were reduced as diffusion weighting increased, thereby bolstering the notion that intravascular spins dominate gradient echo signal changes at 1.5 T. It should be noted that at higher field strengths, intravascular contributions become less important due to the fact that the T_2 of blood is dramatically reduced (T_2 of blood: 109–180 ms at 1.5 T; 6.7 ms at 7 T (15)).

3.2.4. Static Dephasing Regime vs. Motional Narrowing Regime

Blood vessel size and choice of sequence (gradient vs. spin echo) have an impact on the contrast in BOLD images. As mentioned above, extravascular relaxation mechanisms result from the static fields outside of the blood vessel caused by the intravascular deoxyhemoglobin and diffusion of extravascular spins through these fields (20). These mechanisms can largely be broken into static and diffusive regimes.

Initially ignoring diffusion of water, the magnetic fields that extend into the extravascular space give rise to a spatially dependent frequency shift that causes the accumulated phase of one spin to differ from that of its neighbor. The result is a loss of phase coherence in that voxel. Several groups have shown through modeling that this mechanism dominates for larger vessels (≥ 20 μm). It should be noted that this is a static mechanism, can be refocused during a spin echo acquisition, and therefore will not contribute to changes in R_2. This is called the static dephasing regime (SDR) (20).

Diffusion becomes important when the random walk of a spin during the echo time is on the same order as the extent of the extravascular field gradients surrounding the blood vessel, i.e., approximately the radius of the vessel (15). In this situation, spins do not solely experience an increased or decreased field but instead experience a variety of fields throughout the echo time that depends on their particular path. The result is the same, viz., loss of phase coherence and hence a decrease in signal intensity on the NMR images. This mechanism is most important for vessels with radii of 6–8 μm (14). This mechanism is irreversible, will not be refocused during a spin echo acquisition, and therefore will contribute to changes in R_2 and R_2^*.

An additional consideration related to diffusion is when spins interchange position rapidly enough through the field gradients such that each spin, essentially, sees the same spread of frequencies, thereby accumulating approximately the same amount of phase, and maintaining some phase coherence resulting in smaller changes in signal intensity on BOLD images. This is called the motionally narrowed regime (MNR) (20).

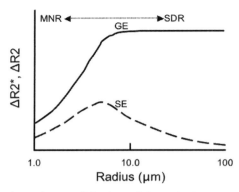

Fig. 2. Summary of the influence of blood vessel size and pulse sequence on BOLD contrast. Gradient echo (GE) data are more sensitive, showing greater changes in relaxation rates across the spectrum of blood vessel sizes. Spin echo (SE) data are specific to the microvasculature. In the static dephasing regime (SDR), local magnetic field gradients generated from intravascular deoxyhemoglobin extend into the extravascular space and result in spins having spatially dependent frequencies. The result is a loss of phase coherence and signal intensity on images. This effect dominates for larger vessels, leads to maximum changes on GE acquisitions, and does not contribute to SE acquisitions because it can be refocused. In the motionally narrowed regime (MNR), diffusion distances are comparable to the extent of the extravascular local magnetic field gradients and spins interchange positions quickly enough that each sees approximately the same environment. Therefore, there is less phase dispersion and loss of signal intensity. This effect dominates for smaller vessels and contributes to changes on GE and SE acquisitions being very similar because it cannot be refocused. (Adapted from Boxerman et al. (14)).

Both vessel size and pulse sequence considerations are referred to in the above discussion. These effects are qualitatively summarized in **Fig. 2**, an adaptation from Boxerman, et al. (14) and Weisskoff et al. (19). One can see from **Fig. 2** that changes in relaxation for gradient echo acquisitions (ΔR_2^*) reach a plateau for vessel sizes greater than $\sim$10 μm. Changes in relaxation for spin echo acquisitions (ΔR_2), however, have a maximum at $\sim$8 μm and decline for vessels smaller than (MNR) or larger than (SDR) this radius. Spin echo acquisitions are a specific reflection of the microvasculature while gradient echo sequences are overall more sensitive (i.e., changes in R_2^* are larger than changes in R_2 across all vessel sizes). As another means to emphasize the fact that BOLD data acquired via a spin echo or gradient echo sequence are weighted by the microvasculature and macrovasculature, respectively, models run varying the blood volume fraction composed of capillaries showed that changes in R_2^* decreased by $\sim$1/3 as the model became solely composed of capillaries while changes in R_2 were relatively unaffected for models composed of 40% capillaries or more (14). The focus on looking for a sequence and/or parameters that reflect changes in the microvasculature stems from the, not unfounded, hypothesis that microvascular changes are "a better reflection of localized neuronal activation" (14). In addition, gradient echo acquisitions depend more heavily on the experimental conditions and regional field properties, the latter of which can be dependent on shimming, for example. Therefore, although gradient echo acquisitions dominate the majority of BOLD investigations, it might be desirable to implement spin echo acquisitions especially at higher field strengths where some sensitivity can be regained.

4. Clinical and Preclinical Application of BOLD Contrast

When the BOLD effect is utilized as the contrast mechanism it is almost always used along with an exogenously applied stimulus to enable the acquisition of images during two (or more) states that differ in the amount of deoxyhemoglobin present in the vasculature of the region of interest. Practically speaking, the various states differ in the amount of oxygen utilization, blood volume, and/or blood flow.

As stated earlier, the most prolific and well-known application of the BOLD effect is in the brain, a.k.a. fMRI. When acquiring fMRI data, the stimulus takes the form of the subject performing some sort of task(s) while in the magnet which causes neuronal activation and an increase in blood flow. A variety of stimulus paradigms have been developed based on the area or function of the brain under investigation. Probably the two most common stimulus strategies are the block design and event-related design. The block design, which was the most frequently used in the early years of fMRI, is the simplest paradigm. It involves switching between two tasks with some periodicity. Typically, there is a stimulus period of relatively long duration (~30 s) followed by a period with no stimulus (21). It can be extrapolated to multiple tasks in which case images acquired during the different tasks can be compared in a number of ways, e.g., subtraction (22). Similarly, the same task could be repeated but at different levels of difficulty/intensity which leads to a parametric comparison. Event-related paradigms involve a quick stimulus being repeated at random intervals. This paradigm is made possible by the temporal resolution of NMR imaging itself.

The physiology (normal and pathological) to which fMRI has been applied clinically is abundant. Perhaps Matthews and Jezzard (21) described fMRI best when they said, "The scope of possibilities is as broad as the range of questions that can be asked." A sentence or two of caution regarding this powerful tool to study neuro-physiology is necessary, however: The changes in signal intensity between states tested in fMRI paradigms are usually on the order of ≤5% and require significant post-acquisition image and statistical analysis to extract (23). This fact combined with the discussion from earlier in the chapter regarding anatomical, physiological, and imaging parameters that influence the BOLD signal requires one to be aware of the "Pitfalls in fMRI" (24) in order to acquire and interpret data effectively.

Table 1 briefly summarizes clinical applications of BOLD contrast. It is by no means complete but seeks to point out practical considerations like reproducibility (25) and more recent investigations such as resting-state fMRI where no stimulus is used (26–29). Also included are references to other organ systems where BOLD has been applied in the clinic (30–34). Muscle in particular provides a tissue bed in which a stimulus such as reactive hyperemia, a physiological reaction where blood supply increases dramatically following a temporary arterial occlusion, is well tolerated and therefore can be used to investigate the BOLD signal in vivo in relation to many of the considerations discussed earlier in this chapter.

Similar to the clinical situation, the most copious use of the BOLD effect in preclinical models has been in the brain. In animal models, however, the breadth of investigation has been much more limited likely due to the requirement to anesthetize the subject while in the magnet. Perhaps one of the easiest paradigms utilized in animal models has been mixed gas breathing. In this approach different gases (O_2, CO_2, N_2) are inhaled by the animal while it lies in the magnet. The differential amount of each

Table 1
Brief list of references related to clinical applications of the BOLD effect

Region of interest and/or pathology	Stimulus
Visual cortex	Flickering checkerboard stimuli (42)
Epilepsy Surgical planning Reproducibility	Cued verbal fluency (25)
Alzheimer's Mild cognitive impairment	Memory tasks (novel, repeated, rest) Face–name paradigm (43)
Brain plasticity after ischemic injury	Motor paradigm (finger tapping, hand tapping) (44)
Multiple sclerosis	Attention task (45) Resting state (27) Review article (46)
Resting state	None (26, 28, 29)
Muscle	Reactive hyperemia (31–34)
Kidney	Pharmacological (30)

gas leads to a change in blood flow and hence BOLD contrast in the images. This approach can provide information regarding vascular reactivity and was one of the first used to provide proof of concept of the BOLD effect in vivo (3).

Since then, the most studied preclinical model has been the rat using paradigms involving paw stimulation, whisker stimulation, and olfactory stimulation (35). Electrical paw stimulation is probably the most commonly used paradigm in rodents due to its robustness and reproducibility. Also, cortical regions related to the forepaw and hindpaw are well mapped out and sizeable. In such work, thin electrodes are inserted subcutaneously in the paw and frequency can be modulated to maximize the BOLD response. Frequency and current should be chosen with care to avoid entering a refractory period or crossing the pain threshold, the latter of which is readily apparent by a concomitant rise in blood pressure. Similarly, whisker stimulation is an attractive approach because each whisker maps to a distinct region in the contralateral cortex called the barrel field. Therefore, stimulation of a particular whisker will result in activation in a distinct region of the cortex. One can stimulate whiskers in the magnet by a focused airstream or vibration of a thin wire attached to a given whisker. Often all other whiskers are clipped close to the skin just before the imaging session. Varying the frequency and intensity of whisker stimulation varies the BOLD response. In olfactory stimulation, an odor is delivered via a nose cone. Here the flow rate, duration of exposure, and amount of odor in the carrier gas is controlled. Similar to frequency and current levels needing to be controlled for paw stimulation to avoid inadvertently altering the BOLD response, prolonged exposure to an odor can reduce the BOLD response after several minutes.

A more recent application of the BOLD effect in the brain preclinically has been in songbirds (36). Although difficult, investigations of the auditory system can be accomplished by using long gradient ramp times to reduce noise generated from the NMR imaging system itself (*see* **Chapter 29**). This animal model is remarkable in that it provided one of the first examples of extensive plasticity in an adult organism, viz., seasonal variation in the size and connectivity of the song control system which is responsible for learned vocalization. Application of BOLD contrast was uniquely able to distinguish subregions of the auditory system based on function as well as demonstrate a larger extent of activation due to structured music or recognizable birdsong as opposed to white noise.

Extension of the BOLD effect into other organ systems in preclinical models parallels what has been done in humans and highlights its translational potential. Changes in BOLD contrast in the murine kidney upon pharmacological application (37) were consistent with work performed in humans. BOLD imaging in the murine hindlimb musculature (38) demonstrated signal intensity patterns very similar to those found in humans and was used to study the contribution of VEGF, age, and atherosclerosis in recovery from an ischemic injury. And finally, returning to where Ogawa began, the mixed gas breathing paradigm has been used for sequence development focusing on the nuances related

Table 2
Summary of concepts related to BOLD contrast

Magnetic susceptibility		
• Diamagnetic	$\chi < 0$	Most tissue
• Paramagnetic	$\chi > 0$	Deoxyhemoglobin
• Ferromagnetic	$\chi \gg 0$	Not naturally found in vivo
Hemoglobin		
• Does not have an evolutionarily conserved sequence but conserved structure and function		
• Increases bloods O_2-carrying capacity approximately 70-fold		
• Fe^{2+} gives deoxyhemoglobin its paramagnetism		
Compartmentalization of deoxyhemoglobin		
• RBCs can contain as many as 280 million hemoglobin molecules		
• Capillaries, venules, and RBCs are ~6–8 μm in diameter		
• Deoxyhemoglobin is spatially distributed along the vascular network		
° arterial hemoglobin: ~97% saturated with oxygen		
° venous hemoglobin: ~75% saturated with oxygen		
Physiological alteration	[Deoxy hemoglobin]	Signal intensity
Increased O_2 utilization	↑	↓
Increased blood volume	↑	↓
Increased blood flow	↓	↑
Increases in blood flow dominate in neuronal activation studies		
Intravascular relaxation		
• Relaxation mechanisms include a dynamic component (two-site exchange or diffusion through field gradients around RBCs)		
• Dominates signal changes at lower field strengths, e.g., 1.5 T		
Extravascular relaxation		
• Static dephasing regime		
° Contributes to gradient echo acquisitions only		
° Dominates at larger vessel radii		
• Diffusion through extravascular field gradients		
° Contributes to gradient echo and spin echo acquisitions		
° Important for capillary- and venule-sized vessels		
• Motionally narrowed regime		
° Dominates at smaller vessel radii		
Sensitivity and specificity		
• Gradient echo is more sensitive, i.e., signal changes are always bigger		
• Spin echo is specific for microvessels		

to imaging diminutive anatomy at high magnetic field strengths (39) as well as to evaluate the maturity and functionality of tumor vasculature (40).

5. Closing Remarks

In many ways, the initial interest in the BOLD effect was driven by a desire to be comparable to positron emission tomography (PET) (4, 41). Although it is highly suspect to think that NMR imaging will ever be as sensitive as PET, per se, it should be noted that the spatial and temporal resolution achievable in BOLD NMR imaging is superior to PET. Furthermore, because the measurement does not require an exogenous contrast agent, blood flow changes can be more closely related to stimuli and studies can be repeated more frequently. Lastly, higher resolution anatomical, connectivity (diffusion tensor imaging), or blood flow-specific (arterial spin labeling) acquisitions can be performed during the same imaging session to provide complimentary data.

In closing, perhaps the best summary is provided by Norris (15): "By any measure the use of BOLD contrast ... is a remarkable success story." And, as pointed out by Ogawa himself (4): "Measurement of the induced signal change is rather simple, requiring only a standard gradient echo sequence"—with the addendum here—while understanding the anatomical (e.g., vessel size), physiological (e.g., blood flow), and imaging (e.g., choice of pulse sequence) parameters which contribute to the induced signal change is another matter entirely. **Table 2** summarizes the concepts discussed within this chapter. The hope is that this chapter has imbued the reader with an intuition related to the various contributions to the BOLD effect and motivates continued education on BOLD contrast as well as application of BOLD imaging in the lab.

References

1. Pauling, L., and Coryell, C. D. (1936) The magnetic properties and structure of hemoglobin, oxyhemoglobin and carbon-monoxyhemoglobin. *Proc Natl Acad Sci USA* **22**, 210–216.
2. Thulborn, K. R., Waterton, J. C., Matthews, P. M., and Radda, G. K. (1982) Oxygenation dependence of the transverse relaxation time of water protons in whole blood at high field. *Biochim Biophys Acta* **714**, 265–270.
3. Ogawa, S., Lee, T. M., Kay, A. R., and Tank, D. W. (1990) Brain magnetic resonance imaging with contrast dependent on blood oxygenation. *Proc Natl Acad Sci USA* **87**, 9868–9872.
4. Ogawa, S., Menon, R. S., Tank, D. W., Kim, S. G., Merkle, H., Ellermann, J. M., and Ugurbil, K. (1993) Functional brain mapping by blood oxygenation level-dependent contrast magnetic resonance imaging. A comparison of signal characteristics

5. Haacke, E. M., Lai, S., Yablonskiy, D. A., and Lin, W. (1995) In vivo validation of the BOLD mechanism: A review of signal changes in gradient echo functional MRI in the presence of flow. *Int J Imaging Syst Technol* **6**, 153–163.
6. van Zijl, P. C., Eleff, S. M., Ulatowski, J. A., Oja, J. M., Ulug, A. M., Traystman, R. J., and Kauppinen, R. A. (1998) Quantitative assessment of blood flow, blood volume and blood oxygenation effects in functional magnetic resonance imaging. *Nat Med* **4**, 159–167.
7. Schenck, J. F. (1996) The role of magnetic susceptibility in magnetic resonance imaging: MRI magnetic compatibility of the first and second kinds. *Med Phys* **23**, 815–850.
8. Haacke, E. M., Brown, R. W., Thompson, M. R., and Venkatesan, R. (1999) *Magnetic Resonance Imaging: Physical Principles and Sequence Design*. Wiley, New York, NY.
9. Wajcman, H., Kiger, L., and Marden, M. C. (2009) Structure and function evolution in the superfamily of globins. *C R Biol* **332**, 273–282.
10. Strandberg, B. (2009) Chapter 1: Building the ground for the first two protein structures: Myoglobin and haemoglobin. *J Mol Biol* **392**, 2–10.
11. Ganong, W. F. (2005) *Review of Medical Physiology*. McGraw-Hill Companies, Columbus.
12. Stuart, J., and Nash, G. B. (1990) Red cell deformability and haematological disorders. *Blood Rev* **4**, 141–147.
13. Berne, R. M., and Levy, M. N. (2001) *Cardiovascular Physiology*, Eighth ed. Mosby, St. Louis.
14. Boxerman, J. L., Hamberg, L. M., Rosen, B. R., and Weisskoff, R. M. (1995) MR contrast due to intravascular magnetic susceptibility perturbations. *Magn Reson Med* **34**, 555–566.
15. Norris, D. G. (2006) Principles of magnetic resonance assessment of brain function. *J Magn Reson Imaging* **23**, 794–807.
16. Boxerman, J. L., Bandettini, P. A., Kwong, K. K., Baker, J. R., Davis, T. L., Rosen, B. R., and Weisskoff, R. M. (1995) The intravascular contribution to fMRI signal change: Monte Carlo modeling and diffusion-weighted studies in vivo. *Magn Reson Med* **34**, 4–10.
17. Buxton, R. B. (2001) The elusive initial dip. *Neuroimage* **13**, 953–8.
18. Stefanovic, B., and Pike, G. B. (2004) Human whole-blood relaxometry at 1.5 T: Assessment of diffusion and exchange models. *Magn Reson Med* **52**, 716–723.
19. Weisskoff, R. M., Zuo, C. S., Boxerman, J. L., and Rosen, B. R. (1994) Microscopic susceptibility variation and transverse relaxation: theory and experiment. *Magn Reson Med* **31**, 601–610.
20. Yablonskiy, D. A., and Haacke, E. M. (1994) Theory of NMR signal behavior in magnetically inhomogeneous tissues: the static dephasing regime. *Magn Reson Med* **32**, 749–763.
21. Matthews, P. M., and Jezzard, P. (2004) Functional magnetic resonance imaging. *J Neurol Neurosurg Psychiatry* **75**, 6–12.
22. Amaro, E., Jr., and Barker, G. J. (2006) Study design in fMRI: Basic principles. *Brain Cogn* **60**, 220–232.
23. Neuroskeptic. (2009) fMRI Gets Slap in the Face with a Dead Fish (September 16, 2009). Retrieved December 5, 2009, from http://neuroskeptic.blogspot.com/2009/09/fmri-gets-slap-in-face-with-dead-fish.html.
24. Haller, S., and Bartsch, A. J. (2009) Pitfalls in FMRI. *Eur Radiol* **19**, 2689–2706.
25. Adcock, J. E., Wise, R. G., Oxbury, J. M., Oxbury, S. M., and Matthews, P. M. (2003) Quantitative fMRI assessment of the differences in lateralization of language-related brain activation in patients with temporal lobe epilepsy. *Neuroimage* **18**, 423–438.
26. Biswal, B., Yetkin, F. Z., Haughton, V. M., and Hyde, J. S. (1995) Functional connectivity in the motor cortex of resting human brain using echo-planar MRI. *Magn Reson Med* **34**, 537–541.
27. Lowe, M. J., Phillips, M. D., Lurito, J. T., Mattson, D., Dzemidzic, M., and Mathews, V. P. (2002) Multiple sclerosis: low-frequency temporal blood oxygen level-dependent fluctuations indicate reduced functional connectivity initial results. *Radiology* **224**, 184–192.
28. Damoiseaux, J. S., Rombouts, S. A., Barkhof, F., Scheltens, P., Stam, C. J., Smith, S. M., and Beckmann, C. F. (2006) Consistent resting-state networks across healthy subjects. *Proc Natl Acad Sci USA* **103**, 13848–13853.
29. Greicius, M. D., Supekar, K., Menon, V., and Dougherty, R. F. (2009) Resting-state functional connectivity reflects structural connectivity in the default mode network. *Cereb Cortex* **19**, 72–78.
30. Gloviczki, M. L., Glockner, J., Gomez, S. I., Romero, J. C., Lerman, L. O., McKusick, M., and Textor, S. C. (2009) Comparison of 1.5 and 3 T BOLD MR to study oxygenation

of kidney cortex and medulla in human renovascular disease. *Invest Radiol* **44**, 566–571.
31. Lebon, V., Brillault-Salvat, C., Bloch, G., Leroy-Willig, A., and Carlier, P. G. (1998) Evidence of muscle BOLD effect revealed by simultaneous interleaved gradient-echo NMRI and myoglobin NMRS during leg ischemia. *Magn Reson Med* **40**, 551–558.
32. Lebon, V., Carlier, P. G., Brillault-Salvat, C., and Leroy-Willig, A. (1998) Simultaneous measurement of perfusion and oxygenation changes using a multiple gradient-echo sequence: application to human muscle study. *Magn Reson Imaging* **16**, 721–729.
33. Donahue, K. M., Van Kylen, J., Guven, S., El-Bershawi, A., Luh, W. M., Bandettini, P. A., Cox, R. W., Hyde, J. S., and Kissebah, A. H. (1998) Simultaneous gradient-echo/spin-echo EPI of graded ischemia in human skeletal muscle. *J Magn Reson Imaging* **8**, 1106–1113.
34. Toussaint, J. F., Kwong, K. K., Mkparu, F. O., Weisskoff, R. M., LaRaia, P. J., Kantor, H. L., and M'Kparu, F. (1996) Perfusion changes in human skeletal muscle during reactive hyperemia measured by echo-planar imaging. *Magn Reson Med* **35**, 62–69.
35. Hoehn, M. (2003) Functional magnetic resonance imaging. In: van Bruggen, N. and Roberts, T. (eds.), *Biomedical Imaging in Experimental Neuroscience*. CRC Press, Boca Raton.
36. Van der Linden, A., Van Meir, V., Boumans, T., Poirier, C., and Balthazart, J. (2009) MRI in small brains displaying extensive plasticity. *Trends Neurosci* **32**, 257–266.
37. Li, L. P., Ji, L., Lindsay, S., and Prasad, P. V. (2007) Evaluation of intrarenal oxygenation in mice by BOLD MRI on a 3.0T human whole-body scanner. *J Magn Reson Imaging* **25**, 635–638.
38. Greve, J. M., Williams, S. P., Bernstein, L. J., Goldman, H., Peale, F. V., Jr., Bunting, S., and van Bruggen, N. (2008) Reactive hyperemia and BOLD MRI demonstrate that VEGF inhibition, age, and atherosclerosis adversely affect functional recovery in a murine model of peripheral artery disease. *J Magn Reson Imaging* **28**, 996–1004.
39. Schneider, J. T., and Faber, C. (2008) BOLD imaging in the mouse brain using a turboCRAZED sequence at high magnetic fields. *Magn Reson Med* **60**, 850–859.
40. Baudelet, C., Cron, G. O., and Gallez, B. (2006) Determination of the maturity and functionality of tumor vasculature by MRI: correlation between BOLD-MRI and DCE-MRI using P792 in experimental fibrosarcoma tumors. *Magn Reson Med* **56**, 1041–1049.
41. Fox, P. T., and Raichle, M. E. (1986) Focal physiological uncoupling of cerebral blood flow and oxidative metabolism during somatosensory stimulation in human subjects. *Proc Natl Acad Sci USA* **83**, 1140–1144.
42. Engel, S. A., Glover, G. H., and Wandell, B. A. (1997) Retinotopic organization in human visual cortex and the spatial precision of functional MRI. *Cereb Cortex* **7**, 181–192.
43. Dickerson, B. C., and Sperling, R. A. (2008) Functional abnormalities of the medial temporal lobe memory system in mild cognitive impairment and Alzheimer's disease: insights from functional MRI studies. *Neuropsychologia* **46**, 1624–1635.
44. Pineiro, R., Pendlebury, S., Johansen-Berg, H., and Matthews, P. M. (2001) Functional MRI detects posterior shifts in primary sensorimotor cortex activation after stroke: evidence of local adaptive reorganization? *Stroke* **32**, 1134–1139.
45. Staffen, W., Mair, A., Zauner, H., Unterrainer, J., Niederhofer, H., Kutzelnigg, A., Ritter, S., Golaszewski, S., Iglseder, B., and Ladurner, G. (2002) Cognitive function and fMRI in patients with multiple sclerosis: evidence for compensatory cortical activation during an attention task. *Brain* **125**, 1275–1282.
46. Filippi, M., and Rocca, M. A. (2009) Functional MR imaging in multiple sclerosis. *Neuroimaging Clin N Am* **19**, 59–70.

Chapter 9

Screening of CEST MR Contrast Agents

Xiaolei Song, Kannie W.Y. Chan, and Michael T. McMahon

Abstract

There has been a tremendous amount of interest in developing new MR contrast agents for cellular and molecular imaging applications such as the visualization of tumors, highlighting areas of angiogenesis, highlighting of contrast agent-labeled therapeutic stem cells, and highlighting of contrast agent-labeled drug delivery vehicles. The contrast properties of paramagnetic and super-paramagnetic relaxation-based agents have allowed MR imaging to be used as a tool for all of the above applications. However, a new class of MR contrast agents, chemical exchange saturation transfer (CEST) agents, provides additional features such as (1) the ability to highlight multiple biological events at once within an image through the distinguishability of the different CEST contrast agents, (2) the ability to toggle the contrast "off-to-on" by applying a saturation pulse, and (3) potentially providing more information about the environment surrounding the contrast agent such as the pH or concentration of metabolites. In this chapter, we will focus on the methods which can be used in terms of acquisition schemes and hardware to screen these agents through MR imaging.

Key words: MRI, CEST, contrast agents, molecular imaging.

1. Introduction

1.1. MR Contrast Agents

Magnetic Resonance Imaging (MRI) is a mature technology which primarily uses the water in tissue of either animals or patients to generate images. This is possible due to the high concentration of water present in tissue, with the overall proton (^{1}H) concentration as high as 110 M. Perturbations in either the concentration or the relaxation properties of the water protons in tissue will produce image contrast, which allows MRI to show excellent soft tissue contrast compared to other imaging modalities. Apart from the endogenous contrast between different imaged soft tissues, the MRI signal intensity can also be perturbed by exogenous materials called MR contrast agents. Through the use of these agents (or probes) the signal in these images, especially where the agents are present, can be altered so that the images are sensitive enough to allow the detection of specific molecular or cellular processes.

To date, there are four major classes of MR molecular probes: (1) paramagnetic agents, such as Gd (1) or Mn (2) complexes or also Mn particles (3) that produce a large positive signal enhancement from decreasing T_1, (2) super-paramagnetic agents, such as iron oxide particles (4), which produce a large negative T_2 contrast. (3) A newer class of probes that has been gaining in popularity are based on nonhydrogen nuclei such as ^{19}F in particles (5–11) or on compounds (12) or drugs (13) or smaller amounts of nuclei using hyperpolarized parahydrogen (14, 15) (*see* **Chapter 11**), hyperpolarized xenon (16–18) (*see* **Chapter 10**), or dynamic nuclear polarization (DNP) (19–23) (*see* **Chapters 11** and **33**). These imaging agents do not interact with water and thus cannot provide anatomical information. Therefore, the images produced by the MR signal from these nuclei are usually co-registered with the water ^1H images to obtain anatomical information. (4) A fourth class, the so-called chemical exchange saturation transfer or CEST agents (24–32), is based on applying a saturation pulse to "switch on" the contrast in water signal through chemical exchange and has now matured with many exciting new features as compared to the other three. This area of research has grown tremendously in the last six years in the number of publications and number of citations of these papers focused on CEST. This review will focus on the unique and interesting properties of CEST contrast agents which have attracted the attention of researchers in this field and the main imaging pulse sequence and acquisition scheme which is now in use to obtain MR images when these contrast agents are present.

1.2. The Basics of CEST Contrast

Magnetic resonance (MR) spectra have been used to study chemical exchange between species for decades, dating back to the first time spectral changes due to exchange were observed in 1951 (33, 34). Chemical exchange processes and their influence on NMR spectra have been presented in many early papers (35–43) in the 1950s and 1960s. One notable study which relates directly to the CEST imaging today was performed by Forsen and Hoffman (37) in which a saturation pulse was placed on resonance with solute protons and then the solvent signal intensity was monitored to study chemical exchange. This experiment, adapted for imaging, is what we call chemical exchange saturation transfer (CEST) but has also been known as magnetization transfer (MT) (*see* **Note 1**). A cartoon depicting the mechanism of this contrast is displayed in **Fig. 1**.

There are two different categories of chemical exchange which can influence MR spectra: *intramolecular exchange*

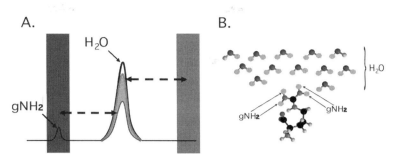

Fig. 1. Cartoon depicting the mechanism of CEST contrast for L-arginine dissolved in water. (**a**) Saturation pulses are placed either on resonance with the exchangeable guanidyl protons or off resonance and on the opposite side of the water resonance. The CEST contrast (*MTR*$_{asym}$) is the difference between the two experiments. (**b**) L-arginine dissolved in water. The guanidyl protons exchange rapidly such that the guanidyl proton saturation spreads through the network of solvent water protons.

(for example helix-coil transitions of nucleic acids or the folding/unfolding processes for proteins) and *intermolecular exchange* (for example protonation/deprotonation processes or binding of small molecules to macromolecules). Intermolecular exchange is most relevant for MR contrast agents, as this includes the interactions of a solute molecule (or well-dispersed particle) with the bulk solvent or the source of image signal. In fact, for CEST, we are particularly interested in describing proton exchange with water because MRI is primarily a water imaging technique in the clinic. The reason that CEST is a viable mechanism for contrast in MRI is that the chemical exchange of protons acts as a "saturation amplifier." Low concentrations of CEST agents can be detected due to the many exchanges between their protons and water protons. These exchanges magnify this signal loss as compared to the relative concentration between agent exchangeable protons and water. Robert Balaban and co-workers were the first to demonstrate this amplification and they coined the term CEST to describe this mechanism (24, 25, 27) when they were studying metabolites such as urea, ammonia, and many others. For a more detailed discussion of the mechanism of CEST, we refer the reader to a recent review of this topic (44).

There are a number of MR pulse sequences developed to acquire data on a system in chemical exchange with the resulting NMR spectra used to quantify the exchange time constants. For CEST imaging, however, the main method used is the basic saturation transfer scheme with a saturation pulse incremented across the NMR spectrum and the heights of the peaks observed to determine which spins are exchanging with the others. How does one describe the contrast mechanism for this sequence? The Bloch equations have long been used to describe the trajectory of the magnetization and can be adapted to include terms describing the physical exchange of spins. Assuming a small pool of solute protons (s) and a large pool of water protons (w) while applying B_1 along the x-axis, the Bloch equations for a two-pool proton exchange model are as follows (45):

$$\frac{dM_{xs}}{dt} = -\Delta\omega_s M_{ys} - R_{2s} M_{xs} - k_{sw} M_{xs} + k_{ws} M_{xw} \quad [1]$$

$$\frac{dM_{ys}}{dt} = \Delta\omega_s M_{xs} + \omega_1 M_{zs} - R_{2s} M_{ys} - k_{sw} M_{ys} + k_{ws} M_{yw} \quad [2]$$

$$\frac{dM_{zs}}{dt} = \Delta\omega_1 M_{ys} - R_{1s}(M_{zs} - M_{0s}) - k_{sw} M_{zs} + k_{ws} M_{zw} \quad [3]$$

$$\frac{dM_{xw}}{dt} = -\Delta\omega_w M_{yw} - R_{2w} M_{xw} + k_{sw} M_{xs} - k_{ws} M_{xw} \quad [4]$$

$$\frac{dM_{yw}}{dt} = \Delta\omega_s M_{xw} + \omega_1 M_{zw} - R_{2w} M_{yw} + k_{sw} M_{ys} - k_{ws} M_{yw} \quad [5]$$

$$\frac{dM_{zw}}{dt} = -\Delta\omega_1 M_{yw} - R_{1w}(M_{zw} - M_{0w}) + k_{sw} M_{zs} - k_{ws} M_{zw} \quad [6]$$

in which $\omega_0 = \gamma B_0$ (Larmor frequency of the static magnetic field) and $\omega_1 = \gamma B_1$ (precession frequency to flip the magnetization during an rf pulse); $\Delta\omega_s$ and $\Delta\omega_w$ are the chemical shift differences between the saturation pulse and the solute and water resonance frequencies, respectively; M_0 is the equilibrium magnetization. Proton exchange between the two pools occurs with rates k_{sw} (solute → water), and k_{ws} (water → solute), and $k_{sw} M_{0s} = k_{ws} M_{0w}$ at equilibrium.

1.3. Types of CEST Agents

For MR imaging using CEST, Balaban and co-workers were the first to come up with the idea of using the saturation transfer experiment for imaging molecules with chemically exchangeable groups (24) and they demonstrated that chemical exchange between protons on metabolites and water could be detected sensitively with MR imaging both ex vivo (25) and in vivo (26) on endogenous metabolites such as urea or ammonia. A cartoon depicting the mechanism for these agents is shown in **Fig. 2a**. In addition van Zijl and colleagues showed that contrast agents could be constructed based on polymers containing many exchangeable protons such as polypeptides (46) or polynucleic acids (47) with the molar sensitivity of these agents greatly enhanced as compared to the small metabolites. In the so-called amide proton transfer (APT) imaging, Zhou et al. (48, 49) then demonstrated that this CEST effect could in fact be used to image the pH effect in the rat brain during ischemia and also highlight regions with large endogenous protein content, such as is the case in tumors in vivo. In addition and concurrently,

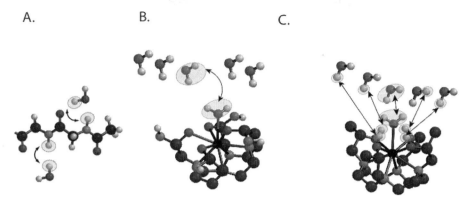

Fig. 2. Cartoon demonstrating the three different types of CEST contrast agents. (**a**) DIACEST N-acetyl-Gly-Gly fragment with chemical exchange occurring between the backbone NH protons and water. (**b**) PARACEST Ln(DOTAm) fragment with chemical exchange occurring between ligated and free water. (**c**) PARACEST Ln (with chemical exchange occurring between ligated and free water and also between ligated alcohol protons and water.

Sherry and co-workers as also Aime and co-workers showed that exogenous agents can also be made up of complexes containing paramagnetic lanthanides (50–52) and termed PARACEST agents. These agents have ligated water or other paramagnetically shifted protons that exchange with bulk water. Two cartoons depicting two types of PARACEST agents are shown in **Fig. 2b** and **c**. More recently, Terreno and co-workers have shown that liposomes which contain paramagnetic shift agents can be used as CEST agents due to the shifting of the water in the lumen of the liposome compared to the bulk, which they have termed LIPOCEST agents (53, 54). A cartoon depicting this type of agent is shown in **Fig. 3a**. Other types of CEST-generating particles have been developed by Winter and co-workers (55), and more recently presented by Liu and co-workers (56), with cartoons depicting the two-exchange behavior of these particles in **Fig. 3b** and **c**.

One exception to the case of using protons as CEST agents has been developed by Pines and colleagues using hyperpolarized xenon gas and molecular cages instead of water and CEST agents which are soluble. They have shown that cryptophane cages can be used to create chemical shift differences between free and encapsulated hyperpolarized xenon, with rapid exchange between the two on the timescale of the T_1 (57, 58). This exchange allows saturation transfer imaging of these cages, which they have termed HYPER-CEST. Since hyperpolarization amplifies the available magnetization by >10^4, the limitation of imaging xenon as opposed to water protons (used for the other agents and abundant in biological tissues) might be overcome. The set of hardware and methods we discuss in this chapter are not suitable for HYPER-CEST experiments/agents.

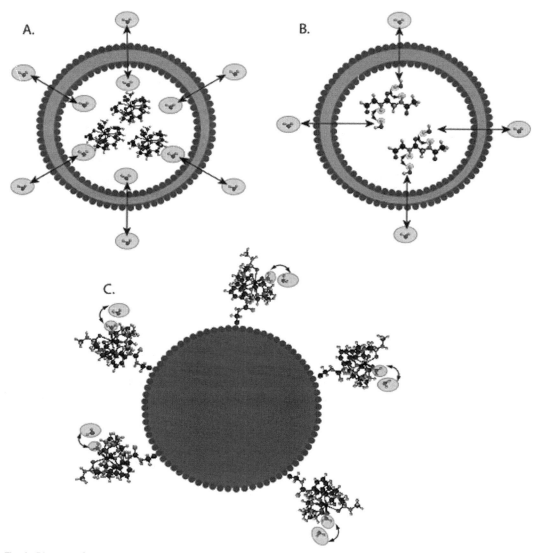

Fig. 3. Diagram of the contrast mechanism for three different types of CEST contrast particles. (a) Traditional LIPOCEST agents, with shift agent enclosed in the liposomal lumen, and chemical exchange occurring between the interior and exterior water. (b) Two-hop LIPOCEST agents, with CEST agent (not shift agent) enclosed in the liposomal lumen, and proton exchange occurring between agent and interior water, then interior water and exterior water. (c) CEST coated particles, where the exchange occurs entirely on the periphery of the particle.

2. Materials

2.1. General Requirements

1. A high-field (3 T or above) MR scanner, which can have a medium bore (89 mm) or higher with a relatively homogenous main magnetic field, fast and reliable gradient coils, and a high signal-to-noise radiofrequency (RF) coil.

2. A set of sample containers which fit within the RF coil, typically 1 mm capillary tubes with a holder (typically plastic) to organize the containers within the coil. We have fashioned the holder from 384-well cell culture plates.

3. For our method the most important component is the post-processing to correct the B_0 inhomogeneities present in the CEST images; as such some sort of image processing software is required and can be one of several commercial software packages, including MATLAB (Mathworks, Natick, MA), IDL(ITT visual information solutions, Boulder,CO), or others.

4. The general lab equipment should include a water bath, pH meter, rotary evaporator (rotovap), fluorescence reader (such as the Victor V, PerkinElmer).

5. DIACEST (such as L-arginine, Poly-L-Lysine or others), PARACEST (such as Eu-DOTA-4AmCE or Tm-DOTA-4AmCE) or shift agent (such as Tm-DOTMA) compounds are needed to insert into the liposomes.

6. The components of the liposomes will be required, which includes egg Phosphatidyl Choline (EggPC), cholesterol, 1,2-distearoyl-*sn*-glycero-3-phosphoethanolamine-N-[maleimide(polyethylene glycol)-2000](DSPE-PEG2000) and a fluorescently labeled lipid such as 1,2-dipalmitoyl-*sn*-glycero-3-phosphoethanolamine-N-(lissamine rhodamine B sulfonyl) (RhodPE). These can be purchased from Avanti Polar Lipids.

7. Saline-based buffers are required for dissolving the CEST agents such as 10 mM phosphate-buffered saline (PBS) or 10 mM tris-buffered saline (TBS) or others.

8. Concentrated solutions of NaOH and HCl will be required for titrating the solutions and chloroform will be required for dissolving the liposome components.

9. A 50 mL round bottom flask with a few hollow glass beads are needed to perform the extended hydration of the liposomes.

10. A liposome extruder, (such as a Liposofast-Basic extruder from Avestin) is needed to adjust the size of the liposomes. Also needed are polycarbonate filters (50–400 nm cutoff) which can be purchased from Northern Lipids Inc. Also a dynamic light scattering instrument such as the ZetaSizer from Malvern Instruments is needed to measure the hydrodynamic size of these nanoparticles.

11. 250 kDa cutoff PVDF dialysis tubing (Spectrum Laboratories, Inc.) and a 2 L beaker are required for separating the liposomes from unencapsulated CEST agent.

2.2. Liposome Sample Influences on CEST Contrast

There are many different ways to prepare DIACEST, PARACEST or LIPOCEST agents. In this chapter, we will describe how to prepare DIACEST liposomes, with the advantage of these being that they can be readily prepared using simple lab equipment, extruded to a variety of sizes (from ~100 nm to ~800 nm) and produce sufficient CEST contrast for detecting these particles in vivo. These particles are depicted in **Fig. 3b**.

1. CEST contrast for these liposomes will vary in part as a function of the permeability of the phospholipid bilayer. This permeability is affected by the lipid/sterol content and also, if cholesterol is added, the relative percentage of cholesterol (mole percentage) to the other components.

2. The surface area–volume (SA–V) ratio of these liposomes also influences the CEST contrast. However the size of the liposome (and resulting SA–V ratio) also affects the biodistribution of the particles so there might not be a choice as to what size liposome should be prepared, even if the maximum contrast is not achieved by this size.

2.3. MRI Pulse Sequences

The following pulse sequences should be available.

1. A multi-slice imaging sequence such as fast spin echo (FSE), or rapid acquisition with relaxation enhancement (RARE) with the ability to place a saturation pulse or pulses in front of the imaging sequence and also with the ability to change the saturation pulse time, field strength, and frequency offset such as that found in the magnetization transfer (MT) module on Bruker scanners.

2. A localized spectroscopy sequence, such as point-resolved Spectroscopy (PRESS).

3. A fast 3D scout image sequence, such as fast low angle shot (FLASH).

2.4. In Vitro B_0 Correction Algorithm

B_0 inhomogeneities, if left uncorrected, can erroneously increase or reduce CEST contrast (59, 60) which presents complications for the practical imaging of these agents. Recently, it was shown that the z-spectra produced by incrementing a low-power, short saturation pulse can be used to map the absolute water frequencies of each voxel in an image, an approach termed water saturation shift reference (WASSR) mapping (61). Since the saturation pulse is short and weak, both CEST and conventional magnetization transfer (MT) contributions to the z-spectra are minimized and only direct water saturation spectra are obtained. The CEST-weighted images can then be corrected in a pixel-by-pixel manner using the WASSR absolute frequency map and the interpolation. These frequency-weighted images can be used to determine the type of CEST agent contained in each voxel in the image, as shown previously (62, 63).

3. Methods

3.1. Liposome Preparation (see Note 2)

1. Dissolve CEST agent (such as L-arginine (Larg)) in 10 mM phosphate-buffered saline (PBS) (or other buffer such as tris-buffered saline (TBS)) at a concentration of 14 mM and titrate to pH 7.4 using HCl or NaOH.

2. The liposome components will have a molar ratio of eggPC:Chol:DSPE-PEG2000:RhodPE of (46.6:46.6:5:1.8). These are all dissolved in chloroform individually, then transferred to a 50 mL round bottom flask using volumes of 0.47 /0.140 /0.27 /1.0 mL respectively, and then stirred manually in a chemical fume hood.

3. The chloroform is evaporated in a rotovap with the water bath set to 50°C. As a result, the lipids form a thin film that coats the side of the flask.

4. Add a few hollow glass beads and hydrate the lipid thin film with 1 mL of the solution containing dissolved CEST agent. Stir the mixture manually until lipids are no longer visible.

5. Anneal for 2 h in a 55°C water bath and stir occasionally.

6. 20 μL of the solution is put aside for fluorescence measurement of the concentration of the liposomes. The remaining solution that contains liposomes is then extruded to reduce the liposome size using 400, 200, 100, and 50 nm cut-off membranes, using two filters. The liposomes are passed through the 400 nm polycarbonate membranes 21 times first, then depending on the desired liposome size, are passed through the 200 nm and 100 nm membranes also using 21 passes through the filters.

7. After the liposomes are extruded, the unencapsulated agent is removed by dialysis overnight in 1× PBS buffer using 250 kDa cutoff PVDF dialysis tubing (Spectrum Laboratories, Inc.).

8. The size and concentration of the liposomes are measured using dynamic light scattering and fluorescence, respectively.

3.2. General MRI Protocol (see Note 3)

1. The phantom should be centered in both the RF coil of the imaging probe and the magnet bore through adjustment of the position of the imaging probe.

2. The RF coil of the imaging probe is then tuned and matched according to the scanner's ^{1}H Larmor frequency (see your scanner manual for details of this procedure).

3. Next, a scout image is acquired (using for example triplot RARE, MSME (multi-slice multi-echo), or FLASH on

Bruker small animal scanners), with a sufficient field of view (FOV) so that three views from the scout (*XZ, YZ, XY*) capture the whole phantom.

4. The pulse sequence parameters should be calibrated including resonance frequency offset, pulse power for 90° and 180° flip angles, and receiver gain; and the magnet can be shimmed (although not entirely necessary with the B_0 correction routine we will employ in **Section 3.3**).

5. The desired slice for the CEST experiment should be chosen, based on considerations from the scout image such as the distance to the center of the coil or position of air bubbles in the sample tubes. Due to the fact that CEST imaging employs a long saturation pulse, a single-slice approach is typically preferred in a CEST acquisition to save time. A single-slice RARE sequence without turning on the saturation pulse is then tested to ensure that the defined single-slice geometry sufficiently covers the region of interest.

3.3. Determining B_0 Inhomogeneity

3.3.1. Shimming

CEST imaging is often (though not always, as demonstrated by Vinogradov and co-workers (64)) a difference imaging technique dependent on the frequency of the solvent and, as such, either good shimming or accurate B_0 maps are crucial to measure CEST contrast. Automatic shimming procedures are available on most scanners (generally restricted to first-order shims), or the shimming can be improved further through manual shimming. The shimming should be performed on a slice with the geometry similar to or equivalent to that used for the CEST imaging and the magnetic field should be shimmed to maximize the signal (*see* **Note 4**).

3.3.2. Estimating B_0 Inhomogeneity Over the CEST Slice

The range of frequencies encompassing the water signal can be determined using the point-resolved spectroscopy (PRESS) sequence without water signal suppression with the voxel geometry approximating the CEST imaging (or up to 50% larger in slice thickness). The typical acquisition parameters used are TR/TE = 2000/20 ms; spectrum acquisition size = 8192; sweep width = 10,080 Hz (25 ppm at 9.4 T); and NA = 1. The full spectral width of the water peak is then used to determine the range of B_0 inhomogeneity, after processing the spectrum (including FFT, baseline correction, apodization, etc.).

3.3.3. WASSR Acquisition

The WASSR images are acquired by adding a saturation pulse (called the magnetization transfer module on Bruker scanners) in front of an FSE or RARE sequence. The saturation pulse we typically use is a 200–500 ms continuous wave (CW) pulse with B_1 = 0.5 µT (21.3 Hz, *see* **Note 5**). The saturation pulse is typically scanned across the entire B_0 inhomogeneity range we determine using PRESS (typically from −1 ppm to +1 ppm with respect to water for twenty 1 mm capillaries in a 20 mm RF coil) using a 0.1 ppm increment between each acquisition. The typical imaging parameters used are acquisition bandwidth = 50 kHz; single slice; 1 mm slice thickness; TE = 6 ms; TR = 1500 ms; RARE factor = 16; FOV = 20 × 20 mm (*see* **Note 6**); and the matrix size is set to 128 frequency-encoding steps and 64 phase-encoding steps. The total acquisition time for generating an absolute B_0 map varies from 3 to 10 min. The error of B_0 estimation is under the hertz level for each pixel at a reasonable signal-to-noise ratio (SNR) (i.e., SNR/pixel > 15), using the parameters listed above.

3.4. High Throughput CEST Imaging

The same saturation transfer (ST) RARE sequences that were used to acquire the WASSR scans can be utilized to acquire CEST images at the relevant saturation frequencies. The typical imaging parameters used are acquisition bandwidth = 50 kHz; single slice; 1 mm slice thickness; TE = 6 ms, TR = 6000 ms; RARE factor = 16; FOV = 20 × 20 mm; 128 × 64 matrix size; 4000 ms CW saturation pulse; B_1 strength = 3.6 µT (153 Hz); and NA = 2. This leads to the acquisition time of approximate 48 s for each image.

3.4.1. CEST Imaging Parameters

3.4.2. High Throughput Collection Scheme

CEST images are collected using a series of CEST-weighted images with the incremented saturation offset to observe the saturation transfer from all the exchangeable protons in the imaging agent, the so-called "z-spectra" approach. For DIACEST liposomes, the frequency can be swept from −5 ppm to +5 ppm in steps of 0.2 ppm. For this sequence of images, the total scan time required will be ~20 min (*see* **Note 7**).

3.5. Image Post-processing and Analysis

3.5.1. Generation WASSR B_0 Maps (*see* **Note 8**)

1. The set of WASSR images are used to generate a B_0 map by writing a MATLAB script (or IDL, etc.) to perform the pixelwise fitting (*see* **Section 2.4**) as follows.

2. Pixel by pixel, the vector of MR signal amplitudes, $S_{\exp}(x, y)$ are fit for the whole set of WASSR images with the saturation offset vector, $\Delta\omega(x, y)$, using the equation

$$S_{\exp}(x,y) = \left\{ \eta^2 + \left[\frac{M_0(x,y)}{1 + \left(\frac{\omega_1(x,y)}{\Delta\omega(x,y) - \delta\omega_0(x,y)}\right)^2 \frac{T_1(x,y)}{T_2(x,y)}} \right]^2 \right\}^{1/2}$$

to estimate $\delta\omega_0(x,y)$. The non-linear fitting function (*lsqcurvefit*) in MATLAB can be used to perform this fitting, with M_0, $\delta\omega_0$, and $\left(\frac{T_1}{T_2}\omega_1^2\right)$ as floating parameters. The experimental noise (η) is estimated from the mean signal of a noise-only region of the image (the air region between the capillaries) and is then fixed.

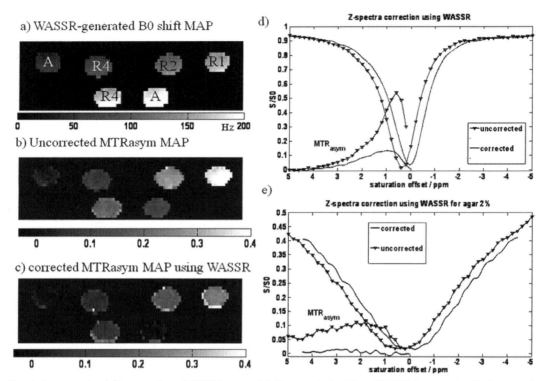

Fig. 4. An example of B_0 correction of CEST images. (a) B_0 map derived from WASSR image set with the sample configuration. A: agar 2%, R1, R2, R4 represent 10 mM, 5 mM, 2.5 mM L-arginine in PBS respectively (the B_0 shift is displayed for this phantom from 0 to 200 Hz from the carrier frequency). (b) Uncorrected MTR_{asym} map at 1.8 ppm (MTR_{asym} is displayed on a scale from 0 to 0.4) and (c) corrected MTR_{asym} map using as described in **Section 3.5.2** (MTR_{asym} is displayed on a scale from 0 to 0.4). (d) Uncorrected and corrected z-spectra and MTR_{asym} curves for the lower R4 sample shown in a–c. (e) Uncorrected and corrected z-spectra and MTR_{asym} curves for the lower agar sample.

3. The absolute B_0 map is constructed as a result of the fits from Step 2 above, by collecting the fit $\delta\omega_0(x,y)$ for each pixel and plotting this, thus providing the water shift information needed to correct the B_0 in the CEST images. The resulting map will look as shown in **Fig. 4a**.

3.5.2. Processing Raw CEST Data

1. The saturation offsets of the CEST images are corrected pixel-by-pixel using the WASSR B_0 map and the expression $\Delta\omega(x,y)_{corrected} = \Delta\omega(x,y) - \Delta\omega_0(x,y)$. In order to perform this correction, the original set of CEST images are interpolated pixelwise to obtain the water signal amplitude, $S_{x,y}(\Delta\omega_{interp})$, at the desired offsets using the cubic-spline fitting (*spline*) function in MATLAB. An example of the resulting MTR_{asym} map for such a phantom is shown in **Fig. 4c** at the shift of the exchangeable guanidyl protons (1.8 ppm).

2. The CEST z-spectrum can then be plotted using the scaled signal intensity, $S^{\Delta\omega}/S_0$ as a function of saturation offset frequency with respect to water. Typically, region of interest (ROI) masks are manually drawn over each tube in the phantom (one way is to use ROI Draw in Matlab) and the mean intensities of the selected tubes are used to plot the z-spectra. Examples of z-spectra correction are shown in **Fig. 4d** for L-arginine and **Fig. 4e** for 2% agar (which should display no CEST contrast).

3. The CEST contrast is quantified for each tube by calculating the asymmetry in the magnetization transfer ratio (MTR_{asym}), as defined by $MTR_{asym} = \left(S^{-\Delta\omega} - S^{\Delta\omega}\right)/S^{-\Delta\omega}$. The MTR_{asym} for each pixel should be calculated and used to construct the parametric map, MTR_{asym} map, which represents the distribution of CEST contrast. An example of the resulting MTR_{asym} map for such a phantom is shown in **Fig. 4c** at the shift of the exchangeable guanidyl protons (1.8 ppm). For a good CEST liposome, the CEST contrast is often in the range of 30% or higher.

4. Notes

1. Magnetization transfer (MT) is more commonly used in the context of signal transfer between immobilized protons and water—in contrast to the transfer of magnetization from exchangeable protons on a CEST agent which possess a sharp resonance line due to the tumbling of the agent.

2. This procedure is modified from the extended hydration method (65) using egg phosphatidyl choline (EggPC), cholesterol (Chol), distearoyl-phosphatidyl-ethanolamine conjugated to polyethylene glycol (molecular weight = 2000 D, DSPE-PEG2000) and rhodamine (Rh)-labeled phosphatidyl-ethanolamine purchased from Avanti Polar Lipids.

3. The MR collection and processing is similar to that described in **Chapter 36**, without the need of equipment for animal anesthesia.

4. For manual shimming, the procedure must be performed in an iterative manner since the shimming coils are coupled to each other and affect all three dimensions (*see also* the handbook of your scanner).

5. For converting tesla to hertz, the equation $f = \gamma\, B_0$ should be used in which γ is the gyromagnetic ratio and is 42.57 MHz/T for ^1H protons.

6. The FOV is determined by the RF coil size and amount of sample tubes enclosed in the phantom. If long capillaries are used, the slice thickness can be increased substantially to improve the SNR.

7. When using a clinical imager, the automated shimming and frequency adjustments must be turned off between the consecutive WASSR-CEST images. This is not an issue for high-resolution spectrometers, where these are not automatically adjusted before each scan.

8. In order to reduce the post-processing time, voxels which contain only noise in the MRI images can be removed through thresholding, for example, requiring $SNR \geq 15$.

References

1. Caravan, P., Ellison, J. J., McMurry, T. J., and Lauffer, R. B. (1999) Gadolinium(III) chelates as MRI contrast agents: structure, dynamics, and applications. *Chemical Reviews*, **99**, 2293–2352.

2. Rocklage, S. M., Cacheris, W. P., Quay, S. C., Hahn, F. E., and Raymond, K. N. (1989) Synthesis and characterization of a paramagnetic chelate for magnetic resonance imaging enhancement. *Inorgnic Chemistry*, **28**, 477–485.

3. Na, H. B., Lee, J. H., An, K., Park, Y. I., Park, M., Lee, I. S., Nam, D. H., Kim, S. T., Kim, S. H., Kim, S. W., Lim, K. H., Kim, K. S., Kim, S. O., and Hyeon, T. (2007) Development of a T_1 contrast agent for magnetic resonance imaging using MnO nanoparticles. *Angewandte Chemie International Edition England*, **46**, 5397–5401.

4. Bjornerud, A. and Johansson, L. (2004) The utility of superparamagnetic contrast agents in MRI: theoretical consideration and applications in the cardiovascular system. *NMR in Biomedicine*, **17**, 465–77.

5. Srinivas, M., Morel, P. A., Ernst, L. A., Laidlaw, D. H., and Ahrens, E. T. (2007) Fluorine-19 MRI for visualization and quantification of cell migration in a diabetes model. *Magnetic Resonance in Medicine*, **58**, 725–734.

6. Janjic, J. M., Srinivas, M., Kadayakkara, D. K. K., and Ahrens, E. T. (2008) Self-delivering nanoemulsions for dual fluorine-19 MRI and fluorescence detection. *Journal of the American Chemical Society*, **130**, 2832–2841.

7. Waters, E. A., Chen, J. J., Yang, X. X., Zhang, H. Y., Neumann, R., Santeford, A., Arbeit, J., Lanza, G. M., and Wickline, S. A. (2008) Detection of targeted perfluorocarbon nanoparticle binding using F-19 diffusion weighted MR spectroscopy. *Magnetic Resonance in Medicine*, **60**, 1232–1236.

8. Waters, E. A., Chen, J. J., Allen, J. S., Zhang, H. Y., Lanza, G. M., and Wickline, S. A. (2008) Detection and quantification of angiogenesis in experimental valve disease with integrin-targeted nanoparticles and 19-fluorine MRI/MRS. *Journal of Cardiovascular Magnetic Resonance*, **10**, 43.

9. Ruiz-Cabello, J., Walczak, P., Kedziorek, D. A., Chacko, V. P., Schmieder, A. H., Wickline, S. A., Lanza, G. M., and Bulte, J. W. M. (2008) In vivo "hot spot" MR imaging of neural stem cells using fluorinated nanoparticles. *Magnetic Resonance in Medicine*, **60**, 1506–1511.

10. Partlow, K. C., Chen, J. J., Brant, J. A., Neubauer, A. M., Meyerrose, T. E., Creer, M. H., Nolta, J. A., Caruthers, S. D., Lanza, G. M., and Wickline, S. A. (2007) F-19 magnetic resonance imaging for stem/progenitor cell tracking with multiple unique perfluorocarbon nanobeacons. *FASEB Journal*, **21**, 1647–1654.

11. Neubauer, A. M., Caruthers, S. D., Hockett, F. D., Cyrus, T., Robertson, J. D., Allen, J. S., Williams, T. D., Fuhrhop, R. W., Lanza, G. M., and Wickline, S. A. (2007) Fluorine cardiovascular magnetic resonance angiography in vivo at 1.5 T with perfluorocarbon nanoparticle contrast agents. *Journal of Cardiovascular Magnetic Resonance*, **9**, 565–573.

12. Hunjan, S., Mason, R. P., Constantinescu, A., Peschke, P., Hahn, E. W., and Antich, P. P. (1998) Regional tumor oximetry: F-19 NMR spectroscopy of hexafluorobenzene. *International Journal of Radiation Oncology, Biology, Physics*, **41**, 161–171.

13. Yu, J. X., Kodibagkar, V. D., Cui, W. N., and Mason, R. P. (2005) F-19: a versatile reporter for non-invasive physiology and pharmacology using magnetic resonance. *Current Medicinal Chemistry*, **12**, 819–848.
14. Adams, R. W., Aguilar, J. A., Atkinson, K. D., Cowley, M. J., Elliott, P. I. P., Duckett, S. B., Green, G. G. R., Khazal, I. G., Lopez-Serrano, J., and Williamson, D. C. (2009) Reversible interactions with para-hydrogen enhance NMR sensitivity by polarization transfer. *Science*, **323**, 1708–1711.
15. Bouchard, L. S., Burt, S. R., Anwar, M. S., Kovtunov, K. V., Koptyug, I. V., and Pines, A. (2008) NMR imaging of catalytic hydrogenation in microreactors with the use of para-hydrogen. *Science*, **319**, 442–445.
16. Navon, G., Song, Y. Q., Room, T., Appelt, S., Taylor, R. E., and Pines, A. (1996) Enhancement of solution NMR and MRI with laser-polarized xenon. *Science*, **271**, 1848–1851.
17. Goodson, B. M., Song, Y. Q., Taylor, R. E., Schepkin, V. D., Brennan, K. M., Chingas, G. C., Budinger, T. F., Navon, G., and Pines, A. (1997) In vivo NMR and MRI using injection delivery of laser-polarized xenon. *Proceedings of the National Academy of Sciences of the United States of America*, **94**, 14725–14729.
18. Song, Y. Q., Gaede, H. C., Pietrass, T., Barrall, G. A., Chingas, G. C., Ayers, M. R., and Pines, A. (1995) Spin-polarized Xe-129 gas imaging of materials. *Journal of Magnetic Resonance Series A*, **115**, 127–130.
19. Ardenkjaer-Larsen, J. H., Fridlund, B., Gram, A., Hansson, G., Hansson, L., Lerche, M. H., Servin, R., Thaning, M., and Golman, K. (2003) Increase in signal-to-noise ratio of > 10,000 times in liquid-state NMR. *Proceedings of the National Academy of Sciences of the United States of America*, **100**, 10158–10163.
20. Golman, K., Ardenaer-Larsen, J. H., Petersson, J. S., Mansson, S., and Leunbach, I. (2003) Molecular imaging with endogenous substances, *Proceedings of the National Academy of Sciences of the United States of America*, **100**, 10435–10439.
21. Day, S. E., Kettunen, M. I., Gallagher, F. A., Hu, D. E., Lerche, M., Wolber, J., Golman, K., Ardenkjaer-Larsen, J. H., and Brindle, K. M. (2007) Detecting tumor response to treatment using hyperpolarized C-13 magnetic resonance imaging and spectroscopy. *Nature Medicine*, **13**, 1382–1387.
22. Golman, K., in't Zandt, R., Lerche, M., Pehrson, R., and Ardenkjaer-Larsen, J. H. (2006) Metabolic imaging by hyperpolarized C-13 magnetic resonance imaging for in vivo tumor diagnosis. *Cancer Research*, **66**, 10855–10860.
23. Gallagher, F. A., Kettunen, M. I., Day, S. E., Hu, D. E., Ardenkjaer-Larsen, J. H., in't Zandt, R., Jensen, P. R., Karlsson, M., Golman, K., Lerche, M. H., and Brindle, K. M. (2008) Magnetic resonance imaging of pH in vivo using hyperpolarized C-13-labelled bicarbonate *Nature*, **453**, 940–U73.
24. Wolff, S. D., and Balaban, R. S. (1990) NMR imaging of labile proton-exchange. *Journal of Magnetic Resonance*, **86**, 164–169.
25. Guivel-Scharen, V., Sinnwell, T., Wolff, S. D., and Balaban, R. S. (1998) Detection of proton chemical exchange between metabolites and water in biological tissues. *Journal of Magnetic Resonance*, **133**, 36–45.
26. Dagher, A. P., Aletras, A., Choyke, P., and Balaban, R. S. (2000) Imaging of urea using chemical exchange-dependent saturation transfer at 1.5 T. *Journal of Magnetic Resonance Imaging*, **12**, 745–748.
27. Ward, K. M., Aletras, A. H., and Balaban, R. S. (2000) A new class of contrast agents for MRI based on proton chemical exchange dependent saturation transfer (CEST). *Journal of Magnetic Resonance*, **143**, 79–87.
28. Ward, K. M. and Balaban, R. S. (2000) Determination of pH using water protons and chemical exchange dependent saturation transfer (CEST). *Magnetic Resonance in Medicine*, **44**, 799–802.
29. Aime, S., Crich, S. G., Gianolio, E., Giovenzana, G. B., Tei, L., and Terreno, E. (2006) High sensitivity lanthanide(III) based probes for MR-medical imaging. *Coordination Chemistry Reviews*, **250**, 1562–1579.
30. Sherry, A. D. and Woods, M. (2008) Chemical exchange saturation transfer contrast agents for magnetic resonance imaging. *Annual Review of Biomedical Engineering*, **10**, 391–411.
31. Zhou, J., and van Zijl, P. C. M. (2006) Chemical exchange saturation transfer imaging and spectroscopy. *Progress in NMR Spectroscopy*, **48**, 109–136.
32. Yoo, B. and Pagel, M. D. (2008) An overview of responsive MRI contrast agents for molecular imaging. *Frontiers in Bioscience*, **13**, 1733–1752.
33. Liddel, U. and Ramsey, N. F. (1951) Temperature dependent magnetic shielding in ethyl alcohol. *Journal of Chemical Physics*, **19**, 1608.
34. Arnold, J. T. and Packard, M. E. (1951) Variations in absolute chemical shift of nuclear

induction signals of hydroxyl groups of methyl and ethyl alcohol. *Journal of Chemical Physics*, **19**, 1608–1609.

35. Gutowsky, H. S. and Holm, C. H. (1956) Rate processes and nuclear magnetic resonance spectra.2. hindered internal rotation of amides. *Journal of Chemical Physics*, **25**, 1228–1234.

36. Gutowsky, H. S. and Saika, A. (1953) Dissociation, chemical exchange, and the proton magnetic resonance in some aqueous electrolytes. *Journal of Chemical Physics*, **21**, 1688–1694.

37. Forsen, S. and Hoffman, R. A. (1963) Study of moderately rapid chemical exchange reactions by means of nuclear magnetic double resonance. *Journal of Chemical Physics*, **39**, 2892–2901.

38. McConnell, H. M. (1958) Reaction rates by nuclear magnetic resonance. *Journal of Chemical Physics*, **28**, 430–431.

39. Arnold, D. L. (1956) Magnetic resonances of protons in ethyl alcohol. *Physical Review*, **102**, 135–150.

40. Allerhand, A. and Gutowsky, H. S. (1964) Spin-echo NMR studies of chemical exchange. 1. some general aspects. *Journal of Chemical Physics*, **41**, 2115–2126.

41. Gutowsky, H. S., Vold, R. L., and Wells, E. J. (1965) Theory of chemical exchange effects in magnetic resonance. *Journal of Chemical Physics*, **43**, 4107.

42. Woessner, D. E. (1961) Nuclear transfer effects in nuclear magnetic resonance pulse experiments. *Journal of Chemical Physics*, **35**, 41–48.

43. McConnell, H. M. and Thompson, D. D. (1957) Molecular transfer of nonequilibrium nuclear spin magnetization. *Journal of Chemical Physics*, **26**, 958–959.

44. Zhou, J. Y. and van Zijl, P. C. M. (2006) Chemical exchange saturation transfer imaging and spectroscopy. *Progress in Nuclear Magnetic Resonance Spectroscopy*, **48**, 109–136.

45. Zhou, J., Wilson, D. A., Sun, P. Z., Klaus, J. A., and van Zijl, P. C. M. (2004) Quantitative description of proton exchange processes between water and endogenous and exogenous agents for WEX, CEST, and APT experiments. *Magnetic Resonance in Medicine*, **51**, 945–952.

46. Goffeney, N., Bulte, J. W. M., Duyn, J., Bryant, L. H., and van Zijl, P. C. M. (2001) Sensitive NMR detection of cationic-polymer-based gene delivery systems using saturation transfer via proton exchange. *Journal of the American Chemical Society*, **123**, 8628–8629.

47. Snoussi, K., Bulte, J. W. M., Gueron, M., and van Zijl, P. C. M. (2003) Sensitive CEST agents based on nucleic acid imino proton exchange: detection of poly(rU) and of a dendrimer-poly(rU) model for nucleic acid delivery and pharmacology. *Magnetic Resonance in Medicine*, **49**, 998–1005.

48. Zhou, J., Lal, B., Wilson, D. A., Laterra, J., and van Zijl, P. C. (2003) Amide proton transfer (APT) contrast for imaging of brain tumors. *Magnetic Resonance in Medicine*, **50**, 1120–1126.

49. Zhou, J., Payen, J.-F., Wilson, D. A., Traystman, R. J., and van Zijl, P. C. (2003) Using the amide proton signals of intracellular proteins and peptides to detect pH effects in MRI. *Nature Medicine*, **9**, 1085–1090.

50. Zhang, S., Merritt, M., Woessner, D. E., Lenkinski, R. E., and Sherry, A. D. (2003) PARACEST agents: modulating MRI contrast via water proton exchange. *Accounts of Chemical Research*, **36**, 783–790.

51. Aime, S., Delli Castelli, D., and Terreno, E. (2002) Novel pH-reporter MRI contrast agents. *Angewandte Chemie-International Edition*, **41**, 4334–4336.

52. Zhang, S., Winter, P., Wu, K., and Sherry, A. D. (2001) A novel europium(III)-based MRI contrast agent. *Journal of the American Chemical Society*, **123**, 1517–1518.

53. Aime, S., Castelli, D. D., and Terreno, E. (2005) Highly sensitive MRI chemical exchange saturation transfer agents using liposomes. *Angewandte Chemie-International Edition*, **44**, 5513–5515.

54. Terreno, E., Cabella, C., Carrera, C., Castelli, D. D., Mazzon, R., Rollet, S., Stancanello, J., Visigalli, M., and Aime, S. (2007) From spherical to osmotically shrunken paramagnetic liposomes: an improved generation of LIPOCEST MRI agents with highly shifted water protons. *Angewandte Chemie-International Edition*, **46**, 966–968.

55. Winter, P. M., Cai, K., Chen, J., Adair, C. R., Kiefer, G. E., Athey, P. S., Gaffney, P. J., Buff, C. E., Robertson, J. D., Caruthers, S. D., Wickline, S. A., and Lanza, G. M. (2006) Targeted PARACEST nanoparticle contrast agent for the detection of fibrin. *Magnetic Resonance in Medicine*, **56**, 1384–1388.

56. Liu, G., Har-el, Y. E., Moake, M., Long, C., Walczak, P., Gilad, A. A., Zhang, J., Cardona, A., Jamil, M., Sgouros, G., Bulte, J. W. M., van Zijl, P. C. M., and McMahon, M. T. (2010) In vivo imaging of lymphatic delivery of multi-color DIACEST liposomes. *Proceedings of ISMRM*, Stockholm, SWE.

57. Schröder, L., Lowery, T. J., Hilty, C., Wemmer, D. E., and Pines, A. (2006)

Molecular imaging using a targeted magnetic resonance hyperpolarized biosensor. *Science*, **314**, 446–449.
58. Schröder, L., Meldrum, T., Smith, M., Lowery, T. J., Wemmer, D. E., and Pines, A. (2008) Temperature response of Xe-129 depolarization transfer and its application for ultrasensitive NMR detection. *Physical Review Letters*, **100**, 257603.
59. Stancanello, J., Terreno, E., Castelli, D. D., Cabella, C., Uggeri, F., and Aime, S. (2008) Development and validation of a smoothing-splines-based correction method for improving the analysis of CEST-MR images. *Contrast Media & Molecular Imaging*, **3**, 136–149.
60. Sun, P. Z., Farrar, C. T., and Sorensen, A. G. (2007) Correction for artifacts induced by B-0 and B-1 field inhomogeneities in pH-Sensitive chemical exchange saturation transfer (CEST) Imaging. *Magnetic Resonance in Medicine*, **58**, 1207–1215.
61. Kim, M., Gillen, J., Landman, B. A., Zhou, J., and van Zijl, P. C. M. (2009) WAter Saturation Shift Referencing (WASSR) for chemical exchange saturation transfer experiments. *Magnetic Resonance in Medicine*, **61**, 1441–1450.
62. McMahon, M. T., Gilad, A. A., DeLiso, M. A., Berman, S. M., Bulte, J. W., and van Zijl, P. C. (2008) New "multicolor" polypeptide diamagnetic chemical exchange saturation transfer (DIACEST) contrast agents for MRI. *Magnetic Resonance in Medicine*, **60**, 803–812.
63. Aime, S., Carrera, C., Delli Castelli, D., Geninatti Crich, S., and Terreno, E. (2005) Tunable imaging of cells labeled with MRI-PARACEST agents. *Angewandte Chemie-International Edition England*, **44**, 1813–1815.
64. Vinogradov, E., Zhang, S., Lubag, A., Balschi, J. A., Sherry, A. D., and Lenkinski, R. E. (2005) On-resonance low $B1$ pulses for imaging of the effects of PARACEST agents. *Journal of Magnetic Resonance*, **176**, 54–63.
65. Castile J. D. and Taylor, K. M. (1999) Factors affecting the size distribution of liposomes produced by freeze–thaw extrusion. *International Journal of Pharmaceutics*, **188**, 87–95.

Chapter 10

Hyperpolarized Noble Gases as Contrast Agents

Xin Zhou

Abstract

Hyperpolarized noble gases (^{3}He and ^{129}Xe) can provide NMR signal enhancements of 10,000 to 100,000 times that of thermally polarized gases and have shown great potential for applications in lung magnetic resonance imaging (MRI) by greatly enhancing the sensitivity and contrast. These gases obtain a highly polarized state by employing a spin exchange optical pumping technique. In this chapter, the underlying physics of spin exchange optical pumping for production of hyperpolarized noble gases is explained and the basic components and procedures for building a polarizer are described. The storage and delivery strategies of hyperpolarized gases for in vivo imaging are discussed. Many of the problems that are likely to be encountered in practical experiments and the corresponding detailed approaches to overcome them are also discussed.

Key words: Hyperpolarized ^{3}He, hyperpolarized ^{129}Xe, spin exchange optical pumping (SEOP), production, storage, and delivery of hyperpolarized noble gases.

1. Introduction

1.1. NMR Sensitivity

Conventional MRI mainly focuses on the nuclear spins of protons because of the large abundance of water and fat tissue in the body. However, certain organs like the lungs have a low proton spin density attributable to the large volume of air dispersed throughout the issue. The low sensitivity of traditional magnetic resonance in the airspace due to the low spin density has motivated the development of techniques using hyperpolarized noble gases for nuclear magnetic resonance (NMR) spectroscopy and biological MRI (1, 2). The spin polarizations of noble gases, like ^{3}He and ^{129}Xe, can be increased by four or five orders of magnitude over thermal equilibrium via spin exchange optical pumping (3–9). As such, noble gases are referred to as "hyperpolarized" or "laser-polarized" gases. Hyperpolarized gases make it possible to obtain magnetic resonance images of the lung airspace with unprecedented spatial resolution and sensitivity (10–19).

Generally, the macroscopic magnetization of a sample containing N nuclear spins can be written as (20),

$$M_0 = \frac{N\gamma\hbar P}{2}, \quad [1]$$

where γ is the gyromagnetic ratio; $\hbar$ is Planck's constant divided by 2π; P is the nuclear spin polarization. Therefore, it is clear that a higher spin concentration and a higher polarization of a sample give rise to a better signal. For a spin system with $I = 1/2$ (I is the nuclear spin angular momentum), the polarization, P, is defined as

$$P = \frac{N_- - N_+}{N_- + N_+}, \quad [2]$$

where N_+ and N_- are the number of spins in the $M_J = 1/2$ and $M_J = -1/2$ states, i.e., the lower and upper Zeeman energy sublevels, respectively.

For the conventional NMR or MRI, i.e., acquiring the nuclear magnetic resonance signal at thermal polarization, the nuclear spin polarization is quite small, and in the high-temperature approximation, it is given by (20)

$$P_0(B_0, T) \approx \frac{\gamma\hbar B_0}{2kT}, \quad [3]$$

where k is Boltzmann's constant; B_0 and T are the magnetic inductivity field strength and the temperature. For example, the proton thermal polarization is only 4×10^{-5} in a magnetic field of 11.7 T at 300 K. It requires the sample's spin density high enough to obtain an observable magnetic resonance signal. That is the reason why the conventional NMR and MRI are only available to liquid, solid samples, or biological tissues, in which the proton spin densities are about three orders of magnitude higher than that in the gas phase samples. Traditionally, increasing the magnetic field (B_0) and lowering temperature (T) are employed to improve the thermal polarization. However, it becomes extremely costly to go to higher and higher magnetic fields with superconducting magnets and the cryogenic temperatures needed for enhanced polarization are not practical for in vivo applications. In order to obtain an observable magnetization in the gas phase, hyperpolarization is needed in order to compensate for the lack of spin density.

A number of methods have been developed to enhance the polarization of samples beyond thermal equilibrium including dynamic nuclear polarization (DNP) (21–26) (*see* **Chapter 11**), parahydrogen-induced polarization (27–29) (see **Chapter 11**), and, in the case of noble gases, spin exchange optical pumping (SEOP). Currently, the spin polarizations of ^{3}He and ^{129}Xe can be routinely increased up to 35–40% (30, 31) and 50–64% (32–34) for liter quantities, respectively. It should be noted that the spin polarization of ^{3}He can also been enhanced by another technology called the metastability exchange (35–39). Since SEOP can be used for hyperpolarizing both ^{3}He and ^{129}Xe gases, in this chapter, we will focus on the technology of SEOP.

1.2. Hyperpolarized ^{3}He and ^{129}Xe

^{3}He and ^{129}Xe are the nuclei normally selected for hyperpolarized noble gases MRI for lungs, but in principle, any nucleus with non-zero spin (like ^{1}H, ^{2}H, ^{3}He, ^{13}C, ^{14}N, ^{15}N, ^{19}F, ^{31}P, ^{129}Xe, ^{131}Xe...) can be polarized above thermal equilibrium. However, the longitudinal relaxation rate of these nuclei can be rapid, causing significant decay of the spin polarization. The monatomic structure of noble gases, such as ^{3}He, ^{19}Ne, ^{21}Ne, ^{37}Ar, ^{83}Kr, ^{129}Xe, and ^{131}Xe, enable long polarization half-lives compared to other nuclei. Among the noble gas isotopes, ^{3}He and ^{129}Xe are the only nuclei with nuclear spin 1/2, resulting in the T_1 (longitudinal relaxation time) of many hours or even days at the standard temperature and pressure. Hyperpolarized ^{83}Kr ($I = 9/2$) has also been reported as a third noble gas probe for lung imaging, but quadrupolar interaction and the presence of a breathable concentration of molecular oxygen heavily affect the T_1 (40). Therefore, hyperpolarized ^{3}He and ^{129}Xe have generally been used for the medical lung imaging so far.

For comparison, the basic physical properties of ^{3}He and ^{129}Xe are listed in **Table 1**. For direct imaging of gas spaces, like lung MRI, ^{3}He has an advantage over ^{129}Xe, because the gyromagnetic ratio of ^{3}He is 2.8 times larger than that of ^{129}Xe. However, the diffusion coefficient of ^{129}Xe is 30 times lower than that of ^{3}He, and the lower diffusion constant would result in less signal attenuation during high-resolution imaging. ^{3}He and ^{129}Xe cover the spectrum from insoluble to highly soluble. ^{3}He is much less soluble than ^{129}Xe due to its smaller size. Since xenon is lipophilic, it has a higher solubility in blood-, oil-, and lipid-containing tissues. ^{129}Xe is also extremely sensitive to its environment, which results in a chemical shift greater than 200 ppm in the body and up to 7500 ppm in different chemical compounds (41). Due to the relatively high solubility and the large chemical shift of xenon in biological tissues, such as in the lungs and the brain (42, 43), dissolved phase hyperpolarized xenon MRI and molecular imaging have shown a great potential in the study of gas exchange in the lungs, brain perfusion, and potential detection of diseases with high sensitivity (44–54). Such solubility and chemical shift properties make ^{129}Xe a great probe for a variety of biomedical and material science applications.

Table 1
The basic properties of ^{3}He and ^{129}Xe

	Gyromagnetic ratio (MHz/T)	Natural abundance (%)	Ostwald solubility in water (37°C)	Ostwald solubility in blood (37°C)	Self-diffusion coefficient (cm^{2}s^{-1})
3He	−32.4	0.00013	0.0098	00099	1.8
129Xe	−11.8	26.4	0.083	0.14	0.06
References	(57)	(57)	(58)	(58)	(59, 60)

1.3. The Principle of Spin-Exchange Optical Pumping

For SEOP, the first step is the transfer of circularly polarized laser light into electronic spin polarization, i.e., optical pumping. **Figure 1** illustrates the principle of optically pumping Rb. In principal, any alkali metal vapor can be optically pumped. Rb is commonly used since the corresponding pumping laser diode arrays (LDA) are routinely manufactured in high power configurations. When the Rb pumping cell is placed in an oven at a temperature up to 160°C, the Rb is vaporized. Since the Rb vapor experiences an externally applied magnetic field (25 G) produced by a pair of Helmholtz coils, the ground state $5^2S_{1/2}$ is split into two Zeeman sublevels, $M_J = -1/2$ and $M_J = 1/2$, i.e., electronic spin "down" or "up." These two Zeeman states have a roughly equal population at room temperature. When the Rb vapor absorbs the circularly polarized laser light ($\sigma+$) centered at 795 nm, a D$_1$ transition of the Rb is observed, i.e., $5^2S_{1/2} \rightarrow 5^2P_{1/2}$. The circularly polarized light is right handed ($\sigma+$), and therefore only the ground state with a sublevel $M_J = -1/2$ is pumped into the

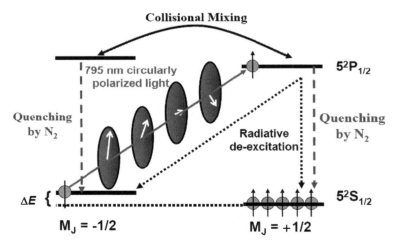

Fig. 1. Schematic diagram of optical pumping of Rb.

excited state, $M_J = +1/2$. Once the valence electron has been optically pumped to the excited state $5^2P_{1/2}$, collisional mixing equalizes the excited state populations. N_2 gas molecules have many degrees of freedom in terms of rotational and vibrational levels. Thus, nitrogen gas will quench the two sublevels of the excited state back to the ground state. As the $M_J = -1/2$ sublevel continues to absorb the circularly polarized light ($\sigma+$) an excess of Rb atoms are optically pumped into the Zeeman sublevel $M_J = 1/2$ while the other sublevel $M_J = -1/2$ is depleted. Therefore, an electronic spin polarization of nearly 100% can be achieved for Rb.

The second step of SEOP is spin exchange between the polarized electronic spin of Rb and the xenon nucleus, as shown in **Fig. 2**. In the high-temperature and high-pressure optical pumping cell, the collision between polarized Rb atoms and xenon atoms results in the transfer of angular momentum from the electronic spin to the Xe nuclear spin. During this collision, the electron wave function of the Rb overlaps the nuclear wave function of xenon which causes spin exchange between the electronic spin and nuclear spin. Binary collisions dominate the spin exchange at high pressure, while three-body collision (by forming a Rb/Xe van der Waals molecule) dominate at low pressure (a few tens of torr) (5).

Generally, the polarization build-up time for ^{3}He is much longer than that of ^{129}Xe, and accordingly, the spin-exchange optical pumping for ^{3}He is normally done in a batch mode, while hyperpolarized ^{129}Xe can be produced in a continuous flow

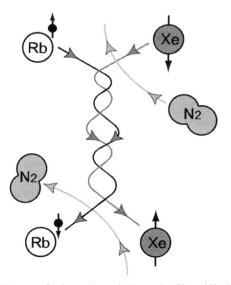

Fig. 2. Schematic diagram of spin exchange between the Rb and Xe. (Courtesy of Brian Patton. Reprinted from Ref. (9) with permission.)

mode. A hybrid optical pumping technique has been developed for ^{3}He to shorten the time needed to polarize the gas (55). In this chapter, we will focus on the building of a xenon polarizer in a continuous flow mode.

2. Materials

2.1. Non-magnetic Platform

1. Aluminum alloy frame.
2. Non-magnetic wheels.
3. Stable optical platform for optical support.

2.2. Vacuum Pipeline System

1. High performance vacuum system (Pfeiffer Vacuum HiPace 400 turbopump).
2. Vacuum gauge.
3. Copper and Teflon gas tubing, valves, fittings (Swagelok, Fremont, CA).
4. Pyrex glass tubes for fabrication of optical pumping cell.
5. Pyrex glass valves and fittings (Chemglass, Vineland, NJ).
6. Glass coating fluid (Surfasil, Thermo scientific, Rockford, IL).
7. Oxygen trap (with option to be regenerated (NuPure) or to be replaced (Agilent))
8. Cold finger.

2.3. Optical Pumping System

1. Litz wire for Helmholtz coils or 40 cm premade Helmholtz coils from Phywe Systeme GmbH & Co KG, Germany.
2. High stability DC power supply (0–60 V).
3. Laser safety glasses designed for a wavelength of 795 nm.
4. 60 W high power tunable laser system with a built-in air-cooled chiller; the central wavelength of laser is 795 nm (Coherent Duo FAP system, Santa Clara, CA).
5. Beam expander, beam splitter (BS), reflector, and two quarter-wave plates with a central wavelength of 795 nm (Coherent, Santa Clara, CA).
6. Optical spectrometer (Ocean Optics USB 2000+, Dunedin, FL).
7. Personal computer (PC).

2.4. Flow Controlling System

1. Mass flow controller (Aalborg DFC-26S, Orangeburg, NY).
2. Mass flowmeter (Aalborg GFM17A Orangeburg, NY).

3. Power supply (+15 V, −15 V) and accessories.
4. Pressure transducer (Swagelok, Oakland, CA).
5. Data Acquisition (DAQ) board (National Instruments PC6251, Austin, TX).
6. DAQ software (National Instruments LabVIEW, Austin, TX).

2.5. Temperature Controlling System

1. Insulated oven box (home built).
2. Air heater, thermocouple, and accessories (Chromalox, Pittsburgh, PA).
3. Flow shutoff safety (Dwyer, Michigan City, IN).
4. Temperature controller (Chromalox, Pittsburgh, PA).

2.6. Cryogenics and Storage

1. Ice water.
2. Liquid nitrogen and dewar.
3. Permanent magnet (2500 G).
4. Cold finger.
5. Non-magnetic lab jack.

2.7. Gases and Alkali Metal

1. Xenon gas mixture (1% Xe, 1–10% N_2, balanced He) (*see* **Notes 1(d)** and **2(a)**).
2. Rubidium ingot, ca. 1 g (Sigma Aldrich, St. Louis, MO).
3. High pressure air (70 psi).
4. High purified nitrogen gas (25–30 psi).
5. Heat gun.
6. Glass Pasteur pipettes.

2.8. Components for In Vivo Delivery

1. Tedlar bag (Jensen Inert, Coral Springs, FL).
2. Hot water.
3. Air ventilator for inspiration.
4. Pneumatic valves.
5. High pressure helium gas (70 psi).

3. Methods

3.1. Xenon Polarizer Platform Fabrication

1. Using aluminum alloy materials and non-magnetic wheels, fabricate a platform for the whole xenon polarizer system, which can be easily moved.

2. The optical system should then be mounted securely on the xenon polarizer platform to minimize the displacement of the laser optics.

3.2. Pipeline System and Vacuum

1. Assemble a pipeline system using copper and Teflon tubing, valves, and fittings according to the schematic in **Fig. 3**.
2. Using Pyrex glass tubes, valves, and fittings, fabricate a cylindrical optical pumping cell, in which the laser beam can illuminate it along the cylindrical axis.
3. The optical pumping cell needs to be washed three times using distilled water and acetone, and then baked in an oven with a temperature of 80°C for more than 24 h.
4. Next, dilute SurfaSil siliconizing fluid in acetone to a concentration of 10% by volume.
5. Flood the optical pumping cell in the diluted SurfaSil solution for at least 10 s.
6. Agitate the solution to ensure a uniform coating, then rinse the object with the acetone and methanol and bake it in an oven at a temperature of 100°C for 1 h.
7. Connect the optical pumping cell, vacuum system, vacuum gauge, and oxygen trap to the pipeline system, and then

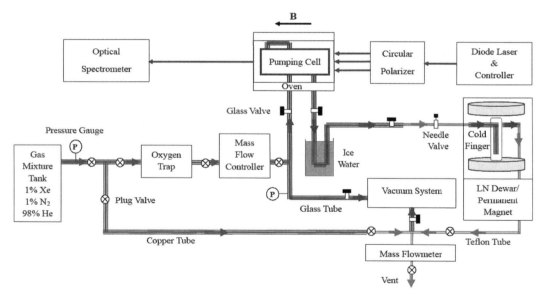

Fig. 3. Schematic diagram of xenon polarizer in a continuous flow mode. The arrows indicate the gas flow paths. Xenon gas mixture initially flows through oxygen trap and mass flow controller into the Rb optical pumping cell, in which SEOP occurs and hyperpolarized xenon is produced. Hyperpolarized xenon gas mixture then flows to a cold finger in ice water in order to remove the residual Rb vapor. Finally, only hyperpolarized xenon is trapped and stored in the cold finger, which is immersed in liquid nitrogen (77 K) and located in a permanent magnetic field of 2500 G, while nitrogen and helium gases in the xenon gas mixture are vented out.

test the vacuum to make sure it is able to reach a vacuum of at least 10 millitorr (*see* **Notes 3(e), 4(d)/(e),** and **1(a)/(b)**).

8. Disconnect the pumping cell from the vacuum system, and bring it to a glove box with highly purified nitrogen environment.
9. Inside the glove box, open the Rb ampoule and use the heat gun to warm the Rb to the liquid state.
10. Then, transfer the liquid Rb to the optical pumping cell with a Pasteur pipette.
11. Finally, close all the valves of optical pumping cell and connect it back to the pipeline system on the polarizer (*see* **Notes 4(a)** and **1(c)**).

3.3. Optical Pumping System

1. Construct a pair of Helmholtz coils with litz wire, with diameters of 40–60 cm.
2. Be sure to center the magnetic field around the optical pumping cell. Connect the high stability DC power supply to the Helmholtz coil, and adjust the current so that the magnetic field strength is 25 G in terms of the formula,

$$B = \left(\frac{4}{5}\right)^{3/2} \frac{\mu_0 n I}{R}, \qquad [4]$$

where n is the number of turns in each Helmholtz coil; I is the current flowing through the coils, and R is the radius (*see* **Note 3(a)**).

3. Wearing laser safety glasses, carefully assemble the laser diode, laser chiller, and controller together (*see* **Note 3(c)**).
4. With the aid of an aiming beam from the laser system, put the beam expander at the output of the laser fiber, and then let the expanded laser beam to pass through a beam splitter (BS) (*see* **Note 4(c)**).
5. The laser beam will become two beams: one is horizontally polarized, and the other is vertically polarized. These two beams will then pass two quarter-wave plates in order to produce two beams with circular polarization in the same direction.
6. Collimate the resulting beams to ensure they are illuminating the entire pumping cell. Place the detector of the optical spectrometer at the back of the pumping cell; connect the USB cable of the optical spectrometer to the computer (PC) so that the laser absorption profile of the Rb vapor can be monitored (*see* **Notes 3(d)** and **4(b)**).

3.4. Flow Controlling System

1. Connect the mass flow controller between the oxygen trap and the gas inlet of optical pumping cell and connect the mass flowmeter before the gas vent valve, as shown in **Fig. 3**.

2. Place one pressure gauge at the outlet of the gas mixture tank and the other one at the gas inlet of optical pumping cell, so that the pressure before and after the mass flow controller can be monitored.

3. The pressure and mass flow controller signal are input into the PC via a National Instruments (NI) PCI6251 data acquisition board. With the NI software, LabVIEW, the gas flow and pressure can be automatically controlled (*see* **Note 2(d)**).

4. For an accurate reading of the flow rate, the mass flow controller should be calibrated based on the composition of the gas mixture (*see* **Note 5(a)**).

5. The room temperature N_2 gas with a pressure of 25–30 psi has been connected to the cold finger, in order to avoid condensation of nitrogen in the xenon gas mixture (*see* **Note 5(c)**).

3.5. Temperature Controlling System

1. Based on the size of optical pumping cell, design an insulated oven which is capable of reaching 160°C.

2. The thermocouple sticks on the optical pumping cell to monitor the Rb temperature.

3. High pressure air (70 psi) is regulated to 25 psi before flowing through the flow shutoff safety (FSS), which is used to turn off the heater in case the airflow ceases. After the air passes the FSS, it flows into the oven through the air heater (*see* **Note 3(b)**).

4. The heated air will keep the oven at the desired temperature by the temperature controller, which is programmed to use the proportional-integral-differential (PID) algorithm.

3.6. Cryogenics and Storage of Hyperpolarized Xenon

1. Cryogenics are needed for purifying xenon from the gas mixture for in vivo applications. The hyperpolarized xenon gas mixture goes from the optical pumping cell and flows through an ice water bath in order to trap the residual Rb vapor (*see* **Note 2(e)**).

2. The gas mixture subsequently flows into a cold finger, which is immersed in liquid nitrogen (77 K) and located in a permanent magnet of 2500 G field strength. Because xenon's melting point is 161 K, xenon solidifies while the nitrogen and helium flow out. Hence, the xenon is separated from the gas mixture. The T_1 of solid xenon can be as long as a few hours in a 2500 G magnetic field and liquid nitrogen temperatures (*see* **Note 5(b)**).

3.7. Delivery for In Vivo Applications

1. Based on the volume of gas needed for either human or animal experiments, the collection time for hyperpolarized solid xenon can be calculated.
2. The hyperpolarized xenon gas can be thawed in a Tedlar bag using hot water.
3. For human experiments, mix hyperpolarized xenon gas with highly purified nitrogen gas to a desired composition (Xe:N_2 =1:2) and then the subject simply needs to inhale the xenon directly.
4. For animal experiments, a hyperpolarized xenon delivery system needs to be designed: The Tedlar bag with the hyperpolarized xenon gas sample is placed in a plastic box, which is pressurized with helium gas. This compresses the hyperpolarized xenon allowing it to flow out.
5. A ventilator can then be constructed with computer-controlled pneumatic valves so that the animal receives oxygen and hyperpolarized xenon in alternating breaths (56).

4. Notes

1. Problem: Rb in the optical pumping cell is often oxidized.
   Solutions:
   (a) Make sure all valves are holding pressure and replace any abnormal or corroded o-rings.
   (b) Double check that the vacuum in the pipeline can be brought down to at least 10 millitorr.
   (c) If the polarizer will not be used for more than 24 h, it should be pressurized above atmospheric pressure by either the xenon gas mixture, or with highly purified nitrogen.
   (d) Check for impurities in the xenon gas mixture and consider replacing tank.
   (e) Ensure the oxygen trap is functioning properly as described in **Note 4(e)**.
2. Problem: The gas flow rate is restricted or not flowing at all.
   Solutions:
   (a) Check the pressure of xenon tank. It should be at least more than 100 psi.
   (b) Make sure that mass flow controller works properly. If not, double check the power supply (+15 V, −15 V, GND). Ensure the flow controller is communicating with the PC by using a terminal communication tool.

(c) Verify there are no clogs in flow path which could restrict the gas flow rate.

(d) Make sure the Labview software that controls the gas flow rate is operating.

(e) Clean the trap located in ice water once in a while and make sure nothing in the trap restricts the hyperpolarized xenon gas flow.

3. Problem: No hyperpolarized Xe NMR signal.
Solutions:
(a) Make sure the power supply of the Helmholtz coil has been turned on. Use a gauss meter to check the magnetic field strength at the center of Helmholtz coil. Double check to ensure the current in each Helmholtz coil produces a magnetic field in the same direction.

(b) Make sure the temperature controller for the oven works properly, and temperature is high enough to vaporize the Rb. Verify that the high-pressure air is flowing. Calibrate the temperature readout to guarantee the accuracy of the feedback loop.

(c) Make sure the laser emission is normal. Check the laser control panel to see if any error indication lights are on and ensure the power supply for the laser is working properly. Double check that there is nothing blocking the air vent for the chiller or any clogs in the coolant flow. If the laser still has problems, follow the manual of the laser diode, chiller, and power supply.

(d) Check the laser absorption profile of the Rb vapor. The profile should show a strong absorbance at 795 nm. If the Rb absorption is not apparent or the profile seems abnormal, please follow **Note 4**.

(e) If all the individual components are operating properly but the xenon NMR signal is still weak, check for areas which might be causing depolarization of the gas. For example, paramagnetic impurities that are contaminating the pipeline for the gas flow, the metal needle valve, leaks, or any zero field spots in the flow path of the hyperpolarized xenon gas mixture.

4. Problem: Rb absorption disappears or is not fully saturated.
Solutions:
(a) Open the oven to check if the rubidium in the optical pumping cell has been oxidized (looks less shiny). If the rubidium oxidizes often, please follow **Note 1**.

(b) Make sure the optical spectrometer works properly by measuring the laser intensity, without heating the oven.

(c) Make sure the laser beams have been aligned properly, i.e., the laser beams pass almost parallel through the optical pumping cell, and converge on the sensor of the optical spectrometer.

(d) Check the pipeline system for leaks and ensure the xenon gas mixture from the tank is not contaminated with oxygen.

(e) Ensure the oxygen trap is working properly. If it has been used for an extended period of time (several months), consider regenerating it by baking it at high temperatures with highly purified nitrogen gas flowing through it for 6 h or replace it.

5. Problem: Little or no accumulation of hyperpolarized solid xenon.
Solutions:

(a) Make sure that hyperpolarized xenon flow rate is stable.

(b) The liquid nitrogen level should be high enough to immerse the cold finger completely in order to freeze the hyperpolarized xenon gas.

(c) Ensure the nitrogen gas flow through the jacket of the cold finger is not clogged. If there is no nitrogen gas flow, hyperpolarized xenon solid might accumulate at the inlet of the cold finger and clog the system.

Acknowledgments

The author thanks Dominic Graziani and Xianping Sun for proof reading and helpful suggestions for the manuscript. The work was supported by the 100 talents program of the Chinese Academy of Sciences, the innovative methods program of the Ministry of Science and Technology of China, and the Director, Office of Science, Office of Basic Energy Sciences, Materials Sciences Division, of the U.S. Department of Energy under Contract DE-AC02-05CH11231.

References

1. Raftery, D., Long, H., Meersmann, T., Grandinetti, P. J., Reven, L., Pines, A. (1991) High-field NMR of adsorbed xenon polarized by laser pumping. *Phys. Rev. Lett.* **66**, 584–587.

2. Albert, M. S., Cates, G. D., Driehuys, B., Happer, W., Saam, B., Springer, C. S., Jr., Wishnia, A. (1994) Biological magnetic resonance imaging using laser-polarized ^{129}Xe. *Nature* **370**, 199–201.

3. Happer, W. (1972) Optical pumping. *Rev. Mod. Phys.* **44**, 169–249.

4. Kastler, A. (1950) Quelques suggestions concernant la production optique et la détection

optique d'une inégalité de population des niveaux de quantifigation spatiale des atomes. Application à l'expérience de Stern et Gerlach et à la résonance magnétique. *J. Phys. Radium* **11**, 255–265.
5. Walker, T. G., Happer, W. (1997) Spin-exchange optical pumping of noble-gas nuclei. *Rev. Mod. Phys.* **69**, 629–642.
6. Zeng, X., Wu, Z., Call, T., Miron, E., Schreiber, D., Happer, W. (1985) Experimental determination of the rate constants for spin exchange between optically pumped K, Rb, and Cs atoms and ^{129}Xe nuclei in alkali-metal-noble-gas van der Waals molecules. *Phys. Rev. A* **31**, 260–278.
7. Appelt, S., Ben-Amar Baranga, A., Erickson, C. J., Romalis, M. V., Young, A. R., Happer, W. (1998) Theory of spin-exchange optical pumping of ^{3}He and ^{129}Xe. *Phys. Rev. A* **58**, 1412–1439.
8. Zhou, X., Sun, X., Luo, J., Zeng, X., Liu, M., Zhan, M. (2004) Production of hyperpolarized ^{129}Xe gas without nitrogen by optical pumping at ^{133}Cs D$_2$ line in flow system. *Chin. Phys. Lett.* **21**, 1501–1503.
9. Patton, B. (2007) *NMR Studies of Angular Momentum Transfer and Nuclear Spin Relaxation*. Ph.D. dissertation, Princeton University.
10. Moller, H. E., Chen, X. J., Saam, B., Hagspiel, K. D., Johnson, G. A., Altes, T. A., de Lange, E. E., Kauczor, H. U. (2002) MRI of the lungs using hyperpolarized noble gases. *Magn. Reson. Med.* **47**, 1029–1051.
11. Tooker, A. C., Hong, K. S., McKinstry, E. L., Costello, P., Jolesz, F. A., Albert, M. S. (2003) Distal airways in humans: dynamic hyperpolarized ^{3}He MR imaging – feasibility. *Radiology* **227**, 575–579.
12. Wild, J. M., Paley, M. N., Kasuboski, L., Swift, A., Fichele, S., Woodhouse, N., Griffiths, P. D., van Beek, E. J. (2003) Dynamic radial projection MRI of inhaled hyperpolarized ^{3}He gas. *Magn. Reson. Med.* **49**, 991–997.
13. Fain, S. B., Panth, S. R., Evans, M. D., Wentland, A. L., Holmes, J. H., Korosec, F. R., O'Brien, M. J., Fountaine, H., Grist, T. M. (2006) Early emphysematous changes in asymptomatic smokers: detection with ^{3}He MR imaging. *Radiology* **239**, 875–883.
14. Conradi, M. S., Saam, B. T., Yablonskiy, D. A., Woods, J. C. (2006) Hyperpolarized He-3 and perfluorocarbon gas diffusion MRI of lungs. *Prog. Nucl. Magn. Reson. Spectr.* **48**, 63–83.
15. Driehuys, B., Walker, J., Pollaro, J., Cofer, G. P., Mistry, N., Schwartz, D., Johnson, G. A. (2007) ^{3}He MRI in mouse models of asthma. *Magn. Reson. Med.* **58**, 893–900.
16. Stupar, V., Canet-Soulas, E., Gaillard, S., Alsaid, H., Beckmann, N., Cremillieux, Y. (2007) Retrospective cine He-3 ventilation imaging under spontaneous breathing conditions: a non-invasive protocol for small-animal lung function imaging. *NMR Biomed.* **20**, 104–112.
17. Hopkins, S. R., Levin, D. L., Emami, K., Kadlecek, S., Yu, J., Ishii, M., Rizi, R. R. (2007) Advances in magnetic resonance imaging of lung physiology. *J. Appl. Physiol.* **102**, 1244–1254.
18. Patz, S., Muradian, I., Hrovat, M. I., Ruset, I. C., Topulos, G., Covrig, S. D., Frederick, E., Hatabu, H., Hersman, F. W., Butler, J. P. (2008) Human pulmonary imaging and spectroscopy with hyperpolarized ^{129}Xe at 0.2 T. *Acad. Radiol.* **15**, 713–727.
19. Matsuoka, S., Patz, S., Albert, M. S., Sun, Y., Rizi, R. R., Gefter, W. B., Hatabu, H. (2009) Hyperpolarized gas MR Imaging of the lung: current status as a research tool. *J. Thorac. Imaging* **24**, 181–188.
20. Abragam, A. (1961) *The Principles of Nuclear Magnetism*. Oxford University Press, London.
21. Hausser, K. H., Stehlik, D. (1968) Dynamic nuclear polarization in liquids. *Adv. Magn. Reson.* **3**, 79–139.
22. Muller-Wamuth, W., Meise-Gresch, K. (1983) Molecular motions and interactions as studied by dynamic nuclear polarization (DNP) in free radical solutions. *Adv. Magn. Reson.* **11**, 1–45.
23. Wind, R. A., Duijvestijn, M. J., Vanderlugt, C., Manenschijn, A., Vriend, J. (1985) Applications of dynamic nuclear polarization in C-13 NMR in solids. *Prog. Nucl. Magn. Reson. Spectr.* **17**, 33–67.
24. rdenkjaer-Larsen, J. H., Fridlund, B., Gram, A., Hansson, G., Hansson, L., Lerche, M. H., Servin, R., Thaning, M., Golman, K. (2003) Increase in signal-to-noise ratio of >10,000 times in liquid-state NMR. *Proc. Natl. Acad. Sci. USA* **100**, 10158–10163.
25. Golman, K., Ardenkjaer-Larsen, J. H., Petersson, J. S., Mansson, S., Leunbach, I. (2003) Molecular imaging with endogenous substances. *Proc. Natl. Acad. Sci. USA* **100**, 10435–10439.
26. Maly, T., Debelouchina, G. T., Bajaj, V. S., Hu, K. N., Joo, C. G., Mak-Jurkauskas, M. L., Sirigiri, J. R., van der Wel, P. C. A., Herzfeld, J., Temkin, R. J., et al. (2008) Dynamic nuclear polarization at high magnetic fields. *J. Chem. Phys.* **128**, 052211.

27. Bowers, C. R., Weitekamp, D. P. (1986) Transformation of symmetrization order to nuclear-spin magnetization by chemical reaction and nuclear magnetic resonance. *Phys. Rev. Lett.* **57**, 2645–2648.
28. Natterer, J., Bargon, J. (1997) Parahydrogen induced polarization. *Prog. Nucl. Magn. Reson. Spectr.* **31**, 293–315.
29. Bouchard, L. S., Burt, S. R., Anwar, M. S., Kovtunov, K. V., Koptyug, I. V., Pines, A. (2008) NMR imaging of catalytic hydrogenation in microreactors with the use of para-hydrogen. *Science* **319**, 442–445.
30. Altes, T. A., Powers, P. L., Knight-Scott, J., Rakes, G., Platts-Mills, T. A., de Lange, E. E., Alford, B. A., Mugler, J. P., III, Brookeman, J. R. (2001) Hyperpolarized ^{3}He MR lung ventilation imaging in asthmatics: preliminary findings. *J. Magn. Reson. Imaging* **13**, 378–384.
31. Jacob, R. E., Morgan, S. W., Saam, B. (2002) He-3 spin exchange cells for magnetic resonance imaging. *J. Appl. Phys.* **92**, 1588–1597.
32. Driehuys, B., Cates, G. D., Miron, E., Sauer, K., Walter, D. K., Happer, W. (1996) High-volume production of laser-polarized Xe-129. *Appl. Phys. Lett.* **69**, 1668–1670.
33. Ruset, I. C., Ketel, S., Hersman, F. W. (2006) Optical pumping system design for large production of hyperpolarization ^{129}Xe. *Phys. Rev. Lett.* **96**, 053002-1-053002-4.
34. Schrank, G., Ma, Z., Schoeck, A., Saam, B. (2009) Characterization of a low-pressure high-capacity ^{129}Xe flow-through polarizer. *Phys. Rev. A* **80**, 063424.
35. Colegrove, F. D., Schearer, L. D., Walters, G. K. (1963) Polarization of He^3 gas by optical pumping. *Phys. Rev.* **132**, 2561–2572.
36. Eckert, G., Heil, W., Meyerhoff, M., Otten, E. W., Surkau, R., Werner, M., Leduc, M., Nacher, P. J., Schearer, L. D. (1992) A dense polarized He-3 target based on compression of optically pumped gas. *Nucl. Instr. Meth. A* **320**, 53–65.
37. Becker, J., Heil, W., Krug, B., Leduc, M., Meyerhoff, M., Nacher, P. J., Otten, E. W., Prokscha, T., Schearer, L. D., Surkau, R. (1994) Study of Mechanical Compression of Spin-Polarized He-3 Gas. *Nucl. Instr. Meth. A* **346**, 45–51.
38. Stoltz, E., Meyerhoff, M., Bigelow, N., Leduc, M., Nacher, P. J., Tastevin G. (1996) High nuclear polarizatoin in ^{3}He and ^{3}He-^{4}He gas mixture by optical pumping with a laser diode. *Appl. Phys. B* **63**, 629–633.
39. Becker, J., Bermuth, J., Ebert, M., Grossmann, T., Heil, W., Hofmann, D., Humblot, H., Leduc, M., Otten, E. W., Rohe, D., et al. (1998) Interdisciplinary experiments with polarized He-3. *Nucl. Instr. Meth. A* **402**, 327–336.
40. Pavlovskaya, G. E., Cleveland, Z. I., Stupic, K. F., Basaraba, R. J., Meersmann, T. (2005) Hyperpolarized krypton-83 as a contrast agent for magnetic resonance imaging. *Proc. Natl. Acad. Sci. USA.* **102**, 18275–18279.
41. Goodson, B. M. (2002) Nuclear magnetic resonance of laser-polarized noble gases in molecules, materials, and organisms. *J. Magn. Reson.* **155**, 157–216.
42. Cherubinia, A., Bifone, A. (2003) Hyperpolarised xenon in biology. *Prog. Nucl. Magn. Reson. Spectrosc.* **42**, 1–30.
43. Oros, A. M., Shah, N. J. (2004) Hyperpolarized xenon in NMR and MRI. *Phys. Med. Biol.* **49**, R105–R153.
44. Swanson, S. D., Rosen, M. S., Coulter, K. P., Welsh, R. C., Chupp, T. E. (1999) Distribution and dynamics of laser-polarized ^{129}Xe magnetization in vivo. *Magn. Reson. Med.* **42**, 1137–1145.
45. Ruppert, K., Brookeman, J. R., Hagspiel, K. D., Mugler, J. P., III (2000) Probing lung physiology with xenon polarization transfer contrast (XTC). *Magn. Reson. Med.* **44**, 349–357.
46. Driehuys, B., Cofer, G. P., Pollaro, J., Mackel, J. B., Hedlund, L. W., Johnson, G. A. (2006) Imaging alveolar-capillary gas transfer using hyperpolarized ^{129}Xe MRI. *Proc. Natl. Acad. Sci. USA.* **103**, 18278–18283.
47. Swanson, S. D., Rosen, M. S., Agranoff, B. W., Coulter, K. P., Welsh, R. C., Chupp, T. E. (1997) Brain MRI with laser-polarized ^{129}Xe. *Magn. Reson. Med.* **38**, 695–698.
48. Duhamel, G., Choquet, P., Grillon, E., Leviel, J. L., Decorps, M., Ziegler, A., Constantinesco, A. (2002) Global and regional cerebral blood flow measurements using NMR of injected hyperpolarized xenon-129. *Acad. Radiol.* **9**, S498–S500.
49. Kilian, W., Seifert, F., Rinneberg, H. (2004) Dynamic NMR spectroscopy of hyperpolarized ^{129}Xe in human brain analyzed by an uptake model. *Magn. Reson. Med.* **51**, 843–847.
50. Schroder, L., Lowery, T. J., Hilty, C., Wemmer, D. E., Pines, A. (2006) Molecular imaging using a targeted magnetic resonance hyperpolarized biosensor. *Science* **314**, 446–449.
51. Zhou, X., Mazzanti, M. L., Chen, J. J., Tzeng, Y. S., Mansour, J. K., Gereige, J. D.,

Venkatesh, A. K., Sun, Y., Mulkern, R. V., Albert, M. S. (2008) Reinvestigating hyperpolarized ^{129}Xe longitudinal relaxation time in the rat brain with noise considerations. *NMR Biomed.* **21**, 217–225.

52. Zhou, X., Graziani, D., Pines, A. (2009) Hyperpolarized xenon NMR and MRI signal amplification by gas extraction. *Proc. Natl. Acad. Sci. USA.* **106**, 16903–16906.

53. Driehuys, B., Moller, H. E., Cleveland, Z. I., Pollaro, J., Hedlund, L. W. (2009) Pulmonary perfusion and xenon gas exchange in rats: MR imaging with intravenous injection of hyperpolarized ^{129}Xe. *Radiology* **252**, 386–393.

54. Zhou, X., Sun, Y., Mazzanti, M. L., Henninger, N., Mansour, J. K., Fisher, M., Albert, M. S. (2011) MRI of stroke using hyperpolarized ^{129}Xe. *NMR Biomed.* **24**, 170–175.

55. Babcock, E., Nelson, I., Kadlecek, S., Driehuys, B., Anderson, L. W., Hersman, F. W., Walker, T. G. (2003) Hybrid spin-exchange optical pumping ^{3}He. *Phys. Rev. Lett.* **91**, 123003.

56. Hedlund, L. W., Cofer, G. P., Owen, S. J., Allan, J. G. (2000) MR-compatible ventilator for small animals: computer-controlled ventilation for proton and noble gas imaging. *Magn. Reson. Imaging* **18**, 753–759.

57. Harris, R. K. (1996) Nuclear spin properties and conventions for chemical shifts. *Encyclopedia of Nuclear Magnetic Resonance*, eds. Grant, D. M., Harris, R. K. pp. 3301–3314. Wiley, Chichester, UK.

58. Abraham, M. H., Kamlet, M. J., Taft, R. W., Doherty, R. M., Weathersby, P. K. (1985) Solubility properties in polymers and biological media. 2. The correlation and prediction of the solubilities of nonelectrolytes in biological tissues and fluids. *J. Med. Chem.* **28**, 865–870.

59. Patyal, B. R., Gao, J. H., Williams, R. F., Roby, J., Saam, B., Rockwell, B. A., Thomas, R. J., Stolarski, D. J., Fox, P. T. (1997) Longitudinal relaxation and diffusion measurements using magnetic resonance signals from laser-hyperpolarized ^{129}Xe nuclei. *J. Magn Reson.* **126**, 58–65.

60. Bock, M. (1997) Simultaneous T_2^* and diffusion measurements with ^{3}He. *Magn Reson. Med.* **38**, 890–895.

Chapter 11

Hyperpolarized Molecules in Solution

Jan Henrik Ardenkjaer-Larsen, Haukur Jóhannesson,
J. Stefan Petersson, and Jan Wolber

Abstract

Hyperpolarization is a technique to enhance the nuclear polarization and thereby increase the available signal in magnetic resonance (MR). This chapter provides an introduction to the concept of hyperpolarization as well as an overview of dynamic nuclear polarization (DNP) and para-hydrogen induced polarization (PHIP), two methods used to generate hyperpolarized molecules in aqueous solution.

Key words: Dynamic nuclear polarization, para-hydrogen induced polarization, hyperpolarization, magnetic resonance.

1. Introduction

For a population of nuclei with spin quantum number I, and thus $2I + 1$ energy states, the distribution of spins in each state is, at thermal equilibrium, governed by Boltzmann's law:

$$N_m = N_0 \frac{e^{-\frac{E_m}{k_B T}}}{\sum_{n=-I}^{n=I} e^{-\frac{E_n}{k_B T}}} \quad [1]$$

where N_m is the number of spins in the state m, N_0 is the total number of spins, E_m is the energy of the state m, T is the absolute temperature, and k_B is Boltzmann's constant. E_m is given by

$$E_m = -m\gamma \hbar B_0 \quad [2]$$

where $m = \pm\frac{1}{2}$, γ is the gyromagnetic ratio for the nucleus in question, and $\hbar$ is the reduced Planck's constant. Nuclei with spin quantum number $I = \frac{1}{2}$ (e.g., ^{1}H, ^{3}He, ^{13}C, ^{129}Xe) have two eigenstates in a magnetic field. The difference in population between the two possible energy states is described by the polarization, P, and it is defined accordingly:

$$P = \frac{N^+ - N^-}{N^+ + N^-} = \tanh\left(\frac{\gamma \hbar B_0}{2 k_B T}\right) \qquad [3]$$

where N^+ and N^- denote the number of spins parallel (spin up) and anti-parallel (spin down) to the external magnetic field, respectively. According to Eqs. [1] and [3], a temperature (spin temperature T_S) can be assigned to the spins for any polarization, e.g., $P = 0$, complete saturation, corresponds to an infinite spin temperature. Depending on the sign of the polarization, the spin temperature can be either negative or positive and will approach zero (cooling) as the polarization approaches unity.

For experiments performed at room temperature and with the highest magnetic field available, E_m is small compared to $k_B T$ and Eq. [3] can be reduced to:

$$P = \frac{\gamma \hbar B_0}{2 k_B T} \qquad [4]$$

From these expressions, the polarization for protons in a 1.5 T MR scanner becomes $\sim 5.0 \times 10^{-6}$ and the polarization for ^{13}C is only $\sim 1.3 \times 10^{-6}$. The polarization may therefore be enhanced 200,000 and 800,000 times respectively if polarization of unity can be achieved.

In an MR scanner operating at field strength of a few megahertz or higher, the dominating source of noise is the imaged object (e.g., the patient) and the signal-to-noise ratio, SNR, may be expressed as (1)

$$\text{SNR} \propto \gamma P c \propto \gamma^2 B_0 c \qquad [5]$$

where c is the concentration of the nucleus in question. The polarization, and thereby the strength of the NMR signal, increases proportionally with the magnetic field (1), which has been the motivation for developing higher field MR systems. The concentration of a hyperpolarized imaging agent may be 0.5 M in the injection syringe, but will decrease to 1–20 mM in vivo due to dilution in the vascular system. In hyperpolarized metabolic mapping applications, where the signals from intracellular metabolites are being visualized, the available concentration may be one to two orders of magnitude lower. This should be compared to the proton concentration of ~ 80 M in the body.

The term hyperpolarization is used to indicate that the available polarization is no longer determined by the main magnetic field of the MR scanner. Instead it is higher than the thermal equilibrium polarization given by Eq. [4]. The polarization for a hyperpolarized agent is created outside the imaging system by means of a polarizer. The hyperpolarization method can be based on several principles (2–6) of which two have successfully been applied to molecules in solution: para-hydrogen induced polarization (PHIP) and dynamic nuclear polarization (DNP) followed by dissolution.

The right-hand side of Eq. [5] is not valid for hyperpolarized molecules; instead the middle part of the expression should be used: the image quality is no longer a function of B_0. However, the choice of nucleus will still influence the SNR through the gyromagnetic ratio.

2. Dynamic Nuclear Polarization (DNP)

2.1. Introduction

In recent years, a novel method for polarizing nuclear spins in molecules in solution has been developed. The method takes advantage of dynamic nuclear polarization (DNP) in the solid state followed by rapid dissolution in a suitable solvent (6–8). The polarization is retained almost completely in the dissolution step creating a solution with a non-thermal nuclear polarization approaching unity.

DNP was first described theoretically by Overhauser in 1953 (9) and a few months later demonstrated by Carver and Slichter (10) in metallic lithium. Overhauser predicted that saturating the conduction electrons of a metal would lead to a dynamic polarization of the nuclear spins. This was a fundamental discovery causing disbelief at the time: that heating of one spin system could lead to the cooling of another. The prediction by Overhauser for metals was soon extended to electron spins in solution by Abragam (11) and most NMR spectroscopists are today familiar with the nuclear and electronic Overhauser effect. However, this effect is limited to solutions where relaxation processes couple the spin systems via the lattice (thermal motions). Soon after, the solid effect was described for spins in the solid state coupled by dipolar interactions (12). Later, DNP in the solid state was extended mechanistically to processes involving several electron spins and thermal mixing (13). The theory of DNP in the solid state, however, has failed to provide a quantitative description in general. In the solid state, the high electron spin polarization is in part transferred to the nuclear spins by microwave irradiation close to the resonance frequency of the electron spin. The efficiency of this process depends on several parameters characterizing the various spin systems, but also on technical factors such as microwave frequency and power.

DNP has mainly been applied to the generation of polarized targets for neutron scattering experiments and it has been demonstrated that the nuclear polarizations of ^{1}H and ^{13}C could be increased to almost 100% and to ~50%, respectively in the solid state by means of DNP at low temperature (14, 15). The mechanism requires the presence of unpaired electrons, which are added to the sample as, for example, an organic radical. The magnetic moment of the electron is 658 times higher than that of the proton. This means that the electron spin will reach polarization of unity at a moderate magnetic field strength and liquid helium temperature. Equation [3] gives an electron spin polarization of 98% at 3.35 T and 1 K. The efficiency of the transfer of polarization from the electron to the nuclear spins depends on several factors.

2.2. DNP Sample Preparation

The formulation of a new compound as a hyperpolarized imaging agent with the DNP method imposes several challenges. In order for the DNP process to be effective, the electron paramagnetic agent must be homogeneously distributed within the sample. Many compounds will be crystalline and have a tendency to crystallize as saturated aqueous solutions. This typically leads to poor polarization. To prevent crystallization and to produce an amorphous solid upon cooling of the sample, glass formers such as glycerol or DMSO can be added to the mixture of the compound and the electron paramagnetic agent. Since it is often a requirement for in vivo studies to achieve a high concentration of the compound after dissolution, it is necessary to be able to formulate the compound in a concentrated form. A solvent mixture with high solubility for the compound and good glassing properties is therefore chosen. Secondly, the solvent mixture has to be either biologically compatible or removed from the solution after polarization.

2.3. Electron Paramagnetic Agent

The source of the unpaired electron is typically an organic radical, but a few metal ions have been employed successfully for DNP, Cr(V) in particular (14–16). The choice of electron paramagnetic agent will depend on a number of factors. First, the electron paramagnetic agent needs to be chemically stable and dissolve readily in the matrix of interest. Second, the electron paramagnetic resonance (EPR) spectrum of the radical should have a width that allows DNP to be effective for the nucleus of interest, i.e., a linewidth that exceeds the Larmor frequency of the nuclear spin. In practice, the above criteria means that two classes of electron paramagnetic agents are available, namely, nitroxides (17, 18) and trityls (19–21). The nitroxides belong to a class of molecules that have been studied extensively by EPR, and which have been used for DNP in many systems. Nitroxides are characterized by having a broad EPR spectrum ($\Delta g/g \approx 4.0 \times 10^{-3}$) (16), covering the Larmor frequency of all nuclear spins. Some of them have reasonable chemical stability and come with different degrees of hydrophilicity. Another class of electron paramagnetic agents with superior properties for direct polarization of low-γ nuclei such as ^{13}C, ^{15}N, and ^{2}H is the trityls ($\Delta g/g \approx 0.25-0.75 \times 10^{-3}$) (16, 22, 23). The trityls also exist with a range of hydrophilicities and some of them are chemically very stable. It was recently discovered that gadolinium can positively affect the solid-state DNP enhancement (24). Other paramagnetic ions and molecules (Mn^{2+} and O_2) can in part have the same effect. The physics is not understood, but Hu et. al. (23) have shown that the longitudinal relaxation time of the electron paramagnetic agent is shortened by the presence of the Gd ions. The effect of adding 1–2 mmol/L Gd^{3+} is a 50–100% improvement of the DNP. The effect seems to be general to most samples, but has to be optimized for each sample similarly to the concentration of the electron paramagnetic agent. There is no direct DNP effect of the Gd^{3+} by itself. Finally, Gd^{3+} may enhance the solid-state polarization by DNP, but care should be taken in eliminating relaxation in the liquid state. Free Gd-ions would cause detrimental liquid-state relaxation and pose an in vivo safety risk.

2.4. DNP Instrumentation

The method and instrumentation for hyperpolarization by DNP was first described by Ardenkjaer-Larsen et al. (6). The polarizer, which is depicted in **Fig. 1**, is built to a large extent on principles established in the polarized target community. Most solid-state DNP has been performed at magnetic fields between 0.35 T (25) and 16.5 T (26, 27), and at temperatures from below 1 K to room temperature. At temperatures below a few kelvin and magnetic field strengths above a few tesla, electron spins are almost fully polarized, and large nuclear polarizations can be obtained. Unlike in NMR spectroscopy applications where non-equilibrium polarizations can be regenerated by repeating the microwave irradiation and NMR acquisition (26), the polarization generated for in vivo applications will decay irreversibly after dissolution. Hence, the goal is to generate polarizations close to unity. It is therefore important to choose initial operating conditions that have been proven to provide high nuclear polarization, but, at the same time, that are easily achievable using standard instrumentation. Temperatures of ∼1 K can be achieved by pumping in liquid helium. In the original work, the liquid helium was supplied to the sample space through a needle valve from the magnet cryostat, but in a recent publication, an alternative arrangement that used a separate helium dewar was described (28). A magnetic field strength of 3.35 T was chosen since microwave sources are readily

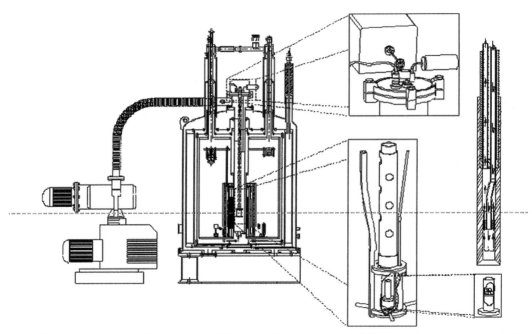

Fig. 1. DNP prototype polarizer design and principle as used in most hyperpolarization experiments for in vivo studies. A superconducting magnet charged to 3.35 T, mechanical pumps (*left-hand side*) to reduce the temperature of liquid helium to ~1.2 K, a microwave source to irradiate the sample at ~94 GHz, and to the *right* the dissolution stick, which is introduced at the end of the experiment to dissolve the frozen material in the sample container in lower *right* corner. Reprinted with permission of John Wiley & Sons, Inc. (6).

available at 94 GHz for irradiation of the electron spin. However, recently it has been demonstrated that for both nitroxides and trityls a significant improvement in polarization can be obtained by increasing the magnetic field strength (29, 30) or lowering the temperature (31). For the compound [1-^{13}C]pyruvic acid the ^{13}C polarization improved from 27% at 3.35 T to 60% at 4.64 T in the solid state. Typically, the literature has been limited and disagreeing on the magnetic field dependence of DNP (32). On the other hand, it has been well established that lowering the temperature is beneficial for DNP. Another recent development in DNP instrumentation for in vivo applications is the closed-cycle sorption pump cryostat (31). This system achieves a base temperature of less than 0.8 K for a pumped ^{4}He bath.

2.5. Dissolution

Once adequate solid-state polarization has been obtained, the sample needs to be dissolved in a suitable buffer. The dissolution may involve neutralization of the agent with acid or base, depending on the solid sample preparation. Buffering of the solution may be required to maintain control of pH and achieve a physiologically acceptable formulation. Physiological buffers such as tris(hydroxymethyl)aminomethane(TRIS) or 4-(2-hydroxyethyl)piperazine-1-ethanesulfonate (HEPES) are commonly used. The dissolution has to be efficient and fast compared to nuclear T_1 in order to preserve the nuclear polarization in this process. Formulating the solid sample as beads or powder may improve the dissolution (in terms of polarization and recovery of the solid sample), but understanding and optimizing the fluid dynamics (33) as well as providing the necessary heat is essential for optimal performance of more difficult agents. Relaxation during the dissolution process can depend on several factors. To minimize relaxation, dissolution is performed in the high field of the polarizer (e.g., 3 T in the case of a 3.35 T polarizer), but above the liquid helium surface. Any paramagnetic ions that could increase the relaxation rate are chelated by adding, for example, EDTA to the dissolution medium. To illustrate the severity of relaxation during dissolution [1-^{13}C]pyruvic acid is chosen as an example. This molecule has been well studied with DNP and has high biological relevance. We assume a best case, i.e., relaxation by paramagnetic impurities is eliminated by having pure samples or chelating any remaining impurities. Chemical shift anisotropy is negligible due to the moderate field of the polarizer. Dipolar relaxation to the protons is unavoidable unless the sample is partially deuterated. In **Fig. 2** the T_1 of the C-1 of neat [1-^{13}C]pyruvic acid at 9.4 T is given as a function of temperature (unpublished data). It can be seen that the minimum T_1 is ~1.6 s. With the trityl radical present (20 mmol/L), there will be an additional (dipolar) relaxation contribution. It can be seen that the contribution from the trityl is marginal and shifts the minimum to a different temperature (correlation time). The values in **Fig. 2** agree well with theory based on the known

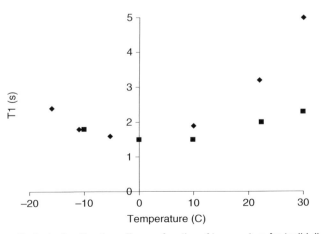

Fig. 2. Longitudinal relaxation time, T_1, as a function of temperature for (*solid diamond*) neat [1-13C]pyruvic acid and (*solid square*) [1-13C]pyruvic acid with 20 mmol/L trityl.

distances between the spins. According to dipolar relaxation theory the minimum T_1 scales approximately as B_0, which means that a minimum T_1 of 0.7 s should be expected during the dissolution (in the 3 T polarizer field). The data illustrates that the nuclear spin during the dissolution should pass through this T_1 minimum on a much faster timescale to avoid a loss of polarization. The example illustrates that it is not unreasonable to expect that the loss of relaxation during dissolution can be overcome. The severity of the problem will depend on the target spin and sample properties, but several parameters can be controlled, e.g., the distance to other spins (labeling position), the abundance of other spins (full or partial deuteration), and the concentration of the electron paramagnetic agent (sacrificing DNP for longer T_1).

Finally, the solution may undergo a filtration or chromatography step to remove the electron paramagnetic agent involved in the DNP process. In case a Gd chelate has been added, this agent may be removed as well. In most cases, the electron paramagnetic agent or Gd chelate do not cause significant relaxation after dissolution and may also be safe to inject into animals, but can often simply be removed by ion exchange or reverse phase chromatography. The filtration can either be in-line with the dissolution process or a subsequent step. In either case, the filtration is completed in a matter of a few seconds with insignificant loss of polarization or target molecule (unpublished work). However, significant research into optimal filtration materials will be required for new combinations of agents.

2.6. Nuclei and Compounds

Many nuclei have been hyperpolarized by the DNP method: ^{1}H (34), ^{6}Li (35), ^{19}F (36), ^{13}C, ^{15}N (37), ^{31}P (38), ^{89}Y (39), and ^{129}Xe (40). Nuclei with spin greater than $^1/_2$ other than ^{6}Li will have too short relaxation times and are unattractive for hyperpolarization studies in vivo. High gyromagnetic ratio nuclei such as ^{1}H, ^{19}F, and ^{31}P will also typically have relatively short T_1. The obvious nucleus of choice is ^{13}C, since many cellular metabolites can be labeled at positions in which the nucleus has a long relaxation time. The signal (polarization) that is created in the polarization process decays to thermal equilibrium with a rate (relaxation time) that depends on a number of factors. Some factors can be controlled in the dissolution process e.g., temperature, magnetic field, pH, and paramagnetic impurities. Others, such as intramolecular dipolar, chemical shift and spin rotation contributions depend on the ^{13}C position in a specific molecule and cannot be controlled as easily (with the possible exception of deuteration in some cases to avoid ^{13}C–^{1}H dipolar relaxation).

The range of compounds that have been polarized by DNP is extensive and demonstrates the generality of the method: [^{13}C]urea (6), [1-^{13}C]pyruvate (41), [2-^{13}C]pyruvate

(42), [1-^{13}C]lactate (43), [^{13}C]bicarbonate (44), [1,4-^{13}C$_2$]fumarate (45), [1-^{13}C]acetate (46), [5-^{13}C]glutamine (47), [1-^{13}C]glutamate (48), amino acids in general (49), [1,1-^{13}C]acetic anhydride (50), and [2-^{13}C]fructose (51). The choice of compound is motivated not only by formulation and relaxation considerations but also by the imaging application; examples of which will be reviewed in **Chapter 33**. The list above is not exhaustive, but illustrates the versatility of the DNP dissolution method. The listed compounds have been imaged in vivo and are examples of potentially useful biomarkers for the hyperpolarized MR.

2.7. Summary

The hyperpolarization of small organic molecules by DNP and dissolution for use as MR agents have been successfully demonstrated for a range of compounds of biological interest. Whereas pre-clinical research applications are now pursued at a number of academic institutions, the development of the technology to clinical routine application is a challenge. Partly, this is due to the need for a relatively complex instrumentation in a hospital environment, but also due to the fact that pharmaceutical requirements of sterility and quality control need to be assured.

3. PHIP

Hydrogenation of an organic molecule with para-hydrogen creates a highly ordered spin state which results in the observation of large anti-phase NMR signals. This was discovered, both theoretically and experimentally, by Bower and Weitekamp in 1986 (52, 53). This phenomenon was initially given the names of PASADENA (53) and ALTADENA (54) for two different implementations, using respectively either a high or low magnetic field during hydrogenation. A more widespread acronym, relating to all different implementations of this phenomenon, PHIP is derived from "para-hydrogen induced polarization." PHIP can be used for magnetic resonance imaging (5, 55–60), where different methods have been developed for transforming the initial spin order into a net polarization. No net polarization is created from the hydrogenation step itself, although large NMR peaks are observed. A further treatment of the system is necessary for conversion into polarization suitable for imaging.

3.1. Properties of Para-hydrogen

The nuclear spin states of the hydrogen molecule exist in two varieties; a triplet state of total spin $S = 1$ called ortho-hydrogen, and a singlet state of total spin $S = 0$ called para-hydrogen. The spin states of ortho- and para-hydrogen are symmetric and anti-symmetric respectively, and are given by

$$\text{triplet:} \quad \begin{aligned} |-1\rangle &= |--\rangle \\ |0\rangle &= \tfrac{1}{\sqrt{2}}(|+-\rangle + |-+\rangle) \\ |1\rangle &= |++\rangle \end{aligned}$$

$$\text{singlet:} \quad |s\rangle = \tfrac{1}{\sqrt{2}}(|+-\rangle - |-+\rangle)$$

where + and − refer to the sign of the two protons spin values along a given axis, i.e., "spin up" and "spin down."

According to the Pauli principle, the total wave function of two identical fermions must be anti-symmetric. The implication of this for the hydrogen molecule is that symmetric rotational quantum states with even rotational quantum numbers are only allowed for the nuclear singlet state and that the anti-symmetric rotational quantum states with odd quantum numbers are only allowed for the triplet states. In the high-temperature limit all four nuclear states are equally populated resulting in $1/4$ of the states being para-hydrogen, whereas in the low temperature limit only the ground state, the singlet state with $J=0$, is populated and we have 100% para-hydrogen. For intermediate temperatures the fraction of para-hydrogen at thermal equilibrium is given by

$$x_p \equiv \frac{N_{\text{para}}}{N_{\text{ortho}} + N_{\text{para}}}$$

$$= \frac{\sum\limits_{J \text{ even}} (2J+1)\,e^{-E(J)/k_BT}}{\sum\limits_{J \text{ even}} (2J+1)\,e^{-E(J)/k_BT} + 3\sum\limits_{J \text{ odd}} (2J+1)\,e^{-E(J)/k_BT}}$$

[6]

where $E(J) = \Theta_r k_B J(J+1)$ is the energy of the rotational state with quantum number J, k_B is the Boltzmann constant, and $\Theta_r = 85.3$ K. At liquid hydrogen temperatures ($\approx$14–20 K), we have essentially pure para-hydrogen, and at room temperature, we are in the high-temperature regime (**Fig. 3**). The conversion between the ortho and para states is however very slow, so for practical purposes an efficient catalyst must be used.

The spin density operators for ortho- and para-hydrogen are respectively equal to

$\sigma_{\text{para}} = \tfrac{1}{4}(1 - 4\mathbf{I}_1 \cdot \mathbf{I}_2)$ and $\sigma_{\text{ortho}} = \tfrac{1}{4}(1 + \tfrac{4}{3}\mathbf{I}_1 \cdot \mathbf{I}_2)$, where $\mathbf{I}_1$ and $\mathbf{I}_2$ are the spin operators ($\mathbf{I}_k = (I_{kx}, I_{ky}, I_{ky})$) of the two protons. For a mixture of ortho- and para-hydrogen, with a molar fraction x_p of para-hydrogen, the total spin density operator is $\sigma_{H_2} = \tfrac{1}{4}(1 - f\mathbf{I}_1 \cdot \mathbf{I}_2)$, with $f = \tfrac{4}{3}(4x_p - 1)$.

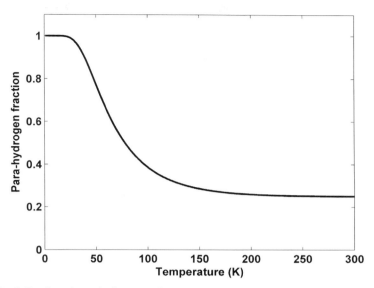

Fig. 3. Fraction of para-hydrogen vs. temperature at thermal equilibrium.

3.2. Hydrogenation with Para-hydrogen

The basic principle behind the PHIP phenomenon is the transfer of order from the para-hydrogen proton spin state to nuclear spins of other molecules. This requires a coupling between the para-hydrogen protons and the other nuclear spins. The most common method to obtain this coupling is to hydrogenate the molecule of interest with para-hydrogen, but also the possibility of using a transient coupling, without hydrogenation, has been demonstrated (61). The hydrogenation requires a catalyst that preserves the spin correlation between the protons, i.e., the hydrogen molecules are transferred to the substrate without scrambling of protons between different para-hydrogen molecules. It has recently been demonstrated that not only homogeneous catalysts can be used but also heterogeneous ones (62, 63).

Hydrogen, enriched in the para form, is used for hydrogenation of a small organic molecule with a double or triple C–C bond close to a ^{13}C spin. A 100% enrichment of para-hydrogen is assumed and the presence of other nuclear spins in the molecule is neglected except for a third spin. However, the general case is straightforward. We denote the spin operators of the protons originating from para-hydrogen by $\mathbf{I}_1$ and $\mathbf{I}_2$ as before, and the spin operator of the carbon spin by $\mathbf{S}$. Since the spin density operator for carbon can be approximated by its high temperature limit, the total spin density operator immediately after the hydrogenation will be given by

$$\sigma_i = \frac{1}{8}\left(\mathbf{1} - 4\mathbf{I}_1 \cdot \mathbf{I}_2\right) \qquad [7]$$

Its subsequent evolution will depend on the method used for order conversion. In the following, spin–lattice relaxation will be neglected because the duration of the order conversion is much shorter than the relaxation time. For a detailed analysis of the evolution of the density operator, the spin Hamiltonian and hence the scalar couplings between the nuclear spins needs to be known.

The spin Hamiltonian (in a doubly rotating proton–carbon frame) for the three-spin system is given by

$$H = J_{12}\mathbf{I}_1 \cdot \mathbf{I}_2 + J_{1S}\mathbf{I}_1 \cdot \mathbf{S} + J_{2S}\mathbf{I}_2 \cdot \mathbf{S} \qquad [8]$$

At fields where the difference between the proton and carbon Larmor frequencies is much larger than the magnitude of the proton–carbon scalar couplings Eq. [8] can be reduced to

$$H = J_{12}\mathbf{I}_1 \cdot \mathbf{I}_2 + J_{1S}I_{1z}S_z + J_{2S}I_{2z}S_z \qquad [9]$$

3.3. Order Transfer

Although hydrogenation with para-hydrogen creates a spin system that is far from thermal equilibrium, the polarization is not enhanced. After all, the para-hydrogen molecule itself has no polarization. However, the high spin order (low entropy) can be converted into polarization. When used as contrast agents for MR, it is preferable to transfer the polarization to a nuclear spin with a longer spin–lattice relaxation time, such as ^{13}C or ^{15}N.

Below we present two methods from our group that have successfully created ^{13}C polarization useful for MR: the field-cycling and the pulse sequence methods.

3.4. The Field-Cycling Method

The field-cycling method is the one that is easiest to implement, but it also is the least efficient. Hydrogenation is performed at a static magnetic field strength where the high-field approximation of the Hamiltonian expression [9] is valid. Expressed in the base of the eigenfunctions of the Hamiltonian, the density operator (Eq. [7]) is not diagonal and the off-diagonal elements will oscillate in time. Averaged over all molecules that are hydrogenated at slightly different times, the off-diagonal elements will vanish and the resulting spin density operator will instead be given by

$$\sigma_{av} = \frac{1}{8}(\mathbf{1} - 4I_{1z}I_{2z}) \qquad [10]$$

The loss of the off-diagonal elements is fast (sub-second) and does amount to a partial loss in spin order, but nevertheless the diagonalized density matrix is still far from thermal equilibrium spin order. Next, the static magnetic field is suddenly reduced to a value where the all scalar product terms in the Hamiltonian (Eq. [8]) must be used. Typically the field should be reduced to the nanotesla range in a millisecond. The averaged density

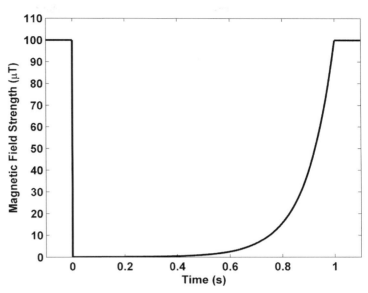

Fig. 4. A typical field-cycling profile with diabatic decrease followed by an adiabatic increase of the field.

operator (Eq. [10]) is now no longer diagonal in the eigenbase of the new Hamiltonian (Eq. [8]), resulting in the creation of oscillating off-diagonal elements corresponding to carbon–proton flip-flops. Immediately after the sudden drop, the field strength is raised adiabatically, in about a second, to the original high-field value (**Fig. 4**). This creates a net polarization of a magnitude that only depends on the scalar couplings. For the three-spin system, the polarization following the field-cycling procedure can be calculated by an analytical expression, but in a realistic case, where other spins cannot be neglected, a numerical simulation has to be performed. The inclusion of additional proton spins reduces the total carbon polarization and deuteration is not an option because its only effect is to reduce the polarization. The reason for this is that the inclusion of more spins, with a non-negligible scalar coupling to the pure three-spin system, creates a number of "surplus" states that will be populated, resulting in a reduced population of states that contribute to the carbon polarization. It can be found from computer simulations that in some cases it is possible to obtain a larger polarization if the increase in the magnetic field is not truly adiabatic, but instead follows an optimized profile (unpublished results).

3.5. The Pulse Sequence Method

In the pulse sequence method, the order transfer is performed at a constant magnetic field strength, but in the presence of radio frequency (rf) pulses that are applied according to a certain sequence. All protons of the organic molecule, other than those originating from the para-hydrogen, should be replaced by deuterons. Only protons with a negligible scalar coupling to the carbon and the two former para-hydrogen protons are allowed in the molecule. This is in contrast to the field-cycling method, where deuterons should be avoided. The reason for this is that additional protons interfere via their scalar coupling and that this cannot be compensated for by using refocusing pulses, whereas for deuterons such compensation is possible. Additional protons will thus reduce the polarization that can be achieved.

By using pulses it is possible to avoid the loss of order due to averaging, i.e., the full initial density operator (Eq. [7]) can be retained. This is done by performing the hydrogenation in the presence of rf irradiation at the proton Larmor frequency, by applying alternating 180° pulses, akin to an ordinary NMR decoupling sequence. During the irradiation, there is effectively no evolution of the initial density operator and at the end of the irradiation, all molecules are in the singlet state, without loss of order (neglecting relaxation). Typically, the duration of this initial part of the pulse sequence is a few seconds, i.e., the time needed for complete hydrogenation and settling of the liquid. After this initial part, a short sequence of pulses on both proton and carbon are applied, interleaved with evolution time periods. The duration of this second part is typically 100–200 ms and the design of the sequence requires knowledge of the proton–proton as well as the proton–carbon scalar couplings. The number of pulses and evolution periods, as well as the durations of the evolution periods, will in general be different for different molecules. During each evolution period, two double spin echo pulses, on both proton and carbon, are applied. A recipe of how to construct pulse sequences, using a geometric picture together with a fictitious spin formalism, has been given by Goldman et al. (59, 60). **Figure 5** shows the pulse sequence used for hydroxyethylpropionate.

3.6. Instrumentation

The instrumentation used by the Malmö group (5, 55–60) for producing para-hydrogen, as well as two polarizers based on the methods above, is presented below.

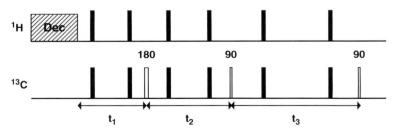

Fig. 5. The pulse sequence used for hydroxyethylpropionate. The white pulses are the "useful" pulses and the black pulses are echo pulses at one quarter and three quarters of each evolution period.

3.6.1. Production of Para-hydrogen

In order to reach the required temperatures for obtaining a large fraction of para-hydrogen (*see* **Fig. 3**), a 3 kW closed-loop helium cryo-cooler was used. By allowing normal hydrogen gas to pass through a commercially available catalyst (C*CHEM, Lafayette, CO) at a temperature of 14 K, high yields of para-hydrogen were obtained. At these conditions, the production rate was one 7 L cylinder filled to a pressure of 30 bar in 1 h. The resulting gas consisted of more than 95% para-hydrogen. The para-hydrogen gas can be stored in aluminum cylinders for several days without losing too much of the para-hydrogen state. Typically, the para-hydrogen fraction decreased almost linearly to 50% during a time period of 25 days. This demonstrates the slow conversion rate between the ortho and para states in the absence of a catalyst.

3.6.2. The Polarizer

3.6.2.1. Field-Cycling Method

A reactor chamber is filled with para-hydrogen gas to a pressure of approximately 10 bar and its temperature is kept at 333 K, an empirically optimized temperature with respect to the final ^{13}C polarization. A mixture of the substrate and the catalyst is injected into the reactor as a narrow jet, a method that was found to give better results than spraying. The liquid is subsequently transferred, after a few seconds, into the low-field chamber where the field is initially set to 100 μT. After a further delay of 0.5 s, the field is reduced to about 30 nT in 1 ms. The field is then increased exponentially to the original field strength of 100 μT in typically 1–2 s, whereupon the sample is ejected and collected in a syringe. The time constant used for the field increase is a compromise between polarization loss due to relaxation and the requirement of adiabaticity.

The low-field chamber is made of glass and it is placed inside a set of two concentric coils, which in turn sit inside a set of three concentric layers of mu-metal. The mu-metal shields are routinely demagnetized in a 50 Hz AC field to give a residual field of 10–30 nT inside the chamber. The two concentric coils have currents running in opposite directions to give as close to zero field as possible on the outside. This is necessary to allow for the rapid change of the field inside the conductive mu-metal shields.

3.6.2.2. Pulse Sequence Method

The reactor is placed inside a solenoid coil with a magnetic field strength of 1.76 mT. The carbon and proton Larmor frequencies are 19 and 75 kHz respectively at this field. Between the solenoid coil and the reactor is an untuned saddle coil for transmitting linearly polarized rf pulses. The B_1 intensity corresponded to a 90° proton (square) pulse of 0.0879 ms and a corresponding carbon pulse of 0.1724 ms. All pulses were generated in MATLAB and were sampled at a frequency of 300 kHz.

The pulse sequence is initiated, as soon as the hydrogenation process starts, with a proton decoupling that lasts for 4 s. The purpose of the proton decoupling is to preserve the spin correlation between the protons, thus preventing the cancellation of the off-diagonal terms in the density operator. The gain is twofold. Firstly, the protons are forced to remain in the singlet state so their mutual dipolar relaxation is reduced. Secondly, this state has a lower entropy which makes it possible to obtain a larger polarization for the carbon nuclei.

The refocusing pulses, simultaneous on both proton and carbon to allow for evolution of scalar couplings, were simple square pulses, whereas the proton decoupling pulses and the remaining carbon pulses were broadband pulses with both a homogeneous excitation and good phase control. All equipment was built into a single temperature-controlled cabinet, divided into compartments where the different steps in the process take place. The valves and the rf pulses were controlled by computer, using software written in LabVIEW.

3.7. Summary

The PHIP method provides a relatively simple and inexpensive means of producing hyperpolarized agents that can be used for MR. Due to the long-lived singlet state of the hydrogen molecule, the production of para-hydrogen does not need to be in the vicinity of the polarizer. There is a limitation concerning the kind of molecules that can be used by the PHIP method, since it is based on molecules that are the product of a hydrogenation reaction. Applications of this method include angiography (5, 55), perfusion (64), catheter tracking (65), and metabolic studies (66).

4. Relaxation Phenomena

4.1. Introduction

The time window for transfer of the sample from the DNP or PHIP polarizer to the MR system is severely limited by the nuclear relaxation time T_1. It is therefore desirable to try to maximize T_1 relaxation time of the hyperpolarized nuclei. First, the sample is subject to some field cycling during transport that may adversely affect relaxation rates. Secondly, the Zeeman T_1 may not be the most efficient way to store non-equilibrium spin order. And thirdly, it may be advantageous to transfer hyperpolarization from a "storage" nucleus with long T_1 to a more desirable nucleus for MR imaging. Some considerations in relation to these concepts are discussed here.

4.2. Relaxation During Transport

After retrieval from the polarizer, the sample needs to be transferred to the NMR spectrometer or MR scanner. During that time, the sample is exposed to varying magnetic fields. The lowest field strength experienced by the sample could be the earth's field or possibly lower, depending on the individual configuration of polarizer and MR system. Most small organic molecules that are hyperpolarized using the PHIP or DNP methods do not show any marked ^{13}C T_1 dependence on B_0 to low fields, hence transport in earth's field is relatively uncritical. As an example, [1-^{13}C]pyruvate is considered again. The ^{13}C T_1 was measured to be 56 s in the background field in our lab (using a low-field NMR spectrometer) and similarly a value of 56 s was measured at 9.4 T and 295 K.

In the case of [1-^{13}C]pyruvate, it was possible to determine the approximate contributions of ^{13}C-^{1}H dipolar and CSA contributions to ^{13}C T_1 from the following measurements:

$T_1 = 56$ s (9.4 T in H$_2$O)   (1)
$T_1 = 84$ s (9.4 T in D$_2$O)   (2)
$T_1 = 107$ s (7.0 T in D$_2$O)   (3)

The decrease of T_1 at 9.4 T compared to 7.0 T can be attributed to chemical shift anisotropy (CSA), which increases as the square of the field. Using the relation $R_1 = R_1^{\text{dipolar}} + B_0^2 \xi$, the dipolar and the CSA contributions can be distinguished from the experimental results (2) and (3) above. We then obtain $T_1^{\text{dipolar}} = 1/R_1^{\text{dipolar}} = 16$ s and $\xi = 6.74 \cdot 10^{-5}\text{s}^{-1}\text{T}^{-2}$. The CSA contribution at 9.4 T is then $T_1^{\text{CSA}} = 168$ s, and at 7.0 T it is $T_1^{\text{CSA}} = 300$ s.

Incidentally, two issues have been observed when transferring samples from our DNP polarizer to a nearby NMR spectrometer, namely, exposure of the sample to zero magnetic field and the coupling of quadrupolar nuclei, e.g., ^{14}N, to ^{13}C at low field during transport. The former issue may exist in a laboratory with several superconducting magnets or due to distortion of the fringe field by magnetic components. When the Larmor frequency of the nucleus in question approaches zero, field changes will become non-adiabatic and can lead to almost complete loss of the hyperpolarization. Mapping of the local field around the polarizer and along the transport path is recommended in order to detect any planes of zero field or non-smoothly varying background fields.

The effect of ^{13}C–^{14}N coupling during transport has not been studied systematically in our laboratory. However, it has been observed that molecules with directly bonded nitrogen to a hyperpolarized ^{13}C site may require transport inside a permanent magnet assembly that ensures a field of several tens of millitesla over the volume of the liquid sample. The most intuitive explanation for this observation is scalar relaxation of the second kind.

4.3. Singlet States

The desire to extend hyperpolarization lifetimes has led to some research around singlet states. The idea is to convert the Zeeman order of the ^{13}C spin into a singlet state just after the retrieval from the polarizer. A second pulse sequence will have to be executed in order to convert this state back to useable ^{13}C hyperpolarization prior to any MR measurement. This concept was pioneered by Carravatta et al. (67). These authors demonstrated that the low-field nuclear spin singlets are unaffected by intramolecular dipole–dipole relaxation. As a result, storage of nuclear spin order for more than 10 times the measured value of T_1 was demonstrated in a two-spin system. However, the generation of a singlet state is contingent on the existence of two ^{13}C sites in the molecule, which limits the application of this concept. Furthermore, the resulting pseudo singlet state may have a longer lifetime than the (Zeeman) T_1, but residual intramolecular scalar couplings exist in this state and cause it to have a finite lifetime. The practical applications of the singlet state to in vivo MR may be limited in several respects, but may well be of advantage in high-resolution NMR spectroscopy. The theory of long-lived singlet states in solution is further described in references (68, 69).

4.4. The "Spin Bank" Concept and Hyperpolarization Transfer

Generally, ^{13}C and ^{15}N relaxation times (T_1) in organic molecules are much longer than ^{1}H relaxation times. Hence, most hyperpolarized MR studies to date have been performed with these low-γ nuclei. However, it is not always desirable to perform the imaging experiment with the same nucleus. In order to achieve a certain spatial resolution, the required gradient integral in the MR imaging sequence is inversely proportional to γ. This is particularly challenging for ^{15}N whose γ is approximately 10 times lower compared to ^{1}H. As a result, it could be desirable to store nonequilibrium polarization on a long-T_1 low-γ nucleus and to incrementally transfer some of this polarization onto another nucleus on the same molecule. This concept has been termed "spin bank" (70). Its implementation requires knowledge of scalar coupling constants in order to optimize the timing of a polarization transfer preparation of the applied MR pulse sequence.

The spin bank concept has not been used extensively in MR imaging applications to date and most of its implementations are proof-of-concept studies that are intended to demonstrate the feasibility of e.g., ^{13}C–^{1}H hyperpolarization transfer, rather than practical applications for in vivo imaging. The efficiency of the transfer depends on various factors; amongst them are the coupling constants in the molecule of choice and the relaxation times of the nuclei in the rotating frame. The situation in a clinical MR scanner is further complicated by the need to have as homogeneous B_0 field as possible.

The feasibility of ^{13}C–^{1}H polarization transfer following ^{13}C hyperpolarization by the PHIP method has been demonstrated by Chekmenev et al. (71). These authors polarized [1-^{13}C]succinate-d$_{2,3}$ and transferred polarization to the two

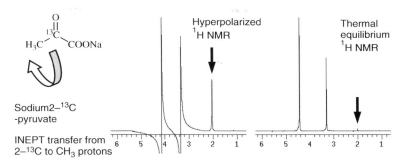

Fig. 6. Schematics of $^{13}C-^{1}H$ polarization transfer in sodium pyruvate-2-^{13}C to methyl protons (*left*), ^{1}H NMR spectrum after the transfer (*middle*), and thermal equilibrium NMR spectrum from the same sample (*right*). The ^{1}H polarization enhancement was 748 for each proton compared to thermal equilibrium at 9.4 T.

remaining ^{1}H sites using an inverse refocused INEPT pulse sequence on a 4.7 T animal scanner. A similar experiment was performed with 2,2,3,3-tetrafluoropropyl[1-^{13}C]propionate-d$_3$ (TFPP). The efficiency of their polarization transfer experiments was 41 and 51%, respectively.

Very similar experiments based on a refocused $^{13}C-^{1}H$ INEPT pulse sequence in a 9.4 T NMR spectrometer were performed in our laboratory, except that the molecules had been polarized using the DNP method and were not deuterated. It was possible to show virtually 100% efficient polarization transfer to the directly ^{13}C-bound ^{1}H in [1-^{13}C]alanine and to the methyl protons in [2-^{13}C]pyruvate. The latter results are shown in **Fig. 6**.

In a third experiment, [1,4-$^{13}C_2$]fumarate was polarized by DNP using the TEMPO radical. The broad EPR line width of this radical causes all nuclear spins—including ^{1}H—to be hyperpolarized. Furthermore, the long ^{1}H T_1 of fumarate enabled dissolution from the DNP polarizer with approximately 6% ^{1}H polarization remaining in the liquid state immediately prior to transfer. Subsequently, a $^{1}H-^{13}C$ refocused INEPT pulse sequence was run, with delay times optimized for transfer of the ^{1}H hyperpolarization to the two ^{13}C sites. The resulting ^{13}C enhancement was significantly higher than what would have been expected from the polarization transfer experiment alone; the ^{13}C polarization is a combination of the direct ^{13}C polarization enhancement by DNP and the polarization transfer from ^{1}H.

4.5. Summary

This section has provided a brief and qualitative overview of relaxation phenomena and concepts that help to prolong the lifetime of hyperpolarization during and after transfer from the polarizer. Furthermore, the feasibility of intramolecular hyperpolarization transfer has been shown in simple model systems.

References

1. Edelstein, W.A., Glover, G.H., Hardy C.J., and Redington, R.W. (1986) The intrinsic signal-to-noise ratio in NMR imaging. *Magn. Reson. Med.* **3**, 604–618.
2. Albert, M.S., Cates, G.D., Driehuys, B., et al. (1994) Biological magnetic resonance imaging using laser-polarized (129)Xe. *Nature* **370**, 199–201.
3. Middleton, H., Black, R.D., Saam, B., et al. (1995) MR imaging with hyperpolarized (3)He gas. *Magn. Reson. Med.* **33**, 271–275.
4. Kauczor, H.U., Hofmann, D., Kreitner, K.F., et al. (1996) Normal and abnormal pulmonary ventilation: Visualization at hyperpolarized He-3 MR imaging. *Radiology* **201**, 564–568.
5. Golman, K., Axelsson, O., Jóhannesson, H., Månsson, S., Olofsson, C., and Petersson, J.S. (2001) Parahydrogen-induced polarization in imaging: Subsecond (13)C angiography. *Magn. Reson. Med.* **46**, 1–5.
6. Ardenkjær-Larsen, J.H., Fridlund, B., Gram, A., et al. (2003) Increase in signal-to-noise ration of >10,000 times in liquid-state NMR. *Proc. Natl. Acad. Sci. USA* **100**, 10158–10163.
7. Ardenkjaer-Larsen, J.H., Axelsson, O., Golman, K., Wistrand, L.-G., Hansson, G., Leunbach, I., Petersson, S. (1998) PCT WO99/35508, priority date 05.01.1998.
8. Wolber, J., Ellner, F., Fridlund, B., et al. (2004) Generating highly polarized nuclear spins in solution using dynamic nuclear polarization, *Nucl. Instr. Meth. Phys. Res.* **526**, 173–181.
9. Overhauser, A. (1953) Polarization of nuclei in metals. *Phys. Rev.* **92**, 411–415.
10. Carver, T.R. and Slichter, C.P. (1953) Polarization of nuclear spins in metals. *Phys. Rev.* **92**, 212–213.
11. Abragam, A. (1955) Overhauser effect in non-metals. *Phys. Rev.* **98**, 1729–1735.
12. Jeffries, C.D. (1957) Polarization of nuclei by resonance saturation in paramagnetic crystals. *Phys. Rev.* **108**, 164–165.
13. Abragam, A. and Goldman, M. (1982) *Nuclear Magnetism: Order and Disorder.* Oxford, UK, Clarendon Press.
14. de Boer, W.D., Borghini, M., Morimoto, K., Niinikoski, T.O., and Udo, F. (1974) Dynamic polarization of protons, deuterons, and C-13 nuclei – thermal contact between nuclear spins and an electron spin-spin interaction reservoir. *J. Low Temp. Phys.* **15**, 249–267.
15. de Boer, W.D. and Niinikoski, T.O. (1974) Dynamic proton polarization in propanediol below 0.5 K. *Nucl. Instr. Meth.* **114**, 495–498.
16. Heckmann, J., Meyer, W., Radtke, E., and Reicherz, G. (2006) Electron spin resonance and its implication on the maximum nuclear polarization of deuterated solid target materials. *Phys. Rev.* **74**, Art. No. 134418.
17. Benjamin, P.S., Fuminori H., Matsumoto, K.-I., Simone, N.L., Cook, J.A., Krishna, M.C., and Mitchell, J.B. (2007) The chemistry and biology of nitroxide compounds. *Free Radical Biol. Med.* **42**, 1632–1650.
18. Eaton, S.S., Eaton, G.R., and Berliner, L. (2005), Part A: Free radicals, metals, medicine, and physiology. Part B: Methodology, instrumentation, and dynamics, Series: Biological magnetic resonance. *J. Biomed. EPR.* **23/24**.
19. Thaning, M. (2004) PCT WO2006/011811, priority date 30.07.2004.
20. Andersson, S., Radner, F., Rydbeck, A., Servin, R., and Wistrand, L.-G. (1995) United States patent US5728370, June 6, 1995.
21. Jagadeeswar Reddy, T., Iwama, T., Halpern, H.J., and Rawal, V.H. (1998) General synthesis of persistent trityl radicals for EPR imaging of biological systems. *J. Org. Chem.* **67**, 4635–4639.
22. Bowman, M.K., Mailer, C., and Halpern, H.J (2005) The solution conformation of triarylmethyl radicals. *J. Magn. Reson.* **172**, 254–267.
23. Hu, K.-N., Bajaj, V.S., Rosay, M., and Griffin R.G. (2007) High-frequency dynamic nuclear polarization using mixtures of TEMPO and trityl radicals. *J. Chem. Phys.* **126**, 044512.
24. Ardenkjaer-Larsen, J.H., Macholl, S., and Johannesson, H. (2008) Dynamic nuclear polarization with trityls at 1.2 K. *Appl. Magn. Reson.* **34**, 509–522.
25. Duijvestijn, M.J., Wind, R.A., and Smidt, J. (1986) Quantitative investigation of the dynamic nuclear polarization effect by fixed paramagnetic centra of abundant and rare spins in solids at room temperature. *Physica B.&C.* **138**, 147–170.
26. Bajaj, V.S., Hornstein, M.K., Kreischer, K.E., et al. (2007) 250 GHz CW gyrotron oscillator for dynamic nuclear polarization in biological solid state NMR. *J. Mag. Res.* **189**, 251–279.
27. Hornstein, M.K., Bajaj, V.S., Kreischer, K.E., Griffin, R.G., and Temkin R.J. (2005) CW second harmonic results at 460 GHz of a gyrotron oscillator – For sensitivity

enhanced NMR. *The Joint 30th International Conference on Infrared and Millimeter Waves and 13th International Conference on Terahertz Electronics.* **2**, 437–438.

28. Comment, A., van den Brandt, B., Uffmann, K., et al. (2007) Design and performance of a DNP prepolarizer coupled to a rodent MRI scanner. *Conc. Magn. Reson.* **31**, 255–269.

29. Jannin, S., Comment, A., Kurdzesau, F., Konter, J.A., Hautle, P., van den Brandt, B., and van der Klink, J.J. (2008) A 140 GHz prepolarizer for dissolution dynamic nuclear polarization. *J. Chem. Phys.* **128**, 241102.

30. Jóhannesson, H., Macholl, S., and Ardenkjaer-Larsen, J.H., (2008) Dynamic nuclear polarization of [1-13C]pyruvic acid at 4.6 tesla. *J. Magn. Reson.* **197**, 167–175.

31. Urbahn J., Ardenkjaer-Larsen J.H., Leach A., Stautner W., Zhang T., and Clarke N., (2008) A closed cycle helium sorption pump system and its use in making hyperpolarized compounds for MR Imaging. *International Cryogenic Engineering Conference 22 and International Cryogenic Materials Conference 2008*, Seoul, Korea.

32. Goertz, S. (2002) Spintemperatur und magnetische Resonanz verdünnter elektronischer Systeme – ein Weg zur Optimierung polarisierter Festkörper-Targetmaterialien, Ruhr-Universität Bochum, Habilitationsschrift, April 15, 2002.

33. Jam, J.A., Dey, S., Muralidharan, L., Leach, A.M., and Ardenkjaer-Larsen, J.H. (2009) Jet impingment melting with vaporization: A numerical study. (2009) *Proceedings of the ASME Summer Heat Transfer Conference 2008*, **2**, 559–567.

34. Mishkovsky, M., Eliav, U., Navon, G., and Frydman, L. (2009) Nearly 10^6-fold enhancements in intermolecular 1H double-quantum NMR experiments by nuclear hyperpolarization. *J. Magn. Reson.* **200**, 142–146.

35. Van Heeswijk, R.B., Uffmann, K., Kurdzesau, F., et al. (2007) Towards detection of sub-micromolar contrast agent concentration with hyperpolarized 6-lithium. *Abstract 1318, Proceedings of International Society for Magnetic Resonance in Medicine*, Berlin, Germany.

36. Kuhn, L.T., Bommerich, U., and Bargon, J. (2006) Transfer of parahydrogen-induced hyperpolarization to (19)F. *J. Phys. Chem.* **110**, 3521–3526.

37. Eykyn, T.R., Reynolds, S., Gabellieri, C., and Leach, M.O. (2007) Hyperpolarised N-15 of choline – potential for observing phospholipid metabolism in cancer. *Proc. Intl. Soc. Mag. Reson. Med.* **15**, 1319.

38. Reynolds, S. and Patel, H. (2008) Monitoring the solid-state polarization of ^{13}C, ^{15}N, ^{2}H, ^{29}Si and ^{31}P. *Appl. Magn. Reson.* **34**, 495–508.

39. Merritt, M.E., Harrison, C., Kovacs, Z., Kshirsagar, P., Malloy, C.R., and Sherry, A.D. (2007) Hyperpolarized (89)Y offers the potential of direct imaging of metal ions in biological systems by magnetic resonance. *J. Am. Chem. Soc.* **129**, 12942–12943.

40. Ardenkjær-Larsen, J.H., Hansson, L., Johannesson, H., Servin, R., and Wistrand, L.-G. (2002) PCT WO2004/037296, priority date 25.10.2002.

41. Golman, K., in't Zandt, R., and Thaning, M., (2006) Real-time metabolic imaging. *PNAS.* **103**, 11270–11275.

42. Schroeder, M.A., Atherton, H.J., Ball, D.R., Cole, M.A., Heather, L.C., Griffin, J.L., Clarke, K., Radda, G.K., and Tyler, D.J. (2009) Real-time assessment of Krebs cycle metabolism using hyperpolarized ^{13}C magnetic resonance spectroscopy, *The FASEB J.* **23**.

43. Chen, A.P., Kurhanewicz, J., Bok, R., Xu, D., Joun, D., Zhang, V., Nelson, S.J., Hurd, R.E., and Vigneron, D.B. (2008) Feasibility of using hyperpolarized [1-13C]lactate as a substrate for in vivo metabolic ^{13}C MRSI studies. *Magn. Reson. Imaging* **26**, 721–726.

44. Gallagher, F.A., Kettunen, M.I., Day, S.E., Hu, D.-E., Ardenkjær-Larsen, J.H., in't Zandt, R., Jensen, P.R., Karlsson, M., Golman, K., Lerche, M.H., and Brindle, K.M. (2008) Magnetic resonance imaging of pH in vivo using hyperpolarized ^{13}C-labelled bicarbonate. *Nature* **453**, 940–943.

45. Karlsson, M., in't Zandt, R., Jensen, P.R., Hansson, G., Månsson, S., Gisselsson, A., and Lerche, M. (2008) Metabolic reactions studied with 13C-DNP-NMR at physiologically relevant conditions in vitro and in vivo. *Experimental Nuclear Magnetic Resonance Conference*, Pacific Grove, CA, USA.

46. Jensen, P.R., in't Zandt, R., Karlsson, M., Hansson, G., Månsson, S., Gisselsson, A., and Lerche, M. (2008) Acetyl-CoA and acetyl-carnitine show organ specific distribution in mice after injection of DNP hyperpolarized 13C1-acetate. *Abstract 892 Proceedings of the International Society for Magnetic Resonance in Medicine*, Toronto, Canada.

47. Gallagher, F.A., Kettunen, M.I., Day, S.E., Lerche, M., and Brindle K.M. (2008) ^{13}C MR Spectroscopy measurements of glutaminase activity in human hepatocellular carcinoma cells using hyperpolarized ^{13}C-labeled glutamine. *Magn. Reson. Med.* **60**, 253–257.

48. Gallagher, F.A., Kettunen, M.I., Day, S.E., Hu, D., in't Zandt, R., Jensen, P.R., Karlsson, M., Golman, K. Lerche, M., and Brindle, K.M. (2007) Real-time visualization of hyperpolarized ^{13}C-labeled glutamine metabolism in human hepatoma cells using magnetic resonance spectroscopy. *Proc. RSNA.* 357.
49. Jensen, P.R., Karlsson, M., Meier, S., Duus, J.Ø., and Lerche, M.H. (2009) Hyperpolarized amino acids for in vivo assays of transaminase activity. *Chem. Eur. J.* **15**, 10010–10012.
50. Wilson, D.M., Hurd, R.E., Keshari, K., Van Criekinge, M., Chen, A.P., Nelson, S.J., Vigneron, D.B., and Kurhanewicz, J. (2009) Generation of hyperpolarized substrates by secondary labeling with [1,1-13C] acetic anhydride. *PNAS* **106**, 5503–5507.
51. Keshari, K.R., Wilson, D.M., Chen, A.P., Bok, R., Larson, P.E.Z., Hu, S., Van Criekinge, M., Macdonald, J.M., Vigneron, D.B., and Kurhanewicz, J. (2009) Hyperpolarized [2-13C]-Fructose: A hemiketal DNP substrate for in vivo metabolic imaging. *J. Am. Chem. Soc.* online Oct 27.
52. Bowers, C.R. and Weitekamp, D.P. (1986) Transformation of symmetrization order to nuclear-spin magnetization by chemical reaction and nuclear magnetic resonance. *Phys. Rev. Lett.* **57**, 2645–2648.
53. Bowers, C.R. and Weitekamp, D.P. (1987) Parahydrogen and synthesis allow dramatically enhanced nuclear alignment. *J. Am. Chem. Soc.* **109**, 5541–5542.
54. Pravica, M.G. and Weitekamp D.P. (1998) Net NMR alignment by adiabatic transport of para-hydrogen addition products to high magnetic field. *Chem. Phys. Lett.* **145**, 255–258.
55. Golman, K., Ardenkjær-Larsen, J.H., Svensson, J., Axelsson, O., Hansson, G., Johannesson, H., Leunbach, I., Månsson, S., Petersson, J.S., Pettersson, G., Servin, R., and Wistrand, L.G. (2002) (13) C-angiography. *Acad. Radiol.* **9**, 507–510.
56. Golman, K, Olsson, L.E., Axelsson, O., Månsson, S., Karlsson, M., and Petersson, J.S. (2003) Molecular imaging using hyperpolarized (13)C. *The Br. J. Radiol.* **76**, 118–127.
57. Jóhannesson, H., Axelsson, O., and Karlsson, M, (2004) Transfer of para-hydrogen spin order into polarization by diabatic field cycling. *C. R. Physique* **5**, 315–324.
58. Goldman, M., Jóhannesson, H., Axelsson, O., and Karlsson, M. (2005) Hyperpolarization of ^{13}C through order transfer from para-hydrogen: A new contrast agent for MRI. *Magn. Reson. Imaging* **23**, 153–157.
59. Goldman, M. and Jóhannesson, H. (2005) Conversion of a proton pair para order into ^{13}C polarization by rf irradiation, for use in MRI. *C. R. Physique* **6**, 575–581.
60. Goldman, M., Jóhannesson, H., Axelsson, O., and Karlsson M. (2006) Design and implementation of ^{13}C hyperpolarization from para-hydrogen, for new MRI contrast agents. *C.R. Chimie* **9**, 357–363.
61. Atkinson, K.D., Cowley, M.J., Duckett, S.B., et al. (2009) Para-Hydrogen induced polarization without incorporation of para-hydrogen into the analyte. *Inorg. Chem.* **48**, 663–670.
62. Koptyug, I.V., Kovtunov, K.V., Burt, S.R., et al. (2007) para-Hydrogen-induced polarization in heterogeneous hydrogenation reactions. *J. Am. Chem. Soc.* **129**, 5580–5586.
63. Balu, A.M., Duckett, S.B., and Luque, R. (2009) Para-hydrogen induced polarization effects in liquid phase hydrogenations catalyzed by supported metal nanoparticles. *Dalton Trans.* **26**, 5074–5076.
64. Johansson, E., Olsson, L.E., Månsson, S., Petersson, J.S., et al. (2004) Perfusion assessment with bolus differentiation: A technique applicable to hyperpolarized tracers. *Magn. Reson. Med.* **52**, 1043–1051.
65. Magnusson, P., Johansson, E., Månsson, S., Petersson, J.S., et al. (2007) Passive Catheter tracking during interventional MRI using hyperpolarized (13)C. *Magn. Reson. Med.* **57**, 1140–1147.
66. Bhattacharya, P., Chekmenev, E.Y., Perman, W.H., et al. (2007) Towards hyperpolarized ^{13}C-succinate imaging of brain cancer. *J. Magn. Reson.* **186**, 150–155.
67. Carravatta, M., Johannessen, O.G., and Levitt, M.H. (2004) Beyond the T_1 limit: Singlet nuclear spin states in low magnetic fields. *Phys. Rev. Lett.* **92**, 153003–153011.
68. Carravatta, M. and Levitt, M.H. (2005) Theory of long-lived nuclear spin states in solution nuclear magnetic resonance. I. Singlet states in low magnetic field. *J. Chem. Phys.* **122**, 1–14.
69. Pileio, G. and Levitt, M.H. (2009) Theory of long-lived nuclear spin states in solution nuclear magnetic resonance. II. Singlet spin locking. *J. Chem. Phys.* **130**, art. no. 214501.
70. Petersson, S., Axelsson, O., and Jóhannesson, H. (2003) Patent US7346384.
71. Chekmenev, E.Y., Norton, A.V., Weitekamp, D.P., and Bhattacharya, P. (2009) Hyperpolarized ^{1}H NMR employing low γ nucleus for spin polarization storage. *J. Am. Chem. Soc.* **131**, 3164–3165.

Chapter 12

MR Oximetry

Jeff F. Dunn

Abstract

MR oximetry includes methods for assessing tissue oxygenation. This chapter focuses on direct measurements of oxygenation. These can be divided into three methods. The first and most common has been termed BOLD MRI and relates to the quantification of deoxyhemoglobin. The second method uses an injected fluorinated agent which has a T_1 that is sensitive to tissue oxygen levels. The third is a direct measurement of T_1 under conditions where the variation in T_1 can be limited to that caused by changes in pO_2. These conditions can be met in the vitreous of the eye or the cerebrospinal fluids. Such changes in the eye have been called the retinal oxygenation response.

Key words: MRI, oximetry, quantification, BOLD, pO_2, oxygen, ^{19}F, retinal oxygenation response.

1. Introduction

Magnetic resonance (MR) oximetry involves obtaining data on tissue oxygenation using MRI. Oxygen plays a role in many pathophysiologies as well as in normal brain function. One of the most well-known applications is that of functional MRI or blood oxygen level-dependent (BOLD) MRI (1) where changes in function are reflected in changes in hemoglobin saturation (oxygenation). Tumor oxygenation is important to understand as increased oxygen imparts radiation sensitivity while hypoxic tumors are radiation insensitive. The list of disorders where oxygenation impacts outcome is too long to mention, but includes stroke, heart attack, traumatic brain injury, Alzheimer's disease, and sudden infant death syndrome. There are a range of potential methods, which vary in complexity and in their capacity to obtain a quantifiable value. There are many which provide only an indirect assessment of oxygenation. These would include measurements such as lactate concentration, whereby under specific conditions the "anaerobic" end product of glycolysis—lactate—may provide an index of tissue oxygenation. A review of direct and indirect methods has recently been published (2).

This chapter will describe three methods: the most widely used method, that of BOLD MRI, one of the most quantifiable methods; that of quantification of the T_1 of ^{19}F and fluorinated hydrocarbons; and finally quantification of T_1 in fluids such as in the eye or cerebrospinal fluid.

BOLD (1) or fMRI has the advantage that it can be undertaken, in some form, on any MRI system. The principle involves the fact that deoxygenated hemoglobin (deoxyHb) is paramagnetic while oxyhemoglobin is diamagnetic. Paramagnetic compounds influence the magnetic field and so will alter the relaxation rates (which are variables in signal intensity). The transverse relaxation rate or R_2 is determined largely by intrinsic properties of the compound while R_2^* has an additional component related to field inhomogeneity. The concentration of a paramagnetic compound can have a linear influence on R_2^* or R_2. Thus, changes in deoxyHb can also induce quantifiable changes in R_2 or R_2^*.

A key point to take into account is that these changes relate only to deoxyHb content. The total deoxyHb can vary by changing blood volume, hemoglobin saturation, or a combination of both (3). Pauling described the paramagnetic properties of deoxyHb (4). Thulborn showed the relationship between R_2, R_2^* and deoxyHb concentration (5). Ogawa coined the term blood oxygen level-dependent or BOLD MRI in studies which he showed that one can use R_2^* as a marker of brain activation (1). In order to use BOLD in MR oximetry, one has to assume that total Hb either does not change, or changes very little (see **Note 1**). In the case of brain, blood volume will increase with increased flow and so has to be controlled.

In the case of T_1 quantification, there are methods for ^{1}H and ^{19}F MRI available whereby the T_1 itself changes linearly with changes in oxygen content of the fluids. For an extended discussion on the impact of O_2 on T_1, see (6). In the case of ^{1}H MRI, this can be undertaken in any fluid where the major change is the oxygen concentration vs. fluids where other compounds such as hemoglobin have a greater impact on the relaxation rates with changes in oxygenation. The main examples in the human body are the vitreous of the eye (7) or in cerebrospinal fluid (8).

2. Materials

2.1. MRI System

1. For imaging, an NMR system with imaging gradients is needed. However, bulk measurements can be obtained without imaging gradients, using spectroscopic methods (*see* **Note 2**).

2. RF coils with optimum sample-to-volume and signal/noise ratios such that the appropriate data can be obtained (*see* **Note 3**).

3. For ^{19}F oximetry, an RF coil and MRI system with a broad frequency response or a dual nuclei capability is needed. The gyromagnetic ratio of ^{1}H is 42.58 MHz/T and of ^{19}F is 40.05 MHz/T; so for a 9.4 T study, one needs frequencies of 400 MHz for ^{1}H and 380.7 MHz for ^{19}F.

4. For BOLD oximetry, MRI sequences are needed that are either weighted largely to T_2 or T_2^*, or are capable of quantification of T_2 or T_2^* ($1/R_2$ or $1/R_2^*$) (*see* **Note 4**). Echo trains can be contaminated by stimulated echoes (*see* **Note 5**). The sequences are often optimized for speed to enable imaging of temporal responses (*see* **Note 6**).

5. For ^{19}F oximetry, sequences are required for quantification of T_1. A standard inversion recovery T_1 map is very slow but accurate. Acquisition time can be reduced by using a range of options including, for example, echo planar imaging (9) or "pulse-burst" saturation recovery (10). In order to improve signal/noise ratio, one can forgo mapping and use a spectroscopy method with a single large voxel (11) (*see* **Note 7**).

6. When being undertaken with animals, one requires appropriate anesthesia and monitoring systems for maintaining animals (*see* **Section 3.1**).

2.2. Chemicals for ^{19}F Oximetry

The following compounds have been used as pO_2 sensitive fluorocarbons.

1. 2-Nitro-alpha-[(2,2,2-trifluoroethoxy)methyl]-imidazole-1-ethanol (TF-MISO). This can be obtained from SynChem OHG Laboratories (CAS no. 21787-91-7) as noted in (11). The chemical formula is $C_8H_{10}F_3N_3O_4$ and the molecular weight is 269.178 g/mol.

2. Hexafluorobenzene HFB, 50 μl, 99.9% (Lancaster Co, Pelham, NH (12); or PCR Inc., Gainesville, FL (13)).

3. 1,3,5-Tris (trifluoromethyl) benzene (PCR Inc., Gainesville, FL).

4. Perfluoro-15-crown-5-ether (Matrix Scientific, Columbia, SC (9, 14)).

2.3. Software for Data Post-processing

Processing methods for fitting multiexponential T_2 decay curves have often been based on a non-negative least squares fitting routine such as that available in MATLAB (The Mathworks Inc.). Simpler models, such as a single or multiexponential function with fewer parameters can be fit with standard graphing packages such as Sigmaplot (Systat Software Inc.) (*see* **Note 9**). There are many good curve fitting routines available.

3. Methods

3.1. Animal Handling

Care is needed to maintain an anesthetized subject in as natural as possible a condition. This is because tissue oxygenation is sensitive to physiological parameters such as inspired O_2 tension and blood flow. Thus, anesthesia paradigms are a key parameter in a study design.

Injectable anesthetics provide a stable imaging condition for up to 2 h in rodents. Inhaled anesthesia can be used for a longer time but requires additional equipment such as the vaporizer and ventilator. Longer time periods may also cause some degree of dehydration. Animals can be maintained with a nose-cone and spontaneously ventilated. More precise control of blood gasses is obtained with ventilation system and either a tracheotomy (usually a non-recovery study) or intubation for delivery of inspired gas. Injectable anesthesia tends to result in lower tissue pO_2 values while isoflurane with proper ventilation tends to result in a pO_2 more consistent with the awake animal (15). The percentage of inspired O_2 will greatly influence tissue pO_2, but can be controlled to some extent by monitoring arterial hemoglobin saturation with a pulse oximeter. For obtaining a strong BOLD response, although a range of anesthetics have been used, the strongest response tends to come using α-chloralose with a paralyzing agent (16, 17). For a review of anesthesia protocols, *see* a general anesthesia reference such as Flecknell (18) (*see* **Chapters 5** and **17**).

1. Anesthetize with either injectable or inhaled anesthesia. A standard inhaled protocol would be isoflurane 3–5% induction and 1–2% maintenance. Monitor respiration and/or blood pressure and reduce the maintenance amount if needed during long anesthesia paradigms if there is a decline in physiological status. A normal hemoglobin saturation of approximately 95–96% is usually achieved with an inspired O_2 of 30% and N_2 constituting the remainder of the gas.

2. Monitor physiological status. The most invasive method is an indwelling catheter for monitoring blood pressure and for periodic sampling of blood gasses. The minimum assessment would be monitoring respiration. This can be done with a custom balloon around the chest connected to a pressure transducer. An MRI-compatible pulse oximeter is useful for continuous monitoring of arterial oxygen saturation. It is possible to purchase an MRI-compatible tail cuff blood pressure monitor. Data can be stored into any digitization system such as the Biopac (Biopac Systems Inc.).

3. Body temperature tends to decline under anesthesia and should be maintained. A rectal probe with a thermocouple is an absolute requirement for physiological monitoring. Maintenance of temperature is more problematic if the bore of the MRI is cold, such as is often the case if the MRI has not been used for a few days, or if the animal is subjected to acute hypoxia, or if the anesthesia is prolonged. The two most common methods of warming are to blow warm air over the subject or to place the subject on a bed of circulating liquid with variable temperature. Circulating liquids can pose a problem for imaging, if the imaging volume includes the waterbed. If needed, rotate the phase-encoding plane away from the water to minimize flow artifacts. It can also be useful to wrap or cover the subject. Rectal probes have been used to provide feedback to temperature regulation devices.

3.2. BOLD MRI

The method involves quantification of R_2, R_2^*, or changes in R_2 and R_2^*. This can be done in spectroscopic mode where the localization is either non-existent or related to the sensitive volume of the RF coil. More commonly, it is done in imaging mode where relaxation data are obtained on a voxel-by-voxel basis. The sensitivity to changes in oxygenation is increased by increasing the weighting to R_2 or R_2^*. In a standard spin or gradient echo sequence, this involves increasing the echo time or increasing the magnetic field strength (*see* **Note 2**). In general, setting the TE to a value close to the T_2 or T_2^* will provide the optimum sensitivity with acceptable loss of signal due to relaxation effects. When quantifying with multi-echo, it is possible to use a wide range of TR (time between repeat pulses) values as long as signal is sufficient. However, if quantifying by measuring delta signal intensity (ΔSI) then a longer TR will minimize the T_1 weighting. This can also be minimized in gradient echo sequences by using a pulse angle approximating the Ernst angle—that is, the angle that provides maximum signal for a given TR (*see* **Note 8**).

Methods based on intravascular contrast agents such as hemoglobin are often not precise in their localization within the vasculature. As BOLD MRI is more sensitive to changes in total deoxyhemoglobin, it follows that changes in large draining vessels may dominate the signal (19)—especially when using gradient echo methods that are sensitive to large vessels. This may influence the interpretation of the results.

The subject is placed in the MRI, and the region of interest (ROI) identified. One optimizes the magnet homogeneity (shim) and calibrates the RF power with the appropriate scanner procedures. A standard study paradigm would be to obtain a series of control images, introduce a change that would change the oxygenation in the region of interest, and collect images after the change.

1. Multi-echo gradient echo or spin echo data can be obtained over a time course of data collection. The TE is in the range of the T_2^* for GE or T_2 for SE. This would be about 15 ms for GE and 40 ms for SE at 7 T. TR in a GE sequence tends to be relatively short in order to gain the temporal resolution required. It is often in the order of 50–200 ms. In such cases, the RF pulse angle needs to be reduced to minimize T_1 saturation. One can start with using the Ernst angle (*see* **Note 8**). The field of view usually exceeds the cross section of the subject to prevent artifacts from image folding. The slice thickness for animal work is often 1 mm. The matrix for fMRI is often 64×64 to minimize acquisition time. If longer times are acceptable, then 128×128 provides a better image.

2. Although oximetry relies on quantification of relaxation parameters, it is often sufficient to measure the change or delta (Δ) in the relaxation parameter. Such a study is technically much simpler to obtain as one can quantify the change in signal intensity using the equation:
$\Delta R_2^* = -\ln(SI_n/SI_0)/TE$
where SI_n is the signal intensity at each time point, SI_0 is the initial signal intensity, and TE is the echo time. One needs to keep the units consistent. The same equation holds for ΔR_2 if one obtains data using a spin echo sequence.

3. For quantification of multi-echo data, one can fit a single exponential function where:
$SI_n = SI_0 \, e^{(-TE/T2)}$
where SI_n is the signal intensity at a given echo time, TE, and SI_0 is the signal intensity at TE = 0. One can also transform the *y*-axis into a log function and plot as a semi-logarithmic plot (**Fig. 1**). The author is not aware of BOLD studies using multiexponential fits, but recognizes that these would be appropriate if the number of echoes is sufficient. Examples of multi-echo analysis of T_2, which the reader can use as an entry into the issues involved with quantifying multi-echo data, can be found in (20, 21) (*see* **Note 9**).

4. Key variables to control for are systemic oxygenation, motion, and blood flow.

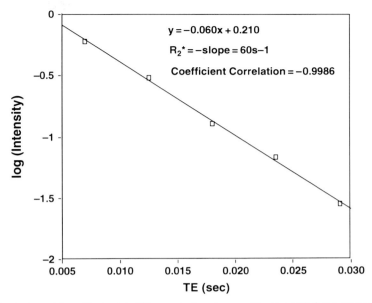

Fig. 1. A semi-logarithmic plot of SI vs. echo time for quantification of R_2^* in a rat brain. As this is a standard mathematical function for a single exponential decay, if the plot results in a linear fit then one can assume that the decay curve is adequately described by a single exponential function. (Reproduced from (43) with permission from Wiley-Liss Inc.).

5. Quantification can be done on a voxel-by-voxel basis or on an ROI basis. The analysis pattern, or workflow, can be undertaken by measuring SI within the ROI and calculating average values or one can undertake the quantification on a voxel-by-voxel basis. After such voxel-based quantification, one can obtain statistical information over the ROI (22).

6. The changes in SI relate to changes in R_2 or R_2^*. Both of these parameters change linearly with the deoxyhemoglobin content (3) which, in turn is a reflection of tissue oxygenation (*see* **Note 1**) (**Fig. 2**).

3.3. ^{19}F, T_1 MRI

This method relies on two key points. One is that there is no inherent ^{19}F signal in living systems under standard conditions and the other is that there are fluorinated hydrocarbons which have a ^{19}F resonance with a T_1 relaxation time that is related to the surrounding oxygen concentration.

In these studies, one injects the ^{19}F tracer and undertakes either ^{19}F-spectroscopy or MRI. Imaging can also include dual nuclei MRI which, when carefully done to ensure similar RF homogeneity between nuclei, allows the oxygen images to be overlaid on an anatomical ^1H MRI.

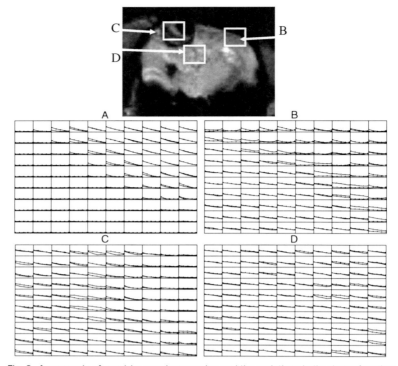

Fig. 2. An example of voxel-by-voxel processing and the variations in the decay function in vivo. A rat with a glial tumor was studied. A multi-echo gradient echo sequence was used. The upper image shows a cross section of the brain. Below are voxel-by-voxel plots of SI vs. TE. Due to the small size of the voxels in the figure, the axis labels have been omitted. SI is in arbitrary units and the TEs are 9, 15.5, 22, 28.5, and 35 ms. **Plot A** shows data from the *lower left* of the head, out of the field of view, that contains noise and muscle. **B** shows normal brain in the *lower left*. **C** is from tumor. **D** is a second tumor region. There are *two lines* in each voxel, one while breathing control gasses and the other while breathing hyperoxygenated gas of 95% O_2 and 5% CO_2. The key point to this figure is that, on close examination of voxels, there are clearly voxels that do not show a simple exponential decay function within the tumor. Such deviation from an exponential function is relatively common and care is needed to prevent any misinterpretation that may arise. (Reproduced from (43) with permission from Wiley-Liss Inc.).

Before injection, one should calibrate the relaxation rate R_1 by undertaking phantom studies. Under controlled conditions, measure the R_1 vs. oxygen in solution. Oxygen can be measured as pO_2 (mmHg, torr, or kPa), or concentration (% O_2, mM, or ppm). If one is using percentage (i.e., equilibrating with 10% O_2, be sure to control for atmospheric pressure. For instance, 10% O_2 is 75.6 mmHg at sea level but only 67.8 mmHg at 1000 m of altitude (23).

The compounds to be used vary in the number of fluorine nuclei adding to the signal, cost, and availability. The options are listed in **Section 2**.

Material can be injected intravenously or directly into the tissue. If the data collection is done in a short period after injection, then substantial ^{19}F signal may arise from the blood after an intravenous injection. It is possible to inject the material and then image hours after injection. This has been done successfully with implanted tumors (10). Under these conditions, the signal may arise from within cells. A confounding issue is that it is not clear which cells contain the ^{19}F.

It is useful to undertake a pilot study to determine an optimum time for imaging whereby there remains sufficient material in the tissue.

Imaging involves quantification of T_1. There are many methods available and one needs to assess the fastest and most sensitive method available on the particular scanner (see **Note 7**).

1. Prepare the animal for direct injection or intravenous injection. For preparation of an injection of TF-MISO, dissolve in sterile saline solution (5 mg/mL) (11). HFB has been injected undiluted at 3 μl per injection (24).

2. As most tissues have low pO_2, reduce the pO_2 in the perfluorocarbon by bubbling with N_2 before use (24).

3. Inject using a small gauge (29G) in multiple tracks to improve penetration.

4. Position in the MRI, shim, and tune the RF coils.

5. Undertake localizing MRI using ^{1}H, T_2w multislice to determine the ROI. This is usually fast, low resolution MRI often with a matrix of 64 × 64 to identify anatomy. A common sequence is a FLASH-type sequence (25) with a relatively low flip angle to reduce T_1 weighting (40–60°) combined with a short TR to minimize acquisition time (100–300 ms). The field of view and orientation to be used is determined at this stage by optimizing the scout image.

6. Change to the ^{19}F frequency and do adjustments like tuning, matching, and flip angle calibration.

7. A range of sequences may be used. The goal is high sensitivity and rapid acquisition time. Options include snapshot flash inversion recovery (24), a driven equilibrium sequence (13, 26), and EPI-inversion recovery (9).

8. Quantify T_1 and compare with pre-determined calibration curves using the same sequence. The equation used to quantify T_1 depends on the acquisition method. Standard saturation or inversion recovery data are fit to a single exponential recovery curve where MRI signal intensities are plotted against a range of recovery times from either an inversion or saturation pulse (see **Note 10**). Faster imaging can be done with a range of methods including echo planar and fast low angle imaging (25, 27). State-free precession data are fit to

3.4. ^{1}H, T_1 MRI

more complex models (28). It is accepted to threshold to remove voxels with random noise (12).

Oxygen in solution influences the T_1 relaxation time of the solution. In living systems, there is a unique opportunity to undertake MRI oximetry within fluids. This has been done in the eye solely by quantifying the T_1 of the solution within the eye—the vitreous. A method termed the retinal oxygenation response (ROR) was developed whereby changes in T_1 within the vitreous are monitored during a step function in inspired O_2 (29).

A major potential source of artifact is the blinking response. This can be reduced or eliminated by a paradigm where the subject holds as still as possible for 10–15 s without blinking during which the MRI can be obtained with fast imaging methods. As animals rarely blink during anesthesia, movement is less of a problem with animal studies. One can cover the eyes with ointment to keep the eyes moist and comfortable during the study.

1. Position subject in the MRI. Subject is instructed to visually target a spot when requested. It is helpful to place gauze over the alternate eye.
2. Place a face mask for administering breathing gases. It is usually adequate to initiate with room air.
3. Place a phantom of water near the eye within the field of view. Isolate the phantom from the skin as much as possible to prevent heat transfer from the subject, which would warm the phantom and so alter the T_1.
4. Practice a blink/no blink cycle with a human subject before imaging.
5. Initiate MRI. Obtain a series of baseline T1w images (at least three). One of the simplest methods would be quantifying changes in SI collected using a snapshot FLASH sequence (TR = 22 ms, TE = 7 ms, flip angle 12° at 1.5 T (30); *see* **Note 7**).
6. Change the breathing gas to 100% O_2. Breath the new gas for at least 5 min. The goal is a PaO_2 of greater than 350 mmHg and a $PaCO_2$ of 35–45 mmHg (29).
7. Obtain a new localizer image to assist with coregistration of image plane.
8. Collect hyperoxia images as needed (at least three).
9. Calculate ΔSI either on an ROI or a voxel-by-voxel basis.
10. Model against published values of ΔSI vs. pO_2 in saline. A relaxivity of 1.75×10^{-4} s^{-1} mmHg^{-1} has been used (31) and modeling has been undertaken to transform data to mmHg of O_2 (29).

4. Notes

1. If total hemoglobin is constant, then ΔdeoxyHb will change with the saturation of blood. The link between saturation and pO_2 is non-linear. One needs to be aware of the shape of the oxyhemoglobin dissociation curve if one wants to make comments on pO_2. Many factors influence the magnitude of the ΔR_2^* effect although the result is that ΔR_2^* is linear in brain over a wide range of hemoglobin saturations. Confounding effects such as vessel orientation, Hb content, flow, and red cell geometry will change the ΔR_2^* vs. deoxyHb content relationship (32, 33).

2. Field sensitivity is important. BOLD methods are more sensitive at higher fields. The balance is that artifacts due to susceptibility are also higher. The R_2 is relatively more sensitive to microvasculature than R_2^* but has lower sensitivity. When using high fields (7 T or more), it may be desirable to use R_2 instead of R_2^* as the susceptibility artifacts are less with R_2 and the sensitivity at high field makes ΔR_2 quantification possible. For an introduction to the literature on field dependence, *see* (34, 35). For a discussion on vessel size dependency, *see* (36–38) (*see* also **Chapter 8**).

3. A volume coil is optimum for quantification, as this geometry provides the most uniform RF transmission. Although it is possible to use a surface coil for measuring R_2, this is not recommended. There will only be a small region of depth where the optimum pulse angle will occur. A surface coil is generally sufficient for quantifying R_2^* if one is using a sequence that does not have RF refocusing pulses (i.e., a standard gradient echo).

4. In a multi-echo spin echo sequence, improvements in the refocusing pulse will greatly improve the accuracy of the measurement. A simple composite pulse has been effective but is limited to single slice measurements (39–41). 3-D quantification methods are under development and should be available soon.

5. Stimulated echoes are a problem and will influence the signal intensity decay curve. Much work has been done to reduce the impact of stimulated echoes (42). An optimized sequence for collecting a high number of echoes is not usually supplied. Some possible modifications can be found in (42).

6. Fast methods of quantifying ΔSI, such as EPI, are commonplace in systems using functional MRI to study brain activation.

7. The sensitivity (change in T_1 per change in oxygen concentration) in ^{1}H studies on fluids is not particularly high and so attention to reducing noise is important.

8. The Ernst angle is the RF pulse angle that provides the maximum signal/noise per unit time of acquisition for a given T_1 and TR. The RF pulse angle (α) is defined as cos $\alpha = \exp(-TR/T_1)$.

9. The method of processing multi-echo data for R_2 or R_2^* is a significant variable. The most common method is to fit a single exponential function to an average signal intensity obtained over a region of interest. It is also possible to quantify on a voxel-by-voxel basis and so obtain a map. The latter provides more data on regional heterogeneity but, especially in R_2^* data, the decay curve may be far from that which would fit a single exponential function (**Fig. 2**). If this occurs, one might consider a totally different metric such as the area under the curve. In multi-echo data one might also consider using a multiexponential fitting routine. For examples, *see* (21, 22).

10. Calculation of T_1 from an inversion recovery sequence can be done using: $M_0[1 - 2\exp(-t/T_1)]$.

Acknowledgments

This work was partially supported by an NIH RO1 EB002085, by the Canadian Institutes of Health Research FIN 79260, the Canadian Foundation for Innovation, and the Alberta Heritage Foundation.

References

1. Ogawa, S., Lee, T. M., Kay, A. R., and Tank, D. W. (1990) Brain magnetic resonance imaging with contrast dependent on blood oxygenation. *Proc Natl Acad Sci USA* **87**, 9868–9872.
2. Dunn, J. F. (2007) Measuring oxygenation in vivo with MRS/MRI—from gas exchange to the cell. *Antioxid Redox Signal* **9**, 1157–1168.
3. Dunn, J. F., Zaim Wadghiri, Y., Pogue, B. W., and Kida, I. (1998) BOLD MRI vs. NIR spectrophotometry: Will the best technique come forward? *Adv Exp Biol Med* **454**, 103–113.
4. Pauling, L., and Coryell, C. D. (1936) The magnetic properties and structure of hemoglobin, oxyhemoglobin and carbonmonoxyhemoglobin. *Proc Natl Acad Sci USA* **22**, 210.
5. Thulborn, K. R., Waterton, J. C., Mathews, P. M., and Radda, G. K. (1982) Oxygenation dependence of the transverse relaxation time of water protons in whole blood at high field. *Biochim Biophys Acta* **714**, 265–270.
6. Prosser, R. S., and Luchette, P. A. (2004) An NMR study of the origin of dioxygen-induced spin-lattice relaxation enhancement and chemical shift perturbation. *J Magn Reson* **171**, 225–232.
7. Berkowitz, B. A., Wilson, C. A., Hatchell, D. L., and London, R. E. (1991) Quantitative determination of the partial oxygen pressure

in the vitrectomized rabbit eye in vivo using ^{19}F NMR. *Magn Reson Med* **21**, 233–241.

8. Zaharchuk, G., Martin, A. J., Rosenthal, G., Manley, G. T., and Dillon, W. P. (2005) Measurement of cerebrospinal fluid oxygen partial pressure in humans using MRI. *Magn Reson Med* **54**, 113–121.

9. Dardzinski, B. J., and Sotak, C. H. (1994) Rapid tissue oxygen tension mapping using ^{19}F inversion-recovery echo-planar imaging of perfluoro-15-crown-5-ether. *Magn Reson Med* **32**, 88–97.

10. Kodibagkar, V. D., Cui, W., Merritt, M. E., and Mason, R. P. (2006) Novel ^{1}H NMR approach to quantitative tissue oximetry using hexamethyldisiloxane. *Magn Reson Med* **55**, 743–748.

11. Procissi, D., Claus, F., Burgman, P., Koziorowski, J., Chapman, J. D., Thakur, S. B., Matei, C., Ling, C. C., and Koutcher, J. A. (2007) In vivo ^{19}F magnetic resonance spectroscopy and chemical shift imaging of tri-fluoro-nitroimidazole as a potential hypoxia reporter in solid tumors. *Clin Cancer Res* **13**, 3738–3747.

12. Xia, M., Kodibagkar, V., Liu, H., and Mason, R. P. (2006) Tumour oxygen dynamics measured simultaneously by near-infrared spectroscopy and ^{19}F magnetic resonance imaging in rats. *Phys Med Biol* **51**, 45–60.

13. Mason, R. P., Rodbumrung, W., and Antich, P. P. (1996) Hexafluorobenzene: A sensitive ^{19}F NMR indicator of tumor oxygenation. *NMR Biomed* **9**, 125–134.

14. Bartusik, D., Tomanek, B., Siluk, D., Kaliszan, R., and Fallone, G. (2009) The application of ^{19}F magnetic resonance ex vivo imaging of three-dimensional cultured breast cancer cells to study the effect of delta-tocopherol. *Anal Biochem* **387**, 315–317.

15. Lei, H., Grinberg, O., Nwaigwe, C. I., Hou, H. G., Williams, H., Swartz, H. M., and Dunn, J. F. (2001) The effects of ketamine-xylazine anesthesia on cerebral blood flow and oxygenation observed using nuclear magnetic resonance perfusion imaging and electron paramagnetic resonance oximetry. *Brain Res* **913**, 174–179.

16. Qiao, M., Rushforth, D., Wang, R., Shaw, R. A., Tomanek, B., Dunn, J. F., and Tuor, U. I. (2007) Blood-oxygen-level-dependent magnetic resonance signal and cerebral oxygenation responses to brain activation are enhanced by concurrent transient hypertension in rats. *J Cereb Blood Flow Metab* **27**, 1280–1289.

17. Liu, Z. M., Schmidt, K. F., Sicard, K. M., and Duong, T. Q. (2004) Imaging oxygen consumption in forepaw somatosensory stimulation in rats under isoflurane anesthesia. *Magn Reson Med* **52**, 277–285.

18. Flecknell, P. (1996) *Laboratory Animal Anesthesia*, 2 ed., Academic, London.

19. Frahm, J., Klaus-Dietmar, M., Hanicke, W., Kleinschmidt, A., and Boecker, H. (1994) Brain or vein-oxygenation or flow? On signal physiology in functional MRI of human brain activation. *NMR in Biomed* **7**, 45–53.

20. Whittall, K. P., MacKay, A. L., Graeb, D. A., Nugent, R. A., Li, D. K., and Paty, D. W. (1997) In vivo measurement of T_2 distributions and water contents in normal human brain. *Magn Reson Med* **37**, 34–43.

21. McCreary, C. R., Bjarnason, T. A., Skihar, V., Mitchell, J. R., Yong, V. W., and Dunn, J. F. (2009) Multiexponential T_2 and magnetization transfer MRI of demyelination and remyelination in murine spinal cord. *Neuroimage* **45**, 1173–1182.

22. Bjarnason, T. A., McCreary, C. R., Dunn, J. F., and Mitchell, J. R. (2010) Quantitative T_2 analysis: The effects of noise, regularization, and multi-voxel approaches. *Magn Reson Med* **63**, 212–217.

23. West, J. B. (1999) Barometric pressures on Mt. Everest: New data and physiological significance. *J Appl Physiol* **86**, 1062–1066.

24. Jordan, B. F., Cron, G. O., and Gallez, B. (2009) Rapid monitoring of oxygenation by ^{19}F magnetic resonance imaging: Simultaneous comparison with fluorescence quenching. *Magn Reson Med* **61**, 634–638.

25. Haase, A. (1990) Snapshot FLASH MRI. Applications to T_1, T_2, and chemical-shift imaging. *Magn Reson Med* **13**, 77–89.

26. Hunjan, S., Zhao, D., Constantinescu, A., Hahn, E. W., Antich, P. P., and Mason, R. P. (2001) Tumor oximetry: Demonstration of an enhanced dynamic mapping procedure using fluorine-19 echo planar magnetic resonance imaging in the Dunning prostate R3327-AT1 rat tumor. *Int J Radiat Oncol Biol Phys* **49**, 1097–1108.

27. Scheffler, K. and Hennig, J. (2001) T(1) quantification with inversion recovery TrueFISP. *Magn Reson Med* **45**, 720–723.

28. Scheffler, K. and Lehnhardt, S. (2003) Principles and applications of balanced SSFP techniques. *Eur Radiol* **13**, 2409–2418.

29. Trick, G. L. and Berkowitz, B. A. (2005) Retinal oxygenation response and retinopathy. *Prog Retin Eye Res* **24**, 259–274.

30. Berkowitz, B. A., McDonald, C., Ito, Y., Tofts, P. S., Latif, Z., and Gross, J. (2001) Measuring the human retinal oxygenation response to a hyperoxic challenge using MRI: Eliminating blinking artifacts and

demonstrating proof of concept. *Magn Reson Med* **46**, 412–416.

31. Berkowitz, B. A. (1996) Adult and newborn rat inner retinal oxygenation during carbogen and 100% oxygen breathing. Comparison using magnetic resonance imaging delta pO_2 mapping. *Invest Ophthalmol Vis Sci* **37**, 2089–2098.

32. Dunn, J. F., and Swartz, H. M. (1997) Blood oxygenation. Heterogeneity of hypoxic tissues monitored using bold MR imaging. *Adv Exp Med Biol* **428**, 645–650.

33. Dunn, J. F., Wadghiri, Y. Z., and Meyerand, M. E. (1999) Regional heterogeneity in the brain's response to hypoxia measured using BOLD MR imaging. *Magn Reson Med* **41**, 850–854.

34. Silvennoinen, M. J., Clingman, C. S., Golay, X., Kauppinen, R. A., and van Zijl, P. C. (2003) Comparison of the dependence of blood R_2 and R_2^* on oxygen saturation at 1.5 and 4.7 Tesla. *Magn Reson Med* **49**, 47–60.

35. Gati, J. S., Menon, R. S., Ugurbil, K., and Rutt, B. K. (1997) Experimental determination of the BOLD field strength dependence in vessels and tissue. *Magn Reson Med* **38**, 296–302.

36. Zhang, N., Yacoub, E., Zhu, X. H., Ugurbil, K., and Chen, W. (2009) Linearity of blood-oxygenation-level dependent signal at microvasculature. *Neuroimage* **48**, 313–318.

37. Boxerman, J. L., Hamberg, L. M., Rosen, B. R., and Weisskoff, R. M. (1995) MR contrast due to intravascular magnetic susceptibility perturbations. *Magn Reson Med* **34**, 555–566.

38. Tropres, I., Grimault, S., Vaeth, A., Grillon, E., Julien, C., Payen, J. F., Lamalle, L., and Decorps, M. (2001) Vessel size imaging. *Magn Reson Med* **45**, 397–408.

39. Levitt, M. H., and Freeman, R. (1981) Compensation for pulse imperfections in NMR spin-echo experiments. *J Magn Reson* **43**, 65–80.

40. Cremillieux, Y., Ding, S., and Dunn, J. F. (1998) High-resolution in vivo measurements of transverse relaxation times in rats at 7 Tesla. *Magn Reson Med* **39**, 285–290.

41. MacKay, A., Laule, C., Vavasour, I., Bjarnason, T., Kolind, S., and Madler, B. (2006) Insights into brain microstructure from the T_2 distribution. *Magn Reson Imaging* **24**, 515–525.

42. Poon, C. S., and Henkelman, R. M. (1992) Practical T_2 quantification for clinical applications. *JMRI* **2**, 541–553.

43. Dunn, J. F., O'Hara, J. A., Zaim-Wadghiri, Y., Zhu, H., Lei, H., Meyerand, M. E., Grinberg, O. Y., Hou, H., Hoopes, P. J., Demidenko, E., and Swartz, H. M. (2002) Changes in oxygenation of intracranial tumors with carbogen, a BOLD MRI and EPR oximetry study. *J Magn Res Imaging* **16**, 511–521.

Chapter 13

MRI Using Intermolecular Multiple-Quantum Coherences

Rosa Tamara Branca

Abstract

Intermolecular multiple-quantum coherences (iMQCs) can generate NMR signals from exceedingly small dipolar interactions between distant spins in solutions. In the last few years, these signals have been used for a wide range of applications in imaging and high-resolution spectroscopy. Recent applications include MRI contrast enhancement, suppression of inhomogeneous broadening in NMR experiments, and more recently, in vivo temperature measurement. In this chapter, we describe how basic iMQC pulse sequences work and how to select the sequence parameters to optimize iMQC signals and to overcome signal contamination.

Key words: Dipolar field, intermolecular multiple-quantum coherences, correlation gradient, correlation distance.

1. Introduction

iMQCs arise from simultaneous spin flips on distant molecules in a solution. Specifically, a resonance corresponding to raising the spin state of a proton on a macromolecule and simultaneously lowering the spin state of a proton in its aqueous solvent is called "intermolecular zero-quantum coherence" or iZQC, because the net number of spin flips (one up, one down) is zero. A resonance corresponding to two simultaneous upward spin flips is called a + double-quantum coherence (iDQC) and a resonance corresponding to two simultaneous downward spin flips is called a − double-quantum coherence. The conventional theoretical framework of NMR predicts that signals from such transitions are impossible to observe. Indeed, they were not evident in the early days of low-field magnetic resonance. Over the last few years, with the increase in magnetic field strengths, iMQC signals nearly as large as the conventional magnetization signals have been detected (1–5). This is because the detection of these signals is made possible by the long-range dipolar interaction, which is proportional to the magnetic field strength (6, 7). This interaction can be ignored even in high magnetic field strength if the magnetization is isotropic, but if the isotropy is broken with odd-shaped samples or gradient pulses, the dipole–dipole coupling can increase to produce noticeable effects.

In the last few years, iMQCs have progressed from academic curiosities into wide usage in imaging and high-resolution spectroscopy. Recent applications include contrast enhancement in magnetic resonance imaging (8–10), functional imaging (11–14), suppression of inhomogeneous broadening (15–17), measurements of local magnetization structure (18, 19), and more recently, measurements of temperature in vivo (20).

The iMQC signal is proportional to the local dipolar field at the correlation distance, $d_c = 1/(\gamma GT)$, that can be tuned by changing the strength GT of the magnetic field gradient pulses used in the iMQC experiments. This distance is usually chosen to be smaller than the typical voxel size, between tens and hundred of micrometers, making the iMQC signal intrinsically sensitive to local dipolar field, altering magnetization and susceptibility structure over sub-voxel distances.

The prototype iMQC pulse sequence is the "CRAZED" (correlated spectroscopy revamped by asymmetric z-gradient echo detection) (5) sequence, shown in **Fig. 1**. The CRAZED pulse sequence is similar to a COSY [COrrelation SpectroscopY] sequence, but differs in that the second CRAZED RF pulse is bracketed by two magnetic field gradient pulses, the correlation gradient pulses, with amplitude G and length T, used to select the signal from a specific order of coherence (iDQC, iZQC, etc.). These gradient pulses break the spatial isotropy and preserve the dipolar field interactions which are not averaged out by molecular diffusion at large separation, and which are responsible for the refocusing of transverse magnetization during the final time period t_2 (6). The refocusing rate of the signal during t_2 is $1/\tau_d$, where $\tau_d = (\gamma \mu_0 M_0)^{-1}$. τ_d is the dipolar demagnetization time and depends on the strength of the dipolar field, which, for this sequence, modulates the magnetization along a well-defined direction and is directly proportional to the local longitudinal magnetization (7). For standard samples in standard magnetic field gradients, the refocusing is very slow (hundreds of milliseconds) and the signal usually relaxes before a complete refocusing (21). In other words, the refocusing is partially hampered by the transverse relaxation time, rendering the available signal much smaller than its theoretical value. For this reason, it is particularly important to optimize the parameters of a CRAZED sequence to yield maximum image contrast over a noisy background signal. The reduced signal also makes such sequences very

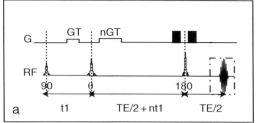 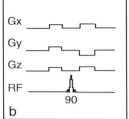

Fig. 1. (a) Standard CRAZED pulse sequence used to observe iMQC signal. The first RF pulse (90°), which excites the equilibrium magnetization, is followed by a delay t_1 during which a gradient pulse of strength G and duration T, dephasing the transverse magnetization, is applied. A second RF pulse θ transfers part of this modulated magnetization along the z-axis creating the correct dipolar field to refocus the transverse magnetization. The ratio $(nGT)/(GT)$ determines the selected coherence order. A third RF pulse (180°), at a time $TE/2 + nt_1$ after the second RF pulse, is used to refocus inhomogeneous broadening during the evolution time t_2 and allow the refocusing of the iMQC signal at TE/2 after the pulse. (b) Crusher module used to suppress signal contamination from stimulated echoes when the magnetization is not completely allowed to relax toward thermal equilibrium between scans. The 90° pulse is surrounded by crusher gradients along the magic angle. The combination of the gradients and the broadband 90° pulse dephases both the transverse and the longitudinal magnetization. The crusher module is usually put at the end of the sequence, provided that dummy scans are performed as well.

sensitive to interference by signals that follow alternate coherence pathways. It is therefore necessary to simultaneously adopt experimental sequence modifications, such as phase cycling to eliminate undesired coherences while preserving the dipolar coupling signal.

Below, we explain how to perform a standard iDQC experiment and how to select all the parameters in a CRAZED pulse sequence.

2. Materials

1. A high-field superconductive magnet.
2. Shim gradient coils.
3. Imaging gradient coils.
4. A scanner equipped with a radio frequency (RF) synthesizer and a power amplifier.
5. A ^{1}H transmitter and receiver coil.
6. A sample tube filled with 1% agarose, 0.5% Magnevist, and 10 Mm isopropanol containing some structured sample (in our case, Legos).
7. A pulse program for the CRAZED sequence (*see* **Note 1**).

3. Methods

1. Position the sample in the center of the RF coil and in the center of the magnet.
2. Connect the coil to the preamplifier.
3. Match and tune the RF coil.
4. Shim the magnetic field using the first- and second-order shims.
5. Find the resonance frequency and select it as the basic transmitter and receiver frequency.
6. Calibrate the RF pulses.
7. Run a preliminary scan to see if the sample is positioned correctly in the center of the magnet.
8. Select the CRAZED sequence parameters, RF pulses, gradient pulses, and delays as follows:
    a. Select the first RF pulse to be a BIR-4 pulse with a 90° flip angle and with 1 ms duration (see **Note 2**).
    b. Select the second RF pulse to be a BIR-4 pulse with a 120° flip angle and with 1 ms duration (see **Note 3**).
    c. Select the two refocusing pulses to be adiabatic full passage hyperbolic secant refocusing pulses with 4 ms duration (see **Note 4**).
    d. Select an axial slice with 2 mm thickness.
    e. Select the direction of the correlation gradient along the z-direction ($G_c = G_z$)
    f. Select the strength of both the correlation gradient pulses to be 12 G/cm (this will select a correlation distance of 90 μm) (see **Note 5**).
    g. Select the duration of the first gradient pulse to 1 ms and the duration of the second gradient pulse to 2 ms (see **Note 6**).
    h. Select the crusher gradients around the refocusing pulses along the z-direction with a strength of 15 G/cm strength and a duration of 1 ms (see **Note 7**).
    i. Select the following delays between the pulses: 90–120°, 2.5 ms; 120–180°, 15 ms; 180–180°, 20 ms; 180° center acquisition window, 10 ms (see **Note 8**).
    j. Select a field of view of 40 mm and a matrix size of 128 × 128 so that the in-plane spatial resolution is about 300 μm/pixel (see **Note 9**).
    k. Select a repetition time of 5 s.

l. Acquire each line of k-space four times by phase cycling the first 90° pulse $(x, -x, y, -y)$ and the receiver $(x, x, -x, -x)$ (see **Notes 10** and **11**).

m. Fourier transform the acquired data.

9. Repeat the same experiment while changing the direction of the correlation gradient along the x-direction.

10. Repeat the same experiment while changing the direction of the correlation gradient along the y-direction.

11. Subtract the magnitude of the images $|G_z| - |G_x| - |G_y|$ to resolve the magnetization density anisotropy map (**Fig. 2**) (see **Note 12**).

12. Repeat the same experiment by using correlation gradients along the z-direction with strengths of 4.2 G/cm and 25 G/cm, and subtract the magnitude of the images (**Fig. 3**).

13. Acquire an image using the standard SPIN-ECHO sequence available on the scanner with an echo time of

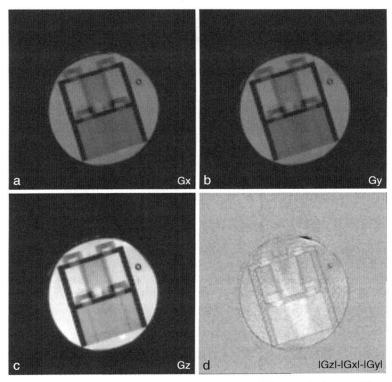

Fig. 2. iDQC images acquired with a correlation distance of 90 μm selected along the x- (**a**), y- (**b**) and z- (**c**) directions (images with the same intensity scale). The signal intensity of the images acquired with the correlation gradient along the x- and y- directions is half that of the image acquired with the correlation gradient along the z-direction. The subtraction image ($|G_z| - |G_x| - |G_y|$) highlights the structural features of the imaged sample (**d**).

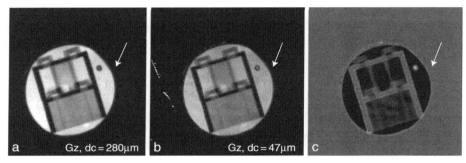

Fig. 3. iDQC images acquired with the correlation gradient along the z-direction, at two different strengths: $G_c = 25$ G/cm (**b**) to select a correlation distance of 47 μm; and $G_c = 4$ G/cm (**a**) to select a correlation distance of 280 μm. The subtraction image (**c**) shows bigger intensity overall for the image acquired with smaller correlation gradients (due to a smaller diffusion attenuation) and a different contrast when the correlation gradient is comparable to structural sample features (*arrow* in **c**).

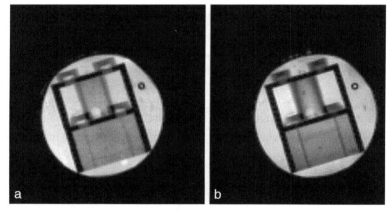

Fig. 4. Comparison between the iDQC image with $G_c = G_z = 12$ G/cm (**a**) and a spin-echo image (**b**) of the same phantom acquired with the same repetition time and the same echo time.

45 ms, using the same image parameters used in the previous experiments (*see* **Note 13**).

14. Compare the spin-echo image with the iDQC image (**Fig. 4**).

4. Notes

1. The CRAZED sequence is typically not available on commercial spectrometers and must therefore be programmed by the user. The sequence is relatively simple and is illustrated in **Fig. 1a**. The main components are the three RF

pulses and the two magnetic field gradient pulses. The correct selection of the sequence delays and the RF and gradient pulses is very important for signal optimization. Bad sequence parameter choice can drastically reduce image SNR and/or produce strong signal contaminations.

2. The first RF pulse is 90° and is intended to excite all the spins in the sample. Therefore, it is a broadband pulse. The pulse shape can be that of a standard amplitude-modulated pulse, like a Hermitian, sinc, or Gaussian pulse. A superior excitation is achieved through the use of adiabatic pulses that ensure uniform sample excitation, such as the adiabatic half passage pulse (AHP) or the B_1-insensitive rotation (BIR-4) pulse. For imaging experiments, this pulse is not slice selective to produce a modulation of the magnetization in the entire sample.

3. The second pulse, also called the mixing pulse, is used to map a portion of the modulated magnetization onto the z-axis. This modulated M_z magnetization will eventually produce the dipolar field necessary to refocus the transverse magnetization. This pulse, along with the first excitation pulse, is non slice-selective: uniform modulation has to be created throughout the sample. A simple broadband amplitude-modulated pulse achieves this aim. For optimal uniformity, an adiabatic pulse should be used.

   The flip angle of the second RF pulse typically determines the strength of the signal. The optimal flip angles are 60, 120, and 45 or 135° for the selection of the +iDQC, −iDQC, and iZQC signals, respectively.

4. In order to maximize CRAZED signal intensity, a spin-echo type acquisition of the DQC signal is usually performed. The refocusing pulse rephases inhomogeneous broadening of the signal during t_2 evolution. In imaging experiments, this is the only slice-selective pulse in the sequence. It can either be a standard amplitude-modulated pulse or an adiabatic pulse. For an adiabatic pulse such as a hyperbolic secant pulse, the spin-echo pulse should be always accompanied by another identical 180° slice-selective pulse to prevent phase rolling across the image. The phase roll in this case come from temporal inversion offset of spins with different resonance frequencies, so that the symmetry of refocusing experiments is perturbed by appreciable amounts (milliseconds), and a serious phase dispersion is introduced.

   The 180° pulse is usually flanked by crusher gradients, whose amplitude is usually different from that of the correlation gradient to preclude the refocusing of spurious signals. The crusher is usually applied along the three orthogonal directions (x, y and z) to crash unwanted magnetization that is not refocused by the pulse.

5. The gradient pulse is used to reintroduce the effect of the dipolar field through spatial modulation of the sample magnetization. The sinusoidal modulation produced by the magnetic field gradient produces a strong dipolar field between spins, which are a pitch, or a correlation distance, apart. Both the direction and the strength of the correlation distance, chosen to select a correlation distance between 30 and 500 µm for imaging experiments, affect the image contrast in structured samples. If the correlation distance is too small, the modulation can be easily wiped out by spin diffusion, while if too long, spurious signals can easily contaminate image contrast. Some of these spurious signals can easily be spotted during image acquisition, since they appear as extra time-shifted echoes (**Fig. 5**).

6. The second gradient pulse should have an area which is n times the area of the first gradient pulse to select the nth-order iMQC coherence. For example, the +DQC requires a 2GT gradient, a −DQC a −2GT gradient, and a ZQC no gradient. To overcome gradient non-linearity, n gradient pulses identical to the first pulse can be applied. The position of the second correlation gradient pulse is usually as close as possible to the mixing pulse, to minimize signal attenuation due to diffusion (**Fig. 3**).

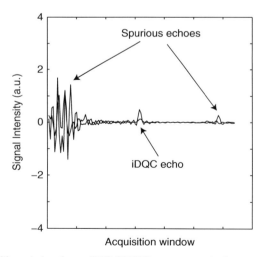

Fig. 5. Acquisition window for an iDQC CRAZED sequence selecting a weak correlation gradient ($G_c = 4$ G/cm) along the read encoding direction. When the correlation gradient is selected along one of the image-encoding directions and the correlation distance is comparable to the image resolution (in this case $d_c = 280$ µm and the image resolution is 300 µm), the correlation gradient couples with the encoding gradients and produces spurious echoes in the acquisition window (*see arrows*) and banding artifacts in the acquired image.

7. A crusher module can be applied at the beginning of the sequence, before the scan–scan pad time, to prevent signal contamination by stimulated echoes. This contamination is a serious problem especially when a very short repetition time is used. The crusher module consists of a 90° pulse sandwiched between a pair of two different gradient pulses oriented along the magic angle (x, y, and z all positive for the first pulse; x inverted for the second pulse) at the beginning of the scan as shown in **Fig. 1b**.

8. Pulse sequencing time delays have a strong effect on signal optimization and the image contrast. IMQC signal evolves during the first delay between the excitation and the mixing pulses. This delay is usually chosen to be few milliseconds to overcome losses due to transverse relaxation, which in the case of iMQC is faster than the standard transverse spin relaxation. Longer delays are usually applied to highlight T_2 inhomogeneities in the sample. In general, the gradient pulse is moved closer to the mixing pulse and a padding time between the end of the gradient pulse and the mixing pulse is used to avoid leakage of the gradient during the pulses.

   The second pulse field gradient is initiated close after the mixing pulse to minimize diffusion attenuation of the signal. The position of the refocusing pulse is usually TE/2 + nt after the mixing pulse, which leads to the refocusing of the signal at TE/2 after the refocusing pulse.

   The total echo time TE + nt is usually chosen to maximize the signal strength. Unlike standard NMR signal, the iMQC signal grows with time after the mixing pulse, reaches a maximum, and then decays. The maximum, in absence of relaxation, is achieved at 2.6 τ_d for the iZQC signal (8) and 2.2 τ_d for the iDQC signal (22). However, in presence of relaxation, the maximum is shifted toward smaller t_2 (21). A good approximation, in this case, is that $t_2 \sim T_2$ in the sample (23).

9. Special attention must be placed on selecting the image resolution, especially if the correlation gradient has been selected along one of the image-encoding directions. In this case, if the voxel size is comparable to the correlation distance, spurious echoes will be refocused in the acquisition window (**Fig. 5**) which will produce banding artifacts in the image.

10. The sequence can use a Cartesian acquisition scheme or an echo-planar imaging (EPI) acquisition scheme. The EPI acquisition scheme, although much faster than the single line scheme, usually introduces artifacts in reconstructed images, including geometric distortion and ghosting due

to a variety of sources, including off resonance, in-plane flow, and echo misalignment.

11. Usually, a phase cycle is used to suppress unwanted signals from different coherent pathways. For +/− DQC signal, a four-step phase-cycling scheme is usually applied to the first excitation pulse $(x, -x, y, -y)$, while the phase of the mixing pulse, (x), is held constant. At the same time, the receiver phase is cycled $(x, x, -x, -x)$. Excellent suppression of spurious signal contributions from unwanted coherence-transfer pathways can also be achieved with a simple two-step phase cycle, first by pulse $(x, -x)$, and then receiver (x,x), provided the repetition time is much longer than the longitudinal relaxation time of the spin system (TR $> 5T_1$). For selection of the ZQC signal, the first RF pulse has to be cycled through $(x, y, -x, -y)$, while the receiver and other pulse phases should be held constant. However, in this case, the phase cycle of the excitation pulse does not suppress unwanted coherences that are excited for the first time by the mixing pulse. In this case, to suppress the standard signal excited by the mixing pulse, the direction of the correlation gradient pulse should be alternated between z and y or x, while the receiver phase is alternated between x and $-x$ (8).

12. In order to prove that the collected signal is purely iMQC, The user can simply look at the phase of the signal as the direction of the correlation gradient is changed from z to either x or y. If the signal is purely iMQC, the phase of the signal should change and the strength should be reduced more or less by half (**Fig. 6a, b**). Moreover if the direction of the correlation gradient is chosen along the magic angle, the signal should be considerably diminished (**Fig. 6c**).

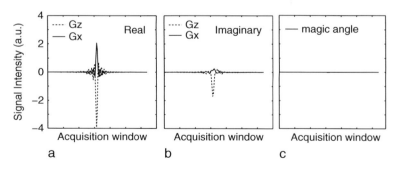

Fig. 6. Comparison of the real (**a**) and imaginary (**b**) parts of the signals obtained from the iDQC CRAZED sequence, selecting the correlation gradient along the z- or x-direction ($G_c = G_x$ and $G_c = G_z$). With respect to the signal obtained with $G_c = G_x$, the signal obtained with $G_c = G_z$ is twice as strong and with an opposite phase. If the correlation gradient is chosen along the direction of the magic angle ($G_z = G_x = G_y$), the iDQC signal disappears (**c**).

These few tests should be done each time a new sequence is programmed into a spectrometer.

13. Theoretically, the iMQC signal should comprise a large fraction of the standard SQC signal.

    When relaxation is included, however, the maximum achievable signal from a typical CRAZED sequence, in the linear regime, is proportional to T_2/τ_d. This means that for samples with a short T_2, as encountered in vivo, signals from intermolecular multiple-quantum coherences (iMQCs) are greatly diminished. This makes it very difficult to perform experiments on regions with T_2 on the order of few milliseconds (liver tissue, for example) unless τ_d is decreased by hyperpolarizing the sample.

Acknowledgments

This work was supported by NIH grant EB 02122.

References

1. Deville, G., Bernier, M., and Delrieux, J. M. (1979) NMR multiple echoes observed in solid He-3. *Phys. Rev. B* **19**, 5666.
2. Einzel, D., Eska, G., Hirayoshi, Y., Kopp, T., and Wolfle, P. (1984) Multiple spin echoes in a normal Fermi liquid. *Phys. Rev. Lett.* **53**, 2312.
3. Bowtell, R., Bowley, R. M., and Glover, P. (1990) Multiple spin echoes in liquids in a high magnetic field. *J. Magn. Reson.* **88**, 643–651.
4. He, Q., Richter, W., Vathyam, S., and Warren, W. S. (1993) Intermolecular multiple-quantum coherences and cross correlations in solution nuclear magnetic resonance. *J. Chem. Phys.* **98**, 6779–6800.
5. Warren, W. S., Richter, W., Andreotti, A. H., and Farmer, B. T. (1993) Generation of impossible cross-peaks between bulk water and biomolecules in solution NMR. *Science* **262**, 2005–2009.
6. Warren, W. S., Lee, S., Richter, W., and Vathyam, S. (1995) Correcting the classical dipolar demagnetizing field in solution NMR. *Chem. Phys. Lett.* **247**, 207–214.
7. Levitt, M. H. (1996) Demagnetization field effects in two-dimensional solution NMR. *Concepts Magn. Reson.* **8**, 77–103.
8. Warren, W. S., Ahn, S., Mescher, M., Garwood, M., Ugurbil, K., Richter, W., Rizi, R. R., Hopkins, J., and Leigh, J. S. (1998) MR imaging contrast enhancement based on intermolecular zero quantum coherences. *Science* **281**, 247–251.
9. Zhong, J., Chen, Z., and Kwok, E. (2000) In vivo intermolecular double-quantum imaging on a clinical 1.5 T MR scanner. *Magn. Reson. Med.* **43**, 335–341.
10. Faber, C., Zahneisen, B., Tippmann, F., Schroeder, A., and Fahrenholz, F. (2007) Gradient-echo and CRAZED imaging for minute detection of Alzheimer plaques in an APPV717I × ADAM10-dn mouse model. *Magn. Reson. Med.* **57**, 696–703.
11. Richter, W., Richter, M., Warren, W. S., Merkle, H., Andersen, P., Adriany, G., and Ugurbil, K. (2000) Functional magnetic resonance imaging with intermolecular multiple-quantum coherences. *Magn. Reson. Imaging* **18**, 489–494.
12. Zhong, J., Kwok, E., and Chen, Z. (2001) fMRI of auditory stimulation with intermolecular double-quantum coherences (iDQCs) at 1.5T. *Magn. Reson. Med.* **45**, 356–364.
13. Gu, T., Kennedy, S., Chen, Z., Schneider, K., and Zhong, J. (2007) Functional MRI at 3T using intermolecular double-quantum coherence (iDQC) with spin-echo (SE)

acquisitions. *Magn. Reson. Mater. Phys.* **20**, 255–264.
14. Schneider, J. T., and Faber, C. (2008) BOLD imaging in the mouse brain using a turboCRAZED sequence at high magnetic fields. *Magn. Reson. Med.* **60**, 850–859.
15. Lin, Y. Y., Ahn, S., Murali, N., Brey, W., Bowers, C. R., and Warren, W. S. (2000) High-resolution, >1 GHz NMR in unstable magnetic fields. *Phys. Rev. Lett.* **85**, 3732.
16. Vathyam, S., Lee, S., and Warren, W. S. (1996) Homogeneous NMR spectra in inhomogeneous fields. *Science* **272**, 92–96.
17. Balla, D. Z., Melkus, G., and Faber, C. (2006) Spatially localized intermolecular zero-quantum coherence spectroscopy for in vivo applications. *Magn. Reson. Med.* **56**, 745–753.
18. Bowtell, R., and Robyr, P. (1996) Structural investigations with the dipolar demagnetizing field in solution NMR. *Phys. Rev. Lett.* **76**, 4971.
19. Ramanathan, C., and Bowtell, R. W. (2002) NMR imaging and structure measurements using the long-range dipolar field in liquids, *Phys. Rev. E* **66**, 041201.
20. Galiana, G., Branca, R. T., Jenista, E. R., and Warren, W. S. (2008) Accurate temperature imaging based on intermolecular coherences in magnetic resonance. *Science* **322**, 421–424.
21. Branca, R. T., Galiana, G., and Warren, W. S. (2007) Signal enhancement in CRAZED experiments. *J. Magn. Reson.* **187**, 38–43.
22. Richter, W., and Warren, S. W. (2000) Intermolecular multiple quantum coherences in liquids. *Concepts Magn. Reson.* **12**, 396–409.
23. Marques, J. P., and Bowtell, R. (2004) Optimizing the sequence parameters for double-quantum CRAZED imaging. *Magn. Reson. Med.* **51**, 148–157.

Section III

Brain Applications

Chapter 14

Experimental Stroke Research: The Contributions of In Vivo MRI

Therése Kallur and Mathias Hoehn

Abstract

Stroke is a disease that develops from the very acute time point of first symptoms during the next several hours and further to a chronic time period of days or even weeks. During this evolution process, a whole series of pathophysiological events takes place. Therefore, the disease is characterized by a continuously changing pathophysiological pattern. In consequence, as the disease develops over time, different imaging modalities must be chosen to accurately describe the status of stroke. In the present chapter, we have divided the evolution of stroke into various dominant steps of the cascade of events, with corresponding time windows. Choice of MRI variables for depiction of the most important aspects during these time windows are presented and their information content is discussed for diagnosis and for investigations into a better understanding of the underlying mechanisms for the disease as well as the relevance of these imaging tools in success assessments for therapeutic strategies.

Key words: Stroke, cerebral blood flow, cytotoxic edema, vasogenic edema, perfusion-weighted MRI, diffusion-weighted MRI, manganese-enhanced MRI (MEMRI), functional brain activation, fMRI.

1. Introduction

The application of magnetic resonance imaging (MRI) to studies of the brain is widely accepted and used in many experimental studies today because of its fantastic battery of measurement sequences leading to information about structural, hemodynamic, metabolic, and even functional aspects of the brain. However, in order to make optimal use of this wide choice of sequences, we believe that the investigator must be closely familiar with the (patho-)physiology of the organ of interest to select the right sequence for the desired information. We, therefore, have split up the chapter into the major blocks of events occurring during cerebral ischemia. We will discuss the relevant MRI application and expected information content for each such pathophysiological time window. For the transparency of the presentation, we have limited the discussion to focal cerebral ischemia, i.e., stroke, as the pathophysiology and, consequently, the selection of optimal sequences may be different for global cerebral ischemia.

Stroke is a leading cause of death and permanent disability worldwide today (1, 2). During an episode of ischemic stroke either an embolic or a thrombotic clot occludes a vessel in the brain, and, thus, deprives the brain territory lying distally from the occluded vessel of its supply of oxygen and glucose. Reduced blood flow is a threat to the survival of cells and, due to the high oxygen and glucose demands of the brain, the cells in the brain are particularly sensitive (3).

The loss of blood flow is more severe in the center and gradually milder in the periphery of the region supplied by the occluded artery. Commonly, the center of the infarction is termed the *core* and the ischemic periphery is termed the *penumbra* (4, 5). Within the ischemic core, during a relatively short period of time, massive cell necrosis occurs, as a consequence of the almost complete disruption of blood flow. The ischemic penumbra is still partially perfused through collateral, non-occluded vessels. Thus, there is a gradual disruption of homeostasis, during which cells are exposed to harmful levels of excitatory neurotransmitters and inflammatory processes and start to undergo apoptosis. Unless interventions to restore blood flow are made, the ischemic penumbra will gradually transform into unrecoverable, necrotic infarction core (6, 7). The time course of major events after an ischemic stroke are schematically depicted in **Fig. 1**.

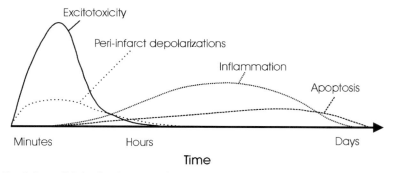

Fig. 1. Immediately after the onset of a focal loss of perfusion, a cascade of detrimental events is initiated. Excitotoxicity lethally damages neurons and glia, and triggers a number of other phenomena, such as peri-infarct depolarizations, inflammation, and apoptosis, all of which contribute to tissue death in the end. Reproduced from (6) with permission.

Stroke can lead to many different symptoms depending on the location of the occlusion, thus brain area affected, commonly resulting in transient or permanent symptoms like complete or partial loss of somatosensory and sensory motor functions, paralysis, aphasia, nausea, and headaches (8). Except for administering recombinant tissue plasminogen activator (rt-PA) as a thrombolytic agent (9) to resolve and/ or to mechanically remove the clot obscuring the blood flow with a MERCI retriever, an embolectomy device (10), there is no other effective treatment for patients suffering a stroke. Thus, not only does stroke tragically affect many people but it also has a severe impact on the society as a whole through hospital care costs, rehabilitation costs, and lost work force capacity from both individuals having a stroke and their families caring and nursing from home (11). Some of the risk factors of suffering a stroke are hypertension, diabetes mellitus, smoking, alcohol abuse, and age. Since the average age of the population is increasing, and the stroke risk doubles with every decade over the age of 50, it is expected that the incidence of stroke also will increase concomitantly (8).

There are several animal models used experimentally in order to study stroke (for an overview see (11)). The most widely used model of stroke is unilateral middle cerebral artery occlusion (MCAO). MCAO can be induced in many ways, such as direct occlusion by permanent ligation after craniotomy (12) and by indirect occlusion by injecting fibrin clots into the common carotid artery (13). However, the most common way is the intraluminal filament occlusion technique, in which a filament is inserted in the common carotid artery, advanced into the circle of Willis and further until reaching the origin of the middle cerebral artery (14). This model, compared to its permanent counterparts, allows for reperfusion at any given time point after occlusion and, thus, the severity of stroke and extent of damage can be adjusted.

Stroke can be described as having an acute phase, during which a massive and rapid cell death occurs, leading to symptoms of functional deficits. Thereafter follows a subacute phase in which slower cell death and secondary phenomena take place, and, finally, a chronic phase which is recognized by tissue reorganization and functional improvement (**Fig. 1**). In the experimental setting, we can use MRI as a tool to study the mechanisms of stroke. The chain of events following an ischemic stroke is very complex and this chapter will not cover an in-depth description of the pathophysiology of stroke. However, we will give a brief overview of the main detrimental events occurring at each of the three above-defined time periods and discuss the ways in which MRI can be used as an instructive and valuable tool in stroke and stroke-related research.

2. MRI Contributions to Stroke Research

2.1. The Acute Phase

2.1.1. Events Occurring Within Minutes and Hours

The reduction in blood flow, due to vascular occlusion, leads to decreased delivery of energy metabolites, such as glucose and oxygen, to the cells in the affected tissue. Rapidly after onset of ischemia, cell respiration fails and the ATP pool is depleted (15). The cells depolarize, since they no longer can maintain the ionic gradients across the membranes (16), and voltage-sensitive ion channels are opened for calcium, sodium, and chloride to enter the cells. Together with these ions, water will stream into the cytoplasm, which causes a swelling of the cells called *cytotoxic edema* (17).

The intracellular calcium overload causes an accumulation of reactive oxygen species, which can degrade DNA and disrupt the electron transport chain in the mitochondria, leading to mitochondrial permeability transition (18). The increased permeability can trigger mitochondria matrix swelling, disruption of the outer membranes and, as a consequence, release of cytochrome C, known to be directly involved in apoptosis (20, 21). Mitochondrial dysfunction is also associated with activation of cyclooxygenase (COX) and nitric oxide synthase (NOS), both of which can be sources of highly reactive free radicals causing biochemical reactions that damage the cells even more (19, 20).

Due to anoxic depolarizations and the presynaptic uptake of calcium, transmitters will be released from intracellular stores. Within minutes after stroke onset, this results in a marked increase in the levels of the excitatory amino acid glutamate in all affected tissue (3, 6). The excess of glutamate in the extracellular space, and the subsequent activation of its ionotropic and metabotropic receptors, produces excitotoxicity since it further increases the intracellular calcium overload (21). This high calcium content within the cells will induce a wide range of detrimental calcium-dependent intracellular signaling cascades, including the activation of endonucleases, proteases, and phospholipases, which degrade the DNA, cytoskeleton, and proteins in the extracellular matrix, and result in cell death (22, 23).

The cells in the core region of the stroke will not be able to repolarize and are irreversibly damaged. However, the cells in the penumbra are still able to repolarize and, thus, undergo cycles of depolarization and repolarization (24). These waves of peri-infarct depolarizations, called spreading depressions, are mediated by increased potassium and glutamate levels, and are initiated by the infarct core but occur in the penumbra where they can travel from their point of origin all over the hemisphere. These spreading depression-like waves are thought to contribute to increased glutamate levels in the penumbra with low oxygen supply, by increased metabolic demand during the repolarizations. This again will, by metabolic exhaustion, recruit the penumbra into the growing infarct core (7, 25–27).

2.1.2. Angiography

Flow through larger vessels can be measured with magnetic resonance angiography (MRA). Techniques of MRA are based on the separation between moving and stationary spins, which can be achieved by applying radiofrequency (rf) pulses to saturate stationary tissue and, thus, enhance the flow-related signal contribution. Since MRA has a relatively low spatial resolution, it has mostly been confined to determine the existence and location of an occlusion or an aneurysm, or the degree of arterial stenosis in humans, having large vessels in the brain.

MRA can, however, be used to visualize the circle of Willis and even smaller branches of the middle cerebral artery (MCA) in both rats and mice (28–30) in three-dimensional datasets within a few minutes scan time. This is obtained without the use of contrast agent employing the time-of-flight (TOF) or inflow angiography. With this method, it is important to use a short echo time, at short repetition time, to make flowing blood much brighter than stationary tissue. It is a fast method providing both information of the potential for recanalization and enabling evaluation of therapeutic success of lysis of obstructing blood clots in experimental studies (31). **Figure 2** illustrates the highly resolved architecture of perfused vessels in the rat brain, both before occlusion, during occlusion by clot emboli, and after treatment with the thrombolytic agent rt-PA.

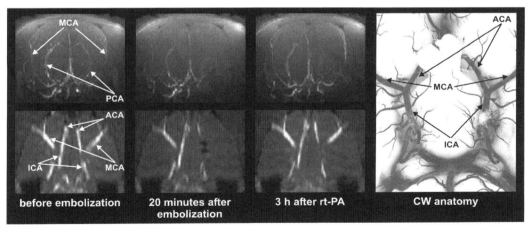

Fig. 2. Maximum intensity projection maps of time-of-flight MR angiograms showing the architecture of perfused intracranial vessels of a Wistar rat. *Upper row*: Coronal projection of the whole brain. *Lower row*: Horizontal projection of the same data set focusing on the circle of Willis. *At right*: Low power photography of the circle of Willis following intravascular perfusion with latex. Angiograms were acquired before (*left column*) and after embolization with autologous clots (*center column*), and 3 h after intraarterial infusion of rtPA. Note the embolic occlusion of the right middle cerebral artery and posterior cerebral artery, followed by full thrombolytic recanalization. Reproduced from (31) with permission.

2.1.3. Perfusion-Weighted Imaging

Together with diffusion-weighted imaging (DWI; *see* **Section 2.1.4**) perfusion-weighted MR imaging (PWI) is the key imaging modality in studies of early stroke evolution, but continues to be important also during later phases of stroke evolution (*see* **Chapter 6**). While MRA depicts the patent or occluded vessel of interest, perfusion-weighted imaging demonstrates the effect of the vessel status on nutrient supply through the microvascular network to the tissue of concern. The ability to image the hemodynamic state of the brain after ischemic stroke is important, since the estimation of regional blood perfusion levels can provide critical information on the development of tissue damage.

The most widely applied MR approaches to assess cerebral microcirculation are, on the one hand, dynamic susceptibility contrast-enhanced (DSC) MRI and steady state susceptibility contrast-enhanced MRI, which measure signal changes after injection of contrast agent, and, on the other hand, arterial spin labeling (ASL) techniques, which use the arterial water as an intrinsic perfusion tracer (32).

After baseline acquisition, DSC, also called bolus track, MRI makes use of rapidly acquired MR images after an intravenous bolus injection of a paramagnetic contrast agent with T_2 and T_2^* susceptibility effects. During the passage of the contrast agent through the vasculature, T_2 and T_2^* are transiently reduced, reflected in a hypointense signal in the vessels and surrounding tissue (33). Thus, the relative loss of signal can be used to quantify cerebral blood flow (CBF) or cerebral blood volume (CBV). Quantification to convert the signal into CBF values requires determination of the arterial input function which sometimes can be difficult when large feeding arteries are not included in the image planes. Alternatively, in experimental context, the technique has been shown to be calibrated, under stable and defined conditions, by CBF autoradiography (34). This perfusion method is, however, less common in pre-clinical research, since repetition of the measurement is limited due to the required intermittent wash out of contrast agent.

The second method, arterial spin labeling (ASL), is completely noninvasive because it uses endogenous blood protons as contrast agent. This technique uses continuous inversion or saturation of inflowing arterial spins entering the region of interest (ROI) (35). In a different approach, it employs a single pulse, such as in a flow-sensitive alternating inversion recovery (FAIR) sequence, to tag spins flowing into the image plane (36). But it must, however, be noted here, that FAIR does not distinguish between arterial inflow and venous flow, but detects both contributions entering the image plane. Comparing the image of tissue perfused by labeled blood protons to the appropriate (untagged) control image results in a T_1-weighted signal difference which is proportional to the amount of inflowing protons or CBF

(38, 40). This method has been used to detect differential perfusion deficits between the ischemic core and penumbral region in rats with focal ischemic stroke induced by an embolic model of MCAO (37). However, ASL is hampered by low signal-to-noise ratios. In addition, the CBF quantification requires knowledge of the apparent T_1 relaxation time, which steadily increases in the acute phase after ischemic stroke. However, recently, a multiecho continuous arterial spin labeling (CASL) sequence and T_2 of the ASL signal were used as an indicator of intra- versus extravascular ratio to incorporate, in the CBF calculations (38), an approach which may improve the accuracy of standard CBF quantification methods (39).

Steady state susceptibility contrast-enhanced MRI is another method with which tissue hemodynamics can be measured continuously (40). By employing T_2- and T_2^*-weighted MRI together with a contrast agent that is not rapidly washed out, but instead remains in the blood for a prolonged time, such as ultrasmall superparamagnetic iron oxide (USPIO) monocrystalline nanoparticles, relative CBV changes can be calculated. This technique has been used to serially measure, by gradient echo and spin echo, the total CBV and microvascular CBV in rats with unilateral MCAO (41). The authors demonstrated that between 1 and 3 h after ischemia there was a steady decrease in total CBV and, in areas destined for infarction, a progressive decline in microvascular CBV was observed. Moreover, gradient and spin echo steady state susceptibility contrast-enhanced MRI are sensitive to different sizes of blood vessels and have shown the potential to detect growth and (re-)distribution of old and newly formed vessels in ROIs (42).

Quantifications using any of the three methods face MRI-intrinsic challenges, since tissue regions with reduced blood flow also exhibit reduced signal intensity. Thus, the greater the perfusion deficit, the more compromised is the tissue perfusion detectability. Even though there are inherent methodological problems in perfusion quantifications, particular strength lies in combining PWI with other modalities such as DWI. A perfusion deficit larger than the lesion delineated by DWI, so-called PWI > DWI mismatch, is used in the clinic and clinical trials to define tissue at risk and selecting patients considered suitable for thrombolysis (43, 44). Essentially, the information given by DWI provides an estimation of tissue with irreversible damage, the core, and PWI provides an estimation of tissue with compromised tissue perfusion, reflecting possibly still reversible damage, with still normal diffusion (45).

2.1.4. Diffusion-Weighted Imaging

Rapidly after vascular occlusion, a cytotoxic edema develops due to the influx of water molecules together with calcium ions into the cells. This leads to a swelling of the cells and shrinkage of the extracellular compartment, but without change in the net water content in the tissue. MRI allows the measurement of incoherent Brownian motion, i.e., self-diffusion of molecules detectable by MR, using diffusion-weighted (DW) MRI. By employing strong magnetic gradient pulses in the rf pulse sequences, DWI can detect diffusion of water in tissue (46). Water diffusion in tissue is influenced by many factors, e.g., cell membranes acting as diffusion barriers. Therefore, the incoherent motion inside a voxel, detected by DWI, is called the *apparent diffusion*. Upon quantification of the effect, one speaks of the apparent diffusion coefficient (ADC).

The shift of water from extracellular to intracellular compartments occurring during the development of cytotoxic edema, results in signal intensity increase on DW MR images, which reflects a reduction in the ADC of brain tissue water (47, 48). Although the exact mechanism is not completely understood, the general assumption is that the ADC will be reduced when water moves to the intracellular compartment with more restricted diffusivity compared to the relatively unrestricted diffusion in the extracellular space (48–50).

A reduction in the tissue water ADC, and a concomitant rise in signal intensity on DWI in the rat brain, can be detected when cerebral blood flow decreases below a threshold of around 30 mL/(100 g/min) (51). This threshold is significantly higher than that required for brain vasogenic (*see* **Section 2.2**) edema, which is close to 10 mL/(100 g/min) and corresponds to the threshold of anoxic depolarization (52). Thus, DWI is a very sensitive tool for detecting the earliest steps of an ischemic event and mild ischemic changes occurring in the penumbral region by recording a gradual decrease of the tissue ADC.

Within the first 15 min following focal brain ischemia in rats, the ADC of brain water declines to around 75% of its normal value. Thereafter, the ADC value further drops to around 60% where it stabilizes in the ensuing next few hours (53). Then, at about 2 days after stroke onset, a "pseudo-normalization" of the ADC value occurs, which can be attributed to the balance of cytotoxic and subsequently developing vasogenic edema, due to an increased extracellular water content (47, 57).

Reductions in ADC can also occur without concomitant ischemia or energy depletion. Diffusion measurements are also sensitive to the abnormal release of neurotransmitters and ions, which takes place during an ischemic brain attack (54). Thus, the measurement of spreading depression-like depolarizations occurring in the ipsilateral cortex in rats subjected to stroke (55), which are accompanied by temporary cell swelling, can be detected as waves of ADC transients by using DWI (29, 30, 36, 60–62).

Whether DWI can provide any prognostic information, such as on the development of the lesion, and differentiate between reversible and irreversible tissue damage or not, remains unsure. The stroke severity is reflected in the degree of ADC decrease. Thus, a comparison of the amount of ADC decrease with breakdown of energy metabolism and with the milder event of tissue acidosis (during anaerobic glycolysis at preserved energy levels) has shown a very stable ADC threshold of 77% of normal ADC for the ischemic core and of 90% for the acidic tissue of the penumbral area (51). But complicating factors are the time for occlusion and reperfusion as well as localization of the lesion, strongly influencing absolute ADC values (56, 57). Therefore, reliable prediction of lesion progression cannot fully be made with DWI or calculated ADC maps alone (58) and whether this is possible at all with only MRI is still debated (52, 59).

2.1.5. Manganese-Enhanced Magnetic Resonance Imaging

Manganese, an essential heavy metal, is an important cofactor in a number of key enzymes in our bodies. The divalent manganese ion, Mn^{2+}, is paramagnetic and is a particularly useful contrast agent for imaging the brain, since it causes a strong reduction of both T_1 and T_2 relaxation times of water in tissue (60) (*see* **Chapter 28**). Mn^{2+} can enter cells in the brain via transport mechanisms of calcium, bind intracellularly to calcium and magnesium binding sites, be transported in axonal tracts, and, in addition, be released into the synaptic cleft along with glutamate (67, 68). As a result of anoxic depolarization caused by ischemic stroke, calcium channels open and calcium and potassium enter the cells, leading to a series of events increasing the calcium influx even further. Therefore, it is expected that anoxic depolarizations will lead to enhanced influx and accumulation of Mn^{2+} in cells when excessive amounts of Mn^{2+} are present in the extracellular space (61).

Manganese-enhanced MRI (MEMRI) has been applied to detect episodes of anoxic depolarizations (62) and cortical spreading depression (63, 64). Aoki and colleagues (62) infused $MnCl_2$ into the carotid artery of rats, permanently occluded their MCA, and recorded multislice T_1-weighted coronal images. Compared to animals in a sham group, they found that the signal intensity was markedly increased in the lateral caudate-putamen and in the lateroventral cortex (areas supplied by the MCA in rats) in animals with stroke. However, the high-intensity area on MEMRI was much smaller than both the measured DWI hyperintensity and ADC map decrease (**Fig. 3**). Therefore, the use of MEMRI appears to detect even earlier events in the acute ischemic phase than DWI, thus focusing on the early depolarized cells while DWI, at this early point, also will encompass all gradual increase of cytotoxic edema in cells not necessarily depolarized although hampered. Aoki and colleagues suggest that MEMRI may be a useful tool in the "super-acute" phase, to define the true ischemic core in animal models of stroke.

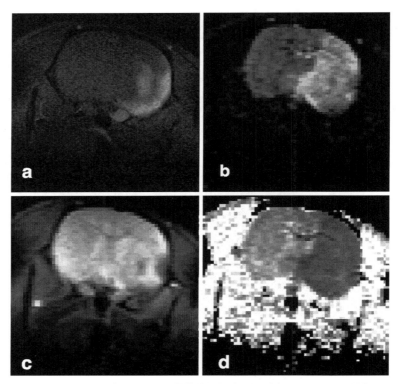

Fig. 3. A comparison of MEMRI (**a**), DWI (**b**), T_2-weighted image (**c**), and ADC map (**d**) using the same model of experimental stroke. The lesion, localized in the lateral caudate-putamen and the lateral cortex, appears as a high-intensity area on MEMRI (**a**) and that area is comparatively smaller than both the DWI hyperintensity (**b**) and decreased ADC (**d**), whereas only a slight decrease in signal intensity was observed on the T_2-weighted image (**c**). Reproduced from Ref. (61) with permission.

One of the major drawbacks of using Mn^{2+} as a contrast agent is its cellular toxicity, known to produce neurological disorders, hepatic failure, and cardiac complications upon chronic exposure. On the other hand, in order to produce detectable and stable contrast, significant amount of Mn^{2+} is a requirement, since MR relaxation rates are proportional to the concentration of Mn^{2+} in the tissue. Therefore, a careful titration of Mn^{2+} concentration must be performed, to achieve best possible contrast without compromising safety (64) or tissue integrity, in the case of local administration (65).

2.2. The Subacute and Chronic Phases

2.2.1. Events Occurring Within Days and Weeks

Initially after a focal ischemic stroke, cells die mainly through necrosis. Subsequently, the blood–brain barrier (BBB) breaks down and plasma proteins such as albumin infiltrate the brain parenchyma, leading to *vasogenic edema* (74, 75). Vasogenic edema increases the tissue water content in the extracellular compartment and peaks at about 1–2 days after the ischemic event.

The volume of water in the tissue is related to the extent of damage and to the degree of BBB impairment, and can lead to a dangerous compression of the midbrain, which calls for urgent craniectomy in the clinic (66). However, at 1 or 2 days following ischemic stroke, more and more cells, particularly in the boundary zones of the lesion, exhibit fragmented DNA typical for apoptosis (67). Caspases are known to be instrumental for the effectuation of apoptosis, and when activated they initiate a series of events, called programmed cell death, that will break down the cell components, shrink the cell, and, in the end, result in phagocytosis of the cell (68).

In addition, during the hours and days following an ischemic insult, the lesioned tissue will react to the insult by gene expression alterations, gliosis, and various inflammatory responses originating from either the central or the peripheral nervous system. These reactions are all highly connected with each other and will influence not only cell death and survival but also tissue reorganization and plasticity, and functional improvement.

Activation of glial cells, gliosis, occurs after many types of lesion in the brain and involves changes of morphology and gene expression, as well as increased proliferation of both astrocytes and microglia (69). Reactive astrocytes are primarily located at the border of the lesion and contribute to the formation of a gliotic scar. Microglia become activated, transform into phagocytes, and produce toxic metabolites and pro-inflammatory cytokines and chemokines such as tumor necrosis factor α (TNFα) and interleukin 1β (IL-1β). However, microglia play a dual role in brain damage by also having a positive effect on cell survival and neurogenesis, depending on the activation mechanism of microglia (70, 71).

Inflammatory mediators, e.g., TNFα and IL-1β, in combination with the enhanced expression of adhesion molecules on the vasculature, facilitate the invasion of inflammatory leukocytes from the periphery into the ischemic brain (72). Invading inflammatory leukocytes are cytotoxic, exert phagocytosis, and produce both antibodies and pro-inflammatory cytokines, all of which may cause increased cell death (69).

Functional, spontaneous recovery after ischemic stroke is frequently observed during the first weeks, but can progress during 6 months or possibly more (73). Reperfusion can explain recovery during the first days, when reversibly damaged regions are re-activated (6). Later achieved recovery is better explained by tissue reorganization and plasticity, although the mechanisms are not fully known. Intact areas close to the lesion may take over the function of the damaged areas or may be mediated by growing and sprouting of new connections during longer periods.

2.2.2. Relaxometry

The most commonly used MRI methods for observing stroke in the subacute and chronic phases are proton density, T_1-, and T_2-weighted MRI. An increase in proton density and in T_1- and T_2-weighted images in pathologic tissue is attributed primarily to an increase in interstitial water associated with the development of vasogenic edema (74). Proton density and T_2-weighted images become hyperintense within a few hours after stroke onset in rats. Depending on stroke severity and reperfusion conditions, the hyperintensity reaches maximal value at around 24 h (75, 76). As T_1 also tends to increase early after stroke onset (but only mildly compared to the increase in T_2), this will lead to a slight hypointensity in T_1-weighted images. The contrast between ischemic lesion and normal tissue is, however, only weak on T_1-weighted images. The varying relaxation-dependent contrast situation of T_1-, T_2-, and T_2^*-weighting, together with the BBB tracer dimegluminegadopentetate (Gd-DTPA) (see below) is depicted in **Fig. 4**.

Due to its sensitivity, T_2-weighted MRI is extensively applied in both clinical and experimental studies to delineate lesions and, thus, estimate lesion size or infarct volume. Within the first days after stroke, T_2 can be used to detect edema and edema-related anatomical changes, such as ventricular compression and midline shift, and ventricular enlargement in the chronic phase. With

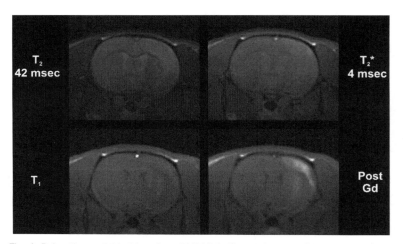

Fig. 4. Relaxation-weighted imaging of MCAO in the rat. In coronal images during the chronic period, six days after stroke induction, the hyperintensity on the T_2-weighted image (top left) indicates the spread of the vasogenic edema and the beginning cystic transformation. This is also well reflected in the T_2^*-weighted image (top right), while there is very little contrast noticeable on the native T_1-weighted image (bottom left). After systemic Gd-DTPA injection, the hyperintensity of the T_1-weighted image allows detection of a disturbed BBB, generated by the leakage of the contrast agent from the vascular compartment into the parenchyma (reproduction courtesy of Tracy D. Farr, Cologne, unpublished).

time, the T_2-weighted signal intensity usually either stabilizes or even increases, when the damaged cells are gradually removed by macrophages and replaced by cerebrospinal fluid.

Although increase and prolongation of T_1 and T_2 are primarily associated with late and irreversible damage, early T_1 and T_2 changes have been observed in the (hyper-) acute stages of experimental ischemic stroke. By using quantitative MR relaxometry, i.e., T_1 and T_2 mapping, it is possible to detect damage earlier than previously anticipated with these parameters (56, 57, 77). In addition, Wegener and colleagues (88) could discriminate between pannecrosis and selective neuronal death in subcortical tissues after transient MCAO, based on T_2 changes over time. They compared T_2 values for animals with either cortical and subcortical or only subcortical damage. Interestingly, the T_2 values returned to normal within about 2 weeks in animals with only subcortical lesions, which was associated with selective neuronal death according to histological analysis. In the larger lesions, on the contrary, T_2 normalization was only transient, followed by secondary T_2 increase. This was related to pannecrosis and cystic transformation of the subcortical lesion. Thus, T_2 mapping can be a useful tool in both diagnosing early determined lesions and in predicting tissue fate at later stages.

Gadolinium-based compounds such as Gd-DTPA are paramagnetic contrast agents, whose dominant effect is observed on T_1 shortening. Normally, the extravasation of injected contrast agent into the parenchyma is prevented by the BBB. But after stroke, the BBB is disrupted and therefore the contrast agent can leak from the vasculature to the brain parenchyma, leading to a local signal enhancement in T_1-weigthed MRI (**Fig. 4**, bottom right). The use of exogenous contrast agents to achieve T_1-weighted contrast has, ever since it was first reported (78), gained widespread use for the study of BBB integrity.

T_2^* imaging also plays an important role in imaging of cerebral ischemia, in particular following the inflammatory response occurring after an ischemic insult to the brain. In order to image the inflammatory response with MRI, contrast agents that shorten longitudinal and transversal relaxation times, are applied. Superparamagnetic iron oxide (SPIO) polycrystalline nanoparticles and USPIOs use magnetite (Fe_3O_4) for iron core and have different hydrophilic coatings for solubility (79). Inflammatory mature leukocytes derived from and activated in the periphery, e.g., blood-borne macrophages, enter the brain via the disrupted BBB. The macrophages are phagocytotic in nature and will, thereby, take up intravenously delivered contrast agent and carry it into the leaky stroke-damaged brain for a macrophage-based contrast. This contrast will be primarily negative but can also become positive depending on the choice of sequence (80–82). However, interpretation of data must be made with caution, since erythrocyte accumulation, i.e., vessels and microbleeds, also produce similar T_2^* effects (83). Moreover, there is still an ongoing debate about the specificity of this effect and whether systemically injected contrast agent can accumulate independent of macrophage infiltration, due to a disrupted BBB or contaminated cerebrospinal fluid (81).

2.2.3. Functional Magnetic Resonance Imaging

Functional magnetic resonance imaging (fMRI) techniques measure the hemodynamic response to neural activity in the active animal brain responding to a stimulus, e.g., during sensory stimulation. Animal fMRI studies allow the assessment of spatial and temporal dynamics of brain recovery, plasticity, and functional reorganization in the chronic phase following MCAO.

Changes in blood flow and blood oxygenation are closely linked to neuronal activity. Upon neuronal activation, an elevation in the metabolic rate results in an increased demand of oxygen, carried by hemoglobin in erythrocytes, and, thus, in the local CBF. However, the increased supply of oxygen exceeds the consumption of oxygen in the activated area. Consequently, the local blood oxygenation increases, leading to an increase of blood oxygenation level-dependent (BOLD) MR signal. The basis for BOLD fMRI is the magnetic property of blood, which is dependent upon the oxygenation state of hemoglobin (84). Deoxygenated hemoglobin is paramagnetic while oxygenated hemoglobin is diamagnetic. Deoxygenation results in increased local magnetic susceptibility differences between intra- and extravascular compartments, thereby causing signal reduction on T_2- and T_2^*-weighted MR images. Although very similar to the $T_2^{(*)}$-shortening seen with exogenously delivered contrast agent, the susceptibility changes caused by deoxyhemoglobin are much smaller.

Upon brain activation, the regional CBF and CBV increases (neurovascular coupling) and this activation-induced increase in perfusion can be measured with CBF- or CBV-weighted MRI. Steady state contrast-enhanced CBV-weighted fMRI studies have described both extensive activity on the side contralateral to the stroke-damaged hemisphere and perilesional activation upon contralateral forelimb stimulation, after both transient and permanent MCAO in rats, thus, providing some evidence for poststroke brain reorganization (96–98). Unfortunately, these studies did not monitor the temporal changes in the same animals, due to the use of the anesthetic α-chloralose, which supposedly preserves vascular–metabolic coupling (85), but is not well suited for survival protocols because of its severe side effects. Instead, the authors have investigated separate animal groups at different survival times after MCAO, thus, eliminating the advantage of longitudinal noninvasive imaging as a method. Therefore, an anesthesia protocol has recently been established, utilizing the α2-adrenoreceptor medetomidine, which permits uncompromised longitudinal fMRI studies after experimental stroke (86, 87).

Although BOLD fMRI is sensitive to small activation-induced changes, it is also sensitive to other factors, such as macrovascular inflow, and is dependent on multiple hemodynamic and metabolic factors, i.e., oxygen metabolism, CBF, and CBV. Despite those drawbacks, it is the most frequently applied fMRI method, especially in human studies, because of the relative ease with which data can be obtained. By employing BOLD fMRI, Weber and colleagues (88) were able, with the medetomidine protocol, to repetitively measure rats subjected to stroke. They found that a subcortical (striatal) lesion induced by unilateral, transient MCAO, initially obstructed activation of the overlying somatosensory cortex upon forepaw stimulation contralateral to the ischemic hemisphere. However, after a few weeks, as the hyperintensity on the T_2 map in the striatum had almost resolved, the activation in the somatosensory cortex reappeared in several animals. This was in full accord with electrophysiological measurements of evoked potentials showing that the coupling between neuronal activity and hemodynamic response remains preserved under those pathophysiological conditions. The functional recovery observed with fMRI was, furthermore, supported by functional improvements assessed with behavioral tests. Thus, fMRI has been proven a very valuable tool to investigate restoration of normal brain function in the somatosensory cortex after transient stroke, be it spontaneous or therapeutically induced (88).

2.3. General Remarks

Various MRI sequences are very helpful in elucidating particular mechanisms and in detecting individual pathophysiological conditions during the (spontaneous) development of the lesion. It should, however, be noted that the description given above, was kept somewhat simplified due to the difficulty of describing the complexity of the whole disease evolution. It goes without saying that meticulous MRI execution is a necessary but not sufficient criterion for a successful characterization of the lesion status and the disease evolution. In order to avoid ambiguous or even faulty interpretation of the MRI data, extensive and continuous monitoring of the physiological status of the animal during the MRI experiment is indispensible. Thus, e.g., impairment of the vascular reactivity (89) or vascular reserve capacity may influence results obtained from perfusion measurements but also from functional brain activation studies (90).

Another issue that can become of importance in generating high contrast of the ischemic tissue against the healthy brain is the interplay of different contrast mechanisms. These may in some cases lead to countering the intensity change of one parameter

Table 1
Selection of optimal MRI conditions for image-based presentation of pathophysiological mechanisms at different phases of ischemic lesion evolution

Acute phase (minutes to hours after onset)	Vascular obstruction Vessel patency Localization of blood clot	MR angiography
	Tissue perfusion status Microvascular perfusion Perfusion deficit	Perfusion-weighted MRI: (a) dynamic contrast enhanced (DCE) MRI (b) arterial spin labeling (ASL) MRI
	Ion and water homeostasis Energy metabolic integrity	Diffusion-weighted MRI, ADC
	Glutamate toxicity Ion channel integrity	MEMRI
	Vasogenic edema	T_2-weighted MRI
	Integrity or impairment of blood–brain barrier (BBB)	T_1-weighted MRI with Gd-DTPA
Subacute phase (days after onset)	Brain swelling Vasogenic edema	T_2-weighted MRI
	Reperfusion	Perfusion-weighted MRI (a) DCE-MRI (b) ASL-MRI
	Severity of lesion; loss of function	fMRI
Chronic phase (weeks after onset)	Tissue perfusion status	Perfusion-weighted MRI (a) DCE-MRI (b) ASL-MRI
	Pannecrosis Selective neuronal death	Temporal profile of T_2 changes
	Functional recovery Brain plasticity Reorganization	fMRI

by another. Thus, as an example, it must be noted that for optimal contrast in diffusion-weighted MRI of stroke, one should avoid reducing TR too far because the T_1 increase in the ischemic territory will lead to signal reduction of T_1-weighted diffusion-weighting, i.e., the hyperintensity generated by the diffusion-weighting will be counteracted by the T_1-weighting. This can always be avoided completely by relying on quantitative parameter maps, where potential negative influence of other MRI variables on the contrast of the variable of choice is excluded.

Also, despite the merits of many of the described sequences for characterizing particular pathophysiological conditions, individual sequences will often fail when used for prognosis of outcome. It has many times been tried to derive prognostic values based on MRI parameter alteration at an early time point. But as shown in many animal experimental studies, it requires combinations of parameters at minimum to allow some outcome prediction with a minimal requested reliability (91).

3. Summary

As has been demonstrated above, the evolving pathophysiological, hemodynamic, and metabolic characteristic patterns will change with the development of the ischemic lesion over time. Therefore, both, the description of the stroke at different phases and the selection of the particular patho-mechanisms of investigation will determine the optimal choice of measurement sequences for the most reliable images with the best contrast. The essential steps of the stroke development, the most intriguing parameters to study at the corresponding time phases, and the consequent choice of MRI variables for characterization are therefore compiled in **Table 1**.

Acknowledgments

The authors gratefully acknowledge the support from the European Union FP6 and FP7 programs (LSHB-CT-2006-037526, StemStroke) and (HEALTH-F5-2008-201842, ENCITE).

References

1. Latchaw, R. E., Alberts, M. J., Lev, M. H., Connors, J. J., Harbaugh, R. E., Higashida, R. T., Hobson, R., Kidwell, C. S., Koroshetz, W. J., Mathews, V., Villablanca, P., Warach, S. and Walters, B. (2009) Recommendations for imaging of acute ischemic stroke. A scientific statement from the American Heart Association. *Stroke* **40**, 3646–3678.
2. Murray, C. J. and Lopez, A. D. (1997) Global mortality, disability, and the contribution of risk factors: Global burden of disease study. *Lancet* **349**, 1436–1442.
3. Lipton, P. (1999) Ischemic cell death in brain neurons. *Physiol. Rev.* **79**, 1431–1568.
4. Astrup, J., Siesjo, B. K. and Symon, L. (1981) Thresholds in cerebral ischemia – the ischemic penumbra. *Stroke* **12**, 723–725.
5. Astrup, J., Symon, L., Branston, N. M. and Lassen, N. A. (1977) Cortical evoked potential and extracellular K+ and H+ at critical levels of brain ischemia. *Stroke* **8**, 51–57.
6. Dirnagl, U., Iadecola, C. and Moskowitz, M. A. (1999) Pathobiology of ischaemic stroke: An integrated view. *Trends Neurosci.* **22**, 391–397.
7. Hossmann, K.-A. (1998) Thresholds of ischemic injury. In: M. D. Ginsberg and J. Bogousslavsky (eds.), Cerebrovascular Disease: Pathophysiology, Diagnosis and Management, pp. 193–204. Blackwell Science, Cambridge, MA.
8. Fatahzadeh, M. and Glick, M. (2006) Stroke: Epidemiology, classification, risk factors, complications, diagnosis, prevention,

and medical and dental management. *Oral Surg., Oral Med., Oral Pathol., Oral Radiol. and Endod.* **102**, 180–191.
9. National Institute of Neurological Disorders and Stroke, r.-P. S. S. G. (1995) Tissue plasminogen activator for acute ischemic stroke. *New Engl. J. Med.* **333**, 1581–1587.
10. Smith, W. S., Sung, G., Starkman, S., Saver, J. L., Kidwell, C. S., Gobin, Y. P., Lutsep, H. L., Nesbit, G. M., Grobelny, T., Rymer, M. M., Silverman, I. E., Higashida, R. T., Budzik, R. F. and Marks, M. P. (2005) Safety and efficacy of mechanical embolectomy in acute ischemic stroke: Results of the MERCI trial. *Stroke* **36**, 1432–1438.
11. Hossmann, K.-A. (2008) Cerebral ischemia: models, methods and outcomes. *Neuropharmacology* **55**, 257–270.
12. Tamura, A., Graham, D. I., McCulloch, J. and Teasdale, G. M. (1981) Focal cerebral ischaemia in the rat: 1. Description of technique and early neuropathological consequences following middle cerebral artery occlusion. *J. Cerebr. Blood Flow Metab.* **1**, 53–60.
13. Kudo, M., Aoyama, A., Ichimori, S. and Fukunaga, N. (1982) An animal model of cerebral infarction. Homologous blood clot emboli in rats. *Stroke* **13**, 505–508.
14. Koizumi, J., Yoshida, Y., Nakazawa, T. and Ooneda, G. (1986) Experimental studies of ischemic brain edema. 1. A new experimental model of cerebral embolism in rats in wich recirculation can be introduced in the ischemic area. *Japanese J. Stroke* **8**, 1–8.
15. Siesjö, B., Kristian, T. and Katsura, K. (1998) Overview of bioenergetic failure and metabolic cascades in brain ischemia. In: M. D. Ginsberg and J. Bogousslavsky (eds.), Cerebrovascular Disease: Pathophysiology, Diagnosis and Management, pp. 1–13. Blackwell Science, Cambridge, MA.
16. Wu, A. and Fujikawa, D. G. (2002) Effects of AMPA-receptor and voltage-sensitive sodium channel blockade on high potassium-induced glutamate release and neuronal death in vivo. *Brain Res.* **946**, 119–129.
17. Hara, M. R. and Snyder, S. H. (2007) Cell signaling and neuronal death. *Annu. Rev. Pharm. Tox.* **47**, 117–141.
18. Siesjo, B. K., Hu, B. and Kristian, T. (1999) Is the cell death pathway triggered by the mitochondrion or the endoplasmic reticulum? *J. Cerebr. Blood Flow Metab.* **19**, 19–26.
19. Samdani, A. F., Dawson, T. M. and Dawson, V. L. (1997) Nitric oxide synthase in models of focal ischemia. *Stroke* **28**, 1283–1288.
20. White, B. C., Sullivan, J. M., DeGracia, D. J., O'Neil, B. J., Neumar, R. W., Grossman, L. I., Rafols, J. A. and Krause, G. S. (2000) Brain ischemia and reperfusion: Molecular mechanisms of neuronal injury. *J. Neurol. Sci.* **179**, 1–33.
21. Choi, D. W. (1994) Calcium and excitotoxic neuronal injury. *Ann. N. Y. Acad. Sci.* **747**, 162–171.
22. Artal-Sanz, M. and Tavernarakis, N. (2005) Proteolytic mechanisms in necrotic cell death and neurodegeneration. *FEBS Lett.* **579**, 3287–3296.
23. Leker, R. R. and Shohami, E. (2002) Cerebral ischemia and trauma – different etiologies yet similar mechanisms: Neuroprotective opportunities. *Brain Res. and Brain Res. Rev.* **39**, 55–73.
24. Nedergaard, M. and Hansen, A. J. (1993) Characterization of cortical depolarizations evoked in focal cerebral ischemia. *J. Cerebr. Blood Flow Metabol.* **13**, 568–574.
25. Mies, G., Iijima, T. and Hossmann, K.-A. (1993) Correlation between peri-infarct DC shifts and ischaemic neuronal damage in rat. *Neuroreport* **4**, 709–711.
26. Busch, E., Gyngell, M. L., Eis, M., Hoehn-Berlage, M. and Hossmann, K.-A. (1996) Potassium induced cortical spreading depressions during focal cerebral ischemia in rats: Contribution to lesion growth assessed by diffusion-weighted NMR and biochemical imaging. *J. Cerebr. Blood Flow Metabol.* **16**, 1090–1099.
27. Busch, E., Hoehn-Berlage, M., Eis, M., Gyngell, M. L. and Hossmann, K.-A. (1995) Simultaneous recording of EEG, DC potential and diffusion-weighted NMR imaging during potassium induced cortical spreading depression in rats. *NMR Biomed.* **8**, 59–64.
28. Beckmann, N., Stirnimann, R. and Bochelen, D. (1999) High-resolution magnetic resonance angiography of the mouse brain: Application to murine focal cerebral ischemia models. *J. Magn. Reson.* **140**, 442–450.
29. Besselmann, M., Liu, M., Diedenhofen, M., Franke, C. and Hoehn, M. (2001) MR angiographic investigation of transient focal cerebral ischemia in rat. *NMR Biomed.* **14**, 289–296.
30. Reese, T., Bochelen, D., Sauter, A., Beckmann, N. and Rudin, M. (1999) Magnetic resonance angiography of the rat cerebrovascular system without the use of contrast agents. *NMR Biomed.* **12**, 189–196.
31. Hilger, T., Niessen, F., Diedenhofen, M., Hossmann, K.-A. and Hoehn, M. (2002) Magnetic resonance angiography of thromboembolic stroke in rats: Indicator of

recanalization probability and tissue survival after recombinant tissue plasminogen activator treatment. *J. Cerebr. Blood Flow Metabol.* **22**, 652–662.

32. Calamante, F., Thomas, D. L., Pell, G. S., Wiersma, J. and Turner, R. (1999) Measuring cerebral blood flow using magnetic resonance imaging techniques. *J. Cerebr. Blood Flow Metabol.* **19**, 701–735.

33. Dijkhuizen, R. M. and Nicolay, K. (2003) Magnetic resonance imaging in experimental models of brain disorders. *J. Cerebr. Blood Flow Metabol.* **23**, 1383–1402.

34. Wittlich, F., Kohno, K., Mies, G., Norris, D. G. and Hoehn-Berlage, M. (1995) Quantitative measurement of regional blood flow with gadolinium diethylenetriaminepentaacetate bolus track NMR imaging in cerebral infarcts in rats: Validation with the iodo[^{14}C]antipyrine technique. *Proc. Natl. Acad. Sci. USA* **92**, 1846–1850.

35. Detre, J. A., Leigh, J. S., Williams, D. S. and Koretsky, A. P. (1992) Perfusion imaging. *Magn. Reson. Med.* **23**, 37–45.

36. Kim, S. G. (1995) Quantification of relative cerebral blood flow change by flow-sensitive alternating inversion recovery (FAIR) technique: application to functional mapping. *Magn. Reson. Med.* **34**, 293–301.

37. Busch, E., Krüger, K., Allegrini, P. R., Kerskens, C. M., Gyngell, M. L., Hoehn-Berlage, M. and Hossmann, K.-A. (1998) Reperfusion after thrombolytic therapy of embolic stroke in the rat: Magnetic resonance and biochemical imaging. *J. Cerebr. Blood Flow Metabol.* **18**, 407–418.

38. Wells, J. A., Lythgoe, M. F., Choy, M., Gadian, D. G., Ordidge, R. J. and Thomas, D. L. (2009) Characterizing the origin of the arterial spin labelling signal in MRI using a multiecho acquisition approach. *J. Cerebr. Blood Flow Metabol.* **29**, 1836–1845.

39. Buxton, R. B., Frank, L. R., Wong, E. C., Siewert, B., Warach, S. and Edelman, R. R. (1998) A general kinetic model for quantitative perfusion imaging with arterial spin labeling. *Magn. Reson. Med.* **40**, 383–396.

40. Hamberg, L. M., Boccalini, P., Stranjalis, G., Hunter, G. J., Huang, Z., Halpern, E., Weisskoff, R. M., Moskowitz, M. A. and Rosen, B. R. (1996) Continuous assessment of relative cerebral blood volume in transient ischemia using steady state susceptibility-contrast MRI. *Magn. Reson. Med.* **35**, 168–173.

41. Zaharchuk, G., Yamada, M., Sasamata, M., Jenkins, B. G., Moskowitz, M. A. and Rosen, B. R. (2000) Is all perfusion-weighted magnetic resonance imaging for stroke equal? The temporal evolution of multiple hemodynamic parameters after focal ischemia in rats correlated with evidence of infarction. *J. Cerebr. Blood Flow Metabol.* **20**, 1341–1351.

42. Dennie, J., Mandeville, J. B., Boxerman, J. L., Packard, S. D., Rosen, B. R. and Weisskoff, R. M. (1998) NMR imaging of changes in vascular morphology due to tumor angiogenesis. *Magn. Reson. Med.* **40**, 793–799.

43. Hjort, N., Butcher, K., Davis, S. M., Kidwell, C. S., Koroshetz, W. J., Rother, J., Schellinger, P. D., Warach, S. and Ostergaard, L. (2005) Magnetic resonance imaging criteria for thrombolysis in acute cerebral infarct. *Stroke* **36**, 388–397.

44. Pantano, P., Totaro, P. and Raz, E. (2008) Cerebrovascular diseases. *Neurol. Sci.* **29**(Suppl 3), 314–318.

45. Baird, A. E., Benfield, A., Schlaug, G., Siewert, B., Lovblad, K. O., Edelman, R. R. and Warach, S. (1997) Enlargement of human cerebral ischemic lesion volumes measured by diffusion-weighted magnetic resonance imaging. *Ann. Neurol.* **41**, 581–589.

46. Le Bihan, D., Breton, E., Lallemand, D., Aubin, M. L., Vignaud, J. and Laval-Jeantet, M. (1988) Separation of diffusion and perfusion in intravoxel incoherent motion MR imaging. *Radiology* **168**, 497–505.

47. Benveniste, H., Hedlund, L. W. and Johnson, G. A. (1992) Mechanism of detection of acute cerebral ischemia in rats by diffusion-weighted magnetic resonance microscopy. *Stroke* **23**, 746–754.

48. Moseley, M. E., Cohen, Y., Mintorovitch, J., Chileuitt, L., Shimizu, H., Kucharczyk, J., Wendland, M. F. and Weinstein, P. R. (1990) Early detection of regional cerebral ischemia in cats: Comparison of diffusion- and T_2-weighted MRI and spectroscopy. *Magn. Reson. Med.* **14**, 330–346.

49. Hoehn-Berlage, M. (1995) Diffusion-weighted NMR imaging: application to experimental focal cerebral ischemia. *NMR Biomed.* **8**, 345–358.

50. Norris, D. G. (2001) The effects of microscopic tissue parameters on the diffusion weighted magnetic resonance imaging experiment. *NMR Biomed.* **14**, 77–93.

51. Hoehn-Berlage, M., Norris, D. G., Kohno, K., Mies, G., Leibfritz, D. and Hossmann, K.-A. (1995) Evolution of regional changes in apparent diffusion coefficient during focal ischemia of rat brain: The relationship of quantitative diffusion NMR imaging to reduction in cerebral blood flow and metabolic disturbances. *J. Cerebr. Blood Flow Metabol.* **15**, 1002–1011.

52. Hossmann, K.-A. (2006) Pathophysiology and therapy of experimental stroke. *Cell. Mol. Neurobiol.* **26**, 1057–1083.
53. Sotak, C. H. (2004) Nuclear magnetic resonance (NMR) measurement of the apparent diffusion coefficient (ADC) of tissue water and its relationship to cell volume changes in pathological states. *Neurochem. Int.* **45**, 569–582.
54. Zoli, M., Jansson, A., Sykova, E., Agnati, L. F. and Fuxe, K. (1999) Volume transmission in the CNS and its relevance for neuropsychopharmacology. *Trends Pharmacol. Sci.* **20**, 142–150.
55. Mies, G., Kohno, K. and Hossmann, K.-A. (1994) Prevention of periinfarct direct current shifts with glutamate antagonist NBQX following occlusion of the middle cerebral artery in the rat. *J. Cerebr. Blood Flow Metabol.* **14**, 802–807.
56. Olah, L., Wecker, S. and Hoehn, M. (2000) Secondary deterioration of apparent diffusion coefficient after 1-hour transient focal cerebral ischemia in rats. *J. Cerebr. Blood Flow Metabol.* **20**, 1474–1482.
57. Olah, L., Wecker, S. and Hoehn, M. (2001) Relation of apparent diffusion coefficient changes and metabolic disturbances after 1 hour of focal cerebral ischemia and at different reperfusion phases in rats. *J. Cerebr. Blood Flow Metabol.* **21**, 430–439.
58. Pillekamp, F., Grüne, M., Brinker, G., Franke, C., Uhlenküken, U., Hoehn, M. and Hossmann, K.-A. (2001) Magnetic resonance prediction of outcome after thrombolytic treatment. *Magn. Reson. Imaging* **19**, 143–152.
59. Fiehler, J., Fiebach, J. B., Gass, A., Hoehn, M., Kucinski, T., Neumann-Haefelin, T., Schellinger, P. D., Siebler, M., Villringer, A. and Rother, J. (2002) Diffusion-weighted imaging in acute stroke – a tool of uncertain value? *Cerebrovasc. Dis.* **14**, 187–196.
60. Silva, A. C. and Bock, N. A. (2008) Manganese-enhanced MRI: An exceptional tool in translational neuroimaging. *Schizophr. Bull.* **34**, 595–604.
61. Aoki, I., Naruse, S. and Tanaka, C. (2004) Manganese-enhanced magnetic resonance imaging (MEMRI) of brain activity and applications to early detection of brain ischemia. *NMR Biomed.* **17**, 569–580.
62. Aoki, I., Ebisu, T., Tanaka, C., Katsuta, K., Fujikawa, A., Umeda, M., Fukunaga, M., Takegami, T., Shapiro, E. M. and Naruse, S. (2003) Detection of the anoxic depolarization of focal ischemia using manganese-enhanced MRI. *Magn. Reson. Med.* **50**, 7–12.
63. Henning, E. C., Meng, X., Fisher, M. and Sotak, C. H. (2005) Visualization of cortical spreading depression using manganese-enhanced magnetic resonance imaging. *Magn. Reson. Med.* **53**, 851–857.
64. Silva, A. C., Lee, J. H., Aoki, I. and Koretsky, A. P. (2004) Manganese-enhanced magnetic resonance imaging (MEMRI): Methodological and practical considerations. *NMR Biomed.* **17**, 532–543.
65. Soria, G., Wiedermann, D., Justicia, C., Ramos-Cabrer, P. and Hoehn, M. (2008) Reproducible imaging of rat corticothalamic pathway by longitudinal manganese-enhanced MRI (L-MEMRI). *Neuroimage* **41**, 668–674.
66. Walz, B., Zimmermann, C., Bottger, S. and Haberl, R. L. (2002) Prognosis of patients after hemicraniectomy in malignant middle cerebral artery infarction. *J. Neurol.* **249**, 1183–1190.
67. Yuan, J., Lipinski, M. and Degterev, A. (2003) Diversity in the mechanisms of neuronal cell death. *Neuron* **40**, 401–413.
68. Honig, L. S. and Rosenberg, R. N. (2000) Apoptosis and neurologic disease. *Am. J. Med.* **108**, 317–330.
69. Stoll, G., Jander, S. and Schroeter, M. (1998) Inflammation and glial responses in ischemic brain lesions. *Prog. Neurobiol.* **56**, 149–171.
70. Hanisch, U. K. and Kettenmann, H. (2007) Microglia: Active sensor and versatile effector cells in the normal and pathologic brain. *Nat. Neurosci.* **10**, 1387–1394.
71. Stoll, G., Jander, S. and Schroeter, M. (2002) Detrimental and beneficial effects of injury-induced inflammation and cytokine expression in the nervous system. *Adv. Exp. Med. Biol.* **513**, 87–113.
72. Huang, J., Upadhyay, U. M. and Tamargo, R. J. (2006) Inflammation in stroke and focal cerebral ischemia. *Surg. Neurol.* **66**, 232–245.
73. Barak, S. and Duncan, P. W. (2006) Issues in selecting outcome measures to assess functional recovery after stroke. *NeuroRx* **3**, 505–524.
74. van Bruggen, N., Roberts, T. P. and Cremer, J. E. (1994) The application of magnetic resonance imaging to the study of experimental cerebral ischaemia. *Cereb. Brain Metab. Rev.* **6**, 180–210.
75. Hoehn-Berlage, M., Eis, M., Back, T., Kohno, K. and Yamashita, K. (1995) Changes of relaxation times (T_1, T_2) and apparent diffusion coefficient after permanent middle cerebral artery occlusion in the rat: Temporal evolution, regional extent, and

comparison with histology. *Magn. Reson. Med.* **34**, 824–834.

76. Neumann-Haefelin, T., Kastrup, A., de Crespigny, A., Yenari, M. A., Ringer, T., Sun, G. H. and Moseley, M. E. (2000) Serial MRI after transient focal cerebral ischemia in rats: Dynamics of tissue injury, blood-brain barrier damage, and edema formation. *Stroke* **31**, 1965–1972; discussion 1972–1973.

77. Calamante, F., Lythgoe, M. F., Pell, G. S., Thomas, D. L., King, M. D., Busza, A. L., Sotak, C. H., Williams, S. R., Ordidge, R. J. and Gadian, D. G. (1999) Early changes in water diffusion, perfusion, T_1 and T_2 during focal cerebral ischemia in the rat studied at 8.5 T. *Magn. Reson. Med.* **41**, 479–485.

78. Runge, V. M., Price, A. C., Wehr, C. J., Atkinson, J. B. and Tweedle, M. F. (1985) Contrast enhanced MRI. Evaluation of a canine model of osmotic blood–brain barrier disruption. *Invest. Radiol.* **20**, 830–844.

79. Bonnemain, B. (1998) Superparamagnetic agents in magnetic resonance imaging: Physicochemical characteristics and clinical applications. A review. *J. Drug Target.* **6**, 167–174.

80. Chambon, C., Clement, O., Le Blanche, A., Schouman-Claeys, E. and Frija, G. (1993) Superparamagnetic iron oxides as positive MR contrast agents: In vitro and in vivo evidence. *Magn. Reson. Med.* **11**, 509–519.

81. Stoll, G. and Bendszus, M. (2009) Imaging of inflammation in the peripheral and central nervous system by magnetic resonance imaging. *Neurosci.* **158**, 1151–1160.

82. Stuber, M., Gilson, W. D., Schar, M., Kedziorek, D. A., Hofmann, L. V., Shah, S., Vonken, E. J., Bulte, J. W. and Kraitchman, D. L. (2007) Positive contrast visualization of iron oxide-labeled stem cells using inversion-recovery with ON-resonant water suppression (IRON). *Magn. Reson. Med.* **58**, 1072–1077.

83. Weber, R., Wegener, S., Ramos-Cabrer, P., Wiedermann, D. and Hoehn, M. (2005) MRI detection of macrophage activity after new experimental stroke in rats: New indicators for late appearance of vascular degradation? *Magn. Reson. Med.* **54**, 59–66.

84. Ogawa, S., Lee, T. M., Kay, A. R. and Tank, D. W. (1990) Brain magnetic resonance imaging with contrast dependent on blood oxygenation. *Proc. Natl. Acad. Sci. USA* **87**, 9868–9872.

85. Ueki, M., Mies, G. and Hossmann, K.-A. (1992) Effect of alpha-chloralose, halothane, pentobarbital and nitrous oxide anesthesia on metabolic coupling in somatosensory cortex of rat. *Acta Anaesthesiol. Scand.* **36**, 318–322.

86. Ramos-Cabrer, P., Weber, R., Wiedermann, D. and Hoehn, M. (2005) Continuous noninvasive monitoring of transcutaneous blood gases for a stable and persistent BOLD contrast in fMRI studies in the rat. *NMR Biomed.* **18**, 440–446.

87. Weber, R., Ramos-Cabrer, P., Wiedermann, D., van Camp, N. and Hoehn, M. (2006) A fully noninvasive and robust experimental protocol for longitudinal fMRI studies in the rat. *Neuroimage* **29**, 1303–1310.

88. Weber, R., Ramos-Cabrer, P., Justicia, C., Wiedermann, D., Strecker, C., Sprenger, C. and Hoehn, M. (2008) Early prediction of functional recovery after experimental stroke: Functional magnetic resonance imaging, electrophysiology, and behavioral testing in rats. *J. Neurosci.* **28**, 1022–1029.

89. Olah, L., Franke, C., Schwindt, W. and Hoehn, M. (2000) CO(2) reactivity measured by perfusion MRI during transient focal cerebral ischemia in rats. *Stroke* **31**, 2236–2244.

90. Bock, C., Schmitz, B., Kerskens, C. M., Gyngell, M. L., Hossmann, K.-A. and Hoehn-Berlage, M. (1998) Functional MRI of somatosensory activation in rat: Effect of hypercapnic up-regulation on perfusion- and BOLD-imaging. *Magn. Reson. Med.* **39**, 457–461.

91. Pillekamp, F., Grüne, M., Brinker, G., Franke, C., Uhlenküken, U., Hoehn, M. and Hossmann, K. (2001) Magnetic resonance prediction of outcome after thrombolytic treatment. *Magn. Reson. Imaging* **19**, 143–152.

Chapter 15

Volumetry and Other Quantitative Measurements to Assess the Rodent Brain

Alize Scheenstra, Jouke Dijkstra, and Louise van der Weerd

Abstract

Morphometry is defined as studying variations in and changes of shapes. Evaluation of shape changes in the brain is a key step in the development of new mouse models, the monitoring of different pathologies, and measuring environmental influences. Traditional morphometry was performed by manual shape delineation, so-called volumetry. Currently, automated methods have been developed that can be roughly divided into three groups: voxel-based morphometry, deformation-based morphometry, and shape-based morphometry. In this chapter, we describe the different approaches for quantitative morphometry and how they can be applied to the quantitative analysis of the rodent brain.

Key words: Image processing, morphometry, volumetry, deformation-based morphometry, shape-based morphometry, automated analysis.

1. Introduction

MRI of the brain is increasingly used for standard phenotyping of transgenic mouse models or for non-invasive monitoring of disease progression and treatment response. Quantitative analysis of the brain images is also referred to as brain morphometry, which is defined as studying variations and changes of different shapes in the brain. The main research question in brain morphometry is how to determine significant differences between two groups of rodents, for example, a diseased and a healthy group or one group of rodents followed over time and measured at multiple time points.

Therefore, the present chapter is intended as a guide for researchers who wish to perform quantitative morphometry on rodent brains. As there are several morphometry methods available that can be used, we review current methods of brain morphometry and how these methods can be used for quantitative group comparison. This chapter contains an image processing section (**Section 2**), a statistical section (**Section 3**), a methods section (**Section 5**), and a notes section (**Section 6**). The image processing and the statistical sections explain terms and techniques that need to be understood for automated morphometry. The methods section describes the several morphometry methods available, including traditional volumetry and automated methods. The notes section indicates the pros and cons of each method and gives some guidance in what method to choose.

The intended audience for this chapter is researchers who have some background in MRI and perhaps some prior knowledge of brain morphometry, but have not (yet) applied this technology in practice. Although, we focus here on MR images, the techniques and protocols are described in such a general way that the same procedures can easily applied on other 3D anatomical images, such as CT or 3D PET.

2. Image Processing and Quantitative Morphology

2.1. Definitions

Image processing is the process of transforming, improving, and extracting information from a digital image, e.g., an MR or CT scan. In this section, the most employed terms and techniques in automated morphometry are briefly explained.

3D digital image: An image is built up by small cubes, so-called voxels. The image size (the matrix) indicates the number of voxels in each direction. Each voxel has a spatial size (in millimeters or micrometers) in each dimension, which is called the resolution. A 3D MR image can be based on a full 3D volume, or alternatively on 2D slices for which the resolution is defined by the in-plane resolution (x- and y-axis) and slice thickness (z-axis). Sometimes, the 2D slices are acquired with some space between them; this is called the slice gap. For example, **Fig. 1** shows an image with a matrix of 6×4 (0.06×0.06 mm^2 in-plane resolution) and five 0.1 mm thick slices with a 0.05 mm slice gap.

Image registration: Image registration is the process of transforming one image, the source image, in such a way that it can be compared to another image, the target image.

The comparison of the images is done by a similarity measure, which assesses how alike the images are. The transformation of an image can be done either affine or nonlinear: Affine transformation allows only global shifts in location, rotation, and scaling, so the transformation is equal for each point in the image. Nonlinear

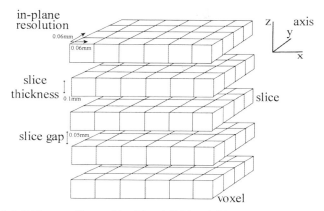

Fig. 1. A digital image with a matrix of 6 × 4 (0.06 × 0.06 mm in-plane resolution) and five 0.1 mm thick slices with a 0.05 mm slice gap.

transformation allows for a different transformation for each voxel in the image. A nonlinear registration might potentially change the shape of the image completely depending on the accuracy of the registration. The result of a nonlinear registration is a deformation field, which is an image with the same dimension as the original image, where each voxel contains a 3D vector indicating the local change of that voxel to go from the source image to the target image. **Figure 2** shows an example of a 3D deformation field overlaid, as a result from warping synthetically deformed image (**b**) (the source), onto its original, image (**a**) (the target).

Image Normalization: Normalization is the process of standardizing images, so that all images have the same orientation, size, and resolution. It places the homologous regions of the several brain images as closely into spatial alignment as possible so that the various brain structures can be compared and analyzed automatically.

The normalization is done by affine or nonlinear registration to a standard reference image. A standard reference image is defined as an average MR image, constructed from brain images

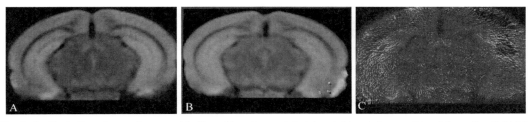

Fig. 2. An illustration of a nonlinear registration of two time points (**a**) and (**b**) and the resulting deformation field (**c**), which is displayed on *top* of the warped image **b**. Each vector indicates to the corresponding voxel in image a. Color coding can be used to indicate the amount of displacement in voxels or in millimeters. The images were acquired in vivo with T_2-weighted protocol on a Bruker 9.4 T system, with the parameters: TE = 12.76 ms, TR = 4000 ms, RARE factor = 4, FOV = 18 mm^2, matrix = 200 × 200, thirty 0.25 mm slices, 12 averages.

of a representative population, which is spatially oriented in a standard viewing direction and has an equal or preferably higher resolution than the group of MR images, which are compared to the reference. The higher resolution of the reference image decreases the effect of partial volume effects in a later stage of the analysis.

A standard reference image should be representative for the population to which it is compared. Therefore, the choice for a standard reference image can be a single mouse brain, or a group average (1). A number of standard mouse brain MRI atlases are freely available (2–4) but as scan parameters and mouse strains vary widely, many people prefer to develop their own reference brain atlases (*see* **Note 1**). A good tool to develop in-house mouse reference brains is developed by Mouse BIRN (5). **Figure 3** shows an example standard reference image for in vivo and ex vivo MRI.

Image segmentation: The segmentation of an image is the manual or automated delineation of structures in the image (6, 7). These segmentations can be visualized in 2D or 3D and might used as input for volumetry.

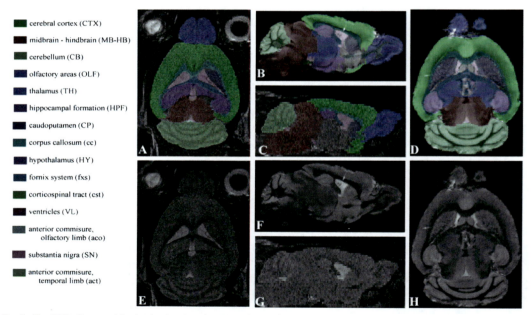

Fig. 3. The MRI atlas used for in vivo (**e**, **g**) and ex vivo (**f**, **h**) MR images with their manual segmentations (**a**, **b**, **c**, and **d**) and corresponding names and abbreviations. For a better understanding, the abbreviations for the brain structures are in upper case where the abbreviations for brain tracts are in lower case. All images are acquired on a Bruker 9.4 T scanner. For the in vivo images a T_2-weighted multi-slice spin echo sequence with TR/TE = 6000/35 ms (four averages) was used with a scanning time of 102 min (matrix size of 256 × 256, 40 slices and a resolution of 97.6 × 97.6 × 200 μm³ per voxel). Ex vivo imaging was performed using a T_1-weighted 3D gradient echo protocol, with TR/TE = 17/7.6 ms, flip angle 25°, and a total scan time of 10 h. The ex vivo volume had a matrix size of 256 × 256 × 256 and an isotropic resolution of 78.1 μm per voxel. (Reproduced from Ref. (6))

3. Statistics in Quantitative Morphology

Although the choice of a particular statistical test is dependent on the measurements which are performed on the brain, the procedure of a statistical test is similar for all morphometry methods. In this section, we explain the basic principles of statistical testing and how these statistical tests are incorporated in morphometry methods.

Producing and interpreting a p-value: First a *null hypothesis* and an *alternative hypothesis* have to be defined. In our experiment, we try to collect as much evidence as possible to prove the null hypothesis is wrong, so we can reject our H_0 and can state that the alternative hypothesis is true with a certain reliability. After the formation of the hypotheses, the experiment is performed, including the measurements on the brain. These measurements are than used as input for a statistical test. The outcome of the test is a *p*-value, which is an indication of the likelihood that the situation that is observed in the experiment, has occurred by chance. As statistical tests work with chances, there is always a probability margin that an error is made, either H_0 is rejected when it should not be (type 1 error) or H_0 is accepted although this is not true (type 2 error). Therefore, this error margin has to be reported; usually only the type 1 error is reported. Reporting a type 1 error can be done in two ways, either by giving the *p*-value directly or by thresholding the results at a predefined significance level α. With this significance level, a statement is made about the maximal error that is allowed when rejecting the H_0. In clinical experiments, this level is usually set to 0.01, which indicates that if the same experiment is repeated 100 times, there is on average one case for which the H_0 is rejected falsely.

Multiple-test correction: In most automated quantitative morphometry methods, a certain hypothesis about group difference is tested for each voxel separately, as explained in the methods section. This results in a *p*-value for each voxel. These *p*-values are displayed in a so called *statistical parametrical map* (SPM), which indicates the significant differences per voxel represented by some color scale usually with the significance overlaid on an MR image to indicate the locations of the significant differences. An example of a SPM is given in **Fig. 4**. All these tests are, unless otherwise specified, independent tests: We need multiple-test correction if we want to draw a general conclusion on the brain instead of the individual voxels (8), i.e., the hypothesis that the brains of the two groups are significantly different instead of a single voxel.

Multiple-testing refers to the testing of more than one hypothesis at the time, where each test has its own error margin.

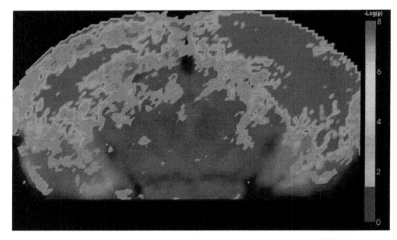

Fig. 4. A statistical parametric shows a *statistical parametrical map* (SPM) for uncorrected *p*-values. The probability scale is overlaid on a MRI image for a better interpretation of the data (unpublished data). The images were acquired in vivo with T_2 weighted protocol on a Bruker 9.4 T system, with the parameters: TE = 12.760 ms, TR = 4000 ms, RARE factor = 4, FOV = 18 mm^2, matrix = 200 × 200, thirty 0.25 mm slices, 12 averages.

For example, two groups of brain MR images from the same population are tested for group difference. The MR images have a 256 × 256 × 128 volume with 8,388,608 voxels. If all hypotheses are tested with $\alpha = 0.01$, on average, 83,886 incorrect rejections of the H_0 are done and thus 83,886 significant voxels might appear to be different just by chance! If we do not correct for this effect, we might draw the conclusion that the groups from the same populations are significantly different.

Multiple-test correction can be done in several ways, of which the following are advised for multiple-test correction in morphometry (8–10).

Bonferroni correction: This is the most stringent and most straightforward correction. This test is the best method for truly independent voxels; however, as the voxels in the brain are correlated, Bonferroni correction is usually too conservative: it avoids false rejections of H_0, but thereby severely increases the chance of a type 2 error. To correct for multiple test with Bonferroni, H_0 per voxel should be rejected if $\alpha/n \leq 0.01$, where n is the number of tests, thus the number of voxels in the volume.

Random field theory: As we should only perform Bonferroni correction on independent voxels, random field theory is used to determine clusters of dependent voxels, so multiple-test correction is only applied on the clusters instead of the voxels. This method requires a smooth SPM, which means that its values change gradually and it has no sharp transitions of probabilities. If a smooth SPM cannot be obtained, the resampling method is advised.

Resampling: The resampling method uses permutation tests to determine the corrected *p*-values. A permutation test iteratively randomizes the two groups and tests if the original situation is significantly different from the randomized groups (11). In general, this method has a high accuracy—higher than the random field theory. But this method is computationally very expensive and much slower than the Bonferroni and random field theory corrections. Therefore, in practice, it is less applied.

4. Materials

1. 3D MR images of two groups of mice, a test and control group, acquired according to the same protocol and with sufficient quality (*see* **Notes 2, 3**, and **4**).
2. A computer with a RAM of minimal 2 GB, preferably 4 GB and CPU power of around 3.0 GHz. (*see* **Note 5**).
3. Image processing software, such as SPM, ImageJ, or Amira (*see* **Note 6**).
4. Statistical software package, such as SPSS, MATLAB, or R (*see* **Note 6**).

5. Methods

All quantitative morphometry methods described in this chapter follow the same basic protocol (*see* **Note 7**):
1. Acquire MR images of the two groups of mice.
2. Find summarizing properties for the structure of interest, the so-called features, which may discriminate between the control group and the test group. Good features might be the volume in mm^3, the location of anatomical landmarks, the outer boundary of the structure indicated by landmarks or segmentation (*see* **Note 8**).
3. Use your software package to extract these features for all images in your dataset.
4. Use your statistical package to test the extracted features to detect a significant difference between the groups.
5. Based on the results, check the original data if some of the signficiant differences can be explained by any artifacts, misalignments, or misregistrations (*see* **Note 9**).
6. Report your result (*see* **Note 10**).

5.1. Manual Volumetry

Volumetry is considered the standard manual approach. This section describes the process of quantification of the brain image by measuring the volume of the structures of interest. The volumetry protocol is based on three steps:

1. *Segmentation:* Use the segmentation tool in your image processing package to segment the structure of interest and ask a second expert to check all segmentations (*see* **Note 11**).

2. *Volume calculation*: Most packages providing segmentation tools also give the volume in mm^3 of the segmented structures. If only the number of voxels is given, the volume can be calculated with the following formula: $V = N \times S = N \times (R_x \times R_y \times R_z)$, where V is the volume of each segmented structure, N is the number of voxels in the structure. S represents the volume of a single voxel, which can be found by the resolution of the x-, y-, and z-directions (R_x, R_y, and R_z) (*see* **Note 12**).

3. *Statistical analysis*: Use your statistical software package to test the data (the volumes of the two different groups), for
   a. Normality with a Shapiro–Wilk test, Kolmogorov–Smirnov test or the Pearson's chi-square test.
   b. Equal variances with a Levene's test. The output of this test is required for all statistical tests on normal data.
   Based on the outcome of (a), (b), and the setup of your test (*see* **Note 13**), test for a significant group difference in your data. Use **Table 1** as guideline for which test to select.

5.2. Automated Volumetry

Based on their feature selection, the automated morphometry methods can be roughly divided into three groups (*see* **Note 14**): voxel-based morphometry (VBM), which calculates for each voxel

Table 1
A guideline for the selection of a statistical test in volumetry

Situation	Normal data		Not normal data	
	Multiple measurements	Independent measurements	Multiple measurements	Independent measurements
2 Groups	Paired *t*-test	Unpaired *t*-test	Wilcoxon matched pairs signed-rank test	Mann–Whitney U test; Wilcoxon rank-sum test
2 Groups	Repeated measures ANOVA	One way ANOVA	Friedman test	Kruskal–Wallis test

the gray and white matter density and uses that for further analysis; deformation-based morphometry (DBM), which warps all images to a standard reference and uses the resulting deformation fields for their analysis; and shape-based morphometry (SBM), which defines the shape based on the contour or landmarks.

5.2.1. Voxel-Based Morphometry

The free SPM software package (12, 13) is specially designed for voxel-based morphometry of human brains and has lately been extended with a special module for rodent brains (14). Here we give a summary of the protocol of VBM as provided by the software package SPM.

1. *Normalization*: Use the registration tool in the software package to coregister your images, so all are brought into the same space and corrected for global brain size differences and brain orientations. If needed, correct them manually.

2. *Segmentation*: Use the segmentation tool in the software package to automatically generate probability maps, where each voxel is labeled with an a posteriori probability of belonging to gray matter (GM), white matter (WM), or cerebrospinal fluid (CSF). If needed, parameters can be tuned to improve the automated segmentation.

3. *Statistics*: Based on the sample setup, select the appropriate test to detect which voxels show a significant group difference. The software package SPM allows correction for multiple comparisons, but it depends on your manuscripts conclusion if you want to present the corrected or the uncorrected p-values (*see* **Section 2.2**).

5.2.2. Deformation-Based Morphometry

The name DBM refers to the nonlinear (deformable) registration which is applied before the morphometry. Another term which is used within this framework is Tensor-based morphometry (TBM). The difference between TBM and DBM lies in the statistical analysis: deformation-based morphometry (DBM) uses deformation vectors directly as they are obtained from the nonlinear registration of brain images to look for local differences in brain volume or shape. TBM in contrast examines the Jacobian determinant (the spatial derivative of the deformation fields) and uses that for the statistical analysis. Many methods have been developed for deformation-based morphometry, but "no ready to use" software packages are available yet, although some authors have published the code of their method on their website (15), and we will describe the protocol according to this method.

1. *Normalization*: Use the registration tool in your software package to register your images to a reference image, so all are brought into the same space and corrected for global

brain size differences and brain orientations. If needed, correct them manually.

2. *Nonlinear registration*: Use the nonlinear registration tool in your software package to register your images (the sources) to a reference image (the target). Check if the images have been warped correctly, without the introduction of artifacts or biologically impossible solutions (*see* **Note 9**).

3. *Statistical analysis*: Download the free statistical software R and the R-package for deformation-based morphometry (15). Run the Moore–Rayleigh test on the deformation vectors per voxel to obtain a significance value for the group difference.

    The *p*-values can be presented with and without multiple comparison corrections, but it depends on your manuscripts conclusion if you want to present the corrected or the uncorrected *p*-values (*see* **Section 2.2**).

5.2.3. Shape-Based Morphometry

SBM is currently mainly applied to human brains (16–18) and is mainly added to this chapter to complete the overview of possible methods. This method is especially useful to assess local changes *within* structures of interest, e.g., in the case of enlarged ventricles, which part of the ventricles is most affected.

6. Notes

1. The standard reference image has to be representative for the control group. For example, in a study for the comparison of mouse models for Alzheimer to their controls, an average should be chosen of the same age group, gender, and genotype as the controls.

2. The groups which are to be compared should be as uniform as possible beside the effect to be measured, such as treatment or specific gene. As many biological factors as possible should be kept equal: e.g., group differences in age and gender might influence the brain morphology.

3. Sufficient image quality may be defined as images having enough detail to visually see the (edges of) structure of interest. Image quality controls the success of the morphology studies; images with a high noise level are difficult to interpret for humans as well as computers. Images with a low resolution suffer from partial volume effects and the analysis is less accurate. But, image quality is not the only constraint to the imaging protocol; a too long scanning

protocol might cause dehydration, which results in shrinking of the ventricles and a changing morphometry. Therefore, it is wise to have a couple of test scans to optimize the MRI protocol for the specific study. The most important is to acquire images with sufficient SNR and resolution within the time constraints of the experiment. A general rule of thumb is to strive for an SNR of at least 20 (19).

4. In morphological studies, the main goal is detecting a shape-related effect in a treatment group or a morphological difference in transgenic mice. This is done by comparison to a normal standard. In comparison studies, it is most important to avoid a bias or external influence, which might be the source for an effect. Many of these effects can be minimized in the design of the study. Therefore, it is worth investigating possible biases (such as selection bias or confounding factors), checking the scanning protocol (see **Note 3** for image quality, but check also external factors, as a simple scanner software update during your experiment can cause an effect on your images), and discussing the complete study setup beforehand with someone from your statistics department.

5. Computer requirements are heavily time dependent, as each year, computers become more powerful. As a result, software packages will use more computer power. Therefore, check the computer requirements of the software package for the latest computer requirements.

6. New image processing software is being placed on the market each year. Therefore, it might be worth looking further for other software packages, which might be more suitable for your study.

7. The methods presented in **Section 5** are all suitable for quantitative morphometry. Depending on the structure of interest and the available options, a quantitative morphometry method can be selected: If there are only a few structures of interest, one can consider shape-based morphometry or volumetry. Volumetry is a good option if one is interested in only the volumetric difference, however, if a local effect in the brain structure is expected, SBM can be considered, as it returns significant differences for each location on the surface of the structure. VBM and DBM are both capable of analyzing the whole brain, which is very suitable for general phenotyping. The choice between VBM and DBM is more subtle and is dependent on the method available in the lab and the preference of the researcher: Both methods produce a statistical parametric map, need smoothing to handle noise in the images, need a perfect

normalization to avoid improper conclusions, and allow the usage of general linear models for the incorporation of global parameters.

Whereas VBM is based on a segmentation which defines composition of brain tissue in amounts of gray and white matter and cerebrospinal fluids, DBM uses the voxel intensity range as input for the nonlinear registration. The usage of deformation vectors allows DBM to perform multi-variate statistics per voxel, whereas VBM applies uni-variate statistics. Uni-variate statistics are less realistic, as they consider only one voxel at the time and not the interaction with their neighbors. However, VBM is available as ready to use software package, where DBM and SBM are only available as free code or through collaborations with the developers. An overview of the different methods is given in **Table 2** below.

8. The key in morphometry is to extract good features to discriminate between two groups. For example, if a structure shows shrinkage, but the overall shape of the structure remains the same, the volume feature will show a significant difference between the two groups. But, if only the shape of the structure is compared, a significant difference might not show.

9. The interpretation of intensity-based morphometry methods such as DBM and VBM is not straightforward. A significant difference implies a significant difference in intensity between the two groups, this can be due to a morphological difference between the structures of the two groups, or it can be caused by one or more errors during the image

Table 2
An overview of the properties of the presented quantitative morphometry methods

Property	Volumetry	VBM	SBM	DBM
Automated	No	Yes	Yes	Yes
Analysis per structure	Yes	No	Yes	No
Full brain analysis	No	Yes	No	Yes
Statistical parametrical map	No	Yes	Yes	Yes
Normalization required	No	Yes	Yes	Yes
Segmentation required	Yes	Yes	Yes	No
Multi-variate analysis	No	No	Yes	Yes
Ready to use software package available	Yes	Yes	No	No

processing. Therefore, if a significant difference is detected, the raw data has to be cross-checked carefully to determine whether the significant effect can be explained by causes other than the detected morphometry differences. Each of the pre-processing steps in the automated morphometry method, such as normalization, segmentation, and/or nonlinear registration to a reference image might introduce errors leading to a significant result (20). **Figure 5** illustrates the effect of a misregistration and its influence on the average reference image. If significant effects for a voxel cannot be explained by any source of error, it can be considered a morphological difference.

10. In voxel-based morphometry studies, normally an SPM is reported, including the uncorrected p-values and the corrected p-values. If the corrected p-values are reported, the correction method should be mentioned. A complete guideline for reporting voxel-based morphometry studies has recently been published (21).

11. Much research has been performed on the variability between the segmentations of one and/or multiple observers. The quality of the manual segmentations is dependent on the accuracy and expertise of the observer, but when many structures have to be delineated; even the best observer may become weary and produce less optimal results. To assure quality, it is best to have multiple observers for each structure or at least one expert for which all segmentations are checked by another expert. Also automated segmentations by a software package need to be

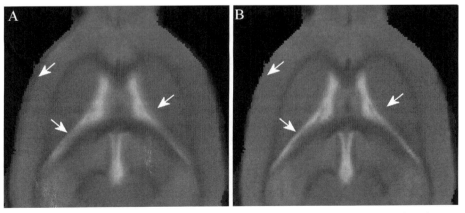

Fig. 5. Two averages of the same group of five mice, where (**a**) shows the effect of a misregistration of one mouse on the average and (**b**) shows accurate alignment of the MRI scans for all mice. The arrows indicate the areas where the misalignment is visible by blurring of the image or by the introductions of new edges. Imaging was performed using a T_2-weighted multi-slice spin echo sequence (102 min) with parameters: TR/TE = 6000/35 ms (four averages), matrix size of 256×256, 40 slices, and a resolution of 97.6 × 97.6 × 200 μm^3 per voxel.

checked, for image artifacts or abnormalities may cause erroneous segmentations.

12. In digital images, partial volume effects will occur and therefore, volume calculation of small structures is less accurate than volumetry of large structures.

13. There are many pitfalls and subtleties in the selection of a statistical test, e.g., when the variance from the different groups is not equal, a t-test with unequal variances needs to be used instead of a normal t-test. Furthermore, a t-test, F-test, or ANOVA are all parametric tests: they assume the data follows some statistical distribution. However, in practice this might be incorrect: the sample size is too small, there are many outliers, or the data follows a different or no distribution. Therefore a nonparametric test needs to be selected. For almost all parametric tests, a nonparametric equivalent is available. As the test depends on the data and the experiment, all recommendations for the statistical tests given in this chapter are guidelines. It is advised to discuss the study design and the statistical method with the statistical department of your institution before the start of the study.

14. Automated image processing methods are developed to save time, to have a more robust overall performance, i.e., reducing observer variability, or to perform more and complicated analyses. These methods allow analyses which are impossible to perform by hand. But for automation to work properly, a fixed protocol is required, as most automated registration methods are designed to compare anatomically similar brains. Unexpected artifacts may influence the results of the automated method. For instance, the excision of a mouse brain in ex vivo brain MRI causes extremely large deformations in the brain. The variations and amount of deformations of the brains shapes caused by the excision are many times larger than the expected in vivo shape variations between subjects (3).

Furthermore, automated methods rely on reference images and are optimized for a certain contrast (e.g., T_1-weighted or T_2-weighted scans). Unexpected inputs, such as a different orientation, or a slightly different scan protocol, or an update in the MRI scanner software may seriously hamper automated analysis. This is one of the reasons automated analysis is not as readily used in small animal research as in clinical settings; in a research environment, there is a tendency to update experimental protocols frequently based on new findings.

References

1. Joshi, S., Davis, B., Jomier, M., and Gerig, G. (2004) Unbiased diffeomorphic atlas construction for computational anatomy. *NeuroImage* **23**, S151–60.
2. Ma, Y., Smith, D., Hof, P.R., Foerster, B., Hamilton, S., Blackband, S.J., Yu, M., and Benveniste, H. (2008) In vivo 3D digital atlas database of the adult C57BL/6 J mouse brain by magnetic resonance microscopy. *Front. Neuroanat.* **2**, 1.
3. Kovacević, N., Henderson, J.T., Chan, E., Lifshitz, N., Bishop, J., Evans, A.C., Henkelman, R.M., and Chen, X.J. (2005) A three-dimensional MRI atlas of the mouse brain with estimates of the average and variability. *Cereb. Cortex* **15**, 639–45.
4. MacKenzie-Graham, A., Lee, E.F., Dinov, I.D., Bota, M., Shattuck, D.W., Ruffins, S., Yuan, H., Konstantinidis, F., Pitiot, A., Ding, Y., Hu, G., Jacobs, R.E., and Toga, A.W. (2004) A multimodal, multidimensional atlas of the C57BL/6 J mouse brain. *J. Anat.* **204**, 93–102.
5. http://nbirn.net/research/mouse/index.shtm, checked December 16th 2009.
6. Scheenstra, A.E., van de Ven, R.C., van der Weerd, L., van den Maagdenberg, A.M., Dijkstra, J., and Reiber, J.H. (2009) Automated segmentation of in vivo and ex vivo mouse brain magnetic resonance images. *Mol. Imaging* **8**, 35–44.
7. Bankman, I.N. (2009) Segmentation. In Bankman, I.N. (eds.), *Handbook of Medical Image Processing and Analysis*, pp. 71–258, 2nd ed. San Diego, CA: Academic.
8. Shaffer, J.P. (1995) Multiple hypothesis testing. *Ann. Rev. Psychol.* **46**, 561–84.
9. Friston, K.J., Holmes, A.P. Worsley, K.J., Poline, J.B., Frith, C.D., and Frackowiak, R.S.J. (1995) Statistical parametric maps in functional imaging: A general linear approach. *Hum. Brain Mapping* **2**, 189–210.
10. Pantazis, D., Nichols, T.E., Baillet, S., and Leahy, R.M. (2005) A comparison of random field theory and permutation methods for the statistical analysis of MEG data. *NeuroImage* **25**, 383–94.
11. Holmes, A.P., Blair, R.C., Watson, J.D., and Ford, I. (1996) Nonparametric analysis of statistic images from functional mapping experiments. *J. Cereb. Blood Flow Metab.* **16**, 7–22.
12. Ashburner, J. and Friston, K.J. (2000) Voxel-based morphometry: The methods. *NeuroImage* **11**, 805–21.
13. http://www.fil.ion.ucl.ac.uk/spm/, checked December 16th 2009.
14. Sawiak, S.J., Wood, N.I., Williams, G.B., Morton, A.J., and Carpenter, T.A. (2009) Voxel-based morphometry in the R6/2 transgenic mouse reveals differences between genotypes not seen with manual 2D morphometry. *Neurobiol. Dis.* **33**, 20–7. (http://www.wbic.cam.ac.uk/~sjs80/spmmouse.html, checked December 16th 2009).
15. Scheenstra, A.E.H., Muskulus, M., Staring, M., van den Maagdenberg, A.M., Lunel, S.V., Reiber, J.H., van der Weerd, L., and Dijkstra, J (2009) The 3D Moore–Rayleigh test for the quantitative groupwise comparison of MR brain images. In *Proc. Inf. Process Med. Imaging* **21**, 564–75. (http://folk.ntnu.no/muskulus/, checked December 16th 2009).
16. Thompson, P.M. and Toga, A.W. (2002) A framework for computational anatomy. *Comput. Vis Sci.* **5**, 13–34.
17. Cootes, T.F., Taylor, C.J. Cooper, D.H., and Graham, J. (1995) Active shape models—their training and application. *Comput. Vis. Image Understanding* **61**, 38–59.
18. Ferrarini, L., Palm, W.M., Olofsen, H., van der Landen, R., van Buchem, M.A., Reiber, J.H., and Admiraal-Behloul, F. (2008) Ventricular shape biomarkers for Alzheimer's disease in clinical MR images. *Magn. Reson. Med.* **59**, 260–7.
19. Kale, S.C., Lerch, J.P., Henkelman, R.M., and Chen, X.J. (2008) Optimization of the SNR-resolution tradeoff for registration of magnetic resonance images. *Hum. Brain Mapp.* **29**, 1147–58.
20. Bookstein, F.L. (2001) "Voxel-based morphometry" should not be used with imperfectly registered images. *NeuroImage* **14**, 1454–62.
21. Ridgway, G.R., Henley, S.M., Rohrer, J.D., Scahill, R.I., Warren, J.D., and Fox, N.C. (2008) Ten simple rules for reporting voxel-based morphometry studies. *NeuroImage* **40**, 1429–35.

Chapter 16

Models of Neurodegenerative Disease – Alzheimer's Anatomical and Amyloid Plaque Imaging

Alexandra Petiet, Benoit Delatour, and Marc Dhenain

Abstract

Alzheimer's disease (AD) is an important social and economic issue for our societies. The development of therapeutics against this severe dementia requires assessing the effects of new drugs in animal models thanks to dedicated biomarkers. According to the amyloid cascade hypothesis, β-amyloid deposits are at the origin of most of the lesions associated with AD. These extracellular deposits are therefore one of the main targets in therapeutical strategies. Aβ peptides can be revealed histologically with specific dyes or antibodies, or by magnetic resonance microscopy (μMRI) that uses their association with iron as a source of signal. The microscopic size of the lesions necessitates the development of specific imaging protocols. Most protocols use T_2-weighted sequences that reveal the aggregates as hypointense spots. This chapter describes histological methods that reveal amyloid plaques with specific stains and MR-imaging protocols for in vivo and ex vivo MR imaging of AD mice.

Key words: Animal model, mouse, amyloid, APP, PS1, imaging, MRI.

1. Introduction

Alzheimer's disease (AD) is a severe dementia with critical social and economic consequences. Senile plaques are one of the hallmarks of this disease. They are microscopic lesions that measure less than 20 μm in humans (1). These lesions are constituted of aggregated extracellular deposition of β-amyloid (Aβ) peptides. Amyloid deposits are believed to occur in the brain a long time, maybe decades, before the occurrence of clinical AD (2) and according to the amyloid cascade hypothesis, amyloid is at the origin of most of the pathological processes associated with AD (3).

To date, there is no curative treatment against AD, but many disease-modifying treatments are under investigation (4). The development of these treatments relies on the use of animals such as transgenic mouse models of amyloidosis (5). Most of these models are based on the overexpression of mutated forms of APP alone or with an additional mutation of presenilin (PS1 or PS2) genes (6). The generation and use of these models require the ability to phenotype these animals and to evaluate AD-like pathologies.

MRI can play a critical role to follow up these models. First, MRI can be used to follow up cerebral atrophy in transgenic mouse models of AD (**Fig. 1**) (7). This biomarker is critical because in humans, cerebral atrophy appears progressively during the evolution of AD (8) and it is associated with disease progression in clinical trials (9). Imaging amyloid plaques would also be critical to follow up the AD pathology in animals. Today, in humans, amyloid plaque imaging relies mainly on positron emission tomography and on specific ligands such as PIB (2). Such ligands are however difficult to use in small rodents (10).

MR methods using MRI are under development to either quantify some parameters that reflect amyloid load (11, 12) or to directly detect amyloid plaques (13, 14).

This chapter describes methods to detect amyloid plaques in vivo and ex vivo in the brains of transgenic mouse models of AD.

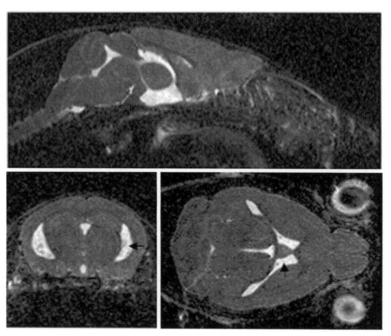

Fig. 1. In vivo 3D acquisition of a mouse brain: sagittal (*top*), coronal (*left*), and axial (*right*) views. Cerebral ventricles filled with CSF appear hyperintense on these images (*arrows*). The measure of their volumes can be used as an index of atrophy.

2. Material

2.1. Mouse Models

APP/PS1dE9 mouse models of AD can be purchased from the Jackson Laboratory (http://www.jax.org/; 600 Main Street Bar Harbor, Maine 04609 USA) (*see* **Note 1**).

2.2. MR Imaging Systems for In Vivo Imaging

1. MRI spectrometer and MR probes: 7-T spectrometer (Pharmascan, Bruker Biospin GmbH) equipped with a 9-cm inner diameter gradient system (760 mT/m strength and 6836 T/m/s slew rate) and interfaced to a console running Paravision 5.0. A birdcage coil (Bruker) of 38 mm diameter can be used for power transmission–reception. Other high-field MRI systems can also be used for in vivo imaging of mouse models of AD.

2. Mouse holder and monitoring (**Fig. 2**): the head of the animal is stabilized in a head holder using ear bars and a bite bar built in a dedicated cradle (available from RAPID Biomedical GmbH, Germany). The head holder is inserted into the radio frequency (rf) coils during the imaging session. This setup prevents movements of the animals during the long imaging acquisitions. Monitoring devices, such as the MR-compatible small animal monitoring and gating system (respiration/IBP module) from SA Instruments Inc (Stony Brook, NY 11790, USA), are used to follow the animal's physiological parameters. The animals are warmed with a water-filled heating blanket connected to a thermoregulated water bath (circulating thermostat system from Bruker).

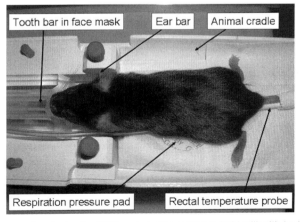

Fig. 2. Mouse setup for in vivo MR imaging. The animal is held still with tooth and ear bars. The isoflurane anesthesia is delivered via a face mask. Respiration is recorded through a pressure pad and body temperature is recorded through a rectal probe.

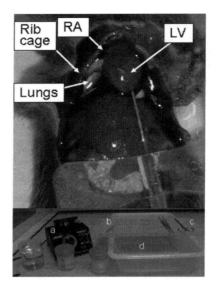

Fig. 3. Mouse intracardiac perfusion. The *top picture* shows the rib cage cut off and the butterfly needle inserted into the left ventricle (LV). The *bottom picture* shows the material used: formalin beakers and PBS beaker, peristaltic pump (**a**), butterfly needle (**b**), forceps, scissors (**c**), surgical board and draining bucket (**d**). RA, right atrium.

3. Anesthetic devices: isoflurane vaporizer connected to several flowmeters for $O_2/N_2O/Air$ and one cage dedicated to anesthesia induction (Minerve, Esternay, France).
4. Isoflurane gas (Belamont, Paris, France) is used for anesthesia.
5. Air, O_2, N_2O tanks.

2.3. Animal Sacrifice

1. Peristaltic pump for intracardiac perfusion: Mini-peristaltic pump II by Harvard Apparatus (**Fig. 3**).
2. Surgical supplies such as scissors (available from WPI Europe (http://www.wpi-europe.com), dressing forceps (10 cm long), micro bulldog clamp (3 cm long), butterfly needle (25 gauge).
3. Perfusion fluids: phosphate-buffered saline, and 10% buffered formalin at room temperature.
4. Surgery board and draining bucket.

2.4. MR Imaging Systems for Ex Vivo Imaging

1. Clinical 7-T spectrometer (Syngo MR, VB15, Siemens), equipped with an AC84 head gradient set with 36-cm available bore (80 mT/m strength and slew rate of 333 mT/m/s). Birdcage coils can be used (inner diameter = 24 mm) for signal transmit–receive. Other MR systems can also be used for ex vivo MRI. For example, we performed

previous studies on a 4.7-T spectrometer (Bruker) (13) or on the 7-T spectrometer used for in vivo imaging.

2. 10-mL syringes can be used to make containers that allow keeping the brain still in place and soaked.

3. Fluorinert™ Electronic Liquid FC-40(3M™, Cergy-Pontoise, France), a fully fluorinated liquid is used to embed the sample before MRI and to remove any background signal.

2.5. Histological Analysis

2.5.1. General Histology

1. Sliding freezing microtome (e.g., LEICA SM2400).
2. Homemade baskets to rinse tissue. Commercial systems (15-mm Netwell insert with 74-μm mesh size polyester membrane from Corning Life Sciences) are available as an alternative.
3. Slow orbital agitator and routine small equipment for histology lab.
4. Phosphate buffer (PB).
5. Dimethyl sulfoxide.
6. Glycerol.
7. Dry ice.
8. Superfrost+ glass slides.

2.5.2. β-Amyloid Staining with BAM10 Antibody

1. Usual glassware and small equipment for histology lab.
2. Phosphate-buffered saline (PBS).
3. Hydrogen peroxide 30% (Sigma H 0904).
4. Octylphenol ethylene oxide condensate 0.2% (Triton X-100™, Sigma).
5. Normal rabbit serum, can be aliquoted at −20°C (Vector® Labs S 5000).
6. Monoclonal BAM10 clone A3981 (Sigma).
7. Biotinylated IgG anti-mouse, BA-9200 (Vector®).
8. Sodium azide 8%, stored at room temperature (Sigma S 8032).
9. Avidin–biotin complex (ABC VECTASTAIN kit, Elite PK 6100).
10. Tyramin biotin reagent, stored at 4°C (Blast PC 2815-0897)
11. Peroxidase substrate kit (VIP SK 4600 substrate kit for peroxidase, Vector®).

2.5.3. β-Amyloid Staining with Congo Red

Labeling of amyloid deposits is done by standard Congo red staining (adapted from Ref. (15)).

1. Usual glassware and small equipment for histology lab.
2. Solution S1: 80° ethanol saturated with NaCl.
3. Solution S2: Saturated solution of Congo red (Fluka, Ref. 60910) made in saline ethanol.
   S1 and S2 solutions are stable for weeks/months but it is recommended to prepare them immediately before staining.
4. Sodium hydroxide.

2.5.4. Iron Staining with Perls-DAB Method

Staining of iron deposits is performed by means of the Perls' method with diaminobenzidine intensification (16).

1. Usual glassware and small equipment for histology lab.
2. Potassium ferrocyanide (Sigma ref. P 9387).
3. Tris buffer.
4. Diaminobenzidine (DAB) dissolved in distilled water (1 g/1000 ml). DAB solution can be prepared and aliquoted at $-20°C$ before use.
5. Methanol.
6. Hydrogen peroxide 35%.
7. Hydrochloric acid 35%.

2.5.5. Analysis and Quantification of Histological Stainings

1. Slide scanner with high optical resolution (e.g., Super CoolScan 8000 ED scanner, Nikon, Champigny sur Marne, France) (*see* **Note 2**). Use a scanner that has a 4000 dpi in-plane digitization resolution (pixel size 6.35 μm^2) to allow quantification of large objects (e.g., plaques and big focal iron deposits). Work under calibrated and constant illumination conditions.
2. Optical microscope equipped with a digital camera (optional).
3. ImageJ freeware (Rasband, W.S., ImageJ, U. S. National Institutes of Health, Bethesda, Maryland, USA, http://rsb.info.nih.gov/ij/, 1997–2005).
4. Adobe Photoshop® software or any image manipulation program (e.g., The Gimp, http://www.gimp.org/, can be a good freeware alternative).

3. Methods

3.1. Mouse Preparation for In Vivo MRI

1. Place the mouse in the induction cage.
2. Turn the oxygen or air tank on.
3. Turn the flowmeter up to 1–1.5 L/min.

Table 1
Example of acquisition parameters for the fast spin echo sequence used to record T_2-weighted MR images in vivo

Parameter		Value	Unit
Repetition time	TR	2500	ms
Echo time	TE	92.3	ms
Field of view	FOVx	15	mm
	FOVy	30	
	FOVz	15	
Matrix	MATx	128	
	MATy	256	
	MATz	128	
Rare factor		16	

4. Induce anesthesia by turning the isoflurane level to 5% until the animal is in lateral decubitus for 2 min.
5. Maintain anesthesia at a concentration of 1.0–1.5% (*see* **Note 3**).
6. Place the mouse prone in the animal cradle of the MR scanner; insert the teeth into the tooth bar and the ear bars into the ear canal; put the respiration and temperature probes in place; cover the animal with the warming blanket.
7. Insert the head of the animal in the rf coil and slide the animal into the magnet for imaging.

3.2. In Vivo MR Brain Imaging

The parameters for a typical T_2-weighted spin echo sequence are presented in **Table 1**. The resolution with these parameters is about 117 μm isotropic and the acquisition time is about 42 min.

3.3. Mouse Sacrifice and Preparation for Ex Vivo MRI

3.3.1. Mouse Sacrifice and Fixation

The entire perfusion fixation procedure should be performed under a hood, possibly equipped with a sink connected to ad hoc disposal to evacuate perfused fluids.

1. Prepare the pump: pour about 500 mL of PBS in a beaker and 500 mL of formalin in another beaker. Connect a butterfly needle to the outflow end of the peristaltic pump and drop the inflow end of the pump into the PBS beaker. Run the pump until no more air bubbles are visible in the perfusion line.
2. Anesthetize the mouse with an intraperitoneal injection of pentobarbital sodium (at a dose of 100 mg/kg).
3. When the animal is deeply anesthetized and all reflexes are lost (toe- and tail-pinch checks), place the animal in supine

position on a dissecting board placed over a draining bucket. Tape or pushpin the limbs away from the body to hold the animal still. Proceed fairly quickly to begin the perfusion before the heart stops beating.

4. Perform a bilateral thoracotomy: reach the sternum with the tooth forceps and, just under the ribs, make a bilateral incision into the skin and through the pleural cavity. Use scissors to cut the ribs towards the shoulders and remove them to expose the heart cavity. Cut through the diaphragm then through the pericardium to expose the heart. Be careful not to puncture any organ. Make sure you can clearly identify the left and right ventricles (that are colored with slightly contrasted red nuances) and the right atrium.

5. Hold the heart with smooth forceps or thumb–index soft pinch and insert the butterfly needle into the left ventricle from the apex up, without piercing into the right ventricle. When the needle is in place (**Fig. 3**), turn the pump on (flow rate ~2 mL/min), and immediately make an incision into the right atrium with a pair of fine scissors to let the blood flow out. Flush the blood out with PBS until the perfusate runs clear. If the perfusion is properly performed, all organs should turn white and the tail should briefly stiffen up.

6. When all the blood is flushed out, turn the pump off and switch the perfusate line to formalin. Turn the pump back on and perfuse the fixative for about 5 min or until the mouse limbs are stiff.

7. Upon completion of the fixation, turn the pump off and remove the needle from the heart.

8. Release the mouse, cut its head off, dispose the carcass in a biological hazard bag, and then proceed to brain dissection.

3.3.2. Brain Dissection

After the mouse has been perfusion fixed and the head has been cut off, quickly proceed to brain dissection to limit tissue degradation.

1. Make a medial incision through the skin of the head from the base of the neck to the nose to expose the skull.

2. Localize the olfactory bulbs through the skull, between the two eyes. Insert the tip of sharp scissors through the skull on the medial line at the tip of the olfactory bulbs. Open the inserted scissors to crack open the skull and to separate it in two halves.

3. With tooth forceps, pull away the two skull halves. Then remove the skull from around the cerebellum and other pieces that might still be covering the brain. Be very careful not to pull apart, squeeze, or slice the brain while removing the skull (*see* **Note 4**).

4. The brain should now be free from the skull. With a small spatula, reach under the brain and gently release it from its cup; the cranial nerves should easily break.

5. Drop the brain in a container with formalin for a 24-h post-fixation at 4°C, and dispose of the rest of the head carcass in a biological hazard bag.

3.3.3. Passive Staining

We adapted and optimized previously published "staining" protocols (13, 17) to Gd-stain the fixed brains.

1. After a post-fixation of at least 24 h, soak the brain sample in a solution of phosphate buffered saline (PBS) and 0.5 M gadoterate meglumine at a dilution of 1:200 (2.5 M) (*see* **Note 5**).

2. Store at 4°C for at least 24 h prior to imaging.

3.3.4. Imaging Holder

Imaging holders can easily be home built with 10-mL syringes (**Fig. 4**).

1. Use two syringe pistons to close each end of the holder made with the syringe. You can use plastic pieces to hold the brain tight in place in the holder but be careful not to squeeze it (*see* **Note 6**).

2. Place the brain in the imaging holder and fill it half way (up to the beginning of the brain) with Fluorinert®.

3. Close the holder and remove all air bubbles. To do so, insert a 26-gauge needle filled with Fluorinert® between the cap and the wall of the holder. Push in some fluid; the air bubbles will exit from the small gap created by the needle. Slowly remove the needle while still pushing in some fluid. All air bubbles should be gone. If not, insert the needle again and repeat Step 3.

3.4. Ex Vivo MR Brain Imaging

A 3D gradient echo sequence can be used (FLASH) to acquire T_2^*-weighted images. **Table 2** gives typical parameters to acquire 72 images in about 14 h at a resolution of about $23 \times 23 \times 90$ μm^3.

Fig. 4. Mouse brain holder for ex vivo imaging. The brain sample is held still in a 10-mL syringe filled with Fluorinert.

Table 2
Example of acquisition parameters for the 3D gradient echo sequence used to record T_2^*-weighted MR images ex vivo

Parameter		Value	Unit
Repetition time	TR	100	ms
Echo time	TE	21	ms
Field of view	FOVx	24	mm
	FOVy	20.25	
Matrix	MATx	1024	
	MATy	864	
Slice thickness		0.09	mm
Number of slices		72	
Flip angle	FA	25	degree

3.5. Histological Studies

3.5.1. General Histology

1. Section whole brains or single hemispheres (frontal 40 μm-thick sections) on a freezing microtome after a 1-week fixation in 10% formalin (see **Note 7**) and subsequent cryoprotection in 20% glycerin and 2% DMSO in 0.1 M PB.

2. Collect 12 batches of serial sections (ranging from the frontal pole to the end of the caudal part of the hippocampus). Immediately rinse in 0.1 M PB the series of free floating sections to be stained and mount them on Superfrost+ glass slides before drying them overnight at room temperature (or in an oven at 40°C). Remaining tissue can be stored at −20°C in cryoprotectant as backup material.

3. For each mouse, it is suggested to perform a Nissl (thionin) stain to control for tissue quality and/or cytoarchitectonic anomalies before processing Congo red and Perls-DAB stains.

3.5.2. β-Amyloid Staining with BAM10 Antibody

Follow these steps for free-floating sections (total solution volume of 5 mL) (**Fig. 5**)

Day 1
1. Rinse the sections six times in PBS for 5 min.
2. Incubate in H_2O_2 (0.3%) for 20 min.
3. Rinse three times in PBS for 10 min.
4. To the (PBS + 0.2% Triton) solution, add 4.5% of normal rabbit serum. Incubate for 30 min.
5. To the (PBS + 0.2% Triton) solution, add 3% of normal rabbit serum and 0.1% of the primary antibody. Incubate for 48 h at room temperature (or 3 days at 4°C).

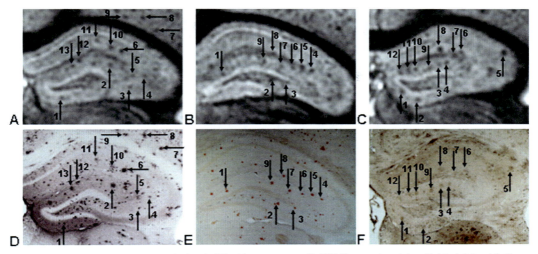

Fig. 5. Hypointense spots detected at the level of the hippocampus with MRI (*top*, **a–c**) match anti-Aβ staining (**d**), Congo red staining (**e**), and Perls' staining (**f**). This indicates that, in the mouse model that we used, these spots correspond to amyloid plaques and that they are loaded with iron.

Day 3
1. Rinse three times in PBS for 10 min.
2. To the (PBS + 0.2% Triton) solution, add 3% of normal rabbit serum and 0.1% of the secondary antibody and incubate for 1 h.
3. Rinse three times in PBS for 10 min.
4. To the (PBS + 0.2% Triton) solution, add kit reagent A (avidin DH) at dilution 1/250 and kit reagent B (biotinylated enzyme) at dilution 1/250. Incubate in this avidin–biotin complex for 1 h.
5. Rinse three times in PBS for 10 min.
6. Revelation with VIP substrate: to 5 mL of PBS, add three drops of each of the kit reagents (reagents 1, 2, 3, and hydrogen peroxide) and mix well. Incubate the free-floating sections in this solution for about 2 min.

3.5.3. β-Amyloid Staining with Congo Red

Labeling of amyloid deposits is done by standard Congo red staining (adapted from Ref. (*15*)) (**Fig. 5**).

1. Prepare S1 and S2; filter S2 before use.
2. Put slides under running tap water for 20 min.
3. Add NaOH $[10^{-4}]$ to S1 and incubate the slides for 30 min.
4. Add NaOH $[10^{-4}]$ to S2 and incubate the slides for 30 min.
5. Rinse under tap water.
6. Dehydrate in alcohols, clear in xylene, and coverslip with Eukitt.

3.5.4. Iron Staining with Perls-DAB Method

Staining of iron deposits is performed by means of the Perls' method with diaminobenzidine intensification (16) (**Fig. 5**).

1. Rehydrate the slides under running tap water for 20 min.
2. Inactivate endogenous peroxidase activity by immersing the tissue in a methanol (20%)/H_2O_2 (3%) solution made in distilled water for 10 min.
3. Rinse in distilled water twice for 5 min.
4. Incubate in acid potassium ferrocyanide solution (for 100 mL of distilled water, add 1 g of potassium ferrocyanide and 1 mL of 35% HCl) for 20 min.
5. Rinse in distilled water for 5 min and then twice in 0.1 M Tris buffer for 5 min.
6. Dilute DAB 2X in 0.2 M Tris; add 30–40 µL of H_2O_2 per 100 mL of final volume just before reaction.
7. Incubate the slides in DAB until a good signal-to-noise ratio is obtained (reaction is monitored under the microscope).
8. Rinse in distilled water. Store DAB in waste container or detoxify it.
9. Dehydrate in alcohols, clear in xylene and coverslip with Eukitt.
(*see* **Notes 8** and **9**).

3.5.5. Analysis and Quantification of Histological Stainings

3.5.5.1. Amyloid Deposits

If the purpose of the study is to register MR images with histologically assessed topography of plaques, no additional processing is required (13).

For quantitative analysis, amyloid loads are evaluated using computer-based thresholding methods. Scans are prepared using Photoshop software to outline selected regions of interest (ROI). Images are then processed with ImageJ freeware using a dedicated macrocommand that extracts amyloid deposits from background tissue (18, 19). Briefly, image processing to detect plaques relies on RGB color component adjustment, global-automated threshold based on entropy criterion, and morphometric filtering according to Feret's diameter. Macro is available on demand.

The step-by-step instructions are as follows:
1. Run ImageJ and open the macrocommand.
2. Select the folder containing the images to be analyzed (tiff format).
3. Automatic processing of all images is then initiated (this can take few minutes in case numerous files have to be processed).
4. After batch processing, a new subdirectory named "processed" is automatically created in the parent directory and contains (1) thresholded images allowing quick visual inspection of the results of plaque segmentation and (2) a

text file that can be imported (tab format) in a spreadsheet program and that contains all morphological data (e.g., total surface, thresholded surface) required to calculate amyloid loads for each image.

Regional amyloid loads are expressed as percent of tissue surface stained by the Congo red dye that corresponds to the proportion of plaque volume according to Delesse's principle (20). Evaluation of amyloid loads can be performed in multiple ROIs. Quantitative analyses are usually performed on several serial sections to sample the whole rostro-caudal extent of each ROI.

3.5.5.2. Iron Deposits

If the purpose of the study is to register MR images with histologically assessed topography of iron deposits, no additional processing is required (14).

Considering now more quantitative aspects (e.g., measuring iron loads), it is known that there is a relationship between "true" tissue iron content (as assessed for instance by atomic absorption spectroscopy) and intensity of Perls' staining (21, 22). We and others perform an analysis of Perls-stained brain tissue by means of optical densitometry, OD (11, 23, 24).

OD determines levels of iron deposition on the basis of transmitted light in the stained tissue. OD can be automatically calculated in selected ROIs using ImageJ or Photoshop. The optical density of each pixel is derived from its gray level and the mean iron load is defined as the mean OD from all pixels of the ROI. Standards to assess absolute iron quantities are not easily available for the analysis of Perls-stained material; therefore only relative quantities of iron deposition can be calculated, still allowing inter-group comparisons.

The step-by-step instructions are as follows:
1. Run image J and set measurements to include the mean gray level variable as output.
2. Open image to be analyzed.
3. Outline the region of interest using the polygon tool.
4. Run the measure command (shortcut: ctrl+M).
5. Uncalibrated optical density of Perls staining is calculated using the following formula: $OD = \log_{10}(255 - M_{GL})$ with M_{GL} as mean gray level of the region of interest.

4. Notes

1. The APP/PS1dE9 double transgenic mice express a chimeric mouse/human amyloid precursor protein (Mo/HuAPP695swe) and a mutant human presenilin 1

(PS1-dE9) both directed to CNS neurons. Both mutations are associated with early onset of AD. The "humanized" Mo/HuAPP695swe transgene allows the mice to secrete a human Aβ peptide. The included Swedish mutations (K595N/M596L) elevate the amount of Aβ produced from the transgene by favoring processing through the β-secretase pathway. The PS1 mutation elevates the amount of Aβ produced from the transgene by favoring processing through the γ-secretase pathway. We used the strain referenced as "Stock Number: 004462." This strain is maintained as a hemizygote line by crossing transgenic mice to B6C3F1/J mice. The strain referenced as "Stock Number: 005864" can also be used. These mice are based on the 004462 strain but were backcrossed to C57BL/6 J for at least eight generations.

2. If the resolution of the scanner appears to be insufficient (e.g., tiny deposits are not detected or with very blurred edges), it is recommended to digitize material at higher resolution using a microscope coupled to a camera. For optimal results, images should be stored as tiff files (jpeg compression has detrimental effects when images are further processed).

3. If O_2 is used as a pushing gas for isoflurane, the percentage of isoflurane will have to be increased to achieve the same level of anesthesia.

4. To avoid pressing against the brain and damaging the tissue during dissection, you can fold the skin around the inferior (ventral) part of the head and hold it between your fingers; this way you should not touch the brain itself at all.

5. By soaking the brain sample in a PBS solution (and 0.5 M gadoterate meglumine) for at least 24 h prior to MR imaging, you will regain some signal due to tissue rehydration.

6. The length of the holder should be at least four times as long as the brain so that the brain sits away from the ends to avoid imaging artifacts that might arise from the pistons. It is also recommended to wipe off the syringe graduation marks with ethanol as they might also cause imaging artifacts.

7. As a general rule, if the animal is fixed via a perimortem intracardiac perfusion, no further tissue fixation is necessary. For fresh brains or brains perfused with saline/PBS only, a 1-week fixation should be done in 10% formalin. However, in the ex vivo protocol for MR imaging, the brains are soaked for days in the staining solution containing PBS instead of formalin. It is therefore recommended to soak the brains again in formalin before sectioning them for histology evaluation.

8. Perls-DAB staining can alternatively be performed on free-floating sections to maximize penetration of reagents.

9. Before dehydrating tissue, a nuclear counterstain can be applied using thionin, nuclear red, or Harris hematoxylin. Do not counterstain when quantification of iron deposition has to be performed by optical densitometry.

Acknowledgments

We thank Fanny Petit from CEA for her assistance in histology processing for the figures presented. Our work was supported by the France-Alzheimer association, the National Foundation for Alzheimer's Disease and Related Disorders, and the NIH (R01-AG020197).

References

1. Hyman, B. T., Marzloff, K., and Arriagada, P. V. (1993) The lack of accumulation of senile plaques or amyloid burden in Alzheimer's disease suggests a dynamic balance between amyloid deposition and resolution. *J. Neuropath. Exp. Neur.* **52**, 594–600.
2. Jack, C. R., Jr., Lowe, V. J., Weigand, S. D., Wiste, H. J., Senjem, M. L., Knopman, D. S., Shiung, M. M., Gunter, J. L., Boeve, B. F., Kemp, B. J., Weiner, M., and Petersen, R. C. (2009) Serial PIB and MRI in normal, mild cognitive impairment and Alzheimer's disease: Implications for sequence of pathological events in Alzheimer's disease. *Brain* **132**, 1355–65.
3. Hardy, J., and Selkoe, D. J. (2002) The amyloid hypothesis of Alzheimer's disease: Progress and problems on the road to therapeutics. *Science* **297**, 353–6.
4. Blennow, K., de Leon, M. J., and Zetterberg, H. (2006) Alzheimer's disease. *Lancet* **368**, 387–403.
5. Duyckaerts, C., Potier, M. C., and Delatour, B. (2008) Alzheimer disease models and human neuropathology: Similarities and differences. *Acta Neuropathol.* **115**, 5–38.
6. Delatour, B., Le Cudennec, C., El Tannir-El Tayara, N., and Dhenain, M. (2006) Transgenic models of Alzheimer's pathology: Success and caveat. In Welsh, E. M. (Ed.), *Topics in Alzheimer's Disease*, pp. 1–34. Nova Publishers, Happauge NY.
7. Delatour, B., Guegan, M., Volk, A., and Dhenain, M. (2006) In vivo MRI and histological evaluation of brain atrophy in APP/PS1 transgenic mice. *Neurobiol. Aging* **27**, 835–47.
8. Valk, J., Barkhof, F., and Scheltens, P. (2002) *Magnetic Resonance in Dementia*. Springer, Heidelberg, Berlin, New York.
9. Albert, M., DeCarli, C., DeKosky, S., de Leon, M., Foster, N. L., Fox, N., Frank, R., Frackowiak, R., Jack, C., Jagust, W. J., Knopman, D., Morris, J. C., Petersen, R. C., Reiman, E., Scheltens, P., Small, G., Soininen, H., Thal, L., Wahlund, L.-O., Thies, W., Weiner, M. W., and Khachaturian, Z. (2005), Report of the neuroimaging work group of the Alzheimer's Association. Available from http://www.alz.org/national/documents/Imaging_consensus_report.pdf.
10. Klunk, W. E., Lopresti, B. J., Ikonomovic, M. D., Lefterov, I. M., Koldamova, R. P., Abrahamson, E. E., Debnath, M. L., Holt, D. P., Huang, G. F., Shao, L., DeKosky, S. T., Price, J. C., and Mathis, C. A. (2005) Binding of the positron emission tomography tracer Pittsburgh compound-B reflects the amount of amyloid-beta in Alzheimer's disease brain but not in transgenic mouse brain. *J. Neurosci.* **25**, 10598–606.
11. El Tannir El Tayara, N., Delatour, B., Le Cudennec, C., Guegan, M., Volk, A., and Dhenain, M. (2006) Age-related evolution

of amyloid burden, iron load, and MR relaxation times in a transgenic mouse model of Alzheimer's disease. *Neurobiol. Dis.* **22**, 199–208.

12. El Tayara Nel, T., Volk, A., Dhenain, M., and Delatour, B. (2007) Transverse relaxation time reflects brain amyloidosis in young APP/PS1 transgenic mice. *Magn. Reson. Med.* **58**, 179–84.

13. Dhenain, M., Delatour, B., Walczak, C., and Volk, A. (2006) Passive staining: A novel ex vivo MRI protocol to detect amyloid deposits in mouse models of Alzheimer's disease. *Magn. Reson. Med.* **55**, 687–93.

14. Dhenain, M., El Tannir El Tayara, N., Wu, T.-D., Guegand, M., Volk, A., Quintana, C., and Delatour, B. (2009) Characterization of in vivo MRI detectable thalamic amyloid plaques from APP/PS1 mice. *Neurobiol. Aging* **30**, 41–53.

15. Puchtler, H., Sweat, F., and Levine, M. (1962) On the binding of Congo red by amyloid. *J. Histochem. Cytochem.* **10**, 355–64.

16. Nguyen-Legros, J., Bizot, J., Bolesse, M., and Pulicani, J.-P. (1980) "Noir de diaminobenzidine": une nouvelle méthode histochimique de révélation du fer exogène ("Diaminobenzidine black": A new histochemical method for the visualization of exogenous iron). *Histochemistry* **66**, 239–44.

17. Petiet, A., Hedlund, L., and Johnson, G. A. (2007) Staining methods for magnetic resonance microscopy of the rat fetus. *J. Magn. Reson. Imaging* **25**, 1192–8.

18. Faure, A., Verret, L., Bozon, B., El Tannir El Tayara, N., Ly, M., Kober, F., Dhenain, M., Rampon, C., and Delatour, B. (2011) Impaired neurogenesis, neuronal loss, and brain functional deficits in the APPxPS1-Ki mouse model of Alzheimer's disease. *Neurobiol. Aging* **32**(3), 407–418.

19. Le Cudennec, C., Faure, A., Ly, M., and Delatour, B. (2008) One-year longitudinal evaluation of sensorimotor functions in APP751SL transgenic mice. *Genes Brain Behav.* 7(Suppl 1), 83–91.

20. Delesse, M. A. (1847) Procédé mécanique pour déterminer la composition des roches [Mechanical process to evaluate the coomposition of rocks]. *Comptes rendus de l'Académie des Sciences (Paris)* **25**, 544–45.

21. Masuda, T., Kasai, T., and Satodate, R. (1993) Quantitative measurement of hemosiderin deposition in tissue sections of the liver by image analysis. *Anal. Quant. Cytol. Histol.* **15**, 379–82.

22. Turlin, B., Loreal, O., Moirand, R., Brissot, P., Deugnier, Y., and Ramee, M. P. (1992) Détection histochimique du fer hépatique (Histochemical detection of hepatic iron. A comparative study of four stains). *Ann. Pathol.* **12**, 371–73.

23. Bizzi, A., Brooks, R. A., Brunetti, A., Hill, J. M., Alger, J. R., Miletich, R. S., Francavilla, T. L., and Di Chiro, G. (1990) Role of iron and ferritin in MR imaging of the brain: A study in primates at different field strengths. *Radiology* **177**, 59–65.

24. Hill, J. M., and Switzer, R. C. (1984) The regional distribution and cellular localisation of iron in the rat brain. *Neuroscience* **11**, 595–603.

Chapter 17

MRI in Animal Models of Psychiatric Disorders

Dana S. Poole, Melly S. Oitzl, and Louise van der Weerd

Abstract

Here we describe MRI and ^{1}H MRS protocols for the investigation of animal models (mainly mice and rats) of psychiatric disorders. The *introduction* provides general findings from brain imaging studies in patients with psychiatric diseases and refers to general rules regarding the use of animal models in research. The *methods* section includes a selection of basic 9.4 T MRI and MRS protocols applicable for the investigation of animal models of psychiatric disorders (T1W, T2W, FLAIR, ^{1}H MRS). The *notes* section discusses in detail a series of factors that can influence the outcome of the experiment: from animal handling, stress-triggering aspects, and experimental design-related factors to technical aspects that affect T_1 and T_2 measurements.

Key words: Animal MRI, ^{1}H NMR spectroscopy methods, imaging methods, magnetic resonance imaging, magnetic resonance spectroscopy, methodological considerations, mood disorders, MRI protocols, stress.

1. Introduction

Over the past four decades, psychiatrists and psychologists have become more and more interested in magnetic resonance imaging and spectroscopic techniques (MRI and MRS, respectively). As a result, more insight has been gained into the anatomy, pathophysiology, and biochemistry of the brain of patients with psychiatric disorders. This chapter is intended for researchers that have a background in MRI or MRS, and perhaps some prior experience with these techniques, but have not yet worked with animal models (of psychiatric disorders).

First, we provide an overview of changes that occur in the brains of humans that have been affected by psychiatric disorders, followed by MRI and MRS findings in animal models of psychiatric diseases such as schizophrenia, attention deficit/hyperactivity disorder (ADHD), and stress-related affective disorders (bipolar disorder, depression, anxiety, and post-traumatic stress disorder (PTSD)). Second, several guidelines for experiments with animal models are presented. The methods section provides detailed information on basic MRI and MRS techniques, as well as short explanations and suggestions for more complex magnetic resonance techniques that can be employed in the study of psychiatric disorders. The shortcomings and possible confounding factors are included in the notes section.

1.1. Main MRI and ^{1}H MRS Results in Humans

1.1.1. Schizophrenia

Volumetric changes of several brain areas have been reported such as reduced volume of the prefrontal and medial temporal cortical regions (1–4), the amygdala (5–7) and the hippocampus (5). Even a reduction in the total brain volume has been described (8). MRS studies have revealed a decrease of N-acetyl-aspartate (NAA) (9). These effects were more pronounced in patients with chronic schizophrenia (10–12).

1.1.2. ADHD

Attention deficit/hyperactivity disorder (ADHD) has been linked with a decrease in the volume of corpus callosum (13), structural abnormalities in striatum (4), and a reduced volume of various compartments of the prefrontal cortex (14). MRS studies have shown increased striatal creatine, glutamate, and glutamate/glutamine in ADHD (15).

1.1.3. Affective Disorders

In major depression, anxiety, and post-traumatic stress disorder, most studies agree on an association with a decrease in the volume of the hippocampus (16–19), frontal cortex (20–22), striatum (22, 23), and subgenual cingulated cortex (22, 24, 25), compared to the healthy individuals. Unipolar depression has been reported to be associated with a smaller frontal lobe, cerebellum, caudate, and putamen (26), while bipolar depression has been shown to be associated with a larger third ventricle (27, 28), reduced amygdala (5–7), and reduced hippocampal volumes (5). Lower hippocampal NAA levels have been reported in patients with hypercortisolaemia (19) and patients diagnosed with post-traumatic stress disorder (29). The decrease of NAA has been shown to be stronger in the left hippocampus compared to the right (30). This suggests the presence of a more complex mechanism, since the mere chronic exposure to stress hormones leads to an elevated NAA content (31). Patients diagnosed with social anxiety disorder display higher whole brain levels of glutamate and glutamine, and higher local glutamine and lower local GABA levels in the thalamus (32). Patients suffering from diabetes mellitus I-associated depression display a higher prefrontal glutamate–glutamine–gamma-aminobutyric acid (Glx) levels compared to healthy controls (33). Higher levels of myo-inositol and choline-containing compounds are found in the hippocampus of patients suffering from depression episodes (34).

1.2. Main MRI and ^{1}H MRS Results in Animal Models

The results presented above are based on comparisons between healthy and diseased individuals. Thus, cause and consequences of a disorder or the influence of pharmacological treatment cannot be disentangled. The advantage of using animal models in research is that the same animal can be investigated before and after the experimental manipulation, and when applicable, also during the treatment or recovery periods. This longitudinal approach also allows monitoring the biochemical changes (using MRS) within a single individual. Animal models allow differentiating between cause(s) and consequences of a disorder. The design of animal models is dedicated to mimic human-specific disorders based on certain validity criteria (face, predictive, and construct validities) with the goal to result in MRI and MRS changes comparable to the human disorders. It is therefore important for the researcher working with animal models of psychiatric disorders to be acquainted with the changes found in the brain of human subjects afflicted with a psychiatric disorder.

MRI and MRS studies on animal models of psychiatric disorder are less numerous and raise of course the question as to what extent they mimic the human disorder. Usually distinct symptoms are singled out of the complex syndrome of the psychiatric disorder in humans. Then, comparable conditions are designed in animals, for example by pharmacological manipulations, alteration of the genetic make-up, or inclusion of life events. This approach has to adhere to certain validity criteria, but is practical and useful in getting insight into the mechanism and the development of a disorder. The most common psychiatric disorders investigated with the aid of MRI and MRS using animal models are schizophrenia and the affective disorders.

1.2.1. Schizophrenia

This is a rather complex disorder and modeling schizophrenia exhaustively in animals by genetic manipulation is therefore a challenging task. The chakragati mouse has been shown to mimic certain symptoms of schizophrenia, like reduced social interaction, psychotic activity (persistent circling, in this case), and hyperactivity, which are inhibited by anti-psychotic drugs (35). An MRI and MRS study of this schizophrenia mouse model has been shown to display a significantly higher Cho/NAA ratio and enlargement of the ventricles, compared to the wild type mice (36).

An alternative way of modeling schizophrenia is by administration of psychosis-inducing drugs to healthy animals. Phencyclidine has been shown to generate a psychosis which closely resembles schizophrenia in both humans and animals, in that it mimics both the negative and the positive symptoms; also, it aggravates the disorder in already established patients (37–39). Reynolds et al. have successfully shown a reduction of NAA in the temporal and medial frontal cortex of rat models induced with phencyclidine (40).

1.2.2. Affective Disorders

It is generally accepted that stress is either the cause of or an important contributor to anxiety disorders, depression, and post-traumatic stress disorder. Animal models use manipulations of the stress system in diverse ways, from alterations of molecular mechanisms to introduction of acute and chronic stressful life events. In rats bred especially to display high anxiety and low anxiety characteristics, the hippocampal volume has been found to correlate inversely with the anxiety-like behaviour, while a direct correlation has been described in an unselected population of rats. Interestingly, the depression-like behaviour does not correlate with the hippocampal volume, suggesting a more complex relationship between these factors (41). The removal of the adrenal glands, combined with the controlled application of high corticosteroids in rats has shown that hypercorticism leads to a decrease in myo-inositol levels, a significant elevation of the glutamate levels, and a trend towards reduced NAA levels. The total brain volume but not the hippocampal volume has been found to be significantly reduced as a consequence of hypercorticism (42). In rats bred for learned helplessness, the glutamate/GABA ratio has been shown ex vivo to be significantly higher than in wild type rats (43).

While using animal models opens many possibilities for investigating psychiatric disorders, it is important to consider which stressors are included in the experimental procedures, e.g. by animal handling, and how this may affect the outcome of the investigation. For this reason, we present a series of guidelines and rules in order to minimize the detrimental influences on the animal and to allow lab-to-lab comparisons to be drawn. The legal aspect of working with animals is also addressed.

1.3. Guidelines for Working with Animals

International and national guidelines secure proper use of animals. We want to highlight several aspects:

- *Biological background*: The species, strain, gender, and age of the animal needs to be well documented and justified, not only to obtain legal approval, but also to ensure the validity of the data produced by the experiment. *See* **Note 2** for details on this aspect.

- *Handling*: Animals do not submit voluntarily to experimental procedures. This means that every handling attempt, from mere grabbing to the most painful procedure, constitutes a stressor uncontrollable and inescapable events elicit to a various degree a general adaptation response (also known as a stress response) from the animal. The stress reaction occurs for both physical and psychological stimuli and is translated into a cascade of biochemical responses, culminating with the secretion of stress hormones (corticosteroids, catecholamines) from the adrenal glands. The consequences of increased corticosteroid levels are changes in the release/uptake of many other hormones and neurotransmitters concentrations in the periphery and the brain, consequently influencing numerous physiological parameters (e.g. heart rate, blood pressure). More details on the stress system are given in **Fig. 1**. A series of adaptive response effects and their interference with an MRI/MRS experiment are discussed in more detail in the **Notes 1** and **2**.

Normal circumstances

Corticosteroid hormones are involved in the regulation of many physiological processes and their strict regulation is vital for health. Corticosteroids (CORT) are produced by the adrenals in a precise balance and are regulated by the pituitary gland and the centers in the brain. Corticosteroids are secreted in a circadian pattern, next to their well known response to physical and psychological stressors. The hypothalamus modulates the activity of the pituitary gland via corticotropin-releasing hormone (CRH). Following these signals, the pituitary gland secretes hormones that regulate cell growth, thyroid and adrenal functions, and the production of reproduction hormones. Adrenocorticotropin (ACTH) from the pituitary stimulates the adrenal glands and thus the production of corticosteroids (cortisol, corticosterone). This system is known as the hypothalamic–pituitary–adrenal (HPA) axis. Two types of CORT receptors in the brain constitute a negative feedback mechanism for the activity of the HPA axis. The main function of corticosteroids in normal circumstances is to regulate and maintain balance (homeostasis) in:
- The inflammatory response, by moderating it/slowing it down
- The metabolism and the insulin cascade
- The cardiovascular functions and blood pressure.

Exceptional circumstances

When a stimulus is *novel, unpredictable, perceived as threatening, or perceived as a loss of control*, a signal is relayed from the cortical areas to the hypothalamus and brain stem. An area in the brain stem known as the locus coeruleus becomes activated and increases its noradrenergic activity. This abundance of catecholamines predisposes the individual to instinctive impulsive actions, oriented mainly towards fight, flight or freeze, via two major routes:
- The catecholaminergic route – Acetylcholine released by the pre-ganglionic sympathetic nerves triggers the production of epinephrine (adrenaline); supplementary norepinephrine (noradrenalin) is also released, activating the sympathetic system, suppressing the para-sympathetic system and thus preparing the body for intense muscular impact.
- The HPA-axis route – The activation of the HPA-axis results in increased production of corticosteroids. Corticosteroid receptors are present in all the organs and mediate a change in each organ's activity (in order to offer maximum support to the fight-or-flight response), in the following manner:
  - Cell growth, digestion, reproduction, and immune activity are suppressed.
  - The metabolism is intensified and anabolism suppressed.
  - The blood flow is redirected from viscera towards the muscles.

The brain CORT receptors allow regaining homeostasis. A dysfunction of this system is detrimental for physical and mental health.
Levels of adrenaline and noradrenaline in the blood increase and decrease quickly, while the levels of corticosteroids remain higher for much longer periods. Therefore, it is common practice to quantify the severity of stress by measuring the blood corticosteroid level.

Fig. 1. Stress hormones and the onset of the stress reaction.

- *Legal aspect*: There are laws in every country that regulate the use of animals in research. Prior to the actual experiments, formal approval needs to be obtained from the appointed organization. Experimenters require a licence that gives them the legal right to experiment on animals. Also, the experimental protocol needs to be well documented and justified. The following three items are of relevance: (1) the number of animals employed needs to be minimal but allow statistical verification; MRI and MRS studies offer the great advantage of a longitudinal approach, resulting in a considerable reduction of the number of animals per experiment. Even after termination of the experiment and the sacrifice of the animal, numerous MRI experiments can be performed on the skull or on the excised brain; (2) the amount of suffering that the animals undergo needs to be as limited as possible. The suffering of an animal model of a psychiatric disorder subjected to a scanner is commonly estimated as moderate, as it combines the unpleasantness and stress of the long-time anaesthesia with the discomforts of the induced disorder. If extra procedures (e.g. surgical) are intended to be included in the study, the severity of the discomfort needs to be re-evaluated; (3) the use of animals should be essential to the investigation, in that it is not possible to replace the animal by in vitro methods or other technology. Modeling psychiatric disorders and especially their development require a living organism, in most cases a mammal, although vertebrates like zebrafish appear to be up-coming animal models. These three items, combined with the necessity to unravel the cause and mechanisms of psychiatric disorders to allow prevention and therapy, make a solid argument for using animal models to investigate psychiatric disorders.

2. Materials

2.1. Anaesthesia: the animal is usually kept under anaesthesia for the entire MRI experiment to prevent animal discomfort and motion artifacts. We prefer inhalation anaesthetics, as these are easily regulated. Alternatively, for relatively short MRI protocols, injectable anaesthetics may be used.
 1. Isoflurane and isoflurane vaporizer.
 2. Oxygen/air or oxygen/nitrogen carrier gas. This may be bottled or dispensed via a wall outlet. It is advisable to use dedicated flowmeters with a dosage range suitable for small animals (0.1–2 L/min). Some laboratories routinely use nitrous oxide as part of the gas

mixture because of the mild analgesic properties; however, be aware that overexposure to nitrous oxide is a severe health risk to the researcher.

3. Anaesthesia mask and tubing. These may be integrated into the animal bed of the MRI scanner.

4. Rodent induction box.

5. Scavenger unit. If you are working in a closed space, it is advisable to scavenge the anaesthetic gasses, e.g. with a carbon filter.

All supplies may be purchased via laboratory or veterinary supply companies (e.g. AgnTho's AB, VetTech).

2.2. Monitoring equipment required to assess the physical well-being of the animal:
1. Device for monitoring respiratory rate, e.g. a pressure-sensitive cushion.
2. Device for monitoring cardiac activity (optional).
3. Heating system with rectal temperature probe (or homeothermic pad).
4. Monitoring device for blood gases (optional).

Monitoring equipment often can be acquired through manufacturers of MRI systems, or alternatively at suppliers of laboratory or veterinary equipment (e.g. SA Instruments, Rapid Biomedical). All devices need to be MRI compatible.

2.3. Imaging equipment:

An animal MRI system (the most commonly used are the Bruker and Varian systems) with animal cradle and a resonator adjusted to the size of the head, usually about 15 mm diameter for a surface coil and 20–25 mm for a volume coil. For different magnetic field strengths or manufacturers, the protocols described in the Methods section need adjusting. Usually the manufacturer provides a number of standard protocols that are good starting points for your own experiment.

2.4. Data processing software:

MRI data: ImageJ is free software that allows the manual delineation of the regions of interest within the scans. If the obtained data are not in DICOM format, a plug-in specific for the type of data acquired by your system needs to be installed. The plug-ins are also freely available.

MRS data: The LCModel/LCMgui package produced by Stephen Provencher Inc., is currently the standard software for processing 1D MRS data. Besides the package, the user also needs a library of individual metabolite spectra.

3. Methods

MRI and MRS are techniques that allow in vivo investigation of tissue composition and structure. *See* **Note 3** for a few considerations on their limits. The present section includes a detailed description of basic MRI and MRS techniques performed on mice and rats under isoflurane anaesthesia. The MR protocols depend on the animal model (*see* **Note 4** for a discussion on the choice of animal model). Brief descriptions of several more advanced techniques that can be used in order to acquire more specific data are given in **Notes 5–7**.

3.1. Data Acquisition

The MRI sequences provided here are based on a vertical 9.4 T Micro-imaging system (Bruker Biospin, Rheinstetten, Germany). ParaVision 5.0 was used for image acquisition. The protocol is suitable for other systems and other manufacturers, but the parameters and routines may have different names; please check the manufacturer's manuals.

3.1.1. Prepare Scan Protocol

1. Decide on a scanning protocol, depending on the desired scan (*see* **Fig. 2** for examples).
   a. Use a T2W protocol for a normal anatomic scan (*see* **Table 1**)
   b. Use a T1W protocol if using a contrast agent (*see* **Table 2**)
   c. Use a FLAIR protocol to acquire more anatomical detail and suppress the bright signal from the ventricles (*see* **Table 3**)

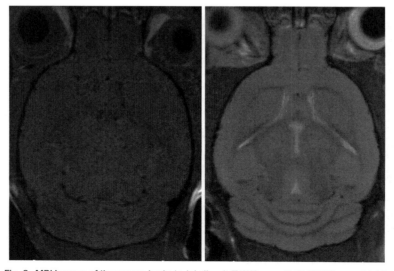

Fig. 2. MRI images of the mouse brain (axial slices): T1W image (*left*); T2W image (*right*).

Table 1
Low- and high-resolution T2W MRI settings for a 9.4 T spectrometer

T2W, 9.4 T	Mice		Rats	
Parameters	Low resolution	High resolution	Low resolution	High resolution
Spin echo/TURBO factor	Fast spin echo/8	Fast spin echo/8	Fast spin echo/8	Fast spin echo/8
TE/TEeff/TR (ms)	9/28/4000	9/28/4000	9/28/4000	9/28/4000
Res/FOV (mm^2)/matrix	$0.100^2/25.6^2/256$	$0.078^2/19.2^2/256^2$	$0.156^2/40^2/256^2$	$0.089^2/35^2/392^2$
Slice thickness (mm)/no. of slices	1/10	0.3/20	1/20	0.5/20
Averages/scan time (min)	1/2′8″	12/25′36″	1/2′8″	6/19′36″

Table 2
Low- and high-resolution T1W MRI settings for a 9.4 T spectrometer

T1W, 9.4 T	Mice		Rats	
Parameters	Low resolution	High resolution	Low resolution	High resolution
Spin echo/TURBO factor	Fast spin echo/4	Fast spin echo/4	Fast spin echo/4	Fast spin echo/4
TE/TEeff/TR (ms)	9/9/1300	9/9/1300	9/9/1300	9/9/1300
Res/FOV (mm^2)/matrix	$0.100^2/25.6^2/256$	$0.078^2/19.2^2/256^2$	$0.156^2/40^2/256^2$	$0.089^2/35^2/392^2$
Slice thickness (mm)/no. of slices	1/10	0.3/20	1/20	0.5/20
Averages/scan time (min)	2/2′4″	12/12′28″	1/1′2″	8/12′40″

2. Decide on a 3D or 2D geometry, depending on the amount of time available for scanning.
    a. Acquire a 3D scan, if enough time available. This will provide a high resolution image that allows multi-planar delineation of the region of interest, resulting in quite an accurate measure of its volume.
    b. Acquire a multi-slice 2D scan. In this case, the accuracy of the results can be increased by acquiring two data sets

Table 3
Low- and high-resolution FLAIR MRI settings for a 9.4 T spectrometer

FLAIR, 9.4 T	Mice		Rats	
Parameters	Low resolution	High resolution	Low resolution	High resolution
Spin echo/TURBO factor	FLAIR/4	FLAIR/4	FLAIR/4	FLAIR/4
TE/TEeff/TI/TR (ms)	10/10/825/3000	10/10/825/3000	10/10/825/3000	10/10/825/3000
Res/FOV (mm^2)/matrix	$0.133^2/25.6^2/192^2$	$0.100^2/19.2^2/256^2$	$0.156^2/40^2/256^2$	$0.089^2/35^2/392^2$
Slice thickness (mm)/no. of slices	1/10	0.5/20	1/20	0.5/20
Averages/scan time (min)	2/4'12"	8/22'24"	1/2'24"	6/22'3"

Table 4
^{1}H MRS settings for a 9.4 T spectrometer

^{1}H MRS, 9.4 T	Mice	Rats
Sequence	PRESS	PRESS
TE/TR (ms)	15/3500	15/3500
No. of points	2048	2048
No. of averages	512	256
Voxel size (μL)[a]	10	20
Scan time (min)	30	15

[a]Given for orientation purposes, it depends strictly on the region of interest.

in different orientations for each sample: one sagittal and one coronal.

3. If desired, include an MRS scan. The most commonly used MRS sequence is the point-resolved spectroscopy (PRESS) sequence (see **Table 4**).

4. If the standard parameters as given in **Tables 2–4** do not satisfy your needs or if you are working at different field strength, you may adjust the parameters (see **Note 8** for more details).

5. Repeat Steps 1–4 if several different scans are desired. Keep the total scan time preferably within 2 h; above 2 h of scanning, consider the effects of dehydration and physiological instability (see **Note 9**).

3.1.2. Animal Preparation

6. Decide on the most appropriate time of day for performing the scans. *See* **Note 10** for the interference of the circadian rhythm of rats and mice with the MR measurements.

7. Bring the animal into the scanning area (*see* **Note 11**) and allow habituation for about 60–120 min. Take care in handling and place the animal in an area of mild sensorial to excitation (thus preventing the animal from becoming stressed by bright light, loud noises, strong smells, etc.) (*see* **Notes 12** and **1**).

8. Direct the gas anaesthetics towards the induction chamber.

9. Turn on oxygen/air mixture at 0.3/0.3 L/min and set the isoflurane to 3.5%.
   a. For alternative anaesthetics, *see* **Note 2**.
   b. For measuring on awake animals, *see* **Note 13**.

10. Place the animal inside the induction chamber. Allow 3 min to achieve deep anaesthesia. Depth of anaesthesia can be checked by pinching the paws – no pain reflex should be present.

11. Redirect the isoflurane flow towards the MRI cradle. Place animal in the cradle in supine position and immediately place the nose of the animal in the isoflurane mask, to prevent awakening.

12. Turn on the heating system. Insert the rectal temperature probe, if available. Make sure that the animal remains under constant temperature (*see* **Note 14** for implications).

13. Place the pressure sensitive cushion under the thorax of the animal. Ensure close contact with the animal body (for example by taping a piece of tissue tightly over the back of the animal).

14. Reduce the isoflurane dosage to 2%. During the entire MR experiment, adjust the isoflurane dose in order to keep the breathing rate at a constant of 50 ± 15 bpm, in order to avoid moving artifacts. Usually the dose gradually decreases to about 1% for a 2 h scanning protocol.

15. Place the surface coil on the animal's head, or drag the animal into the volume coil.

16. Place an external reference tube (2 mM $CuSO_4$ in H_2O) near the head, inside the coil.

17. Fixate the head of the animal. Most animal beds have a tooth bar to hook behind the front teeth. Some animal beds have ear bars to prevent head motion. These should be inserted carefully to avoid rupturing the ear drums. A more animal-friendly method is to fixate the head and the coil with memory foam (e.g. standard noise-reducing ear plugs).

3.1.3. MRI

18. If your model requires the administration of certain drugs or contrast agents by i.v. injection immediately prior to the scanning, administer the i.v. injection now.

19. Place the cradle in the centre of the magnet and connect the coil to the system.

20. Create a new scan protocol ("new patient").

21. Load a pilot scan #1 provided by your system ("Pilot_GE" for a Bruker system) and tune the coil to resonance ("wobble").

22. Run the pilot scan to check the position of the brain in the centre of the coil: 2D gradient echo with TR = 100 ms, TE = 4 ms, one average, rf excitation = sinc7H; 1 ms; FA = 15°, three orthogonal slices of 1 mm with FOV = 30 × 30 mm, matrix 128 × 128, scan time 13 s. When you run scans using automatic acquisitions ("traffic light"), shimming, pulse calibration, and receiver gain adjustment will be done automatically.

23. Run a second pilot scan and edit the geometry based on the first scout scan. Aim to place three perpendicular fields that intersect on the median lines of the brain.

24. If necessary, repeat Step 23 after re-editing the geometry based on each previous scout scan, until the positioning on the median lines is acquired.

25. Open the predefined MRI protocols as determined in Steps 1–5, adjust the geometry based on the last acquired pilot scan, and run the required protocols. If no MRS scans will be performed at this point, continue with Step 32.

3.1.4. MRS

26. Open the PRESS protocol without water suppression, define the voxel geometry and place it such that the centre of the voxel coincides with the centre of the area of interest.

27. Ensure that the voxel homogeneity is optimal, using localized shim protocols. Bruker has several options:
    a. Run "method-specific adjustment" of the field homogeneity for the PRESS protocol. Ensure that the voxel is placed in the correct area. Consult the manufacturer's manual for more details.
    b. Run a Fastmap protocol according to the manufacturer's instructions.
    c. Run a Fieldmap protocol according to manufacturer's instructions.
    Some tips can be found in **Note 15**.

28. Record a water spectrum (water suppression off) with the same TE and TR as the PRESS protocol. Only one average is needed.

29. Determine the line width of the water peak at full-width-half-height (FWHH). At a Bruker system you can use the macro "CalcLinewidth" to do this automatically. FWHH should be below 30 Hz at 9.4 T. With proper shimming, line widths of about 15 Hz can be obtained in homogeneous brain areas. If necessary, repeat Steps 26–28.

30. Open the PRESS sequence and run it on the same voxel, with the water suppression and the outer volume suppression ON, according to your imaging system. An example of a ^{1}H MRS spectrum at 9.4 T has been acquired by Tkac et al. (44) in the mouse brain (*see* **Fig. 3** and **Table 5**). Note that this spectrum was acquired using a very sophisticated shim set-up and you will usually obtain results that have broader lineshapes.

3.1.5. Recovery After MRI

31. Remove the animal from the magnet and allow recovery in a well-ventilated dark warm chamber, e.g. by placing a heating lamp above the cage.

32. Check for signs of recovery (the pedal reflex should be noticeable after 1 min).

33. Check for signs of activity (the animal should be completely recovered within 15 min). Avoid transport to the facility before the animal is completely recovered.

3.2. MRI Data Processing

3.2.1. Manual Delineation and Classification of Images

1. The experienced observer who will delineate the areas of interest should be blinded to the experimental set-up, in order to avoid bias. Rename the measurement files if necessary.

2. Load an individual scan into the analysis software of choice. A commonly used package is the free-ware image processing package ImageJ (http://rsbweb.nih.gov/ij/).

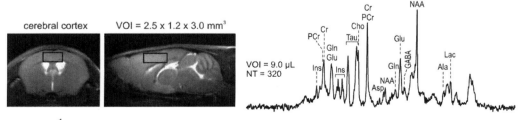

Fig. 3. In vivo ^{1}H NMR spectrum of the mouse brain (C57BL/6) with the volumes of interest (VOIs) centred in the cerebral cortex. Ins, myo-inositol; PCr, phosphocreatine; Cr, creatine; Gln, glutamine; Glu, glutamate; Tau, taurine; Cho, choline; Asp, aspartate; NAA, N-acetyl-aspartate; GABA, gamma-aminobutyric acid; Ala, alanine; Lac, lactate; VOI, volume of the voxel; NT, number of averages. Reproduced with permission from Ref. (82).

Table 5
Most common brain metabolites involved in psychiatric disorders and detectable in mice and rats

Neuro-chemical compound	Peak position (ppm)	Biological interpretation
NAA	2.02	Neuronal marker, the decrease may reflect neuronal loss, decreased neuronal viability, impaired neuronal function
GABA	2.3	Inhibitory neurotransmitter
Glutamate	2.3	Neurotransmitter, high levels may reflect neuronal damage as effect of anoxia or brain injury, may indicate dopaminergic dysfunction
Glutamine	2.3	Amino acid, serves as substrate for glutamate in the glutamate–glutamine cycle
Myo-inositol	3.56	Sugar, involved in neuronal signalling systems; neuronal marker and astrocyte marker, involved in cell membrane metabolism and osmoregulation
Choline	3.23	Neuronal constituent, sometimes used as reference for quantifying other peaks, may be involved in modulating synaptic activity, was linked with membrane turnover and biosynthesis of membrane-bound phosphatidylcholine
Creatine/ phosphocreatine	3.02	Brain metabolite considered stable, presently used as internal standard

3. Find a slice that can be associated with most certainty with its correspondent histological section in a mouse brain atlas (e.g. www.mbl.org);
4. Open the ROI Manager.
5. Draw the contour of the region of interest in the particular slice, using the histology atlas as a reference guide.
6. If the contour is sufficiently precise, save the region of interest in the ROI Manager.
7. Proceed to the next slice and repeat Steps 5–6.
8. When all areas of interest have been delimited, the collection of ROIs can be saved.
9. Calculate the area of delimited regions using the "Measure" function in the ROI manager. It will produce a table that can be copied to an Excel file. For quantitative volume measurements, you need to multiply the

number of voxels with the voxel volume, which is (in-plane resolution)^2 × slice thickness (including slice gap).

10. The table produced by ImageJ contains also the average intensity of the delineated regions; to assess signal intensity changes, this average intensity should be normalized by an internal or external reference. The reference tube can be used as an external reference. If this is not available, an area of tissue that is assumed to remain constant can be used, such as muscle.

11. More details on image processing and volumetry can be found in **Chapter 15**.

3.2.2. MRS Data Processing

1. Load an individual scan into the LCMgui.
2. Load the water peak corresponding to the same MRS scan.
3. LCModel will open a window in which attenuation factors can be provided. These can be calculated from the T_1 and T_2 decay times of the water in the region of interest, in order to have the correct estimation of the metabolites concentrations. A formula for the calculation can be found in the LCModel manual (http://s-provencher.com/pub/LCModel/manual/manual.pdf). T_1, T_2, and proton density values have to be obtained experimentally, or literature values can be used.
4. Run the processing.
5. The output will contain concentrations of metabolites accompanied by the Cramer–Rao lower bound as an estimate of variance. Commonly, a threshold of 20% is used to discard fits with insufficient accuracy and therefore of the estimated concentration.
6. If the fitting procedure does not produce reliable results (e.g. too many parameters require exclusion, the baseline is not fitted properly), please refer to the LCModel manual for a more detailed explanation of the procedure.

4. Notes

1. *Handling*: Proper handling is necessary for the safety of both the experimenter and the animal. Improper handling can lead to stress reaction in the animals and potential injury (45). The stress reaction generated by handling can be partially avoided by allowing the animals to get used to the experimental environment, for example by placing the cage in the experimental area at least 1 h before starting the experiment. When more animals are measured on the same

day, the last animal will be more habituated to the new environment of the lab than the first animal. Therefore, it is necessary to rotate the animal measurements between days, to avoid measuring repeatedly the same animal at the same time of the day.

It is important to add that every animal responds differently to different researchers, as every experimenter has its own manner of handling the animals. When the animals are handled by different persons, this can artificially introduce variations in the obtained results. Ideally, only one person should handle the animals during all the experiments.

Repeated handling in longitudinal studies: Repeated handling can lead to either habituation or chronic stress. When the handling is done carefully, habituation occurs, which means that initial stress responses decline or vanish. The animal gets used to the procedure. Moreover, the amplitude of the stress response to the same stressor can vary between animals. This can induce uncontrolled and undesired changes in time, and thus may lead to larger variation of results. It is usually helpful to allow the animals a fixed habituation period before actively starting the experiment.

On the other hand, when the handling is experienced by the animal as very stressful, the animal can develop a chronic stress symptomatology. For example, repeated restraint has been reported to lead to suppressed neurogenesis (46). In order to avoid these problems, it is recommended that the animals be handled according to the existing standard protocols, designed to minimize/avoid inducing a chronic stress condition unintentionally.

To control the effects of circadian rhythm, the animals should all be tested at the same time of day. Allow the animals a minimum of 2 weeks to adjust to the circadian cycle of the laboratory, as the provider might have used different day/night settings. If the number of animals is too large and imposes several measurements per day, the animals should be submitted to experiments in a randomized manner, in which both the control and the experimental groups are measured at different time points during the day.

2. *Effects of anaesthesia, sedation, and other pharmaca*: To ensure the reliability of data, animals have to be totally immobile during an MRI or MRS experiment. This is usually achieved by anaesthesia or heavy sedation. Also, certain pharmaca are used to induce a disorder. Anaesthetics and other drugs have been shown to alter the proton relaxation times and may thus interfere with the experiment (47–49)

if they are still present in the organism when the experiments are performed. In some cases, the variation is also tissue dependent. A number of compounds have been investigated with respect to their ability to interfere with MRI or MRS experiments. All these characteristics make the choice of the anaesthetic a delicate aspect and will be discussed in more detail in this section. The alternative – using no anaesthetics, will be addressed briefly thereafter.

It has been known since the 1980s that inhalational anaesthetics (halothane, enflurane, and isoflurane) can change both the T_1 and T_2 relaxation times, while injectable anaesthetics (such as Innovar, pentobarbital, ketamine, and alpha-chloralose) do not (48). While injectable anaesthetics might appear more convenient, they are only suitable for rather short experiments due to the duration of action, though some researchers have development protocols for maintaining anaethesia in the magnet. Continuous subcutaneous infusion of a sedative has also been successfully performed, enabling researchers to use non-interfering injectable compounds to keep the animals immobile while avoiding the side effects typical for alpha-chloralose. The method was developed by Weber et al. (50), who used medetomidine antagonized at the end of experiment by atipamezole administered intraperitoneally (50).

Nevertheless, the depth of anaesthesia is more easily controllable for inhalation anaesthetics and is therefore associated with fewer fluctuations in the MRI signal (49). Isoflurane has nowadays replaced halothane as the anaesthetic of choice, because of the shorter recovery period and the reduced health risk to the researcher compared to halothane (51). Anaesthetics and other drugs, can influence the outcome of an MR experiment also indirectly. For example, they can decrease the blood pressure by inducing blood withdrawal and suppressing cerebro-vascular reactivity (52). The result is a baseline with lower variations than in the case of a conscious animal, and with a lower signal-to-noise ratio as well (53). Differences in cerebrovascular reactivity have been found between the awake and anaesthetized animals when certain anaesthetics, such as urethane, are used (54, 55). Other anaesthetics, such as isoflurane, appear to have a less negative impact (56, 57).

The choice of the anaesthetic is equally important. The anaesthetic equithesin was reported to generally inhibit the brain response to stimuli, while isoflurane allowed a robust neural activation (58). Some anaesthetics are less appropriate for longitudinal measurements. Alpha-chloralose can only be employed in terminal experiments, due to its slow elimination, numerous side effects and the fact that it induces peritonitis and adynamic ileus (59). Other anaesthetics, such as isoflurane, allow fast and complete recovery and are suitable for longitudinal studies (60).

3. There are various internal and external factors that can influence the outcome of an MRI/MRS experiment. For didactic purposes, they can be divided into three categories: homeostasis-related factors (a stress response is triggered), magnetic property-related factors (T_1 and T_2 are affected), and experiment-dependent factors (certain aspects in the experimental set-up condition the outcome of the measurements). For example, a certain anaesthetic or drug can trigger a stress response due to its toxicity, can affect T_1 or T_2, or/and can make a longitudinal experiment impossible due to the insufficient recovery of the animal after the anaesthesia.

   MRI volumetric studies are affected mainly by chronically present factors, whilst the metabolic profiles detected by MRS are more susceptible to alterations due to the acute condition of the animal at the time of measurement. Consequently, strict mathematical corrections should be applied and the experimental set-up needs to be very carefully designed in order to avoid bias.

   While it is relatively straightforward to follow a protocol describing how to carry out an MR measurement for rodent models for psychiatric disorders, the design of the overall experimental set-up is crucial, as all aspects of animal handling may trigger stress responses and thereby may directly influence the experimental outcome.

4. *Type, strain, and gender of the animal*: Rats and mice are commonly used for MRI assessment of psychiatric models. In general, rats are more "cooperative" and less aggressive, and their bigger size enables the researcher to handle and investigate them more easily. On the other hand, mice offer the advantage of having a completely mapped genome – making transgenic models readily available. Moreover, for several disorders only mice models exist. These aspects are of consequence and need to be considered when the optimal model is chosen for a study.

   After deciding on the species, the peculiarities of the gender and strain need to be taken into consideration. Strain-related differences in the size of brain areas exist. For example, Flinders Sensitive-Line and Wistar-Kyoto rats, while both used as models for depression, have been shown to display opposite behavioral patterns under the influence of stressors (61). Gender-related dimorphism of the brain has also been shown (62). Thus, results obtained with different strains or gender should be carefully interpreted. It is important to decide prior to the experiment which published works will serve as background, and employ the same strain and gender for comparison purposes.

5. When the anatomical details are not sufficient, manganese-enhanced MRI (MEMRI) can be employed. In this case, manganese ions enter cells through voltage-gated calcium channels after neuronal activation. As not all brain areas are equally active, this improves the image contrast. Manganese enhancement can be detected by T1W imaging or FLAIR and usually a 3D sequence is used to obtain an isotropic full brain image.

6. Apart from 1D ^{1}H MRS, there are several other methods that have not been used extensively, but may have great potential. The following paragraphs describe them briefly. In contrast to ^{1}H, ^{13}C MRS is very insensitive, mainly due to its natural abundance of only 1.1%. However, if a substrate selectively enriched with ^{13}C is administered to the animal, this insensitivity turns into an advantage, since it then becomes possible to track the activity of the metabolic pathways specifically for the enriched substrate. This is known as ^{13}C cycling. Various substrates can be used, depending on which process is to be monitored. For the study of psychiatric disorders, the most relevant cycle is the glutamate–glutamine cycle (63, 64), as glutamate is the most important excitatory neurotransmitter in the mammalian brain (64–67). Abnormalities in the glutamate–glutamine cycle, involving both neurons and astrocytes, are thought to participate in many psychiatric disorders. A great tool to study metabolic pathways and metabolic interactions between the neurons and the astrocytes is MRS after simultaneous injection of [1,2-^{13}C] acetate and [1-^{13}C] glucose (68, 69). The most important resonances detected this way represent glutamate, glutamine, GABA, NAA, inositol, and glucose (70).

7. While some cerebral metabolites can be quantified reliably in absolute values from 1D ^{1}H MRS (e.g. NAA, creatine (Cr), and choline (Ch)), other metabolites (such as glutamate/glutamine, myo-inositol, aspartate, and GABA) can overlap, making the interpretation of the in vivo spectra difficult. In contrast to localized 1D MRS, localized two-dimensional (2D) MRS overcomes the problem of spectral overlap considerably because the resonances are spread over a two-dimensional surface rather than along a single frequency dimension. The most important drawback of these methods is the long acquisition time compared to one-dimensional MRS. More information about 2D

methods for mouse brain can be found in the papers of Meric et al. (71) and Braakman et al. (72).

8. Any protocols can be adjusted for various purposes, keeping in mind two constrictive aspects: the time required for scanning and the signal to noise ratio (SNR). SNR is proportional to the volume of the voxel and to the square root of the number of averages and phase-encoding steps. Decreasing the FOV, increasing the resolution, or decreasing the slice thickness will decrease the SNR. Likewise, increasing the FOV, decreasing the resolution, and increasing the slice thickness will improve the SNR.

9. *Dehydration*: When the animal is submitted to a prolonged MRI measurement, dehydration may occur. Dehydration can affect the MRI results as it diminishes the proton density in the tissue. Also, it has been shown in rabbits that a decrease in the relaxation times of the tissue can be expected as a consequence of a drop in tissue water content (73). The time length of the MR measurement is therefore an important aspect. When planning a long measurement, an intra-peritoneal infusion with a solution of sodium chloride (and optionally glucose) can prevent dehydration. As the water intake takes place during the dark period, usually at the same time with food intake, it is recommended to plan the experiments outside the animal's feeding and drinking time.

10. *Circadian rhythm*: In the lab, rats and mice are housed on a 12:12 h light/dark cycle. They are nocturnal animals and their circadian peak of activity occurs around the time the light is switched off (74). In most labs, lights are on between 7:00 and 19:00 h – which is the inactive (sleeping) period of mice and rats, and thus, parallels our human diurnal activity rhythm. The corticosteroid concentrations follow a circadian pattern: in nocturnal animals corticosteroid secretion is very low in the morning and the first hours of the light (inactive) period and increase during the course of the afternoon. Circadian rhythm is therefore another aspect to be considered for each procedure that is being performed on the animal. Anaesthesia, for instance, takes longer to have an effect and requires a higher amount of drug when corticosteroids are elevated – either due to the circadian or stress-induced corticosteroid peaks. MRI and MRS measurements around circadian peak times are best avoided. Since day/night conditions can be set artificially in the lab, the "day" can be fortunately programmed to overlap with the period of measurement. For obvious reasons, the day/night settings should be kept constant

during the entire experiment, unless changing the cycle is specifically dictated by the experimental set-up.

11. *Transport*: To avoid the stress caused by transport, the animals can be anaesthetized in their housing facility. The option should however be used with moderation, and only if the transport-related distress is estimated to be higher than the anaesthesia-related distress.

12. *Stress*: The stress response is coordinated partly by the hypothalamic–pituitary–adrenal (HPA) axis which constitutes a major component of the neuro-endocrine system; this controls and regulates many processes in the organism, such as digestion, metabolism and anabolism, blood circulation, immune response, cell growth, sexual response, moods, and emotions, etc. As mentioned in the introductory section, stress resulting from the experimental procedure is an important confounder in all animal experiments; also, vulnerability to stress is critical for certain psychiatric disorders such as depression and post-traumatic stress disorders. Systemic and local corticosteroid concentrations are generally accepted as markers for HPA axis activity, expressing the experienced level of stress. **Figure 1** describes schematically the physiological mechanisms of the onset of stress reaction.

Therefore, researchers have to be aware of stressors in the lab due to experimental procedures, to avoid or at least control them. Animal–researcher interactions can activate the HPA axis. Stress responses are species and strain dependent and depend on the history of the animal. Thus, the intensity of the stress response depends on the genetic make-up, the age, the perceived strength of a threat, and the duration and frequency of previous exposures to stress. The time and manner of occurrence of a stressor in relation to the experimental procedure, its duration, and the intensity of the stress response should be assessed prior to the experiment.

For example, prolonged or repeated stress can delay recovery or affect the therapeutic effect of drugs. Acute stress due to improper handling can delay the action of an anaesthetic and can lead to unstable heart rate/blood pressure/breathing under anaesthesia. Also, it can render the animal highly "uncooperative", which can lead to escape, and/or injuries to either the animal or the researcher. Chronic stress due to long-term exposure to unpredictable, uncontrollable conditions can render the whole adaptive system less flexible to new stressors and generate inadequate responses during the experiment.

13. *Measuring on awake/aware animal*: Recently, mechanical restraining of the animal has been used as an alternative to anaesthesia. Several groups have reported being able to perform the investigations on restrained awake rats placed in especially designed devices (52, 75). In order to perform such experiments, the animals need to be previously trained for these measurements. The training is rather focused on accustoming the animal to the restraining device in order to prevent struggle attempts, rather than a "volunteer" immobility, and needs a skilled practitioner with long experience in working with animals. The procedure is more easily implemented with rats; mice are less likely to "cooperate" than rats and are therefore more difficult to train, though MRI experiments have been performed with a certain degree of success on restrained awake mice as well (76). Quite important here is that restraint is a very powerful stressor.

    It is interesting to note that for partial body restraint (head), a degree of adaptation was achieved after 14 days of daily identical restraint, both for rats and mice. Gradually increased restraint however prevented adaptation (77). The achieved adaptation is partial and it is associated with exaggerated stress responses to novel stimuli (78). Changes in the type of restraint are therefore not advisable and measures need to be taken to keep the whole protocol as identical as possible for each measurement. It is also important to bear in mind that different strains react differently to acute and chronic stressful events. The effects of the restraint in rats and mice influence brain morphology and come with a variety of physiological and behavioral effects (79, 80).

14. Temperature: Due to immobility and anaesthesia, the animal is prone to hypothermia. A drop in, or fluctuations of the body temperature affects the intensity of the signal, due to a temperature-dependent change in the magnetization of the protons ($0.32\%/°C$) and in water density ($\sim 0.03\%/°C$) (81). Hypothermia is rather easy to deal with: the animal can be warmed abdominally with the aid of an isothermal heating pad that is controlled by a rectal thermometer during the MR experiment. Alternatively, warm air can be used, though this can increase the rate of dehydration. After the MR experiment, the animal needs to be returned to its cage under an infrared lamp until it recovers from anaesthesia, with care taken not to overheat the animal.

15. Shimming of a small voxel for MRS can be quite challenging. On a Bruker system, several tools are available

to improve the line width. A proper shimming procedure should have the following steps:

1. Load a good shim file. This can be a stored shim setting of a previous experiment. However, in this case, you need to be sure that the positioning of your animal is exactly the same as in that experiment. Alternatively, a shim set for a water phantom is usually a good start. You can determine the line width for different shim files to see which one gives the best results.
2. Do a global shim using first- and second-order shims.
3. Shim locally using one of the shim tools mentioned in the methods.

If you are using the method-specific shim tool, it usually helps to include second-order shims in your shim procedure. You can also increase the number of iterations to improve the shimming. If you are using Fastmap, the Bruker manual gives detailed instructions on the procedure. For mouse models, you standardly have to decrease the stick dimensions, but increase the number of averages to improve the SNR. If the Fastmap procedure does not yield useable line widths, you may run the procedure again to further improve the shimming.

References

1. Abou-Saleh, M. T. (2006) Neuroimaging in psychiatry: an update. *J. Psychosom. Res.* **61**, 289–293.
2. Barkley, R. A., Grodzinsky, G., and DuPaul, G. J. (1992) Frontal lobe functions in attention deficit disorder with and without hyperactivity: a review and research report. *J. Abnorm. Child Psychol.* **20**, 163–188.
3. Mattes, J. A. (1980) The role of frontal lobe dysfunction in childhood hyperkinesis. *Compr. Psychiatry* **21**, 358–369.
4. Bush, G., Valera, E. M., and Seidman, L. J. (2005) Functional neuroimaging of attention-deficit/hyperactivity disorder: a review and suggested future directions. *Biol. Psychiatry* **57**, 1273–1284.
5. Blumberg, H. P., Kaufman, J., Martin, A., Whiteman, R., Zhang, J. H., Gore, J. C., Charney, D. S., Krystal, J. H., and Peterson, B. S. (2003) Amygdala and hippocampal volumes in adolescents and adults with bipolar disorder. *Arch. Gen. Psychiatry* **60**, 1201–1208.
6. Chang, K., Karchemskiy, A., Barnea-Goraly, N., Garrett, A., Simeonova, D. I., and Reiss, A. (2005) Reduced amygdalar gray matter volume in familial pediatric bipolar disorder. *J. Am. Acad. Child Adolesc. Psychiatry* **44**, 565–573.
7. Rosso, I. M., Killgore, W. D., Cintron, C. M., Gruber, S. A., Tohen, M., and Yurgelun-Todd, D. A. (2007) Reduced amygdala volumes in first-episode bipolar disorder and correlation with cerebral white matter. *Biol. Psychiatry* **61**, 743–749.
8. Andreone, N., Tansella, M., Cerini, R., Rambaldelli, G., Versace, A., Marrella, G., Perlini, C., Dusi, N., Pelizza, L., Balestrieri, M., Barbui, C., Nose, M., Gasparini, A., and Brambilla, P. (2007) Cerebral atrophy and white matter disruption in chronic schizophrenia. *Eur. Arch. Psychiatry Clin. Neurosci.* **257**, 3–11.
9. Steen, R. G., Hamer, R. M., and Lieberman, J. A. (2005) Measurement of brain metabolites by ^{1}H magnetic resonance spectroscopy in patients with schizophrenia: a systematic review and meta-analysis. *Neuropsychopharmacology* **30**, 1949–1962.
10. Molina, V., Sanchez, J., Reig, S., Sanz, J., Benito, C., Santamarta, C., Pascau, J., Sarramea, F., Gispert, J. D., Misiego, J. M., Palomo, T., and Desco, M. (2005) N-acetyl-aspartate levels in the dorsolateral prefrontal

cortex in the early years of schizophrenia are inversely related to disease duration. *Schizophr. Res.* **73**, 209–219.
11. Ohrmann, P., Siegmund, A., Suslow, T., Spitzberg, K., Kersting, A., Arolt, V., Heindel, W., and Pfleiderer, B. (2005) Evidence for glutamatergic neuronal dysfunction in the prefrontal cortex in chronic but not in first-episode patients with schizophrenia: a proton magnetic resonance spectroscopy study. *Schizophr. Res.* **73**, 153–157.
12. Abbott, C., and Bustillo, J. (2006) What have we learned from proton magnetic resonance spectroscopy about schizophrenia? A critical update. *Curr. Opin. Psychiatry* **19**, 135–139.
13. Luders, E., Narr, K. L., Hamilton, L. S., Phillips, O. R., Thompson, P. M., Valle, J. S., Del'Homme, M., Strickland, T., McCracken, J. T., Toga, A. W., and Levitt, J. G. (2009) Decreased callosal thickness in attention-deficit/hyperactivity disorder. *Biol. Psychiatry* **65**, 84–88.
14. Seidman, L. J., Valera, E. M., and Makris, N. (2005) Structural brain imaging of attention-deficit/hyperactivity disorder. *Biol. Psychiatry* **57**, 1263–1272.
15. Carrey, N. J., MacMaster, F. P., Gaudet, L., and Schmidt, M. H. (2007) Striatal creatine and glutamate/glutamine in attention-deficit/hyperactivity disorder. *J. Child. Adolesc. Psychopharmacol.* **17**, 11–17.
16. Sheline, Y. I., Mittler, B. L., and Mintun, M. A. (2002) The hippocampus and depression. *Eur. Psychiatry* **17**(Suppl 3), 300–305.
17. Mervaala, E., Fohr, J., Kononen, M., Valkonen-Korhonen, M., Vainio, P., Partanen, K., Partanen, J., Tiihonen, J., Viinamaki, H., Karjalainen, A. K., and Lehtonen, J. (2000) Quantitative MRI of the hippocampus and amygdala in severe depression. *Psychol. Med.* **30**, 117–125.
18. Vakili, K., Pillay, S. S., Lafer, B., Fava, M., Renshaw, P. F., Bonello-Cintron, C. M., and Yurgelun-Todd, D. A. (2000) Hippocampal volume in primary unipolar major depression: a magnetic resonance imaging study. *Biol. Psychiatry* **47**, 1087–1090.
19. Brown, E. S., D, J. W., Frol, A., Bobadilla, L., Khan, D. A., Hanczyc, M., Rush, A. J., Fleckenstein, J., Babcock, E., and Cullum, C. M. (2004) Hippocampal volume, spectroscopy, cognition, and mood in patients receiving corticosteroid therapy. *Biol. Psychiatry* **55**, 538–545.
20. Bremner, J. D. (2002) Structural changes in the brain in depression and relationship to symptom recurrence. *CNS Spectr.* **7**, 129–130, 135–129.
21. Bremner, J. D., Vythilingam, M., Vermetten, E., Nazeer, A., Adil, J., Khan, S., Staib, L. H., and Charney, D. S. (2002) Reduced volume of orbitofrontal cortex in major depression. *Biol. Psychiatry* **51**, 273–279.
22. Bremner, J. D., Narayan, M., Anderson, E. R., Staib, L. H., Miller, H. L., and Charney, D. S. (2000) Hippocampal volume reduction in major depression. *Am. J. Psychiatry* **157**, 115–118.
23. Krishnan, K. R., McDonald, W. M., Escalona, P. R., Doraiswamy, P. M., Na, C., Husain, M. M., Figiel, G. S., Boyko, O. B., Ellinwood, E. H., and Nemeroff, C. B. (1992) Magnetic resonance imaging of the caudate nuclei in depression. Preliminary observations. *Arch. Gen. Psychiatry* **49**, 553–557.
24. Drevets, W. C., Price, J. L., Simpson, J. R., Jr., Todd, R. D., Reich, T., Vannier, M., and Raichle, M. E. (1997) Subgenual prefrontal cortex abnormalities in mood disorders. *Nature* **386**, 824–827.
25. Hirayasu, Y., Shenton, M. E., Salisbury, D. F., Kwon, J. S., Wible, C. G., Fischer, I. A., Yurgelun-Todd, D., Zarate, C., Kikinis, R., Jolesz, F. A., and McCarley, R. W. (1999) Subgenual cingulate cortex volume in first-episode psychosis. *Am. J. Psychiatry* **156**, 1091–1093.
26. Drevets, W. C. (2000) Neuroimaging studies of mood disorders. *Biol. Psychiatry* **48**, 813–829.
27. Soares, J. C., and Mann, J. J. (1997) The anatomy of mood disorders – review of structural neuroimaging studies. *Biol. Psychiatry* **41**, 86–106.
28. Soares, J. C., and Mann, J. J. (1997) The functional neuroanatomy of mood disorders. *J. Psychiatr. Res.* **31**, 393–432.
29. Schuff, N., Neylan, T. C., Fox-Bosetti, S., Lenoci, M., Samuelson, K. W., Studholme, C., Kornak, J., Marmar, C. R., and Weiner, M. W. (2008) Abnormal N-acetylaspartate in hippocampus and anterior cingulate in post-traumatic stress disorder. *Psychiatry Res.* **162**, 147–157.
30. Mohanakrishnan Menon, P., Nasrallah, H. A., Lyons, J. A., Scott, M. F., and Liberto, V. (2003) Single-voxel proton MR spectroscopy of right versus left hippocampi in PTSD. *Psychiatry Res.* **123**, 101–108.
31. Neylan, T. C., Schuff, N., Lenoci, M., Yehuda, R., Weiner, M. W., and Marmar, C. R. (2003) Cortisol levels are positively correlated with hippocampal N-acetylaspartate. *Biol. Psychiatry* **54**, 1118–1121.
32. Pollack, M. H., Jensen, J. E., Simon, N. M., Kaufman, R. E., and Renshaw, P. F.

(2008) High-field MRS study of GABA, glutamate and glutamine in social anxiety disorder: response to treatment with levetiracetam. *Prog. Neuropsychopharmacol. Biol. Psychiatry* **32**, 739–743.
33. Lyoo, I. K., Yoon, S. J., Musen, G., Simonson, D. C., Weinger, K., Bolo, N., Ryan, C. M., Kim, J. E., Renshaw, P. F., and Jacobson, A. M. (2009) Altered prefrontal glutamate-glutamine-gamma-aminobutyric acid levels and relation to low cognitive performance and depressive symptoms in type 1 diabetes mellitus. *Arch. Gen. Psychiatry* **66**, 878–887.
34. Milne, A., MacQueen, G. M., Yucel, K., Soreni, N., and Hall, G. B. (2009) Hippocampal metabolic abnormalities at first onset and with recurrent episodes of a major depressive disorder: a proton magnetic resonance spectroscopy study. *Neuroimage* **47**, 36–41.
35. Dawe, G. S., and Ratty, A. K. (2007) The chakragati mouse: a mouse model for rapid in vivo screening of antipsychotic drug candidates. *Biotechnol. J.* **2**, 1344–1352.
36. Torres, G., Hallas, B. H., Gross, K. W., Spernyak, J. A., and Horowitz, J. M. (2008) Magnetic resonance imaging and spectroscopy in a mouse model of schizophrenia. *Brain Res. Bull.* **75**, 556–561.
37. Luby, E. D., Cohen, B. D., Rosenbaum, G., Gottlieb, J. S., and Kelley, R. (1959) Study of a new schizophrenomimetic drug; sernyl. *AMA Arch. Neurol. Psychiatry* **81**, 363–369.
38. Allen, R. M., and Young, S. J. (1978) Phencyclidine-induced psychosis. *Am. J. Psychiatry* **135**, 1081–1084.
39. Javitt, D. C., and Zukin, S. R. (1991) Recent advances in the phencyclidine model of schizophrenia. *Am. J. Psychiatry* **148**, 1301–1308.
40. Reynolds, L. M., Cochran, S. M., Morris, B. J., Pratt, J. A., and Reynolds, G. P. (2005) Chronic phencyclidine administration induces schizophrenia-like changes in N-acetylaspartate and N-acetylaspartylglutamate in rat brain. *Schizophr. Res.* **73**, 147–152.
41. Kalisch, R., Schubert, M., Jacob, W., Kessler, M. S., Hemauer, R., Wigger, A., Landgraf, R., and Auer, D. P. (2006) Anxiety and hippocampus volume in the rat. *Neuropsychopharmacology* **31**, 925–932.
42. Schubert, M. I., Kalisch, R., Sotiropoulos, I., Catania, C., Sousa, N., Almeida, O. F., and Auer, D. P. (2008) Effects of altered corticosteroid milieu on rat hippocampal neurochemistry and structure – an in vivo magnetic resonance spectroscopy and imaging study. *J. Psychiatr. Res.* **42**, 902–912.
43. Sartorius, A., Mahlstedt, M. M., Vollmayr, B., Henn, F. A., and Ende, G. (2007) Elevated spectroscopic glutamate/gamma-amino butyric acid in rats bred for learned helplessness. *Neuroreport* **18**, 1469–1473.
44. Tkac, I., Henry, P. G., Andersen, P., Keene, C. D., Low, W. C., and Gruetter, R. (2004) Highly resolved in vivo ^{1}H NMR spectroscopy of the mouse brain at 9.4 T. *Magn. Reson. Med.* **52**, 478–484.
45. Donovan, J., and Brown, P. (2004) Handling and restraint. *Curr. Protoc. Neurosci.* Appendix 4, Appendix 4D.
46. Pham, K., Nacher, J., Hof, P. R., and McEwen, B. S. (2003) Repeated restraint stress suppresses neurogenesis and induces biphasic PSA-NCAM expression in the adult rat dentate gyrus. *Eur. J. Neurosci.* **17**, 879–886.
47. Chen, Y. C., Galpern, W. R., Brownell, A. L., Matthews, R. T., Bogdanov, M., Isacson, O., Keltner, J. R., Beal, M. F., Rosen, B. R., and Jenkins, B. G. (1997) Detection of dopaminergic neurotransmitter activity using pharmacologic MRI: correlation with PET, microdialysis, and behavioral data. *Magn. Reson. Med.* **38**, 389–398.
48. Karlik, S. J., Fuller, J., and Gelb, A. W. (1986) Anesthetics change tissue proton NMR relaxation. *Acta Radiol. Suppl.* **369**, 500–502.
49. Willis, C. K., Quinn, R. P., McDonell, W. M., Gati, J., Parent, J., and Nicolle, D. (2001) Functional MRI as a tool to assess vision in dogs: the optimal anesthetic. *Vet. Ophthalmol.* **4**, 243–253.
50. Weber, R., Ramos-Cabrer, P., Wiedermann, D., van Camp, N., and Hoehn, M. (2006) A fully noninvasive and robust experimental protocol for longitudinal fMRI studies in the rat. *Neuroimage* **29**, 1303–1310.
51. Pawson, P., and Forsyth, S. (2008) Anaesthetic agents. In Maddison, J. E., Page, S. W., and Church, D. B. (Eds.), *Small Animal Clinical Pharmacology*, 2nd ed., pp. 83–112. Saunders, Edinburgh.
52. Sicard, K., Shen, Q., Brevard, M. E., Sullivan, R., Ferris, C. F., King, J. A., and Duong, T. Q. (2003) Regional cerebral blood flow and BOLD responses in conscious and anesthetized rats under basal and hypercapnic conditions: implications for functional MRI studies. *J. Cereb. Blood Flow Metab.* **23**, 472–481.
53. Brevard, M. E., Duong, T. Q., King, J. A., and Ferris, C. F. (2003) Changes in MRI signal intensity during hypercapnic challenge under conscious and anesthetized

conditions. *Magn. Reson. Imaging.* **21**, 995–1001.

54. Martin, C., Jones, M., Martindale, J., and Mayhew, J. (2006) Haemodynamic and neural responses to hypercapnia in the awake rat. *Eur. J. Neurosci.* **24**, 2601–2610.

55. Martin, C., Martindale, J., Berwick, J., and Mayhew, J. (2006) Investigating neural-hemodynamic coupling and the hemodynamic response function in the awake rat. *Neuroimage* **32**, 33–48.

56. Zhao, F., Jin, T., Wang, P., and Kim, S. G. (2007) Improved spatial localization of poststimulus BOLD undershoot relative to positive BOLD. *Neuroimage* **34**, 1084–1092.

57. Zhao, F., Jin, T., Wang, P., and Kim, S. G. (2007) Isoflurane anesthesia effect in functional imaging studies. *Neuroimage* **38**, 3–4.

58. Dashti, M., Geso, M., and Williams, J. (2005) The effects of anaesthesia on cortical stimulation in rats: a functional MRI study. *Australas. Phys. Eng. Sci. Med.* **28**, 21–25.

59. Silverman, J., and Muir, W. W., 3rd. (1993) A review of laboratory animal anesthesia with chloral hydrate and chloralose. *Lab. Anim. Sci.* **43**, 210–216.

60. Sommers, M. G., van Egmond, J., Booij, L. H., and Heerschap, A. (2009) Isoflurane anesthesia is a valuable alternative for alpha-chloralose anesthesia in the forepaw stimulation model in rats. *NMR Biomed.* **22**, 414–418.

61. Braw, Y., Malkesman, O., Merenlender, A., Dagan, M., Bercovich, A., Lavi-Avnon, Y., and Weller, A. (2009) Divergent maternal behavioral patterns in two genetic animal models of depression. *Physiol. Behav.* **96**, 209–217.

62. Spring, S., Lerch, J. P., and Henkelman, R. M. (2007) Sexual dimorphism revealed in the structure of the mouse brain using three-dimensional magnetic resonance imaging. *Neuroimage* **35**, 1424–1433.

63. Berl, S., Nunez, R., Colon, A. D., and Clarke, D. D. (1983) Acetylation of synaptosomal protein: inhibition by veratridine. *J. Neurochem.* **40**, 176–183.

64. Brenner, E., Sonnewald, U., Schweitzer, A., Andrieux, A., and Nehlig, A. (2007) Hypoglutamatergic activity in the STOP knockout mouse: a potential model for chronic untreated schizophrenia. *J. Neurosci. Res.* **85**, 3487–3493.

65. Sibson, N. R., Mason, G. F., Shen, J., Cline, G. W., Herskovits, A. Z., Wall, J. E., Behar, K. L., Rothman, D. L., and Shulman, R. G. (2001) In vivo (13)C NMR measurement of neurotransmitter glutamate cycling, anaplerosis and TCA cycle flux in rat brain during. *J. Neurochem.* **76**, 975–989.

66. Lebon, V., Petersen, K. F., Cline, G. W., Shen, J., Mason, G. F., Dufour, S., Behar, K. L., Shulman, G. I., and Rothman, D. L. (2002) Astroglial contribution to brain energy metabolism in humans revealed by ^{13}C nuclear magnetic resonance spectroscopy: elucidation of the dominant pathway for neurotransmitter glutamate repletion and measurement of astrocytic oxidative metabolism. *J. Neurosci.* **22**, 1523–1531.

67. Shulman, R. G., Rothman, D. L., Behar, K. L., and Hyder, F. (2004) Energetic basis of brain activity: implications for neuroimaging. *Trends. Neurosci.* **27**, 489–495.

68. Sonnewald, U., and Kondziella, D. (2003) Neuronal glial interaction in different neurological diseases studied by ex vivo ^{13}C NMR spectroscopy. *NMR Biomed.* **16**, 424–429.

69. Taylor, A., McLean, M., Morris, P., and Bachelard, H. (1996) Approaches to studies on neuronal/glial relationships by ^{13}C-MRS analysis. *Dev. Neurosci.* **18**, 434–442.

70. Rodrigues, T. B., and Cerdan, S. (2005) A fast and sensitive ^{1}H NMR method to measure the turnover of the H_2 hydrogen of lactate. *Magn. Reson. Med.* **54**, 1014–1019.

71. Meric, P., Autret, G., Doan, B. T., Gillet, B., Sebrie, C., and Beloeil, J. C. (2004) In vivo 2D magnetic resonance spectroscopy of small animals. *MAGMA* **17**, 317–338.

72. Braakman, N., Oerther, T., de Groot, H. J., and Alia, A. (2008) High resolution localized two-dimensional MR spectroscopy in mouse brain in vivo. *Magn. Reson. Med.* **60**, 449–456.

73. Kundel, H. L., Schlakman, B., Joseph, P. M., Fishman, J. E., and Summers, R. (1986) Water content and NMR relaxation time gradients in the rabbit kidney. *Invest. Radiol.* **21**, 12–17.

74. Ottenweller, J. E., Meier, A. H., Russo, A. C., and Frenzke, M. E. (1979) Circadian rhythms of plasma corticosterone binding activity in the rat and the mouse. *Acta Endocrinol. (Copenh)* **91**, 150–157.

75. Duong, T. Q. (2007) Cerebral blood flow and BOLD fMRI responses to hypoxia in awake and anesthetized rats. *Brain Res.* **1135**, 186–194.

76. Kiryu, S., Inoue, Y., Watanabe, M., Izawa, K., Shimada, M., Tojo, A., Yoshikawa, K., and Ohtomo, K. (2009) Evaluation of gadoxetate disodium as a contrast agent for mouse liver imaging: comparison with gadobenate dimeglumine. *Magn. Reson. Imaging.* **27**, 101–107.

77. Narciso, S. P., Nadziejko, E., Chen, L. C., Gordon, T., and Nadziejko, C. (2003) Adaptation to stress induced by restraining rats and mice in nose-only inhalation holders. *Inhal. Toxicol.* **15**, 1133–1143.
78. Harris, R. B., Gu, H., Mitchell, T. D., Endale, L., Russo, M., and Ryan, D. H. (2004) Increased glucocorticoid response to a novel stress in rats that have been restrained. *Physiol. Behav.* **81**, 557–568.
79. Dunn, A. J., and Swiergiel, A. H. (2008) Effects of acute and chronic stressors and CRF in rat and mouse tests for depression. *Ann. N. Y. Acad. Sci.* **1148**, 118–126.
80. Swiergiel, A. H., Leskov, I. L., and Dunn, A. J. (2008) Effects of chronic and acute stressors and CRF on depression-like behavior in mice. *Behav. Brain Res.* **186**, 32–40.
81. Venkatesan, R., Lin, W., Gurleyik, K., He, Y. Y., Paczynski, R. P., Powers, W. J., and Hsu, C. Y. (2000) Absolute measurements of water content using magnetic resonance imaging: preliminary findings in an in vivo focal ischemic rat model. *Magn. Reson. Med.* **43**, 146–150.
82. Tkác, I., Henry, P. G., Andersen, P., Keene, C. D., Low, W. C., and Gruetter, R. (2004) Highly resolved in vivo ^{1}H NMR spectroscopy of the mouse brain at 9.4 T. *Magn. Reson. Med.* **52**, 478–484.

Chapter 18

Spectroscopic Imaging of the Mouse Brain

Dennis W.J. Klomp and W. Klaas Jan Renema

Abstract

Magnetic resonance spectroscopic imaging (MRSI) of the mouse brain reveals a wealth of metabolic information, not only from a single region of interest (single voxel), but spatially mapped over potentially the entire brain. However, MRSI requires challenging methods before the data can be obtained accurately. When applied in vivo, MRSI is generally combined with volume-selective spin perturbation to exclude artifact originating from outside the volume of interest. To obtain good magnetic field (B_0) uniformity at this volume, accurate B_0 shimming is required. Finally, the immensely large signals originating from water spins need to be suppressed to prevent sidebands that contaminate the spectra, or even saturate the dynamic range of the MR receiver. This chapter describes solutions for these challenges and ends with a rationale between single-voxel MRS versus MRSI.

Key words: Spectroscopic imaging, chemical shift imaging, mouse brain metabolism, MRSI, CSI.

1. Introduction

Magnetic resonance imaging (MRI) is widely used to study anatomy and function using contrast parameters of T_1, T_2, diffusion, perfusion, magnetization transfer, and T_2^* of water and lipid tissue in humans and animal models. A wealth of information can be added to these parameters using MR spectroscopy (MRS) since levels of metabolites can in vivo be detected with this technique. In fact, for mice brain, levels of up to 17 metabolites can be detected using proton MRS (*see* **Note 1**), which can be used for understanding metabolism in fundamental research or used as diagnostic markers (1). Mice are of special interest because numerous (transgenic) mouse models are available to investigate metabolism or to mimic pathology (2). **Figure 1** shows an

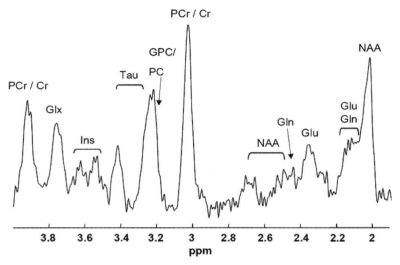

Fig. 1. MR spectrum obtained from a 4 μL volume in the cortex/hippocampus of a mouse brain at 7 T (23). Clear resonances can be depicted of NAA, glutamate (Glu), glutamine (Gln), (phospho-) creatine (PCr/Cr), (glycero-)phospocholine (GPC/PC), taurine (Tau), and myo-inositol (Ins).

example of a proton MR spectrum obtained from a 4 μL volume in the cortex/hippocampus of a mouse brain, illustrating a number of detectable metabolites.

Spatial mapping of metabolism benefits research in brain metabolism even further (3); however, most of the published proton MR spectra of mice brain are obtained from single volume elements (voxel). Limiting the detection to a single voxel allows focused optimization of the MR system in maximizing the quality of the MR spectrum, whereas spatial mapping of metabolism over a larger field of view (FOV) requires more technical challenges to overcome. Especially in the mouse brain, this is challenging due to its small size.

MRSI data of good quality has to meet quite a few prerequisites (1). First of all, the signal-to-noise ratio (SNR) needs to be sufficient to detect metabolic alterations with respect to controls. Secondly, good B_0 shim needs to be obtained to minimize linewidths of the peaks in order to reduce signal overlap, thereby enabling distinct detection of metabolites (4). Thirdly, the signals from the highly concentrated water content have to be accurately suppressed not only to overcome dynamic range limitations of the MR receiver but also to reduce spectral contamination caused by potential vibrations, eddy currents, and (physiologic) motion (5). Fourthly, contaminating signals originating from areas outside the volume of interest need to be suppressed to below noise level as these may leak into the voxels of interest due to the point spread function in spatial Fourier transform. Fifthly, the actual spatial resolution should be sufficient to obtain metabolic information from distinct anatomical areas. Finally, nonuniformities in spin excitation, spin relaxation, and signal reception both in spatial and spectral domain should be known and accounted for when analyzing the data.

Although MRSI of the mouse brain is feasible, artifacts in spectra can easily occur if the acquisition methods are not optimized properly (6). In MRSI, spectra are obtained from a relatively large area, therefore B_0 shimming has to be optimized for the entire region of interest. However, if a small part of this area includes tissue transitions with substantial differences in susceptibility (e.g., brain–air transition in ear cavities in mice results in inhomogeneities), the B_0 shimming using conventional shim systems (up to third-order shim coils) will be poor, leading to broad and overlapping peaks in the spectra.

When regions of interest are selected that exclude large susceptibility transitions (e.g., the centre of the brain), B_0 shimming can be excellent over the entire region of interest. As such, spectral resolution will be optimal and chemical shift-selective water suppression can be obtained relatively easily. Outside the region of interest, however, the B_0 shim is not optimal, leading to off resonances of the spins. Water suppression will therefore be insufficient leading to high remaining signals of water with a high probability that these will shift into the spectral range of the metabolites of interest. These signals can contaminate the spectra in the region of interest due to voxel bleeding caused by spatial Fourier transform.

Another potential artifact originates from highly concentrated lipids at the skull, generally outside the region of interest. With only chemical shift-selective water suppression techniques, these signals will not be attenuated and may therefore leak into the voxels of interest by spatial Fourier transform.

Both hardware and pulse sequences have been developed to meet the challenges in MRSI of the mouse brain, overcoming most of the limitations described above. In this chapter, the basics of spectroscopic imaging are described including application of filtering to reduce voxel bleeding. In addition, pulse sequences are discussed that use spatially selective spin perturbation, thereby excluding unwanted signals outside the region of interest. Finally, dedicated hardware for mice MRSI in terms of RF coils and B_0 shims are briefly discussed.

2. Materials

2.1. Scanner and Animal Handling

Compared to the rat or the human brain, the mouse brain is relatively small; thus the number of spins that contribute to the signal is relatively low, potentially resulting in a low SNR. The strength of the magnet that is used to obtain MR spectra of the mouse brain is therefore generally at least 7 T, but preferably even stronger. The bore size must be at least the size of the mouse, including space for the RF coils and gradient system. For example, a magnet bore size of 12 cm can be used with a gradient set of 8 cm and an RF coil of 6 cm, still enabling the positioning of the mouse.

Although a larger bore size may be beneficial for positioning the mouse, it may result in higher costs and reduced gradient strength performance. For mouse MRS, small voxel sizes are required that can only be effectively obtained with strong gradients. Therefore at least 200 mT/m of gradient strength is recommended for MRS(I) of the mouse brain (*see* **Note 2**).

The ideal RF coil setup for mouse MRSI would be a volume transmit coil, which can generate a uniform B_1 field for spin excitation/manipulation, with an array of local receiver coils for maximum SNR (7). It should be considered, however, that whole body volume coils are not very efficient in obtaining strong B_1 fields resulting in a low bandwidth with severe chemical shift displacement artifacts (as described below). The use of RF amplifiers with more peak power can overcome this issue, but also a dedicated (small-size) mouse headcoil would help. On the signal reception side, it should also be pointed out that an array may not be that efficient, even compared to a single loop coil. In mouse brain imaging, the tissue losses do not dominate the coil losses; adding multiple elements will therefore add more noise to the data (8). However, when parallel imaging in MRSI is required, multi-element arrays are essential. Nevertheless, when SNR is crucial, the receiver coil should be positioned as close as possible to the tissue of interest.

Finally, the mouse has to be positioned in the coil setup, while being kept under anesthesia for the duration of the experiment. As these experiments may take hours, the temperature of the mouse has to be regulated as well, since mice are unable to maintain body temperature under these conditions. **Figure 2** shows an illustration of a setup that can be used for MRSI of the mouse brain using a surface coil and anesthetics system mounted on an animal tray that can be shifted inside the MR system. A gas mixture of 1–2% of isoflurane with a 50–50% mixture of O_2 and N_2 gas may be applied to the mouse using a gas mask. Hot water can be pumped through a mouse bed to warm the mouse, while rectum thermometry is used to control the temperature. All cables, water tubes and gas lines are guided through a filter plate at the end of the bore of the magnet, to prevent RF noise sources entering the bore.

2.2. Basic Scanning Considerations

In analogy with phase encoding in MRI, spatial information in spectroscopic imaging can be obtained by phase encoding of the

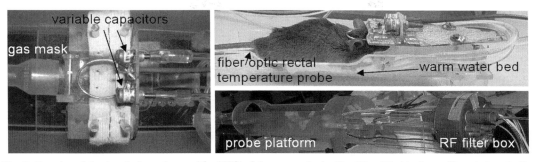

Fig. 2. Experimental setup that can be used for MRSI of the mouse brain. The RF coil is ideally positioned close to the mouse brain while the mouse is kept under anesthesia using a gas mask. The temperature of the mouse is controlled by circulating warm water through a dedicated mouse bed and monitored using fiber optic rectal thermometry. The setup is positioned in the magnet via a probe platform, the cables and tubes of which are interfaced through an RF filter box.

excited spins using an array of B_0 gradient settings (9). For each spatial dimension (i.e., x-, y-, and z-axes), a unique number of N phase encoding steps are required to encode the data in that dimension in N spatial segments. The dataset is stored into an array called k-space. The nominal dimensions of the segments (voxel size) and field of view (FOV) of the encoded spatial domain are related as voxel size = FOV/N. The actual dimensions are related to the gradient strength and length used to encode the spins. In particular, the step size of the gradient (dGt) defines the FOV, whereas the maximum gradient change (i.e., dGt × N) defines the pixel size with dGt = $1/\gamma$ FOV, where Gt is the integral of the gradient over its duration t and γ the gyromagnetic ratio of the nuclear spin. Spatial information can be decoded from the dataset in k-space by fast Fourier transform (FFT) of each spatial dimension. Since the number of gradient steps is limited, FFT will result in point spread artifacts; i.e., signals originating from a voxel will leak into other voxels (voxel bleeding, **Fig. 3**). Such leakage may be reduced by application of filters like Hamming, but will result in broadening of the voxel dimension (i.e., for a 3D Hamming filtered SI dataset with a nominal voxel size of 1 mm in all dimensions, this would lead to a voxel of a sphere with a diameter of 1.78 mm) (10).

As a unique phase-encoding step is required to encode spins in spatial domain, a three-dimensionally encoded dataset consists of a k-space matrix of $N_x N_y N_z$ data lines. Each line is obtained by acquisition of the signals from the excited spins. The acquisition will be repeated $N_x N_y N_z$ times leading to a scan duration of $N_x N_y N_z$TR (with TR being the repetition time that elapses between two acquisitions). In case spectra are required from the entire mouse brain with a FOV of 16 × 16 × 8 mm^3 at a resolution of 1 mm^3 with a TR of 5 s, a typical 3D MRSI sequence would take 16 × 16 × 8 × 5 s = 3 h.

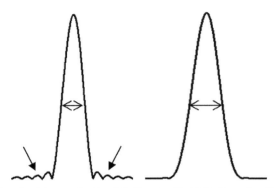

Fig. 3. Effects of point spread without (*left*) and with (*right*) Hamming filtering. Application of a Hamming filter results in a broader voxel size; however, contamination of signals from outside this voxel (voxel bleeding) is substantially reduced (*see arrows*).

Scan times can be reduced by decreasing TR; however, the signal intensities acquired will have a stronger dependence on the T_1 of the specific spins making quantification more challenging. Proton echo planar spectroscopic imaging (PEPSI) can be applied to reduce scan time by one dimension (11). In this method, as in MRI, a readout gradient is applied during the acquisition of the data lines. To maintain spectral information, each data point in the time domain is obtained with the same balanced readout gradient.

FOV can be reduced to reduce scan time, but care must be taken not to excite spins outside the FOV for these will fold into the region of interest as a consequence of phase wrap. Using an array of receiver coils, these wraps can still be unfolded by means of SENSE reconstruction (12). Slice-selective excitation can also be applied to reduce scan times, as spatial encoding can be omitted in one dimension.

2.3. Volume Selection

As a consequence of voxel bleeding by spatial FFT, signals outside the region of interest need to be excluded to prevent contamination of the spectra of interest. The most effective methods combine spectroscopic imaging with volume selection and refocusing techniques, like stimulated echo acquisition method (STEAM) (13) or point resolved spectroscopy (PRESS) (14). With STEAM, a combination of three orthogonally oriented slice selective radio frequency (RF) pulses are applied. Stimulated echoes from these three RF pulses result in a detectable echo signal that originates from the cross section of the three slices. With PRESS, a slice-selective RF pulse is succeeded with two orthogonally oriented slice-selective refocusing pulses resulting in two echoes of which the second detectable echo originates from the cross section of the three slices. Intrinsically PRESS can give twice the SNR compared to STEAM (*see* **Note 3**) but may come with potentially more artifacts.

A slice-selective RF pulse is obtained with a B_0 gradient applied in conjunction with an RF pulse. The bandwidth of the RF pulse (BW, in hertz) combined with the gradient strength (G, in mT/m) defines the thickness (d, in mm) of the slice by $d = BW/(\gamma \cdot G)$, where γ is the gyromagnetic ratio of the nuclear spin (i.e., 42.56 MHz/T for ^{1}H spins). The slice position (p, in mm) with respect to the isocenter of the magnet can be controlled by setting the offset frequency (Δf, in Hz) of the RF pulse to $\Delta f = BW \cdot p/d$.

Spins of metabolites with different chemical shifts will experience a different offset frequency between their Larmor frequency and the selected frequency of the RF pulse. The position of the slice will therefore be different for metabolites with different chemical shifts, which is called chemical shift displacement artifact (CSDA). This is expressed as the relative mismatch of the slice position over the entire chemical shift range ($\Delta \sigma$, in ppm) of the metabolites of interest as $CSDA = \Delta \sigma \cdot \gamma \cdot B_0 / BW$. By maximizing the bandwidth of the RF pulse (*see* **Note 2**), the chemical shift displacement artifact can be minimized (15).

Apart from a minimum CSDA, the slice profile needs to be appropriate to not only uniformly perturb the spins within the slice, but also exclude excitation of spins outside the selected slice. Inverse FFT can be applied to calculate the RF pulse shape for small tip angles. As can be predicted by the nonlinear behavior of the Bloch equations, Shinnar–le Roux optimizations need to be applied to calculate these RF pulses for the larger tip angles used in MR spectroscopy (16). When assuming a uniform B_1 field over the entire subject of investigation, the maximum bandwidth of these RF pulses is related to the maximum B_1 field that can be obtained with the MR system, whereas the transition from full perturbation to no excitation can be minimized by maximizing the length of the RF pulse (**Fig. 4a** and **b**). This

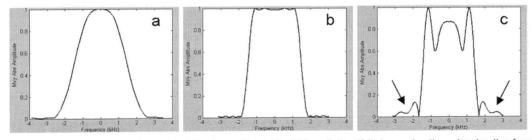

Fig. 4. Frequency (or slice) profiles of an RF pulse, optimized for 90° excitation (24). Increasing the pulse duration from 0.67 to 2 ms results in a better slice definition (**a** and **b** respectively). When the actual B_1 level is 50% more than the nominal B_1 level, the slice profile degrades substantially, not only within the slice, but also outside the slice (*see arrows*, **c**).

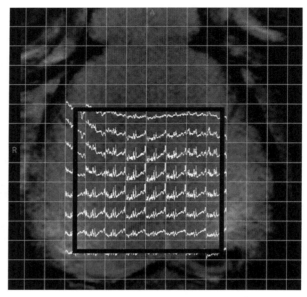

Fig. 5. 2D proton MRSI data acquired from a mouse brain at 7 T. A STEAM sequence was applied as volume-selective (*see black square*) spin perturbation, preceded by a VAPOR water suppression scheme and followed by phase encoding for spatial mapping.

will have a consequence for the minimum echo time that can be obtained. When the B_1 field over the entire subject is not uniform or does not match the expected B_1 field, the slice profile itself becomes a function of the actual B_1 strength (**Fig. 4c**). Not only do the spins inside the selected slice have a deviated perturbation, but also spins outside the slice are partially being excited, giving rise to unwanted signals or artifacts. At lower flip angles, e.g., 90°, in a STEAM sequence, this effect is less severe than in a PRESS sequence that uses two 180° pulses. Another option would be to adapt the PRESS sequence into an adiabatic version like a LASER sequence (localization by adiabatic selective refocusing), which is not sensitive to B_1 nonuniformities. This method, however, requires substantially longer and more RF pulses which has a consequence for the TE (17). **Figure 5** shows an MRSI dataset obtained with STEAM selection of the mouse brain.

When chemical shift displacement artifacts are still severe, and the slice-selective RF pulses are not sufficient to exclude excited spins outside the slice, additional slice-selective RF pulses can be applied prior to the STEAM or PRESS sequence to crush the magnetization of the spins outside the selected volume (outer volume suppression, OVS) (18).

2.4. Water Suppression

Having excited the spins in the volume of interest, while excluding signals outside this region, artifacts may still arise originating from spins of the highly concentrated water molecules. The four orders in magnitude difference in concentration of water versus the metabolites of interest not only stresses a strong demand on the dynamic range of the MR receiver but also on the stability of the setup. Any coherence in vibration, for instance, caused by the application of gradients, by breathing of the mouse, or by the cardiac cycle, can result in sidebands of the water signal that contaminates the spectral region of interest (5). Suppression of the signals from water can therefore exclude these potential artifacts.

Many methods for water suppression have been described, all of which use chemical shift-selective RF pulses to selectively perturb the spins of water. The bandwidths of these pulses define to what extent the spectral region of interest remains unaffected. Most methods apply water suppression prior to excitation of spins of interest by dephasing the selectively excited spins of water with B_0 gradients (crushers). Using a number of these pulses with optimized flip angles and inter-pulse delays, the suppression can become more effective considering the differences in T_1 of water as well as potential nonuniformities in B_1 (i.e., WET (19) or VAPOR (20)). Another way of effective suppression can be applied during refocusing of the spins of interest. Using chemical shift-selective refocusing pulses with crushers (BASING (21), MEGA (22)), water suppression can be obtained independent of T_1, but with a consequence of lengthening the TE (*see* **Note 4**).

2.5. B_0 Shimming

Since spectral resolution and spectral signal-to-noise ratio are both affected by B_0 nonuniformities, optimization of B_0 shimming is crucial in MR spectroscopy. Although the minimum spectral linewidth is defined by the inherent T_2 of the spins of the metabolite, the true linewidth is generally defined by the B_0 shim. B_0 shimming can be adjusted by changing the static electrical currents through external coils, which affect the B_0 field spatially. Typically, up to three orders of spatial harmonics in B_0 shim can be applied in animal MR systems, which can be optimized using MR sequences that are able to detect the nonuniformities in B_0. B_0 maps are acquired using MR sequences obtained with at least two different TEs, either from the entire volume, or from several rods through the volume of interest (FASTMAP) (4). These maps are then fitted to the known harmonics of the shim fields, from which the required current can be calculated. As in MRSI a larger volume needs to be shimmed compared to single voxel, the linewidth of the metabolites in MRSI may be worse. Recent studies, however, have shown that use of additional shim coils or passive shims can improve the B_0 shim over the entire mouse brain (25).

3. Methods

MRSI of the mouse brain provides a wealth of information but can be challenging. As an alternative, multiple single-voxel MRS experiments may be performed. The method of choice depends on many variables like the application of interest and availability of sequences. This section briefly describes typical methods that can be applied in single-voxel MRS and in MRSI.

3.1. Typical Single-Voxel MRS

1. The mouse has to be positioned in the coil setup, while being kept under anesthesia for the duration of the experiment (using, for instance, a gas mask with a flow of 1% isoflurane with a 50%/50% gas mixture of O_2 and N_2). The temperature of the mouse can be regulated using a heated mouse bed (i.e., using a flow of warm water through tubes positioned in close contact with the mouse) and rectal temperature control.

2. Conventional MR images of the mouse brain are needed to guide the position of the region of interest (voxel). As metabolite concentrations are substantially different between gray and white matter, the location of the voxel needs to be accurate. Generally, spin echo images acquired at relatively long echo times provide good contrast between gray and white matter (for instance, TE = 50 ms, TR = 2000 ms, slice thickness 0.5 mm, matrix size 512×256).

3. To obtain a localized single voxel MR spectrum, a STEAM or PRESS method can be used. These methods require adjustments of the B_1 field in order to yield the correct flip angles. Multiple B_1 adjustment methods are available in literature, but if there is none present at the MR system, one may alter the RF power of the RF pulses until a maximum SNR is obtained in the MR spectra (*see* **Note 5**).

4. B_0 shimming can be focused on the region of interest. If B_0 mapping techniques are used, e.g., FASTMAP, B_0 maps of sufficient spatial resolution need to be acquired to correctly fit the shim fields to the B_0 maps (i.e., at least three data points within the voxel per spatial dimension for second-order shims). Detailed adjustments (fine-tuning) of the shim currents may be applied iteratively using for instance manual shimming. In this method, a single-shot MR spectrum of water is obtained and visualized repeatedly, while shim currents can be adjusted in real time to maximize the length of the free induction decay (FID) or minimize the linewidth of the water peak in the MR spectrum (*see* **Note 6**).

5. Even after careful adjustment of the RF power settings of the experiment, fine adjustments of the RF power may still be needed for effective water suppression, particularly when a relatively short TR is used with respect to the T_1 of water. This can be optimized by monitoring the residual water signal in repeated single-shot acquisitions while adjusting the RF pulse amplitudes of the suppression pulses. In addition, as the bandwidth of the RF pulses used for water suppression is narrow, the carrier frequency of these pulses has to be set to the exact resonance frequency of the spins of water. Use of a VAPOR water suppression scheme is relatively insensitive to B_1 variances and effective for a range of T_1 values.

6. Finally the number of averages has to be set, in which a compromise has to be made between scan time, voxel volume, and SNR. One may choose to record the data of each acquisition separately to enable in postprocessing data corrections for potential B_0 drifts and phase errors due to movements. In cases of large voxel sizes (i.e., covering almost all brain), MR spectra may be obtained in a single shot (i.e., acquisition time in the order of 1 s), whereas if voxel sizes of approximately 1 mm^3 are used, acquisition times may be in the order of a few minutes.

7. The acquired data can be processed into an MR spectrum with FFT using either the scanner software or external software (e.g., MRUI (26) or LCModel (27)). Particularly when the T_2^* of the metabolites is substantially less than the acquisition time, time domain filtering (for instance using an exponential decay filter that matches the decay caused by T_2^*) will enhance the SNR in the spectrum. Finally, the data can be fitted to model spectra obtained using simulations or phantom measurements, which allows quantification of metabolite levels (27).

3.2. Typical MRSI

1. As in single voxel MRS, anatomical images of the mouse brain can be used to guide the location of the MRSI grid and to register the voxels to its anatomic area (for acquisition parameters of scout images, see **Section 3.1**, Step 2). More importantly, these images are used to position the selection box (i.e., STEAM or PRESS) for volume-selective spin perturbation to exclude signal contamination from spins outside the selected volume. Generally, the box size is maximized to include the brain without including any skull, subcutaneous fat, or areas with strong susceptibility differences (like air cavities in ears or nose).

2. Although RF power adjustments or flip angle calibrations are identical for MRSI as for single voxel MRS, B_0 optimizations may be more challenging in MRSI. To optimize the B_0

shims in the entire volume of interest, the B_0 variance over the spatial domain needs to be mapped and fitted to spatial shim fields of the B_0 shim setup to set the corresponding current levels (see **Note 7**). As the variance in B_0 field and therefore the resonance frequency of the water spins is more than in single voxel MRS, the bandwidth of the water suppression RF pulse has to increase to ensure proper suppression of all signals from water. The performance of the water suppression can be adjusted in repeated single-shot acquisitions by switching off the phase-encoding gradients, while varying the amplitude of the water suppression RF pulses.

3. Finally, the FOV and number of data points are set to meet the desired spatial resolution, which in turn defines the minimum acquisition time. As an example, a 2D slice-selective slice thickness (1 mm) MRSI with a FOV of 16 × 16 mm obtained with a 16 × 16 matrix results in a spatial resolution of 1 mm^3 which, at a TR of 5 s, takes about 20 min (see **Note 8**).

4. The data can be decoded into multiple single voxels with spatial FFT using the scanner software or external software (such as 3DiCSI (28)). This can be applied with spatial filtering (like Hamming filtering) to reduce point spread artifacts. Similar to single-voxel spectroscopy, this reconstructed multiple voxel data can be processed and quantified in metabolite levels for each voxel (see **Section 3.1**, Step 7).

4. Notes

1. The maximum number of metabolites that can distinctly be detected depends on the B_0 field strength. With proton MR spectroscopy at 9.4 T, levels of 17 different metabolites can be distinguished (1).

2. The required gradient strength for volume-selective spin perturbation can be calculated as: $G = BW/\gamma/\text{thickness}$, where BW is the bandwidth of the RF pulse. For example, when a chemical shift displacement artifact (CSDA) of 10% between water and lipids is considered acceptable, and the field strength of 7 T is used, then for a slice-selective RF pulse of 1 mm, a gradient strength (G) of (4.7 ppm–0.9 ppm)7 T42.5 MHz/T/10%/42.5 MHz/T/1 mm = 266 mT/m is required.

3. A STEAM sequence is generally used for its well-defined voxel definition and the ability to acquire MRS data at short echo times. However, as the method is based on a stimulated

echo rather than on a full echo, the intrinsic sensitivity is half of that of a pulse acquire or spin echo sequences, like PRESS.

4. BASING or MEGA water and/or lipid suppression will lengthen the minimum echo time of a sequence, as with these methods, chemical shift selective refocusing pulses are applied while the spins of interest are in the transverse plain of rotation. Due to the required low bandwidth of these pulses (i.e., only refocus spins of water and/or lipids without affecting the spins of interest), the pulses have to be long (i.e., in the order of 10 ms). In addition, crusher gradients are required during this state, adding another few ms of time to the TE of the sequence. In order to acquire MR spectra without phase errors of the signals of interest due to the application of MEGA of BASING pulses, these pulses need to be applied in pairs (prior to and after refocusing). Typically, the minimum TE will be lengthened by at least 25 ms when BASING or MEGA water/lipid suppression is applied.

5. A relatively fast adjustment of the B_1 field to get the correct flip angle is to use a STEAM or PRESS sequence without water suppression and run this sequence with a number of different RF power settings. The RF power setting that coincides with a maximum signal is by approximation the correct power calibration for a 90° pulse for the selected region of interest (voxel). Be aware that the slice profile at different power settings may also alter; therefore, this method is biased toward a higher actual flip angle. Also ensure that the TR of the sequence is substantially longer than the T_1 of water in the voxel; particularly if contributions of CSF are included in the voxel.

6. Although manual B_0 shimming based on minimizing the linewidth of the water resonance can be an effective method for single-voxel MRS, it should be noted that the shimming may not get the most optimal B_0 shim. Particularly when the T_2 relaxation of water is lower than the T_2 for the metabolites of interest and approaching the T_2^* caused by macroscopic B_0 nonuniformity. In these cases, B_0 phase mapping is the method of choice.

7. Spectral resolution: If very high spectral resolution is required to avoid overlap of resonances of metabolites of interest, single voxel with local B_0 shimming may be considered, as in MRSI the B_0 shim is a compromise to obtain reasonable shim over the entire volume.

8. Single-voxel MRS may take a few seconds, whereas MRSI takes several minutes. If the intention of the experiment is to monitor fast dynamic responses, MRSI is not the most optimal method. However, SNR per unit of time remains

approximately equal, thus if SNR is the limiting factor, MRSI may be used as well. In case SNR is not the limiting factor, MR spectra of multiple regions can be obtained with multiple single-voxel MRS. The SNR of the each of the single voxels per unit of total examination time is less (by approximately a factor of (number of voxels)) than in MRSI.

Acknowledgments

The authors would like to thank Professor Arend Heerschap and Professor Peter Luijten for initiating and financing this work.

References

1. Tkac, I., Dubinsky, J.M., Keene, C.D., Gruetter, R., Low, W.C. (2007) Neurochemical changes in Huntington R6/2 mouse striatum detected by in vivo ^{1}H NMR spectroscopy. J Neurochem 100(5), 1397–1406.
2. Heerschap, A., Kan, H.E., Nabuurs, C.I., Renema, W.K., Isbrandt, D., Wieringa, B. (2007) In vivo magnetic resonance spectroscopy of transgenic mice with altered expression of guanidinoacetate methyltransferase and creatine kinase isoenzymes. Subcell Biochem 46, 119–148. Review.
3. Maudsley, A.A., Domenig, C., Govind, V., Darkazanli, A., Studholme, C., Arheart, K., Bloomer, C. (2009) Mapping of brain metabolite distributions by volumetric proton MR spectroscopic imaging (MRSI). Magn Reson Med 61(3), 548–559.
4. Gruetter, R. (1993) Automatic, localized in vivo adjustment of all first- and second-order shim coils. Magn Reson Med 29(6), 804–811.
5. Nixon, T.W., McIntyre, S., Rothman, D.L., de Graaf, R.A. (2008) Compensation of gradient-induced magnetic field perturbations. J Magn Reson 192(2), 209–217.
6. Kreis, R. (2004) Issues of spectral quality in clinical ^{1}H-magnetic resonance spectroscopy and a gallery of artifacts. NMR Biomed 17(6), 361–381. Review.
7. Sutton, B.P., Ciobanu, L., Zhang, X., Webb, A. (2005) Parallel imaging for NMR microscopy at 14.1 Tesla. Magn Reson Med 54(1), 9–13.
8. Kumar, A., Edelstein, W.A., Bottomley, P.A. (2009) Noise figure limits for circular loop MR coils. Magn Reson Med 61(5), 1201–1209.
9. Buchthal, S.D., Thoma, W.J., Taylor, J.S., Nelson, S.J., Brown, T.R. (1989) In vivo T_1 values of phosphorus metabolites in human liver and muscle determined at 1.5 T by chemical shift imaging. NMR Biomed 2(5–6), 298–304.
10. Pohmann, R., von Kienlin, M. (2001) Accurate phosphorus metabolite images of the human heart by 3D acquisition-weighted CSI. Magn Reson Med 45(5), 817–826.
11. Posse, S., DeCarli, C., Le Bihan, D. (1994) Three-dimensional echo-planar MR spectroscopic imaging at short echo times in the human brain. Radiology 192(3), 733–738.
12. Otazo, R., Tsai, S.Y., Lin, F.H., Posse, S. (2007) Accelerated short-TE 3D proton echo-planar spectroscopic imaging using 2D-SENSE with a 32-channel array coil. Magn Reson Med 58(6), 1107–1116.
13. Haase, A., Frahm, J., Matthaei, D., Hänicke, W., Bomsdorf, H., Kunz, D., Tischler, R. (1986) MR imaging using stimulated echoes (STEAM). Radiology 160(3), 787–790.
14. Bottomley, P.A. (1984) U.S.-Patent, 4, 480, 228:0.
15. Klomp, D.W., van der Graaf, M., Willemsen, M.A., van der Meulen, Y.M., Kentgens, A.P., Heerschap, A. (2004) Transmit/receive headcoil for optimal ^{1}H MR spectroscopy of the brain in paediatric patients at 3T. MAGMA 17(1), 1–4.
16. Shinnar, M., Leigh, J.S. (1993) Inversion of the Bloch equation. J Chem Phys 98(8), 6121–6128.

17. Garwood, M., DelaBarre, L. (2001) The return of the frequency sweep: designing adiabatic pulses for contemporary NMR. J Magn Reson 153(2), 155–177.
18. Henning, A., Fuchs, A., Murdoch, J.B., Boesiger, P. (2009) Slice-selective FID acquisition, localized by outer volume suppression (FIDLOVS) for (1)H-MRSI of the human brain at 7 T with minimal signal loss. NMR Biomed 22(7), 683–696.
19. Ogg, R.J., Kingsley, P.B., Taylor, J.S. (1994) WET, a T_1- and B_1-insensitive water-suppression method for in vivo localized 1H NMR spectroscopy. J Magn Reson B 104(1), 1–10.
20. Tkác, I., Starcuk, Z., Choi, I.Y., Gruetter, R. (1999) In vivo ^{1}H NMR spectroscopy of rat brain at 1 ms echo time. Magn Reson Med 41(4), 649–656.
21. Star-Lack, J., Nelson, S.J., Kurhanewicz, J., Huang, L.R., Vigneron, D.B. (1997) Improved water and lipid suppression for 3D PRESS CSI using RF band selective inversion with gradient dephasing (BASING). Magn Reson Med 38(2), 311–321.
22. Mescher, M., Tannus, A., Johnson, M.O., Garwood, M. (1996) Solvent suppression using selective echo dephasing. J Magn Res A 123, 226–229.
23. in 't Zandt, H.J., Renema, W.K., Streijger, F., Jost, C., Klomp, D.W., Oerlemans, F., Van der Zee, C.E., Wieringa, B., Heerschap, A. (2004) Cerebral creatine kinase deficiency influences metabolite levels and morphology in the mouse brain: a quantitative in vivo ^{1}H and ^{31}P magnetic resonance study. J Neurochem 90(6), 1321–1330.
24. Matson, G.B. (1994) An integrated program for amplitude-modulated RF pulse generation and re-mapping with shaped gradients. Magn Reson Imaging 12, 1205–1225.
25. Koch, K.M., Brown, P.B., Rothman, D.L., de Graaf, R.A. (2006) Sample-specific diamagnetic and paramagnetic passive shimming. J Magn Reson 182(1), 66–74.
26. Naressi, A., Couturier, C., Devos, J.M., Janssen, M., Mangeat, C., de Beer, R., Graveron-Demilly, D. (2001) Java-based graphical user interface for the MRUI quantitation package. MAGMA 12, 141–152.
27. Provencher, S.W. (2001) Automatic quantitation of localized in vivo 1H spectra with LCModel. NMR Biomed 14(4), 260–264.
28. http://mrs.cpmc.columbia.edu/3dicsi.html.

Section IV

Spine Applications

Chapter 19

Spinal Cord – MR of Rodent Models

Virginie Callot, Guillaume Duhamel, and Frank Kober

Abstract

Different MR techniques, such as relaxation times, diffusion, perfusion, and spectroscopy have been employed to study rodent spinal cord. In this chapter, a description of these methods is given, along with examples of normal metrics that can be derived from the MR acquisitions, as well as examples of applications to pathology.

Key words: Spinal cord, rodent, MRI, diffusion, perfusion, spectroscopy.

1. Introduction

A growing number of experimental rodent models of spinal cord injury (SCI) and multiple sclerosis (experimental autoimmune encephalomyelitis, EAE) are studied to better understand physiological consequences of these pathologies and to develop and test new therapies that enhance repair or functional recovery.

Histological studies provide excellent insight into the understanding of pathologies. A non-invasive modality capable of depicting the evolution of the disease and its response to therapeutic intervention is, however, an important adjunct to assess behavioral outcomes and restoration of function.

Magnetic resonance (MR) is among the appropriate tools for longitudinally and non-invasively studying animals. An MR evaluation of the pathologic spinal cord (SC) permits recording the time course of the pathology and monitoring of the effects of neuro-protecting (1) or regenerative drugs. A qualitative description in terms of volumetry, tissue properties, and potentially inflammatory territories can be obtained using conventional

T_1- and T_2-weighted MRI. In addition to qualitative assessments, MR techniques permit reliable measurements of quantitative parameters directly related to the pathophysiological state of the spinal cord in vivo. Attention has particularly been paid to diffusion MRI, which conveys information on tissue integrity and structure, relevant for determining the extent of axonal loss or demyelination. Perfusion MRI, which has recently been proven applicable for SC studies, may also play an important role in SC examinations, especially for the characterization of tissue alteration and regeneration following spinal cord injury. Other experimental MR techniques could provide complementary information. Among them, MR spectroscopy may be of particular interest for the diagnosis of neurodegenerative disease and for monitoring tissue repair following SCI.

After a brief presentation of the anatomy of the spinal cord, different MR protocols dedicated to spinal cord investigation including morphology, diffusion, and perfusion will be described and illustrated for both healthy and pathologic cases. MR spectroscopy, magnetization-transfer (MT) contrast, and functional MRI (fMRI), which may prove valuable in the future, are still scarcely used MR modalities, and will only be briefly approached.

2. Materials

2.1. Gross Anatomy of the Spinal Cord

The spinal cord is a highly organized and complex part of the central nervous system, which plays an important role in sensation, autonomic, and motor control functions. The spinal cord is protected by the body structures of the vertebral column, which can be divided into five groups of vertebrae (**Fig. 1**): cervical (C1 to C7), thoracic (T1 to T13), lumbar (L1 to L6), sacral (S1 to S4), and coccygeal (tail) segments.

The spinal cord is composed of gray matter (GM) and white matter (WM). The gray matter is made up of neuronal cell bodies, dendrites, axons, and glial cells and it can be divided into dorsal, intermediate, and ventral horns, and central gray matter around the central canal (**Fig. 2c**). Gray matter is bordered by white matter at its circumference (**Fig. 2d**), except at locations where the dorsal horns touch the margin of the spinal cord. The white matter consists mostly of longitudinally running axons and glial cells and it is conventionally divided into the dorsal, dorsolateral, lateral, ventral, and ventrolateral funiculi. The spinal cord is enclosed by a tube of cerebrospinal fluid (CSF) and protected by three membranes called the spinal meninges (**Fig. 2b**).

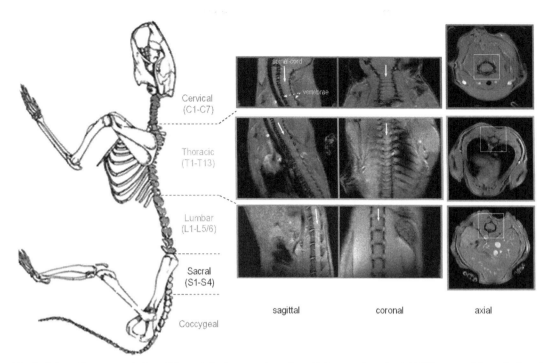

Fig. 1. Schematic representation of the vertebral segments and sagittal, coronal, and axial MR images of the spinal cord at the cervical, thoracic, and lumbar levels. The *arrows* (sagittal and coronal views) and *squares* (axial view) indicate the spinal cord.

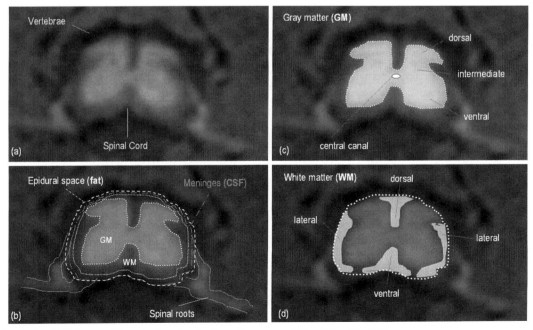

Fig. 2. Axial MR image of the spinal cord and schematic representation of the structures.

In rodents, the overall diameter of the spinal cord ranges from 2 to 3 mm. The small size of the spine and its surrounding bony structure render mouse or rat SC MRI comparatively challenging.

For further details on the structures and nomenclature, the reader can refer to available atlases of rat and mouse spinal cord (2).

2.2. MR Systems and rf Coils

1. High-field micro-imaging systems are adapted to SC investigation since they offer both high MR signal compared to clinical scanner fields (1.5 T or 3 T) and strong gradient performance. Such dedicated small animal systems are therefore of great help for achieving adequate resolution and signal-to-noise ratio (SNR). Some studies carried out on clinical scanners have been reported (3–5) and the readers can refer to these earlier studies to obtain more information on protocols using those MR systems. However, most of the rodent studies have been performed with dedicated small animal MR systems at 4.7 T (6–14), 7 T (15–20), and 9.4 T (21–30). Although, very high field systems are not yet widely present in labs, studies at 11.75 T (31–36) and 17.6 T (37–39) have been reported successful.

2. Since MR microscopy requires strong field gradients, actively shielded gradient systems with integrated shim coils and high performances, including high gradient strength, short gradient slew rates, good gradient duty cycle, optimal gradient linearity, and high gradient shielding quality, are recommended, especially when using rapid imaging techniques such as echo planar imaging (EPI) sequences. The combination of diffusion gradients with EPI readouts requires a reliable gradient pre-emphasis unit.

3. Beyond high-field magnets, well-adapted radio frequency (rf) coils should be considered for an optimized MR SC investigation. Although many studies reported in the literature were carried out using conventional small animal transmit–receive volume coils (17, 32), transmit–receive surface coils (38), and separate transmit and receive coils (geometric or active decoupling) (8, 27), much effort has been put into investigation of highly specialized implantable coils (21, 40), phased arrays (18) or cryogenic probes (41). All coil systems have proven to be applicable in SC studies, but their advantages and disadvantages may lead an investigator to choose one system over the others. Implantable coils provide excellent SNR, but the procedure is invasive and surgery may be difficult in the deeper cervical region. Surface coils offer limited spatial coverage but generally good SNR. Volume coils are suitable for large screening purposes offering increased spatial coverage and good field homogeneity but

lower SNR as compared to surface coils. Phased arrays offer both high SNR and spatial coverage and they provide parallel imaging capabilities, which can be helpful for a better motion robustness of various imaging techniques.

2.3. Animal Handling

1. *Anesthesia*: In most of the studies, animals are kept anesthetized by spontaneous respiration of a mixture of air and isoflurane through a nose cone. Flow rate (250–300 mL/min) and isoflurane concentration (1.5–2%) have to be optimized to obtain a regular breathing. A mixture of intraperitoneal ketamine (80 mg/kg) and xylazine (10 mg/kg) can also be used.

2. *Temperature regulation*: In small-bore systems, the physiological temperature can be maintained by heating the gradient cooling system to appropriate temperatures. In larger bore systems, a warming pad must be used.

3. *Animal positioning*: Animals are positioned on an animal bed usually in prone position for mice and supine for rats. A MRI-compatible device designed to immobilize the spinal column and suspend the lower body of the animal may eventually be used to isolate respiratory motion from the cord (8). In vertical MR systems, the animals are positioned head up and maintained using a teeth holder and by taping the neck and pelvis to the cradle. No other restraining device is necessary (32).

4. *Physiologic monitoring*: In order to minimize motion artifacts in the MR images, the respiratory signal is taken from a pressure sensor connected to an air-filled balloon placed under the abdomen of the animal (Rapid Biomedical (Rimpar, Germany) or SA instruments (Stony Brook, NY, USA)) and acquisitions are synchronized with breath motion.

2.4. Post-processing Software

1. For relaxation time map calculation (T_1, T_2, or T_2^*), home-built programs capable of performing pixel-by-pixel fitting of the function $S(t) = f(T_1, T_2,$ or $T_2^*)$ to the serial images data acquired with the MR system, are needed. Programming toolkits such IDL (Interactive Data Language, ITT Visual Solutions, Boulder, CO, USA) or MATLAB (Mathworks, Natick, MA, USA) are well adapted. These toolkits also provide helpful libraries for analyzing the metrics in different regions of interest.

2. The DTI metrics can usually be calculated directly on the MR system (e.g., the "DTI_proc_tensor" macro on Bruker ParaVision 3.02) or by using dedicated software such as FSL (42) (www.fmrib.ox.ac.uk/fsl/).

3. The MRS data may be analyzed using the magnetic resonance user interface software package (MRUI, http://sermn02.uab.cat/mrui/) or the LCModel software (http://s-provencher.com/pages/lcmodel.shtml).

3. Methods

This paragraph provides an overview of the principal MR applications useful to study spinal cord in rodent models. They mainly include T_1 and T_2-w imaging for the morphology, diffusion tensor imaging for the tissue structure, and perfusion imaging for the hemodynamics.

These MRI techniques, which can be performed in most of the MR systems, allow obtaining significant in vivo information on the pathophysiology and recovery mechanisms involved in spinal cord diseases. Advances in MR technology, such as more powerful gradient systems, refined rf coil design, and new sequence designs, will provide increased sensitivity and faster acquisition. Therefore, in the near future, MRI studies, in combination with other imaging modalities and histology, are expected to play an increasing role in the delineation of pathophysiologic mechanisms, the characterization of novel therapeutic strategies, and the longitudinal assessment of recovery mechanisms.

3.1. Preparation Scans

1. Scout images are acquired to check the animal position inside the coil and to provide a localization reference. The SC segments of interest have to be positioned in the magnet center.

2. Spectrometer adjustments such as tuning and matching of the rf coil, base frequency, reference rf gain, and shimming are performed at a level of automation depending on the MR scanner.

3. A preliminary sagittal gradient echo image of high spatial resolution may be acquired as further localization reference. A strict sagittal image with clear delineation of the disks and segments can be obtained when the sagittal slice crosses the spinal canal and the spinous process.

3.2. Anatomic MRI

3.2.1. Spin Echo Sequences – T_1 and T_2 Measurements

– *T_1- and T_2-weighted imaging*: Conventional spin echo (c-SE) or EPI-based spin echo sequences (SE-EPI, *see* **Notes 1–3**) can be used with appropriate repetition time (TR) and echo time (TE) to obtain different useful image contrasts (*see* **Table 1**). T_1- and T_2-weighted images allow size and volumetric evaluation and provide complementary

Table 1
Relation between TR, TE, and image contrast for a spin echo sequence

MR parameters	Short TR	Long TR
Short TE	T_1	PD
Long TE	T_1, T_2	T_2

information on the pathology. T_1-weighted images optimally highlight normal soft-tissue anatomy and fat, whereas T_2-weighted images optimally highlight the presence of fluid and pathology (e.g., tumors, inflammation, trauma, and edema). Examples of T_1 and T_2-weighted images are given in **Fig. 3**.

- T_1 *calculation*: T_1 values can be calculated by using a saturation or an inversion recovery SE technique, with increasing TR or inversion time (TI), respectively. Examples of protocol are given in **Table 2**.

- T_2 *calculation*: T_2 values can be measured by using a multi-spin echo (MSE) sequence, but one major problem of this approach is potential contamination of the spin echo train by stimulated echoes if the 180° refocusing pulse is imperfect. To limit this contamination, acquisitions are often performed separately for each TE (TE-step method). Examples of T_2 protocols are given in **Table 2**.

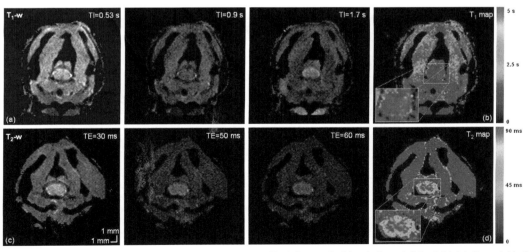

Fig. 3. Healthy mouse T_1 and T_2-weighted imaging at 11.75 T, at the cervical level. *From left to right*: (**a**) T_1-weighted images for different TI and (**b**) resulting T_1 map. (**c**) T_2-weighted images for different TE and (**d**) resulting T_2 map. T_1 values are in the order of 1870 and 1820 ms in the GM and WM, respectively. T_2 values are around 28 ms in GM and 40 ms in WM.

Table 2
Protocols for T_1, T_2, and PD measurements in the axial plane. *See* **Note 3 for additional comments on the EPI acquisitions**

Spin echo sequence – T_1 and T_2 relaxometry				
Animal model	**Rat**		**Mouse**	
References	(40)	(17)	(32)	(24)
Readout/sequence	SE-EPI	c-SE	SE-EPI	c-SE
rf Coil	Implanted coil	Birdcage coil	Birdcage coil	Inductively-coupled surface coil
Magnetic field (T)	2	7	11.75	9.4
FOV (mm^2)	25 × 25	15 × 30	17 × 17	12 × 8
Matrix	128 × 128	64 × 128	128 × 128	64 × 64
Slice thickness (mm)	2	3	0.75	2
T_1 measurements	Saturation recovery	–	Inversion recovery	Saturation recovery
TR (s)	0.3[a], 0.5[a], 0.7[a], 1.0, 2.0, 3.0, 5.0		10	0.25, 0.50, 0.75, 2.0, 3.5, 6.0
TI (s)			0.13, 0.53, 0.9, 1.4, 1.7, 4, 9.0	
TE (ms)	25		10.7	12
# of averages	2 or 4[a]		9	2
T_2 measurements	TE-step	TE-step	TE-step	TE-step
TR (s)	1.0	2.5	10.0	4.0
TE (ms)	25, 30, 35, 40, 60[a], 80[a]	10, 20, 40, 60, 80, 100, 120	20, 30, 50, 60, 70	12, 24, 36, 48
# of averages	2 or 4[a]	2	5	2
PD measurements				
TR (s)/TE (ms)	5/25			6/12

[a]Acquisitions performed with four signal averages, otherwise two averages.

T_1 and T_2 values of gray and white matter at different magnetic fields are summarized in **Table 3**. Examples of T_1 and T_2 maps are given in **Fig. 3b** and **d**.

– *Proton density calculation*: Although not frequently used, proton density measurements can also be performed. They are in some cases necessary for calibration purposes when quantitative MR parameter information is required. The protocol and typical values are indicated in **Tables 2** and **3**, respectively.

Table 3
T_1, T_2, and PD values measured at different magnetic fields

	Normal values (healthy rodents)					
	T_1 (ms)		T_2 (ms)		PD (%)	
Magnetic field	WM	GM	WM	GM	WM	GM
2 T (40), *	1089	1021	79	64	40	60
7 T (17), *			57	43		
9.4 T (24), +	1690	1730	38	33	44	56
11.75 T (32), +	1820	1870	40	28		

(*) and (+) indicate rat and mouse studies, respectively.

3.2.2. Gradient Echo Sequences – T_2^* Measurements

– T_2^*-*weighted imaging:* Gradient echo sequences may be advantageous when using surface coils for transmission because they are less sensitive to rf field inhomogeneities. T_2^*-weighted images may also help in lesion detection. An inflammatory lesion will appear as a hyper-intense area in the normal-appearing white matter and hemorrhage will present as hypo-intense area on the images. Examples of gradient echo images are given in **Fig. 4a** (healthy SC) and **4b** (contused SC).

– T_2^* *calculation:* T_2^* relaxation measurement may, for example, be exploited when iron-based contrast agents are employed, i.e., when labeling cells with small iron oxide particles.

– T_2^* maps are calculated by using multi-gradient echo sequences. An example of a protocol is given in **Table 4**. An illustration of the derived T_2^* map is given in **Fig. 4c**.

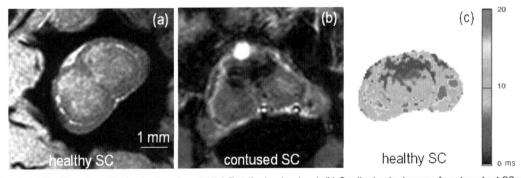

Fig. 4. (**a**) Healthy rat gradient echo imaging at 17.6 T at the lumbar level. (**b**) Gradient echo image of contused rat SC at mid-thoracic level. The hypo-intense structure, which is due to hemorrhage, represents the lesion center, with disruption of the normal GM/WM. (**c**) Healthy rat T_2^* map of the spinal cord. Values range from 5 to 20 ms. A ventral to dorsal T_2^*-gradient resulting from distortion of the magnetic field by the nearby bony structures is visible. Modified from Ref. (38), copyright [2004], with permission from ESMRMB and acknowledgment to V. Behr (Wuerzburg, Germany).

Table 4
Protocol for T_2^* measurement and gradient echo imaging

Gradient echo sequence – T_2^* relaxometry	
Animal model	Rat
Ref.	(38)
rf coil	Transmit–receive surface coil
Magnetic field (T)	17.6
FOV (mm^2)	24 × 24
Matrix	192 × 192
Slice thickness (mm)	0.5
# of echoes	8
TR/TE/inter-echo spacing (ms)	200 /2.75 /3.9
# of averages	32

3.3. Diffusion MRI

In addition to the standard contrasts (T_1, T_2, and PD), diffusion MRI provides excellent differentiation between gray and white matter due to the differences in mobility that the water molecules experience during their random walk in the microscopic tissue organization (43, 44). Diffusion-weighted imaging (DWI) and diffusion tensor imaging (DTI) (43) thus offer helpful information about tissue micro-architecture. The changes in the metrics that can be derived from such experiments (fractional anisotropy and directional diffusivities, *see* **Note 4**) usually correlate with pathophysiologic changes and the histopathology. DTI metrics may especially be seen as a complement to routine imaging, since variations may show tissue alteration or disruption in regions that appear normal on the conventional MRI (16, 45), suggesting the potential of the technique in detecting subtle pathology.

- *DWI* : The diffusion techniques in rodent studies are usually based on c-SE, however SE-EPI techniques have also been proven to work for both rat and mice studies (18, 33, 36, 46) (*see* **Note 1**). Diffusion contrast is obtained by adding diffusion gradients during the preparatory phase of an imaging sequence. Many of the reported diffusion measures were determined from the diffusion-weighted images acquired with gradients applied parallel and perpendicular to the axis of the magnet (13, 15, 17, 26, 30, 47) rather than on the complete determination of the diffusion tensor. Resulting contrasts are illustrated in **Fig. 5**.

- *DTI* : DTI is widely used to explore the brain and its application to SC investigation has considerably increased during the last 5 years. DTI is particularly useful for assessing alterations in WM tracts occurring in multiple or amyotrophic lateral sclerosis (7, 8, 31, 48, 49), as well as those occurring after SCI or infarction (7, 9, 16, 26, 50–52).

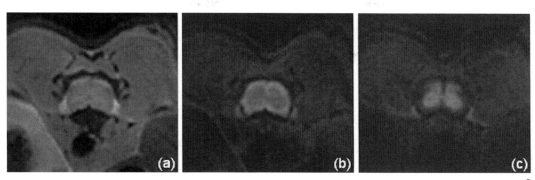

Fig. 5. Mouse spinal cord transverse images at 11.75 T, without diffusion gradient (**a**) and with gradient ($b = 800$ s/mm^2) applied perpendicular (**b**) and parallel (**c**) to the spinal cord axis. The observed contrast between WM and GM (**b, c**) is in agreement with the preferential alignment of the WM fibers along the SC axis.

DTI could also help to characterize glial processes in the GM adjacent to the site of injury (51). In general, it has been demonstrated that decreased axial diffusivity is associated with axonal injury and dysfunction and increased radial diffusivity with myelin injury (11, 53).

– To acquire DTI data, use parameters as given for different protocols in **Table 5**. The DTI metrics that can be derived from such measurements are presented in **Table 6**.

– *DTI and MS*: In multiple sclerosis models, decrease in axial diffusivity ($\lambda_{//}$) has been demonstrated to be associated with altered axon and increase in radial diffusivity ($\lambda_\perp$) with loss of integrity of myelin (8).

– *DTI and SCI*: In SCI models, DTI metrics are also essential for quantifying the direct mechanical damage of the cord (7, 9, 13, 16, 19, 26, 39, 50, 53–56), evaluating the secondary injury, predicting functional outcomes (57), and to test regenerative strategies (1, 28, 29, 58). At the lesion site, increased radial diffusivity in both GM and WM reflects the disruption of cell membranes and myelin sheaths. Increase in the axial diffusivity for GM shows disintegration with hemorrhages and disappearance of normally discernable cellular elements, whereas decrease in WM is associated with mechanical disruption of neural tissues. DTI also allows quantification of the asymmetric rostral and caudal variations associated with the regional extent of the injury, swelling processes, cyst formation, and secondary excitotoxic processes (59). An illustration of DTI applied to SCI model is given in **Fig. 6**.

3.4. Vascular and Perfusion MRI

The assessment of spinal cord (SC) hemodynamics, and especially SC blood flow (SCBF), may provide additional and complementary information useful in the description of SC diseases and may help to discover new mechanistic insights or therapeutic

Table 5
Protocols for DTI acquisitions in the axial plane. Additional comments are given in Notes 3 (EPI sequence), 5 (bulk motion), 6 (SNR consideration), and 7 (diffusion-encoding directions)

Diffusion tensor imaging

Animal species	Rat		Mouse	
References	(46)	(27)	(33)	(8)
Readout/sequence	EPI / SE (8-shots)	c-SE	EPI / SE (4-shots) echo position 18%	c-SE
rf coil	Implanted coil inductively coupled to an external coil	Volume transmit coil and surface receive coil	Volume coil	Surface coil inductively coupled to an Helmholtz coil
Magnetic field (T)	2	9.4	11.75	9.4
FOV (mm²)	25 × 25	38.4 × 38.4	12.8 × 12.8 + 2 OVS	10 × 10
Matrix	12 8 × 80	128 × 128	128 × 128	128 × 128
Slice thickness (mm)	2	2	0.75	0.75
TR (ms)	2000	500	3500	1200
TE (ms)	60	31	15.25	38
# of directions	6 (1,0,0),(0,1,0),(0,0,1),(2/2,2/2,0),(2/2,0,2/2),(0,2/2,2/2)	6 (±0.33,0.67,0.67),(0.67,±0.33,0.67),(0.67,0.67,±0.33)	6 (1,±1,0),(1,0,±1),(0,1,±1)	6 (±1,1,0),(1,0,±1),(0,±1,1)
b-Values (s/mm²)	{0, 750–1000}	{0, 1000}	{0, 700}	{0, 785}
δ/Δ (ms)	13/26	3/15	2.3/6.8	7/20
# of averages	4	1	18	8
Synchronization	Respiration	Respiration	Respiration	Respiration

opportunities. As a matter of fact, the relationship between axonal guidance molecules, vessel path-finding, and network formation have been described (60–62) and demonstrated to contribute to important diseases when dysregulated. It has also been shown that vasculature and blood flow are highly involved in the physiopathology of SC damages such as ischemia or spinal cord injury and are related to the severity of the injury (63–65). Vasculature and blood flow are also involved in the spinal cord recovery

Table 6
Examples of DTI metrics derived from DTI measurements on healthy rodents. Additional comments are given in Notes 8 (ROI location), 9 (in vivo vs. ex vivo), and 10 (fast and slow components)

Normal values			FA		$\lambda_{//}$ (10^{-3} mm^2/s)		$\lambda_{\perp}$ (10^{-3} mm^2/s)	
Species	Level		WM	GM	WM	GM	WM	GM
Rat	Cervical	C3:C7 (18)	0.82 ±0.05	0.36 ±0.04	2.15 ±0.26	1.54 ±0.20	0.33 ±0.05	0.83 ±0.06
		C3 (27)	0.70 ±0.01	0.59 ±0.01	0.85 ±0.01	0.72 ±0.01	0.23 ±0.01	0.26 ±0.01
	Thoracic	T7:T12 (18)	0.83 ±0.05	0.38 ±0.04	1.94 ±0.12	1.49 ±0.11	0.29 ±0.07	0.77 ±0.12
		T8 (27)	0.62 ±0.01	0.57 ±0.01	0.72 ±0.01	0.71 ±0.01	0.25 ±0.01	0.28 ±0.01
	Caudal	Equine (27)	0.56 ±0.01		0.97 ±0.02		0.38 ±0.01	
Mouse	Cervical	C1:C6 (36)	0.83 ±0.02	0.27 ±0.02	2.03 ±0.12	1.03 ±0.04	0.31 ±0.01	0.69 ±0.02
		T6–T10 (28)	0.80 ±0.08	0.36 ±0.11	1.37 ±0.12	0.72 ±0.12	0.26 ±0.11	0.42 ±0.04
	Thoracic/ lumbar	T8–T12 (24)	0.76 ±0.09	0.61 ±0.07	1.95 ±0.13	1.25 ±0.11	0.46	0.48
		T10–L2 (33)	0.71 ±0.03	0.30 ±0.02	1.82 ±0.08	1.08 ±0.07	0.48 ±0.05	0.69 ±0.04

through increased blood vessel density and angiogenesis (66–70).

- *MR Angiography*: One attempt to non-invasively image the vascular anatomy of the rat spinal cord has been performed using 3D time-of-flight MR angiography (25). The reported angiograms were clearly able to depict the arterial supply and surface arteries, but smaller intra-medullary arteries, which are of major interest, could not be visualized. This protocol is therefore not further described in this chapter. Future studies should focus on improving the sensitivity of detection in terms of spatial resolution and SNR.

- *MR perfusion*: Spinal cord blood flow (SCBF) has been previously measured in animals with invasive techniques such as autoradiography or labeled microspheres but a non-invasive MR technique has only been proposed very recently (34). The technique is based on arterial-spin labeling (ASL) (71), which is particularly well suited for animal studies performed at high field due to long T_1 relaxation times and high MR signal amplitudes encountered. The technique provides absolute quantification of the tissue blood flow expressed in

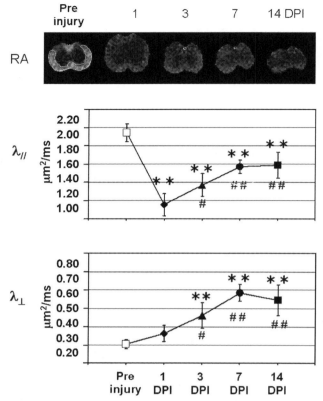

Fig. 6. (*Top*) Pre and post-injury relative anisotropy (RA) maps at the injury site (T12), demonstrating the temporal evolution of acute spinal cord injury. (*Bottom*) Axial (//) and radial (⊥) mean diffusivity variations, calculated in ROI outlined in white, for the pre-injury control and 1, 3, 7, and 14 days post-injury (DPI). The observed variations at the epicenter, the existing rostral–caudal asymmetry (data not shown), and the transient changes, suggest that DTI provides a sensitive tool for monitoring the dynamic pathophysiologic processes that occurs after SCI. Modified from Ref. (9), copyright [2007], with permission from Wiley and acknowldgements to S-K Song (St-Louis, USA).

mL/100 g/min. It has been extensively used for brain perfusion measurement and its feasibility for SCFB measurements has been demonstrated for both cervical and thoraco-lumbar levels (35). The ASL protocol proposed in **Table 7** is based on a flow-sensitive alternating inversion recovery (FAIR) magnetization preparation (72) and EPI readout. The SCBF values that can be derived from ASL experiments are presented in **Table 8**. Illustrations of SCBF maps are given in **Figs. 7** and **8**. Further comments and advises on ASL protocols and SCBF calculations are given in **Chapter 6** of this book.

– *Perfusion MRI and pathology*: Perfusion MRI bears great potential to understand the pathophysiology of SC diseases such as SC compression or contusion, involving alteration

Table 7
Protocol for perfusion measurements at cervical (C) and lumbar (L) levels

ASL-based perfusion MRI

Animal model	Mouse
Reference	(35)
Readout/sequence	EPI/SE (4-shots)
rf coil	Volume coil
Magnetic field (T)	11.75
FOV (mm^2)	17 × 17
Matrix	128 × 128
Slice thickness (mm)	0.75
TE (ms)	10.6
Magnetization preparation	Presaturated FAIR (94)
Saturation pulse	Adiabatic hyperbolic secant (0.8 ms)/12 repetitions
Inversion pulse	Adiabatic hyperbolic secant (4 ms)
TI (s)	1.3
τ (s)	3.4 (C)/10 (L)
# of averages	32 (C)/24 (L)
Inversion slice/Image slice ratio	3:2
Respiratory synchronization	Off (C)/On (L)
Slice-selective IR prescan	
TI (s)	0.13, 1.9, 10
TR (s)	10
# of averages	3
Respiratory synchronization	Off (C)/On (L)

of the vessel network or perfusion defect such as chronic or permanent ischemia. Reported studies are very scarce (73) because of the novelty of the technique; however, non-invasive longitudinal follow-ups should allow evaluating the tissue at risk, as well as tissue and vascular recovery. Several assumptions such as increased perfusion and vascular recruitment around the lesion, angiogenesis in relation with improved support for white matter and axonal tracts, or asymmetry between caudal and rostral areas should be verified. **Figure 8** illustrates the potential of the technique.

– *Dynamic contrast-enhanced MRI (DCE-MRI)*: DCE-MRI following administration of gadolinium-based contrast agent has been reported to evaluate quantitatively the permeability of the blood–spinal cord barrier (BSCB) whose structure and biochemical integrity may be altered when the SC is

Table 8
Normal perfusion values for mouse spinal cord. Brain perfusion values acquired with the same experimental setup and same ASL sequence are indicated for comparison

Normal perfusion values for mouse (34, 35)

SCBF (mL/100 g/min)	Cervical (C3)	WM	121 ± 23
		GM	310 ± 17
	Lumbar (L1)	WM	100 ± 32
		GM	285 ± 27
CBF (mL/100 g/min)	Brain	Thalamus	295 ± 22
		Cortex	226 ± 26

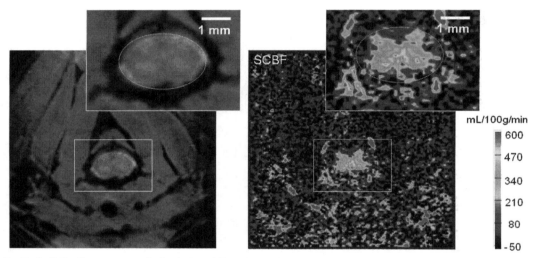

Fig. 7. (*Left*) Healthy mouse cervical spinal cord T_2-weighted SE-EPI image at 11.75 T, (*Right*) corresponding SCBF map demonstrating high perfusion values in the gray matter and low perfusion values in the white matter.

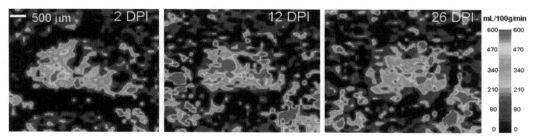

Fig. 8. Perfusion maps of an injured spinal cord 2, 12, and 26 days post-injury (73), at 11.75 T. The contusion initially caused a dramatic decrease of perfusion in the gray matter dorsal horn (*left*). Then, perfusion increased in the perilesioned area, suggesting a vascular recruitment or neovasculature (*middle*). Normalization of the perfusion slowly reappears in correlation with behavioral recovery (*right*).

Table 9
Protocol for DCE-MRI of the spinal cord

DCE-MRI Animal species	Rat	Mouse
References	(54)	(74)
Readout/sequence	c-SE	c-SE
rf coil	Implanted coil inductively coupled to an external coil	Inductively couple surface coil
Magnetic field (T)	2	9.4
FOV (mm^2) Matrix Slice thickness	25 × 25 80 × 128^a/192 × 256^b 2 mm (12 slices)	12 × 8 128 × 128 1 mm (14 slices)
TR/TE (ms)	500/25	1000/12
Baseline scan	2^c	
# of averages	2	4
Contrast agent	Magnevist	Magnevist
Concentration/bolus	0.1 mmol/kg (bolus <5 s)	0.1 mmol/kg (bolus <5 s)
Temporal resolution (min)	1.5	10
Acquisition repetition	Up to 1 h	Up to 2 h

^{a}The first acquisition can be acquired with a small matrix size for improved temporal resolution, ^{b}whereas subsequent acquisitions may have a better spatial resolution.
^{c}Baseline scan or pre-contrast acquisitions need to be acquired (one per spatial resolution, ie. 80 × 128 and 192 × 256).

injured. Correlations between changes in BSCB permeability and neurofunctional improvements have been related during endogenous repair processes (54, 74). Assessment of BSCB permeability may thus provide prognostic information on the progression of SCI and DCE-MRI may assist interventions aiming at restoring the barrier integrity. An example of a protocol is presented in **Table 9**. Post-processing is based on a two-compartment pharmacokinetic model (plasma/spinal cord tissue) (75).

3.5. Spectroscopy

Magnetic resonance spectroscopy (MRS) allows non-invasive and in vivo exploration of the tissue metabolism under normal and pathological conditions. The use of MRS has been reported in healthy (37, 76, 77) and experimental autoimmune encephalomyelitis (78) rat models, and more recently, in mouse experiments (79). Typical protocols are proposed in **Table 10**. Additional comments are given in **Note 11**. Principal metabolites that can be derived from such acquisitions are summarized in **Note 12**.

Table 10
Protocols for single voxel MRS acquisitions on rats and mice

MR spectroscopy Animal species	Rat	Mouse
References	(37)	(79)
Sequence	PRESS	PRESS
rf coil	Transmit–receive surface coil	Volume coil
Magnetic field (T)	17.6	11.75
TR (s)	3	4
TE (ms)	60	10
Spectral width (ppm)	8	10
Water suppression	VAPOR	VAPOR
Voxel size	2 × 2 × 2 to 4 mm^3 (8–16 µl)	2 × 8 × 1.1 mm^3 (4 µl)
# of averages	16–512	256
# of points	4096	512
Synchronization		Respiratory gating

- *MRS and pathology*: The spatial selection of the spinal cord tissue and the relatively low measurement sensitivity still limit the application of ^{1}H MRS. One particular obstacle is the strong magnetic field distortion resulting from deformation of the spinal cord or hemorrhages in case of spinal cord injury, for example. However, MRS may help to increase the understanding of the SC pathologies by offering markers of disease progression or tissue repair. For instance, the *N*-acetylaspartate peak is thought to be a neuronal marker indicative of axonal integrity and it has been found reduced in EAE lesions (78). Increase in choline-containing compounds is usually considered to reflect inflammation, demyelination, and remyelination and myo-inositol is a possible indicator of gliosis or glial cell proliferation. Follow-ups of spatial and temporal variations of the metabolites may thus provide additional useful information that may help to evaluate therapeutics and management of SC disease.

3.6. Other Techniques

3.6.1. Magnetization Transfer

MT techniques are based on the interaction between mobile protons of free water and protons with restricted motions in proteins and macromolecules. A selective irradiation of the immobile protons causes a partial saturation of these protons and a decrease in the intensity of the mobile proton pool, thus creating an

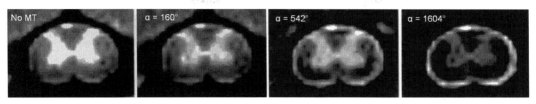

Fig. 9. MT-weighted images of a cervical rat spinal cord. MT efficiency is dependent on the amplitude, offset, and pattern of the saturation pulses applied to generate the contrast. For $\alpha = 160°$, MTR of WM, GM, and CSF equal 7.6, 5.5, and 2.9, respectively. For $\alpha = 1604°$, MTR of WM, GM, and CSF equal 51.4, 50.9, and 20.5, respectively. Modified from Ref. (17), copyright [2000], with permission from Wiley and acknowledgments to L. Lemaire (Angers, France).

additional source of tissue contrast for imaging, particularly used in the brain. Spinal cord applications are scarce (17) and MT imaging appears especially useful for the detection of the thin CSF layer surrounding the cord and to unambiguously distinguish it from the WM. Illustration is provided in **Fig. 9**. An example protocol is given in **Table 11**. The magnetization transfer ratio (MTR) is calculated by subtraction of image intensities obtained with and without a saturation pulse.

Table 11
Protocol for magnetization transfer imaging experiments on rats

Magnetization transfer MRI

Animal model	Rat
Reference	(17)
Readout/sequence	c-SE
rf coil	Birdcage coil
Magnetic field (T)	7
FOV (mm^2)	15 × 30
Matrix	64 × 128
Slice thickness (mm)	3
TR/TE (ms)	2500/7.8
# of averages	2
MT preparation pulses	Gaussian
# of RF pulse/s	20
Pulse duration (ms)	7.5
Interpulse delay (μs)	300
Peak RF amplitude (μT)	0.55, 1.89, 3.81, 5.56
Frequency offset (Hz)	1500
Corresponding flip angle (α)	160, 542, 1094, 1604

Table 12
Protocol functional imaging experiments on rats

fMRI Readout/sequence	Fast spin echo	GE-EPI
Animal species	Rat	Rat
References	(83)	(84)
rf coil	Surface coil	Volume transmit coil and surface receive coil
Magnetic field (T)	9.4	4.7
TR/TE (ms)	1800/85	1000/12.4
FOV (cm^2)	2 × 2	4 × 2
Matrix	128 × 64	64 × 64
Slice thickness	2 mm, 6 axial slices	1 mm, 4 sagittal slices
Stimulus	Capsaicin injection	Electrical stimulation (on/off) (0.3 ms, 1.5 mA, 40 Hz)
# of repetitions	[10 (baseline)–40 (inj.)]×3	20 (off)-20(on)-40(off)

3.6.2. Functional MRI (fMRI)

fMRI (80) has been extensively used to study the brain function and the extension of the method to SC, which is the communication channel between the body and the brain, is very motivating. As a matter of fact, voluntary, involuntary, and sensory functions, such as walking, breathing, and feeling pain, all rely on communication between the brain and the spinal cord. SC fMRI may thus help to better assess and understand defects following SC injury and disease or to test and discover new pharmacological agents.

So far the number of published studies is limited; however SC fMRI has been demonstrated to have sufficient sensitivity to noninvasively detect regions of the SC that are functionally active due to a given stimulus (81–85), such as active agent injection, light touch stimulation, or electrical stimulation. Examples of protocol are proposed in **Table 12**.

4. Notes

1. *Spin Echo-Echo Planar Imaging (SE-EPI) technique*: Single-shot EPI techniques, which are largely used in human studies, cannot be used in rodents because of increased ghosting and susceptibility-induced artifacts at higher magnetic

field strengths. However, recent technological improvements, including accurate gradients systems (*see* **Note 2**) and improved pre-emphasis correction, combined with the use of multi-shot EPI readout, now allow for acquisition of good-quality images. The use of four to eight shots appears to be optimal for rodent studies (32, 36, 46, 52, 86). By using such SE-EPI techniques, a gain factor of 3–4 in the acquisition time can be obtained allowing inclusion of other imaging modalities in the protocol while keeping identical and reasonable duration of anesthesia for the animal (87).

2. *SE-EPI and gradient performance*: Recent gradients are usually actively shielded and present high performance (1 T/m gradient strength, 9 kT/m/s slew rate). These strong and fast gradients enable reducing the time required to form an EPI image. They also enable strong diffusion weighting in a short period of time. All together, this permits to acquire EPI images and diffusion-weighted images at short TE, which improves SNR and reduces susceptibility effects.

    Note that residual magnetic fields induced by gradient switching may persist after the gradients are turned off (eddy currents). This may cause scaling, shift, and shear artifacts (88).

3. *Additional MR parameters for SE-EPI sequences*: In order to further ensure a limitation of ghost and distortion artifacts, EPI adjustments should include automatic shimming, pre-emphasis correction, B_0 shift compensation, and automatic ghost correction. The sampling bandwidth (BW) has to be large (~400 kHz) as a compromise between reduction of distortions and loss of SNR. To further reduce the echo time, the readout module can be shifted in the sampling window (e.g., echo shift 18%). Off-resonance artifacts are eliminated by using a fat-suppression module (e.g., Gaussian pulse, 1.5 ms, 1750 Hz BW, 2 ms spoiler). When reduced FOV are used, outer volume suppression (OVS) may be applied on each side of the body.

4. *DTI metrics*: The mean diffusivity (MD) and the fractional anisotropy (FA), which denotes the degree of directionality of the water diffusion and ranges from 0 (isotropy) to 1 (maximum anisotropy), are defined as:

$$\mathrm{MD} = \frac{\lambda_1 + \lambda_2 + \lambda_3}{3} \text{ and}$$

$$\mathrm{FA} = \frac{\sqrt{(\lambda_1 - \lambda_2)^2 + (\lambda_2 - \lambda_3)^2 + (\lambda_3 - \lambda_1)^2}}{\sqrt{2 \times (\lambda_1^2 + \lambda_2^2 + \lambda_3^2)}},$$

with λ_1, λ_2, and λ_3 the main diffusivities (eigenvalues). $\lambda_{//} = \lambda_1$ additionally represents the axial or longitudinal diffusivity and $\lambda_\perp = (\lambda_2 + \lambda_3)/2$ the radial or transverse diffusivity. Additional explanations may be found in the "High Field Diffusion Tensor Imaging in Small Animals and Excised Tissue" chapter of this book (**Chapter 7**).

In spinal cord, the direction of maximum diffusivity coincides with the white matter (WM) tract orientation (i.e., along the spinal cord axis).

5. *Diffusion MRI and bulk motion*: Diffusion-weighted images are sensitive to microscopic water self-diffusion but also to macroscopic movements such as bulk subject motion, respiration, and cardiac pulsation, which cause additional dephasing of the magnetization and thus more attenuated diffusion-weighted signals.

    Acquisitions, especially at the thoracic and lumbar levels, therefore need to be synchronized with respiration in order to minimize the bulk motion contribution to DTI metrics measurements. Although not dominant (∼4%, i.e., lower than the in-ROI variation (33)), bulk motion still exists and interferes with the motion-sensitive diffusion-encoding processes. This may be revealed by slightly lower in vivo FA values as compared to postmortem values.

    Respiration motion being controlled, cardiac motion and cerebrospinal fluid (CSF) pulsatility are the main residual causes of bulk motion in the spinal cord. These effects have already been described in human studies (89), in which cardiac gating and acquisition during the quiescent phase have been demonstrated to be key points. No similar studies have been reported on rodents. Given the rather small volume of the surrounding CSF in rodent (∼1 or 2 pixels), the spread of the pulsatility effect may be reduced as compared to human studies.

6. *Diffusion MRI and SNR*: Insufficient SNR is undesirable because weak diffusion-weighted signals close to the background noise level bias the estimated diffusion tensor parameters. SNR considerations thus limit the choice of the maximum b values used for the diffusion experiments. Values around 700 s/mm^2 are well adapted to SC investigations.

7. *DTI and number of diffusion-encoding directions*: To estimate the diffusion tensor, diffusion weighting with high b-value along at least six noncollinear directions are needed, in addition to the $b = 0$ s/mm^2 image. In order to boost SNR, a common practice is to repeat the same acquisition (signal averaging, with increasing NEX). However, diffusion tensor estimation may benefit from increased

diffusion-encoding directions rather than repeated scanning of the six directions.

Most of the reported literature on rodent SC investigation deal with six directions; however, studies with higher number of diffusion-encoding directions have also been reported (16, 52).

8. *DTI analyses and region of interest (ROI) location*: DTI analyses are commonly performed in the ventral, lateral, and dorsal WM (v-WM, l-WM, and d-WM, respectively (*see* **Fig. 2**) as well as in the ventral and dorso-lateral GM (v-GM and d-GM, respectively). Since differences have been reported between the substructures of the white and gray matter, it is important to keep in mind the ROI delineation and location of each particular study when comparing the results with other studies.

    For instance, in rat studies (14, 30), differences in longitudinal diffusivities have been reported between the different substructures of the white matter: the reticulospinal and the vestibulospinal tracts (both in v-WM), the dorsal cortical spinal tract and the fasciculus gracilis (in d-WM), or the rubrospinal (in l-WM) and the dorsal cortical spinal tracts.

    Similarly, dorsal gray matter is known, from fixed rat SC experiments (90), to have a more anisotropic structure compared to the ventral gray matter. However, in vivo studies failed to differentiate DTI metrics in ventral and dorsal GM. This lack of differentiation may be attributed to the difficulties in delineating the substructures (ventral, intermediate, dorsal, and substantia gelatinosal), as well as the spatial resolutions that were used.

9. *In vivo, ex vivo, and perfusion-fixed DTI*: If in vivo DTI metrics have to be compared with ex vivo or perfusion-fixed tissue values, the reader has to keep in mind that alteration of the biophysical environment of the tissue upon excision and fixation has an effect on the diffusivities (10, 30). Decrease in water diffusivity is observed in postmortem conditions. However, it has been demonstrated (10) that water diffusion anisotropy is equivalent in vivo, in situ after death (up to 10 h before fixation), and ex vivo (15 weeks after immersion fixation).

10. *Biexponential DTI – fast and slow component analysis*: Non-monoexponential behavior of diffusion in spinal cord at high b values (e.g., using increasing b-values, up to 16 per direction, with a maximum of 6200 s/mm^2) that can reasonably be modeled in terms of fast and slow diffusion components have been demonstrated (91, 92). The presence of these components in the gray matter and white matter could either imply the existence of two

compartments or restricted diffusion within a single compartment in the cord tissue. However, there is no consensus so far about the cellular origin of these two components.

11. *¹H MRS of the spinal cord – calibration and shimming:* If measurements are performed with a transmit–receive surface coil, excitation profiles have to be mapped with gradient echo images acquired with a pre-pulse to account for the B_1-profile of the surface coil. The strength of the pre-pulse has to be adjusted to achieve full saturation over the largest possible region of the spinal cord.

    Then, for good spectral quality, voxels extending into the surrounding bone have to be avoided. To further minimize field variations, shimming is necessary, e.g., using FASTMAP (fast automatic shimming technique by mapping along projections (93)) on a $(3\ mm)^3$ voxel (for rat) or $(2\ mm)^3$ voxel (for mouse) positioned inside the spinal cord.

    Optimization of the voxel geometry to the shape of the cord should also greatly help.

12. *¹H MRS metabolites:* The principal metabolites that can be analyzed are *N*-acetyl aspartate (NAA, at 2.02 ppm), whose concentration decrease is associated to neuronal injury; choline-containing compounds (Cho, at 3.2 ppm), which are mostly involved in cell membrane metabolism (inflammation, demyelination); creatine/phosphocreatine (Cr, at 3.02 ppm), which is a marker of overall cellular density; glutamate/glutamine (Glx, at 2.05/2.5 ppm), which is involved in excitatory and inhibitory neurotransmission and excitotoxicity; myo-inositol (mI, at 3.56 ppm), whose concentration increases in cases of glial activation or proliferation; lactate (Lac, at 1.35 ppm), which indicates hypoxic conditions; and macromolecules and lipids (Lip, at 0.9–1.3 ppm), which correlate to the extent of tissue necrosis.

References

1. Banasik, T., Jasinski, A., Pilc, A., Majcher, K., and Brzegowy, P. (2005) Application of magnetic resonance diffusion anisotropy imaging for the assessment neuroprotecting effects of MPEP, a selective mGluR5 antagonist, on the rat spinal cord injury in vivo. *Pharmacol Rep.* 57, 861–866.
2. Watson, C., Paxinos, G., and Kayalioglu, G. (2009) *The Spinal Cord. A Christopher and Dana Reeve Foundation Text and Atlas*, Elsevier ed., Academic Press, Elsevier, London.
3. Sandner, B., Pillai, D. R., Heidemann, R. M., Schuierer, G., Mueller, M. F., Bogdahn, U., Schlachetzki, F., and Weidner, N. (2009) In vivo high-resolution imaging of the injured rat spinal cord using a 3.0 T clinical MR scanner. *J Magn Reson Imaging.* 29, 725–730.
4. Dunn, E. A., Weaver, L. C., Dekaban, G. A., and Foster, P. J. (2005) Cellular imaging of inflammation after experimental spinal cord injury. *Mol Imaging.* 4, 53–62.
5. Heckl, S., Nagele, T., Herrmann, M., Gartner, S., Klose, U., Schick, F., Weissert, R., and Kuker, W. (2004) Experimental autoimmune encephalomyelitis (EAE): lesion

visualization on a 3 Tesla clinical whole-body system after intraperitoneal contrast injection. *Rofo.* **176**, 1549–1554.

6. Budde, M. D., Kim, J. H., Liang, H. F., Russell, J. H., Cross, A. H., and Song, S. K. (2008) Axonal injury detected by in vivo diffusion tensor imaging correlates with neurological disability in a mouse model of multiple sclerosis. *NMR Biomed.* **21**, 589–597.

7. Budde, M. D., Kim, J. H., Liang, H. F., Schmidt, R. E., Russell, J. H., Cross, A. H., and Song, S. K. (2007) Toward accurate diagnosis of white matter pathology using diffusion tensor imaging. *Magn Reson Med.* **57**, 688–695.

8. Kim, J. H., Budde, M. D., Liang, H. F., Klein, R. S., Russell, J. H., Cross, A. H., and Song, S. K. (2006) Detecting axon damage in spinal cord from a mouse model of multiple sclerosis. *Neurobiol Dis.* **21**, 626–632.

9. Kim, J. H., Loy, D. N., Liang, H. F., Trinkaus, K., Schmidt, R. E., and Song, S. K. (2007) Noninvasive diffusion tensor imaging of evolving white matter pathology in a mouse model of acute spinal cord injury. *Magn Reson Med.* **58**, 253–260.

10. Kim, J. H., Trinkaus, K., Ozcan, A., Budde, M. D., and Song, S. K. (2007) Postmortem delay does not change regional diffusion anisotropy characteristics in mouse spinal cord white matter. *NMR Biomed.* **20**, 352–359.

11. Song, S. K., Sun, S. W., Ju, W. K., Lin, S. J., Cross, A. H., and Neufeld, A. H. (2003) Diffusion tensor imaging detects and differentiates axon and myelin degeneration in mouse optic nerve after retinal ischemia. *Neuroimage.* **20**, 1714–1722.

12. Song, S. K., Sun, S. W., Ramsbottom, M. J., Chang, C., Russell, J., and Cross, A. H. (2002) Dysmyelination revealed through MRI as increased radial (but unchanged axial) diffusion of water. *Neuroimage.* **17**, 1429–1436.

13. Fraidakis, M., Klason, T., Cheng, H., Olson, L., and Spenger, C. (1998) High-resolution MRI of intact and transected rat spinal cord. *Exp Neurol.* **153**, 299–312.

14. Gullapalli, J., Krejza, J., and Schwartz, E. D. (2006) In vivo DTI evaluation of white matter tracts in rat spinal cord. *J Magn Reson Imaging.* **24**, 231–234.

15. Benveniste, H., Qui, H., Hedlund, L. W., D'Ercole, F., and Johnson, G. A. (1998) Spinal cord neural anatomy in rats examined by in vivo magnetic resonance microscopy. *Reg Anesth Pain Med.* **23**, 589–599.

16. Deo, A. A., Grill, R. J., Hasan, K. M., and Narayana, P. A. (2006) In vivo serial diffusion tensor imaging of experimental spinal cord injury. *J Neurosci Res.* **83**, 801–810.

17. Franconi, F., Lemaire, L., Marescaux, L., Jallet, P., and Le Jeune, J. J. (2000) In vivo quantitative microimaging of rat spinal cord at 7 T. *Magn Reson Med.* **44**, 893–898.

18. Mogatadakala, K. V., and Narayana, P. A. (2009) In vivo diffusion tensor imaging of thoracic and cervical rat spinal cord at 7 T. *Magn Reson Imaging* **27**, 1236–1241.

19. Narayana, P. A., Grill, R. J., Chacko, T., and Vang, R. (2004) Endogenous recovery of injured spinal cord: longitudinal in vivo magnetic resonance imaging. *J Neurosci Res.* **78**, 749–759.

20. Pirko, I., Ciric, B., Gamez, J., Bieber, A. J., Warrington, A. E., Johnson, A. J., Hanson, D. P., Pease, L. R., Macura, S. I., and Rodriguez, M. (2004) A human antibody that promotes remyelination enters the CNS and decreases lesion load as detected by T_2-weighted spinal cord MRI in a virus-induced murine model of MS. *Faseb J.* **18**, 1577–1579.

21. Bilgen, M. (2004) Simple, low-cost multipurpose RF coil for MR microscopy at 9.4 T. *Magn Reson Med.* **52**, 937–940.

22. Bilgen, M., Abbe, R., Liu, S. J., and Narayana, P. A. (2000) Spatial and temporal evolution of hemorrhage in the hyperacute phase of experimental spinal cord injury: in vivo magnetic resonance imaging. *Magn Reson Med.* **43**, 594–600.

23. Bilgen, M., Abbe, R., and Narayana, P. A. (2001) Dynamic contrast-enhanced MRI of experimental spinal cord injury: in vivo serial studies. *Magn Reson Med.* **45**, 614–622.

24. Bilgen, M., Al-Hafez, B., Berman, N. E., and Festoff, B. W. (2005) Magnetic resonance imaging of mouse spinal cord. *Magn Reson Med.* **54**, 1226–1231.

25. Bilgen, M., Al-Hafez, B., He, Y. Y., and Brooks, W. M. (2005) Magnetic resonance angiography of rat spinal cord at 9.4 T: a feasibility study. *Magn Reson Med.* **53**, 1459–1461.

26. Bonny, J. M., Gaviria, M., Donnat, J. P., Jean, B., Privat, A., and Renou, J. P. (2004) Nuclear magnetic resonance microimaging of mouse spinal cord in vivo. *Neurobiol Dis.* **15**, 474–482.

27. Ellingson, B. M., Kurpad, S. N., Li, S. J., and Schmit, B. D. (2008) In vivo diffusion tensor imaging of the rat spinal cord at 9.4 T. *J Magn Reson Imaging.* **27**, 634–642.

28. Gaviria, M., Bonny, J. M., Haton, H., Jean, B., Teigell, M., Renou, J. P., and Privat, A. (2006) Time course of acute phase in mouse spinal cord injury monitored by

ex vivo quantitative MRI. *Neurobiol Dis.* **22**, 694–701.

29. Schwartz, E. D., Chin, C. L., Shumsky, J. S., Jawad, A. F., Brown, B. K., Wehrli, S., Tessler, A., Murray, M., and Hackney, D. B. (2005) Apparent diffusion coefficients in spinal cord transplants and surrounding white matter correlate with degree of axonal dieback after injury in rats. *AJNR Am J Neuroradiol.* **26**, 7–18.

30. Schwartz, E. D., Cooper, E. T., Chin, C. L., Wehrli, S., Tessler, A., and Hackney, D. B. (2005) Ex vivo evaluation of ADC values within spinal cord white matter tracts. *AJNR Am J Neuroradiol.* **26**, 390–397.

31. Ahrens, E. T., Laidlaw, D. H., Readhead, C., Brosnan, C. F., Fraser, S. E., and Jacobs, R. E. (1998) MR microscopy of transgenic mice that spontaneously acquire experimental allergic encephalomyelitis. *Magn Reson Med.* **40**, 119–132.

32. Callot, V., Duhamel, G., and Cozzone, P. J. (2007) In vivo mouse spinal cord (SC) imaging using Echo-Planar Imaging (EPI) at 11.75T. *Magn Reson Mater Phy.* **20**, 169–173.

33. Callot, V., Duhamel, G., Le Fur, Y., Decherchi, P., Marqueste, T., Kober, F., and Cozzone, P. J. (2010) Echo planar diffusion tensor imaging of the mouse spinal cord at thoracic and lumbar levels: a feasibility study. *Magn Reson Med.* **63**, 1125–1134.

34. Duhamel, G., Callot, V., Cozzone, P. J., and Kober, F. (2008) Spinal cord blood flow measurement by arterial spin labeling. *Magn Reson Med.* **59**, 846–854.

35. Duhamel, G., Callot, V., Decherchi, P., Le Fur, Y., Marqueste, T., Cozzone, P. J., and Kober, F. (2009) Mouse lumbar and cervical spinal cord blood flow measurements by arterial spin labeling: sensitivity optimization and first application. *Magn Reson Med.* **62**(2), 430–439.

36. Callot, V., Duhamel, G., Cozzone, P. J., and Kober, F. (2008) Short scan-time multi-slice diffusion MR Imaging of the mouse cervical spinal cord using echo planar imaging. *NMR Biomed.* **21**, 868–877.

37. Balla, D. Z., and Faber, C. (2007) In vivo intermolecular zero-quantum coherence MR spectroscopy in the rat spinal cord at 17.6 T: a feasibility study. *Magma.* **20**, 183–191.

38. Behr, V. C., Weber, T., Neuberger, T., Vroemen, M., Weidner, N., Bogdahn, U., Haase, A., Jakob, P. M., and Faber, C. (2004) High-resolution MR imaging of the rat spinal cord in vivo in a wide-bore magnet at 17.6 Tesla. *Magma, Magn Reson Mater Phys.* **17**, 353–358.

39. Weber, T., Vroemen, M., Behr, V., Neuberger, T., Jakob, P., Haase, A., Schuierer, G., Bogdahn, U., Faber, C., and Weidner, N. (2006) In vivo high-resolution MR imaging of neuropathologic changes in the injured rat spinal cord. *AJNR Am J Neuroradiol.* **27**, 598–604.

40. Narayana, P., Fenyes, D., and Zacharopoulos, N. (1999) In vivo relaxation times of gray matter and white matter in spinal cord. *Magn Reson Imaging.* **17**, 623–626.

41. Ginefri, J. C., Poirier-Quinot, M., Girard, O., and Darrasse, L. (2007) Technical aspects: development, manufacture and installation of a cryo-cooled HTS coil system for high-resolution in-vivo imaging of the mouse at 1.5 T. *Methods.* **43**, 54–67.

42. Smith, S. M., Jenkinson, M., Woolrich, M. W., Beckmann, C. F., Behrens, T. E., Johansen-Berg, H., Bannister, P. R., De Luca, M., Drobnjak, I., Flitney, D. E., Niazy, R. K., Saunders, J., Vickers, J., Zhang, Y., De Stefano, N., Brady, J. M., and Matthews, P. M. (2004) Advances in functional and structural MR image analysis and implementation as FSL. *Neuroimage.* **23**(Suppl 1), S208–219.

43. Basser, P. J., and Jones, D. K. (2002) Diffusion-tensor MRI: theory, experimental design and data analysis – a technical review. *NMR Biomed.* **15**, 456–467.

44. Beaulieu, C. (2002) The basis of anisotropic water diffusion in the nervous system – a technical review. *NMR Biomed.* **15**, 435–455.

45. Ford, J. C., Hackney, D. B., Alsop, D. C., Jara, H., Joseph, P. M., Hand, C. M., and Black, P. (1994) MRI characterization of diffusion coefficients in a rat spinal cord injury model. *Magn Reson Med.* **31**, 488–494.

46. Fenyes, D. A., and Narayana, P. A. (1999) In vivo diffusion tensor imaging of rat spinal cord with echo planar imaging. *Magn Reson Med.* **42**, 300–306.

47. Tu, T. W., Kim, J. H., Wang, J., and Song, S. K. (2010) Full tensor diffusion imaging is not required to assess the white matter integrity in mouse contusion spinal cord injury. *J Neurotrauma.* **27**, 253–262.

48. Niessen, H. G., Angenstein, F., Sander, K., Kunz, W. S., Teuchert, M., Ludolph, A. C., Heinze, H. J., Scheich, H., and Vielhaber, S. (2006) In vivo quantification of spinal and bulbar motor neuron degeneration in the G93A-SOD1 transgenic mouse model of ALS by T_2 relaxation time and apparent diffusion coefficient. *Exp Neurol.* **201**, 293–300.

49. Hesseltine, S. M., Law, M., Babb, J., Rad, M., Lopez, S., Ge, Y., Johnson, G., and Grossman, R. I. (2006) Diffusion tensor imaging in multiple sclerosis: assessment of regional differences in the axial plane within normal-appearing cervical spinal cord. *AJNR Am J Neuroradiol.* **27**, 1189–1193.

50. Bilgen, M. (2006) Imaging corticospinal tract connectivity in injured rat spinal cord using manganese-enhanced MRI. *BMC Med Imaging.* **6**, 15.

51. Schwartz, E. D., Duda, J., Shumsky, J. S., Cooper, E. T., and Gee, J. (2005) Spinal cord diffusion tensor imaging and fiber tracking can identify white matter tract disruption and glial scar orientation following lateral funiculotomy. *J Neurotrauma.* **22**, 1388–1398.

52. Madi, S., Hasan, K. M., and Narayana, P. A. (2005) Diffusion tensor imaging of in vivo and excised rat spinal cord at 7 T with an icosahedral encoding scheme. *Magn Reson Med.* **53**, 118–125.

53. Schwartz, E. D., Cooper, E. T., Fan, Y., Jawad, A. F., Chin, C. L., Nissanov, J., and Hackney, D. B. (2005) MRI diffusion coefficients in spinal cord correlate with axon morphometry. *Neuroreport.* **16**, 73–76.

54. Bilgen, M., Dogan, B., and Narayana, P. A. (2002) In vivo assessment of blood-spinal cord barrier permeability: serial dynamic contrast enhanced MRI of spinal cord injury. *Magn Reson Imaging.* **20**, 337–341.

55. Loy, D. N., Kim, J. H., Xie, M., Schmidt, R. E., Trinkaus, K., and Song, S. K. (2007) Diffusion tensor imaging predicts hyperacute spinal cord injury severity. *J Neurotrauma.* **24**, 979–990.

56. Bilgen, M. (2007) Magnetic resonance microscopy of spinal cord injury in mouse using a miniaturized implantable RF coil. *J Neurosci Methods.* **159**, 93–97.

57. Ma, M., Basso, D. M., Walters, P., Stokes, B. T., and Jakeman, L. B. (2001) Behavioral and histological outcomes following graded spinal cord contusion injury in the C57Bl/6 mouse. *Exp Neurol.* **169**, 239–254.

58. Fraidakis, M. J., Spenger, C., and Olson, L. (2004) Partial recovery after treatment of chronic paraplegia in rat. *Exp Neurol.* **188**, 33–42.

59. Schwartz, E. D., Yezierski, R. P., Pattany, P. M., Quencer, R. M., and Weaver, R. G. (1999) Diffusion-weighted MR imaging in a rat model of syringomyelia after excitotoxic spinal cord injury. *AJNR Am J Neuroradiol.* **20**, 1422–1428.

60. Carmeliet, P. (2003) Blood vessels and nerves: common signals, pathways and diseases. *Nat Rev Genet.* **4**, 710–720.

61. Eichmann, A., Le Noble, F., Autiero, M., and Carmeliet, P. (2005) Guidance of vascular and neural network formation. *Curr Opin Neurobiol.* **15**, 108–115.

62. Park, K. W., Crouse, D., Lee, M., Karnik, S. K., Sorensen, L. K., Murphy, K. J., Kuo, C. J., and Li, D. Y. (2004) The axonal attractant Netrin-1 is an angiogenic factor. *Proc Natl Acad Sci USA.* **101**, 16210–16215.

63. Koyanagi, I., Tator, C. H., and Lea, P. J. (1993) Three-dimensional analysis of the vascular system in the rat spinal cord with scanning electron microscopy of vascular corrosion casts. Part 2: Acute spinal cord injury. *Neurosurgery.* **33**, 285–291; discussion 292.

64. Guha, A., Tator, C. H., and Rochon, J. (1989) Spinal cord blood flow and systemic blood pressure after experimental spinal cord injury in rats. *Stroke.* **20**, 372–377.

65. Whetstone, W. D., Hsu, J. Y., Eisenberg, M., Werb, Z., and Noble-Haeusslein, L. J. (2003) Blood-spinal cord barrier after spinal cord injury: relation to revascularization and wound healing. *J Neurosci Res.* **74**, 227–239.

66. Beggs, J. L., and Waggener, J. D. (1979) Microvascular regeneration following spinal cord injury: the growth sequence and permeability properties of new vessels. *Adv Neurol.* **22**, 191–206.

67. Blight, A. R. (1991) Morphometric analysis of blood vessels in chronic experimental spinal cord injury: hypervascularity and recovery of function. *J Neurol Sci.* **106**, 158–174.

68. Imperato-Kalmar, E. L., McKinney, R. A., Schnell, L., Rubin, B. P., and Schwab, M. E. (1997) Local changes in vascular architecture following partial spinal cord lesion in the rat. *Exp Neurol.* **145**, 322–328.

69. Hamamoto, Y., Ogata, T., Morino, T., Hino, M., and Yamamoto, H. (2007) Real-time direct measurement of spinal cord blood flow at the site of compression: relationship between blood flow recovery and motor deficiency in spinal cord injury. *Spine.* **32**, 1955–1962.

70. Dray, C., Rougon, G., and Debarbieux, F. (2009) Quantitative analysis by in vivo imaging of the dynamics of vascular and axonal networks in injured mouse spinal cord. *Proc Natl Acad Sci USA.* **106**, 9459–9464.

71. Detre, J. A., Leigh, J. S., Williams, D. S., and Koretsky, A. P. (1992) Perfusion imaging. *Magn Reson Med.* **23**, 37–45.

72. Kim, S. G. (1995) Quantification of relative cerebral blood flow change by flow-sensitive alternating inversion recovery (FAIR) technique: application to functional mapping. *Magn Reson Med.* **34**, 293–301.

73. Duhamel, G., Decherchi, P., Marqueste, T., Cozzone, P. J., and Callot, V. (2009) Vascular modifications occuring during spinal cord injury (SCI) recovery using high-resolution arterial spin labeling (ASL). *Annual meeting of ISMRM*. p. 1298.
74. Tatar, I., Chou, P. C., Desouki, M. M., El Sayed, H., and Bilgen, M. (2009) Evaluating regional blood spinal cord barrier dysfunction following spinal cord injury using longitudinal dynamic contrast-enhanced MRI. *BMC Med Imaging*. **9**, 10.
75. Bilgen, M., and Narayana, P. A. (2001) A pharmacokinetic model for quantitative evaluation of spinal cord injury with dynamic contrast-enhanced magnetic resonance imaging. *Magn Reson Med*. **46**, 1099–1106.
76. Bilgen, M., Elshafiey, I., and Narayana, P. A. (2001) In vivo magnetic resonance microscopy of rat spinal cord at 7 T using implantable RF coils. *Magn Reson Med*. **46**, 1250–1253.
77. Silver, X., Ni, W. X., Mercer, E. V., Beck, B. L., Bossart, E. L., Inglis, B., and Mareci, T. H. (2001) In vivo ^{1}H magnetic resonance imaging and spectroscopy of the rat spinal cord using an inductively-coupled chronically implanted RF coil. *Magn Reson Med*. **46**, 1216–1222.
78. Zelaya, F. O., Chalk, J. B., Mullins, P., Brereton, I. M., and Doddrell, D. M. (1996) Localized ^{1}H NMR spectroscopy of rat spinal cord in vivo. *Magn Reson Med*. **35**, 443–448.
79. Tachrount, M., Duhamel, G., Maues de Paula, A., Laurin, J., Marqueste, T., Decherchi, P., and Cozzone, P. J. (2011) Medullar and thalamic metabolic alterations following spinal cord injury (SCI): a preliminary mice study, combining early and longitudinal follow-ups using high-spatially resolved MRS and DTI at high field. *Annual meeting of ISMRM*. p. 401.
80. Ogawa, S., Lee, T. M., Kay, A. R., and Tank, D. W. (1990) Brain magnetic resonance imaging with contrast dependent on blood oxygenation. *Proc Natl Acad Sci USA*. **87**, 9868–9872.
81. Lawrence, J., Stroman, P. W., and Malisza, K. L. (2007) Comparison of functional activity in the rat cervical spinal cord during alpha-chloralose and halothane anesthesia. *Neuroimage*. **34**, 1665–1672.
82. Majcher, K., Tomanek, B., Tuor, U. I., Jasinski, A., Foniok, T., Rushforth, D., and Hess, G. (2007) Functional magnetic resonance imaging within the rat spinal cord following peripheral nerve injury. *Neuroimage*. **38**, 669–676.
83. Malisza, K. L., Stroman, P. W., Turner, A., Gregorash, L., Foniok, T., and Wright, A. (2003) Functional MRI of the rat lumbar spinal cord involving painful stimulation and the effect of peripheral joint mobilization. *J Magn Reson Imaging*. **18**, 152–159.
84. Zhao, F., Williams, M., Meng, X., Welsh, D. C., Coimbra, A., Crown, E. D., Cook, J. J., Urban, M. O., Hargreaves, R., and Williams, D. S. (2008) BOLD and blood volume-weighted fMRI of rat lumbar spinal cord during non-noxious and noxious electrical hindpaw stimulation. *Neuroimage*. **40**, 133–147.
85. Zhao, F., Williams, M., Meng, X., Welsh, D. C., Grachev, I. D., Hargreaves, R., and Williams, D. S. (2009) Pain fMRI in rat cervical spinal cord: an echo planar imaging evaluation of sensitivity of BOLD and blood volume-weighted fMRI. *Neuroimage*. **44**, 349–362.
86. Fenyes, D. A., and Narayana, P. A. (1998) In vivo echo-planar imaging of rat spinal cord. *Magn Reson Imaging*. **16**, 1249–1255.
87. Duhamel, G., Decherchi, P., Marqueste, T., Cozzone, P. J., and Callot, V. (2009) Assessment of pathologic mouse spinal cord recovery using high-resolution diffusion and ASL-based perfusion imaging, *Annual meeting of ISMRM*. p. 1295.
88. Le Bihan, D., Poupon, C., Amadon, A., and Lethimonnier, F. (2006) Artifacts and pitfalls in diffusion MRI. *J Magn Reson Imaging*. **24**, 478–488.
89. Summers, P., Staempfli, P., Jaermann, T., Kwiecinski, S., and Kollias, S. (2006) A preliminary study of the effects of trigger timing on diffusion tensor imaging of the human spinal cord. *AJNR Am J Neuroradiol*. **27**, 1952–1961.
90. Inglis, B. A., Yang, L., Wirth, E. D., 3rd, Plant, D., and Mareci, T. H. (1997) Diffusion anisotropy in excised normal rat spinal cord measured by NMR microscopy. *Magn Reson Imaging*. **15**, 441–450.
91. Elshafiey, I., Bilgen, M., He, R., and Narayana, P. A. (2002) In vivo diffusion tensor imaging of rat spinal cord at 7 T. *Magn Reson Imaging*. **20**, 243–247.
92. Inglis, B. A., Bossart, E. L., Buckley, D. L., Wirth, E. D., 3rd, and Mareci, T. H. (2001) Visualization of neural tissue

water compartments using biexponential diffusion tensor MRI. *Magn Reson Med.* **45**, 580–587.

93. Gruetter, R. (1993) Automatic, localized in vivo adjustment of all first- and second-order shim coils. *Magn Reson Med.* **29**, 804–811.

94. Pell, G. S., Thomas, D. L., Lythgoe, M. F., Calamante, F., Howseman, A. M., Gadian, D. G., and Ordidge, R. J. (1999) Implementation of quantitative FAIR perfusion imaging with a short repetition time in time-course studies. *Magn Reson Med.* **41**, 829–840.

Section V

Cardiovascular Applications

Chapter 20

Assessment of Global Cardiac Function

Jürgen E. Schneider

Abstract

High-resolution magnetic resonance cine imaging (cine-MRI) allows for a non-invasive assessment of ventricular function and mass in normal mice and in genetically and surgically modified mouse models of cardiac disease. The assessment of myocardial mass and function by cine-MRI does not rely on geometric assumptions, as the hearts are covered from the base to the apex, typically by a stack of two-dimensional images. The MR data acquisition is then followed by image segmentation of specific cine frames in each slice to obtain geometric and functional parameters, such as end-diastolic volume (EDV), end-systolic volume (ESV) or ejection fraction (EF). This technique has been well established in clinical routine application and it is now also becoming the reference method in experimental cardiovascular MRI. The cine images are typically acquired in short- and long-axis orientations of the heart to facilitate an accurate assessment of cardiac functional parameters. These views can be difficult to identify, particularly in animals with diseased hearts. Furthermore, data analysis can be the source of a systematic error, mainly for myocardial mass measurement. We have established protocols that allow for a quick and reproducible way of obtaining the relevant cardiac views for cine-MRI, and for accurate image analysis.

Key words: Magnetic resonance imaging, cine-MRI, rodent models, cardiac function, heart failure.

1. Introduction

Mice (and rats) have become the predominant animal models in cardiovascular research, allowing for investigation of physiological and pathophysiological conditions of the heart in much more detail than it would be possible in humans. Developments in transgenic techniques, i.e. deletion, over-expression or mutation of specific genes, have lead to an enormous increase in genetically modified mouse models, enabling study of the influence of certain genes on cardiac morphology and function. Furthermore, dedicated surgical techniques, such as ligation of the left coronary artery (LCA) or constriction of the transverse aorta (TAC), can generate conditions that closely resemble those found in patients with heart disease. Specifically, LCA induces myocardial infarction and subsequently heart failure, whereas TAC reduces the cross-sectional area of the aorta to cause pressure overload of the heart. This results in progressive hypertrophy of the left ventricle as a compensatory mechanism, eventually leading to heart failure.

Magnetic resonance imaging (MRI) is the most sophisticated, non-invasive imaging modality to investigate the functional and structural effects of gene alteration or the consequences of surgically induced myocardial stress on the hearts of rodents. MRI has become the method of choice in many laboratories around the world, as it is non-invasive, uses intrinsic contrast and is capable of obtaining true 3D information on the heart and the vascular system (e.g. (1–9)).

Ventricular volumes, which are measured to characterize global function of the heart, and ventricular mass are frequently obtained at ultra-high magnetic fields using a fast, 2D spoiled gradient echo (GE)-type sequences applied in multi-frame ('cine') mode continuously throughout the cardiac cycle. The heart is then covered from base to apex with multiple, contiguous slices. The contrast of such a sequence is based on spins flowing into the imaging slice (i.e. blood) while saturating spins that are stationary relative to the imaging slice. Thus, myocardial and skeletal muscle appears dark, whereas blood shows bright contrast ('bright-blood contrast', e.g. (1, 9, 10)). Alternatively, 2D murine black-blood cine images have been shown to accurately assess left-ventricular mass and function (11). The blood signal is suppressed by a double inversion recovery technique applied at the beginning of a cardiac cycle prior to the cine train. Although this approach is more time consuming compared to the more common bright-blood imaging, the black-blood preparation may reduce inter-observer variability (11, and could be beneficial for regional cardiac function analysis (11, where the velocity (using phase contrast MRI (12, 13)) or the displacement (by means of tagging (2, 14, 15) or displacement encoding –'DENSE' (16–18)) of the myocardial wall is measured. 2D sequences for global or regional cardiac function assessment are typically applied in an orientation defined by the geometry of the heart, i.e. the short- and long-axis orientations. Multi-frame sequences applied in 3D have also been reported for global cardiac function analysis (19–21). While they eliminate the need for special planning of the imaging slab and provide better spatial resolution particularly in long-axis orientation of the heart, they typically require long scan times (19) or the application of dedicated contrast agents (20).

Although sequence design and post-processing are different for global and regional cardiac function analysis, they are identical in animal preparation and scanning procedure. The focus of this chapter therefore is to demonstrate how the relevant information on global cardiac function can be obtained in mice (the procedure in rats is identical). Besides the general requirements, we explain in particular, how the relevant cardiac views (i.e. short- and long-axis) can be found reproducibly in normal and diseased hearts, and how an accurate data analysis can be performed. The description follows closely our recent 'How to …' publication in the *Journal of Cardiovascular Magnetic Resonance* (22).

2. Materials

2.1. MR Hardware and Software Requirements

1. MR system equipped with a horizontal magnet and preferably providing a field strength of ≥ 7 T (*see* **Notes 1 and 2**).
2. Water-cooled gradient system with maximum strength of ≥ 500 mT/m and switching times of 100–200 μs (from 0 to maximum gradient strength) (*see* **Note 3**).
3. Dedicated RF coils optimized in length and diameter and adjusted to physiological loading conditions (*see* **Note 4**).
4. Calibration sequences for frequency and power adjustments.
5. Sequence (preferably slice selective and double gated) to correct for magnetic field inhomogeneities ('shimming').
6. 2D multi-frame gradient echo sequence with echo times (TE) <2 ms and repetition times (TR) <8 ms, cardiac and, depending on magnetic field strength and required number of averages, also respiratory gated (*see* **Notes 5–7**). A graphical illustration of the sequence design is shown in **Fig. 1**.

2.2. Animal Handling

1. Standard anaesthesia set-up, comprising oxygen supply, vaporizer, anaesthetic chamber and scavenging unit with absorbers for isoflurane (*see* **Note 8**).
2. Dedicated animal cradles, optimized in diameter and length, to maintain stable animal physiology throughout an experiment. The animal cradle should be comprised of a heating blanket to maintain the core body temperature at 37°C, a nose cone for continuous delivery of the anaesthetic and a scavenging line for anaesthetic gas recovery. Additional lines may be required if cardiac-stimulating drugs such as dobutamine, or MR contrast agents need to be administered.
3. Ophthalmic lubrication (e.g. Viscotears, Novartis) in order to keep the eyes moist throughout the procedure as the eye closure reflex is suppressed during anaesthesia.

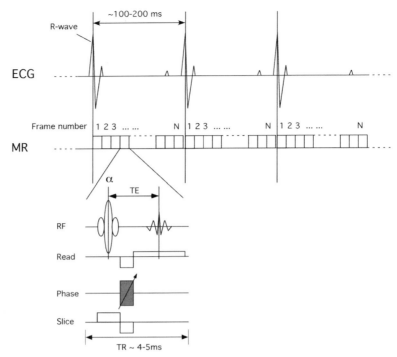

Fig. 1. Diagram of a spoiled, bright-blood gradient echo cine sequence frequently used in murine MRI: after the detection of the R-wave in the ECG, the same k-space line is acquired repeatedly with a constant value for the phase-encoding gradient. The number of frames N per cardiac cycle depends on the sequence timing and the heart rate of the animal and ranges between 15–30 frames. The illustrated scheme is repeated in the next cardiac cycle with a different value for the phase-encoding gradient. Thus, the product of number of phase-encoding steps times number of averages is the number of cardiac cycles that are required in total to obtain a full cine data set for one slice. If respiratory gating is employed, the scheme is interrupted during respiration (with or without steady-state maintenance) and the imaging time is prolonged. (Taken from Ref. (22), originally published by BioMed Central).

4. Surface-mounted, or needle electrodes (~25 gauge) inserted subcutaneously into the forelimbs to derive cardiac signal (*see* **Note 9**).
5. Pressure pad to detect respiratory motion.
6. Gating and monitoring device (*see* **Note 5**).
7. Tools, such as a pair of tweezers and scissors.
8. Consumables, such as gauze, (surgical) tape, gloves, isoflurane and isoflurane absorbers, saline.
9. Heat pad for recovery.

2.3. Post-processing

1. Computer to perform off-line data analysis.
2. Off-line analysis may require additional software (typically MATLAB (The MathWorks, Natick, US) or IDL (ITT

Visual Information Solutions, Boulder, US) to read MR data in the specific scanner format (raw or image data), process and/or convert them into a suitable image format (such as TIFF or JPEG).

3. Analysis software, for example Amira (licensed – http://www.amiravis.com/) or ImageJ (free – http://rsb.info.nih.gov/ij/).

3. Methods

The overall procedure for assessing cardiac function can be divided into the following six steps:

1. Preparation of the animal and positioning in the magnet with the heart in the isocentre of the RF-coil/gradients.
2. Scan preparations, including tuning and matching the coil (if applicable), (slice-selective) shimming and RF-power calibration.
3. Scouting for the short- and long-axis views.
4. Multi-slice cine-MRI in short-axis orientation.
5. Post-experimental procedures including animal recovery.
6. (Off-line) data analysis.

3.1. Preparing the Set-Up

1. Check isoflurane level at the vaporizer and the O_2-level at the gas cylinder and refill/replace if required.
2. Turn on the monitoring and recording device.
3. Put some tissue into the anaesthetic chamber and open the valve in the gas line for filling the anaesthetic chamber.
4. Connect animal handling with gas lines and electrical connections (i.e. ECG etc.).
5. Mount appropriate cradle to animal handling system.
6. Verify that ECG electrodes and pressure pad are sensitive.
7. Switch on the heater and pre-heat the animal cradle.
8. Cut sufficient number of tape strips of different lengths required during preparation and wet a small piece of gauze.
9. Insert the corresponding RF probe into the magnet and connect it up.
10. Set up experiment at the console computer by creating new study, selecting correct gradient and RF coil (where applicable) and provide scan-specific information, such as orientation, animal id, gender and purpose of experiment. Load the protocol for the first scan (*see* **Section 3.3**).

3.2. Animal Preparation

1. Fill the anaesthetic chamber (O_2 flow 4 L/min, isoflurane 4% for 4 min). Operate scavenging unit connected to the anaesthetic chamber in order to provide a homogeneous concentration of isoflurane in the chamber.
2. Put the animal into the chamber until it is deep under anaesthesia (*see* **Note 10**).
3. Reduce the oxygen flow to 1.5 l/min and the isoflurane concentration to 2.5% and start flushing the gas lines to the animal cradle.
4. Weigh the animal and place it prone into the cradle with the nose into the cone. If tooth bars are available, hook it up to the front tooth and pull it back so that the nose is secured in the nose cone.
5. Turn off the gas flow to the anaesthetic chamber.
6. Connect the animal to the ECG electrodes; in case of needle electrodes carefully insert the needles subcutaneously into the front limbs and fix them with tape.
7. Start recording the ECG trace. It is important to get optimal coupling between the animal and the electrodes. Sensitivity that is lost in first place can never be gained back!
8. If applicable, ensure that the upper torso is resting on the surface coil(s).
9. Place the respiratory pad below the abdomen of the animal (*see* **Note 11**).
10. Tape down the front limbs on the side or the top of the nose cone; try to minimize the loop area that is formed by the limbs and the electrodes (this will minimize future problems related to gradient interference with the ECG) (*see* **Note 12**).
11. Put ointment on both eyes, place a wet piece of gauze over the eyes and secure it with tape.
12. Put a strip of tape loosely across the lower part of the back.
13. Cover the animal with a lid and secure the lid.
14. Reduce the isoflurane concentration to 1–1.5%.

3.3. Preparing the MR Scans

3.3.1. General

1. Adjust and set the frequency.
2. Adjust amplification of physiological signals (i.e. ECG and respiration) at the monitoring device and set trigger levels appropriately. Where applicable, adjust the window length for respiratory gating.
3. SCAN_1: Run the scouting sequence consisting of three orthogonal planes, i.e. axial, sagittal and coronal orientations, respectively (**Fig. 2**).

Assessment of Global Cardiac Function 393

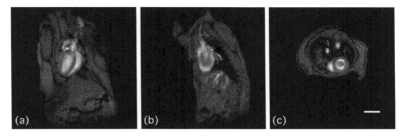

Fig. 2. Three orthogonal scout images. (**a**) Coronal, (**b**) sagittal and (**c**) axial orientation. Scale bar: 5 mm.

4. Reposition the animal and re-run if required.
5. SCAN_2: If the heart is approximately in the centre of the coil(s), load and run the second scout consisting of four axial slices (gap 1–2 mm – **Fig. 3**).
6. Tune and match the RF coil(s).
7. Perform the shim procedure.

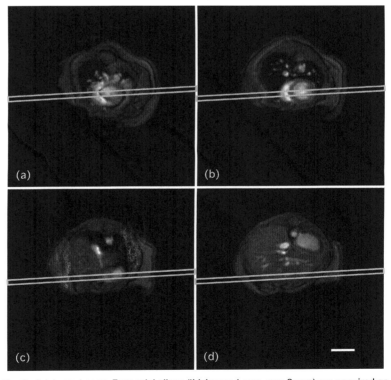

Fig. 3. Axial scout scan. Four axial slices (thickness 1 mm, gap 2 mm) are acquired as the second scout scan, using a segmented GE sequence. The *grey boxes* in (**a**)–(**d**) show the orientation of the next, coronally placed scout scan (referred to as SCAN_3). Scale bar: 5 mm.

394   Schneider

8. Re-adjust the frequency.
9. Perform power calibration.

3.3.2. Scouting for Short-Axis

1. SCAN_3: Run single-slice scout with the slice placed perpendicular to 'SCAN_2' (i.e. read-out in head–foot direction of the animal) through the left and right ventricle as indicated in **Fig. 3**. The result is shown in **Fig. 4a** (*see* **Note 13**).
2. SCAN_4: Place the slice perpendicular to 'SCAN_3' through the apex and outflow tract of the left ventricle as

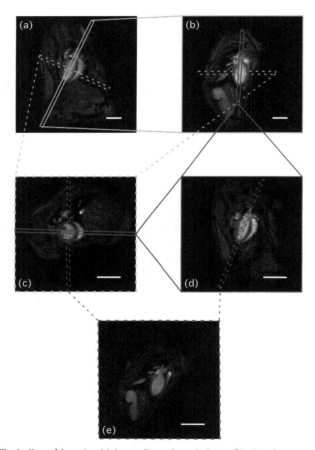

Fig. 4. Illustration of how to obtain cardiac relevant views. Starting from (**a**) coronally orientated slice (SCAN_3), the next, more sagittally orientated view (SCAN_4 – shown in (**b**)) is obtained by cutting orthogonally through the outflow tract and the apex of the left ventricle of SCAN_3 as indicated by the *solid light grey box* in (a). (**c**) Short-axis scan (SCAN_5), obtained by placing the slice perpendicular to SCAN_3 and SCAN_4, respectively (*light grey dashed boxes* in (a, b)). The *dark grey box* in (b, c) indicates the slice position for obtaining the long-axis four-chamber view (SCAN_LA4c). (**d**) Longitudinal scan SCAN_LA4c. The *dashed dark grey box* in (c, d) illustrates the orientation to obtain the long-axis two-chamber view (SCAN_LA2c), which is depicted in (**e**). Note that FOV for images (c–e) is reduced (i.e. 25.6 × 25.6 mm versus 40 × 40 mm in (a, b)). Scale bars: 5 mm.

indicated by the light grey rectangle in **Fig. 4a**. The result is shown in **Fig. 4b**.

3. SCAN_5 – short-axis orientation: Starting from 'SCAN_3', rotate the imaging slice by 90° (dashed light grey box in **Fig. 4a**); load 'SCAN_4' as reference and turn the imaging slice roughly perpendicular to the long-axis of the heart (dashed light grey box in **Fig. 4b**). It is recommended to set the correct field of view prior to running this scan (i.e. typical FOV 2.56–3 cm). Shift slice in plane so that the animal is fully contained. Swap read-out and phase encoding directions if required. The result (i.e. 'SCAN_5') is shown in **Fig. 4c**.

3.3.3. Running the Functional Study

1. Load the cine protocol, import the slice orientation from the scan 'SCAN_5'.
2. Specify the number of frames according to the heart rate and the repetition time (take a delay of 10–20 ms at the end of the cine train into consideration to allow for variations in heart rate during the scan).
3. Modify the slice offset for each scan to cover the heart from base to apex as shown in **Fig. 5**. Where applicable, update heart rate in the software and respiratory window length on the gating device.

3.3.4. Scouting for Long-Axis

1. SCAN_LA4c – four-chamber long-axis: Starting from 'SCAN_3', place the imaging slice perpendicular to 'SCAN_4' through the base and the apex of the heart (dark grey rectangle in **Fig. 4b**) to cut through both, right and left ventricle in the short-axis scan ('SCAN_5') as indicated by the dark grey box in **Fig. 4c**. The resulting long-axis four-chamber (i.e. 'SCAN_LA4c') is shown in **Fig. 4d**.

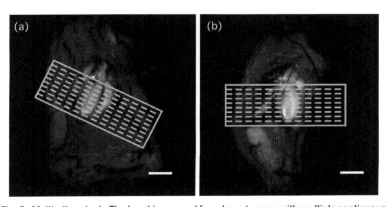

Fig. 5. Multi-slice stack. The heart is covered from base to apex with multiple contiguous slices as indicated on (**a**) SCAN_3 and (**b**) SCAN_4, respectively. Scale bars: 5 mm.

2. SCAN_LA2c – two-chamber long-axis: Starting from 'SCAN_LA4c', rotate the imaging slice by 90° on the reference image 'SCAN_5' and angulate it to section through the apex and outflow tract on 'SCAN_LA4c' (dark grey dashed boxes in **Fig. 4c, d**). The resulting two-axis four-chamber (i.e. 'SCAN_LA2c') is shown in **Fig. 4e**.

3.3.5. Scouting for Right-Ventricular Function Analysis

Due to the difference in geometry between the right and left ventricles, a slightly modified procedure is required to assess right ventricular function:

1. Follow description in **Section 3.3.2** up to 'SCAN_4'.
2. Position the short-axis slices orthogonal to the septum rather than the long-axis of the left ventricle (23), as shown in **Fig. 6**.

3.4. Post-experimental Procedures

1. Take the animal handling unit out of the magnet.
2. Carefully remove all tapes and electrodes.
3. Turn off isoflurane but keep oxygen flowing to aid recovery from anaesthesia.
4. Provide heat support until the animal is active and fully recovered from the anaesthesia. In particular, animals have to be placed alone inside a recovery cage and supplied with easily accessible water and food. Their recovery process has to be monitored. Once fully recovered they can be returned to their main cage.

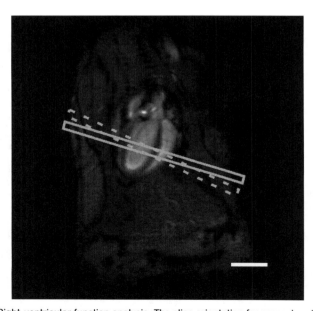

Fig. 6. Right-ventricular function analysis. The slice orientation for assessing right ventricular function is orthogonal to the septum (*solid box*) rather than the long-axis of the left ventricle (*dashed box*). Scale bar: 5 mm.

5. Disinfect all tools, cradle parts, probes, surfaces and the anaesthetic chamber that have been in contact with animal.

6. Refill isoflourane at vaporizer and weigh the absorber of the scavenging unit.

3.5. Post-processing

3.5.1. Manual or Semi-automated Analysis of Cardiac Function

1. Export the reconstructed MR data into a suitable image format (for example TIFF or JPEG) that can be read by post-processing software (*see* **Note 14**).

2. Load all frames of a slice into the segmentation software.

3. For each slice, select the frames with maximal and minimal ventricular volume, corresponding to end-diastole and end-systole. Viewing the frames in movie mode may aid the decision on which frame to segment.

4. Manually or semi-automatically outline the epicardial border first using a *Spline*-, *Lasso*- or *AutoTrace* function (dark grey trace in **Fig. 7**).

5. The ventricular cavity can subsequently be segmented using a threshold tool. (The endocardial border is depicted as the light grey trace in **Fig. 7**) (*see* **Note 15**).

6. In order to obtain ventricular mass, multiply the myocardial volume by the density of myocardial tissue (1.05 g/cm^3) (24).

7. In basal slices where the LV cavity is not fully surrounded by myocardium, the epicardial contour should be connected at both ends by a straight line through the blood pool (25).

8. Complete segmentation for all the slices and add up the results to obtain ventricular volumes and mass (*see* **Note 16**).

9. Calculate the cardiac functional parameters as listed in **Table 1** (*see* **Note 17**).

3.5.2. Infarct Size Measurements from Cine Images

The degree of heart failure (as indicated by the ejection fraction; hearts with an EF < 45% may be defined as failing) can be assessed with cine-MRI in animal models of chronic myocardial infarction (3–5, 26). To stratify the hearts according to their infarct size (*see* **Note 18**):

1. Trace and measure the endo- and epicardial circumferences in the end-diastolic frame (solid and dashed white lines in **Fig. 8**).

2. Trace and measure the length of the akinetic region (i.e. infarct) in the same frame (solid and dashed grey lines in **Fig. 8**).

3. Repeat these measurements for all slices. The infarct size IS can be calculated according to:

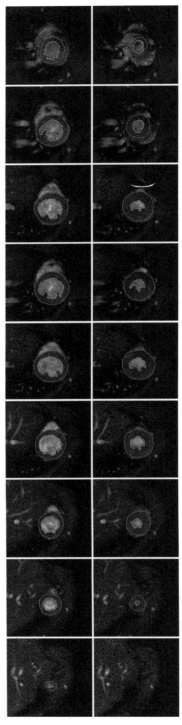

Fig. 7. Segmented cine data. End-diastolic (*left column*) and end-systolic frames are shown for all slices covering the heart from base (*top row*) to apex (*bottom row*). The *dark grey* trace indicates the epicardial border and is segmented first, while the *light grey line* depicts the endocardial border. Note that the apical slice is only segmented in end-diastole due to the through-plane motion of the heart during contraction.

Table 1
Cardiac functional parameters

Description	Acronym	Definition	Unit
Heart rate	HR	Beats per minute	bpm
End-systolic volume	ESV		μL
End-diastolic volume	EDV		μL
End-diastolic left-ventricular mass	EDM		mg
End-systolic left-ventricular mass	ESM		mg
Left-ventricular mass	LVM	$0.5 \cdot (EDM + ESM)$	mg
Stroke volume	SV	$EDV - ESV$	μL
Ejection fraction	EF	$100\% \cdot SV/EDV$	%
Cardiac output	CO	$SV \cdot HR$	mL/min

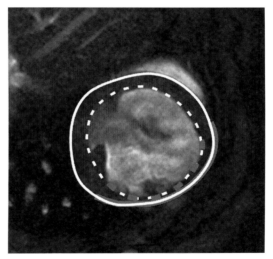

Fig. 8. Infarct size measurement. Mid-ventricular end-diastolic frame in the short-axis orientation of an infarcted mouse heart as used for measuring infarct size. Traces of the endo- and epicardial circumferences are shown in *white* and correspond to $T_{\text{epi/endo}}$. The epi-/endocardial akinetic sections are highlighted in *grey*, and correspond to $I_{\text{epi/endo}}$.

$$\text{IS} = \frac{1}{\#\text{SLICES}} \sum_{i=1}^{\#\text{SLICES}} \frac{1}{2} \left(\frac{I_{\text{epi}}^i}{T_{\text{epi}}^i} + \frac{I_{\text{endo}}^i}{T_{\text{endo}}^i} \right) \cdot 100\%. \quad [1]$$

(With I_{epi}, I_{endo} – epi- and endocardial lengths of infarcted tissue; T_{epi}, T_{endo} – total epi- and endocardial circumferences of left ventricle, respectively).

4. Notes

1. While cardiac functional imaging can in principle be performed at any magnetic field strength including clinical MR systems (e.g. (27)), ultra-high field magnets with $B_0 \geq 7$ T benefit the signal-to-noise ratio, which can be exploited to improve on temporal and spatial resolution and to reduce the overall scan time.

2. Horizontal magnets, compared to vertical magnets, eliminate any potential orthostatic effect (*see* for example Ref. (28). We have shown that normal mice can be studied in vertical position up to 3 h with only small effects on left-ventricular volumes and function (29). However, changes in volumes, ejection fraction and cardiac output were larger in failing hearts, although statistical significance was not reached in these hearts due to variability in infarct sizes. Therefore, it cannot automatically be assumed that all new genetically modified mouse models of cardiac dysfunction show long-term stability in the vertical position.

3. The stronger and faster the gradient system can be switched, the shorter echo and repetition times that can be achieved. Furthermore, the gradient coils should be cooled efficiently in order to minimize temperature changes being transferred to the animal.

4. Birdcage coils (6 provide an excellent homogeneity of the RF-field and can be applied in 'quadrature mode', resulting in an additional increase in SNR by up to a factor of ~2 (30). Surface coils suffer from an intensity gradient orthogonal to the coil, which may impact on the analysis. Cardiac-tailored correction algorithms have been proposed to overcome this problem (31).

5. Self- or retrospectively gated multi-frame sequences have been reported for small animal imaging (21, 32–35). While they overcome the need for utilizing the physiological signals to synchronize the MR scan with heart and respiratory rates of the animal, they do not eliminate the requirement for monitoring the animal during the procedure. It is an ethical (and in various countries also legal) duty of any researcher conducting animal experiments to ensure animal welfare. While the physiological signals particularly in models of heart failure may be too weak to derive a reproducible trigger signal, they are sufficient to monitor the animal.

6. Respiratory gating may not be required at lower magnetic field strengths such as 4.7 T (4), but is crucial at 11.7 T (9) in order to obtain virtually artefact-free images, as

motion artefacts become more pronounced with increasing magnetic field strength (36). Interrupting acquisition during respiration may lead to an intensity modulation in the images, or to additional – non-motion related – artefacts, caused by T_1-modulation of the amplitude and depending on the phase-encoding scheme used. These artefacts can be minimized by maintaining steady state during respiration without data acquisition (9, 37).

7. The echo time of the cine sequence is B_0-field dependent and should be chosen such that lipid and water protons have an opposite phase to enhance the contrast between various tissue types (38). The flip angle of the sequence needs to be optimized according to the respective repetition and relaxation times in order to maximize the contrast between blood and myocardium. Specifically, repetition times of less than 5 ms per frame freeze cardiac motion and should be kept constant. Hence, the number of available frames per RR interval (typically 15–30 frames in mice) is adapted to the respective heart rate. Typical sequence parameters at 9.4 T with a 33 mm quadrature-driven birdcage coil are TE/TR = 1.79/4.60 ms, matrix size 256 × 256, field of view: 25.6 × 25.6 mm, slice thickness 1 mm, flip angle 15°, bandwidth 147 kHz, two averages, ECG and respiratory gated with steady state maintenance during respiration. Scan time per slice: 2–2.5 min.

8. Isoflurane is the preferred anaesthetic reagent as the dose is easily titrated, it is easy to administer and has the least impact on cardiac function (39). Nevertheless, a dose-dependent effect of isoflurane on myocardial blood flow has been demonstrated (40). Hence, it is advised to keep the dose as low as possible.

9. In our experience needle ECG electrodes provide a better coupling, and therefore a more robust and sensitive approach, particularly in diseased animals or mice with ECG distortion.

10. The depth of anaesthesia is characterized in mice by snatch breathing with a low respiration rate and a lack pedal reflex.

11. If a respiratory loop is used to derive respiratory information, it should be placed and fixed over the back. Note: a signal can only derived in the magnetic field.

12. ECG acquisition may be prone to MR gradient breakthrough problems (41). Specifically, rapid and strong gradient switching may cause mechanical vibrations or induce a voltage in the loop formed between the electrodes and the animal. This signal superimposes the intrinsic ECG trace, leading to false triggering, which can substantially

deteriorate the image contrast and quality. In order to minimize the impact on the sequence timing or to avoid mistriggering, appropriate measures have to be taken such as damping of cradle (to minimize vibrations), minimizing loop area formed by mouse/ECG electrodes and/or suppression of gradient signal on physiological trace.

13. In order to take into consideration the geometry of the heart and to minimize partial volume effects, it is recommended to analyze the heart in a coordinate system that is defined by the symmetry of the heart itself. In the laboratory (i.e. patient/gradient) coordinate system, the three orthogonal planes have a transverse (axial), sagittal and coronal orientation (**Fig. 2**), while short-axis-, long-axis two- and long-axis four-chambers define the three orthogonal views in the heart (**Fig. 4c–e**). The cardiac-relevant views are typically double oblique relative to the coordinate system of the patient. In the short-axis view, the left ventricle has a characteristic doughnut shape, with the crescent-shaped right ventricle attached to it (**Fig. 4c**). The papillary muscles typically do not appear connected to the ventricular muscle in end-diastolic frames in mid-ventricular slices and can be seen as dark areas inside the bright ventricular cavity (**Fig. 4c**).

14. Exporting MR data into TIFF or JPEG may require some offline post-processing using appropriate software. We perform offline reconstruction of the raw MR data using purpose-written IDL routines and export the data into TIFF format.

15. While ventricular cavities are easy to segment, left-ventricular mass measurements are much more prone to systematic errors, particularly in the most basal and apical slices due to reduced image contrast or structural complexity of the heart. Thus, special care needs to be taken to only include structures for LV mass in normal hearts that are actively contracting and moving towards the LV cavity centre.

16. Since the heart is fully relaxed in the stack of end-diastolic frames and maximally contracted in the end-systolic ones, fewer slices may need to be segmented in the latter case (*see* **Fig. 7**). Moreover, the number of pixels for LV mass in the end-systolic frame is not comparable to the pixel number in the end-diastolic frame of the same slice, but the sum for all slices, i.e. end-diastolic mass and end-systolic mass should agree well with an accuracy of 5% or better. This provides an inherent quality control for the data analysis.

17. It is recommended to validate the segmentation process against autopsy first and also to establish inter- and intra-observer variability.

18. In order to distinguish between normal and infarcted tissue, the multi-frame capability is essential for non-contrast-enhanced infarct size measurements. Thickening in systole indicates viable myocardium. In the mouse chronic CAL model, akinetic areas are considered scarred, and therefore non-viable; long-term hibernation does not occur in this model. Because the infarction is almost invariably transmural in the mouse, the infarcted area will also appear significantly thinner.

Acknowledgements

This work was funded by the British Heart Foundation (BHF, Grant No. BS/06/001). We thank Ms Hannah Barnes for critical reading of the manuscript and segmentation of the cine data.

References

1. Ruff, J., Wiesmann, F., Hiller, K. H., Voll, S., von Kienlin, M., Bauer, W. R., Rommel, E., Neubauer, S., and Haase, A. (1998) Magnetic resonance microimaging for noninvasive quantification of myocardial function and mass in the mouse. *Magn. Reson. Med.* **40**, 43–48.
2. Zhou, R., Pickup, S., Glickson, J. D., Scott, C. H., and Ferrari, V. A. (2003) Assessment of global and regional myocardial function in the mouse using cine and tagged MRI. *Magn. Reson. Med.* **49**, 760–764.
3. Wiesmann, F., Ruff, J., Engelhardt, S., Hein, L., Dienesch, C., Leupold, A., Illinger, R., Frydrychowicz, A., Hiller, K. H., Rommel, E., Haase, A., Lohse, M. J., and Neubauer, S. (2001) Dobutamine-stress magnetic resonance microimaging in mice: Acute changes of cardiac geometry and function in normal and failing murine hearts. *Circ. Res.* **88**, 563–559.
4. Ross, A. J., Yang, Z., Berr, S. S., Gilson, W. D., Petersen, W. C., Oshinski, J. N., and French, B. A. (2002) Serial MRI evaluation of cardiac structure and function in mice after reperfused myocardial infarction. *Magn. Reson. Med.* **47**, 1158–1168.
5. Yang, Z., Bove, C. M., French, B. A., Epstein, F. H., Berr, S. S., DiMaria, J. M., Gibson, J. J., Carey, R. M., and Kramer, C. M. (2002) Angiotensin II type 2 receptor overexpression preserves left ventricular function after myocardial infarction. *Circulation* **106**, 106–111.
6. Hayes, C. E., Edelstein, W. A., Schenck, J. F., Mueller, O. M., and Eash, M. (1985) An efficient, highly homogeneous radio-frequency coil for whole-body NMR imaging at 1.5 T. *J. Magn. Reson.* **63**, 622–628.
7. Franco, F., Dubois, S. K., Peshock, R. M., and Shohet, R. V. (1998) Magnetic resonance imaging accurately estimates LV mass in a transgenic mouse model of cardiac hypertrophy. *Am. J. Physiol.* **274**, H679–683.
8. Kubota, T., McTiernan, C. F., Frye, C. S., Slawson, S. E., Lemster, B. H., Koretsky, A. P., Demetris, A. J., and Feldman, A. M. (1997) Dilated cardiomyopathy in transgenic mice with cardiac-specific overexpression of tumor necrosis factor-alpha. *Circ. Res.* **81**, 627–635.
9. Schneider, J. E., Cassidy, P. J., Lygate, C., Tyler, D. J., Wiesmann, F., Grieve, S. M., Hulbert, K., Clarke, K., and Neubauer, S. (2003) Fast, high-resolution in vivo cine magnetic resonance imaging in normal and failing mouse hearts on a vertical 11.7 T system. *J. Magn. Reson. Imaging* **18**, 691–701.
10. Epstein, F. H., Yang, Z., Gilson, W. D., Berr, S. S., Kramer, C. M., and French, B. A. (2002) MR tagging early after myocardial infarction in mice demonstrates contractile

dysfunction in adjacent and remote regions. *Magn. Reson. Med.* **48**, 399–403.

11. Berr, S. S., Roy, R. J., French, B. A., Yang, Z., Gilson, W., Kramer, C. M., and Epstein, F. H. (2005) Black blood gradient echo cine magnetic resonance imaging of the mouse heart. *Magn. Reson. Med.* **53**, 1074–1079.

12. Streif, J. U., Herold, V., Szimtenings, M., Lanz, T. E., Nahrendorf, M., Wiesmann, F., Rommel, E., and Haase, A. (2003) In vivo time-resolved quantitative motion mapping of the murine myocardium with phase contrast MRI. *Magn. Reson. Med.* **49**, 315–321.

13. Herold, V., Morchel, P., Faber, C., Rommel, E., Haase, A., and Jakob, P. M. (2006) In vivo quantitative three-dimensional motion mapping of the murine myocardium with PC-MRI at 17.6 T. *Magn. Reson. Med.* **55**, 1058–1064.

14. Henson, R. E., Song, S. K., Pastorek, J. S., Ackerman, J. J., and Lorenz, C. H. (2000) Left ventricular torsion is equal in mice and humans. *Am. J. Physiol. Heart Circ. Physiol.* **278**, H1117–1123.

15. Liu, W., Ashford, M. W., Chen, J., Watkins, M. P., Williams, T. A., Wickline, S. A., and Yu, X. (2006) MR tagging demonstrates quantitative differences in regional ventricular wall motion in mice, rats, and men. *Am. J. Physiol. Heart Circ. Physiol.* **291**, H2515–2521.

16. Gilson, W. D., Yang, Z., French, B. A., and Epstein, F. H. (2004) Complementary displacement-encoded MRI for contrast-enhanced infarct detection and quantification of myocardial function in mice. *Magn. Reson. Med.* **51**, 744–752.

17. Gilson, W. D., Yang, Z., Sureau, F. C., French, B. A., and Epstein, F. H. (2004) Multi-slice DENSE with three dimensional displacement encoding: Development and application in a mouse model of myocardial infarction, in *Proc. Intl. Soc. Mag. Reson. Med.*, p. 1789, Kyoto.

18. Sureau, F. C., Gilson, W. D., Yang, Z., French, B. A., and Epstein, F. H. (2004) Comprehensive assessment of systolic function in the mouse heart using volumetric DENSE MRI, in *Proc. Intl. Soc. Mag. Reson. Med.*, p. 1786, Kyoto.

19. Feintuch, A., Zhu, Y., Bishop, J., Davidson, L., Dazai, J., Bruneau, B. G., and Henkelman, R. M. (2007) 4D cardiac MRI in the mouse. *NMR Biomed.* **20**, 360–365.

20. Bucholz, E., Ghaghada, K., Qi, Y., Mukundan, S., and Johnson, G. A. (2008) Four-dimensional MR microscopy of the mouse heart using radial acquisition and liposomal gadolinium contrast agent. *Magn. Reson. Med.* **60**, 111–118.

21. Nieman, B. J., Szulc, K. U., and Turnbull, D. H. (2009) Three-dimensional, in vivo MRI with self-gating and image coregistration in the mouse. *Magn. Reson. Med.* **61**, 1148–1157.

22. Schneider, J. E., Wiesmann, F., Lygate, C. A., and Neubauer, S. (2006) How to perform an accurate assessment of cardiac function in mice using high-resolution magnetic resonance imaging. *J. Cardiovasc. Magn. Reson.* **8**, 693–701.

23. Wiesmann, F., Frydrychowicz, A., Rautenberg, J., Illinger, R., Rommel, E., Haase, A., and Neubauer, S. (2002) Analysis of right ventricular function in healthy mice and a murine model of heart failure by in vivo MRI. *Am. J. Physiol. Heart Circ. Physiol.* **283**, H1065–1071.

24. Manning, W. J., Wei, J. Y., Katz, S. E., Litwin, S. E., and Douglas, P. S. (1994) In vivo assessment of LV mass in mice using high-frequency cardiac ultrasound: Necropsy validation. *Am. J. Physiol.* **266**, H1672–H1675.

25. Barbier, E. C., Johansson, L., Lind, L., Ahlström, H., and Bjerner, T. (2007) The exactness of left ventricular segmentation in cine magnetic resonance imaging and its impact on systolic function values. *Acta Radiologica* **48**, 285–291.

26. Yang, Z., French, B. A., Gilson, W. D., Ross, A. J., Oshinski, J. N., and Berr, S. S. (2001) Cine magnetic resonance imaging of myocardial ischemia and reperfusion in mice. *Circulation* **103**, E84.

27. Gilson, W. D., and Kraitchman, D. L. (2007) Cardiac magnetic resonance imaging in small rodents using clinical 1.5 T and 3.0 T scanners. *Methods* **43**, 35–45.

28. Loeppky, J. A. (1975) Cardiorespiratory responses to orthostasis and the effects of propranolol. *Aviation, Space, and Environ. Med.* **46**, 1164–1169.

29. Schneider, J. E., Hulbert, K. J., Lygate, C. A., Hove, M. T., Cassidy, P. J., Clarke, K., and Neubauer, S. (2004) Long-term stability of cardiac function in normal and chronically failing mouse hearts in a vertical-bore MR system. *Magma* **17**, 162–169.

30. Chen, C. N., Hoult, D. I., and Sank, V. J. (1983) Quadrature detection coils – A further 2 improvement in sensitivity. *J. Magn. Reson.* **54**, 324–327.

31. Sosnovik, D. E., Dai, G., Nahrendorf, M., Rosen, B. R., and Seethamraju, R. (2007) Cardiac MRI in mice at 9.4 Tesla with a transmit-receive surface coil and a

cardiac-tailored intensity-correction algorithm. *J. Magn. Reson. Imaging* **26**, 279–287.

32. Bishop, J., Feintuch, A., Bock, N. A., Nieman, B., Dazai, J., Davidson, L., and Henkelman, R. M. (2006) Retrospective gating for mouse cardiac MRI. *Magn. Reson. Med.* **55**, 472–477.

33. Hiba, B., Richard, N., Janier, M., and Croisille, P. (2006) Cardiac and respiratory double self-gated cine MRI in the mouse at 7 T. *Magn. Reson. Med.* **55**, 506–513.

34. Hiba, B., Richard, N., Thibault, H., and Janier, M. (2007) Cardiac and respiratory self-gated cine MRI in the mouse: Comparison between radial and rectilinear techniques at 7T. *Magn. Reson. Med.* **58**, 745–753.

35. Heijman, E., de Graaf, W., Niessen, P., Nauerth, A., van Eys, G., de Graaf, L., Nicolay, K., and Strijkers, G. J. (2007) Comparison between prospective and retrospective triggering for mouse cardiac MRI. *NMR Biomed.* **20**, 439–447.

36. Wood, M. L., and Henkelman, R. M. (1986) The magnetic field dependence of the breathing artifact. *Magn. Reson. Imaging* **4**, 387–392.

37. Cassidy, P. J., Schneider, J. E., Grieve, S. M., Lygate, C., Neubauer, S., and Clarke, K. (2004) Assessment of motion gating strategies for mouse magnetic resonance at high magnetic fields. *J. Magn. Reson. Imaging* **19**, 229–237.

38. de Kerviler, E., Leroy-Willig, A., Clement, O., and Frija, J. (1998) Fat suppression techniques in MRI: An update. *Biomed. Pharmacother.* **52**, 69–75.

39. Roth, D. M., Swaney, J. S., Dalton, N. D., Gilpin, E. A., and Ross, J., Jr. (2002) Impact of anesthesia on cardiac function during echocardiography in mice. *Am. J. Physiol. Heart Circ. Physiol.* **282**, H2134–2140.

40. Kober, F., Iltis, I., Cozzone, P. J., and Bernard, M. (2005) Myocardial blood flow mapping in mice using high-resolution spin labeling magnetic resonance imaging: Influence of ketamine/xylazine and isoflurane anesthesia. *Magn. Reson. Med.* **53**, 601–606.

41. Felblinger, J., Slotboom, J., Kreis, R., Jung, B., and Boesch, C. (1999) Restoration of electrophysiological signals distorted by inductive effects of magnetic field gradients during MR sequences. *Magn. Reson. Med.* **41**, 715–721.

Chapter 21

Plaque Imaging in Murine Models of Cardiovascular Disease

Gert Klug, Volker Herold, and Karl-Heinz Hiller

Abstract

Comprehensive imaging of the cardiovascular system of murine models of atherosclerosis requires high spatial and temporal resolution as well as a high soft tissue contrast. High-field (≥ 7 T) experimental magnetic resonance imaging can provide noninvasive, high-resolution images of the murine cardiovascular system. High-field scanners, however, require special equipment and imaging protocols. The aim of this chapter is to provide instructions on how to obtain morphological and functional data on the murine cardiovascular system in animal models of atherosclerotic disease on a very high-field scanner (17.6 T). Equipment requirements are presented, and a comprehensive description of the methods needed to complete a magnetic resonance imaging exam, including the animal preparation, imaging, and image analysis are discussed. In addition, common problems during high-field MRI experiments and methods to validate MRI results are reviewed. The steps can be adopted to other MRI scanners and modification of the imaging parameters might allow for a more individual assessment of cardiovascular diseases in a number of transgenic mouse models.

Key words: Atherosclerosis, mouse models, MRI, high-field, plaque composition, contrast agents.

1. Introduction

In recent years, high-resolution MRI has become the most promising technique for noninvasive characterization of atherosclerotic plaque composition, evolution, and progression. Current MRI strategies are based on multicontrast (non-contrast agent) approaches as well as on the use of commercially available nonspecific contrast agents or targeted molecular imaging probes.

ApoE$^{-/-}$ mice are widely used as popular animal models for atherosclerosis as they develop similar atherosclerotic plaques as observed in humans (1). The advantages of these transgenic murine models are rapid plaque progression, low maintenance costs, and a well-known genome for further genetic engineering.

Besides non-contrast agent MR-imaging techniques, the passive targeting of intraplaque macrophages using commercially available iron oxide particles is a widespread MR approach for identifying, characterizing, and quantifying plaque inflammation.

2. Materials

2.1. MRI Hardware

In order to provide sufficient signal-to-noise ratio, commonly applied MR systems consist of main magnets generating static magnetic fields of 7 T, 9.4 T, 11.7 T, or 17.6 T (2–5) (*see* **Note 1**).

Rapidly switching gradient inserts (rise time <300 μs, field strength > 200 mT/m) are important to enable the use of pulse sequences with short echo times.

Radiofrequency (rf) coils represent a critical part in the MR hardware setup and a careful choice of coils adjusted to the needs of the experiment offers a good way to improve data quality (SNR). Two types of coils are commonly used for small animal imaging (*see* **Note 2**):

1. Volume coils offer a high homogeneity and sensitivity increases with smaller inner diameters. Compared to linear designs (birdcage), quadrature coils increase SNR.

2. Surface coils offer high sensitivity but suffer low penetration depth and homogeneity. Arrays of surface coils may reduce scan time or increase SNR.

While surface coils may provide advantages in cardiac imaging or imaging of the ascending thoracal aorta in rodents, we prefer the use of volume coils. These allow homogeneous data acquisition throughout the transversal imaging plane. To choose the optimal coil setup, one has to consider the main magnetic field strength, the dimensions of the gradient bore, and the dimensions of the rodents:

40-mm gradient bore (17.6 T):
  1. 20-mm Birdcage: Mice up to <20 g bodyweight.
  2. 27-mm Birdcage/TEM: Mice 20–40 g bodyweight.

57-mm gradient bore (17.6 T):
  1. 38-mm Birdcage: Mice >40 g bodyweight.

20-cm gradient bore (7 T):
  1. Birdcage/Volume-transmit–surface-receive/quadrature coils for all rodents.

2.2. Animal Handling

This section describes animal handling as it is performed in our laboratory on a 17.6-T vertical bore magnet (Bruker Avance 750) with a 40-mm gradient bore (field strength 1000 mT/m) (*see* **Fig. 1**). Furthermore, as described above, probeheads with an inner diameter of 20–27 mm are used (*see* **Fig. 2**) (6). Compared to conventional horizontal bore magnets, which often provide sufficient space for animal handling containers and warming pads, space is critical in this setup. Hemodynamics are not influenced by upright position of the mice (7).

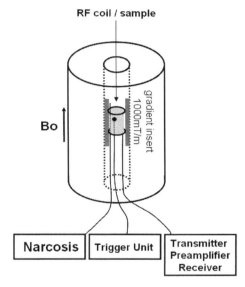

Fig. 1. Schematic setup of the probehead when using a vertical bore main magnet.

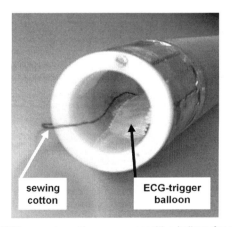

Fig. 2. A 27-mm TEM resonator with pressure-sensitive balloon for respiratory gating and cardiac triggering.

Anesthesia induced by inhalation of isoflurane is the method of choice in our laboratories (*see* **Note 3**).

Required materials:

1. Adjustable oxygen (O_2) supply (cylinders with compressed oxygen).
2. Isoflurane vaporizer.
3. Narcosis container (cage-sized) with cap and hole for gas supply.
4. PVC flexible tube (inner diameter ~0.5 cm) for gas supply with three-way stopcock: O_2 supply → isoflurane-vaporizer → O_2 + isoflurane → switchable three-way stopcock. (a) First output: to the narcosis container/preparation desk (length ~2–3 m). (b) Second output: to the MR coil (~5–10 m, depending on the distance to the MR magnet).
5. Nose cone to be applied on the PVC tube on the preparation desk (e.g., made from a 10-mL syringe (*see* **Fig. 3**)).
6. A cone to cover the upper opening of the coil after positioning the mouse inside it (e.g., lower end of a 50-mL Falcon™ tube).
7. A clock and a scale for protocol.
8. ECG trigger unit (*see* below).

Body temperature is kept constant using a temperature-controlled electrical warming pad designed for use in MR scanners (e.g., RAPID Biomedical GmbH). With experiments

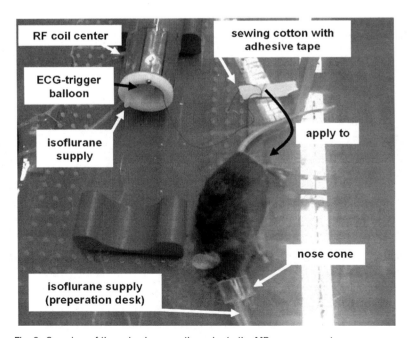

Fig. 3. Overview of the animal preparation prior to the MR measurement.

using coils of a very small inner diameter (20.0–30.0 mm), extra warming pads are not applicable. To stabilize the temperature of the sample volume inside the rf coil, the temperature adjustment of the gradient cooling unit can also be used. However, to find the right adjustments during different gradient duty cycles, preliminary experiments are required to measure the temperature while applying the desired pulse sequences.

Given a constant body temperature during the MR experiments, wildtype mice and ApoE$^{-/-}$ mice can be examined for a measurement time up to 2 h (8). However, when just examining plaque morphology at a limited vascular region, a total measurement time of 20–30 min is usually sufficient.

2.3. Cardiac Triggering and Respiratory Gating

When investigating vascular sections which are highly exposed to the cardiac and respiratory movements, dedicated trigger and gating mechanisms are necessary.

1. Commonly, an electrocardiogram (ECG) signal is used to synchronize cardiac movement and MR data acquisition. Therefore, ECG electrodes are connected to the front limbs of the animal.
2. The trigger signal is generated from the filtered electrical ECG signal with a dedicated trigger unit (RAPID Biomedical, Rimpar, Germany).
3. Respiration is monitored using the pressure signal of a respiration sensor (Graseby Medical Limited, Hertfordshire, UK) transformed into an electrical signal outside the magnet with a pressure transducer (Honeywell Inc., Freeport, IL).
4. In case of strong interferences between the electromagnetic signal of rapid switching gradient pulses and the ECG signal, the respiration sensor can also be applied to monitor cardiac motion. When placed upon the cardiac region (*see* **Fig. 2**), the filtered and amplified signal shows both cardiac and respiratory motion (*see* **Fig. 4**).

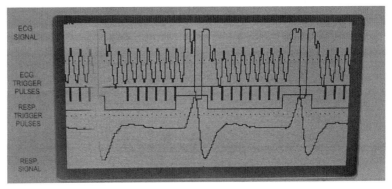

Fig. 4. Signal obtained with the pressure-sensitive balloon for cardiac triggering and respiratory gating.

3. Methods

3.1. Animal Preparation

Mice should be kept in anesthesia shorter than 2 h. A good rule of thumb is that for every hour in anesthesia, an animal should be allowed to recover for 1 day without anesthesia. This is especially important in genetically engineered mice (e.g., ApoE$^{-/-}$) or mice with invasively induced morbidities (e.g., ligature of the vasculature). These facts have impact on mortality during the experiment. If done routinely, the average time between induction of anesthesia and insertion of the probe into the magnet is 10–15 min. So:

Check the equipment *before* inducing anesthesia:
1. Enough oxygen gas available?
2. Isoflurane vaporizer filled? (One filling ~3 mice).
3. Supply lines connected correctly?
4. Warming pad/lamp on preparation desk on?
5. ECG trigger balloon connected correctly? Trigger unit setup adequate?
6. MR scanner ready?
7. Gradient cooling system temperature 33–38°C?
8. Adhesive tape strips ready (two of ~4 cm, one of ~8 cm, one to fix the cap-cone onto the coil)?

Check the coil (all items are inserted from the bottom of the coil toward the probehead, *see* **Figs. 1–3**):
9. Narcosis supply line inserted?
10. ECG trigger balloon positioned within the coil center with leucoplast strips? This depends on the area you would like to investigate. The balloon must be positioned right over the apex of the mouse heart to ensure optimal trigger signal. The region of interest (ROI) should be right in the center of the coil. So if examining the thoracal aorta, position the heart and the balloon in the coil center.
11. If necessary (for volume coils in which the animal should be pulled in), sewing cotton (loose end of ~10 cm on the end of the probehead) thread through the probehead?

Then:
12. Set three-way stopcock to the narcosis container line.
13. Turn on O_2 (flow 2 L/min) and isoflurane (4%) to flood container with gas.
14. After ~3 min, insert the mouse into the container. Record time as "narcosis induced" in measurement protocol.
15. Wait for the mouse to fall asleep (~2 min). Check gasping frequency (should be around 60/min).
16. Weigh mouse while asleep and record weight.

17. Connect the supply line from the narcosis container to the nose cone on the preparation desk (*see* **Fig. 3**).
18. Set isoflurane flow to 2.5%.
19. Apply mouse to nose cone. Make markings on the mouse (tail, ear, ...) if necessary.
20. Fix sewing cotton with adhesive tape on the hind feet of the mouse (**Fig. 3**, "apply to").
21. Fix forefeet with second adhesive tape.
22. Apply 8 cm leucoplast perpendicular to the one applied to the forefeet to allow "hanging" the mouse into the coil with it.
23. Pull mouse into the coil by pulling the lower end of the sewing cotton. Mouse heart's apex beat should be positioned over the ECG trigger balloon.
24. Check trigger signal on the trigger unit.
25. Fix mouse with leucoplast applied to the forefeet inside the coil. If coil is positioned vertically, the mouse should not move downwards.
26. Set three-way stopcock to the gas supply line inserted into the coil.
27. Close upper end of the coil (nose end of the mouse) with the cone to prevent anesthetic gas from leaking.
28. Turn isoflurane flow to its definite value of 1.5–2%.
29. Fix cap-cone with leucoplast.
30. Check ECG trigger signal again.

3.2. MR Imaging

The steps in this section describe the imaging protocol using an MR system that comprises the following components: main magnet, Bruker Avance 750 (17.6 T); gradient system, 1000 mT/m (ID 40 mm, rise time 150 μs); rf coil, homebuilt transverse electromagnetic (TEM) resonator (ID 27 mm); trigger system, RAPID ECG Trigger Unit + respiration sensor (Graseby); and imaging software, ParaVision 4.0 (Bruker).

1. Tune and match the rf coil according to the description in the Bruker manual (A-3-9).
2. Generate a new scan ("New Scan" button) and select the scan protocol "A_TRIPILOT_GE_bas". Adjust the following parameter with the spectrometer control tool ("Edit Method" button): field of view, 40 mm × 40 mm.
3. Start adjustment routines with the "Traffic Light" button. The following standard adjustments are performed: linear-order auto shim to compensate field inhomogeneities, basic frequency adjustment, adjustments for the reference pulse gain (RF gain) and the receiver gain (RG).

4. The tripilot protocol is automatically performed subsequently.

5. Set up a new scan and initialize the parameter measuring method with "FLASH".

6. Change the following PVM parameters to create a customized Cine Flash protocol: effective spectral bandwidth, 75,757.6 Hz; echo position, 25%; duration of slice rephase, 0.6 ms; read spoiler duration, 0.75 ms; read spoiler strength, 20.0%; slice spoiler duration, 2.0 ms; slice spoiler strength, 20.0%; field of view, 27 mm × 27 mm; matrix size, 256 × 256; movie mode, "on"; number of movie cycles, 10; and trigger parameters: trigger mode, "per_phase_step"; trigger delay, 1.0 ms; echo time, 1.815 ms; repetition time, 12 ms.

7. Save this protocol.

8. Open the geometry editor and set the parameter slices to "2".

9. Use the coronal and the sagittal "Tripilot" scans to navigate the two Cine Flash slices.

10. Create two transverse slices, one approximately at the level of the ascending aorta and one at the level of the abdominal aorta.

11. Press GOP to acquire the two transversal image slices.

12. Now three spots of the aortic cross section should mark the course of the ascending and descending aorta as shown in **Fig. 5a**.

13. Navigate the coronal oblique Cine slice according to **Fig. 5b** to acquire a cine data set of the entire aortic arch and the descending aorta. Adjust the trigger delay parameter of the trigger unit so that the trigger pulse coincides with the early systole. This ensures that the data acquisition

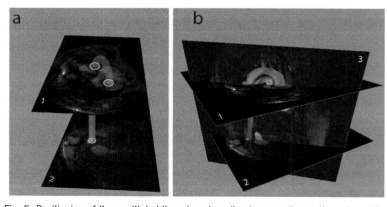

Fig. 5. Positioning of the sagittal oblique imaging slice to cover the aortic arch and the descending aorta.

for the morphologic measurements is gated to occur during systole (i.e., when blood flow velocity is at a maximum) such that excited fast-flowing blood is not refocused during acquisition and thereby generating images with a black blood contrast.

14. Generate a new scan ("New Scan" button) and adjust the parameters given in **Table 1** with the spectrometer control tool ("Edit Method" button).

    This parameter setup allows for the acquisition of 10 slices with an interleaved multi-slice mode. During each cardiac cycle, one k-space line in frequency encoding direction is acquired. The repetition time can be calculated by the formula (when omitting respiratory gating):

    TR = number of slices × RR (duration of one cardiac cycle). Assuming an RR interval of 130 ms, the present parameter setup would thus generate a repetition time

Table 1
Parameter adjustments for the multi-slice spin echo measurements

Echo time	9.0 ms
Slice bandwidth	100%
Excitation pulse choice	"gauss512"
Refocusing pulse choice	"gauss512"
Effective spectral bandwidth	83,333.3 Hz
Duration of read dephase	1 ms
Duration of 2D phase gradient	1 ms
Slice spoiler duration	0.5 ms
Slice spoiler strength	20.0%
Field of view	20.0 mm × 20.0 mm
Matrix size	265 × 256
Slice thickness	0.4 mm
Object ordering mode	"Interlaced"
Number of slices in slice packages	10
Slice distance in slice packages	0.8 mm
Trigger module	"On"
Fat suppression	"On"
Fat suppression pulse list	"sinc3"
Fat suppression bandwidth	1600.103 Hz
Fat suppression spoiler duration	2.0 ms
Fat suppression spoiler strength	40%

of 1.3 s and would lead to mainly T_1-weighted images. The balance between TR and TE is responsible for the image contrast behavior and can be used to identify different plaque compartments. **Table 2** gives some guidelines on how to emphasize the MR signal of different plaque regions with multicontrast imaging when performing high-field imaging. T_1 weighting can be customized by adjusting the slice number of the multi-slice data set.

15. After data acquisition, the MR images can be exported directly from ParaVision by using the export function. **Figure 6** shows an example of one slice of a 2D multi-slice spin echo set. For further analysis *see* **Notes 4** and **5**.

Table 2
Contrast behavior of different atherosclerotic plaque regions

	T_1 weighting	PD weighting	T_2 weighting
Thrombus	+ − +/−	− − +/−	− − +/−
Lipid core	+	+	−
Fibrous tissue	+/−	+	+/− − +
Calcium	−	−	−
TE/TR	10 ms /500 ms	10 ms /2000 ms	30 ms /2000 ms

+: hyperintense; −: hypointense

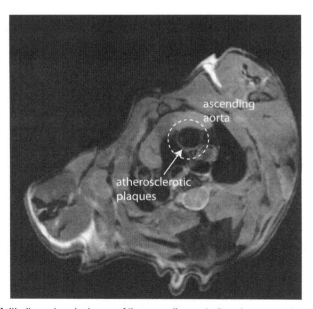

Fig. 6. Multi-slice spin echo image of the ascending aorta (imaging parameters *see* text).

4. Notes

1. *Field Strength and SNR*

    Generally, the signal-to-noise ratio (SNR) can be assumed to increase linearly with the magnetic field strength B_0 when ignoring the effects of relaxation parameters. Due to the convergence of T_1 and T_2, however, relaxation effects with high magnetic field strength reduce the overall gain on SNR. With

    $$\text{SNR} \propto B_0 \sqrt{\frac{T_2}{T_1}},$$

    the SNR increases less than linearly as shown by de Graaf et al. (9). T_1 relaxation times generally increase with increasing field strength while the differences between tissue T_1s decrease. The improved SNR at high field strengths, however usually also leads to an improved contrast-to-noise ratio (CNR). By contrast, the T_2 relaxation time decreases at high magnetic fields while the relative differences between tissue T_2s remain the same or even increase. This makes T_2 weighting a more appropriate imaging strategy at high fields than T_1 weighting. However, for both weightings a gain of CNR can be expected by increasing the B_0 field.

2. *Coils*

    The performance of the coil is of paramount importance to the imaging experiment. Always handle coils and connectors with great care. The signal quality (and finally SNR) achieved during the experiment might be optimized by adjustments to the coil setup. Two major problems may arise while performing small animal scans:

    No signal when performing auto-SF. Then check the following:
    1. Coil/probe in the center of the magnet?
    2. Wiring correct?
    3. Correct frequency selected?
    4. Correct nuclei selected?
    5. Coil tunable/matchable? If not, consider a coil defect.
    6. None of the above? Consider other hardware/software defects.

    Low SNR. Check:
    1. Optimal coil selected?
    2. ROI in coil center?
    3. Optimized tune/match?

4. Correct receiver gain?

5. Correct flip-angle calibration?

6. Signs of interfering signals?

If experiencing a drop in SNR between two scans, clone the first scan (localizer) and perform it again. Compare the image quality. If they are nearly the same, your lack of SNR might originate from the sequence design (e.g., in-plane resolution). If you experience a drop in SNR between two separate experiments (mice), this is mostly due to hardware difficulties or global scan parameters (e.g., shim).

3. *Narcosis*

Narcosis might be induced by intravenous/intraperitoneal injection of agents or via inhalation. As stated above, we prefer inhalation narcosis to injections. One has to consider the following pros and cons when working with isoflurane narcosis:

Pros:

1. Wide therapeutic index and easy management of anesthesia depth.

2. Easy integration into MR hardware (diameter of flexible PVC feed line: ~0.75 cm).

3. Noninvasive approach.

4. A constant and adjustable rate of gasping allows easy breath-hold techniques.

Cons:

1. No analgesia.

2. MR-active ^{19}F-compound with accumulation in fat tissue (only in ^{19}F-MRI).

3. Circulatory and hemodynamic effects (10).

4. *Segmentation*

For morphometric analysis, the pixel size of the MR images should not exceed 100×100 μm^2. Since the healthy murine aortic wall has a diameter of approximately 50 μm, a given resolution of 100×100 μm^2 would only allow for the quantification of a significant increased aortic wall cross section due to atherosclerotic plaque. Computer-aided morphometry can be performed using software for image analysis such as Image-Pro Plus (Media Cybernetics, Silver Spring, MD) or Amira (Visage Imaging GmbH, Berlin, Germany). The cross-sectional area of the vessel wall can be quantified as the area bounded by the outer limits of the adventitia minus the lumen area. To reduce errors for the vessel wall segmentation, adjust the image zoom prior to tracing boundaries.

Changing the image contrast during the segmentation process can lead to significant errors in the area quantification.

5. *Validating Results with Histology*
   Validation of MR results obtained form small animals in vivo is essential for the correct interpretation of in vivo results and is mostly performed by optical procedures ex vivo. Generally, mice are sacrificed under deep anesthesia within a short time (days) after the NMR experiment by blood removal followed by injection of either saline or a fixation agent via the left ventricle. Thereafter the organs of interest are removed. Histological evaluation is either done by bright-field microscopy, fluorescent microscopy or electron microscopy. Further validation might be performed by functional tests (e.g., endothelial function). According to the desired goal of your study on rodent vasculature, the following procedures may be considered:

Morphometry:
1. Use vessels fixed with paraformaldehyde (alternatively frozen sections).
2. Cut sections of 5–10 μm thickness.
3. Stain with hematoxilin–eosin stain.
4. Photograph vessels with a camera attached to the microscope and perform morphometry on a dedicated, calibrated software.

Studies of vessel wall composition:
1. Since you will have to perform different staining procedures use every third to fourth section for one staining.
2. Use vessels fixed with paraformaldehyde for the following: cellular/extracellular matrix (van Gieson stain), elastic fibers (elastica stain), collagen fibers (AZAN stain), calcification (von Kossa stain).
3. Fat: Perform Oil-Red staining on unfixed or frozen sections.

Fluorescent microscopy:
1. Allows detection of multiple cell types, antigens, receptors, or magneto-fluorescent contrast agents.
2. Should be performed on frozen sections since preparation (heat) might otherwise degrade proteins.
3. Nuclei of cells are mostly DAPI-stained (blue) to give orientation.
4. Be aware of auto-fluorescent tissues when interpreting results. Always compare to a non-stained control section.

Electron microscopy:
1. Allows detection of particles ~10 nm (e.g., nanoparticles, intracellular organelles).
2. Samples fixed in glutaraldehyde.
3. Only a close region might be analyzed (samples are typically 1×1 mm^2).

References

1. Nakashima, Y., Plump, A. S., Raines, E. W., Breslow, J. L., and Ross, R. (1994) ApoE-deficient mice develop lesions of all phases of atherosclerosis throughout the arterial tree. *Arterioscler. Thromb.* **14**, 133–140.
2. Wiesmann, F., Szimtenings, M., Frydrychowicz, A., Illinger, R., Hunecke, A., Rommel, E., Neubauer, S., and Haase, A. (2003) High-resolution MRI with cardiac and respiratory gating allows for accurate in vivo atherosclerotic plaque visualization in the murine aortic arch. *Magn. Reson. Med.* **50**, 69–74.
3. Fayad, Z. A., Fallon, J. T., Shinnar, M., Wehrli, S., Dansky, H. M., Poon, M., Badimon, J. J., Charlton, S. A., Fisher, E. A., Breslow, J. L., and Fuster, V. (1998) Noninvasive in vivo high-resolution magnetic resonance imaging of atherosclerotic lesions in genetically engineered mice. *Circulation* **98**, 1541–1547.
4. Schneider, J. E., McAteer, M. A., Tyler, D. J., Clarke, K., Channon, K. M., Choudhury, R. P., and Neubauer, S. (2004) High-resolution, multicontrast three-dimensional-MRI characterizes atherosclerotic plaque composition in ApoE$^{-/-}$ mice ex vivo. *J. Magn. Reson. Imaging* **20**, 981–989.
5. Herold, V., Wellen, J., Ziener, C. H., Weber, T., Hiller, K.-H., Nordbeck, P., Rommel, E., Haase, A., Bauer, W. R., Jakob, P. M., and Sarkar, S. K. (2009) In vivo comparison of atherosclerotic plaque progression with vessel wall strain and blood flow velocity in ApoE(−/−) mice with MR microscopy at 17.6 T. *Magn. Reson. Mater. Phys.* **22**, 159–166.
6. Klug, G., Kampf, T., Ziener, C., Parczyk, M., Bauer, L., Herold, V., Rommel, E., Jakob, P. M., and Bauer, R. W. (2009) Murine atherosclerotic plaque imaging with the USPIO Ferumoxtran-10. *Front. Biosci.* **14**, 2546–2552.
7. Wiesmann, F., Neubauer, S., Haase, A., and Hein, L. (2001) Can we use vertical bore magnetic resonance scanners for murine cardiovascular phenotype characterization? Influence of upright body position on left ventricular hemodynamics in mice. *J. Cardiov. Magn. Reson.* **3**, 311–315.
8. Itskovich, V. V., Choudhury, R. P., Aguinaldo, J. G., Fallon, J. T., Omerhodzic, S., Fisher, E. A., and Fayad, Z. A. (2003) Characterization of aortic root atherosclerosis in ApoE knockout mice: high-resolution in vivo and ex vivo MRM with histological correlation. *Magn. Reson. Med.* **49**, 381–385.
9. de Graaf, R. A., Brown, P. B., McIntyre, S., Nixon, T. W., Behar, K. L., and Rothman, D. L. (2006) High magnetic field water and metabolite proton T1 and T2 relaxation in rat brain in vivo. *Magn. Reson. Med.* **56**, 386–394.
10. Zuurbier, C. J., Emons, V. M., and Ince, C. (2002) Hemodynamics of anesthetized ventilated mouse models: aspects of anesthetics, fluid support, and strain. *Am. J. Physiol. Heart. Circ. Physiol.* **282**, H2099–H2105.

Chapter 22

Interventional MRI in the Cardiovascular System

Harald H. Quick

Abstract

Endovascular stent-graft placement for thoracic aortic disease such as aortic dissection or aortic aneurysms is usually performed under conventional X-ray guidance. The experimental concept of using magnetic resonance imaging (MRI) for image-based guidance of vascular instruments for this specific intervention potentially offers a number of features that – aside from not using ionizing radiation – may provide added diagnostic value to the interventional therapy. It allows not only pre-interventional evaluation and detailed anatomic diagnosis but also permits immediate post-interventional, anatomical, and functional delineation of procedure success that may serve as a baseline for future comparison during follow-up.

Key words: Interventional MRI, cardiovascular interventions, interactive, real-time MR imaging, aortic dissection, aortic stent grafting, vascular device delivery, device visualization, MR compatibility.

1. Introduction

Several attributes render magnetic resonance imaging (MRI) attractive as imaging modality for guidance of intravascular therapeutic procedures, including its excellent soft tissue contrast, imaging in arbitrary oblique planes, lack of ionizing radiation, and the ability to provide functional information, such as flow velocity or flow volume per unit time, in conjunction with morphologic information.

The transition of MR from a purely diagnostic imaging modality into an imaging platform to guide therapeutic cardiovascular interventions requires realization of more technical features than just instrument and device visualization. For an interventional imaging concept to succeed, the setup must account for imaging flexibility and interactivity in order to streamline the workflow required for performing image-guided interventions. Flexible, interactive user interfaces with in-room monitors or large MR-compatible displays as well as fast and interactive real-time user interfaces are important prerequisites to successfully perform MR-guided vascular interventions (1–4).

For pre-clinical experiments featuring MR-guided cardiovascular interventions for development and evaluation of MR-compatible devices, new procedures, as well as workflow issues, large animal models such as pigs have been successfully established over the last two decades (1–10). This is owed to the fact that these large animals provide cardiovascular anatomy and size as well as physiological cardiovascular parameters such as cardiac output volume, heart rates, and blood flow velocities comparable to humans.

This chapter describes the interventional MR scanner hardware setup, interventional vascular devices, animal preparation and anaesthesia, the interventional MR-imaging sequences as well as the experimental workflow for performing MR image-guided aortic stent-graft delivery in a pig model with aortic dissection using commercially available and MR-compatible instruments (11–14). In the opinion of the author of this chapter, this kind of cardiovascular intervention is at the border of the pre-clinical experimental phase and will be among the first clinical applications in humans. It can serve as a general example for other basic MR-guided device delivery interventions.

2. Materials

2.1. Interventional MR Scanner Setup

1. 1.5-T short- and wide-bore MR scanner (e.g., Magnetom Espree; Siemens Healthcare Sector, Erlangen, Germany).
2. Radiofrequency (RF) receiver coils for RF signal reception. Three clusters of the SpineArray coil each featuring three RF channels as well as two BodyMatrix RF coils each with six RF channels were activated for RF signal reception (Tim Technology, Siemens Healthcare Sector, Erlangen, Germany).
3. In-room console with 18" display and keyboard for interactive scanner operation from inside the scanner room (Siemens Healthcare Sector, Erlangen, Germany). Alternatively/additionally:
4. Large format backprojection screen and video beamer contained in an RF-shielded box for in-room display of real-time images (*see* **Fig. 1**).

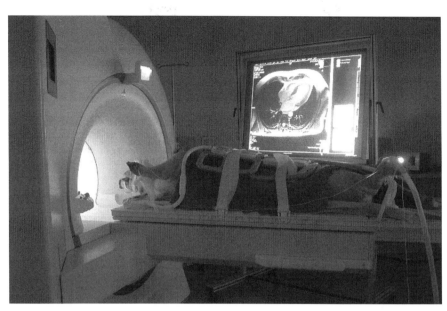

Fig. 1. Photograph of a pig lying head first, supine, on the scanner patient table in front of a 1.5-T MR system. The animal is covered by two BodyMatrix RF surface coils. Large in-room display of the real-time images is provided by a large format backprojection screen in conjunction with a video beamer that is contained in an RF-shielded box.

5. Interactive real-time user interface software (e.g., "Interactive Front End" works in progress; Siemens Healthcare Sector, Erlangen, Germany).

6. Automatic contrast injector (Spectris; Medrad Inc., Indianola, PA, USA) for controlled injection of paramagnetic contrast agent.

7. All non-MR-compatible anaesthesia equipment such as monitors, perfusors, and ventilation devices are placed outside the MR scanner room (*see* **Fig. 2**) (*see* **Note 1**).

8. Intercommunication system for communication in the operation room while the MR scanner is running in real-time mode (*see* **Note 2**).

2.2. Interventional Vascular Devices

1. Guidewire, 0.89 mm diameter, 260 cm length, with hydrophilic coating and nitinol core (Terumo Standard Glidewire; Terumo Medical Corporation, Somerset, NJ, USA) for introduction of the introducer sheath (*see* below) with dilator into the iliac artery (*see* **Notes 3** and **4**).

2. Introducer sheath with silicon pinch valve, inner diameter of 6 mm, free length of 300 mm for introduction into the artery, overall length of 700 mm with dilator (Gore Introducer Sheath; W.L. Gore Inc., Flagstaff, AZ, USA).

3. Thoracic aortic stent-graft endoprosthesis (100 mm length, 26 mm diameter) made from self-expanding nitinol wire

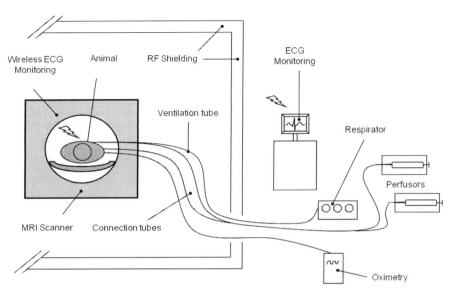

Fig. 2. Schematic diagram for the ventilation and monitoring setup for experiments in the MR scanner room. All equipment for ventilation and monitoring is placed outside the MR scanner room. Connection to the animal inside the scanner requires long ventilation tubing and connection tubing. (Redrawn from (10).)

covered with an expanded polytetrafluoroethylene (ePTFE) membrane (GoreTAG; W.L. Gore Inc., Flagstaff, AZ, USA). The loaded stent-graft is constrained by an implantable ePTFE sleeve on the leading end of an 18-French (equates to 6.0 mm diameter) polyurethane delivery catheter. Middle-to-end deployment is initiated by retraction of a GoreTex filament, which rapidly releases the stent-graft from the ePTFE sleeve (*see* **Fig. 3**).

2.3. Animal Preparation and Anaesthesia

1. Domestic pigs (60–90 kg depending on application) that will be sedated and fully anaesthetized (*see* **Note 5**).
2. Sedation is performed with intramuscular injection of ketamine hydrochloride (30 mg/kg), azaperone (2 mg/kg), and atropine (0.025 mg/kg).
3. Intravenous access in ear vein with 18-gauge cannula.
4. Anaesthesia is maintained by continuous intravenous infusion of propofol (160 mg/h), fentanyl (0.4 mg/h), and midazolam (40 mg/h).
5. Infusion pump for anaesthesia (Perfusor Compact; Braun, Melsungen, Germany) (*see* **Note 1**).
6. Endotracheal tube (5.0–8.0 mm inner diameter) for animal ventilation.
7. Respirator for ventilation (Oxylog; Dräger, Lübeck, Germany) (*see* **Note 1**).

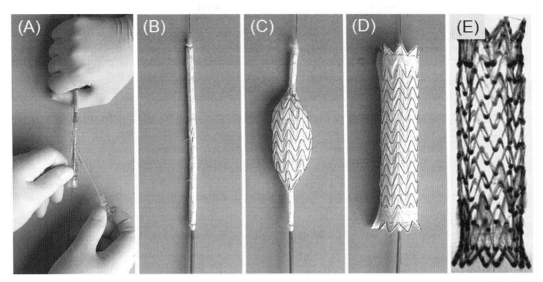

Fig. 3. Photographs of the GoreTAG stent-graft. (**a**) Trailing end of the stent-graft delivery device shows position of the guidewire as well as the Gore filament that can be manually pulled to release the loaded stent-graft at the distal end of the delivery device (*see* b). (**b**) The stent-graft loaded on the 18-Fr polyurethane delivery catheter and covered by the Gore membrane. (**c**) The graft is released in a fraction of a second by manually retracting a Gore filament (*see* a) string that prevents the graft from unfolding. (**d**) The fully expanded nitinol-based, membrane-covered stent-graft. (**e**) Minimal-intensity projection of the fully expanded GoreTAG stent-graft acquired with a high-resolution 3D FLASH sequence in a water phantom. The lumen of the stent is free of artefacts thus allowing for detailed evaluation of the stent lumen and surroundings.

8. Monitoring of heart rate with a four-channel ECG unit of an MR scanner (Siemens Healthcare Sector, Erlangen, Germany).
9. Monitoring of oxygenation with peripheral oxygen sensor (Vet/Ox 4404 pulse oximeter; Heska Animal Health Products, Clayton, Australia) (*see* **Note 1**).
10. Contrast-enhanced MR-angiography (MRA) with Gd-DTPA (Magnevist; Bayer Schering, Berlin, Germany) or Gd-DOTA (Dotarem; Guerbet, Cedex, France).
11. Euthanization with bolus injection of pentobarbital (80 mg/kg).

2.4. Interventional MR Imaging and Sequences

MR imaging is comprised of pre-interventional diagnostic imaging, real-time MR image guidance, and post-interventional imaging of therapeutic success. Image sequences are listed with imaging parameters. The abbreviations used are as follows: FOV, field of view; TR, repetition time; TE, echo time; slice, slice thickness; BW, bandwidth; flip, flip angle; and TA, acquisition time.

1. Localizer imaging (TrueFISP sequence: TR = 4 ms, TE = 2 ms, FOV = 400 × 400 mm, matrix = 192 × 192 pixel, slice = 8 mm, flip = 70°, providing 3 slices for each of

the basic orientations (axial, coronal, sagittal)) provides fast anatomic overview and anatomic landmarks.

2. Cine-TrueFISP retro, ECG-gated, respiration-suspended, retrospective image reconstruction (TrueFISP sequence: TR = 40 ms, TE = 1.1 ms, FOV = 380 × 330 mm, matrix = 192 × 168, slice = 6 mm, BW = 930 Hz/pixel, flip = 70°, TA = 15 s, 20 imaging phases/RR-interval, parasagittal orientation along the aortic arch and axial to the course of the aorta) for cardiac-gated pre- and post-interventional evaluation of aortic pathology.

3. Test bolus timing (Fast gradient echo (GRE) sequence: TR = 34.1 ms, TE = 1.2 ms, FOV = 400 × 400 mm, matrix = 256 × 192, slice = 10 mm, BW = 400 Hz/pixel, flip = 30°, 60 measurements providing 1 image/s) in axial orientation for determination of the contrast arrival time in the vessel segment of interest (aorta) when planning first-pass arterial MR angiography (MRA).

4. Contrast-enhanced (CE) 3D MR angiography (FLASH sequence: TR = 2.5 ms, TE = 1.0 ms, FOV = 400 × 400 mm, matrix = 384 × 246, slab thickness of 154 mm, 128 slices with an interpolated slice thickness of 1.2 mm, BW = 685 Hz/pixel, flip = 15°, TA = 18 s, coronal orientation) for pre- and post-interventional angiographic display of the vascular lumen.

5. Interactive real-time TrueFISP (TrueFISP sequence with projection reconstruction (PR): TR = 3.0 ms, TE = 1.5 ms, FOV = 360 × 360 mm, matrix = 192 × 192, slice = 6 mm, BW = 1530 Hz/pixel, 49 echoes were acquired for each reconstructed image, resulting in a "real-time" frame rate of 7 fps) (*see* **Notes 6, 7**, and **8**).

6. Post-contrast T_1-weighted VIBE (VIBE sequence: TR = 3.1 ms, TE = 1.3 ms, FOV = 400 × 400 mm, matrix 256 × 134, slice = 3 mm, BW = 560 Hz/pixel, flip = 20°, fat saturation, 128 slices acquired in 21 s) to confirm for post-interventional thrombus formation in false lumen of dissection and to perform contrast-enhanced MRA after application of a paramagnetic contrast agent (Gd-DTPA or Gd-DOTA), as well as post-contrast evaluation of thrombus formation in the false lumen.

3. Methods

Before the actual interventional procedure can be performed, the MRI scanner has to be prepared for the experiment (**Section 3.1**). The stent-graft delivery device with its introducer sheath is prepared for vascular introduction (**Section 3.2**). The animal will be sedated and prepared for anaesthesia (**Section 3.3**). Surgical preparation for vascular access is performed (**Section 3.3.1**). Finally, the animal is brought to the interventional MR scanner suite and prepared for MR imaging. The MR imaging section (**Section 3.4**) describes the chronologic order of the imaging protocol with all the sequences used for pre-interventional diagnostics, interventional real-time instrument guidance, and post-interventional evaluation of therapeutic success.

3.1. Interventional MR Scanner Setup

1. The RF coils used in the experiment are prepared for use in an animal intervention. The SpineArray RF coil is placed on the patient table of the MR scanner. The SpineArray coil is covered with paper covers. Two BodyArray RF coils are covered with plastic bags to protect the coils from contact with body fluids.

2. The mobile in-room console is placed next to the scanner such that the operator can run the scanner with keyboard commands from inside and such that the interventionalist simultaneously can monitor the course of the intervention on the 18" monitor.

3. The large format in-room backprojection screen with the RF-shielded video beamer is placed in the scanner room such that everybody in the operation room can monitor and follow the course of the intervention (*see* **Fig. 1**).

4. All imaging sequences are stored in accordance with the imaging parameters provided in **Section 2.4** Imaging sequences are stored in chronological order to form an imaging protocol according to the imaging workflow. The localizer sequences are opened first.

3.2. Interventional Vascular Devices

1. The 18-Fr dilator is introduced with its leading end through the silicone pinch valve into the Gore introducer sheath.

3.3. Animal Preparation and Anaesthesia

1. Initial sedation of the animals is achieved by intramuscular injection of ketamine hydrochloride (30 mg/kg), azaperone (2 mg/kg), and atropine (0.025 mg/kg) vertically into the animal's neck (*see* **Note 9**).

2. Intravenous access is achieved by insertion of an 18-gauge cannula into an ear vein.

3. Anaesthesia is induced by venous bolus injection of propofol (2 mg/kg BW) and fentanyl (0.005 mg/kg BW).

4. Endotracheal intubation is performed in prone position of the animal on the operation table. An assisting person opens the snout by means of a gauze bandage applied to the animal's upper and lower jaws. Swallowing reflex is completely

defunct. Tube sizes range from 5.0 to 8.0 mm depending on the size of the pig. After pre-oxygenation with 100% O_2, the tongue is carefully put into a forward position. When palate and palatal velum become visible, the tube can be lead in under eye contact toward the epiglottis and carefully pushed forward simultaneously rotating it round its axis.

5. The animal is ventilated with a respirator (Oxylog) using 50% oxygen and a ventilation rate between 10 and 12 breaths/min. The tidal volume varies between 600 and 1000 mL depending on the weight of the pig. Pre-oxygenation using 100% oxygen is performed prior to each induction of apnea.

6. Monitoring includes heart rate and ECG via external leads, oxygenation via peripheral oxygen sensor at the pig's tail, and respiration via external chest sensors.

7. Anaesthesia is maintained by continuous intravenous infusion of an initial dose of propofol of 2.29 mg/kg BW/h, midazolam of 1.14 mg/kg BW/h, and fentanyl of 0.009 mg/kg BW/h delivered via infusion pump (Perfusor Compact) (*see* **Note 10**).

8. For MR procedures, the potentially ferromagnetic infusion pumps and the respirator are placed outside the scanner room. For connections to the animal lying in the isocentre of the magnet, a ventilation tube and connector tubing 6 m in length – depending on the room size and setup – have to be used (*see* **Note 1**).

3.3.1. Surgical Animal Preparation

1. A paramedian incision of the left groin is performed to expose the left iliac artery. For that, the iliacal muscles are dissected bluntly and the underlying iliac artery is laid open.

2. The Terumo guidewire is introduced into the iliac artery and pushed forward to be located with its leading end into the abdominal aorta of the animal.

3. The Gore introducer sheath with the 18-Fr dilator in place is gently shifted with its leading end over the trailing end of the guidewire. The catheter sheath is subcutaneously tunnelled up into groin area over the guidewire until the leading end of the catheter sheath is safely positioned within the aorta of the animal.

4. All skin lesions and the bluntly dissected muscles are closed by single-stitch technique. The animal is then transported to the MR scanner unit (*see* **Note 11**) by the interventional team (*see* **Note 12**).

5. The animal is placed head first in a supine position inside the scanner on the paper-covered spine phased-array radio

frequency (RF) coil with three clusters of coil elements activated for signal reception. For ECG electrode placement, the animal's chest is shaved with a disposable shaver and contact gel is used to assure good electrode-to-skin contact. The ECG trigger unit of the MR scanner is connected to the electrodes. Two body flex phased-array RF coils, each consisting of two clusters of coil elements, are placed anteriorly on the pig. The legs of the animal are fixed with bandages to the scanner table in order to result in a stable position (*see* **Fig. 1**).

6. The dilator is retracted from the GORE introducer sheath over the Terumo guidewire. The 18-Fr delivery catheter with the mounted GoreTAG endoprosthesis is pushed over the guidewire through the silicon pinch valve to enter the catheter sheath. The delivery device is pushed forward in order to introduce the catheter-mounted endoprosthesis through the sheath into the iliac artery. The Terumo guidewire is removed before MR imaging is performed. The stent-graft and catheter are left in this position in the iliac artery until instrument guidance with real-time imaging is performed (*see* **Section 3.4.2**).

3.4. Interventional MR Imaging and Sequences

Imaging sequences in the following will be referred to as listed in **Section 2.4**. All necessary imaging parameters are given in **Section 2.4**.

3.4.1. Pre-interventional Diagnostics

1. The patient table of the MR scanner is moved until the chest of the animal is positioned in the laser landmark crosshairs. This position is confirmed with "landmark" in order to position this body region in the isocentre of the magnet.

2. Pre-interventional MRI evaluation is commenced with TrueFISP localizer imaging providing a fast overview of the anatomical situations and landmarks (*see* Step 1, **Section 2.4**).

3. Cine-TrueFISP retro sequences (Step 2, **Section 2.4**) are acquired using ECG gating while animal breathing is suspended. The image orientation is parasagittal in order to cover the aortic arch and the thoracic and abdominal aorta to best advantage. The descending aorta subsequently is covered with multiple imaging slices in an orientation axial to the course of the vessel to achieve transaxial views of the site of aortic dissection.

4. Test bolus timing (Step 3, **Section 2.4**) is performed by injecting 2 mL of contrast agent (Gd-DTPA or Gd-DOTA) at a flow rate of 2 mL/s to be flushed with 30 mL of saline at the same flow rate. The test bolus sequence is started

simultaneously with contrast test bolus injection. The time between start of contrast agent injection and resulting signal enhancement in the vascular region of interest (aorta in cross-sectional orientation) is the time delay necessary to consider between the initiation of actual contrast agent injection and starting the image acquisition for the next step, 3D CE-MRA.

5. Pre-contrast 3D CE-MRA (Step 4, **Section 2.4**) is acquired as a native imaging mask *before* the injection of the contrast agent. For this, the MRA sequence is set up in four identical runs, acquired in coronal orientation with suspended breath holds, with user-defined pauses in between. The first sequence run provides a native, non-contrast, data set for subsequent image data subtraction removing non-vascular background signal from the later following contrast-enhanced MR angiographic phases.

6. A volume of 16–32 mL of contrast agent is injected with the automatic contrast injector into the ear vein access with an injection rate of 2 mL/s followed by a 30-mL saline flush at the same flow rate.

7. Post-contrast 3D CE-MRA (Step 4, **Section 2.4**) is performed as described in Step 5 in this section, with identical imaging parameters. The initialization of the second run of the 3D CE-MRA sequence is timed according to the animal's circulation time (*see* test bolus timing in Step 4 of this section.) in order to result in a first-pass arterial MR angiogram. The third and fourth sequence runs follow in immediate succession in order to provide the arterio-venous and venous contrast phase, respectively. Images of the second (arterial) and third acquisition phases (venous) were post-processed to produce maximum-intensity projections (MIP) enabling comprehensive 3D angiographic visualization of the aortic pathology.

3.4.2. Real-Time Interventional Instrument Guidance

1. For aortic stent-graft implantation, the delivery system with the mounted stent-graft is advanced from the iliac artery to the thoracic aorta under MR fluoroscopy using the Interactive real-time TrueFISP sequence with projection reconstruction (Step 5, **Section 2.4**). Real-time imaging is performed during free breathing and without cardiac triggering. The reconstructed imaging frame rate is 7 fps. Image slice position, orientation, and contrast parameters can be changed and adapted interactively from inside the scanner room using the in-room console while the sequence is running. Imaging position is adjusted to a parasagittal orientation for anatomic display of the descending thoracic aorta as derived from the MRI sequence to visualize the longest

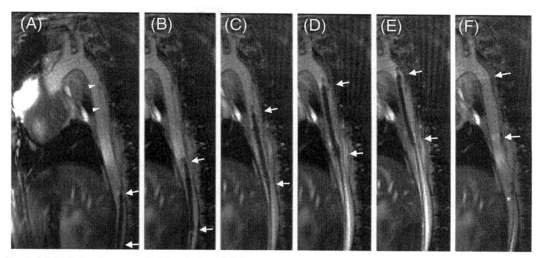

Fig. 4. (**a**)–(**e**) Safe advancement of the stent-graft delivery system (*arrows*) up to the level of the dissection (*arrowheads*) under real-time MRI guidance. (**f**) The correct stent-graft position (*arrows*) is confirmed immediately after stent-graft deployment showing complete coverage of the dissection. (Reproduced from (14) with permission from Oxford University Press).

course of the thoracic aorta including ascending aorta, aortic arch, and descending aorta (*see* **Fig. 4**).

2. While the real-time sequence is running, the catheter is moved forward along the abdominal and thoracic aorta under visual guidance till the distal end with the endoprosthesis reaches a position across the floating dissection membrane that covers the entry tear of the false lumen.

3. When the correct position of the endoprosthesis is confirmed, the interventionalist retracts the Gore filament at the proximal end of the delivery catheter to release the endoprosthesis in the animal's aorta. Stent deployment is monitored with real-time MR imaging.

4. Subsequent to full deployment of the stent, the delivery catheter is retracted under real-time monitoring. Imaging can be stopped when the distal tip of the catheter has reached the introducer sheath and no further imaging is necessary to fully retrieve the catheter through the introducer sheath.

3.4.3.
Post-interventional
Evaluation of
Therapeutic Success

1. After successful stent-graft placement, the pre-interventional MRI protocol as listed in **Section 2.4**, including parasagittal and axial Cine-TrueFISP retro sequences (*see* **Fig. 5**) as well as 3D CE-MRA with FLASH (*see* **Fig. 6**) in three contrast phases (arterial first-pass, arterio-venous, venous), was repeated to evaluate therapeutic procedure success. Additionally, a post-contrast T_1-weighted 3D VIBE sequence (Step 6, **Section 2.4**) is acquired in axial orientation to

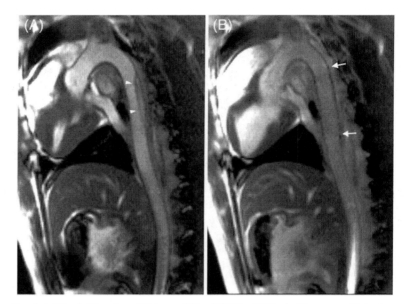

Fig. 5. (a) Pre-interventional high-resolution MRI (TrueFISP retro) in parasagittal orientation showing the dissection flap (*arrowheads*) in the proximal descending thoracic aorta. One of 20 acquired frames per RR-interval is shown. (b) Corresponding post-interventional MRI demonstrating correct position of the stent-graft (*arrows*) with complete coverage of the dissection. (Reproduced from (14) with permission from Oxford University Press).

evaluate and confirm post-interventional thrombus formation in the false lumen of the treated dissection (*see* **Fig. 7**).

2. Following completion of all experiments, the animal is euthanized by intravenous bolus injection of 80 mg/kg BW pentobarbital.

3. Open surgical preparation of the animal's thorax is carried out for removal of aortic specimen with stent-graft in order to evaluate and verify the actual position of the stent-graft. This specimen serves as the standard of reference for macroscopic evaluation and comparisons with imaging (*see* **Fig. 8**).

4. For scientific evaluation of interventional procedure success, similar to pre-interventional evaluation, minimal true lumen diameter is measured on TrueFISP images. Furthermore, images are to be assessed qualitatively for grade of false lumen thrombosis and compared with macroscopic examination of the ex vivo aorta.

5. For statistical analysis, continuous variables are presented as medians and interquartile range. Diameters of true lumen before and after stent-graft placement are compared using the non-parametric Wilcoxon signed-rank test.

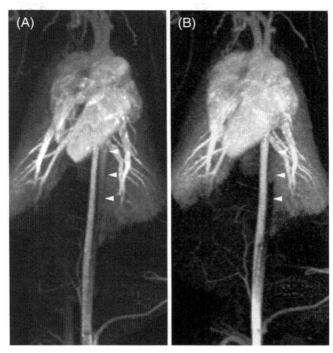

Fig. 6. (a) Pre-interventional contrast-enhanced 3D MR angiography (anterior–posterior projection) showing true and false lumen (*arrowheads*). (b) Corresponding contrast-enhanced 3D MRA after stent-graft placement showing that the false lumen (*arrowheads*) is completely obliterated. (Reproduced from (14) with permission from Oxford University Press).

4. Notes

1. Safety Note: Only use devices and materials in the MR scanner room that are non-ferromagnetic and that have been tested and labelled MR compatible. Since the infusion pumps and respirators described are not explicitly labeled MR compatible they are potentially ferromagnetic and should not be brought closer to the MR scanner than the 5 Gauss line. Aside from the potential hazards due to magnetic forces on ferromagnetic, non-MR-compatible devices, imaging RF artefacts might originate from non-MR-compatible electrical equipment that is used in the MR scanner room.

2. An intercommunication system is helpful for communication among the interventional team while the scanner is running in noisy real-time imaging mode. The system interconnects all staff that are directly involved in the procedure. This is especially important when clinical

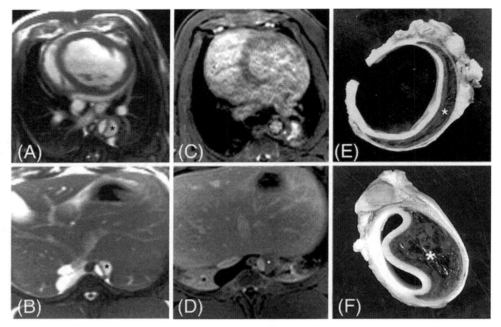

Fig. 7. (**a, b**) Pre-interventional high-resolution MRI (TrueFISP retro) in axial orientation showing dissection of the thoracic (**a**) and abdominal aorta (**b**), with true and false lumen (*) and the undulating dissection flap. (**c, d**) Corresponding post-interventional MRI (post-contrast T_1-weighted 3D VIBE) showing the stent-graft within the true lumen (**c**) while the false lumen is completely thrombosed (*). False lumen thrombosis (*) extends down to the abdominal aorta. (**e, f**) Corresponding macroscopic slices at the level of the in vivo stent-graft position: (**e**) stent-graft removed), and at the level of the abdominal aorta (**f**), confirming complete thrombosis (*) of the false lumen (FL). (Reproduced from (14) with permission from Oxford University Press).

applications in humans are envisioned. Such a system, however, was not available during the described experiments.

3. Safety Note: Be aware that vascular devices (guidewires and catheters) that are used in a conventional X-ray surrounding often come with metal cores (also the Terumo Glidewire) and metal braidings, although from the outside they might look like being made completely from plastic. Such metal cores and braidings are used for device reinforcement and increased instrument push and steerability. These conventional devices with their mechanically supporting, electrically conducting structures might act as antennae in an MRI surrounding thus concentrating RF energy around the device that eventually might heat up. This may result in a safety hazard to the animal (patient), as well as to the interventionalist (15–18).

4. MR-compatible guidewires are currently not commercially available. MR-compatible guidewires made from PEEK as well as from fibre glass are currently under development for safe use in MR-guided vascular interventions (19–23).

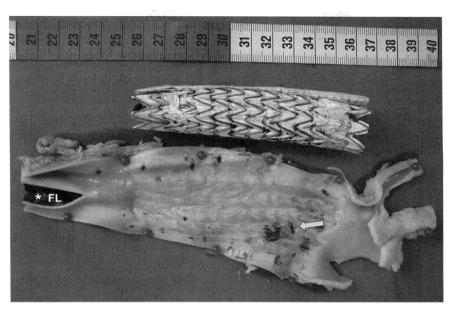

Fig. 8. Ex vivo specimen of the descending thoracic aorta (view onto the true lumen) showing the proximal dissection tear (*arrow*). The intimal imprints of the stent-graft demonstrate the correct in vivo position of the stent-graft. Successful obliteration of flow is evidenced by complete thrombosis (*) of the false lumen (FL). (Reproduced from (14) with permission from Oxford University Press).

5. The animal size should be chosen large enough to fit the relatively large catheter dimension. Otherwise, the iliac artery diameter might be too small to enter the 18-Fr catheter sheath and devices.

6. Projection reconstruction (PR) real-time TrueFISP imaging for MR-guided cardiovascular instrument guidance (in own experiments) has shown to be less susceptible to flow and pulsation artefacts than conventional Cartesian reconstruction imaging.

7. Projection reconstruction TrueFISP imaging allows for high frame rates when the number of new projections chosen is low (undersampling). On the other hand, high undersampling rates for improved temporal resolution come at the cost of increased streak artefacts.

8. Increasing image acquisition speed with parallel imaging while using many individual coil elements (RF channels) increases image reconstruction times, but – depending on the reconstruction computer hardware – a certain limit can be reached where a delay in image reconstruction times can be observed that even adds up during the course of an intervention. Then the "real-time" display is no longer in synchrony with the real intervention.

9. Adequate sedation was assessed when the animal failed to withdraw after its snout was stroked.
10. If necessary, the initial dose was increased according to the reflex status of the pig, resulting in a doubling of the dose in individual cases.
11. Ventilation during the transport from the animal facility (or catheter lab) to the MR suite is maintained by the use of an anaesthesia bag.
12. To perform the described MR-guided vascular interventions on a pig, the interventional team should be comprised (at least) of the following staff:
    - Veterinary medicine doctor (DVM) experienced in large animal care for animal preparation, sedation, anaesthesia, and surgical preparation of the iliac artery.
    - Radiological technologist (RT) experienced in cardiovascular MR for MR scanner preparation and MR scanning.
    - Medical doctor (MD) for assistance in animal preparation and experienced in clinical cardiovascular intervention such as stent-graft placement.
    - PhD with technical background for MR hardware setup, setup and optimization of real-time sequences, integrity of safety issues, and overall logistic workflow.

Acknowledgements

The author acknowledges the help and teamwork of Dr. Stephanie Aker, DVM, Lena Schaefer, RT, Dr. Holger Eggebrecht, MD, Gernot M. Kaiser, MD, Hilmar Kuehl, MD, Frank Breuckmann, MD, Michael O. Zenge, PhD, and Mark E. Ladd, PhD, (all at the University Hospital Essen, Essen, Germany), when performing the actual series of interventional experiments that are described here and that have led to corresponding publications as cited in the reference list. Parts of this work have been published in *Eur. Heart J.* 2006; 27, 613–620.

References

1. Quick H. H. Kuehl H., Kaiser G., Hornscheidt D., Mikolajczyk K. P., Aker S., Debatin J. F., and Ladd M. E. (2003) Interventional MRA using actively visualized catheters, TrueFISP, and real-time image fusion. *Magn. Reson. Med.* **49**, 129–137.

2. Quick H. H., Kuehl H., Kaiser G., Aker S., Bosk S., Debatin J. F., and Ladd M. E. (2003) Interventional MR angiography with a floating table. *Radiology* **229**, 598–602.
3. Bock M., Müller S., Zuehlsdorff S., Speier P., Fink C., Hallscheidt P., Umathum R., and Semmler W. (2006) Active catheter tracking using parallel MRI and real-time image reconstruction. *Magn. Reson. Med.* **55**, 1454–1459.
4. Guttman M. A., Ozturk C., Raval A. N., Raman V. K., Dick A. J., DeSilva R., Karmarkar P., Lederman R. J., and McVeigh E. R. (2007) Interventional cardiovascular procedures guided by real-time MR imaging: an interactive interface using multiple slices, adaptive projection modes and live 3D renderings. *J. Magn. Reson. Imaging* **26**, 1429–1435.
5. Quick H. H., Ladd M. E., Zimmermann-Paul G. G., Erhart P., Hofmann E., von Schulthess G. K., and Debatin J. F. (1999) Single-loop coil concepts for intravascular magnetic resonance imaging. *Magn. Reson. Med.* **41**, 751–758.
6. Spuentrup E., Ruebben A., Schaeffter T., Manning W. J., Günther R. W., and Buecker A. (2002) Magnetic resonance-guided coronary artery stent placement in a swine model. *Circulation* **105**, 874–879.
7. Quick H. H., Kuehl H., Kaiser G., Bosk S., Debatin J. F., and Ladd M. E. (2002) Inductively coupled stent antennas in MRI. *Magn. Reson. Med.* **48**, 781–790.
8. Buecker A., Adam G. B., Neuerburg J. M., Kinzel S., Glowinski A., Schaeffter T., Rasche V., van Vaals J. J., and Guenther R. W. (2002) Simultaneous real-time visualization of the catheter tip and vascular anatomy for MR-guided PTA of iliac arteries in an animal model. *J. Magn. Reson. Imaging* **16**, 201–208.
9. Quick H. H., Zenge M. O., Kuehl H., Kaiser G., Aker S., Massing S., Bosk S., and Ladd M. E. (2005) Interventional magnetic resonance angiography with no strings attached: wireless active catheter visualization. *Magn. Reson. Med.* **53**, 446–455.
10. Kaiser G. M., Breuckmann F., Aker S., Eggebrecht H., Kuehl H., Erbel R., Fruhauf N. R., Broelsch C. E., and Quick H. H. (2007) Anesthesia for cardiovascular interventions and magnetic resonance imaging in pigs. *J. Am. Assoc. Lab. Anim. Sci.* **46**, 30–33.
11. Mahnken A. H., Chalabi K., Jalali F., Günther R. W., and Buecker A. (2004) Magnetic resonance-guided placement of aortic stents grafts: feasibility with real-time magnetic resonance fluoroscopy. *J. Vasc. Interv. Radiol.* **15**, 189–195.
12. Raman V. K., Karmarkar P.V., Guttman M. A., Dick A. J., Peters D. C., Ozturk C., Pessanha B. S., Thompson R. B., Raval A. N., DeSilva R., Aviles R. J., Atalar E., McVeigh E. R., and Lederman R. J. (2005) Real-time magnetic resonance-guided endovascular repair of experimental abdominal aortic aneurysm in swine. *J. Am. Coll. Cardiol.* **45**, 2069–2077.
13. Eggebrecht H., Zenge M., Ladd M. E., Erbel R., and Quick H. H. (2006) In vitro evaluation of current thoracic aortic stent-grafts for real-time MR-guided placement. *J. Endovasc. Ther.* **13**, 62–71.
14. Eggebrecht H., Kühl H., Kaiser G. M., Aker S., Zenge M. O., Stock F., Breuckmann F., Grabellus F., Ladd M. E., Mehta R. H., Erbel R., and Quick H. H. (2006) Feasibility of real-time magnetic resonance-guided stent-graft placement in a swine model of descending aortic dissection. *Eur. Heart J.* **27**, 613–620.
15. Ladd M. E., and Quick H. H. (2000) Reduction of resonant RF heating in intravascular catheters using coaxial chokes. *Magn. Reson. Med.* **43**, 615–619.
16. Konings M. K., Bartels L. W., Smits H. F., and Bakker C. J. (2000) Heating around intravascular guidewires by resonating RF waves. *J. Magn. Reson. Imaging* **12**, 79–85.
17. Nitz W. R., Oppelt A., Renz W., Manke C., Lenhart M., and Link J. (2001) On the heating of linear conductive structures as guide wires and catheters in interventional MRI. *J. Magn. Reson. Imaging* **13**, 105–114.
18. Buecker A., Spuentrup E., Schmitz-Rode T., Kinzel S., Pfeffer J., Hohl C., van Vaals J. J., and Günther R. W. (2004) Use of a nonmetallic guide wire for magnetic resonance-guided coronary artery catheterization. *Invest. Radiol.* **39**, 656–660.
19. Krueger S., Schmitz S., Weiss S., Wirtz D., Linssen M., Schade H., Kraemer N., Spuentrup E., Krombach G., and Buecker A. (2008) An MR guidewire based on micropultruded fiber-reinforced material. *Magn. Reson. Med.* **60**, 1190–1196.
20. Mekle R., Zenge M. O., Ladd M. E., Quick H. H., Hofmann E., Scheffler K., and Bilecen D. (2009) Initial in vivo studies with a polymer-based MR-compatible guide wire. *J. Vasc. Interv. Radiol.* **20**, 1384–1389.
21. Kos S., Huegli R., Hofmann E., Quick H. H., Kuehl H., Aker S., Kaiser G. M., Borm P. J., Jacob A. L., and Bilecen D. (2009) First magnetic resonance imaging-guided

aortic stenting and cava filter placement using a polyetheretherketone-based magnetic resonance imaging-compatible guidewire in swine: proof of concept. *Cardiovasc. Intervent. Radiol.* **32**, 514–521.

22. Kos S., Huegli R., Hofmann E., Quick H. H., Kuehl H., Aker S., Kaiser G. M., Borm P. J., Jacob A. L., and Bilecen D. (2009) Feasibility of real-time magnetic resonance-guided angioplasty and stenting of renal arteries in vitro and in swine, using a new polyetheretherketone-based magnetic resonance-compatible guidewire. *Invest. Radiol.* **44**, 234–241.

23. Kos S., Huegli R., Hofmann E., Quick H. H., Kuehl H., Aker S., Kaiser G. M., Borm P. J., Jacob A. L., and Bilecen D. (2009) MR-compatible polyetheretherketone-based guide wire assisting MR-guided stenting of iliac and supraaortic arteries in swine: feasibility study. *Minim. Invasive. Ther. Allied. Technol.* **1**, 181–188.

Chapter 23

MR for the Investigation of Murine Vasculature

Christoph Jacoby and Ulrich Flögel

Abstract

The investigation of alterations in vessel morphology of transgenic mouse models generally requires time-consuming and laborious planimetry of histological sections. This postmortem analysis is per se restricted to endpoint studies and, furthermore, may reflect the situation in vivo to a limited degree only. For the repetitive and noninvasive monitoring of dynamic changes in the murine vasculature, several protocols for high-resolution 3D MR angiography (MRA) at a vertical 9.4 T system are described. These protocols are based on flow-compensated 3D gradient echo sequences with application-dependent spatial resolution, resulting in voxel sizes between 1 and 13 nL. To ensure constant physiological conditions, particular attention is paid to minimize the acquisition time. All measurements are carried out without a contrast agent to avoid temporal inconstancy of the contrast-to-noise-ratio (CNR) as well as toxic side effects. Moreover, metabolic alterations as a consequence of disturbed vascularization and blood supply are monitored by ^{31}P MR spectroscopy.

Key words: Arteriogenesis, carotid arteries, circle of Willis, femoral arteries, flow velocity, hindlimb ischemia, kidney, muscle energetics, phosphorus MR spectroscopy, stenosis, time-of-flight angiography, vascular remodeling.

1. Introduction

Genetically modified mice are well-established models to study dynamic processes in vascular biology, such as the development of vessel stenosis or the initiation of neovascularization. In general, such animal models are evaluated by ex vivo histopathology providing information on vessel morphology at the time of harvest only. Consequently, fast, noninvasive 3D monitoring is highly desirable for the continuous vessel analysis in transgenic mouse models. Here, we present high-resolution 3D MR angiography (MRA) protocols for serial measurements of vascular blood flow and vessel lumina in the mouse in vivo. Segmentation for volumetric analysis of the 3D MR images is carried out by dedicated software. Quantification of luminal volumes from 3D data yields highly reproducible results with low intra- and interobserver variabilities (1). The simultaneous acquisition of ^{31}P MR spectra provides additional information about the consequences of vascular alterations at the metabolic level.

Implementation of the described protocols allows the repetitive analysis of individual mice lacking or overexpressing relevant target genes to identify the key factors involved in regulation of vascular remodeling. MR angiograms of the murine renal, thoracic, hindlimb, and cerebral vessel systems are presented. Furthermore, pathophysiological applications for the serial assessment of vessel narrowing or growing in response to different vascular injury models and the consequences for the metabolic state of the tissue are given.

2. Materials

2.1. MR Equipment

1. The vertical MR system has to be fitted with a microimaging unit. For the experiments described in this chapter, a Bruker DRX 9.4 T wide-bore (89 mm) NMR spectrometer equipped with actively shielded 40 or 57 mm gradient sets (1 or 0.2 T/m maximum gradient strength, 110 μs rise time at 100% gradient switching) operated by ParaVision 4.0 was used (all from Bruker, Rheinstetten, Germany).
2. With regard to the size of the object to be investigated, adequately sized coils are used: resonators with an inner diameter of 10, 25, 30, and 38 mm attached to microimaging probe heads (Bruker, Rheinstetten, Germany). In order to perform in parallel, ^{31}P MR spectroscopy double-tuned ^{1}H/^{31}P resonators or additional ^{31}P surface coils are required.

2.2. Data Visualization and Quantification

1. Personal computer with adequate processor (≥ 3 GHz) and memory (≥ 2 GB) running Debian Linux (≥ 3.0) as operating system and the Angiotux module within the Eccet platform (www.eccet.de) (*See* **Note 1**).
2. Flow quantification was performed on a standard Windows (XP, Service Pack 2) PC using dedicated software written in LabVIEW (v6, National Instruments, Austin, TX, USA).

2.3. Animal Anesthesia and Handling

1. Isoflurane vaporizer (Dräger, Lübeck, Germany) filled with adequate amounts of isoflurane (*see* **Note 2**).

2. Gas flow control unit with flow meters (V-100 from Voegtlin, Aesch, Switzerland) for O_2 and N_2 (*see* **Note 3**).

3. Gas-washing bottle for saturating the anesthesia gas mixture with water (*see* **Note 4**).

4. Nose cone for mice (*see* **Note 5**).

5. Small animal monitoring system providing rectal thermoprobe (Pt10Rh-Pt from Greisinger, Regenstauf, Germany), pneumatic pillow (Graseby, Watford, UK), and ECG electrodes (Klear-Trace from CAS Medical Systems, Branford, CT, USA) for monitoring of vital functions of the animal (e.g., M1025 system, SA Instruments, Stony Brook, NY, USA), which can be used to synchronize data acquisition with cardiac and respiratory motion.

6. Custom-made animal handling system to insert anesthetized animals into the magnet.

7. Tempering unit for keeping the animals' body temperature at 37°C (UWK 45, Haake, Karlsruhe, Germany; *see* **Note 6**).

3. Methods

3.1. Animal Preparation

1. Mice are anesthetized with 1.5% isoflurane in a water-saturated gas mixture of 30% oxygen in nitrogen applied at a rate of 75 ml/min by manually restraining the animal and placing its head in an in-house-built nose cone. Under these conditions, the animals respire spontaneously at a rate of approximately 100 min^{-1}, if body temperature is kept at 37°C as described in the previous section.

2. If ECG triggering is required, the signal can be derived by sticking fore- and hindpaws to small pediatric electrodes which should be fixed at the animal handling system to minimize moving artifacts in the ECG. The quality of the ECG can significantly be enhanced by usage of electrode gel. Triggering to the QRS complex ensures that excitation takes places during end-diastole, unless an additional (but constant) delay is introduced by the user.

3. A pressure sensor is placed at a body position where the respiratory motion has its maximum amplitude to monitor the vital functions of the mouse. If respiration triggering should be performed, a window is defined within the expiration phase to control the starting point of the individual phase-encoding steps. For combination with ECG

triggering, this starting point is additionally bound to a QRS signal (*see* **Note 7**).

4. The mouse is gently fixed after final positioning within the animal handling system to avoid a shift when the system is introduced into the vertical magnet (*see* **Note 8**).

3.2. Starting the MR Experiment

1. After insertion of the animal into the magnet, the central frequency is set on-resonance and a pilot scan is performed to verify the correct positioning of the animal within the resonator (*see* **Note 9**).

2. Upon final positioning, the coils for ^{1}H and X nuclei are tuned and matched for optimization of the resonance circuits.

3. Thereafter, scout images are again acquired for definite localization of the area of interest.

4. Shimming is started from a phantom-optimized shim file. Large homogeneous objects like the brain can easily be shimmed by automatic shim procedures like FASTMAP (2); otherwise, manual shimming is required. For FASTMAP, a large voxel (e.g., $7 \times 7 \times 7$ mm^3) is placed within the area of interest. To ensure successful performance, take care not to position this voxel too close to susceptibility jumps. Do not forget to readjust the center frequency after shimming.

3.3. Angiographic Methods

1. All angiographic methods used are based on the time-of-flight (TOF) or inflow effect which provides unsaturated spins from outside the selected slice ready to be excited by the next radio frequency (rf) pulse. On the other hand, stationary spins are consecutively saturated and appear dark in the image. As a consequence, the contrast between blood in the vessels and surrounding tissue is maximized with short repetition times and high flip angles. In principle, either 2D or 3D gradient echo sequences can be employed for MR angiographic imaging.

2. From 2D multislice gradient echo experiments, the spatial structure of the vessel system can easily be obtained by stacking the images in the right order. However, to achieve a satisfactory resolution in the third dimension, i.e., a value comparable to the in-plane resolution, the slice thickness has to be chosen very small, which requires strong gradients. Furthermore, to accomplish a sharp slice-selection profile for this, a narrow rf excitation pulse bandwidth is needed which is associated with long excitation pulses. Hence, echo times are considerably increased leading to a loss in signal-to-noise ratio. Moreover, short repetition times are necessary to achieve a maximum contrast between tissue and flowing

blood, so that the number of slices in the multislice sequence and thereby the coverage in the third dimension is limited. To overcome this problem, 2D angiographic methods are typically used with the phase encoding included in the slice loop, i.e., all slice images are acquired sequentially. This acquisition scheme also reduces the loss in contrast due to inflow of partially saturated spins into another slice. Indeed, the total acquisition time in this mode approximates that of 3D methods but provides lower resolution in z-direction and a worse slice-selection profile.

3. In a 3D gradient echo sequence, the introduction of a second phase-encoding direction, usually termed *partition encoding*, allows to achieve an identical (isotropic) resolution in all three dimensions. Therefore, 3D methods are strongly recommended for volumetric analysis of small vessel systems. However, since each excitation is applied to the whole volume of interest, spins are progressively saturated while flowing through the imaged volume which might result in a substantial signal decrease. This effect particularly becomes relevant for slow flows, so that in this case, a 2D method may be preferable. Examples for application of both 2D and 3D methods are given below.

4. To countervail the signal loss due to progressive saturation in a 3D sequence there are, in principle, two methods: first, the FOV in flow direction can be divided into several smaller sections (*slabs*). In this case, a respective number of 3D datasets must be acquired and reassembled afterwards. Second, an excitation pulse with a spatially dependent flip angle can be used, which is small when the blood enters the volume and becomes larger while flowing through the FOV. This leads to an almost constant saturation, since the spins that exhibit more excitations are only slightly tilted into the transverse plane in the beginning. The design of this so-called TONE-pulse (tilted optimized nonsaturating excitation (3)) strongly depends on flow velocity, so that it is mainly optimized to one specific vessel.

5. In order to avoid signal loss or spatial misregistration due to flow effects, it is strongly recommended to use flow-compensated gradient echo methods, even if this leads to longer echo and acquisition times (*see* **Note 10**).

6. As long as the inflow effect results in sufficient signal-to-noise ratio (*see* **Note 11**), all measurements can be carried out without any contrast agent (*see* **Note 12**). Thus, temporal inconstancy of the contrast-to-noise ratio (CNR) as well as toxic side effects can be avoided.

7. Water signal of the background tissue can be reduced using magnetization transfer contrast (MTC) which uses an rf

pulse off-resonant to free water. Since bound water protons exhibit a very broad absorption band, they are saturated by the pulse. Afterwards, magnetization properties can be transferred to free water protons leading to partial saturation of tissue compared to vessel protons.

3.4. Protocols and Parameters

1. Scout images: Three (or more) orthogonal slices in axial, coronal, and sagittal orientation; FLASH, TE = 2.2 ms, TR = 100 ms, flip angle = 10°, FOV = 3 × 3 cm^2, slice thickness = 1 mm, matrix = 128 × 128, acquisition time = 12 s.

2. Shimming with FASTMAP: TR = 1000 ms, FOV ~7 × 7 × 7 mm^3, acquisition size = 128, evolution time = 40 ms, stick size ~1.5 mm, time requirement ~1–2 min.

3. Kidneys: 2D flow-compensated gradient echo, TR = 12 ms, TE = 3.3 ms, flip angle = 75°, FOV = 2.56 × 2.56 cm^2 in sagittal orientation, slice thickness = 0.4 mm with overlapping slices at a center-to-center distance of 0.25 mm, total length of the multislice stack = 2 cm, matrix = 256 × 256 × 80, accumulations = 8, acquisition time = 24.6 min. For surface rendering of the kidneys: RARE, 16 contiguous slices, RARE factor = 8, TR = 5000 ms, TE = 4.4 ms, FOV = 3 × 3 cm^2, slice thickness = 1 mm, accumulations = 4, acquisition time = 5.3 min.

4. Carotid arteries: 3D flow-compensated gradient echo, TR = 30 ms, TE = 2.9 ms, flip angle = 35°, FOV = 3 × 3 × 1.5 cm^3, matrix = 128 × 128 × 64, acquisition time = 4.1 min.

5. Hindlimb: 3D flow-compensated gradient echo, TR = 23 ms, TE = 3.5 ms, flip angle = 35°, five slabs with FOV = 2.56 × 2.56 × 0.64 cm^3, matrix = 256 × 256 × 64, acquisition time = 6.2 min/slab, total length of the overlapping slabs = 1.92 cm, total scan time = 31 min (*see* **Note 13**). For acquisition of ^{31}P MR spectra from ischemic hindlimbs, the affected hind leg is placed in a 10-mm tilt resonator and non-volume-selective spectra are recorded over the entire hindlimb: spectral width = 6460 Hz, transients = 1024, data size = 1024, TR = 250 ms, flip angle = 30°, exponential weighting resulting in a 20-Hz line broadening, acquisition time = 6 min. Chemical shifts are referenced to the phosphocreatine (PCr) resonance at −2.52 ppm and peak areas are obtained by integration after phase and baseline correction.

6. Brain: 3D flow-compensated gradient echo, TR = 23 ms, TE = 2.9 ms, flip angle = 30°, FOV = 3 × 2.56 × 2.56 cm^3, matrix = 256 × 256 × 128, acquisition time = 12.3 min.

3.5. Visualization of 3D Datasets

After acquisition, 3D datasets are reconstructed and displayed like a 2D multislice dataset, i.e., the image volume is divided into N_z slices where N_z is the number of points in the partition-encoding direction. Using multiplanar reconstruction (MPR), a tomographic image can be calculated for any user-defined angled slice. However, to assess the morphology of a vessel system that is three-dimensional by nature, a method providing more spatial impression is needed. The easiest way to do this is to generate a maximum intensity projection (MIP) as shown in **Fig. 1** for the vessel system around the kidneys. It can be performed without any user-defined parameters by simply displaying the brightest pixels along each line of sight perpendicular to the image plane. This projection, usually performed for different perspectives, is sufficient if only the brightest vessels within the dataset are of interest. The main disadvantage is that depth information can be distorted and smaller vessels are covered by more intense arteries, regardless of the true morphology.

A more adequate 3D view is obtained from surface or volume-rendered datasets as carried out for visualization of the kidneys within the angiogram in **Fig. 1**. For both methods, in a first step the user has to define intensity intervals in order to discriminate the background. The spatial impression for a calculated object surface is improved by light effects in the so-called surface-shaded display (SSD) representation. However, surface

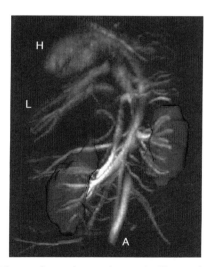

Fig. 1. MIP view of the renal vessels superimposed with a surface-rendered view of the kidneys for anatomical orientation. The angiogram was reconstructed from a 2D multislice dataset which was acquired from an FOV in sagittal orientation to maximize the inflow effect for the vessels running from the aorta (A) to the kidney. Signals from the portal system of the liver (L) and the chambers of the heart (H) are less marked due to cardiac and respiratory motion.

rendering of an object requires rather extensive work from the user and is therefore superseded by volume-rendering reconstruction. For this technique, the whole image volume is kept in the working memory with intensity or color (RGB) values stored together with opacity information. The main advantage of volume compared to surface rendering is that no interior information is discarded, allowing for viewing and volumetrically quantifying the 3D dataset as a whole. Segmentation of adjacent voxels within the pre-defined intensity interval can easily be performed by defining a seed point via mouse click. However, storing and recalculating of this multidimensional vector is very time consuming and necessitates powerful computers. In Angiotux (*see* **Section 3.6**), handling of 3D objects in terms of translation, rotation, and zoom can be accelerated by omitting all vector elements lying outside the defined intensity interval.

3.6. Quantification Methods

1. Tomographic slices can be analyzed by planimetrical determination of luminal areas using, e.g., the ParaVision (Bruker, Ettlingen, Germany) Region of Interest (ROI) Tool or can be automatically derived by an edge-detection algorithm.

2. For analysis of the Fourier-transformed 3D MRI datasets, the Angiotux program based on the freely available package Eccet for processing and visualizing voxel data in 3D and 4D can be used (1). With this software vessel segments can be sectioned with a plane freely adjustable in any orientation. An algorithm for the measurement of vessel lengths is implemented in order to account for curvature and varying diameter. Segmented regions are assigned to defined colors and absolute volumes are quantified by counting the voxels of these colors.

3. Since the intensity in angiographic images is mainly determined by flow velocity which rapidly decreases near the vessel walls due to a laminar flow profile, vessel sizes may be underestimated by 2D or 3D image analysis. Calculation of absolute vessel sizes may further be complicated within longitudinal experimental series of pathological models by alterations in body weight (animal size) or also anesthesia. To account for the latter effects, it is recommended to relate the areas/volumes to a vessel area or segment at the contralateral side that is defined by anatomical landmarks. Using such ratios will cancel out a variety of error sources and greatly reduces the standard deviations of the individual parameters, so that differences between experimental groups can be more sensitively detected.

4. There are different approaches to determine flow velocities and/or volume flows from gradient echo images (4–6).

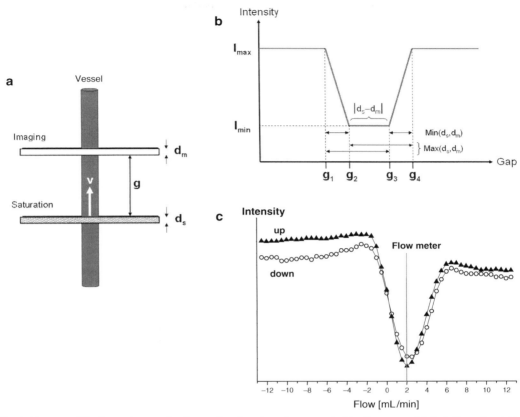

Fig. 2. Flow quantification using a saturation and an imaging slice separated by the gap g. (a) Definition of variables with thicknesses d_m and d_s and velocity v. (b) Theoretical curve for the general case $d_m \neq d_s$. If the interslice gap is continuously varied at fixed Δt between saturation and measurement, significant changes in intensity are expected to occur at the gaps labeled as g_1 to g_4 ($g_1 = v\Delta t - d_s - d_m$; $g_2 = v\Delta t - \text{Max}(d_s,d_m)$; $g_3 = v\Delta t - \text{Min}(d_s,d_m)$; $g_4 = v\Delta t$). (c) A phantom with two tubes (ø 500 μm) of identical velocity but different flow directions was used for flow quantification. In both tubes, the intensity minimum was located at a gap attributed to a flow of ~2 mL/min, identical to the value determined concomitantly by an ultrasonic flow meter.

An alternative and straightforward 2D method is illustrated in **Fig. 2a** showing a vessel and two slices for saturation and imaging, respectively, positioned perpendicular to flow direction. Excitation of both slices is separated by a fixed time delay (Δt) and a variable gap (g) which can be continuously increased, thereby altering the image intensity as shown in **Fig. 2b**. If an identical thickness $d = d_s = d_m$ is used, the general intensity progression simplifies to a curve with only one distinct minimum as illustrated in **Fig. 2c**. This graph shows the signal intensity against the interslice gap for two tubes of diameter D, in which water is flowing with identical velocity but in opposite directions. Both minima nearly coincide at a flow of 2 mL/min as calculated by

$$\dot{V} = \frac{\pi D^2 (g_{\min} + d)}{4\Delta t}$$

and which was confirmed by direct measurement with an ultrasonic flow meter (Transonic Systems Inc., Ithaca, NY, USA).

To assess alteration in vascular architecture from in vivo measurements, the cross section of the vessel has to be determined, which can automatically be performed by edge-detecting the circumference of the vessel lumen. If the tomographic slice is not exactly perpendicular to the flow direction, the calculated flow velocity has to be corrected (*see* below). In contrast, volume flow is not significantly affected, since the decrease in v is nearly compensated by an increasing cross section. If $d_s \neq d_m$ is chosen, the equations above can also be used with d replaced by the maximum (for gap g_2 in **Fig. 2b**) or the minimum (for g_3) of both values d_m and d_s.

5. Generally, positioning of a slice perpendicular to the vessel direction as estimated from the scout images is straightforward. However, in most cases, it is very difficult to realize a true right angle with respect to the third dimension. Assuming a circular vessel cross section, the extent of deviation from the perpendicular position is reflected by the quotient of both radii of the resulting ellipse (**Fig. 3a**). If the vessel is angled by θ, the length of the semimajor axis is given by $r/\cos(\theta)$ with the vessel radius r corresponding to the

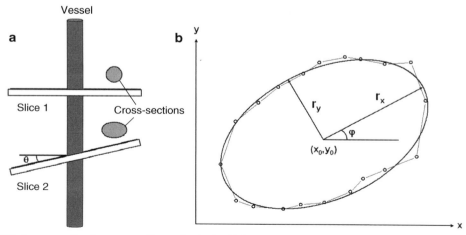

Fig. 3. Elliptical correction for cross section and velocity. (**a**) Slicing the vessel at an angle $\theta \neq 0$ leads to an elliptical cross section in the image. (**b**) ROI points defining the luminal border fitted to a tilted ellipse. The parameters of the fit are the coordinates of the center of the ellipse (x_0, y_0), its semi-axes (r_x, r_y), and the tilt angle φ (for uniqueness: $0 \leq \varphi < 90°$).

semiminor axis length. To correct cross section, vessel diameter, and flow velocity for θ, an analysis of the elliptical shape should be performed. **Figure 3b** shows a fit of a polygon describing the circumference of a vessel lumen to a tilted ellipse (*see* **Note 14**). The points of the polygon were automatically obtained by edge detection. In this example, the angle θ was calculated from the radii to be 54°.

3.7. MRA Applications to Pathophysiological Models

In the following subsections, selected MRA examples are provided representative of the large number of applications in studying the vasculature of mice or small animals in general.

3.7.1. Detection of Vessel Stenosis

A widely used injury model to study the temporal development of stenosis is the denudation of the carotid artery by a flexible guidewire. This procedure may lead to an occlusion of the artery, thereby completely disrupting blood flow. In moderate cases, signal loss is restricted to the location of neointima formation triggered by the intervention as shown in **Fig. 4**. The volumetric quantification of the 3D MRA datasets as described above show a good correlation to planimetric data obtained postmortem from immunohistochemical sections (1). Interestingly, similar images are also obtained after stepwise recanalization of an artery, which was initially occluded by thrombus formation. Here, the advantage of noninvasive MRA clearly becomes evident: only the continuous observation of the same animal can help to clarify the underlying mechanisms resulting in the observed endpoint.

3.7.2. Collateral Vessel Formation

Besides vessel narrowing, vessel formation due to arteriogenesis can also be repetitively monitored. This is demonstrated in the left column of **Fig. 5** which illustrates the continuous restoration

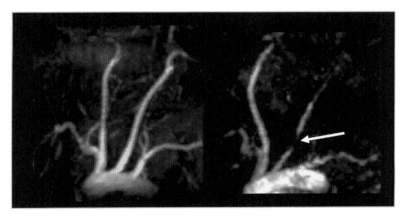

Fig. 4. MIP view of the aortic arch with carotid arteries from the same mouse before (*left*) and after wire injury (*right*). The *arrow* indicates the vessel segment where reduced flow leads to hypointense signal in the angiographic image.

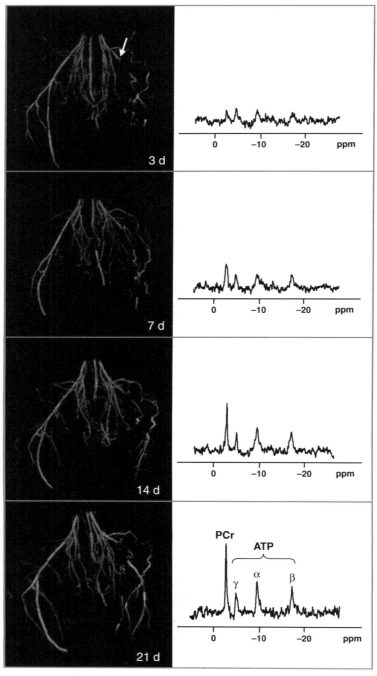

Fig. 5. Time course for collateral vessel formation after hindlimb ischemia induced by occlusion of the right femoral artery (*arrow in the upper left image*). Vessels were segmented with Angiotux from reconstructed MRI data. The increase in collateral vessel density shown for days 3, 7, 14, and 21 after ligation correlates with rising ATP and PCr levels in the ^{31}P MR spectra reflecting the enhanced metabolic recovery due to the restored blood supply into the distal areas of the hindlimb.

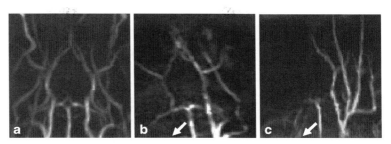

Fig. 6. MIP images (sections) of the circle of Willis in the murine brain. (**a**) Untreated animal. (**b**) Measurement immediately after ligation of the left carotid artery (*arrow*) without any further interventions. (**c**) With pertussis toxin administration preceding ligation.

of blood flow after induction of hindlimb ischemia by ligation of the femoral artery. The formation of arteries from smaller vessels (arterioles) leads to a gradual restoration of blood flow to the ischemic regions. This is also reflected by the ^{31}P MR spectra which were immediately acquired after the angiographic measurements (**Fig. 5**, right). Collateral vessel density and adenosine triphosphate (ATP) and phosphocreatine (PCr) levels are continuously increasing with time demonstrating not only the enhanced blood flow into the distal regions of the hindlimb but also, as a consequence, the metabolic recovery of the tissue.

3.7.3. Cerebral Blood Supply

The circle of Willis is a circular vessel structure within the brain that has the function to ensure full cerebral blood supply even in the case when flow from one of the afferent carotid arteries fails. An example is given in **Fig. 6b** where blood supply is completely maintained within the circle after ligation of the left carotid artery (arrow). **Figure 6a** shows an untreated mouse for comparison. After functional inactivation of G_i proteins using pertussis toxin, this function of the circle is strongly impaired. As illustrated by the angiogram in **Fig. 6c**, no blood flow within the circle of Willis is observable on the side of ligation due to reduction of either flow velocity or vessel diameter. Hence, the ipsilateral side of the brain is undersupplied leading to severe cerebral ischemia within this area.

4. Notes

1. Debian Linux as operating system and the Eccet platform are free software. Alternative solutions capable of handling 3D data stacks may be used. As a commercial product Amira (Mercury Computer Systems, Berlin, Germany) running on Windows, Mac, and Linux is recommended.

2. Inhalation anesthesia is preferred to injection anesthesia, since it can be conveniently controlled throughout the experiment. Furthermore, induction of and recovery from isoflurane anesthesia are rapid, and cardiorespiratory depression is minimal, resulting in physiological heart (~600 bpm) and respiration rates (100/min), respectively. A clinical vaporizer can be used but should be equipped with a gas flow control unit suitable for mice (*see* **Note 3**).

3. Mixing the gas for inhalation anesthesia with N_2 and O_2 has the advantage over the use of air that also experiments with modified O_2 levels can easily be carried out. No standard flow meters attached to clinical inhalation anesthesia devices should be used, since they are generally designed to regulate gas flows in the range of 1–2 L/min and cannot be adjusted to lower flows. However, the tidal volume (i.e., the volume of air that is breathed in) of a mouse is generally assumed to be between 0.1 and 0.2 mL. At a physiological respiration rate of 100/min a maximum required flow of 20 mL/min is calculated. Therefore, a flow rate of 50–75 mL/min is more than adequate to support normal ventilation of the mouse. The application of a flow of 1–2 L/min may have serious consequences for the animals, since such high flow rates will abet dehydration and definitively result in unphysiological conditions. The V-100 units used in our setup are calibrated for gas flows of 10–100 mL/min (N_2) and 5–50 mL/min (O_2).

4. To avoid dehydration of mucous membranes due to the continuous gas flow applied, the anesthesia gas mixture is moisturized by passing the gas flow through a gas-washing bottle with water.

5. Connect not only a gas inlet but also a gas offtake to avoid accumulation of isoflurane within the probe head and also the laboratory.

6. Keeping the animal at normal body temperature is essential for meaningful MRA, since vasoregulation is heavily temperature dependent. Loss of body heat results in significant vasoconstriction, reduced blood flow in small vessels, and underestimation of the vascular architecture when using time-of-flight techniques. In our setup, the temperature is kept at 37°C by regulating the temperature of the gradient cooling device (Haake UWK 45). However, several other possibilities of temperature regulation are conceivable.

7. Respiration and/or ECG gating will not necessarily result in improved image quality. Due to elongated effective repetition times, the contrast-to-noise ratio is considerably decreased, while the total acquisition time will substantially

increase (1) which in turn makes it more difficult to ensure constant physiological conditions.

8. The longitudinal position of the mice in the vertical magnet requires some additional precautions. Of note, we and several others (7–9) have independently shown that vertical-bore MR scanners allow measurements under physiologic conditions, and that in the steady state there are no significant alterations in hemodynamic and functional parameters due to the animal's position. However, it cannot automatically be assumed that all transgenic mouse models show a similar long-term stability in the vertical position, which may become especially relevant for studies of animals in labile hemodynamic conditions, such as infarction or stroke. Therefore, it is of special importance to continuously monitor the vital functions of the animals (ECG, respiration, body temperature) during the experiment. To exclude that vertical positioning of the mice is physiologically problematic, functional parameters can be evaluated in independent experiments by continuous registration of left ventricular hemodynamics with microtip catheters in supine position and after tilting the mouse to a 90° vertical position.

9. The resonators used in our setup exhibit a limited FOV in the xy-plane due to physical restrictions, but also in the z-direction depending on the coil design (usually 25–30 mm). Thus, it is important to position the animal within the resonator in such a way that the area of interest is as close as possible to the center of the system in order to ensure the greatest possible magnetic field homogeneity for the MRA measurements.

10. Vessel flow can lead to artifacts in slice selection, read, and phase-encoding directions. In contrast to effects caused by diffusion or periodic motions (also including pulsatile flow), constant velocity flow artifacts can easily be overcome by use of (first-order) flow-compensated methods. In **Fig. 7a** a phase-evolution diagram is shown for a common readout gradient switching which only in the stationary case (dashed line) leads to a spin rephasing at the time of echo (TE). A velocity component in the read direction introduces a quadratic dependence producing a constant phase shift for all spins flowing with identical velocity (solid line). Non-refocused spins do not contribute to the signal so that there is less contrast in the resulting image. By using a symmetric gradient between 0 and TE, rephasing of all spins can be accomplished if positive and negative areas within this interval are equal (**Fig. 7b**). While this holds also for slice selection, flow in the phase-encoding direction

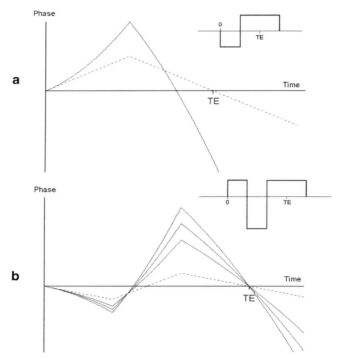

Fig. 7. Phase-evolution diagrams of stationary (*dashed*) and flowing (*solid*) spins for the gradient switchings shown *top right*. (**a**) Common readout gradient leading to velocity-dependent phase shifts at time TE. (**b**) A symmetric gradient between 0 and TE leads to a rephasing of all spins moving with arbitrary but constant velocity. This is shown for three different velocities > 0.

leads to a misregistration. Here, a rephasing gradient lobe of variable amplitude (depending on the phase-encoding gradient) is used to compensate for this effect.

11. Since particular attention is paid to minimize the acquisition time, we used voxel sizes of 1–13 nL in standard applications, depending on the vessel size of interest. However, in further experiments the voxel size could be reduced to 0.125 nL according to an isotropic length of 50 μm with satisfying signal-to-noise. Hence, vessels up to about 80 μm in diameter can readily be imaged using a flow-compensated 3D gradient echo method without any contrast agent (at least at 9.4 T). Smaller vessels only partially contribute to the voxel intensity. Hence, the signal-to-noise ratio strongly depends on the fraction of unsaturated spins and is further diminished by the low flow velocities typically present in small vessels. However, even with contrast agents, imaging of small vascular structures is exceedingly difficult due to the limited spatial resolution in MRI (*see also* **Note 12**).

12. The use of contrast agents for MRA in the mouse is not as straightforward as in larger animals and hampered by several problems: (1) A well-defined intravenous application of the contrast agent within the magnet is complicated by large dead space within the delivering tubes versus small injection volumes. (2) The very fast heart rate on the one hand and the small dimensions on the other hand are an obstacle to first-pass measurements with the required high resolution. (3) The high cardiac output (~20 mL/min) in relation to the small entire blood pool (~2 mL) results in a much faster distribution of the contrast agent from the vessels into the interstitium and the surrounding tissue and thereby in a quick loss of the desired contrast increase from vessel to tissue.

13. Monitor carefully the body temperature of the animal: Heavy-duty cycles over a longer period at high field will result in a significant deposition of energy within the animal. Under certain conditions, we could observe an increase of 1°C in body temperature within 10 min.

14. The ROI was fitted to the following function describing a tilted ellipse with its center in (x_0, y_0) and the semi-axes r_x and r_y:

$$y = y_0 + p \pm \sqrt{p^2 + q}$$

where

$$p \equiv p(x) = cs(x - x_0) \frac{r_x^2 - r_y^2}{c^2 r_x^2 + s^2 r_y^2}$$

$$q \equiv q(x) = \frac{r_x^2 r_y^2 - \left(s^2 r_x^2 + c^2 r_y^2\right)(x - x_0)^2}{c^2 r_x^2 + s^2 r_y^2}$$

and c and s are cosine and sine of the rotation angle φ (cf. **Fig. 3b**), respectively. In the fit algorithm, the y (of the two possible values) nearer to the experimental coordinate was used. Additionally, φ was forced to be between 0 and 90° to ensure a unique definition of r_x and r_y.

Acknowledgments

The authors would like to thank Professor Jürgen Schrader for his continuous support and encouragement as well as Yang Chul Böring and Katja Pexa for their contributions to the studies of hindlimb and cerebral ischemia, respectively. This work was supported by a grant from the Deutsche Forschungsgemeinschaft (DFG) through SFB612, TP Z2.

References

1. Jacoby, C., Böring, Y. C., Beck, A., Zernecke, A., Aurich, V., Weber, C., Schrader, J., and Flögel, U. (2008) Dynamic changes in murine vessel geometry assessed by high-resolution magnetic resonance angiography: a 9.4T study. *J. Magn. Reson. Imaging* **28**, 637–645.
2. Gruetter, R. (1993) Automatic, localized in vivo adjustment of all first- and second-order shim coils. *Magn. Reson. Med.* **29**, 804–811.
3. Roditi, G. H., Smith, F. W., and Redpath, T. W. (1994) Evaluation of tilted, optimized, non-saturating excitation pulses in 3D magnetic resonance angiography of the abdominal aorta and major branches in volunteers. *Br. J. Radiol.* **67**, 11–13.
4. Wagner, S., Helisch, A., Bachmann, G., and Schaper, W. (2004) Time-of-flight quantitative measurements of blood flow in mouse hindlimbs. *J. Magn. Reson. Imaging* **19**, 468–474.
5. Greve, J. M., Les, A. S., Tang, B. T., Draney Blomme, M. T., Wilson, N. M., Dalman, R. L., Pelc, N. J., and Taylor, C. A. (2006) Allometric scaling of wall shear stress from mice to humans: quantification using cine phase-contrast MRI and computational fluid dynamics. *Am. J. Physiol.* **291**, H1700–H1708.
6. Miraux, S., Franconi, J. M., and Thiaudiere, E. (2006) Blood velocity assessment using 3D bright-blood time-resolved magnetic resonance angiography. *Magn. Reson. Med.* **56**, 469–473.
7. Jacoby, C., Molojavyi, A., Flögel, U., Merx, M. W., Ding, Z., and Schrader, J. (2006) Direct comparison of magnetic resonance imaging and conductance microcatheter in the evaluation of left ventricular function in mice. *Basic Res. Cardiol.* **101**, 87–95.
8. Wiesmann, F., Neubauer, S., Haase, A., and Hein, L. (2001) Can we use vertical bore magnetic resonance scanners for murine cardiovascular phenotype characterization? Influence of upright body position on left ventricular hemodynamics in mice. *J. Cardiovasc. Magn. Reson.* **3**, 311–315.
9. Schneider, J. E., Cassidy, P. J., Lygate, C., Tyler, D. J., Wiesmann, F., Grieve, S. M., Hulbert, K., Clarke, K., and Neubauer, S. (2003) Fast, high-resolution in vivo cine magnetic resonance imaging in normal and failing mouse hearts on a vertical 11.7 T system. *J. Magn. Reson. Imaging* **18**, 691–701.

Section VI

Lung Applications

# Chapter 24

MRI of the Lung: Non-invasive Protocols and Applications to Small Animal Models of Lung Disease

Magdalena Zurek and Yannick Crémillieux

Abstract

Magnetic resonance imaging (MRI) can be used in pre-clinical studies as a non-invasive imaging tool for assessing the morphological and functional impact of lung diseases and for evaluating the efficacy of potential treatments for airways diseases. Hyperpolarized gases (^{3}He or ^{129}Xe) MRI provides insight into the lung ventilation function. Lung proton MRI provides information on lung diseases associated with inflammatory activity or with changes in lung tissue density. These imaging techniques can be implemented with non-invasive protocols appropriate for longitudinal investigations in small animal models of lung diseases. This chapter will detail two ^{3}He and proton lung MR imaging protocols applied on two models of lung pathology in rodents.

Key words: Magnetic resonance imaging, MRI, lung, mouse, rat, rodent, hyperpolarized gases, Helium-3, Xenon-129, lung ventilation imaging, bronchoconstriction, methacholine, asthma model, LPS, lung inflammation, spontaneous breathing lung imaging.

1. Introduction

The low proton density of the lungs (approximately 20–30%) combined with susceptibility gradients induced by the air–tissue interfaces and motions within thoracic cavity produces an inherently weak MR signal from the lung parenchyma and makes the lungs the most challenging organ to be imaged by means of MRI.

Despite difficulties associated with lung MRI, there is growing interest in the potential of MR techniques applied to lung diagnostics in patients or to the investigation of lung diseases in animal models. This increase in MR applications for lungs is related to technological breakthrough (use of hyperpolarized gases such as ^{3}He or ^{129}Xe), hardware improvement (MR gradient performance), and methodological progress (for instance short echo time imaging or oxygen-enhanced lung MRI). The application of MRI to lungs is further motivated by its inherent non-invasiveness (of high interest in the case of chronic diseases), its potential for translational research and its complementary readouts as compared to those of CT imaging.

The large NMR signal offered by hyperpolarized (HP) gases allows imaging their distribution in the pulmonary tree and the alveolar spaces. The first NMR biomedical application of HP ^{129}Xe was reported in 1994 with intrapulmonary space imaging of excised mouse lungs (1) (*see* also **Chapter 10**). The first in vivo lung ventilation images obtained in rodents using HP ^{3}He were reported 1 year later (2) followed by the first human lung images (3, 4). The spatial resolution of lung ventilation images obtained by means of ^{3}He exceeds by an order of magnitude the spatial resolution that is routinely obtained with scintigraphy techniques using radioactive gases. In human studies, typical spatial resolutions in the millimeter range are reported while in rodent studies sub-millimetric resolutions are usually reached (5, 6).

Apparent diffusion coefficient (ADC) values of HP gases in airspaces depend on the restriction of gas atoms diffusion by the broncho-alveolar walls. The diffusion length of helium atoms during typical diffusion-sensitizing times (a few milliseconds) exceeds the diameter of alveolar sacks (a few hundreds of micrometers). Hence, in the timescale of MR diffusion acquisition, ^{3}He diffusion in alveolar space takes place in a restricted regime. The dependence of HP gases ADC values upon the dimensions of the alveolar space has been proposed as a non-invasive approach for probing the lung architecture at a sub-pixel level. Indeed, ^{3}He ADC values have been shown to significantly increase in patients with emphysema compared to healthy volunteers (7). These ^{3}He ADC changes in emphysematous lungs are attributed to the morphological changes in alveolar structure and more specifically to the airspace enlargements that characterize emphysema (8–10). Emphysema disease in animal models has been extensively studied using ^{3}He and ^{129}Xe MRI. Elastase-induced emphysema has been investigated in rat (8, 10), mouse (11), and rabbit (12) using ^{3}He or ^{129}Xe diffusion MRI. When measurements were carried out at total lung capacity, ^{3}He ADC values increased from 0.15 cm^2/s in normal rats to 0.18 cm^2/s in elastase-challenged animals; moreover, a significant correlation was found between the ^{3}He ADC values and the alveolar internal area assessed by histology in lungs fixed with formalin at an airway pressure corresponding to the total lung capacity (10). Similarly, ^{3}He ADC values averaged over the entire lungs were found to be approximately 25% higher in emphysema mice than in healthy animals (11).

The relaxation rate R_1 of ^{3}He varies linearly with the partial pressure of oxygen through dipolar interactions of the ^{3}He nucleus with paramagnetic molecular oxygen (13). By measuring the time variation of the relaxation time of HP ^{3}He in the lungs, it is then possible to compute locally the intrapulmonary oxygen concentration and the oxygen consumption rate of oxygen in vivo (14–18). As the intrapulmonary pO_2 distribution is governed by local ventilation, perfusion, and O_2 uptake, pO_2 assessment can be used to evaluate lung function. The rate at which oxygen reaches the alveolus is determined by its ventilation and the inspired pO_2. The rate at which oxygen leaves the alveolus is determined by its perfusion. Consequently, the determination of alveolar pO_2 is an indirect measure of the ventilation/perfusion ratio. The potential of ^{3}He imaging for detecting perfusion abnormalities due to their effect on alveolar pO_2 was demonstrated in an experimental pig model (19). After isolated pulmonary arterial occlusion using a balloon catheter, a focal T_1 reduction corresponding to an abnormally high pO_2 (because of the absence of perfusion) was observed, which normalized upon deflation of the balloon. More recently, pO_2 imaging and oxygen depletion rate imaging were extended to small animal investigation in rat and mouse studies (20, 21).

Chemical shift imaging has been demonstrated using HP ^{129}Xe. The high solubility in blood and tissues and the large chemical shift (several hundreds of ppm) of xenon allow one to differentiate between xenon in alveoli and xenon dissolved in tissue. The so-called xenon polarization transfer contrast (XTC) technique aims to probe the xenon exchange between alveolar space and blood/tissue compartments (22). The method is based on the selective destruction of the xenon polarization in the lung parenchyma. Due to the rapid exchange of xenon between the gas and tissue-dissolved phases, the depolarization of xenon dissolved in tissue affects the xenon signal from the gaseous phase. Using an appropriate pixel-based signal analysis of this effect, the XTC lung images with a contrast related to the tissue and alveolar xenon exchange can be obtained.

The potential of ^{3}He MRI for assessing airways constriction has been investigated in the methacholine-induced bronchoconstriction rat model. Using a Cine MRI approach in which image acquisition was synchronized with the inhalation of the gas mixture (^{3}He with oxygen and nitrogen), a heterogeneously distributed airways constriction resulting in a partition of the lung between ventilated and non-ventilated regions was observed (23). The diameter of the main airways decreased by approximately 11% following methacholine challenge (30 μg). In a methacholine-induced bronchoconstriction rat model (24), dynamic ventilation image series obtained from a single breath were used to generate parametric pixel-by-pixel maps of gas arrival time, filling time constant, inflation rate, and gas volume. Quantitative and regional analysis of gas flow, volume, and arrival times demonstrated statistically significant differences between the baseline and methacholine-constricted states.

In animals, hyperpolarized ^{3}He can be inhaled using either tracheal intubation or invasive tracheotomy. The gas delivery to the animal lungs can be performed using a variety of protocols and apparatus. Small animal respirators compatible with polarized ^{3}He have been developed by several groups (25–27). These respirators allow a fine control of the delivered gas volume and of the lung ventilation timing. Triggering and synchronization of the imaging sequence with the gas delivery can be used for performing lung ventilation. However, acquisition of ^{3}He lung images in spontaneously breathing mice and rats has also been reported recently (28, 29). In this case, gas was administered through a mask.

Apart from the HP gases imaging techniques, many efforts have been made to employ proton lung MRI. The insignificant cost, easy protocol implementation, and versatility of proton lung MR imaging make it an adequate tool in research of lung diseases in animal model studies. Consequently, the development of methodologies and experimental protocols for proton MR pharmacological studies became of interest (30–32). For instance, it was recently demonstrated that conventional gradient echo MR imaging can be efficiently applied to show inflammation hallmark in various models of lung diseases. Considering that a significant contrast is obtained between a dark-appearing lung parenchyma and hyperintense fluids being associated with diverse lung diseases, the gradient echo imaging has been successfully applied to numerous animal models for the detection and quantification of fluid secretion in asthma, chronic obstructive pulmonary diseases (COPD), emphysema, or lung fibrosis (33–36). Furthermore, potential motion artifacts from breathing and cardiac cycles can be reduced by image averaging. For all the above-mentioned applications, the averaging approach based on Cartesian encoding has been recognized as robust, easy to implement, and well adapted to scanning of large cohorts of animals. In order to detect signal from the lung parenchyma, characterized by a very short transverse relaxation time, non-standard imaging techniques are required. For instance a radial ultra-short echo time (UTE) sequence which reduces considerably the echo time (TE) was proposed for this application (37, 38). With the use of a UTE sequence, a TE of the order of 450 μs (with a standard excitation pulse) can be achieved allowing for visualization of lung tissue with a high signal-to-noise ratio (SNR) (39). The improved parenchyma visualization may enable the detection of emphysema and/or fibrosis-like microstructural changes of the lung (40–42) or alternations in ventilation/perfusion ratio (43–45) reflecting different pathophysiological effects in lung injury models. UTE radial imaging has recently also been applied to detect edema and mucus plugging in an acute lipopolysaccharide (LPS)-induced inflammation rat model (46). The radial encoding is less sensitive to motion and blurring artifacts occurring in the thoracic cavity during respiration (47, 48), and thus can provide more accurate results as compared to Cartesian imaging techniques.

There is a great interest in investigating the relationship between anatomical and functional changes in lung disorders. Therefore, proton lung MRI combined with ventilation imaging using HP gases was explored in order to examine the correlation between lesion area detected by proton MRI and the observed ventilation defects (49). To assess the regional pulmonary ventilation, oxygen-enhanced MRI is a technique of choice (50). Molecular oxygen is weakly paramagnetic and shortens the spin–lattice relaxation time (T_1) of the lung parenchyma. The effect of ventilation is thus visualized by signal intensity increase in the T_1-weighted images of the lung acquired with subjects breathing 100% oxygen as compared to room air. Oxygen-enhanced ventilation imaging was applied in human studies to show the ventilation defects in fibrotic, emphysematous, and pneumonic lungs (51, 52). Recently, the investigation of lung ventilation in mouse using oxygen as a contrast agent was reported (43).

Another challenge of proton lung MRI arises from the cardiac and respiratory movements, which can degrade the image quality and make the correct image readout impossible. To address this issue, several acquisition methods have been proposed, including scan-synchronous ventilation combined with electrocardiographic (ECG) synchronization (53, 54), the averaging method (55), and self-gating approaches (56–58). In contrast to conventional synchronization, the averaging and self-gating methods permit to avoid delays associated with the setup of ECG equipment, thus reducing the overall imaging time and the cost of the studies involving a large number of animals. The self-gating methods combined with a radial UTE acquisition allow obtaining highly resolved images of lung parenchyma, which can be synchronized to a specific phase of the cardio-respiratory cycle (39). Alternatively, the combination of UTE acquisition with cardio-respiratory gating can be used to assess and quantify the regional signal changes in the lung parenchyma due to blood perfusion in the lungs (44).

This chapter focuses on two selected lung imaging techniques applied to small animal, namely hyperpolarized (HP) ^{3}He MRI and ultra-short echo time (UTE) proton MRI. The potential of these techniques is illustrated and detailed on two models of lung pathologies in rodents. Both presented approaches are suitable for longitudinal investigations in animal models of lung diseases.

2. Materials

2.1. Breathing of Hyperpolarized ^{3}He

1. 40 mL hyperpolarized ^{3}He.
2. 60-mL syringe.
3. Two- and three-way Luer-lock valves.
4. Rat or mouse.
5. Animal mask (*see* **Note 1**).
6. Latex balloon.
7. Warming pad.
8. Anesthesia (e.g., ketamine/xylazine) (*see* **Note 2**).

2.2. Bronchoconstriction Induction by Injection of Methacholine

1. Methacholine (85 μg in 1 mL of saline).
2. Saline.
3. Heparin.
4. Infusion syringe pump.
5. 2-mL syringe.
6. 22–25 gauge needle (rat) or 27–30 gauge needle (mouse) attached to a 1-mL syringe.
7. Rat or mouse.
8. Anesthesia (e.g., ketamine/xylazine) (*see* **Note 2**).

2.3. Inflammation Induction by the Instillation of Endotoxin Lipopolysaccharide (LPS)

1. LPS (1 mg/kg in 0.2 mL of saline).
2. Balance.
3. 1-mL tube.
4. Vortexer.
5. Rat (e.g., Brown Norway, Wistar).
6. Blunt forceps, light source, 1-mL syringe, 26-G catheter, ventilator.
7. Anesthesia (e.g., ketamine/xylazine) (*see* **Note 2**).
8. Warming pad.
9. A mask, gloves, lab coat, and eye protection.

3. Methods

3.1. Ventilation Imaging of Bronchoconstriction Model Under Spontaneous Breathing Conditions

3.1.1. Animal Preparation

1. Anesthetize the animal (e.g., with intraperitoneal injection of ketamine/xylazine).
2. Insert a 22–25 gauge needle (rat) or 27–30 gauge needle (mouse) into a vein of the tail. Flush the line with saline solution and connect a 2-mL syringe filled with methacholine solution to the catheter.
3. Place the head of the animal within the mask (**Fig. 1**). The homemade mask can be screwed, at one end, to the balloon containing the polarized ^{3}He. On the other end of the mask, a latex sleeve surrounds the neck of the animal. Tighten the latex sleeve around the neck of the animal in order to limit air leakage.
4. Center the chest of the animal in the NMR rf coil. Position the NMR coil within the magnet isocenter for MRI acquisitions.

3.1.2. Free Breathing and Acquisition of Ventilation Images

1. Extract HP ^{3}He from the storage cell to a 60-mL syringe. Transfer the ^{3}He gas from the syringe to the latex balloon. Attach the balloon to the mask positioned on the head of the animal. Allow the animal to breathe spontaneously the gas from the balloon (**Fig. 1**).
2. Launch the MRI sequence. Following MR acquisition (typically 20 s time), disconnect the balloon from the head mask and allow the animal to recover.
3. This protocol can be repeated to investigate bronchoconstrictive effects of methacholine. Trigger intravenous injection of methacholine solution (10 µg/mL) using infusion syringe pump (6 mL/h for rat).
4. Repeat Steps 1 and 2 as many times as required by the follow-up imaging protocol.

3.1.3. Imaging Parameters

MRI field strength is equal to 2 T (optimal magnetic field strength for HP gases ventilation imaging in small animals). To acquire ^{3}He signal at very short echo, a radial-sampling imaging sequence is used.

Repetition time (TR) = 5 ms; TE = 40 µs; field of view (FOV) = 80 mm; number of radial directions (NA) = 200; number of sampled points (NS) = 128; number of experiments (NEX) = 20; no slice thickness; flip angle = 12°; total acquisition time = 20 s.

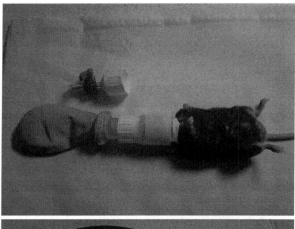
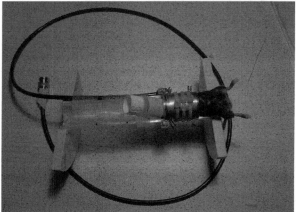

Fig. 1. Illustration of the experimental setup for the lung MR imaging studies using hyperpolarized gas with free breathing. The *top picture* shows the homemade mask positioned on the head of the animal (C57BL/6 mouse). The head mask is connected to a balloon filled with gas. The *bottom picture* represents the animal, with the mask, positioned in the NMR rf coil. The balloon is connected to the mask once the animal is positioned within the magnet.

3.1.4. Reconstruction Parameters

Image reconstruction is performed using a gridding algorithm (regridding of radially acquired samples onto a Cartesian grid). Reconstruction oversampling factor = 2; sampling density compensation using the Jacobian of the transformation; Kaiser–Bessel kernel parameters: shape factor $a = 2.8$, window width $L = 3.0$; final reconstruction matrix 256×256 (*see* **Note 3**).

Images of ventilation synchronized with the breathing cycle of the animal are obtained using the retrospective Cine reconstruction techniques (29). These time-resolved images of ventilation can be processed to generate parametric maps of gas arrival time in the broncho-alveolar space (29).

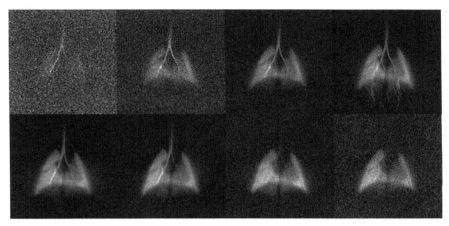

Fig. 2. Examples of lung ventilation images obtained under free-breathing conditions. The images represent the intrapulmonary distribution of polarized ^{3}He for different time windows of the breathing cycle of the animal. In this example, images correspond to consecutive 100-ms time windows. Images are reconstructed using a retrospective Cine algorithm. The total acquisition time is 20 s.

3.1.5. Expected Results

The expected results for ventilation imaging under spontaneous breathing conditions are the visualization of ventilated airspaces at tidal volumes. The Cine reconstruction algorithm generates images of ventilation at different phases of the breathing cycle (**Fig. 2**). Images of ventilation and gas arrival maps can be monitored during slow intravenous injection of methacholine. With increasing injected dose of methacholine, ventilation defects and delayed gas arrival time in the airspaces can be observed (**Fig. 3**).

3.2. Lung Proton MR Imaging in LPS Model of Lung Inflammation

3.2.1. Administration of LPS

1. Dissolve 1 mg/kg of LPS in 0.2 mL of saline.
2. Anesthetize the rat (injection of ketamine/xylazine) (*see* **Note 2**).
3. Wait until the animal loses its toe pinch reflex and position it supine on the board with the head tilted up (*see* **Note 4**).
4. Intubate the animal with a 26-gauge flexible polyethylene catheter attached to the 1-mL syringe (*see* **Note 5**) into a trachea (it is recommended to verify the length of the catheter first) and instill 0.2 mL of the solution (**Fig. 4**).
5. In order to have a homogeneous distribution of the suspension in the lungs, provoke hyperventilation by using a ventilator (*see* **Note 6**).
6. Allow the animal to recover maintaining its temperature on a heating pad.

3.2.2. Imaging Parameters

Two different approaches can be used: the first one uses a gradient echo imaging sequence which permits to obtain a high

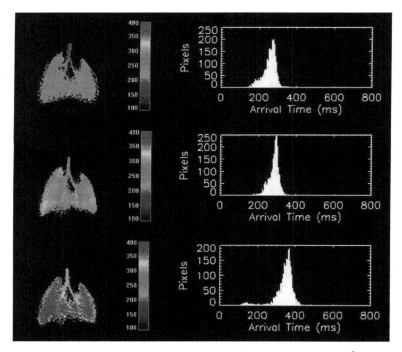

Fig. 3. Examples of parametric ventilation maps obtained in a rat using the ^{3}He free-breathing protocol. Parametric maps represent the local arrival time values in milliseconds. The arrival time is defined as the time delay between the arrival of gas into the trachea and the arrival of gas in a given pixel. The corresponding histogram for arrival time value is shown on the *right*. From *top* to *bottom*, the images correspond to the data acquisition before intravenous injection of methacholine, 11 and 21 min after the start of the injection of methacholine (10 μg/mL, 6 mL/h injection rate), respectively.

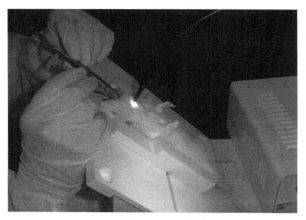

Fig. 4. Example of instillation setup. For instillation of agents, position a rat supine with the head tilted up. Intubate rats perorally with a 26-gauge flexible polyethylene catheter into the trachea. Inject 0.2 mL of solution.

contrast between dark-appearing lung parenchyma and the expected inflammation hallmark; the second, based on a radial ultra-short echo time sequence, is less sensitive to motion and blurring artifacts occurring in the thoracic cavity during respiration, and thus can provide more accurate assessment of image borders. The optimal sequence parameters at 4.7 T are as follows:

- Gradient echo imaging: TR = 5.6 ms, TE = 2.7 ms; bandwidth = 100 kHz, flip angle = 15°, FOV = 6 × 6 cm², matrix size = 256 × 128, and slice thickness = 1.5 mm. A single slice with 60 averages is acquired, resulting in an acquisition time of 75 s (55). To cover the entire lung volume, use 12–28 consecutive axial slices.
- Radial ultra-short echo time imaging: TR = 80 ms, TE = 450 μs; bandwidth = 64 kHz, flip angle = 20°, FOV 6 × 6 cm², 400 radials/image, 128 samples/view, slice thickness = 1.5 mm. A multi-slice acquisition with four averages is applied. Neither cardiac nor respiratory gating is used. The total acquisition time is equal to 4 min with an acquisition covering 12 contiguous axial slices.

3.2.3. Optimal Imaging Times

To observe an inflammation hallmark, the optimal imaging time is 48 h after LPS exposure (34, 49). However, the inflammation can be detected at its earlier stage (6 h after the challenge) as well as at later time points (up to 72–144 h after the challenge), reflecting pleura and mucus hypersecretion, respectively (34, 59). The protocol scheme is presented in **Fig. 5**.

3.2.4. Expected Results

The expected result is a hyperintense signal visualized on the T_1-weighted images. The observed signal is usually characterized by two components: one, of higher intensity, is attributed to the

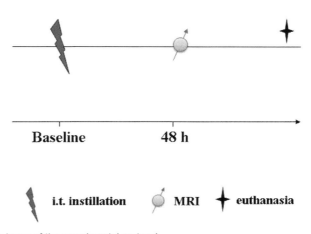

Fig. 5. Scheme of the experimental protocol.

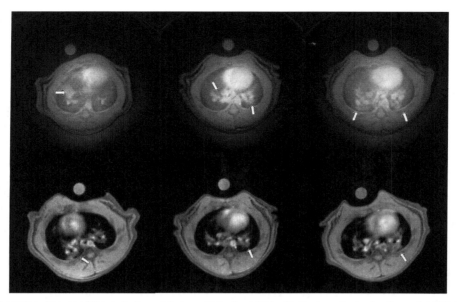

Fig. 6. Exemplary images of thorax in the axial plane 48 h after LPS instillation acquired using the radial UTE (*top*) and the gradient echo imaging sequence (*bottom*). The *arrows* indicate signal from inflammation.

edema; the second patchy-like appearance, with a less intense signal, represents an evidence of mucus secretion (**Fig. 6**).

3.2.5. Application in Pharmacology

The exposure to the endotoxin LPS is a well-established model of acute inflammation in rodents and similar to that observed in human COPD. The proposed protocols have the potential to provide easy implementation and accurate and non-invasive means of monitoring the effect of anti-inflammatory drugs in the experimental research of lung inflammation (31).

4. Notes

1. The homemade mask was realized as follows. The barrel of a standard plastic syringe was cut to the appropriate length to cover the head of the animal. The tip of the syringe barrel was sectioned to fit the nose of the animal. A hole was drilled in the screw cap of a plastic vial. This screw cap was then glued to the tip of the barrel of the syringe. This mask was positioned on the head of the animal. The body of the plastic vial was sectioned. The balloon (later on filled with polarized helium) was tied on the body of the plastic vial. Before launching the ^{3}He MR acquisition, the plastic vial attached to the helium-filled balloon is screwed on the animal mask.

2. Ketamine–xylazine cocktail is a preferred anesthesia over pentobarbital in ventilation studies because it induces less cardio-respiratory depression. Standard dose for intraperitoneal injection in rat and mouse is 3–9 mL/kg body weight of a mixture of 1 mL ketamine (100 mg/mL), 1 mL xylazine (20 mg/mL), and 5 mL saline.

   Alternatively, isoflurane inhalation can be used to anesthetize animals during MR acquisitions (proton studies).

3. Reconstruction of MR images acquired with radial k-space sampling requires dedicated reconstruction algorithms (regridding, i.e., interpolation of radially sampled data onto a Cartesian grid or filtered back-projection). MR manufacturers might provide such software in their reconstruction package. However, it is likely that off-line dedicated reconstruction software will be needed. In our studies, we use a set of homemade programs developed under IDL software (Research Systems Inc., Boulder, CO).

4. Hold the rat's head tight on to the board by a loop string or rubber which passes under the upper incisors.

5. or intubation, put out the rat's tongue and open the rat's mouth 2 cm wide using smooth blunt forceps. While inserting the catheter, lighten the larynx from the exterior using a light source.

6. Ventilate the rat at a frequency of 60 breaths per minute for approximately 1 min.

Acknowledgments

Magdalena Zurek acknowledges a fellowship from the European Network PHELINET (MRTN-CT-2006-36002).

References

1. Albert, M. S., Cates, G. D., Driehuys, B., Happer, W., Saam, B., Springer, C. S. Jr., and Wishnia, A. (1994) Biological magnetic resonance imaging using laser-polarized ^{129}Xe. *Nature* **370**, 199–201.
2. Middleton, H., Black, R. D., Saam, B., Cates, G. D., Cofer, G. P., Guenther, R., Happer, W., Hedlund, L. W., Johnson, G. A., and Juvan, K. (1995) MR imaging with hyperpolarized ^{3}He gas. *Magn. Reson. Med.* **33**, 271–275.
3. Ebert, M., Grossmann, T., Heil, W., Otten, W. E., Surkau, R., Leduc, M., Bachert, P., Knopp, M. V., Schad, L. R., and Thelen, M. (1996) Nuclear magnetic resonance in humans using hyperpolarized helium-3. *Lancet* **347**, 1297–1299.
4. MacFall, J. R., Charles, H. C., Black, R. D., Middleton, H., Swartz, J. C., Saam, B., Driehuys, B., Erickson, C., Happer, W., Cates, G. D., Johnson, G. A., and Ravin, C. E. (1996) Human lung air spaces: potential

5. Viallon, M., Cofer, G. P., Suddarth, S. A., Möller, H. E., Chen, X. J., Chawla, M. S., Hedlund, L. W., Crémillieux, Y., and Johnson, G. A. (1999) Functional MR microscopy of the lung with hyperpolarized ^{3}He. *Magn. Reson. Med.* **41**, 787–792.
6. Chen, B. T., Yordanov, A. T., and Johnson, G. A. (2005) Ventilation-synchronous magnetic resonance microscopy of pulmonary structure and ventilation in mice. *Magn. Reson. Med.* **53**, 69–75.
7. Saam, B. T., Yablonskiy, D. A., Kodibagkar, V. D., Leawoods, J. C., Gierada, D. S., Cooper, J. D., Lefrak, S. S., and Conradi, M. S. (2000) MR imaging of diffusion of (3)He gas in healthy and diseased lungs. *Magn. Reson. Med.* **44**, 174–179.
8. Chen, X. J., Hedlund, L. W., Möller, H. E., Chawla, M. S., Maronpot, R. R., and Johnson, G.A. (2000) Detection of emphysema in rat lungs by using magnetic resonance measurements of ^{3}He diffusion. *Proc. Natl. Acad. Sci. U.S.A.* **10**, 11478–11481.
9. Yablonskiy, D. A., Sukstanskii, A. L., Leawoods, J. C., Gierada, D. S., Bretthorst, G. L., Lefrak, S. S., Cooper, J. D., and Conradi, M. S. (2002) Quantitative in vivo assessment of lung microstructure at the alveolar level with hyperpolarized ^{3}He diffusion MRI. *Proc. Natl. Acad. Sci. U.S.A.* **99**, 3111–3116.
10. Peces-Barba, G., Ruiz-Cabello, J., Crémillieux, Y., Rodríguez, I., Dupuich, D., Callot, V., Ortega, M., Rubio Arbo, M. L., Cortijo, M., and Gonzalez-Mangado, N. (2003) Helium-3 MRI diffusion coefficient: correlation to morphometry in a model of mild emphysema. *Eur. Respir. J.* **22**, 14–19.
11. Dugas, J. P., Garbow, J. R., Kobayashi, D. K., and Conradi, M. S. (2004) Hyperpolarized (3)He MRI of mouse lung. *Magn. Reson. Med.* **52**, 1310–1317.
12. Mata, J. F., Altes, T. A., Cai, J., Ruppert, K., Mitzner, W., Hagspiel, K. D., Patel, B., Salerno, M., Brookeman, J. R., de Lange, E. E., Tobias, W. A., Wang, H. T., Cates, G. D., and Mugler, J. P. 3rd. (2006) Evaluation of emphysema severity and progression in a rabbit model: a comparison of hyperpolarized He-3 and ^{129}Xe diffusion MRI with lung morphometry. *J. Appl. Physio.* **102**, 1273–1280.
13. Saam, B., Happer, W., and Middleton, H. (1995) Nuclear relaxation of ^{3}He in the presence of O_2. *Phys. Rev. A* **52**, 862–865.
14. Eberle, B., Weiler, N., Markstaller, K., Kauczor, H., Deninger, A., Ebert, M., Grossmann, T., Heil, W., Lauer, L. O., Roberts, T. P., Schreiber, W. G., Surkau, R., Dick, W. F., Otten, E. W., and Thelen, M. (1999) Analysis of intrapulmonary O(2) concentration by MR imaging of inhaled hyperpolarized helium-3. *J. Appl. Physiol.* **87**, 2043–2052.
15. Deninger, A. J., Eberle, B., Ebert, M., Grossmann, T., Heil, W., Kauczor, H., Lauer, L., Markstaller, K., Otten, E., Schmiedeskamp, J., Schreiber, W., Surkau, R., Thelen, M., and Weiler, N. (1999) Quantification of regional intrapulmonary oxygen partial pressure evolution during apnea by ^{3}He MRI. *J Mag. Reson.* **141**, 207–216.
16. Deninger, A. J., Eberle, B., Ebert, M., Grossmann, T., Hanisch, G., Heil, W., Kauczor, H.U., Markstaller, K., Otten, E., Schreiber, W., Surkau, R., and Weiler, N. (2000) ^{3}He-MRI-based measurements of intrapulmonary pO2 and its time course during apnea in healthy volunteers: first results, reproducibility, and technical limitations. *NMR Biomed.* **13**, 194–201.
17. Deninger, A. J., Eberle, B., Bermuth, J., Escat, B., Markstaller, K., Schmiedeskamp, J., Schreiber, W. G., Surkau, R., Otten, E., and Kauczor, H-U. (2002) Assessment of a single-acquisition imaging sequence for oxygen-sensitive ^{3}He. *Magn. Reson. Med.* **47**, 105–114.
18. Wild, J. M., Fichele, S., Woodhouse, N., Paley, M. N., Kasuboski, L., and van Beek, E. J. (2005) 3D volume-localized pO2 measurement in the human lung with ^{3}He. MRI *Magn. Reson. Med.* **53**, 1055–1064.
19. Jalali, A., Ishii, M., Edvinsson, J. M., Guan, L., Itkin, M., Lipson, D. A., Baumgardner, J. E., and Rizi, R. R. (2004) Detection of simulated pulmonary embolism in a porcine model using hyperpolarized ^{3}He MRI. *Magn. Reson. Med.* **51**, 291–298.
20. Cieślar, K., Stupar, V., Canet-Soulas, E., Gaillard, S., and Crémillieux, Y. (2007) Alveolar oxygen partial pressure and oxygen depletion rate mapping in rats using (3)He ventilation imaging. *Magn. Reson. Med.* **57**, 423–430.
21. Cieślar, K., Alsaid, H., Stupar, V., Gaillard, S., Canet-Soulas, E., Fissoune, R., and Crémillieux, Y. (2007) Measurement of nonlinear pO2 decay in mice lungs using 3He-MRI. *NMR Biomed.* **20**, 383–391.
22. Ruppert, K., Brookeman, J. R., Hagspiel, K. D., and Mugler, J. P. 3rd. (2000) Probing lung physiology with xenon polarization transfer contrast (XTC). *Magn. Reson. Med.* **44**, 349–357.
23. Chen, B. T. and Johnson, G. A. (2004) Dynamic lung morphology of

methacholine-induced heterogeneous bronchoconstriction. *Magn. Reson. Med.* **52**, 1080–1086.
24. Mosbah, K., Crémillieux, Y., Adeleine, P., Dupuich, D., Stupar, V., Nemoz, C., Canet, E., and Berthezène, Y. (2006) Quantitative measurements of regional lung ventilation using helium-3 MRI in a methacholine-induced bronchoconstriction model. *J. Magn. Reson. Imaging* **24**, 611–616.
25. Hedlund, L. W., Moller, H. E, Chen, X. J., Chawla, M. S., Cofer, G. P., and Johnson, G. A. (2000) Mixing oxygen with hyperpolarized (3)He for small-animal lung studies. *NMR Biomed.* **13**, 202–206.
26. Ramirez, M. P., Sigaloff, K. C., Kubatina, L. V., Donahue, M. A., Venkatesh, A. K., and Albert, M. S. (2000) Physiological response of rats to delivery of helium and xenon: implications for hyperpolarized noble gas imaging. *NMR Biomed.* **13**, 253–264.
27. Chen, B. T., Yordanov, A. T., and Johnson, G. A. (2005) Ventilation-synchronous magnetic resonance microscopy of pulmonary structure and ventilation in mice. *Magn. Reson. Med.* **53**, 69–75.
28. Imai, H., Narazaki, M., Inoshita, H., Kimura, A., and Fujiwara, H. (2006) MR imaging of mouse lung using hyperpolarized ^{3}He: image acquisition and T_1 estimation under spontaneous respiration. *Magn. Reson. Med. Sci.* **5**, 57–64.
29. Stupar, V., Canet-Soulas, E., Gaillard, S., Alsaid, H., Beckmann, N., and Crémillieux, Y. (2007) Retrospective Cine (3)He ventilation imaging under spontaneous breathing conditions: a non-invasive protocol for small-animal lung function imaging. *NMR Biomed.* **20**, 104–112.
30. Beckmann, N., Cannet, C., Karmouty-Quintana, H., Tigani, B., Zurbruegg, S., Blé, F. X., Crémillieux, Y., and Trifilieff, A. (2007) Lung MRI for experimental drug research. *Eur. J. Radiol.* **64**, 381–396.
31. Beckmann, N., Crémillieux, Y., Tigani, B., Karmouty Quintana, H., Blé, F. X., and Fozard, J. R. (2006) Lung MRI in small rodents as a tool for the evaluation of drugs in models of airways diseases. In: Beckmann, N. (ed.), *In Vivo MR Techniques in Drug Discovery and Development*. Taylor & Francis Group, New York, pp. 351–372.
32. Nieman, B. J., Bishop, J., Dazai, J., Bock, N. A., Lerch, J. P., Feintuch, A., Chen, X. J., Sled, J. G., and Henkelman, R. M. (2007) MR technology for biological studies in mice. *NMR Biomed.* **20**, 291–303.
33. Beckmann, N., Tigani, B., Ekatodramis, D., Borer, R., Mazzoni, L., and Fozard, J. R. (2001) Pulmonary edema induced by allergen challenge in the rat: non-invasive assessment by magnetic resonance imaging. *Magn. Reson. Med.* **45**, 88–95.
34. Beckmann, N., Tigani, B., Sugar, R., Jackson, A. D., Jones, G., Mazzoni, L., and Fozard, J. R. (2002) Noninvasive detection of endotoxin-induced mucus hypersecretion in rat lung by MRI. *Am. J. Physiol. Lung Cell Mol. Physiol.* **283**, 22–30.
35. Karmouty-Quintana, H., Cannet, C., Zurbruegg, S., Blé, F. X., Fozard, J. R., Page, C. P., and Beckmann, N. (2006) Proton MRI as a noninvasive tool to assess elastase-induced lung damage in spontaneously breathing rats. *Magn. Reson. Med.* **56**, 1242–1250.
36. Karmouty-Quintana, H., Cannet, C., Zurbruegg, S., Blé, F. X., Fozard, J. R., Page, C. P., and Beckmann, N. (2007) Bleomycin-induced lung injury assessed noninvasively and in spontaneously breathing rats by proton MRI. *J. Magn. Reson. Imaging* **26**, 941–949.
37. Bergin, C. J., Pauly, J. M., and Macovski, A. (1991) Lung parenchyma: projection, reconstruction MR imaging. *Radiology* **179**, 777–781.
38. Gewalt, S. L., Glover, G. H., Hedlund, L. W., Cofer, G. P., MacFall, J. R., and Johnson, G. A. (1993) MR microscopy of the rat lung using projection reconstruction. *Magn. Reson. Med.* **29**, 99–106.
39. Zurek, M., Bessaad, A., Cieslar, K., and Crémillieux, Y. (2010) Validation of simple and robust protocols for high resolution lung proton MR imaging in mice. *Magn. Reson. Med.* **64**, 401–407.
40. Olsson, L. E., Lindahl, M., Onnervik, P. O., Johansson, L. B., Palmér, M., Reimer, M. K., Hultin, L., and Hockings, P. D. (2007) Measurement of MR signal and T2* in lung to characterize a tight skin mouse model of emphysema using single-point imaging. *J. Magn. Reson. Imaging* **25**, 488–494.
41. Takahashi, M., Togao, O., Obara, M., Cauteren, M., Ohno, Y., Malloy, C., and Dimitrov, I. (2009) Ultra-short echo time (UTE) MR imaging of the lung: comparison between normal and emphysematous mice. *Proceedings of the 17th Annual Meeting of ISMRM*, Honolulu, Hawaii, USA, p. 11.
42. Suga, K., Yuan, Y., Ogasawara, N., Tsukuda, T., and Matsunaga, N. (2003) Altered clearance of gadolinium diethylenetriaminepentaacetic acid aerosol from bleomycin-injured dog lungs: initial observations. *Am. J. Respir. Crit. Care Med.* **167**, 1704–1710.

43. Watt, K. N., Bishop, J., Nieman, B. J., Henkelman, R. M., and Chen, X. J. (2008) Oxygen-enhanced MR imaging of mice lungs. *Magn. Reson. Med.* **59**, 1412–21.
44. Zurek, M., Cieslar, K., Sigovan, M., Bessaad, A., Canet-Soulas, E., and Crémillieux, Y. (2009) Perfusion-weighted pulmonary MRI in Mouse—Preliminary Results. *Proceedings of the 17th Annual Meeting of ISMRM*, Honolulu, Hawaii, USA, p. 2005.
45. Mistry, N. N., Pollaro, J. Song, J. Lin, M. D., and Johnson, G. A. (2008) Pulmonary perfusion imaging in the rodent lung using dynamic contrast-enhanced MRI *Magn. Reson. Med.* **59**, 289–297.
46. Zurek, M., Carrero-Gonzalez, L., Bucher, S., Kaulisch, T., Stiller, D., and Crémillieux, Y. (2010) Inflammation assessment in the lungs of LPS-challenged rodents: comparison between radial ultra short echo time (UTE) and Cartesian MR Imaging. *Proceedings of the 18th Annual Meeting of ISMRM*, Stockholm, Sweden.
47. Glover, G. H. and Pauly, J. M. (1992) Projection reconstruction techniques for reduction of motion effects in MRI. *Magn. Reson. Med.* **28**, 275–289.
48. Schäffter, T., Rasche, V., and Carlsen, I. C. (1999) Motion compensated projection reconstruction. *Magn. Reson. Med.* **41**, 954–963.
49. Olsson, L. E., Smailagic, A., Onnervik, P. O., and Hockings, P. D. (2009) (1)H and hyperpolarized (3)He MR imaging of mouse with LPS-induced inflammation. *J. Magn. Reson. Imaging* **29**, 977–981.
50. Edelman, R. R., Hatabu, H., Tadamura, E., Li, W., and Prasad, P. V. (1996) Noninvasive assessment of regional ventilation in the human lung using oxygen-enhanced magnetic resonance imaging. *Nat. Med.* **2**, 1236–1239.
51. Stadler, A., Stiebellehner, L., Jakob, P. M, Arnold, J. F. T., Eisenhuber, E., Katzler, I., and Bankier, A. A. (2007) Quantitative and O(2) enhanced MRI of the pathologic lung: findings in emphysema, fibrosis, and cystic fibrosis. *Int. J. Biomed. Imaging* **23**, 624.
52. Jakob, P. M., Wang, T., Schultz, G., Hebestreit, H., Hebestreit, A., and Hahn, D. (2004) Assessment of human pulmonary function using oxygen-enhanced T(1) imaging in patients with cystic fibrosis. *Magn. Reson. Med.* **51**, 1009–1016.
53. Maï, W., Badea, C. T., Wheeler, C. T., Hedlund, L. W., and Johnson, G. A. (2005) Effects of breathing and cardiac motion on spatial resolution in the microscopic imaging of rodents. *Magn. Reson. Med.* **53**, 858–865.
54. Cassidy, P. J., Schneider, J. E., Grieve, S. M., Lygate, C., Neubauer, S., and Clarke, K. (2004) Assessment of motion gating strategies for mouse magnetic resonance at high magnetic fields. *J. Magn. Reson. Imaging* **19**, 229–237.
55. Beckmann, N., Tigani, B., Mazzoni, L., and Fozard, J. R. (2001) MRI of lung parenchyma in rats and mice using a gradient-echo sequence. *NMR Biomed.* **14**, 297–306.
56. Hiba, B., Richard, N., Janier, M., and Croisille, P. (2006) Cardiac and respiratory double self-gated cine MRI in the mouse at 7 T. *Magn. Reson. Med.* **55**, 506–513.
57. Esparza-Coss, E., Ramirez, M. S., and Bankson, J. A. (2008) Wireless self-gated multiple-mouse cardiac cine MRI. *Magn. Reson. Med.* **59**, 1203–1206.
58. Nieman, B. J., Szulc, K. U., and Turnbull, D. H. (2009) Three-dimensional in vivo MRI with self-gating and image coregistration in the mouse. *Magn. Reson. Med.* **61**, 1148–1157.
59. Karmouty-Quintana, H., Cannet, C., Schaeublin, E., Zurbruegg, S., Sugar, R., Mazzoni, L., Page, C., Fozard, J. R., and Beckmann, N. (2006) Identification with MRI of the pleura as a major site of the acute inflammatory effects induced by ovalbumin and endotoxin challenge in the airways of the rat. *Am. J. Physiol. Lung Cell Mol. Physiol.* **291**, 651–657.

Section VII

Cancer Models

Chapter 25

Characterization of Tumor Vasculature in Mouse Brain by USPIO Contrast-Enhanced MRI

Giulio Gambarota and William Leenders

Abstract

Detailed characterization of the tumor vasculature provides a better understanding of the complex mechanisms associated with tumor development and is especially important to evaluate responses to current therapies which target the tumor vasculature. Magnetic resonance imaging (MRI) studies of tumors have been mostly performed using gadolinium-diethylenetriamine pentaacetic acid (Gd-DTPA) contrast-enhanced imaging, which relies on Gd-DTPA leakage from hyperpermeable tumor vessels and subsequent accumulation in the tumor interstitium. In certain tumor types, especially diffuse glioma in the brain, incorporated tumor vessels are not necessarily leaky, complicating effective diagnosis via Gd-DTPA contrast-enhanced MRI. Another class of contrast agents, based on superparamagnetic ultrasmall iron oxide particles (USPIO), allows for non-invasive assessment of vascular volume within the tumor. Vascular volume can be obtained by calculating the change in water proton transverse relaxation rate (R_2 or R_2^*) following USPIO administration. This allows for an objective comparison between vascular volumes of different tumors and also allows to perform longitudinal studies in order to assess, for example, treatment efficacy. Moreover, since the USPIO T_2 relaxivity is up to 20 times that of Gd-DTPA, USPIO provides a highly sensitive marker for alterations in vascular volume among tissues; this characteristic might be exploited for tumor detection. Thus, USPIO imaging may be a very attractive alternative to the most commonly used Gd-DTPA imaging and will at least have added value, especially for detection and delineation of diffuse infiltrative brain tumors.

Key words: Gd-DTPA, USPIO, VEGF, contrast agent, mouse brain, tumor, vascular volume, magnetic resonance imaging, T_2 relaxation time.

1. Introduction

Tumor vasculature plays a key role in tumor growth (1). Tumors have developed a number of ways to ensure a proper blood supply, such as incorporation of pre-existent vessels (also referred to as vessel co-option), vessel modulation (e.g., dilatation, intussusception), and angiogenesis (2, 3). One of the key features of the angiogenic process is increased permeability of the newly formed vessels, which can be effectively monitored in T_1-weighted MRI by extravasation of small paramagnetic contrast agents such as gadolinium- diethylenetriamine pentaacetic acid (Gd-DTPA). Hyperpermeability is a response to tumor-derived vascular endothelial growth factor-A (VEGF-A), the most important angiogenic factor known. VEGF-A is now regarded as an important therapeutic target and a number of inhibitors of the VEGF pathway have now entered oncology practice (2). One of the profound effects of these inhibitors is normalization of tumor vasculature. The concomitant decrease in permeability creates problems with Gd-DTPA MRI-based tumor diagnosis (4).

Therefore, the detailed characterization of tumor vasculature with respect to both permeability (vascular leakage) and vascular (blood) volume provides essential insight into tumor physiology and is a prerequisite to investigate and evaluate tumor response to anti-angiogenic therapy. For example, glial brain tumors may contain significant areas of diffuse infiltrative growth without a substantial vascular hyperpermeability (2). These areas are not detected via Gd-DTPA-enhanced T_1-weighted MRI but may have altered blood volumes as compared to surrounding normal tissue, making tumor blood volume another important parameter that characterizes the tumor vasculature (5). Several studies have shown that magnetic susceptibility effects, caused by blood pool contrast agents that consist of ultrasmall superparamagnetic particles of iron oxide (USPIO), can be used to assess blood volume and vessel size within tumors (6–10). In particular, the enhancement in the water proton transverse relaxation rates R_2 ($=1/T_2$, where T_2 is the transverse relaxation time) and R_2^* ($=1/T_2^*$, see **Note 1**), following administration of USPIO, provides an index proportional to the blood volume of the microvasculature and macrovasculature, respectively (11, 12). The quantitative assessment of blood volume allows for an objective comparison of vascular volumes of different tumors and also allows to perform longitudinal studies in order to assess, for example, treatment efficacy. This protocol provides an overview of technical aspects of MRI in combination with USPIO administration, for investigations of tumor vasculature in mouse brain.

2. Materials

2.1. Animal Preparation

1. A basic gas anesthesia apparatus, using isoflurane mixed with O_2 and N_2O.

2. U87 and cerebral metastases of human Mel57 metastases. Describing the specific tumor-model preparation is outside the scope of this chapter, and we refer the reader to the appropriate literature (13).
 3. A catheter for tail vein injection (28G, with 23G thin wall needle (Braintree Scientific, Inc., Braintree, MA, USA).
 4. A homebuilt plastic cradle with nose cone, to hold the animal in the scanner.
 5. A fluoroptic probe (Luxtron 712; Luxtron, Santa Clara, CA, USA).
 6. An optical sensor of motion (Siracust 401; Siemens, Erlangen, Germany).
 7. A circulating warm-water (37°C) blanket (homebuilt, with plastic tubing, a water pump, and an electrical heater equipped with a thermostat).

2.2. Scanner Hardware/Software

 1. MR console (Surrey Medical Imaging Systems, Surrey, UK) interfaced with a 7.0 T, 200 mm horizontal bore magnet (Magnex Scientific, Abingdon, UK) and a 150 mT/m gradient insert.
 2. A 10 mm diameter surface transmitter/receiver RF coil.
 3. Gradient-echo imaging sequence and spin-echo imaging sequence. These are standard sequences that are provided with every MR imaging console.
 4. MATLAB (Mathworks, Natick, MA, USA) is used in this study as post-processing imaging software.

2.3. Contrast Agents

 1. Gd-DTPA (Magnevist Schering, Berlin, Germany) is diluted in 0.9% NaCl to 20 mM/L.
 2. USPIO with an iron oxide core of 4–6 nm coated by dextran to yield particles of ~30 nm diameter (Sinerem, Guerbet Laboratories, Aulnay-Sous-Bois, France) is diluted in 0.9% NaCl and injected in mice at a dose of 12.5 mg Fe/kg.

3. Methods

There are a large number of brain tumor models and various ways to inject the tumor cells. For instance, cell injection can be performed transcranially or – in order to closely mimic the development of tumor metastasis in humans – in the internal carotid artery. This part of the experiment is, however, outside the scope of this chapter, and we refer the reader to the appropriate literature (13). Here we focus on the animal preparation, which is a crucial part of the experiment, and the imaging procedure. Data shown in this article refer to glioma xenografts U87 and cerebral metastases of human Mel57 metastases, either engineered or not engineered to express VEGF-A_{165} (14).

3.1. Animal Preparation

1. Mice are anesthetized via inhalation of isoflurane (1.5–2%) in N_2O/O_2.
2. In each mouse, a catheter is inserted in a lateral tail vein to allow administration of the contrast agent (*see* **Note 2**). After catheter insertion, the animal is placed on a custom-made plastic cradle. The cradle includes a plastic cone that is positioned near the mouse nose for anesthesia delivery during the experiment.
3. A 10 mm diameter surface transmitter/receiver RF coil is positioned over the mouse head (*see* **Note 3**).
4. A rectal fluoroptic probe, inserted approximately 1 cm into the mouse rectum, is used to monitor the core temperature of the animal.
5. During the experiment, the temperature of the animal is maintained at $38.0 \pm 0.5°C$ with use of a circulating warm-water (37°C) blanket. The respiration rate is monitored with an optical sensor of motion (*see* **Note 4**).
6. The cradle with the animal is inserted in the magnet bore.

3.2. Magnetic Resonance Imaging

1. All MR images are acquired with the MR console.
2. Three gradient-echo scout images in orthogonal planes, i.e., axial, coronal, and sagittal, are acquired for anatomical localization of the brain. It should be noted that, since the scout images are used only as a simple reference image to position the images for the contrast-enhanced protocol, the imaging parameters of the scout are not particularly crucial. Typical scout imaging parameters are repetition time (TR) = 400 ms, echo time (TE) = 10 ms, image matrix size of 128 × 128, field of view (FOV) of 5.8 cm × 5.8 cm, slice thickness (SLT) of 0.7 mm, and one average.
3. The protocol of Gd-DTPA contrast-enhanced MRI consists of T_1-weighted multislice gradient-echo images. We find it convenient to acquire brain images in the axial direction, since this slice orientation provides an easy anatomical visualization of the brain regions. The imaging parameters are the following: TR/TE = 400/6 ms, voxel size = 136 × 136 × 1000 μm, 16 contiguous slices, receiver bandwidth = 100 kHz, scan time = 1 min 16 s. In **Fig. 1a** (Mel57-VEGF-A_{165} metastases) and **Fig. 2a** (U87 tumor), images prior to administration of Gd-DTPA show virtually no contrast between tumor and healthy brain tissue.

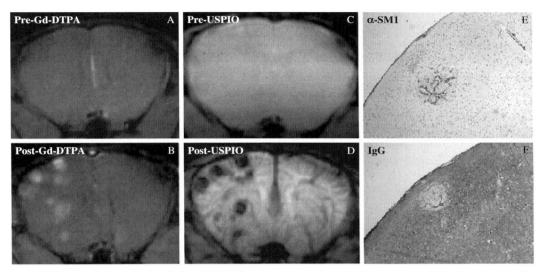

Fig. 1. MR images of mouse brain with Mel57-VEGF-A_{165} tumors, acquired prior to (**a** and **c**) and following administration of Gd-DTPA (**b**) and USPIO (**d**). Immunostainings for pericytes (**e**) and extravasated mouse immunoglobulins (**f**) of a lesion in a section matched with the MRI slice. The immunohistochemical stainings, which show highly dilated tumor vasculature and the high level of extravasated IgG, confirm the contrast-enhanced MRI findings. *See also* (14).

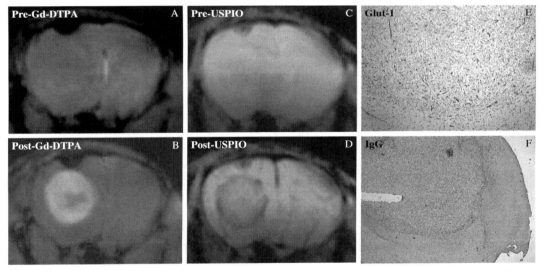

Fig. 2. MR images of mouse brain with a U87 tumor, acquired prior to (**a** and **c**) and following administration of Gd-DTPA (**b**) and USPIO (**d**). Immunohistochemical staining for Glut-1 (**e**), highlighting vessels, and an immunostaining for extravasated mouse immunoglobulins (**f**) confirm the MRI results. *See also* (14).

4. Contrast agent administration. Gd-DTPA contrast agent is administered at a dose of 0.2 mmol/kg in a volume of 200 μL phosphate-buffered saline via injection in a lateral tail vein. Injection is performed slowly, such that it takes about 5 s. The same injection procedure is performed in the

protocol of USPIO contrast-enhanced MRI, with USPIO administered at a dose of 12.5 mg/kg.

5. Post-contrast agent imaging. Tumors are hyperintense on T_1-weighted images acquired 2 min following Gd-DTPA administration, indicating Gd-DTPA extravasation (**Figs. 1b** and **2b**). This correlates well with the high vascular permeability in the tumors, as assessed by anti-IgG staining (**Figs. 1f** and **2f**).

6. Steps for anti-IgG staining. With respect to the details of the staining procedure, since this part of the experiment is outside the main scope of this chapter, we refer the reader to the appropriate literature (13).

7. MR imaging with USPIO is performed 2–3 h after Gd-DTPA imaging. As Gd-DTPA has a biological half-life of 20 min in mouse, this time gap allows near-complete washout of Gd-DTPA, precluding interference of Gd-DTPA with the USPIO-enhanced MRI protocol and enabling a correlation between Gd-DTPA and USPIO enhancement.

8. The protocol of USPIO contrast-enhanced MRI consists of two experiments: multislice gradient-echo imaging (TR/TE=1500/7 ms, 16 slices, voxel size = 136 × 136 × 1000 μm, receiver bandwidth = 100 kHz, scan time = 4 min 54 s); and multislice spin-echo imaging (TR/TE=2000/9 ms, 16 slices, voxel size = 136 × 136 × 1000 μm, receiver bandwidth = 100 kHz, scan time = 6 min 32 s) performed prior to and following USPIO administration (*see* **Note 5**).

9. Similar to T_1-weighted images acquired prior to contrast agent administration, T_2-weighted images show virtually no contrast between tumor and healthy brain tissue (**Fig. 1c**, Mel57-VEGF-A_{165} metastases, and **Fig. 2c**, U87 tumor). Following USPIO administration, tumors are hypointense on T_2-weighted images, indicating a higher concentration of contrast agent and therefore higher blood volume (**Figs. 1d** and **2d**). This correlates well with the presence of highly dilated vessels ((**Fig. 1e**, α-SM1 staining, and **Fig. 2e**, Glut-1). The Mel57-VEGF-A_{165} metastases appear as black spots and show an expansive growth pattern (**Fig. 1d**), whereas the U87 tumors are characterized by a ring-like structure (**Fig. 2d**). It should be noted that MRI in combination with USPIO administration has the potential benefit of improving tumor detection and delineation. In fact, leakage of Gd-DTPA from the tumor vessels into the interstitium can extend into the surrounding normal tissue, thus precluding an accurate determination of tumor boundaries. Since USPIO remains intravascular for a prolonged period of time and generates a high signal contrast between regions

with different blood volumes, due to its high R_2^* relaxivity (R_2^* up to 100 mM^{-1} s^{-1} (10)), a better delineation of tumor is feasible (15–17).

3.3. Data Analysis

1. Pixel-by-pixel ΔR_2^* maps are obtained from the formula: $\Delta R_2^* = (1/TE)\log(S_o^{bef}/S_o^{aft})$, where TE is the echo time and S_o the signal amplitude pre-USPIO (S_o^{bef}) and post-USPIO (S_o^{aft}), in the gradient-echo images (*see* **Note 6**). The same algorithm is used to generate ΔR_2 maps from the spin-echo images. The algorithm for calculating ΔR_2^* (and ΔR_2) from the formula reported above is implemented as a script written in MATLAB (*see* **Note 7**).

2. After generating ΔR_2^* and ΔR_2 maps, ΔR_2^* and ΔR_2 values are measured in manually segmented ROIs within the region of interest. The algorithm for generating ROIs is also implemented as a script written in MATLAB.

3. For each ROI, mean ΔR_2^* and ΔR_2 are calculated by averaging the values of all pixels within the ROI.

4. The Mel57-VEGF-A$_{165}$ metastases show very high values of ΔR_2^* and ΔR_2 ($\Delta R_2^* = 280 \pm 90$ s^{-1} and $\Delta R_2 = 40 \pm 10$ s^{-1}, mean ± standard deviation, assessed in four mice and for each mouse five lesions with the highest ΔR_2^* values are considered).

5. The U87 tumors are characterized by relatively low ΔR_2^* and ΔR_2 values in the core ($\Delta R_2^* = 17 \pm 6$ s^{-1} and $\Delta R_2 = 5 \pm 3$ s^{-1}, assessed in four mice) and moderately high ΔR_2^* and ΔR_2 values in the peritumoral region ($\Delta R_2^* = 54 \pm 15$ s^{-1} and $\Delta R_2 = 13 \pm 5$ s^{-1}).

6. It is also of interest, for reference, to measure the values of ΔR_2^* and ΔR_2 in healthy brain regions (cortex: $\Delta R_2^* = 21 \pm 9$ s^{-1}, $\Delta R_2 = 5 \pm 2$ s^{-1}; striatum: $\Delta R_2^* = 36 \pm 10$ s^{-1}, $\Delta R_2 = 9 \pm 3$ s^{-1}). An example of the macrovascular blood volume map (i.e., ΔR_2^* map) of Mel57-VEGF-A$_{165}$ and U87 tumors is shown in **Fig. 3a** and **c**, respectively. As explained in point 1 of this section, the algorithm for calculating ΔR_2^* (and ΔR_2) is implemented as a script written in MATLAB.

7. In Mel57-VEGF-A$_{165}$ tumors, a high threshold for ΔR_2^* is used in the tumor areas (region inside the box, **Fig. 3a**; note the ΔR_2^* scale up to 300 s^{-1}). Outside the tumor region, a low cutoff threshold (ΔR_2^* scale up to 100 s^{-1}) is used to maximize the contrast between striatum (ROI indicated on the corresponding gradient-echo image by the horizontal arrow, **Fig. 3b**) and cortex (ROI indicated by the vertical arrow, **Fig. 3b**).

8. In **Fig. 3d**, the gradient-echo image corresponding to the U87 tumor macrovascular blood volume map is shown.

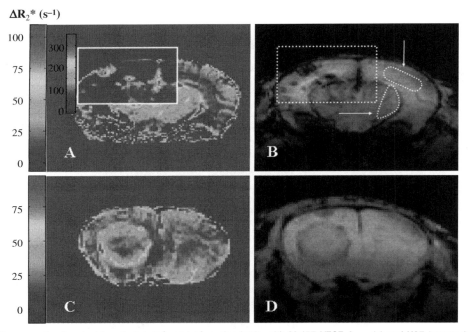

Fig. 3. An example of a pixel-by-pixel ΔR_2^* map of murine brain with Mel57-VEGF-A$_{165}$ (a) and U87 tumors (c). The gradient-echo images post-USPIO are shown in (b) and (d), respectively. In order to highlight the differences in regional blood volume between brain regions that do not include tumor, the ΔR_2^* map of Mel57-VEGF-A$_{165}$ (a) is displayed with a high cutoff threshold in the tumor areas (region inside the *box*) and a low cutoff threshold outside tumor regions. For reference, ΔR_2^* and ΔR_2 values are also calculated in healthy brain regions (striatum, indicated by the *horizontal arrow* and cortex, indicated by the *vertical arrow*).

4. Notes

1. The T_2 relaxation time is the time constant that governs the loss of transverse magnetization (i.e., of signal) in spin-echo imaging. This relaxation process is due to the dephasing of the individual magnetic dipoles because of the existence of non-stationary random local magnetic fields. On the other hand, the loss of transverse magnetization (i.e., of signal) in gradient-echo imaging is characterized by a relaxation time T_2^*. Two factors contribute to the T_2^* decay: the non-stationary random local magnetic fields (which generate the T_2 decay) and the static (i.e., constant in time) inhomogeneities in the magnetic field. Static distortions in the magnetic field are particularly dominant near interfaces between regions with different magnetic susceptibilities (air–tissue interface, for example). It should be noted that the dephasing of spins resulting from static field inhomogeneities is a reversible process, which is accomplished by applying a 180° pulse (which is present in the spin-echo imaging). The

dephasing due to fluctuating magnetic fields, however, is not reversible.

2. Inserting the catheter in a mouse tail vein can be challenging and requires high technical skills. Catheters are held in place by a plaster. In our laboratory, this operation was performed by technicians with years of experience in animal procedures. Once the catheter is inserted, specific care has to be taken to make sure that, during the positioning of the animal in the cradle and then in the magnet bore, the catheter remains well in place. Before attaching a line with contrast agent, we routinely flush minimal amount of heparinized PBS (5 U/mL) to prevent clogging in the catheter.

3. We found this size of coil convenient for spatial coverage of the whole mouse brain, providing at the same time high sensitivity, which resulted in images with a good signal-to-noise ratio.

4. We cannot stress enough the importance of a well-designed setup, including the plastic cradle where the animal is positioned, the supports that keep the coil well in place on the head of the animal, the location of the tubing that delivers the anesthesia, and the positioning of the warm pad. All these aspects are crucial for a successful experiment. It is also important to start the imaging acquisition protocol only when the animal is in a stable condition, so care has to be taken in checking continuously that temperature and respiration rate are stable – if not, change appropriately the flow of anesthetic and/or temperature of circulating water.

5. A number of pilot experiments might be needed to optimize the protocol of blood volume measurements at the given field strength of the magnet. At the dose of USPIO agent used in the current study, we find that a good compromise between signal-to-noise ratio and contrast enhancement is achieved for TE values in the range of 5–10 ms. Longer (shorter) values of TE might be advantageous for blood volume measurements at a lower (higher) field. On the other hand, it should be noted that, for longer TEs, the USPIO-induced signal dephasing tends to extend more beyond the size of the main magnetic field perturbers, i.e., the blood vessels, and this might confound tumor delineation. Thus, when precise tumor delineation is needed, it is preferable to keep TE short.

6. In this study, ΔR_2^* and ΔR_2 maps are obtained by acquiring a T_2^*- and T_2-weighted images, respectively, prior to and following USPIO administration. This approach has the advantage of experimental simplicity, since only T_2^*- and T_2-weighted images are required; furthermore, the data analysis is relatively simple and straightforward.

Another way to obtain ΔR_2^* maps is to map the T_2^*, prior to and following USPIO administration with multi-echo gradient-echo imaging and then use directly the formula: $\Delta R_2^* = (1/T_2^*\text{post}) - (1/T_2^*\text{pre})$ (8, 10). The same applies to ΔR_2 maps, with the only difference being that multi-echo spin-echo (i.e., Carr–Purcell–Meiboom–Gill or CPMG) imaging is required. The latter approaches for ΔR_2^* and ΔR_2 are potentially more robust. However, the pulse sequences are more complex and need to be properly designed. In particular, the T_2 mapping with CPMG could give some problems with incidental magnetization transfer effects and stimulated echoes. Furthermore, a careful choice of the number of echoes and interpulse delay is necessary. Finally, the data processing becomes more intensive, since, for each slice, the multiple echoes need to be fit to an exponential decay to generate T_2^* or T_2 maps.

7. In order to generate the ΔR_2 and ΔR_2^* maps, an image-processing software is required. In the current study, maps were calculated using MATLAB. MATLAB is a matrix-based computer language; as such, it is particularly convenient for processing imaging data. In fact, images are simply matrices, where each value of the matrix represents the intensity of the MR signal in the specific pixel. To generate the ΔR_2 and ΔR_2^* maps, the first operation consists of reading the image data into a matrix. Afterwards, the formula for calculating ΔR_2^* (and ΔR_2), reported in **Section 3.3**, is implemented as a matrix operation. It should be noted that most commands needed for image processing require the additional MATLAB software called Image Processing tool box. For more details regarding the commands for image processing we refer the reader to the MATLAB manual.

References

1. Folkman, J. (1992) The role of angiogenesis in tumor growth. *Semin Cancer Biol* **3**, 65–71.
2. Kusters, B., Leenders, W. P., Wesseling, P., Smits, D., Verrijp, K., Ruiter, D. J., Peters, J. P., van Der Kogel, A. J., and de Waal, R. M. (2002) Vascular endothelial growth factor-A(165) induces progression of melanoma brain metastases without induction of sprouting angiogenesis. *Cancer Res* **62**, 341–345.
3. Leenders, W. P., Kusters, B., Verrijp, K., Maass, C., Wesseling, P., Heerschap, A., Ruiter, D., Ryan, A., and de Waal, R. M. (2004) Antiangiogenic therapy of cerebral melanoma metastases results in sustained tumor progression via vessel co-option. *Clin Cancer Res* **10**, 6222–6230.
4. Leenders, W. P., Kusters, B., Pikkemaat, J., Wesseling, P., Ruiter, D., Heerschap, A., Barentsz, J., and de Waal, R. M. (2003) Vascular endothelial growth factor-A determines detectability of experimental melanoma brain metastasis in GD-DTPA-enhanced MRI. *Int J Cancer* **105**, 437–443.
5. Claes, A., Gambarota, G., Hamans, B., van Tellingen, O., Wesseling, P., Maass, C., Heerschap, A., and Leenders, W. P. (2008) Magnetic resonance imaging-based detection of glial brain tumors in mice after antiangiogenic treatment. *Int J Cancer* **122**, 1981–1986.

6. Dennie, J., Mandeville, J. B., Boxerman, J. L., Packard, S. D., Rosen, B. R., and Weisskoff, R. M. (1998) NMR imaging of changes in vascular morphology due to tumor angiogenesis. *Magn Reson Med* **40**, 793–799.
7. Le Duc, G., Peoc'h, M., Remy, C., Charpy, O., Muller, R. N., Le Bas, J. F., and Decorps, M. (1999) Use of T2-weighted susceptibility contrast MRI for mapping the blood volume in the glioma-bearing rat brain. *Magn Reson Med* **42**, 754–761.
8. Tropres, I., Grimault, S., Vaeth, A., Grillon, E., Julien, C., Payen, J. F., Lamalle, L., and Decorps, M. (2001) Vessel size imaging. *Magn Reson Med* **45**, 397–408.
9. Bremer, C., Mustafa, M., Bogdanov, A., Ntziachristos, V., Petrovsky, A., and Weissleder, R. (2003) Steady-state blood volume measurements in experimental tumors with different angiogenic burdens—a study in mice *Radiology* **226**, 214–206.
10. Gambarota, G., van Laarhoven, H. W., Philippens, M., Lok, J., van der Kogel, A., Punt, C. J., and Heerschap, A. (2006) Assessment of absolute blood volume in carcinoma by USPIO contrast-enhanced MRI. *Magn Reson Imaging* **24**, 279–286.
11. Weisskoff, R. M., Zuo, C. S., Boxerman, J. L., and Rosen, B. R. (1994) Microscopic susceptibility variation and transverse relaxation: theory and experiment. *Magn Reson Med* **31**, 601–610.
12. Kennan, R. P., Zhong, J., and Gore, J. C. (1994) Intravascular susceptibility contrast mechanisms in tissues. *Magn Reson Med* **31**, 9–21.
13. Kusters, B., Westphal, J. R., Smits, D., Ruiter, D. J., Wesseling, P., Keilholz, U., and de Waal R. M. (2001) The pattern of metastasis of human melanoma to the central nervous system is not influenced by integrin alpha(v)beta(3) expression. *Int J Cancer* **92**, 176–180.
14. Gambarota, G., Leenders, W. P., Maass, C., Wesseling, P., van der Kogel, A. J., van Tellingen, O., and Heerschap, A. (2008) Characterisation of tumour vasculature in mouse brain by USPIO contrast-enhanced MRI. *Br J Cancer* **98**, 1784–1789.
15. Enochs, W. S., Harsh, G., Hochberg, F., and Weissleder, R. (1999) Improved delineation of human brain tumors on MR images using a long-circulating, superparamagnetic iron oxide agent. *J Magn Reson Imaging* **9**, 228–232.
16. Taschner, C. A., Wetzel, S. G., Tolnay, M., Froehlich, J., Merlo, A., and Radue, E. W. (2005) Characteristics of ultrasmall superparamagnetic iron oxides in patients with brain tumors. *AJR Am J Roentgenol* **185**, 1477–1486.
17. Varallyay, P., Nesbit, G., Muldoon, L. L., Nixon, R. R., Delashaw, J., Cohen, J. I., Petrillo, A., Rink, D., and Neuwelt, E. A. (2002) Comparison of two superparamagnetic viral-sized iron oxide particles ferumoxides and ferumoxtran-10 with a gadolinium chelate in imaging intracranial tumors. *AJNR Am J Neuroradiol* **23**, 510–519.

Chapter 26

Cancer Models—Multiparametric Applications of Clinical MRI in Rodent Hepatic Tumor Model

Feng Chen, Frederik De Keyzer, and Yicheng Ni

Abstract

Small animal imaging has been a major player in an increasing amount of oncological experiments wherein magnetic resonance imaging (MRI) has become a favorite choice of measures for in vivo small animal imaging due to its advantages of excellent resolution and innocuousness. Based on a clinical MRI scanner, we propose a protocol of multiparametric MRI for noninvasive characterization and therapeutic evaluation of a rat model with implanted liver tumors. This protocol contains six sequences, namely, T_1-weighted image (T1WI), T_2-weighted image (T2WI), diffusion-weighed imaging (DWI), T_1-weighted dynamic contrast-enhanced MRI (DCE-MRI), T_2^*-weighted dynamic susceptibility contrast-enhanced MRI (DSC-MRI), and contrast-enhanced T1WI (CE-T1WI), for acquiring anatomic, diffusion, and perfusion information of tumor models. In this chapter, the details about this complete MRI protocol and the rodent liver tumor model are described in order to facilitate the readers to perform their own translational animal imaging research.

Key words: MRI, contrast enhanced, diffusion imaging, perfusion imaging, tumor models, rodents.

1. Introduction

Advances in the development of new anti-tumor strategies and preclinical evaluation of anti-tumor drugs highlight the need for animal models that are hopefully more reflective of human cancer. Rapidly growing subcutaneous (ectopic) tumor transplants have been the most favorable patterns and the most extensively studied tumor models, mainly because of their experimental convenience (1–4). However, one drawback of such traditional ectopic cancer models is the overly optimistic/exaggerated outcomes obtained due to certain factors, e.g., rapidly growing subcutaneous tumor transplants (5). Furthermore, human malignant tumors are often deeply seated in body organs and their vascular networks and microenvironment are quite different from those of subcutaneous tumors. Therefore, an appropriate and readily available animal model in visceral organs, e.g., in the liver, is crucial in oncological research (5, 6).

With its superb soft tissue contrast, excellent temporal and spatial resolutions, and its nature of noninvasiveness, magnetic resonance imaging (MRI) has been established as a tool for in vivo monitoring of morphological, functional, and metabolic changes in various experimental animal models (7). Compared to dedicated high-field-strength and small-bore animal MRI systems, the main strength of using clinical MRI scanners is the easier translation of the obtained results to the clinical settings with patients, despite their limited spatial, temporal, and spectral resolution. Since the susceptibility artifacts and motions associated with cardiac pulsation, breathing, and peristalsis can cause geometric distortion in MRI pictures and thus degrade image quality, in vivo abdominal or liver MRI are more technically challenging in small rodents (8). Rapid advances in MRI technology have been made in the past decade and many powerful and versatile MRI sequences and methods have been developed and widely applied in preclinical research (7). These developments facilitate an updated and high throughput experimental platform that comprises comprehensive MRI protocols to fully exploit the potential of these new developments, particularly in the application of abdominal visceral organs (6, 9).

We use a multiparametric MRI approach in a variety of translational imaging studies in rats. Besides the conventional anatomical imaging, i.e., T_1-weighted image (T1WI) and T_2-weighted image (T2WI), the protocol also includes functional imaging such as diffusion-weighted imaging (DWI), T_1-weighted dynamic contrast-enhanced MRI (DCE-MRI) T_2^*-weighted dynamic susceptibility contrast-enhanced MRI (DSC-MRI), and contrast-enhanced T1WI (CE-T1WI). The main advantage of using this complete set of MRI sequences is that in addition to morphological information, you can obtain not only organ hemodynamic parameters such as blood flow and blood volume but also capillary permeability parameters, e.g., K_{trans}, at a single acquisition within about 30 min. This approach has been proven feasible, reproducible, and informative by a series of studies including longitudinal monitoring of stroke model (10, 11), characterization of ischemic/reperfusion liver infarction model or liver tumor models (12, 15), and assessment of therapeutic effects of anti-vascular agents for tumor treatment (13, 14).

In this chapter, we introduce a rat liver tumor model with implanted rhabdomyosarcoma, describe the animal handling for MRI imaging and detail the multiparametric MRI protocol with six sequences that are operable at a clinical 1.5-T MRI scanner.

2. Materials

2.1. Animal Model and Anesthesia

1. WAG/Rij rats
2. Pentobarbital (Nembutal; Sanofi Sante Animale, Brussels, Belgium)
3. Isoflurane (Halocarbon, River Edge, NJ, USA)
4. Gas anesthesia system (Harvard Apparatus, Holliston, MA, USA)
5. 14-Gauge core tissue biopsy needle (Bauer Medical, Inc., Clearwater, FL, USA)
6. Tissue glue (Histoacryl; Aesculap, Tuttlingen, Germany)
7. Gauge-27 needle (Vygon, Ecouen, France)
8. Plastic animal holder homemade with an ordinary plastic bottle (**Fig. 1**)
9. Plastic animal mask homemade with an ordinary 50-mL syringe (**Fig. 1**)

2.2. MRI

1. 1.5 T clinical MR scanner (Sonata; Siemens, Erlangen, Germany)
2. Phased array wrist radio frequency coil (MRI Devices Corporation, Waukesha, WI, USA)
3. MR contrast agent: gadoterate meglumine (Dotarem®, Guerbet, France) (*see* **Note 1**)

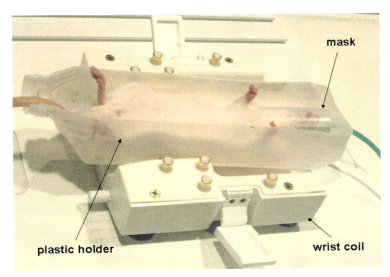

Fig. 1. MR imaging setup for rodents. A rat was placed supinely into a homemade plastic holder covered with a homemade mask connecting to an inhalation anesthesia system via a plastic tube.

3. Methods

3.1. Animal Tumor Model

A subcutaneous rhabdomyosarcoma (R1 tumor) in the flank of an adult WAG/Rij rat is used as donor tissue (*see* **Note 2**). Intrahepatic implantation of R1 is performed in male adult rats of the same strain, which have a mean weight of 250 g ± 20 (standard deviation). Rats are anesthetized with intraperitoneal injection of 40 mg of pentobarbital (Nembutal; Sanofi Sante Animale, Brussels, Belgium) per kilogram of body weight. The viable part of the donor tumor is minced into small cubes (approximately 2 mm^3). Then, a tumor tissue fragment of 2 mm^3 is loaded into a homemade tumor-seeding device or implanter (*see* **Note 3**), which is made by modifying a 14-gauge core tissue biopsy needle (Bauer Medical, Inc., Clearwater, FL, USA). Recipient rats undergo midline laparotomy and the left and right liver lobes are exposed and implanted successively. The loaded implanter should be inserted about 1 cm in the subcapsular liver parenchyma, and the pith, when fully advanced, should be about 1 mm longer than the needle tip to ensure complete discharge of the tumor tissue. Upon withdrawing the needle, a droplet of tissue glue (Histoacryl; Aesculap, Tuttlingen, Germany) is placed over the puncture site to stop bleeding instantaneously. The rats are allowed to recover for 1–2 h after closure of the abdomen with layered sutures and tumors are allowed to grow for about 14–16 days before MRI.

3.2. Animal Anesthesia and Handling

During MRI measurements, two appropriate methods can be used for rodent anesthesia, namely gas anesthesia and intraperitoneal injection of anesthetics.

1. Gas anesthesia: The rat is lying supine in a plastic holder inside a human wrist rf coil. Placed over the rat mouth, a homemade plastic mask is connected via a 10-m long polyethylene tube to a gas anesthesia system (IMS; Harvard Apparatus, Holliston, MA, USA) located outside the MRI room. Rapid initial anesthetic induction is performed with inhalation of 4% isoflurane in a mixture of 70% oxygen and 30% room air and maintenance during imaging with 2% isoflurane in a mixture of 20% oxygen and 80% room air. The flow rate of O_2 is 2 L/min for the initial phase and 0.2 L/min for the maintaining phase. The flow rate of room air is always kept at 0.8 L/min. Besides the superb safety, this method allows both rapid immobilization and quick recovery of the animals and is therefore ideal for experiments in which repeated anesthesia is needed.

2. Intraperitoneal anesthesia: Rats are injected intraperitoneally with Nembutal (sodium pentobarbital; Sanofi Sante Animale, Brussels, Belgium) at 40 mg/kg. This is a very easy

approach and can be used when only a few scanning sessions are conducted on a daily schedule (*see* **Note 4**). When necessary, body temperature can be maintained at 37.5°C and monitored with commercially available MRI-compatible monitoring system (Model 1025; Monitoring & Gating System, SAII, Stony Brook, NY, USA).

3. Vein cannulation: The penile vein or tail vein of rats is cannulated for bolus injection of contrast agent with a gauge-27 needle set (*see* **Note 5**). The cannulation is usually done after acquisition of T1WI, T2WI, and DWI sequences without change of animal's position.

3.3. Multiparametric MRI

1. MR scanner and coil: Rodent MRI is typically performed with a clinical 1.5-T whole-body system (Sonata, Siemens, Erlangen, Germany) with a maximum gradient capability of 40 mT/m. To allow parallel imaging, phased array coils such as the commercially available four-channel phased array wrist coil (MRI Devices Corporation, Waukesha, WI, USA) are used for shortened imaging acquisition time and improved signal homogeneity (*see* **Note 6**). Initially, fast sagittal, coronal and axial gradient-echo-based sequences are obtained as localizer for positioning of the subsequent MRI acquisitions (*see* **Note 7**). All sequences are performed afterwards transversely with the same slice thickness of 2 mm and interval of 0.2 mm as well as the same geometry to maintain comparability between the different imaging sequences.

2. T1WI: Fast spin echo (SE) T1WI is performed with the following parameters: repetition time /echo time (in milliseconds) (TR/TE), 535/9.2 ms; turbo factor (*see* **Note 8**), 7; field of view (FOV), 140 × 70 mm; imaging acquisition matrix, 256 × 256; two concatenations; fat saturation (*see* **Note 9**); four signals acquired; in-plane resolution, 0.5 × 0.3 mm; and total examination time, 1 min 24 s.

3. T2WI: FSE T2WI is acquired with fat saturation and the following parameters: TR/TE, 3860/106 ms; turbo factor, 19; field of view, 140 × 70 mm; acquisition matrix, 256 × 256; one concatenation; three signals acquired; and total examination time, 1 min 25 s.

4. DWI: DWI is performed with a two-dimensional SE echoplanar imaging sequence with a FOV of 140 × 82 mm and an acquisition matrix of 192 × 91 (zero-filled for reconstruction), which lead to an in-plane resolution of 0.7 × 0.9 mm. To reduce susceptibility artifacts and examination time, a parallel imaging technique (*see* **Note 10**) is applied with the following parameters: TR/TE, 1700/83 ms and six signals acquired, including repeated measurements for 10 different

b-values (0, 50, 100, 150, 200, 250, 300, 500, 750, and 1000 s/mm^2) (*see* **Note 11**) and resulting in a total examination time of 4 min 51 s. For DWI, three directions (*x*, *y*, and *z*) are measured and averaged for the calculation of the isotropic apparent diffusion coefficient (ADC) value. Postmortem DWI as an alternative evaluating measure can be done immediately after sacrifice of the animals (*see* **Note 12**).

5. DCE-MRI: DCE-MRI is acquired by using a three-dimensional T1W gradient-echo sequence (volumetric interpolated breath hold examination, VIBE) with fat saturation and the following parameters: TR/TE, 7.02/2.69 ms; matrix, 154 × 192 (zero-filled for reconstruction); acceleration factor two; field of view, 81.3 × 130 mm; voxel size, 0.5 × 0.7 × 2.0 mm; a single signal acquired; and a total acquisition time of 3.7 s for the entire volume of 16 sections This sequence is applied continuously for 80 measurements resulting a total acquisition time of 4 min 58 s. After the first 20 measurements, an intravenous bolus of gadoterate meglumine (Dotarem®; Guerbet, France) with a gadolinium concentration of 0.5 mmol/mL is administered by a manual injection at a dose of 0.04 mmol/kg over a maximum period of 2 s (*see* **Note 13**).

6. DSC-MRI: DSC-PWI is done by using a T_2^*-weighted echo planar imaging (EPI) sequence with the following parameters: TR/TE, 2000/46 ms; matrix, 128 × 128; FOV, 140 × 70 mm; in-plane resolution, 1.1 × 0.5 × 2 mm; and parallel imaging. . The dynamic sequence is repeated for 80 measurements resulting in a total scan time of 2 min 46 s (*see* **Note 14**). During the dynamic series a dose of 0.3 mmol/kg intravenous bolus of Dotarem® is started after the 20th measurement to ensure a sufficient precontrast baseline. The bolus injection is done manually in less than 2 s without saline flush (*see* **Note 15**).

7. CE-T1WI: A postcontrast FSE TSE-T1WI is started immediately after the DSC-MRI sequence. The sequence and parameters are identical to that described for T1WI (*see* **Note 16**).

3.4. Imaging Post-processing

On most current scanner setups, the on-scanner tools are sufficient for initial image analysis, as these allow simple delineations and region of interest values. However, some advanced analyses need to be performed off-line, for instance, on a LINUX workstation using dedicated software (Biomap; Novartis, Basel, Switzerland). T1WI and T2WI usually do not require further post-processing, although these are often used off-line as a template for coregistration and distortion correction of the functional images. This allows better assessment of lesion heterogeneity using the combined information of coregistered anatomical and functional images.

1. DWI
   (a) A typical DWI scan results in two sets of images. The first is the set of diffusion or trace images that present signal intensity. Another is the ADC map which is obtained by a pixel-by-pixel calculation of the ADC in the tissue using all native b-value images. It is in general more reliable to use the ADC for tissue characterization (*see* **Note 17**).
   (b) The averaged ADC map for 10 b-values is generated automatically by the built-in software of the MRI machine. To separate "both diffusion and perfusion contributions to the ADC" or so-called "intravoxel incoherent movement (IVIM)" (*see* **Note 18**), the ADC values of the regions of interest (ROIs) are obtained separately for low b-values ($b = 0$, 50, and 100 s/mm^2; ADC$_{low}$) and high b-values ($b = 500$, 750, and 1000 s/mm^2; ADC$_{high}$) from the average signal intensity (SI) per ROI and per b-value. Each ADC value is calculated by using a least squares solution of the following system of equations:

$$\text{ADC}_{low}: S_i = S_0 \times \exp(-b_i \times \text{ADC}_{low})$$
$$\text{For } i = 0, 50, 100$$
$$\text{ADC}_{high}: S_j = S_0 \times \exp(-b_j \times \text{ADC}_{high})$$
$$\text{for } j = 500, 750, 1000$$

   where S_i and S_j are the signal intensities (SI) measured on the DWI images acquired with the corresponding b-values b_i and b_j, S_0 represents the SI with b equal to 0 s/mm^2 (9).

2. DCE-MRI
   The image analysis needs to be done off-line, for instance, on a LINUX workstation using dedicated software (Biomap; Novartis, Basel, Switzerland). The functional parameters derived from DCE-MRI, i.e., initial slope (IS) and volume transfer constant k (*see* **Note 19**) can be calculated by the following steps.
   (a) Contrast enhancement–time curves (CTC) map generation. For evaluation of the DCE-MRI images, CTC maps are calculated from the perfusion images by using the following formula (14): $\text{CTC}(t_i) = [I(t_i) - I(t_0)]/I(t_0)$, where $\text{CTC}(t_i)$ is the contrast

enhancement–time curve at time point t_i, $I(t_i)$ is the SI at perfusion imaging at time point t_i, and $I(t_0)$ is the SI at baseline perfusion imaging.

(b) The arterial input function (AIF) calculation. AIF is assessed by placing an ROI in the aorta, which provides the measured signal intensity enhancement pattern in the artery. This is then normalized to the baseline value ($Cp(t_i) \sim AIF(t_i) - AIF(t_0)/AIF(t_0)$), for each time point t_i). The plasma concentration of gadoterate meglumine (Cp), needed for further quantitative evaluation of tissue perfusion, is then obtained through a fitting procedure usually according to the formula $Cp(t) = -k/TE * \ln(S(t)/S_0)$, with k a constant, TE the echo time of the sequence and $S(t)$ the signal intensity in the artery at time t, and S_0 the baseline signal intensity. For exact quantification, the constant k should be estimated; however, since the most used calculations result in relative values, this factor is usually set to unity (16).

(c) Volume transfer constant k calculation. For the ROIs in the tissue, the volume transfer constant k (in L/s) is calculated using the Tofts and Kermode model (17) according to the following formula: $dC_t/dt = -(k/V_e) \times C_t \gtrless k(1 + V_p/V_e) \times C_p + V_p (dC_p/dt)$, where C_t is the concentration of contrast agent in the tumor tissue, which is assumed to be represented by the previously determined contrast enhancement–time curve, and C_p is the concentration of contrast agent in the blood plasma, determined from the AIF delineation (*see* Step (b) above). V_e and V_p are the volumes per unit of tumor tissue belonging to the extracellular extravascular space and blood plasma, respectively (*see* **Note 20**).

(d) Mergence of AIF and CTC maps. Copy the ROI for AIF and transfer it to the CTC map, thus, the volume transfer constant k can be generated by the software.

(e) IS is obtained by calculating the slope of the contrast enhancement–time curve at the time point of maximal contrast agent inflow, which was defined as the initial slope (IS): $IS = \max[d(CTC)/dt]$.

3. DSC-MRI
Most providers add the DSC-MRI calculation tools as part of an optional neuroradiology tool package, which can be automatically included in the scanner purchase, or can be added on separately. This software usually allows placement of an AIF and definition of cutoff time points (one for the steady state, one at baseline end, and one at the end of

Table 1
A multiparametric MRI protocol for liver tumor model in rats at 1.5 T

Parameters	T1WI	T2WI	DWI	DCE-MRI	DSC-MRI
Sequence	FSE	FSE	SE-EPI	VIBE	FID-EPI
Contrast agent				Dotarem®	Dotarem®
Dose				0.04 mmol/kg	0.3 mmol/kg
TR/TE (ms)	535/9.2	3860/106	2000/46	7.02/2.69	2000/46
FOV (mm)	140 × 70	140 × 70	140 × 82	130 × 81	140 × 70
Imaging matrix	256 × 256	256 × 256	192 × 91	192 × 154	128 × 128
Slice thickness (mm)	2	2	2	2	2
Intersection gap (mm)	0.4	0.4	0.4	0.4	0.4
Voxel size (mm)	0.5 × 0.3 × 2.0	0.5 × 0.3 × 2.0	0.7 × 0.9 × 2.0	0.7 × 0.5 × 2.0	1.1 × 0.5 × 2.0
No. of averages	4	3	6	1	1
Bandwidth (Hz/pixel)	195	130	790	180	752
Total measured time (min:s)	1:24	1:25	4:51	4:58	2:46
Measurements				80	80
Time resolution (s)				3.7	2.1
b-values (s/mm²)			0, 50, 100, 150, 200, 250, 300, 500, 750, 1000		
Directions measured			Trace in x, y, z		Trace in read-out direction
EPI factor or turbo factor	7	19	91		64
Parallel imaging (GRAPPA)[a]			Yes	Yes	Yes
Acceleration factor			2	2	2
Echo spacing (ms)	9.24	11.8	1.69	1.52	1.52
Applications	Anatomy	Anatomy	ADC	k, IS	rBV, rBF, MTT, TTP, K_2

T1WI T_1-weighted imaging, *T2WI* T_2-weighted imaging, *DWI* diffusion-weighted imaging, *DCE-MRI* dynamic contrast-enhanced MRI, *DSC-MRI* dynamic susceptibility contrast-enhanced MRI, *FSE* fast spin echo, *SE-EPI* echo planar imaging, *VIBE* volumetric interpolated breath hold examination, *FID* free induced decay, *TE* echo time, *TR* repetition time, *FOV* field of view, *GRAPPA* generalized autocalibrating partially parallel acquisition, *ADC* apparent diffusion coefficient, *k* volume transfer constant or permeability surface area product per unit of tissue volume (K_{trans}), *IS* initial slope, *rBV* relative blood volume, *rBF* relative blood flow, *MTT* mean transition time, *TTP* time to peak, K_2 a permeability parameter

[a]Only applicable when phased array wrist coil (channel ≥ 4) is applied

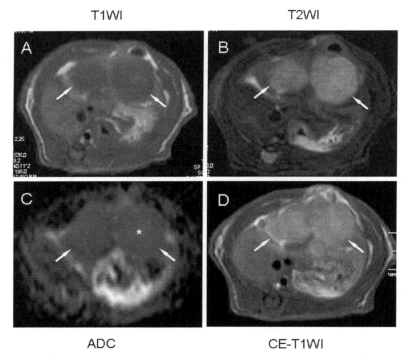

Fig. 2. MRI visualization of rhabdomyosarcoma 16 days after implantation in a Wag/Rij rat. Two tumors (*arrows*) in both right and left liver lobes appear hypointense on T_1-weighted image (T1WI) (**a**) and hyperintense on T_2-weighted image (T2WI) (**b**). Apparent diffusion coefficient map (ADC) (**c**) shows slight hyperintensity in the center of tumors indicating potential necrosis (*). Contrast-enhanced T1WI (CE-T1WI) shows hyperintense tumors after i.v. injection of contrast agent (**d**).

contrast passage) and then automatically calculates pixelwise relative blood volume (rBV), relative blood flow (rBF), time to peak (TTP), and mean transit time (MTT) maps. Further explanation is based on the software provided on the Siemens MRI scanners (Syngo platform), but similar methods are used in most available software programs.

(a) Determination of AIF curve. During post-processing, a 75-mm² ROI—which is further divided into 49 pixels (each 1.24 × 1.24 mm)—is placed at the region covering the aorta and the hepatic artery to measure the AIF. On the panel, pixels that are representative of the aorta or hepatic artery branch are selected as the final defined AIF curve for further calculation of hemodynamic parameters.

(b) Generation of parameter maps. With use of the built-in software, DSC-MRI-derived parameter maps of relative blood flow (rBF), relative blood volume (rBV), time to peak (TTP), and mean transition time (MTT) in both

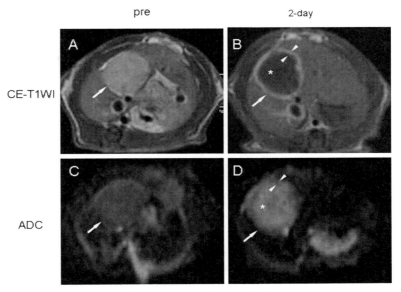

Fig. 3. MR images acquired in an upper section cranially in the same animal as seen in **Fig. 2**. Compared to pre-treatment (**a**, **c**), tumor (*arrow*) shows extensive necrosis (*) with a residual viable rim (*arrow heads*) on both CE-T1WI (**b**) (necrosis: hypointense signal; rim: hyperintense signal) and ADC map (**d**) (necrosis: hyperintense signal; rim: slightly hypointense signal) 2 days after treatment with a vascular disrupting agent (CA-4P).

tumoral and normal hepatic tissue are derived automatically with arbitrary units (*see* **Note 21**). Local dynamic mean SI curves are obtained by placing the ROI at targeted areas.

(c) Generation of permeability-weighted maps. A permeability index, K_2 (*see* **Note 22**), from tumor and normal liver regions can also be obtained during a single DSC-MRI acquisition apart from other parameter maps by using dedicated Biomap software.

3.5. Examples

Some examples of liver R1 tumor model in rats evaluated by multiparametric MRI are shown with imaging acquisition parameters as summarized in **Table 1**. An adult WAG/Rij rat with R1 tumors implanted in the liver is presented as shown in **Figs. 2–6** with multiparametric MRI obtained before and 2 days after a vascular disrupting agent, CA-4P, treatment.

3.6. Conclusion

Animal experiments remain an indispensable part of oncological research, in which small animal imaging is more and more playing an important role (7). Although several imaging modalities exist, MRI becomes the favorite choice of measures in a large amount

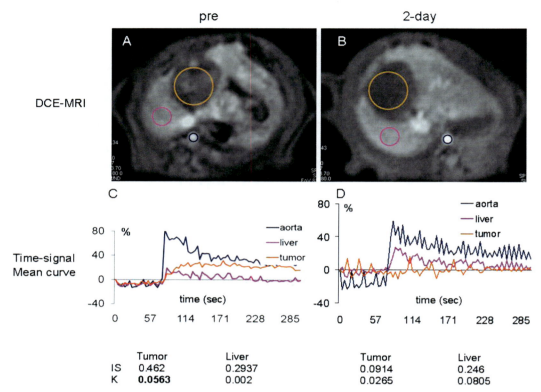

Fig. 4. T_1-weighted dynamic contrast-enhanced MRI (DCE-MRI) acquired in the same animal and same section as showed in **Fig. 2**. Before treatment with vascular disrupting agent CA-4P (**a, c**), the amplitudes of the tumor time–signal curve (*orange line* in **c** and **a**) are higher than that of liver parenchyma (*pink line* in **c** and **a**) indicating an elevated perfusion. After treatment (**b, d**), the tumor curve becomes flat (*orange line* in **d** and **b**) showing little perfusion due to the vascular shutdown. *Blue line*: time–signal curve for aorta in **d**. The indented shape of the curves may be attributed to motion artifacts. The initial slope (IS) and volume transfer constant (*k*) in the tumor are greatly reduced after treatment obtained with an off-line Biomap software.

of in vivo imaging experiments on small animals due to its advantages of excellent spatial and temporal resolutions, superb soft tissue contrast, and the apparent innocuousness. Based on a clinical MRI scanner, we have proposed a comprehensive MRI protocol to noninvasively evaluate liver-implanted tumors in a rat model. Our previous studies indicate that this tumor model has a 100% success rate of implantation and is suitable for MRI studies. MRI characterization and therapeutic evaluation in this liver tumor model involve morphologic, functional, and metabolic information. This proposed protocol has proven feasible and reproducible (6). Thus, the use of this research setup including the liver tumor model and the comprehensive MRI protocol may provide an upgraded platform for animal research in translational oncology. Upon this approach, a series of preclinical studies can be facilitated for the assessment of new diagnostic and therapeutic strategies (12, 13, 15).

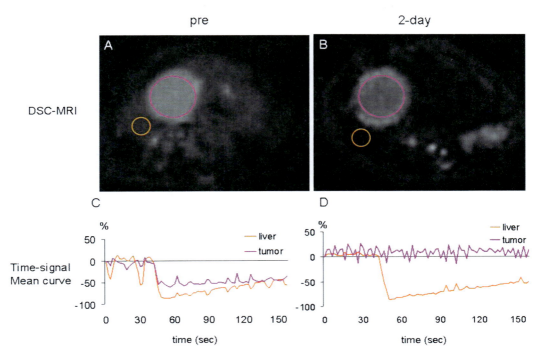

Fig. 5. T_2^*-weighted dynamic susceptibility contrast-enhanced MRI (DSC-MRI) acquired in the same animal and same section as showed in **Fig. 2**. Before treatment with vascular disrupting agent CA-4P (**a, c**), the amplitudes of tumor time–signal curve (*pink line* in **c** and **a**) are smaller than that of liver parenchyma (*orange line* in **c** and **a**) because of more perfusion induced by dual blood supply in the liver. After treatment (**b, d**), the tumor curve becomes flat (*pink line* in **d** and **b**) showing little perfusion due to the vascular shutdown, while the liver curve is virtually unchanged (*orange line* in **d** and **b**).

4. Notes

1. The gadoterate meglumine is one of the representatives of extracellular fluid (ECF) space agents, other common ECF contrast agents include gadopentetate dimeglumine (Magnevist®; Bayer Schering Pharma, Germany) and gadodiamide injection (Omniscan™; GE Healthcare, USA).

2. Rhabdomyosarcoma (R1) in rats is among the most widely used rodent tumor models in cancer research for its biological stability, minimal immunogenicity, low metastatic potential, and responsiveness to various therapeutic interventions (18).

3. The use of the tumor implanter has eased the procedure, reduced the complications such as tumor spreading or liver damage, and raised the success rate (18).

4. Always keep in mind the risk of "deadly easy yet easily dead" of this anesthetic regime using pentobarbital due to

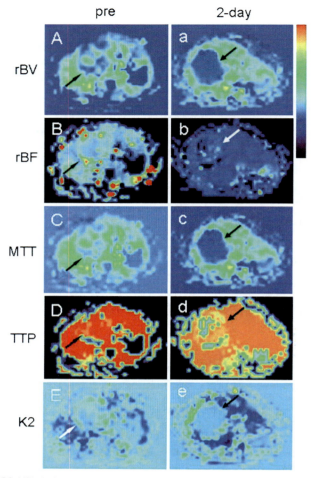

Fig. 6. DSC-MRI-derived parameter maps acquired in the same animal and same section as showed in **Fig. 2**. These maps are generated off-line with a Biomap software before (**A–E**) and after treatment (**a–e**). After CA-4P treatment, all parameters including the relative blood volume (rBV), relative blood flow (rBF), mean transit time (MTT), time to peak (TTP), and permeability parameter K_2 are reduced due to vascular shutdown and extensive necrosis in the tumor (*arrow*) compared to pre-treatment.

the very narrow margin between effective and lethal dose of this drug.

5. It is relatively easier to perform a bolus injection via penile vein compared to tail vein for a better first-pass perfusion effect.
6. Small single element surface loop coils may also be used for a high local signal-to-noise ratio (SNR) and favorable cost-effectiveness. However, the parallel imaging can only be acquired with multichannel phase array coils.
7. In clinical systems, automated scanner protocols are used, with automatic adjustments of receiver bandwidth and

magnetic field homogeneity according to the shim box. Size and positioning of this shim box is very important in small animals. If the shim box is too large, the field inside the shim box cannot be homogenized, completely leading to suboptimal signal homogeneity. If, however, the shim box is made too small, the entire animal values will not be sampled and consequently bad choices in receiver bandwidth are made. Ideally, a shim box for small animals should cover as much of the animal as possible, while at the same time avoiding too much air or areas with susceptibility differences, usually air–tissue boundaries. For small animal imaging, especially when looking at functional imaging, manual shimming is strongly recommended.

8. The turbo factor is the number of echoes acquired after each excitation. This is a measure of the scan time acceleration, e.g., at turbo factor 3 the scan time is three times as fast as an SE sequence with comparable parameters.

9. Any type of fat saturation is allowed in small animal imaging. A first type is spectral fat saturation, which is based on the different Larmor frequencies of fat and water protons. In spectral fat suppression (usual provider acronyms FATSAT, CHEMSAT, CHESS), before the normal excitation pulse, an extra excitation pulse is added at the frequency of fat, after which a strong spoiler gradient is used to destroy the fat signal. A second type is based on using a complete inversion pulse at the start of the sequence. The actual excitation pulse is then applied at the zero-crossing of the fat signal, yielding only the signal of non-fatty tissues. This is called short TI inversion recovery (STIR). Advanced fat saturation prepulses, used especially in higher field strengths combine both ideas into a spectral inversion pulse at the frequency of fat with the actual excitation starting again at the zero-crossing of the fat signal (methods such as SPIR, SPAIR). In small animal imaging research, especially at 1.5 T, with well-designed local coils and with a dedicated shim box (*see* previous note) any type of fat saturation will be sufficient to suppress nearly all fat signal, so we advocate using the spectral version, such as FATSAT, as this requires the least amount of extra scan time.)

10. Any type of parallel imaging technique can be used to speed up acquisition while at the same time reducing susceptibility artifacts. There are two types of parallel imaging used nowadays, both based on acquiring undersampled images with each coil element, depending on the coil sensitivities. The first performs the Fourier transform on these undersampled data, and the resulting images with fold-in artifacts are then combined to create an artifact-free image;

this is called the sensitivity-encoding approach (SENSE or mSENSE (modified SENSE), depending on the provider). The second manner is not based on the reconstructed images, but rather attempts to fill in the skipped lines in the undersampled k-space from the information obtained using the different coil element measurements; this is the approach of generalized autocalibrating partially parallel acquisition (GRAPPA) as used in our protocol.

11. The b-value is a factor of diffusion-weighted sequences. The b-factor summarizes the influence of the gradients on the diffusion-weighted images. The higher the value b, the stronger the diffusion weighting. For accurate assessment of proton diffusion in the tissue, care must be taken when choosing the b-values in the DWI. To reduce influence of noise on the calculations, DWI with at least three different b-values are selected. To separate both diffusion and perfusion contributions to the ADC and reduce the effects of respiratory movement, a large range of b-values is needed (e.g., 10 b-values in the range of 0–1000 s/mm^2) in the same setting (4, 19, 20) (*see also* **Section 2.4** imaging postprocessing).

12. To overcome the problems associated with motion and susceptibility artifacts with in vivo DWI acquisition, postmortem DWI can be performed immediately after animal sacrifice, because studies have shown that the postmortem DWI can reflect the true in vivo diffusion properties in different tissues (21).

13. A previous study showed that low doses (0.04 mmol/kg) of a gadolinium-based contrast agent provide maximal signal enhancement with the same sequence introduced in this chapter (14).

14. The extra 6 s (2 min 46 s minus the 80 timepoints with 2 s per timepoint (TR)) are just preparation of the sequence (such as gradients for timing of external equipment). They are not prescans to get to a steady-state condition. There are in total 20 baseline scans, of which the first two or so induce the steady-state signal, but these are usually discarded afterwards. The real baseline values are then calculated as the average signal intensities of the remaining 18 baseline scans.

15. Exact amount of contrast agent can be injected by loading additional 0.2 mL contrast agent in a 1-mL syringe to fill an equal volume of a Gauge-27 cannulation tube. Thus, saline flush is not needed.

16. Since the DSC-MRI sequence lasts about 3 min, the succeeding CE-T1WI sequence is actually acquired during the equilibrium phase of contrast agent in the liver.

17. Due to the underlying T_2-weighted contrast in native DWI images, lesions with long T_2 relaxation times can also remain bright on higher b-value images, which may make it difficult to know whether they represent an area of restricted diffusion or not. This phenomenon of "T_2 shine-through" (22) map can be easily recognized on the ADC map, since only tissues with true restricted diffusion present a low ADC value on the ADC map.

18. IVIM is used to designate microscopic translational motions that occur in each voxel in MRI (19, 23), which contains both diffusional (reflected on ADC at high b-values) and perfusional (reflected on ADC at low b-values) partitions.

19. The volume transfer constant k is the permeability surface area product per unit of tissue volume. This value is now generally known as K_{trans}; however, because of small deviations in absolute value due to the assumptions made, we continue to use the denomination k (14).

20. This is based on the assumption that the permeability of tumor vessels for contrast agent flow from intravascular to extravascular extracellular space is identical to their permeability for flow from extravascular extracellular to intravascular space. As V_p and V_e have not been used commonly in the literature, only the volume transfer constant k is calculated here (14).

21. For each tissue pixel, a concentration–time curve is established, usually using $C_m(t) = -k/\text{TE} * \ln(S(t)/S_0)$, with k a constant, TE the echo time of the sequence and $S(t)$ the signal intensity at time t, with S_0 the baseline signal intensity. The TTP is defined as the time between the end of baseline and the maximal contrast enhancement (C_{max}). Estimates of the MTT are either based on calculating the full width at half maximum (FWHM) of $C_m(t)$, or by using the area to height relation $\text{MTT}=(\int C_m(t)dt)/C_{max}$. The amount of blood in a given volume of tissue (rBV) is then calculated as the area under the measured concentration curve (C_m), normalized to the AIF as follows: rBV = constant * $(\int C_m(t)dt/\int \text{AIF}(t)dt)$, where the constant is the same for every pixel in the image and is usually disregarded for relative blood volume maps. Finally, the rBF follows from the formula rBF = rBV/MTT. A good overview is provided for brain applications by Rempp et al. (24).

22. The K_2 map (6), the map of a permeability parameter similar to K_{trans} as described in **Note 19**, is generated to process T_2^*-weighted data with the method described by Weisskoff et al. (25) and further validated by Ostergaard et al. (26) and Provenzale et al. (27).

Acknowledgments

This work was partially supported by National Natural Science Foundation of China project 30670603 and an EU higher education project Asia-Link CfP 2006-EuropeAid/123738/C/ACT/Multi-Proposal No. 128-498/111.

This document has been produced with the financial assistance of the European Union. The contents of this document are the sole responsibility of Katholiek Universitat of Leuven and can under no circumstances be regarded as reflecting the position of the European Union.

References

1. Antoine E., Pauwels C., Verrelle P., Lascaux V., and Poupon M. F. (1988) In vivo emergence of a highly metastatic tumour cell line from a rat rhabdomyosarcoma after treatment with an alkylating agent. *Br J Cancer* **57**, 469–474.
2. Hermens A. F. and Barendsen G. W. (1967) Cellular proliferation patterns in an experimental rhabdomyosarcoma in the rat. *Eur J Cancer* **3**, 361–369.
3. Denekamp J. (1992) The choice of experimental models in cancer research: the key to ultimate success or failure? *NMR Biomed* **5**, 234–237.
4. Thoeny H. C., De Keyzer F., Chen F., et al. (2005) Diffusion-weighted MR imaging in monitoring the effect of a vascular targeting agent on rhabdomyosarcoma in rats. *Radiology* **234**, 756–764.
5. Kerbel R. S. (2008) Tumor angiogenesis. *N Engl J Med* **358**, 2039–2049.
6. Chen F., Sun X., De Keyzer F., et al. (2006) Liver tumor model with implanted rhabdomyosarcoma in rats: MR imaging, microangiography, and histopathologic analysis. *Radiology* **239**, 554–562.
7. Brockmann M. A., Kemmling A., and Groden C. (2007) Current issues and perspectives in small rodent magnetic resonance imaging using clinical MRI scanners. *Methods* **43**, 79–87.
8. Inderbitzin D., Stoupis C., Sidler D., Gass M., and Candinas D. (2007) Abdominal magnetic resonance imaging in small rodents using a clinical 1.5 T MR scanner. *Methods* **43**, 46–53.
9. Chen F., De Keyzer F., Wang H., et al. (2007) Diffusion weighted imaging in small rodents using clinical MRI scanners. *Methods* **43**, 12–20.
10. Chen F., Suzuki Y., Nagai N., et al. (2004) Visualization of stroke with clinical MR imagers in rats: a feasibility study. *Radiology* **233**, 905–911.
11. Chen F., Suzuki Y., Nagai N., et al. (2007) Microplasmin and tissue plasminogen activator: comparison of therapeutic effects in rat stroke model at multiparametric MR imaging. *Radiology* **244**, 429–438.
12. Wu X., Wang H., Chen F., et al. (2009) Rat model of reperfused partial liver infarction: characterization with multiparametric magnetic resonance imaging, microangiography, and histomorphology. *Acta Radiol* **50**, 276–287.
13. Wang H., Sun X., Chen F., et al. (2009) Treatment of rodent liver tumor with combretastatin A4 phosphate: noninvasive therapeutic evaluation using multiparametric magnetic resonance imaging in correlation with microangiography and histology. *Invest Radiol* **44**, 44–53.
14. Thoeny H. C., De Keyzer F., Vandecaveye V., et al. (2005) Effect of vascular targeting agent in rat tumor model: dynamic contrast-enhanced versus diffusion-weighted MR imaging. *Radiology* **237**, 492–499.
15. Wang H., Van de Putte M., Chen F., et al. (2008) Murine liver implantation of radiation-induced fibrosarcoma: characterization with MR imaging, microangiography and histopathology. *Eur Radiol* **18**, 1422–1430.
16. Carroll T. J., Rowley H. A., and Haughton V. M. (2003) Automatic calculation of the arterial input function for cerebral perfusion

imaging with MR imaging. *Radiology* **227**, 593–600.
17. Tofts P. S. (1997) Modeling tracer kinetics in dynamic Gd-DTPA MR imaging. *J Magn Reson Imaging* **7**, 91–101.
18. Ni Y., Wang H., Chen F., et al. (2009) Tumor models and specific contrast agents for small animal imaging in oncology. *Methods* **48**, 125–138.
19. Le Bihan D., Breton E., Lallemand D., Aubin M. L., Vignaud J., and Laval-Jeantet M. (1988) Separation of diffusion and perfusion in intravoxel incoherent motion MR imaging. *Radiology* **168**, 497–505.
20. Yamada I., Aung W., Himeno Y., Nakagawa T., and Shibuya H. (1999) Diffusion coefficients in abdominal organs and hepatic lesions: evaluation with intravoxel incoherent motion echo-planar MR imaging. *Radiology* **210**, 617–623.
21. Sun X., Wang H., Chen F., et al. (2009) Diffusion-weighted MRI of hepatic tumor in rats: comparison between in vivo and postmortem imaging acquisitions. *J Magn Reson Imaging* **29**, 621–628.
22. Burdette J. H., Elster A. D., and Ricci P. E. (1999) Acute cerebral infarction: quantification of spin-density and T2 shine-through phenomena on diffusion-weighted MR images. *Radiology* **212**, 333–339.
23. Le Bihan D. (2008) Intravoxel incoherent motion perfusion MR imaging: a wake-up call. *Radiology* **249**, 748–752.
24. Rempp K. A., Brix G., Wenz F., Becker C. R., Gückel F., and Lorenz W. J. (1994) Quantification of regional cerebral blood flow and volume with dynamic susceptibility contrast-enhanced MR imaging. *Radiology* **193**, 637–641.
25. Weisskoff R., Boxerman J. L., Sorensen A. G., Kulke S. M., Campbell T. A., and Rosen B. R. (1994) Simultaneous blood volume and permeability mapping using a single Gd-based contrast injection. In: *Proceedings of the Society for Magnetic Resonance*. Berkeley, CA: Society for Magnetic Resonance, Vol. 1, p. 279.
26. Ostergaard L., Hochberg F. H., Rabinov J. D., et al. (1999) Early changes measured by magnetic resonance imaging in cerebral blood flow, blood volume, and blood-brain barrier permeability following dexamethasone treatment in patients with brain tumors. *J Neurosurg* **90**, 300–305.
27. Provenzale J. M., Wang G. R., Brenner T., Petrella J. R., and Sorensen A. G. (2002) Comparison of permeability in high-grade and low-grade brain tumors using dynamic susceptibility contrast MR imaging. *AJR Am J Roentgenol* **178**, 711–716.

Section VIII

Functional MRI

Chapter 27

BOLD MRI Applied to a Murine Model of Peripheral Artery Disease

Joan M. Greve

Abstract

Peripheral artery disease (PAD) is the narrowing or complete occlusion of vessels due to the progression of atherosclerosis. Ultimately, the reduction in blood supply, due to a reduced lumen diameter, results in a functional deficit, e.g., reduced mobility. Because function is closely tied to blood flow through large-caliber vessels, therapeutic development to treat PAD has recently focused on arteriogenesis rather than angiogenesis. Optimally, the preclinical investigations related to such therapeutic development would take place in murine models of PAD to allow for future studies utilizing transgenic strains. However, it can be challenging to quantify functional recovery of the peripheral vascular network in murine models. The purpose of this work is to provide a protocol of temporally and spatially resolved methods for functional assessment of arteriogenesis in a murine model.

Key words: BOLD, reactive hyperemia, arteriogenesis, VEGF, mouse.

1. Introduction

Adequate blood supply, provided by bulk-flow arterial supply routes connecting to capillary beds that deliver nutrients and oxygen, enables proper function of tissues throughout the body. In the case of peripheral artery disease (PAD), the progression of atherosclerosis results in narrowing or occlusion of arteries, most commonly those carrying blood to the lower limbs (1). Initially, even with a reduction in lumen diameter, no functional deficit is evident while the individual remains at rest. However, upon initiating activity, an imbalance between metabolic demand and blood supply develops and the person will suffer pain in the lower leg, termed intermittent claudication. Critical limb ischemia, inadequate blood supply causing pain even at rest, can occur when the diameter of the artery is reduced to approximately 25% of its original value (1).

In an attempt to restore proper function to a diseased network, two compensatory remodeling mechanisms, angiogenesis and arteriogenesis, proceed spontaneously. Angiogenesis is the de novo formation of capillaries from an existing capillary network, is driven primarily by hypoxia-mediated vascular endothelial growth factor (VEGF) (2), and enhances local delivery of nutrients and oxygen (3). Arteriogenesis is the enlargement of pre-existing arterioles into collateral arteries, is largely influenced by hemodynamic forces, and is required to restore function (3, 4). Although survival of ischemic tissue may be attributed to angiogenesis providing sufficient perfusion acutely, successful therapeutic treatment of PAD, as measured by restoration of original function, is thought to require large-vessel remodeling and re-establishment of adequate bulk flow, i.e., arteriogenesis (3, 4).

The development path of any therapeutic must pass through animal models prior to application in humans. Due to the ease with which its genome can be manipulated to examine the contribution of specific genes or molecules to normal and pathological states, the animal model of choice is the mouse. Ideally, the techniques used to assess functional recovery in a murine model of PAD would satisfy three criteria:

1. Be analogous to exercise yet independent of extrinsic factors, e.g., mouse strain (5).

2. Provide temporally and spatially resolved data at the muscle bed.

3. Hold the potential for translation to the clinic.

Reactive hyperemia (RH) can fulfill priorities 1 and 3. RH is a physiological reaction where blood supply increases dramatically following a temporary arterial occlusion. It represents the capacity of the vascular network to provide additional flow when demand increases. RH has been utilized clinically (6–8) where it has been shown to be reproducible (9, 10) and capable of distinguishing diseased from healthy individuals (11) and localizing disease in the vasculature (6).

Blood-oxygen-level-dependent (BOLD) MRI can fulfill priorities 2 and 3. BOLD MRI results in images whose signal intensity is determined by the intravascular ratio of oxy- to deoxyhemoglobin and extravascular tissue–vessel susceptibility effects (12). Depending on which MR pulse sequence is chosen, the contribution of each component can be emphasized (12). Unlike the more common methods used to quantify RH (strain-gauge plethysmography and Doppler flowmetry which reflect bulk and superficial blood flow, respectively), BOLD MRI can provide temporally and spatially resolved data at the muscle bed of interest. The BOLD effect in muscle has been substantiated in the clinic (13–16), but has yet to be utilized to investigate pathology.

Analogous to the, perhaps, more familiar task-based activation BOLD MRI studies performed in the brain (17, 18), we use RH as a stimulus to increase blood flow to skeletal muscle in order to enlarge the dynamic range of the response to be measured by the signal intensity in our BOLD MR images acquired from the murine hindlimb. This protocol describes how to induce RH while the animal remains in the bore of the magnet by developing a system that allows remote, transient occlusion. Combining RH and BOLD MRI makes it possible to:

- Reproducibly acquire data between animals and studies performed over nearly 1 year.

- Develop an assay sensitive enough to discern differences even between sham-operated 10-week-old C57Bl6 (YNG), 9-month-old C57Bl6 (OLD), and 9-month-old atherosclerotic (DKO; $Ldlr^{-/-}Apobec^{-/-}$ (19)) mice (*see* **Note 1**).

- Quantify differences between the spontaneous functional recovery in YNG, OLD, and DKO mice following creation of an arterial insufficiency, showing that inhibition of VEGF, age, and the presence of atherosclerosis prevent complete functional restoration but likely for different reasons.

- Provide further evidence that VEGF plays a significant role in arteriogenesis and functional recovery (20, 21).

2. Materials

2.1. General Surgery Materials

1. Lab coat

2. Mask (for animal allergen control, preferably, N95 9211/37022; 3 M, St Paul, MN)

3. Surgical microscope (e.g., Leica M651; less expensive benchtop versions work as well)

4. Heating pad set to 37°C (e.g., homeothermic blanket control unit; Harvard Apparatus, Holliston, MA) (*see* **Note 2**)

5. Isoflurane (*see* **Note 3**)

6. Surgical instruments
   a. Large forceps, delicate serration (e.g., Adson dressing forceps, 4-3/4″ (12.1 cm); Miltex part no. 6-118; York, PA)
   b. Large forceps, deep serration (e.g., Brown-Adson tissue forceps, 7 × 7 teeth, 4-3/4″ (12.1 cm); Miltex part no. 6-124; York, PA)

      c. Large surgical scissors, straight, blunt (e.g., Fine Science Tools part no. 14501-14; Foster City, CA)

      d. Small surgical scissors, straight (e.g., Strabismus scissors, 4″ (10.2 cm); Miltex part no. 5-312; York, PA)

      e. Small forceps, half-curved, extra delicate serration (e.g., eye dressing forceps, 4″ (10.2 cm); Miltex part no. 18-781; York, PA)

      f. Small forceps, angled, smooth platform (e.g., McPherson micro iris suturing forceps, 3-1/2" (8.9 cm); Miltex part no. 18-949; York, PA)

      g. Needle holder, fenestrated jaws (e.g., Collier; Miltex part no. 8-2; York, PA)

7. Clippers or depilatory (e.g., Nair)
8. Medium alcohol prep pads (e.g., Professional Disposables International, part no. b33901; Orangeburg, NY)
9. Sterile cotton-tipped applicators (e.g., Puritan, Guilford, ME)
10. Sterile triangular surgical spears (e.g., Ultracell-Aspen Surgical, Caledonia, MI)
11. 5-0 Silk suture (e.g., Ethicon, Inc., Somerville, NJ, USA)
12. Sterile saline.

2.2. Femoral Artery Ligation Surgery

The additional materials listed here are required to ligate the femoral artery.

1. Hair cover
2. Povidone–iodine swabsticks (e.g., Professional Disposables International, part no. s41350; Orangeburg, NY)
3. Ophthalmic ointment (e.g., Phoenix Pharmaceutical, St. Joseph, MO)
4. Sterile surgeon's gloves (sized appropriately; e.g., Henry Schein, Melville, NY)
5. Small sterile drape with aperture (e.g., Steri-Drape, 40 × 40 cm, ID 6.3 cm; 3 M, part no. 1020; St Paul, MN)
6. Syringe (1–3 mL) with sterile saline
7. Wound clip kit (7 mm, RF7-KIT; Braintree Scientific Inc.; Braintree, MA)
8. Buprenorphine or other suitable post-operative analgesic (*see* **Note 4**)

2.3. Reactive Hyperemia Surgery

The additional materials listed here are required to perform the surgery that will enable transient and reversible occlusion of the aorta and vena cava in the magnet.

1. Isoflurane
2. Nylon fishing line (6 lb and 14 lb test)

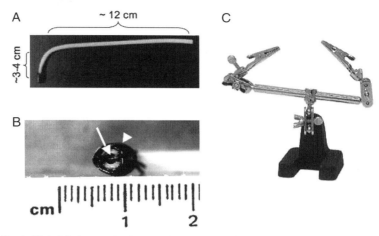

Fig. 1. Materials for reactive hyperemia surgery. (**a**) Occlusion tube. (**b**) Bisected end of occlusion tube through which the 6-lb test fishing line is threaded and which is placed into the intraperitoneal (IP) cavity of the animal (*arrow*: thin strips of electrical tape; *arrow head*: circumferentially wrapped electrical tape covered in epoxy; *see* **Note 14**). (**c**) Soldering stand helper with alligator clips.

    3. Occlusion tube (PE 280, semi-rigid but formable by hand, length ~12 cm, bent portion ~3–4 cm; Intramedic; Becton-Dickinson, Franklin Lakes, NJ; **Fig. 1a** and **b**)

    4. Electrical tape

    5. Waterproof epoxy

    6. Soldering stand helper with alligator clips (**Fig. 1c**).

    7. Occlusion weight, 6–10 g (*see* **Note 5**)

2.4. BOLD MRI

The additional materials listed here are required to induce reactive hyperemia in the lower leg of the mouse while it remains in the magnet and acquire and process imaging data.

1. Physiological monitoring and control (rectal temperature probe at minimum; ideally respiration pad too but may be difficult due to placement of occlusion tube; *see* **Note 6**).

2. Eye ointment not required because of non-survival procedure.

3. Lubricating jelly.

4. 3-cm Inner diameter transmit–receive volume radiofrequency (RF) coil for proton imaging (*see* **Note 7**).

5. Plastic dowel long enough to run the extent of the magnet's bore (**Fig. 3**).

6. Cross bar over which occlusion line and occlusion weight can be run (**Fig. 3**).

7. Scope trigger or some other means to provide a signal to personnel who will be lowering and raising occlusion weight.

8. Fast spin echo (FSE) sequence providing adequate temporal resolution, ≤10 s per image (*see* **Notes 8** and **9**)

9. Image analysis software (MATLAB, including the Image Processing Toolbox; The Mathworks, Natick, MA; or MRVision, Winchester, MA)

3. Methods

All animal work should be carried out *only* upon review and approval of the methods by your institution's Institutional Animal Care and Use Committee (IACUC). For those new to small animal surgery and/or these procedures, instruction should be sought from experts in the field prior to initiating any studies (22).

3.1. Femoral Artery Ligation Surgery

Aseptic technique is required for this procedure (lab coat, mask, hair cover; sterile instruments, sterile gloves, and a sterile field). If not already proficient, obtain training in aseptic technique from experienced personnel.

The following can be done with non-sterile gloves and preferably in an area separate from where the surgery will take place. If the latter is not possible, then be careful to remove fur thoroughly from the surgical area prior to starting the surgery and maintain sterility of instruments and field.

1. Anesthesia should be induced at ~3% isoflurane in a carrier gas flowing at 1 L/min (medical grade air, or 70/30 mixture of N_2O/O_2 will work but are listed here in order of preference).

2. Upon induction, place animal on the heating pad ventral surface up with its nose in a nose cone to continue to deliver isoflurane at ~1.5–2% in 1 L/min of the carrier gas. The animal's head should be pointing away from you.

3. Before proceeding, check for the appropriate plane of anesthesia (e.g., by pinching one of the hindpaws of the animal).

4. Opthalmic ointment should be placed on each eye.

5. The fur over the left femoral artery at the level of the inguinal ligament should be removed by using clippers or depilatory (*see* **Note 10**)—externally the landmark is where the hindlimb attaches to the intraperitoneal (IP) cavity (**Fig. 2**).

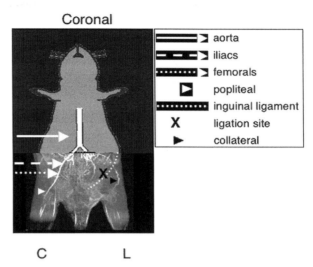

Fig. 2. Murine anatomy following femoral artery ligation as visualized by an MR angiogram. Various levels of the vascular tree are highlighted along with the femoral artery ligation site: ▬▬▷ aorta; ▬■▬▷ iliacs; ●●●●●▷ femorals; ▷ popliteal; ▪▪▪▪▪▪▪ inguinal ligament; X ligation site; ▶ collateral. (Adapted and reproduced from (20) with permission from John Wiley & Sons, Inc.)

6. Prepare the skin over the femoral artery by scrubbing it with three separate povidone–iodine swabsticks followed by one to two alcohol pads. Let the skin dry.

At this point, transfer the animal to the surgery area and put on sterile gloves. Place the sterile drape over the animal so that the surgical aperture is over the furless skin.

7. Using the large, delicate serration forceps and small surgical scissors, make a small incision (~1 cm in length) through the skin, only, over the femoral artery at the level of the inguinal ligament. The incision should be parallel to the inguinal ligament (see **Fig. 2**).

8. You will need to move aside a small portion of subcutaneous fat before visualizing the femoral artery, which will reside between the femoral vein on the left and the femoral nerve on the right.

9. Have a small syringe (1–3 mL) filled with sterile saline available to moisten the surgical site as you are separating the femoral artery from surrounding tissue.

10. Isolate the femoral artery from the vein using the small, angled, smooth platform forceps (see **Note 11**).

11. When the femoral artery has been isolated, run a 5-0 silk suture beneath it and tie off the artery. A drop or two of saline can provide lubrication. Cut the ends of the ligature to appropriate lengths (~2 mm) (see **Note 12**).

12. Close the skin using the needle holders, 5-0 silk suture, and a wound clip. The wound clip should be removed after 7 days if it is still in place.
13. Administer appropriate post-operative analgesic, e.g., buprenorphine (0.05–0.1 mg/kg subcutaneously).
14. Recover the animal in a cage, with no bedding, that is placed half-on/half-off a warm heating pad until ambulatory.

3.2. Inhibition of VEGF

Administration of such injections can be done with non-sterile gloves and without anesthesia. If not already proficient, obtain training in administering injections via the intraperitoneal (IP) cavity from experienced personnel.

1. Gently restrain animal by holding the skin at the nape of the neck between thumb and forefinger with one hand and tail with the other hand.
2. Turn animal so its abdominal surface is facing toward the ceiling.
3. Place the animal's tail between the fourth and fifth fingers (ring and pinky) of the hand that is also holding the skin at the nape of the neck between thumb and forefinger.
4. Inject animal just proximal to the inguinal ligament with needle entering at an approximately 45° angle relative to abdominal surface. The needle should be pointed superiorly (*see* **Note 13**).

3.3. Reactive Hyperemia Surgery

You will need to construct the occlusion tube (*see* **Note 14**) at least 24 h in advance of performing this surgery to allow for the epoxy to dry. This surgery is performed just prior to performing the BOLD MRI. The surgical area needs to be close to the preparation area for imaging as the animal will need to be quickly moved from the surgical bench to imaging preparation area without its regaining consciousness.

Non-sterile technique is allowed because this is a non-survival procedure. However, best practice suggests care should be taken to avoid contaminating the surgical site, for example, with fur.

1. Repeat Steps 1–3 of the femoral artery ligation surgery.
2. The fur on the midline of the abdomen, from approximately the sternum to the level of the inguinal ligament, should be removed using clippers or depilatory.
3. Using the large, delicate serration forceps and small surgical scissors, a small incision (~2–3 cm in length) through the skin only, should be made parallel to the aorta.
4. You will now see the ventral musculature of the intraperitoneal (IP) cavity. Again, make a small incision (~2–3 cm in length) through the muscle wall (*see* **Note 15**).

5. Your field of view will now largely be encompassed by the intestinal tract. You will need to move aside the intestinal tract before visualizing the aorta and vena cava.

6. Have a small syringe (1–3 mL) filled with sterile saline available to moisten the surgical site as you are separating the aorta and vena cava from surrounding tissue.

7. Do not separate the aorta and vena cava from each other. Instead, using the small, angled, smooth platform forceps and/or the triangular surgical spears, separate the pair from the surrounding tissue (*see* **Note 16**).

8. A drop or two of saline should be used to provide lubrication during this procedure. When there is a clear tunnel beneath both the aorta and vena cava, take a length of 6-lb test fishing line (approximately double the length of your occlusion tube) and run it beneath the vessels until half is protruding from each side.

9. Thread both ends of the 6-lb test line through the occlusion tube and tie it off.

10. Secure the occlusion tube in one of the alligator clips on the soldering stand helper (**Fig. 1c**). The long portion of the occlusion tube should be pointing toward the head of the mouse (**Fig. 3**).

11. Gently manipulate the soldering stand arm such that the surface of the occlusion tube is ~5 mm above the aorta and vena cava. Make sure not to put pressure on the vessels thereby blocking flow.

12. Using the 5-0 suture, suture the muscle and skin layer closed around the occlusion tube.

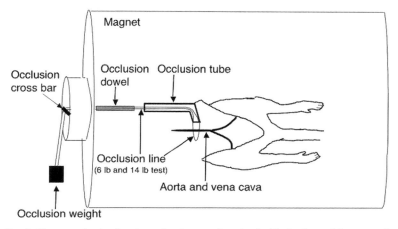

Fig. 3. Diagram of animal and reactive hyperemia setup inside the bore of the magnet. The design allowed transient occlusion and reflow of the aorta and vena cava while acquiring data continuously. (Adapted and reproduced from (21) with permission from John Wiley & Sons, Inc.)

13. When the surgery is complete, elevate the anesthetic to 2–3% for ~3 min in anticipation of transferring the animal to the imaging preparation area where a nose cone should be located with anesthetic already flowing via a carrier gas.

14. Transfer the animal to the imaging preparation area.

3.4. BOLD MRI

Once the animal has been transferred to the imaging preparation area, check once again that the correct plane of anesthesia has been maintained. Once that has been confirmed, you will likely want to return the isoflurane to ~1.5–2%.

1. The rectal temperature probe should be inserted using lubricating jelly and taped in place.

2. Both the animal's hindlimbs should be gently straightened and taped in this position (*see* **Note 17**).

3. The end of the occlusion tube outside of the RF coil should be supported so that it remains parallel to the body of the mouse thereby minimizing the possibility of the occlusion line becoming taut beneath the aorta and vena cava.

4. Tie the 14-lb test line (long enough to reach through the extent of the bore of the magnet) to the 6-lb test line running through the occlusion tube.

5. Place the animal inside the RF coil. If you have an RF coil with its active extent toward one end of the coil, only the animal's hindlimbs up to the occlusion tube will be inside the RF coil (*see* **Note 7**).

6. Tape the 14-lb test line to a plastic dowel long enough so that it can be easily reached from the other end of the magnet. There should be a small extent of occlusion line between the end of the dowel and the end of the occlusion tube. You will need to make sure that this dowel is supported at both ends to keep it in line, i.e., at the same height, with the occlusion tube. This will minimize the possibility of tension being put on the 6-lb test occlusion line that is running beneath the aorta and vena cava (**Fig. 3**).

7. a. When you place the setup (RF coil, animal, fishing line, and dowel) into the magnet, the fishing line should be long enough to be draped over the occlusion cross bar and the occlusion dowel should be long enough to be easily reached inside the bore of the magnet in order to actively push the line back through the occlusion tube and reflow the aorta and vena cava, *see* **Fig. 3**. The occlusion dowel does not need to slide into the occlusion tube to reflow the vessels.

b. Secure the occlusion weight to the end of the fishing line. If you have used a weight made of an MR-compatible material, simply set the weight on the inside of the bore of the magnet until ready to lower the weight and occlude vessels.

8. Proceed with the normal steps to prepare for image acquisition, e.g., tuning/matching of RF coil, shimming, setting the frequency, calibrating the power of the RF pulses.

9. To acquire a single transverse slice located 3 mm below the tibial plateau (**Fig. 4a**), begin by acquiring a set of scout images (*see* **Note 18**). The coronal plane is likely to be the most useful at first. Ideally, you will be able to visualize the knee joint, and hence the tibial plateau in one of these images. If that is not possible, it may be necessary to perform a sagittal, multi-slice acquisition and then plan another set of coronal slices off of the sagittal scout images.

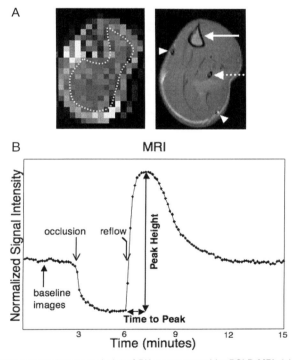

Fig. 4. Representative image and plot of RH as measured by BOLD MRI. (**a**) Representative image and overlaid ROI (*dotted line*) used for quantification (*left*) and image of the anatomy of the lower limb (*right: solid arrow*, tibia; *dashed arrow*, fibula; *arrowheads*, blood vessels). (**b**) Representative reactive hyperemia plot as measured using the fast spin echo sequence. Each RH acquisition included 3 min of baseline, 3 min of occlusion, and 9 min of reflow. The peak height (PH) of the RH response and the time to peak (TTP) were used to quantify the RH response. (Adapted and reproduced from (21) with permission from John Wiley & Sons, Inc.)

10. Once you are able to visualize the knee joint in your scout images and plan an axial slice 3 mm distal from the tibial plateau, you are ready to acquire data with the fast spin echo sequence (*see* **Note 8**).

11. In a single acquisition, you will want to acquire a set of images prior to occlusion (baseline images) and throughout occlusion and reflow (**Fig. 4b**) (*see* **Note 19**).

12. Using a trigger output from the MR scanner and displayed on a scope that can be seen from the end of the magnet where the occlusion weight resides, accurate occlusion and reflow times can be achieved (*see* **Notes 20** and **21**).

13. *Rapid visualization* of changes in signal intensity (SI) post acquisition consisted of: drawing a region of interest (ROI) that encompassed musculature while avoiding the tibia on the first image of the dataset (*see* **Fig. 4a**, dotted line overlaid on *left* image). This ROI is propagated through the dataset resulting in a plot of mean SI (averaged over the ROI) as a function of time (MRvision, Winchester, MA) (*see* **Fig. 4b**). The mean SI of the first 20 baseline images is averaged. Subsequently, the SI at each time-point was normalized to this baseline average (Microsoft Excel, Redmond, WA) (*see* **Note 22**).

14. Post acquisition *analysis* of changes in SI consisted of: a trained operator using the anatomy apparent in the first image together with the function *roipoly* from the MATLAB Image Processing Toolbox to define the ROI (*see* **Fig. 4a**, dotted line overlaid on *left* image). The ROI was saved as a bitmap and each pixel was identified as either "in" or "out" of the ROI. Subsequent Matlab code then read in that ROI bitmap (using *load*) together with the full dataset (using *fopen*) and used it for normalizing and computing the hyperemic curves. Peak height (PH) and time to peak (TTP) of the hyperemic curve were quantified using the *max* function in MATLAB (The Mathworks, Natick, MA) (*see* **Notes 23** and **24**). Various methods for smoothing the hyperemic curves prior to computing the PH and TTP were investigated, but did not produce results substantially different than those obtained by applying the *max* function to the unsmoothed curves.

15. The reproducibility of the RH and BOLD MRI methodology is illustrated in **Fig. 5**. It is highly reproducible between animals (**Fig. 5a**) and between studies that span nearly 1 year (**Fig. 5b**).

16. The sensitivity of the RH and BOLD MRI methodology is highlighted in **Figs. 6** and **7**. Differences are discernible even between sham operated YNG, OLD, and DKO mice.

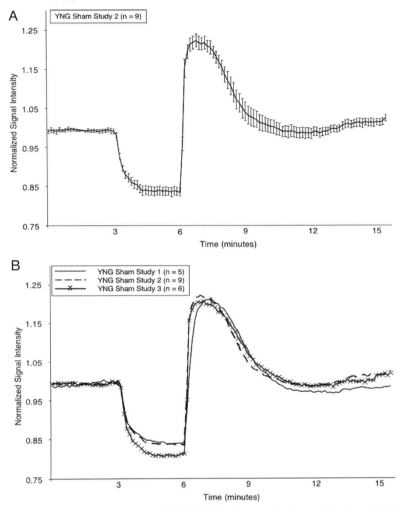

Fig. 5. Reproducibility of RH and BOLD MRI between animals and studies. (**a**) RH response for 10-week-old C57Bl6 mice who underwent sham surgery (mean ± std error). (**b**) Average RH response for 10-week-old C57Bl6 mice that underwent sham surgery in three separate studies that spanned approximately 1 year.

17. The RH and BOLD MRI assay is also able to quantify the differential effects that inhibition of VEGF, age, and the presence of atherosclerosis have on spontaneous functional recovery following creation of an arterial insufficiency. Each results in incomplete functional recovery but likely for different reasons. Inhibition of VEGF (*see* **Note 13**) results in a significant reduction in PH (**Figs. 6a** and **7a**), suggesting a compromised nitric oxide (NO)-dependent response. Age results in a significant increase in TTP (**Figs. 6b** and **7b**), suggesting a residual increase in vascular resistance. The presence of atherosclerosis (*see* **Note 1**) results in a

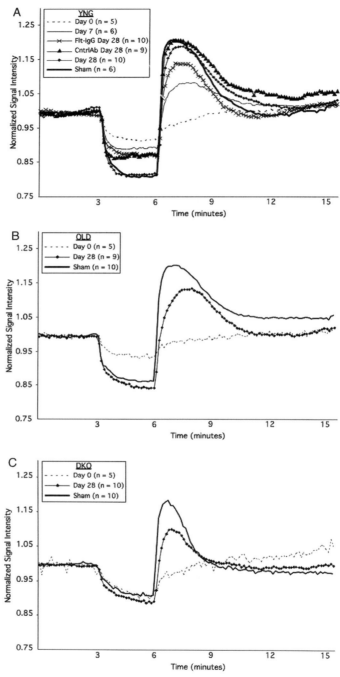

Fig. 6. Functional recovery as measured by RH and BOLD MRI. (**a**) Average RH response for 10-week-old C57Bl6 mice after femoral artery ligation (day 0, day 7, day 28), sham operation (sham), and femoral artery ligation followed by intraperitoneal administration of a soluble VEGF inhibitor or control antibody (Flt-IgG and ContrlAb, respectively). (**b**) Average RH response for 9-month-old C57Bl6 mice after femoral artery ligation (day 0, day 28) and sham operation (sham). (**c**) Average RH response for 9-month-old Ldlr/Apobec/ mice after femoral artery ligation (day 0, day 28) and sham operation (sham). (Reproduced from (21) with permission from John Wiley & Sons, Inc.)

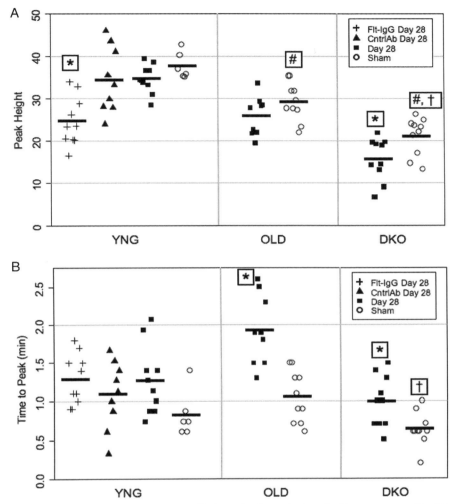

Fig. 7. Statistical analysis of functional recovery at 28 days after femoral artery ligation as measured by RH and BOLD MRI. (a) Peak height: By day 28, no statistically significant difference was detected between YNG C57Bl6 mice and their respective sham controls. However, application of Flt-IgG significantly reduced PH (*) compared with all other groups. By day 28, no statistically significant difference was detected between OLD C57Bl6 mice and their respective sham controls; however, OLD sham animals had a significantly reduced PH compared with the YNG sham group (#). DKO mice 28 days after femoral artery ligation had a significantly reduced PH (*) compared with their respective sham group. DKO mice also had a significantly diminished peak height compared with YNG (#) and OLD (†) sham groups. (b) Time to peak: All groups of young C57Bl6 mice were statistically indistinguishable. Old C57Bl6 mice had a significantly slower TTP (*) at day 28 after femoral artery ligation compared with the respective sham animals. At day 28, DKO mice had a significantly slower time to peak (*) compared with sham operated controls. Furthermore, DKO sham mice had a significantly faster time to peak compared with OLD sham animals (†). *$P < 0.05$ compared with sham animals within the same group; #$P < 0.05$ compared with young sham animals; †$P < 0.05$ compared with old sham animals. (Reproduced from (21) with permission from John Wiley & Sons, Inc.)

reduced PH and increased TTP (**Figs. 6c** and **7a** and **b**), implying both a compromised NO-dependent response and increased vascular resistance. The multi-parametric nature of the RH curve provides insight not attainable by bulk or superficial flow measurements (*see* **Note 24**).

4. Notes

1. Transgenic mice with only a mutation in the LDL receptor Ldlr$^{-/-}$ have only mildly elevated LDL cholesterol and little atherosclerosis. Ldlr$^{-/-}$Apobec$^{-/-}$ mice were developed as a model of human familial hypercholesterolemia and provide a means for studying atherosclerosis. They were bred on a wild-type C57Bl6 background and do not develop extensive atherosclerosis until approximately 6–9 months of age. Hence, the appropriate control group to assess the contribution of age, independent of the presence of atherosclerosis, is an age matched cohort of C57Bl6 mice. A cohort of young C57Bl6 mice was used for assessment of the spontaneous recovery that was known to occur in young animals and to evaluate the contribution of VEGF to this recovery.

2. Consider using a rectal temperature probe, inserted using lubricating jelly, during the femoral artery ligation surgery to confirm that the heating pad set to 37°C results in a similar core temperature.

3. When proficient, the femoral artery ligation surgery will take 20–30 min. Inhalation anesthetic is preferable because it allows quick recovery and avoids inadvertent death due to re-dosing.

4. Buprenorphine is a controlled substance requiring registration with the Drug Enforcement Agency and use must be documented. Possible alternatives which are not controlled substances include carprofen and meloxicam.

5. It is necessary to use a weight that is made of a material that is MR compatible, i.e., will not be pulled into the magnet, e.g., the lead fishing weight. This may at first seem like a lot of weight, but you need to factor in the necessity to keep the occlusion dowel and line from slackening back through the occlusion tube until reflow is desired.

6. At the time this work was performed, we were using a homebuilt heating and monitoring system with no respiratory component. Since then, we have switched to using a commercially available system (SA Instruments Inc., Stony Brook, NY).

7. Ideally, the active region of the RF coil would be toward one end so that the occlusion tube does not have to fit inside the RF coil. A larger ID coil may be necessary if the active region of the coil is not toward one end.

8. Sequence parameters for fast spin echo acquisition: Field of view, $(3 \text{ cm})^2$; slice thickness, 1.5 mm; matrix, 64×64; repetition time (TR) = 1 s; effective echo time (TEeff) = 80 ms; echo train length = 16; echo spacing = 10 ms; number of excitations (NEX) = 2.

9. Alternatively you could use a gradient echo sequence: Field of view, $(3 \text{ cm})^2$; slice thickness, 1 mm; matrix, 128×64; TR = 30 ms; TE = 20 ms; NEX = 4. We have utilized both gradient echo and fast spin echo sequences. Although not explored extensively, results/conclusions were not different between the two acquisitions. When using the gradient echo sequence, we shimmed on the slice of interest until the mean $T2^*$ across the gastrocnemius was ≥ 16 ms. Because reliable field optimization was difficult due to the small area of tissue (which was especially true with older animals that have a higher fat content) and a high-field magnet and long echo times necessary to accentuate the subtle differences under investigation, we ultimately chose the fast spin echo sequence.

10. If using depilatory, immediately after placing depilatory on the animal take cotton-tipped applicators and start gently spreading the depilatory around. Depilatory action will begin to occur almost immediately. To avoid irritation to the skin, it is necessary to be attentive and remove the depilatory expeditiously followed by a rinse with sterile saline.

11. Care is requisite when separating the femoral artery from the vein; the vein is thin-walled and can easily be punctured. Also of note should be the femoral nerve that will be a white shiny string(s) to the right of the artery. Damaging the nerve will result in the animal not recovering ambulation properly following surgery and therefore would need to be euthanized.

12. Sham-operated animals undergo all the same procedures as ligated animals but instead of tying off the ligature after it is threaded below the femoral artery, it is removed from beneath the vessel and Steps 12–14 are completed.

13. At the time this work was performed, we did not have copious amounts of a murine antibody to VEGF (23). Instead, we used Flt(1–3)-IgG (Flt-IgG; constructed in-house). Flt-1 is a transmembrane VEGF receptor found on the surface of endothelial cells. Flt-IgG consists of the first three

extracellular domains of Flt-1, which are responsible for binding VEGF, spliced to an IgG domain. The result is a fusion protein forming a soluble receptor that can sequester endogenous murine VEGF expressed after ischemic injury. YNG mice, which received Flt-IgG, were pre-treated with 25 mg/kg 24 h prior to surgery and daily thereafter with 10 mg/kg via IP injection. Many antibodies can be administered IP and it is an easier route of administration to master compared to the tail vein, for example.

14. The PE 280 tubing is easily formable by hand or with some warmth provided by a hairdryer. Bend the PE 280 tubing so the long end is ~12 cm and the shorter portion is ~3–4 cm. You will need to cut very thin strips of electrical tape (width ~1 mm); this is best achieved using a razor blade. Drape 3 or 4 of these thin strips across the opening of the occlusion tube and secure with electrical tape wrapped around the circumference of the tube. Apply epoxy over the circumferentially wrapped electrical tape to provide some resistance to it loosening due to exposure to biological fluids. The tube can be rinsed thoroughly and re-used without strict sterilization due to this being a non-survival procedure.

15. You will need to gently pull up on the musculature with the forceps before making the incision. This will prevent accidentally cutting into any organs within the IP cavity in which case the animal would need to be euthanized.

16. Care will need to be taken to avoid puncturing small vessels that run dorsally and irritating any surrounding nerves (again these will be white, shiny strings).

17. Due to the diminutive size of the mouse, any bend in the leg causes significant over lap in anatomy when being imaged. Taping both legs extended, and hopefully in the same plane as each other, will allow you to use the contralateral leg as an internal control for how well the RH procedure was performed in the magnet. RH curves from the contralateral leg should appear nearly identical to the sham-operated RH curves in **Figs. 5** and **6**.

18. A spin-echo sequence with a relatively short TR and TE, FOV = $(3 \text{ cm})^2$, slice thickness = 1 mm, number of slices = 5–7, and gap = 0 or 1 mm is a good place to start.

19. Three minutes of occlusion likely results in maximal hemoglobin desaturation and vasodilation (16) and therefore might be expected to be the maximum length of time necessary for occlusion to maximize signal intensity decrease during occlusion and signal intensity increase

during reflow. The reflow period should be at least twice as long as the occlusion time.

20. Since occlusion will not start at the beginning of the imaging sequence (similarly for reflow), you will not be able to simply use the noise produced by the scanner to know when to lower the occlusion weight. Most scanners have an output in the form of a BNC port on the back of the console electronics from which a trigger signal can be relayed to a scope at time of occlusion and reflow. A minor modification to the pulse sequence may be necessary.

21. While learning the RH and BOLD MRI techniques, one way to evaluate how well the transient and remote occlusion and reflow of the aorta and vena cava is being performed is to use 3D time of flight MR angiography. The following parameters should result in an acquisition time of ~4 min: FOV = $(3\text{ cm})^3$, TR = 15 ms, TE = 2.6 ms, matrix 128^3 (this can probably be reduced to reduce acquisition time while retaining enough resolution to see the hindlimb vessels), NEX = 1, slab thickness = 20 mm. Acquire an angiogram prior to occlusion—you should see the vessels of the lower limb similar to the contralateral side in **Fig. 2**; lower the occlusion weight and acquire another angiogram—you should see no vessels; raise the weight and acquire a third angiogram—you should now see the vessels again.

22. SI normalized = SIi/mean SI1–20 (SIi = mean SI, averaged over the ROI, at a given time-point; mean SI1–20 = mean signal intensity of the first 20 images).

23. PH = change in normalized signal intensity between the final image prior to reflow and the image that had the highest signal intensity during the RH response. TTP = the time between these images (**Fig. 4b**).

24. Peak height is thought to be an indication of nitric oxide (NO)-mediated vasodilation (10), whereas the time to peak is weighted toward a combination of total vascular resistance (microvasculature and large arteries) and compliance (11).

References

1. Cooke, J. P. (1997) The pathophysiology of peripheral arterial disease: rational targets for drug intervention. *Vasc Med* **2**, 227–230.
2. Ferrara, N. and Bunting, S. (1996) Vascular endothelial growth factor, a specific regulator of angiogenesis. *Curr Opin Nephrol Hypertens* **5**, 35–44.
3. van Royen, N., Piek, J. J., Buschmann, I., Hoefer, I., Voskuil, M., and Schaper, W. (2001) Stimulation of arteriogenesis; a new concept for the treatment of arterial occlusive disease. *Cardiovasc Res* **49**, 543–553.
4. Scholz, D., Ziegelhoeffer, T., Helisch, A., Wagner, S., Friedrich, C., Podzuweit, T., and

Schaper, W. (2002) Contribution of arteriogenesis and angiogenesis to postocclusive hindlimb perfusion in mice. *J Mol Cell Cardiol* **34**, 775–787.

5. Lerman, I., Harrison, B. C., Freeman, K., Hewett, T. E., Allen, D. L., Robbins, J., and Leinwand, L. A. (2002) Genetic variability in forced and voluntary endurance exercise performance in seven inbred mouse strains. *J Appl Physiol* **92**, 2245–2255.

6. Van De Water, J. M., Indech, C. D., Indech, R. B., and Randall, H. T. (1980) Hyperemic response for accurate diagnosis of arterial insufficiency. *Arch Surg* **115**, 851–856.

7. Van den Brande, P., and Welch, W. (1988) Diagnosis of arterial occlusive disease of the lower extremities by laser Doppler flowmetry. *Int Angiol* **7**, 224–230.

8. Vogelberg, K. H., Helbig, G., and Stork, W. (1988) Doppler sonographic examination of reactive hyperemia in the diagnosis of peripheral vascular disease. *Klin Wochenschr* **66**, 970–975.

9. Kragelj, R., Jarm, T., and Miklavcic, D. (2000) Reproducibility of parameters of postocclusive reactive hyperemia measured by near infrared spectroscopy and transcutaneous oximetry. *Ann Biomed Eng* **28**, 168–173.

10. Dakak, N., Husain, S., Mulcahy, D., Andrews, N. P., Panza, J. A., Waclawiw, M., Schenke, W., and Quyyumi, A. A. (1998) Contribution of nitric oxide to reactive hyperemia: impact of endothelial dysfunction. *Hypertension* **32**, 9–15.

11. Wahlberg, E., Line, P. D., Olofsson, P., and Swedenborg, J. (1994) Correlation between peripheral vascular resistance and time to peak flow during reactive hyperaemia. *Eur J Vasc Surg* **8**, 320–325.

12. Silvennoinen, M. J., Clingman, C. S., Golay, X., Kauppinen, R. A., and van Zijl, P. C. (2003) Comparison of the dependence of blood R2 and R2* on oxygen saturation at 1.5 and 4.7 Tesla. *Magn Reson Med* **49**, 47–60.

13. Donahue, K. M., Van Kylen, J., Guven, S., El-Bershawi, A., Luh, W. M., Bandettini, P. A., Cox, R. W., Hyde, J. S., and Kissebah, A. H. (1998) Simultaneous gradient-echo/spin-echo EPI of graded ischemia in human skeletal muscle. *J Magn Reson Imaging* **8**, 1106–1113.

14. Lebon, V., Brillault-Salvat, C., Bloch, G., Leroy-Willig, A., and Carlier, P. G. (1998) Evidence of muscle BOLD effect revealed by simultaneous interleaved gradient-echo NMRI and myoglobin NMRS during leg ischemia. *Magn Reson Med* **40**, 551–558.

15. Lebon, V., Carlier, P. G., Brillault-Salvat, C., and Leroy-Willig, A. (1998) Simultaneous measurement of perfusion and oxygenation changes using a multiple gradient-echo sequence: application to human muscle study. *Magn Reson Imaging* **16**, 721–729.

16. Toussaint, J. F., Kwong, K. K., Mkparu, F. O., Weisskoff, R. M., LaRaia, P. J., Kantor, H. L., and M'Kparu, F. (1996) Perfusion changes in human skeletal muscle during reactive hyperemia measured by echo-planar imaging. *Magn Reson Med* **35**, 62–69.

17. Norris, D. G. (2006) Principles of magnetic resonance assessment of brain function. *J Magn Reson Imaging* **23**, 794–807.

18. Hoehn, M. (2003) Functional magnetic resonance imaging. In: van Bruggen, N. and Roberts, T. (eds.), *Biomedical Imaging in Experimental Neuroscience*. CRC Press LLC, Boca Raton.

19. Powell-Braxton, L., Veniant, M., Latvala, R. D., Hirano, K. I., Won, W. B., Ross, J., Dybdal, N., Zlot, C. H., Young, S. G., and Davidson, N. O. (1998) A mouse model of human familial hypercholesterolemia: markedly elevated low-density lipoprotein cholesterol levels and severe atherosclerosis on a low-fat chow diet. *Nat Med* **4**, 934–938.

20. Greve, J. M., Chico, T. J., Goldman, H., Bunting, S., Peale, F. V., Jr., Daugherty, A., van Bruggen, N., and Williams, S. P. (2006) Magnetic resonance angiography reveals therapeutic enlargement of collateral vessels induced by VEGF in a murine model of peripheral arterial disease. *J Magn Reson Imaging* **24**, 1124–1132.

21. Greve, J. M., Williams, S. P., Bernstein, L. J., Goldman, H., Peale, F. V., Jr., Bunting, S., and van Bruggen, N. (2008) Reactive hyperemia and BOLD MRI demonstrate that VEGF inhibition, age, and atherosclerosis adversely affect functional recovery in a murine model of peripheral artery disease. *J Magn Reson Imaging* **28**, 996–1004.

22. National Research Council. (1996) *Guide for the Care and Use of Laboratory Animals*. National Academy Press, Washington, D.C.

23. Liang, W. C., Wu, X., Peale, F. V., Lee, C. V., Meng, Y. G., Gutierrez, J., Fu, L., Malik, A. K., Gerber, H. P., Ferrara, N., and Fuh, G. (2006) Cross-species vascular endothelial growth factor (VEGF)-blocking antibodies completely inhibit the growth of human tumor xenografts and measure the contribution of stromal VEGF. *J Biol Chem* **281**, 951–961.

Chapter 28

Manganese-Enhanced Magnetic Resonance Imaging

Susann Boretius and Jens Frahm

Abstract

Manganese-enhanced magnetic resonance imaging (MEMRI) relies on contrasts that are due to the shortening of the T_1 relaxation time of tissue water protons that become exposed to paramagnetic manganese ions. In experimental animals, the technique combines the high spatial resolution achievable by MRI with the biological information gathered by tissue-specific or functionally induced accumulations of manganese. After in vivo administration, manganese ions may enter cells via voltage-gated calcium channels. In the nervous system, manganese ions are actively transported along the axon. Based on these properties, MEMRI is increasingly used to delineate neuroanatomical structures, assess differences in functional brain activity, and unravel neuronal connectivities in both healthy animals and models of neurological disorders. Because of the cellular toxicity of manganese, a major challenge for a successful MEMRI study is to achieve the lowest possible dose for a particular biological question. Moreover, the interpretation of MEMRI findings requires a profound knowledge of the behavior of manganese in complex organ systems under physiological and pathological conditions. Starting with an overview of manganese pharmacokinetics and mechanisms of toxicity, this chapter covers experimental methods and protocols for applications in neuroscience.

Key words: Manganese-enhanced magnetic resonance imaging (MEMRI), manganese, toxicity, activity-induced MEMRI, brain function, cellular layer.

1. Introduction

The bivalent cationic form of manganese (Mn^{2+}) has a long history as a paramagnetic contrast agent in NMR and MRI (1, 2). Mn^{2+} shortens both the T_1 and T_2 relaxation times of nearby protons as, for example, found in water molecules within the intra- and extracellular compartments of biological tissues. In NMR, the range of Mn^{2+} applications comprises studies of water exchange through cell membranes (3), binding to macromolecules and three-dimensional (3D) conformation of proteins (4). In MRI, a chelate complex of Mn^{2+}, $Mn(II) N,N'$-dipyridoxylethylenediamine-N,N'-diacetate-5,5'-bis(phosphate) (Mn-DPDP), finds clinical applications for liver imaging (5–7). Most importantly, however, the application of free (not complex-bound) Mn^{2+} has gained increasing interest in MRI studies of experimental animals, where the approach is usually referred to as manganese-enhanced MRI (MEMRI). MEMRI combines the access to high spatial resolution achievable by MRI with the biological information resulting from tissue-specific or functionally induced accumulations of Mn^{2+} that present as hyperintensities in T_1-weighted images due to the T_1 shortening of affected protons.

Although recent MEMRI applications have expanded into the field of cardiac imaging, for example, see (8), this chapter focuses entirely on the use of Mn^{2+} in neuroscience. The majority of studies of living animals may be classified into three different categories: (1) The differential cellular uptake and cerebral distribution of Mn^{2+} offers in vivo analyses of the neuroaxonal cytoarchitecture. (2) The biochemical similarity of Mn^{2+} to calcium (Ca^{2+}) is exploited for functional assessments of brain activity. (3) The axonal transport of Mn^{2+} is used for the identification of neuronal tracts and the establishment of axonal connectivities.

1.1. Metabolism and Physiological Importance of Manganese

Depending on its concentration and state of oxidation, manganese is both an essential trace element and potential toxicant. As a cofactor of several enzymes, manganese plays a role in the regulation of the cell energy, immune response, blood sugar homeostasis, blood clotting, reproduction, digestion, skeletal system development, and bone growth (9). It is also important for the development and function of the brain. In rats, for example, a deficiency of manganese has been shown to lead to an increased susceptibility to seizures (10).

Several important enzymes employ or even fully depend on manganese. In particular, this applies to manganese superoxide dismutase (Mn-SOD), which is an intracellular antioxidant metallo-enzyme. It is localized in the mitochondrial matrix and converts two superoxide anions into hydrogen peroxide and oxygen (11, 12). Mn-SOD concentrations are threefold higher in gray matter than in white matter (13). Manganese is also a cofactor of the pyruvate carboxylase and the arginase. Furthermore, Mn^{2+} (besides Mg^{2+}) can serve as an activator of glutamine synthetase (GS), which catalyzes the conversion of glutamate (Glu), adenosine triphosphate (ATP), and ammonia into glutamine (Gln). In the brain, this enzyme is localized in astrocytes (14–16), which take up Glu from the synaptic cleft where it has been released from neurons during synaptic transmission. After conversion of Glu into Gln via GS, Gln is channeled back into neurons where it can be used to resynthesize Glu.

Under balanced nutrition, manganese deficiency is rare. Approximately 1–5% of the ingested manganese is absorbed across the intestinal wall. Within the blood, manganese is mainly contained in erythrocytes (17, 18). Within plasma, more than 80% of Mn^{2+} (the primary form of manganese in the blood) are bound to gamma-globulin and albumin (19, 20), whereas Mn^{3+} is bound to the iron-carrying plasma protein transferrin (21, 22). About 6.4% of Mn^{2+} in plasma is assumed to exist as hydrated ions, 5.8% as a complex with HCO^{3-}, 2% bound to citrate, and 1.8% bound to other low-molecular-weight substances (22). In normal rats, the total manganese plasma concentration is about 0.53 μM (23). Manganese is absorbed from the gut and transported to the liver via the portal vein, where the biliary secretion becomes the major pathway for its excretion from the body (19).

There are three ways for manganese to enter the brain: (1) Uptake via the nose is accomplished by the olfactory epithelium, which is then followed by a transport along the olfactory nerve to the olfactory bulb (24, 25). (2) Uptake from the capillaries of the choroid plexus is performed by crossing the blood–cerebrospinal fluid barrier (BCB) into the cerebrospinal fluid (CSF). From the ventricular system, manganese diffuses into the brain, because of the absence of a barrier between CSF and brain parenchyma. (3) Direct uptake from the bloodstream into the brain parenchyma takes place by crossing the blood–brain barrier (BBB). The two pathways from the blood utilize at least partially different mechanisms. While Mn^{2+} transport via the BCB into the ventricle is not saturated up to a plasma concentration of at least 78 mM, the transport via the BBB becomes saturated with a cortical concentration at half of the maximum transport rate of 0.94 μM (26). At physiological plasma concentrations, Mn^{2+} is primarily transported across the endothelial cells of the brain capillaries, whereas at high concentrations, for example after bolus injection, uptake is mainly via the choroid plexus (26, 27).

Transporters responsible for manganese transport across the BBB are still under investigation. So far, no unique mammalian manganese transporter has been identified. Most likely, manganese uptake into the brain utilizes both facilitated diffusion and active transport processes (26–29). **Table 1** summarizes the currently considered transport systems.

After reaching the brain or more precisely the interstitial fluid, manganese can enter neurons via calcium channels (30, 31). Furthermore, the transferrin–transferrin receptor system (32), the divalent metal transporter (DMT1) (33), and the dopamine transporter (DAT) (12) may be involved in neuronal uptake of manganese. In the absence of strongly elevated manganese levels, astrocytes serve as the major storage site and homeostatic regulator in the brain (34, 35). About 80% of the normal brain manganese is associated with cardiac astrocyte-specific GS (15). Intracellular Mn^{2+} is sequestered by the Ca^{2+} uniporter of the

Table 1
Mn^{2+} transport across the BBB and proteins relevant for brain uptake

Transport system	References
Divalent metal transporter (DMT1, also known as NRAMP-2 or DCT-1)	(33, 197–199)
Transferrin–transferrin receptor system	(21, 28, 80, 200)
Voltage-gated calcium channels	(44)
Store-operated calcium channels	(201)
Choline transporter	(202)
Homeric purinoreceptor	(203, 204)

mitochondria, where over 97% of the Mn^{2+} is bound to membrane and matrix proteins (36). Mn^{2+} is transported out of the mitochondria via a very slow Na(+)-independent efflux mechanism (37).

Throughout the brain, manganese is heterogeneously distributed. In normal rats, manganese concentrations in increasing order have been observed in the cortex, striatum, hippocampus, and cerebellum ranging from 7.2 to 10.9 µmol/kg wet weight of tissue (10, 38). Mn^{2+} ions are axonally transported (39) and—similar to Ca^{2+}—they are released into the synaptic cleft upon reaching a synapse and taken up by Ca^{2+} channels on the postsynaptic membrane (40, 41).

While the influx of manganese into the brain is at least partially mediated by carriers, no such way has been found for its efflux from the brain. Perfusion studies have shown that the rate of manganese efflux is consistent with its estimated diffusion rate (42), so that repeated exposure to manganese leads to an accumulation over time. The half-life time of manganese in the brain appears to depend on the applied dose. For low-dose exposures a half-life time of 51–74 days was reported for rats (43) and more than 100 days for monkeys (44). On the other hand, the exposure to high doses was found to be accompanied by a shorter persistence in the brain (45).

1.2. Toxicity of Manganese

Although manganese is essential for the normal functioning of a variety of physiological processes, this is only true for concentrations in the nanomolar range. Acute and chronic exposure to high concentrations is toxic and can result in a permanent neurological disorder known as manganism. Reported symptoms in human and nonhuman primates include Parkinson's disease-like disturbances of motor function such as tremor, rigidity, bradykinesia, and clumsiness as well as cognitive and behavioral dysfunction (29).

At high concentrations manganese rapidly accumulates in myocardial tissues, so that acute manganese intoxication compromises cardiovascular function with the risk of acute heart failure (46). The mechanism of this toxic outcome seems to be related to the interaction of Mn^{2+} with cardiac calcium channels (8). Acute manganese exposure can also lead to irritation of the eyes, the skin, and the respiratory system (material safety data sheets (MSDS), Sigma-Aldrich). **Table 2** summarizes the acute LD50 values for mice and rats for different routes of application,

Table 2
LD 50 (in μmol/kg) for different routes of $MnCl_2$ (acute exposure)

	oral	ip	iv	im	sc
Mouse	1.9	0.73	0.3	2	1.6
Rat	7.5	0.7	0.7	5.5	

Source: MSDS for $MnCl_2$, Sigma Aldrich, USA, MSDS for $MnCl_2$, Allan Chemical Corporation, USA, and (205)
oral oral consumption, *ip* intraperitoneal injection, *iv* intravenous injection, *im* intramuscular injection, *sc* subcutaneous

Table 3
Toxic effects of manganese as a function of local concentration

Mn^{2+}/μM	Effects	References
10	Alteration of electrocardiogram	(89)
10	No observed cytotoxicity in cell culture	(78)
30	Depression of the contractile function of the isolated heart	(91)
100	No significant alteration in neuronal cell culture	(98)
100	Death by cardiac arrest	(8)
200	Cardiac ventricular fibrillation	(89)
500	Increase of F2-isoprostanes in neuronal cell cultures	(98)
>800	Astrogliosis	(99)
900[a]	Depression of visually evoked potentials	(161)
1000	Significant cytotoxicity in cell culture	(78)
>1600	Neuronal death	(99)
>2700[b]	Death of retinal ganglion cells	(170)

[a]Assuming a vitrial volume of 7 μL in mice; [b]Assuming a vitrial volume of 55 μL in rats

Table 4
MEMRI protocols for studies of mice

Admin.	Mn^{2+}/μmol/kg #Local injection/μmol	Solution/mM	Volume (infusion/μL/min)	References
iv	45.5–884.3	120	0.4–7.4 mL/kg (4.2)	(121)
	40	100	0.4 mL/kg (3.3)	(159)
ip	101.0–404.2	30	3.4–13.5 mL/kg	(206)
	333.5	100	3.3 mL/kg	(162)
	200	4	50 mL/kg	(122)
sc	1000	100	10 mL/kg	(126)
	160	320	0.5 mL/kg	(169)
ivc	0.00125–0.0125#	5–50	0.25 μL	(129)
ic	0.005#	5	10 μL	(188)
io	0.8–1.8#	800	1–2 μL	(168)
	0.06#	120	0.5 μL (0.5)	(177)
	0.05#	200	0.25 μL	(161)
	1#	1000	1 μL (0.2)	(181)
	0.0125#	25	0.5 μL (0.1)	(171)
in	0.7#	10	7 μL	(160)
	15.6#	3900	4 μL	(168)

iv intravenous, *ip* intraperitoneal, *ivc* intraventricular, *ic* intracerebral, *in* intranostril, *io* intraocular, *sc* subcutaneous

while **Table 3** lists the toxic effects observed as a function of local manganese concentration. **Tables 4**, **5**, and **6** demonstrate that MEMRI studies of mice, rats, and nonhuman primates often exceed these thresholds.

Chronic exposure to manganese is known to cause neurotoxicity. This process is mediated by oxidative damage to neuronal cells, disturbance of mitochondrial activity (47), changes in neurotransmission and glutamate-mediated excitotoxicity. Manganese accumulates in mitochondria, where it inhibits oxidative phosphorylation and leads to reduced ATP levels (48–50). In cell culture, manganese disturbs the function of astrocytes, hampering their ability to provide neurons with substrates for energy and neurotransmitter metabolism (51). Several studies demonstrate altered levels of glutamate (52–58), GABA (49, 56, 59–66), dopamine (49, 55, 67–74), and choline (75), although to a different extent and often with conflicting results. Part of the controversial findings may result from differences in dose, route of application, and duration of exposure (67). For a better comparison, it has been proposed to estimate the internal cumulative dose (ICD$_{Mn}$), which is the total amount of Mn^{2+} taken up into the circulatory system by the time the endpoint was detected

Table 5
MEMRI protocols for studies of rats

Admin.	Mn^{2+}/μmol/kg #Local injection/μmol	Solution/mM	Volume (infusion/μL/min)	References
iv	2 × 450^a–1 × 910^b	100	4.5–9.1 mL/kg (0.021)	(95)
	884.3	64	2 mL (0.03)	(109)
	158.9–1399	120	1.3–11.7 mL/kg (0.0375)	(45, 173)
ip	6 × 150	100	1.5 mL/kg	(95)
	3 × 300^c		3.0 mL/kg	
ivc	0.45#	30	15 μL (0.25)	(110)
ic	1.25–2.5#	25–50^d	25–50 μL	(123)
	0.002–0.012#	10–60	0.2 μL	(110)
	0.008#	100	80 nL (0.5)	(99)
	0.012–0.06#	60	0.2–1.0 μL	(185)
	0.2#	1000	0.2 μL (0.05)	(187)
	0.04#	1000	0.04 μL	(186)
io	0.1#	1000	0.1 μL (0.2)	(169)
	0.15–0.30#	50–100	3 μL	(170)
in	10#	500	20 μL	(173)
	10#	1000	10 μL	(174)

iv intravenous, *ip* intraperitoneal, *ivc* intraventricular, *ic* intracerebral, *in* intranostril, *io* intraocular
^{a}Occasional tail necrosis. ^{b}Consistent tail necrosis. ^{c}Occasionally abdominal induration. ^{d}5/6 Animals died within 6 h after injection.

Table 6
MEMRI protocols for studies of nonhuman primates

Admin.	Mn^{2+}/μmol/kg #Local injection/μmol	Solution/mM	Volume (infusion/μL/min)	References
iv	4 × 0.15	40	3.8 mL/kg (0.021)	(114)
ic	0.024–0.036	120	0.2–0.3 μL	(113)
	0.4^a	800	0.5 μL	
io	0.08–0.8#	800	0.1–1.0 μL (0.6–2.4)	(111)

iv intravenous, *ic* intracerebral, *io* intraocular
^{a}Areas of complete cell loss

(67). In rodents, ICD$_{Mn}$ values of 80–120 μmol/kg affect the motor activity, while values of ICD$_{Mn}$ ≥140 μmol/kg lead to alterations of neurotransmitter levels. This evidence confirms that the presence of manganese in the brain parenchyma affects the neurotransmitter metabolism in a dose-dependent manner at concentrations that are commonly used or even exceeded by MEMRI (**Tables 4, 5**, and **6**).

2. Materials

2.1. MnCl$_2$ Preparations

See **Table 7**.

2.2. Solvents

1. Deionized water.
2. Isotonic (0.9%) NaCl solution (DeltaSelect, Pfullingen, Germany).
3. Glucose solution (5%) (DeltaSelect, Pfullingen, Germany).

2.3. Chemicals for pH Adjustment

1. Bicine (N,N-bis(2-hydroxyethyl)glycine, $C_6H_{13}NO_4$) (Sigma Aldrich, Taufkirchen, Germany).
2. Sodium hydroxide solution (1 M), NaOH (Sigma Aldrich, Taufkirchen, Germany).
3. Sodium bicarbonate, $NaHCO_3$ (Sigma Aldrich, Taufkirchen, Germany).
4. Tris-HCl (10 mM) (Sigma Aldrich, Taufkirchen, Germany).

2.4. Opening of the Blood–Brain Barrier

D-Mannitol solution ($C_6H_{14}O_6$, 25%) (Intramed, Port Elizabeth, South Africa).

2.5. Pain Treatment

Novaminsulfon (Metamizole) (Winthrop, Mülheim-Kärlich, Germany).

2.6. Pumps and Cannulae

1. Syringe pump: Cole-Parmer Instruments, IL, USA; Model 2681 (Harvard Instruments, MA, USA).
2. Pressure-controlled picospritzer (Parker Hannifin, OH, USA).

Table 7
Commercial preparations of MnCl$_2$ (Sigma Aldrich)

Name	Chemical formula	CAS number	Molecular weight	Product number	Purification
Manganese chloride tetrahydrate	MnCl$_2$ · 4H$_2$O	13446-34-9	197.1 g/mol	63543	>99%
Manganese(II) chloride	MnCl$_2$	7773-01-5	125.84 g/mol	244589 (powder) 450995 (beads) 416479 (flakes)	>99% 99.999% 97%
MnCl$_2$ solution	MnCl$_2$	7773-01-5	1.00 ± 0.01 M	M1787	Prepared in 18 megohm water

3. Micro-osmotic pumps: Alzet (Durect Corporation, Cupertino, CA, USA).

4. MRI-compatible intrabrain cannulae (Plastics One, Inc., Roanoke, VA, USA).

3. Methods

3.1. Administration of Manganese

3.1.1. Manganese Solutions

Although manganese can exist in multiple oxidation states, physiological conditions emphasize Mn^{2+} and Mn^{3+}. Under in vitro conditions, Mn^{2+} can be oxidized to Mn^{3+} by superoxide radicals (76). Mn^{3+} is more reactive and toxic than Mn^{2+} (47, 62). The failure to detect significant amounts of Mn^{3+} within the brain mitochondria by X-ray absorption near edge structure (XANES) spectroscopy suggests that Mn^{3+} only occurs as a transient state (77). Repetitive intraperitoneal exposure of rats to equimolar doses of Mn^{2+} and Mn^{3+} using $Mn(II)Cl_2$ and Mn(III)pyrophosphate, respectively, revealed significant higher blood and brain concentrations after Mn^{3+} than after Mn^{2+} exposure (62). However, little is known about the distribution and possible interconversion of Mn^{2+} and Mn^{3+} in the brain. Because of its lower toxicity and biochemical similarity to Ca^{2+}, only Mn^{2+} has hitherto been utilized for MEMRI.

In most cases, Mn^{2+} is administered as an aqueous solution of $MnCl_2$. The salt or the solutions are provided by many manufacturers. **Table 7** summarizes products from Sigma Aldrich, Germany. It should be noted that the amount of $MnCl_2$ administered to an animal is often given in mass units. To calculate the corresponding load of Mn^{2+} in molar units, the different molecular weights of $MnCl_2$ (125.84 g/mol) and $MnCl_2 \cdot 4\,H_2O$ (197.92 g/mol) have to be taken into account.

In situ perfusion techniques revealed a greater influx rate of ^{54}Mn citrate into the brain than of ^{54}Mn chloride. The use of pH 8.6 buffer-treated $MnCl_2$ which has a higher affinity to transferrin leads to a significantly lower brain manganese concentration in comparison to the use of $MnCl_2$ in Tris-HCl at pH 7.4. Only little is known of how different manganese salts and solvents may affect the MR contrast. So far, MEMRI studies almost exclusively employ $MnCl_2$ dissolved in deionized water, 5% glucose solution, or 0.9% NaCl solution.

For parenteral applications of Mn^{2+}, the injected solution should be biocompatible with the organism or tissue of interest. This refers to parameters such as concentration, chemical characteristics, applied volume, and sterility. For example, the osmolarity of most extracellular body fluids is about 300 mOsM. Because $MnCl_2$ consists of 3 ions, a 100 mM $MnCl_2$ solution would result in an isotonic solution. Manganese solutions in distilled water with a lower (higher) concentration are hypotonic (hypertonic). Although iso-osmolarity should be achieved whenever possible to minimize the occurrence of cell necrosis, minor deviations are less critical for intravenous and intraperitoneal injections, because of the large ratio of the target to the injected volume and the expected rapid mixing. For subcutaneous and intracerebral injections, however, strictly isotonic solutions have to be prepared to avoid pain and tissue damage. For $MnCl_2$ concentrations below 100 mM this condition can be easily achieved by adding sodium chloride. For example, a 10 mM $MnCl_2$ solution that corresponds to 30 mOsM would have to be dissolved in a 270 mOsM solution of NaCl to reach 300 mOsM.

Another aspect is the pH of the solution. Unbuffered aqueous solutions of $MnCl_2$ are slightly acidic. Although this may be compensated by plasma or extracellular physiological buffer systems, the pH should be better controlled by adjustment to values near 7.4 using sodium hydroxide or by preparing manganese solutions with a buffer like bicine or Tris-HCl.

Besides pH and osmolarity the toxicity of manganese depends on the local concentration (**Table 3**). In cell culture, a direct cytotoxic effect of concentrations as low as 1 mM $MnCl_2$ has been reported, but not for $MnCl_2$ concentrations of 10 μM (78). The lowest possible concentration for MEMRI, however, is limited by the amount of manganese required to yield sufficient contrast in the target region and the tolerable solution volume. **Table 8** summarizes the recommended maximum injection volumes for different routes of application. For subcutaneous injections the administration of manganese may be distributed to several locations in order to reduce local concentrations (*see* **Note 1**).

3.1.2. Routes of Manganese Administration

Delivery of Mn^{2+} to the central nervous system may be achieved using three different routes: (1) systemic applications with manganese entering the brain via the blood, (2) direct intracerebral or intraventricular injections, and (3) nostril administration.

Table 8
Recommended maximum injection volumes for mice and rats

	Volume/mL		Volume/mL		Volume/mL/kg bw	
	sc	im	icv	ic	ip	iv
Mouse (30 g)	0.5	0.03	0.01	0.01–0.03	50 (1.5)	5.0 (0.15)
Rat (250 g)	2.0	0.25	0.1		20 (5.0)	5.0 (1.25)

Source: Society for Laboratory Animal Science, recommended maximum volume of injection for experimental animals, 1999

iv intravenous, *ip* intraperitoneal, *ivc* intraventricular, *ic* intracerebral, *im* intramuscular, *sc* subcutaneous

For systemic applications, the common routes are intravenous, intraperitoneal, and subcutaneous injections. In principle, oral applications are also possible, but less well controlled as absorption from the gastrointestinal tract can be influenced by many factors. These include the chemical form (79), the presence of other minerals and dietary constituents (33, 80, 81), and the age of the animal (82–84). Furthermore, adaptive changes to high dietary manganese intake lead to reduced absorption and enhanced hepatic (biliary) or pancreatic excretion of manganese (19, 85–88).

A major advantage of the intravenous application is the fast delivery to the systemic circulation. This is utilized for activity-induced MEMRI where the accumulation of $MnCl_2$ in conjunction with a brief opening of the BBB may be related to brain activity during stimulation. On the other hand, intravenous bolus injections of manganese can lead to a high transient blood concentration and bear the risk of altered heart function. In fact, injections of 0.01 mmol/kg Mn^{2+} over 30 s in dogs and rabbits caused prolongation of PR and QTc intervals in the electrocardiogram, while 0.2 mmol/kg Mn^{2+} led to ventricular fibrillation (89). Intravenous infusion of 3.6 μmol/min Mn^{2+} in rats resulted in a drop in blood pressure (90) and Mn^{2+} concentration of 30 μM caused immediate depression of the contractile function in isolated, perfused rat hearts (91). Considering a volume of manganese distribution of about 1.16 L/kg (23) this local concentration corresponds to a bolus injection of 0.035 mmol/kg Mn^{2+}. In general, the cardiac side effects can be markedly reduced by slow infusion rates, which lower the transiently high extracellular manganese level in the myocardium. Examples for intravenous injections of $MnCl_2$ are given in **Tables 4**, **5**, and **6**. The most often used veins in rodents are the tail vein and the femoral vein with a recommended needle size of 27 G for rats and 30 G for mice.

Only few groups used an arterial route for the administration of manganese (92) without differences from intravenous injections (90). No significant distinctions in temporal and regional T_1 changes were reported for subcutaneous, intraperitoneal, and intravenous injections (93, 94). The best way of administration therefore depends on the desired temporal resolution (e.g., subcutaneous injections significantly delay the peak plasma concentration), the manganese concentration (e.g., intravenous injections allow for higher concentrations, but hold a high risk of heart failure), and other practical considerations (e.g., in mice, intraperitoneal injections are much easier to perform than intravenous injections).

To reduce the acute toxicity of manganese, while retaining its potential as an MRI contrast agent, a fractionated application may be helpful (95). In rats, repeated doses of 0.15 mmol/kg Mn^{2+} every 48 h yielded a better signal enhancement than a single administration of 0.15 mmol/kg, indicating accumulation of manganese in the brain. A single administration of the cumulative dose revealed a stronger T_1 shortening at the expense of higher acute toxicity.

Only few studies attempted to quantitatively determine the manganese concentration in the brain after systemic administration (45, 96, 97). In rats, 1 day after a single 1.4 mmol/kg Mn^{2+} injection, the respective brain concentrations were about 0.11 mM in the olfactory bulb and 0.06 mM in the cortex (45), which correspond to a range in which no toxic effects were seen in cell cultures (98). On the other hand, a single subcutaneous injection of 1.8 mmol/kg Mn^{2+} significantly increased the occurrence of biomarkers for cerebral oxidative damage (98).

A direct and localized injection of manganese into the central nervous system allows for a marked reduction of the total amount of applied manganese (**Tables 4, 5, and 6**). To minimize possible artifacts due to tissue damage, the manipulation should be as minimally disruptive as possible. This includes a proper choice of the needle size, infusion duration, injection volume, pH value, osmolarity, sterility, and concentration of the manganese solution. For rodents, Hamilton syringes (0.5–2.5 μL) or pressurized controlled picospritzers with a needle size of 32 G and smaller are recommended (99). Examples of recent protocols for intracerebral and intraventricular injections are given in **Tables 4, 5, and 6**. Although the total manganese burden for the organism is lower than that of a systemic administration, the induction of local cytotoxic effects should be kept in mind (**Table 3**). Primary cultures of cortical neurons that were exposed to 0.5 mM $MnCl_2$ for 2 h showed significant increases in F2-isoprostanes, which are markers for oxidative stress, while a fivefold lower concentration of 0.1 mM $MnCl_2$ did not cause significant alterations (98). For a range of volumes and concentrations, histological investigations of rats 1 week after $MnCl_2$ injection into the brain revealed thresholds for $MnCl_2$ toxicity at the injection site of 16 nmol for neuronal death and 8 nmol for astrogliosis in a tissue volume of 8–11 μL (99). Notably, most protocols for intracerebral injection exceed these values, so that a reduction of toxic side effects remains an ongoing research goal. One approach may be the use of micro-osmotic pumps that constantly infuse manganese into a targeted brain area for a period of about 48 h (99).

3.2. Magnetic Resonance Imaging

3.2.1. Manganese-Induced Reductions of T_1

As with many other MRI contrast agents, Mn^{2+} ions shorten both the T_1 and T_2 relaxation times of the surrounding water protons. The extent of $T_{1,2}$ reduction depends on the local Mn^{2+} concentration and can be described by

$$R_i([Mn^{2+}]) = R_i(0) + r_i \times [Mn^{2+}], \qquad [1]$$

where $R_i = 1/T_i$ ($i = 1, 2$), $[Mn^{2+}]$ is the Mn^{2+} concentration in mM, $R_i([Mn^{2+}])$ is the R_i at this concentration in s^{-1}, and $R_i(0)$ is the native R_i without manganese ($[Mn^{2+}] = 0$).

The specific Mn^{2+} relaxivities r_i ($i = 1, 2$) in s^{-1} mM^{-1} have extensively been studied (45, 97, 100–105). In water, the very short rotational correlation time of Mn^{2+} ensures an almost constant value for r_1 of about 7–8 s^{-1} mM^{-1} (100, 103), while r_2 increases with field strength from 40 s^{-1} mM^{-1} at 0.47 T to 120 s^{-1} mM^{-1} at 9.4 T (100, 103, 104). Under in vivo conditions the intracellular binding of Mn^{2+} to macromolecules causes a much higher r_1 of about 60 s^{-1} mM^{-1} for magnetic fields below 1 T (100, 102, 103). This effect is lost at higher fields of 4.7 T and above, where the in vivo r_1 approaches the value for aqueous Mn^{2+} (45, 97, 101).

For the range of concentrations used in MEMRI studies and for field strengths greater than 4.7 T where r_1 is similar for intra- and extracellular manganese, the linear proportionality of the R_1 relaxation rate to the manganese concentration may be exploited to estimate the absolute amount of manganese present in tissue (45).

3.2.2. Optimization of T_1 Contrast

Due to the paramagnetism of Mn^{2+} and the fact that the native T_1 relaxation time of tissue water protons is much longer than the T_2 relaxation time, low concentrations of Mn^{2+} are best visualized by MRI sequences yielding T_1 contrast (45). Thus, the majority of MEMRI studies relies on T_1-weighted images and quantitative maps of T_1 relaxation times.

For a conventional spin echo (or gradient echo) MRI sequence, the signal equation is given by

$$S(TR, TE) = \rho_0 \cdot \left(1 - e^{-TR \cdot R_1}\right) \cdot e^{-TE \cdot R_2} \quad [2]$$

with ρ_0 the spin density, TR the repetition time, and TE the echo time (for gradient echo MRI R_2 has to be replaced by R_2^*). Unfortunately, R_1 and R_2 have opposite effects: while a shortening of T_1 leads to a signal enhancement, a shortening of T_2 causes a signal decrease. In order to optimize the contrast, the shortest possible echo time will be the best choice. This depends on experimental parameters such as the receiver bandwidth, desired resolution or matrix size, and maximum gradient strength of the MRI system. If the use of an echo time with $TE \ll T_2$ minimizes the effect of the T_2 signal attenuation, then equation [2] simplifies to

$$S(TR) = \rho_0 \cdot \left(1 - e^{-TR \cdot R_1}\right). \quad [3]$$

Using equations [1] and [2], the signal difference of a region before and after manganese application can be described by

$$S(TR, [Mn^{2+}], R_1(0))$$
$$= \rho_0 \cdot \left(\left(1 - e^{-TR \cdot (r_1 \cdot [Mn^{2+}] + R_1(0))}\right) - \left(1 - e^{-TR \cdot R_1(0)}\right)\right) \quad [4]$$

$$S(\text{TR}, [Mn^{2+}], R_1(0)) = \rho_0 \cdot \left(1 - e^{-\text{TR} \cdot (r_1 \cdot [Mn^{2+}])}\right) \cdot e^{-\text{TR} \cdot R_1(0)}.$$
[5]

Thus, the TR for maximum signal enhancement becomes a function of the manganese concentration and the native T_1 of the tissue water protons: The higher the local manganese concentration for a given $R_1(0)$, the shorter the TR that gives the best signal difference between pre- and post-contrast conditions and the higher the pre-contrast $R_1(0)$, the shorter the optimal TR and the lower the signal gain by a given manganese concentration (**Fig. 1**).

However, there is no general answer to this optimization problem as the choice of parameters always needs to reflect the

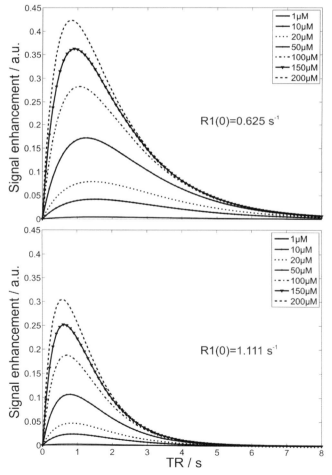

Fig. 1. Simulated MEMRI signal enhancement as a function of repetition time TR for two different native relaxivities $R_1(0)$ and a wide range of Mn^{2+} concentrations (assuming TE $\ll T_2$). For details see text.

actual scientific question and the experimental setting. The latter not only involves the parameters of a particular MRI sequence, for example, TR and flip angle for FLASH, but also depends on the local T_1 change, the applied amount of Mn^{2+}, and the time point of the measurement relative to the Mn^{2+} administration (*see* **Note 2**).

3.3. MEMRI of Brain Anatomy

After systemic application, manganese accumulates in many tissues including liver, pancreas, kidney, heart, and brain (106, 107). Region-specific enrichments in the brain may be used to delineate the neuroarchitecture in rodents (108–110), nonhuman primates (97, 111–114), birds (115–117), and insects (118, 119).

3.3.1. Time Course of MEMRI Signal After Systemic Mn^{2+} Administration

MEMRI findings after systemic administration need to consider modulations by the manganese pharmacokinetics and the transfer into the brain. Manganese exhibits a high blood clearance with an elimination half-life time of 1.83 h after intravenous administration in rats (23). Already 5–10 min after the beginning of a manganese infusion in rats, the MRI signals of structures outside the BBB start to increase (109). This process leads to high manganese concentrations in the choroid plexus, pituitary gland, and pineal gland after about 1 h (97, 107, 120). Additional strong signal enhancements can be found in the veins and subarachnoid space (97, 121). After about 2 h, the signal from the parenchyma surrounding the ventricles presents with an increasingly strong signal enhancement, while the signals from the vasculature and subarachnoid space start to decline (97, 109) (**Fig. 2**). Interestingly, in rats, a Mn^{2+} infusion rate of about 32 µmol/h resulted in a signal enhancement of the entire CSF between 10 min and 2 h after administration (109), whereas a slightly lower rate of 14 µmol/h Mn^{2+} failed to yield a signal increase of the CSF in rats and marmosets (97).

The initial enhancement of the choroid plexus and the subsequent enhancement of the periventricular tissue indicate that the BCB is the main delivery route of manganese into the brain, at least for a systemic administration at higher than physiological doses. Between 2 and 24 h after administration, manganese distributes into other brain areas accompanied by a gradual MRI signal increase (108, 121). The maximum tissue enhancement is observed about 24 h after application (122) (**Fig. 2**) and persists for about 24–48 h followed by a slow recovery to pre-contrast baseline over 2–3 weeks (108, 121, 123).

The clearance of manganese from the brain is rather slow. After intravenous infusion of 1.4 mmol/kg $MnCl_2$ in rats, the value of ΔR_1 and the absolute manganese concentration in the brain peaked at day 1 and subsequently declined to near control levels after 28–35 days with an estimated brain half-life time of 11–12 days (from ΔR_1) and 15–24 days (based on

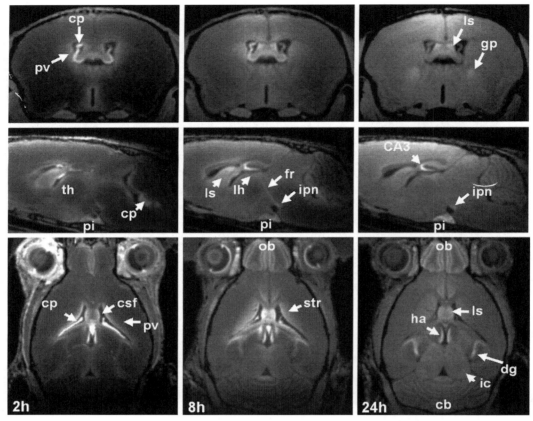

Fig. 2. (*Top*) Axial, (*middle*) sagittal and (*bottom*) horizontal MEMRI sections of a mouse brain in vivo 2, 8, and 24 h after intraperitoneal injection of 320 μmol/kg MnCl$_2$ (9.4 T, 100 μm isotropic resolution). Cb, cerebellum; cp, choroid plexus; CSF, cerebrospinal fluid; dg, dental gyrus; fr, fasciculus retroflexus; gp, globus pallidus; ha, habenula; ic, inferior colliculus; ipn, interpeduncular nucleus; lh, lateral habenula; ls, lateral septum; ob, olfactory bulb; pi, pituitary gland; pv, periventricular tissue; str, striatum; th, thalamus.

manganese concentration as determined by inductively coupled plasma–mass spectrometry) (45). A radioactive ^{54}MnCl$_2$ analysis, however, revealed a threefold longer brain half-life time of 51–74 days (43) indicating a long terminal elimination phase. Moreover, intravenous injection of ^{54}Mn even led to a gradual increase of manganese in some brain areas until day 6 after administration (107). ^{54}Mn studies of rhesus monkeys yielded a half-life time of more than 100 days (44, 124) for very low doses of the radioactive tracer, different from MEMRI, which relies on higher doses. The difference between these two types of studies may indeed be explained by saturation effects of manganese-binding sites (45), which result in a faster clearance rate for high doses. This view is further supported by a study where ^{54}Mn excretion after intramuscular injection was accelerated by increased dietary manganese (125).

For longitudinal MEMRI studies, it is important to consider potentially confounding effects from preceding injections of MnCl$_2$. The presence and time course of a residual signal enhancement strongly depends on the region of interest, the accumulated local amount of manganese, and the sensitivity of the applied MRI sequence. In addition, preceding manganese injections may have caused neurotoxicity with accompanying cellular and/or behavioral abnormalities that may hamper subsequent measurements.

3.3.2. Spatial Distribution of Manganese

Figure 2 shows axial, coronal, and sagittal MEMRI sections of a mouse brain in vivo 24 h after intraperitoneal injection of 0.32 mmol/kg MnCl$_2$. These T_1-weighted images confirm the heterogeneous distribution of manganese in the brain. A closer look at the olfactory system in **Fig. 3** reveals a signal enhancement in the glomerular layer and mitral cell layer, which are well separated from the less-enhanced granular layer and even darker external plexiform layer. This pattern can already be observed 2 h after a subcutaneous administration of 0.07 mmol/kg MnCl$_2$ (121). As shown in **Fig. 4**, the hippocampus presents with a bright signal in the dentate gyrus and CA3 region, whereas CA1 can hardly be separated from the less-enhanced subiculum. In the cerebellum, the granular layer exhibits a pronounced signal enhancement, which becomes even stronger in the layer formed by Purkinje cells (**Fig. 5**). In the developing cerebellum, a correlation between manganese uptake and granule cell density has been reported (122).

The combination of high-resolution MRI, for example, corresponding to $100 \times 100 \times 100$ μm^3 in mice and $50 \times 50 \times 750$ μm^3 in rats, with Mn^{2+} doses equal to or higher than

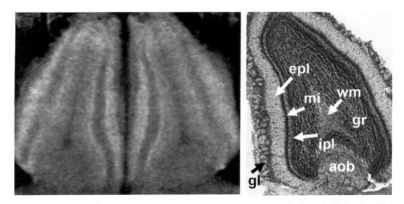

Fig. 3. (*Left*) Magnified horizontal MEMRI section of a mouse olfactory bulb in vivo 24 h after intraperitoneal injection of 320 μmol/kg MnCl$_2$ (9.4 T, $30 \times 30 \times 300$ μm^3 resolution) and (*right*) corresponding Nissl stain (http://www.brainmaps.org/). aob, accessory olfactory bulb; epl, external plexiform layer; gl, glomerular layer; gr, granule layer; ipl, inner plexiform layer; mi, mitral cell layer; wm, white matter.

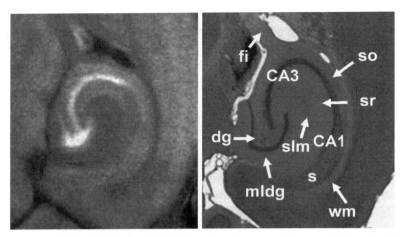

Fig. 4. (*Left*) Magnified horizontal MEMRI section of a mouse hippocampal formation in vivo 24 h after intraperitoneal injection of 320 μmol/kg MnCl$_2$ (30 × 30 × 300 μm^3 resolution) and (*right*) corresponding Giemsa stain (http://www.brainmaps.org/). Dg, dental gyrus; fi, fimbria hippocampi; mldg, molecular layer of the dental gyrus; s, subiculum; slm, stratum lacunosum moleculare; so, stratum oriens; sr, stratum radiatum; wm, white matter.

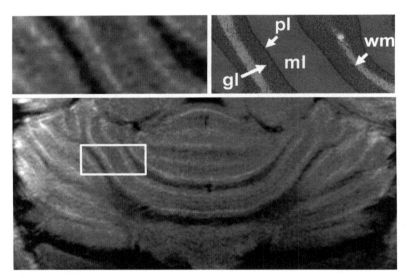

Fig. 5. (*Top*) Magnified view and (*bottom*) overview of a horizontal MEMRI section of a mouse cerebellum in vivo 24 h after intraperitoneal injection of 320 μmol/kg MnCl$_2$ (9.4 T, 30 × 30 × 300 μm^3) and (*right*) corresponding Giemsa stain (http://www.brainmaps.org/). Gl, granular layer; ml, molecular layer; pl, Purkinje cell layer; wm, white matter.

0.7 mmol/kg (109, 121) allows for a detection of cellular layers in the cerebral cortex. In particular, an increased MRI signal is observed in regions corresponding to layer II, the transition region of layers IV and V, and layer VIb, whereas layers I, III, and VIa are much less enhanced (109). This heterogeneous distribution across layers of the cerebral and piriform cortex has been exploited for an in vivo characterization of different mouse mutants (110, 126, 127). Moreover, the more pronounced shortening of T_1 in the primary visual cortex (V1) relative to the secondary visual cortex (V2) in common marmosets enabled the in vivo delineation of the V1/V2 border (114).

The underlying causes for the differential distribution of manganese in the brain are still a matter of debate. Nevertheless, some mechanisms are generally accepted to contribute to the signal pattern. A major aspect is the location of a structure relative to the ventricles. This is because Mn^{2+} predominantly passes the BCB with uptake across the choroid plexus for the high plasma concentrations as used in MEMRI (128). Studies using radioactive ^{54}Mn have shown that the manganese influx in the cerebral cortex is nearly saturated for plasma concentrations of 2–7.5 μM, but the influx into the hippocampus and other regions close to the ventricular system increases with increasing plasma concentration (26). In agreement with these considerations, marmosets showed a strong enhancement in the visual cortex, which in this species is rather close to the lateral ventricles, whereas the frontal cortex due to its distance from the ventricles showed no significant enhancement (*see* **Note 3**).

A second relevant aspect is the axonal transport of manganese. For example, the distinct manganese accumulation in the globus pallidus 24 h after administration is most likely due to the anterograde transport of Mn^{2+} from the striatum (gp and str in **Fig. 2**) (111, 112). Similar evidence for anterograde transport exists for the habenular complex (lh and fr in **Fig. 2**) (129).

Further parameters influencing the cerebral pattern of manganese are the cell density and neuronal activity (130). For example, strong signal enhancements are observed for the mitral cell layer in the olfactory bulb and the Purkinje cell layer in the cerebellum, which possess high neuronal densities. Moreover, the degree of neuronal activity has been shown to alter the manganese accumulation (*see* **Section 3.4**). In addition, the density of voltage-gated Ca^{2+} channels (VGCC) may play a role. Overexpression of N-type VGCC in rats with optic neuritis was accompanied by a strong manganese-induced signal enhancement (131), which could significantly be reduced by treatment with ω-conotoxin GVIA, an N-type specific blocker (132). Furthermore, excitotoxicity in brain ischemia is characterized by enhanced Ca^{2+} influx. Consequently, ischemic lesions exhibited a marked signal enhancement after administration of manganese (133).

The effects supporting MEMRI are also related to the function of cGMP-gated channels (134). Comparison of dark-adapted and light-adapted rats revealed a higher uptake of manganese in the outer retina of the dark-adapted group, which is expected to develop higher cGMP levels resulting in a higher

Ca^{2+} influx. Another factor is the presence of manganese-binding enzymes such as GS and Mn-SOD. For instance, the observation of T_1 hyperintensities after mild ischemia in rats and humans was accompanied by endogenous manganese accumulation and Mn-SOD and GS induction in reactive astrocytes (135). Moreover, in the late phase of mild hypoxic-ischemic injury and following the administration of manganese in rats, MRI revealed enhanced gray matter lesions that correlated with increased immunoactivities of GS and Mn-SOD (136). Manganese-induced signal enhancements were also found to be co-localized with high concentrations of activated microglia (137, 138). Although suggestive, it should be noted that these correlations do not represent a final proof that the MRI findings directly or exclusively reflect the accumulation of manganese-binding proteins.

Attention should also be paid to the fact that the brain uptake of manganese (and respective signal enhancements) may be modulated by other metals. For example, low iron or calcium may result in a higher manganese uptake (80, 139). Conversely, the intravenous administration of ferric hydroxide–dextran complex and a high iron intake may reduce the brain manganese level (80). Furthermore, the amount of manganese in the animal chow may influence its clearance rate, whereas a disturbance of manganese excretion by the liver may lead to manganese deposits in the brain (140). Manganese homeostasis may also be altered in animal models of human brain disorders such as schizophrenia, Alzheimer's disease, or Parkinson's disease (141) (*see* **Notes 4** and **5**). Finally, in humans but not in animals, a gender difference has been reported for manganese absorption (142).

3.4. MEMRI of Brain Function

3.4.1. Physiological Basis

Another application of MEMRI is the mapping of functional brain activity in response to an external stimulus. This ability is based on the fact that Mn^{2+} may behave similar to Ca^{2+}, which accumulates in neurons or astrocytes in an activity-dependent manner. In contrast to other functional MRI approaches that exploit the blood oxygenation level-dependent (BOLD) effect or detect changes of cerebral blood flow (CBF) or volume (CBV), MEMRI does not rely on hemodynamic responses but activity-related changes in calcium flux.

There are three gates for Ca^{2+} uptake into a cell: (1) the VGCC, (2) the ligand-gated nonspecific calcium channels (e.g., the ionotropic glutamate receptors NMDA and AMPA), and (3) the receptor-activated calcium channels (RACC) for which two main types have been described, the store-operated calcium channels and the intracellular messenger-activated nonselective channels (143, 144).

Because Mn^{2+} and Ca^{2+} have the same charge and rather similar ionic radii of 89 and 114 pm, respectively, they behave similarly in many biological systems. In experiments with synaptosomes of rat brain, it was shown that Mn^{2+} can enter the pre-synaptic nerve endings via VGCC and even induce the release of dopamine from the depolarized nerve terminals (30). Moreover, in the absence of extracellular Ca^{2+}, Mn^{2+} increased the frequency of acetylcholine release from parasympathetic nerve endings during stimulation (145). In frog, the influx of Mn^{2+} (and Mg^{2+}) into neuromuscular junctions was reduced by verapamil, an L-type VGCC blocker (31). Diltiazem, another L-type VGCC blocker, attenuated the signal enhancement in several MEMRI experiments (146–148). Besides VGCC, NMDA receptors also seem to play a role in Mn^{2+}-dependent MRI signal changes. Application of NMDA antagonists such as MK-801 resulted in signal suppression, whereas NMDA agonists such as glutamate and NMDA increased the MEMRI signal in rats. On the other hand, NBQX, an AMPA antagonist, did not influence the signal enhancement in MEMRI experiments (94).

Mn^{2+} may also act as a competitive substrate of calcium transporters. $MnCl_2$ reduced the Ca^{2+} uptake and calcium-dependent sodium efflux of intact squid axons (149) and Mn^{2+} accumulated during hyperkalemic hypoxia in perfused rat myocardium via Na^+/Ca^{2+}-exchanger-mediated transport (150). The ability of Mn^{2+} to mimic Ca^{2+} has also been confirmed by the quenching of fluorescent calcium indicators following Mn^{2+} entry into various cell types (151).

Further evidence for the pertinency of MEMRI to measure neuronal activity comes from studies of c-fos expression. C-fos is an intermediate early gene and its mRNA and protein expression is a widely used functional marker of activated neurons in the CNS. MEMRI signals due to hypothalamic activation, which was induced by intracarotid arterial hypertonic NaCl infusion, correlated well with c-fos expression (152). However, the apparent manganese accumulation during osmotic stress may not only be due to calcium channels. For instance, the DMT1 receptor and transferrin iron transport system seem to be further involved. Furthermore, cGMP-gated channels (134) as well as other metal transporters have to be taken into consideration.

Despite their similarities, Ca^{2+} and Mn^{2+} control different biological functions and, in contrast to Ca^{2+}, manganese can also be converted into other oxidation states. It has several similarities to iron including a similar ionic radius, oxidation states 2+ and 3+ under physiological conditions, and similar binding affinities to transferrin. It is therefore of utmost importance to further understand all factors contributing to the cellular uptake of Mn^{2+} before MEMRI may be used as a quantitative marker for Ca^{2+} fluxes (*see* **Notes 4** and **5**).

3.4.2. Short-Term Activity-Induced MEMRI

In order to map functional brain activity by MEMRI within a temporal range of seconds to minutes, $MnCl_2$ is intravenously or intra-arterially infused, while the stimulus of interest is applied. The mechanism relies on local increases of brain activity that lead to corresponding accumulations of Mn^{2+}, which in turn cause a signal increase in affected regions. To achieve a sufficiently high manganese level under these conditions, the BBB needs to be opened. Typically, infusion of a concentrated solution of D-mannitol (~25%) into the bloodstream yields a temporary osmotic disruption of the BBB (92, 153). The potential drawbacks of this technique are the long persistence of manganese inside the cells and the risk of the strong hyperosmolarity of the applied mannitol solution. Thus, initial increases in activation can be visualized, but further variations or deactivations cannot be resolved.

Short-term activity-induced MEMRI is usually performed on anesthetized animals, so that the depth of anesthesia will influence the functional contrast. Deep anesthesia reduces neuronal activity, whereas light anesthesia may lead to competitive brain activity without relation to the stimulus. To control for such confounding effects, it was suggested to perform four sets of measurements (92): The first after starting $MnCl_2$ infusion but before mannitol injection, the second after bolus injection of mannitol to estimate the Mn^{2+} accumulation caused by nonspecific effects, the third during functional stimulation, and the fourth after the end of $MnCl_2$ infusion and stimulation. However, additional complications may arise from different narcotic drugs and their putative modulations of manganese uptake and transport. For example, ketamine, an NMDA receptor antagonist, can reduce the signal enhancement by manganese (94), while isoflurane depresses its transsynaptic transport (41).

So far, short-term activity-induced MEMRI has been applied for mapping activation of distinct brain areas by pharmacological substances such as KCl (154), NaCl (155), glutamate (90), amphetamines, and cocaine (148) as well as in response to electrical forepaw stimulation (148, 156) and barrel activation (157). A comparison of MEMRI, CBF, and BOLD activations due to forepaw stimulation revealed a well co-localization of the respective maps in the somatosensory cortex. Interestingly, the strongest T_1 shortening was observed in cortical layer IV exhibiting a high cell density, while BOLD MRI resulted in stronger activations near the cortical surface close to draining veins (156).

A few brain regions allow for a sufficient manganese accumulation without opening the BBB in a relatively short period of less than 3 h. In mice, these regions are adjacent to the third ventricle and include nuclei of the hypothalamus. Respective MEMRI studies have been performed on fasted and non-fasted mice as well as in response to anorexigenic agents (158, 159). Another well-accessible region is the olfactory bulb. The delivery of $MnCl_2$ directly into the nasal cavity leads to a signal enhancement in the olfactory bulb via uptake by the nasal epithelium and axonal transport. This approach has been used to visualize odorant representation in the glomerular and mitral cell layer (146, 160).

3.4.3. Long-Term Activity-Induced MEMRI

A second approach to activity-induced MEMRI relies on the fast influx and slow efflux of manganese into the brain and utilizes the cumulative signal change over several hours or even days. The assessment of functional brain activation involves a comparison of two groups of animals, one with and the other without stimulation or treatment. Typically, a stimulus is applied over several hours and MEMRI is performed 24 h after systemic Mn^{2+} administration. Differences in the signal enhancement between the groups are interpreted as stimulus response. Because anesthesia is only necessary during MRI, the approach yields maps of accumulated activity from awake and behaving animals. However, the longer the time between Mn^{2+} injection and MEMRI, the more axonal transport may influence the results: areas with high Mn^{2+} uptake may lead to signal enhancements in axonally connected areas independent of their own activity (161). To reduce such transport effects, shorter stimulation periods of only 9 h have been suggested (162). Due to the low manganese accumulation, this strategy is at the expense of sensitivity, especially in areas remote from the ventricles. Therefore longer periods of stimulus exposure have also been employed (163).

As an example, long-term activity-induced MEMRI has successfully been applied for mapping the tonotopic organization of the mouse inferior colliculus (164) and its plasticity (165). In rats, visual stimulation evoked increased signal intensities in layers IV and V of the primary visual cortex and in deeper portions of the superficial superior colliculus when the MEMRI effects were compared to a control group kept in darkness (162).

3.5. MEMRI of Axonal Connectivity

3.5.1. Physiological Basis

The fact that manganese is axonally transported has turned MEMRI into a widely used tool for neuronal tract tracing in living animals. In general, there are two possible directions of axonal transport: anterograde transport from the cell body to the synapse and retrograde transport from the synapse back to the cell body. A further distinction can be made between slow and fast axonal transport processes. Slow axonal transport with a velocity of $0.002–0.3$ $\mu m/s$ only occurs in anterograde direction and is mainly used for proteins. The fast transport process at $2–5$ $\mu m/s$ is bidirectional and mediated by the motor protein kinesin in anterograde direction. Fast retrograde transport is accomplished by dynein and vesicular transport along the microtubuli.

Autoradiographic analyses revealed an anterograde axonal transport of ^{54}Mn after injection into the striatum (39, 166) and after administration to the nostril (24, 25). Treatment with colchicines, an inhibitor of microtubule formation, reduced the manganese transport in a dose-dependent manner and supported the involvement of the microtubules (39, 146, 167). Subsequently, Pautler et al. (168) demonstrated that topical administration of MnCl$_2$ to the nostrils and the retinal ganglion cells in mice allowed for a visualization of respective pathways by MEMRI. After neuronal uptake, manganese is sequestered in the endoplasmatic reticulum, packed into vesicles, and anterogradely transported along the microtubules (146, 168). The transport velocity was estimated by MEMRI in the range 3.6–21.6 μm/s (161, 169, 170). These values are compatible with the fast axonal transport mechanism and seem to depend on Mn^{2+} concentration (111) and temperature (167). In comparison, the anterograde transport of mitochondria of about 1 μm/s is clearly slower than that of manganese observed by MEMRI. Moreover, kinesin contributes to but is not essential for the neuronal transport of Mn^{2+} (161). Interestingly, electrical activity is not necessary for the uptake and transport of Mn^{2+} in the optic nerve as shown in retinal blind CBA mice (161). Furthermore, blockage of the retinal ganglion cell activity by intravitrial co-injection of the sodium channel blocker tetrodotoxin failed to reduce the manganese-induced signal enhancement in the superior colliculus (171). However, electrical activity seems to be required for synaptic transmission (161), where manganese is released together with neurotransmitters as shown by ^{54}Mn (172). Transsynaptic manganese transport is reduced by isoflurane, which is known to depress the action potential and vesicular release at the pre-synaptic site, and by a blockage of the post-synaptic NMDA-receptor (41). On the other hand, mice that lack the voltage-gated potassium channel Kv4.2 and therefore exhibit an increase in post-synaptic Ca^{2+} influx, showed elevated transsynaptic manganese transport (41).

Neuronal pathways that cross up to a maximum of three synapses have hitherto been visualized by local Mn^{2+} injections (112). Further support for Mn^{2+} as a tract tracer stems a combined injection of Mn^{2+} and the histological tracer wheat germ agglutinin conjugated to horseradish peroxidase (WGA-HRP) (111, 112) as both compounds result in a similar projection pattern (111). In general, however, unspecific enhancements due to a possible leakage of manganese into the blood vessels should be considered, particularly after intranasal administration (173).

3.5.2. Olfactory System

Under general anesthesia, Mn^{2+} is administrated via a polyethylene tube or humidifier unilaterally or bilaterally into the nostrils (for commonly used volumes and doses *see* **Tables 4, 5, and 6**). From the olfactory epithelium, Mn^{2+} is taken up by the olfactory receptor neurons and transported anterograde to the olfactory bulb. About 1–1.5 h after nasal administration, an enhancement of the olfactory is detected (146, 174). Crossing a synapse, manganese is further transported by the mitral cells to the lateral olfactory tract and piriform cortex with corresponding signal enhancements 24–36 h after administration (146). After 36–48 h, the enhancement reaches the anterior commissure and amygdalae (174). Unilateral application of manganese further enhances the contralateral olfactory epithelium, bulb, and cortex, albeit to lower extent and in line with ^{54}Mn studies (175). Radiotracer studies in rats demonstrated a saturation effect of manganese transport in the olfactory receptor neurons at a dose of 1.5 μM. Thus, higher doses will only increase the local toxicity without improving the MEMRI signal. Using state-of-the-art MRI sequences, the manganese enhancement in the olfactory bulb is detectable for 3–5.5 days (168, 174). Olfactory tract tracing has also been applied to quantitatively assess axonal dysfunction, recovery, and reorganization in rats and mice (41, 167, 176). Such applications may offer wide opportunities for a characterization of animal models dealing with neurodegenerative diseases.

3.5.3. Visual System

Intraocular injections of Mn^{2+} highlight the optic nerve and tract and connected parts of the visual system in mice (161, 177), rats (169, 170, 178), and monkeys (112) under normal and pathological conditions (178–180). The retinal ganglion cells that are responsible for manganese uptake are located in the innermost layer of the retina. Their axons form the optic nerve. In rodents almost 100% of these axons cross the optic chiasm and project into the contralateral superior colliculus and the lateral geniculate nucleus which relays visual information to the primary visual cortex. About 0.5–1 h after intraocular Mn^{2+} injection the MEMRI signal of the optic nerve begins to increase (161). At 24–48 h, the images reveal a continuous pattern of anterograde labeling of the visual pathways up to the superior colliculus and lateral geniculate nucleus (**Fig. 6**) (169, 170). In mice, a significant signal increase was even seen in the contralateral primary visual cortex and the lateral visual cortex (181). In rhesus monkeys unilateral intravitrial Mn^{2+} injection revealed a clear signal enhancement of the entire optical system. Remarkably, even the layer-specific projection of the retinal ganglion cells to the lateral geniculate nucleus could be resolved by MEMRI. Furthermore, the rate of Mn^{2+} transport in the magnocellular layers was higher than in the parvocellular layers, which most likely reflects the known differences of respective axonal sizes (112).

In mice with an estimated vitreous volume of about 7 μL (182), toxicity studies after intraocular Mn^{2+} injection found that a volume of 0.5 μL saline depressed visually evoked potentials

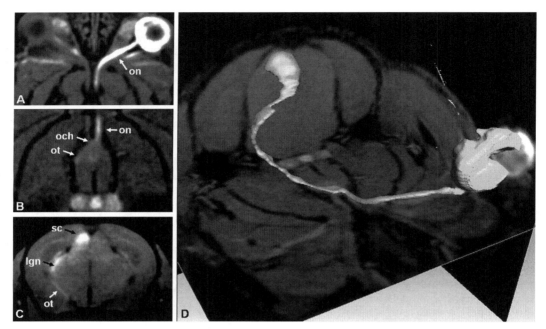

Fig. 6. (**a, b**) Horizontal and (**c**) axial MEMRI sections and (**d**) 3D volume rendering of a mouse optical system in vivo 24 h after intravitrial administration of 0.06 μmol MnCl$_2$ (2.35 T, 117 μm isotropic resolution). lgn, lateral geniculate nucleus; och, optic chiasm; on, optic nerve; ot, optic tract; sc, superior colliculus.

(VEP), whereas 0.25 μL did not (182). However, a volume of 0.125 μL of 50 mM MnCl$_2$ (corresponding to 6.25 nmol Mn^{2+}) depressed the VEP response at 4 h, but allowed for a full recovery at 24 h (161). Unfortunately, because this low amount of Mn^{2+} failed to generate an MRI signal enhancement, the suggested injection was 0.25 μL of 200 mM MnCl$_2$ (50 nmol Mn^{2+}) (161). It is worth noting that 4 months after this dose, a loss of 10–20% of the optic nerve axons was detected (161). In another study, a maximum T_1 shortening in the superior colliculus was obtained after the injection of 0.5 μL of 25 mM MnCl$_2$ (12.5 nmol Mn^{2+}) into the eye, while higher concentrations degraded the result (171).

In rats, an intravitreal dose of 150 nmol MnCl$_2$ (3 μL of 50 mM MnCl$_2$ solution) emerged as a threshold below which there was no significant reduction of retinal ganglion cells, whereas higher doses led to increasingly lower cell densities (170). The use of a second injection of 3 μL of 50 mM MnCl$_2$ 14–20 days after the first administration revealed a similar contrast-to-noise ratio on T_1-weighted MRI (170, 178).

A return to basal signal intensities in the visual system was reported 3–14 days after intraocular injection of Mn^{2+} (168, 181) (*see* **Note 6**).

3.5.4. Axonal Projections After Intracerebral Injection

In vivo tract tracing by intracerebral manganese injections was used to delineate cortical projections in mice (183), rats (99, 184, 185), and monkeys (113) under normal and pathological conditions (184, 186, 187). Moreover, several studies addressed the transport of manganese into the striatum, amygdala (111, 112, 188), and hippocampus (189). In birds, the song control circuit could be visualized by Mn^{2+} injection into the high vocal center (HVL) (115).

All intracerebral applications acquire a stereotactic injection of Mn^{2+} into the brain. In view of the manganese toxicity, it is therefore mandatory to minimize the local injury, potential osmotic and pressure effects, the injection volume and needle size, and the concentration of the manganese solution (*see* **Section 3.1.2**) (*see* **Note 7**). Examples for volumes and concentrations are given in **Tables 4**, **5**, and **6**.

3.5.5. Spinal Cord

Tract tracing by MEMRI using stereotactic injections of Mn^{2+} in the spinal cord were performed in healthy animals and animals with spinal cord injuries (190–194). The findings correlate well with locomotor ratings (190). Improved locomotion by treatment was accompanied by an increase of manganese uptake into the injured spinal cord (191).

3.6. Quantitative Analyses

3.6.1. MEMRI Signal

The signal intensity of T_1-weighted images depends on a number of experimental parameters and therefore is not directly comparable. To overcome some of these problems for quantitative assessments, normalization of signal intensities (174) in selected regions-of-interest (ROI) with the signal of a presumably unaffected region is a common procedure. For that purpose, both external standards such as Gd-DTPA or manganese solutions of known concentration and internal standards such as muscle tissue are applied (167, 181). The latter may also partly account for nonspecific signal enhancements or unexpected differences in the administered dose.

The comparison between pre-contrast and post-contrast images may be accomplished by ROI analyses or on a voxel-by-voxel basis. The latter approach requires a co-registration of respective MRI data sets and an intensity normalization is recommended (174). Voxel-wise statistical assessments of significant signal enhancements result in thresholded statistical maps that may be superimposed onto anatomical images (112, 195).

In principle, quantitative determinations of regional T_1 relaxation times allow for a more accurate analysis of the manganese accumulation and distribution. However, because other factors such as field strength and MRI sequence design may play a role, it is common practice to use relative T_1 changes due to manganese. Moreover, because the respective MRI measurements are much more time-consuming than the acquisition of a

3.6.2. Axonal Transport

T_1-weighted image, only few attempts to T_1 mapping have been performed and in many cases with lower spatial resolution.

A simple way to quantify the axonal transport is to analyze the normalized signal change in a target area as a function of the time after Mn^{2+} administration. The slope of the curve may be taken as an indicator for the transport rate. Depending on the relative locations of administration and target areas, this value reflects both the axonal transport and the properties of any intermediate transsynaptic transfer. To address the latter the ratio of the signal intensities in the pre- and post-synaptic ROI 24 h after local Mn^{2+} administration was suggested as a magnetization transfer index (41). Alternatively, a connectivity index was introduced (99), which is the difference in signal intensity between the baseline image and an image 10 h after local Mn^{2+} injection relative to the signal increase at the injection site 30–40 min after Mn^{2+} administration. In principle, all these methods exploit (temporal) changes of the signal intensity in the destination area. Alternatively, one may obtain the signal intensities at specific distances along a single tract as a function of time (196).

4. Notes

For biomedical research involving experimental animals, MEMRI has developed into an exciting tool with unique tissue contrasts that offer structural and functional information complementary to plain MRI techniques. However, manganese is a potential toxicant, which requires careful administration. In addition, its complex chemical and biological behavior in the living organism should be taken into account to avoid misinterpretation of MEMRI findings.

1. In general, the lowest possible manganese concentration should be used in the form of isotonic solutions buffered at physiological pH. A slow manganese release by micropumps or the use of manganese complexes that slowly dissociate in biological media may be alternative approaches to reduce the acute manganese toxicity.

2. Attempts should be made to increase the sensitivity of MEMRI by optimizing the MRI hardware, MRI sequences, and data analysis to allow for using the lowest possible manganese concentration for a specific biological question.

3. The disruption of the BBB is a matter of concern, in particular, in MEMRI studies of disease-associated enhancements in animal models of brain disorders. If not accounted for,

the accumulation of manganese may be caused by a higher passage of manganese through a BBB leakage rather than reflect a specific pathology, which in turn may be accompanied by an increased cellular calcium influx or elevated levels of manganese-binding enzymes.

4. Many applications rely on a presumed uptake of manganese via voltage gated calcium channels (VGCCs). However, this is only one way of how manganese may enter into cells, while other mechanisms may prevail under specific pathological conditions. The general identification of Mn^{2+} accumulation with Ca^{2+} influx is an unwarranted and oversimplified assumption. Further studies are necessary to understand the molecular basis of manganese turnover and possible alterations by pathology and across species.

5. The manganese homeostasis (influx, binding, efflux) may be disturbed in disease models and observed changes in MEMRI may be related to differences in manganese uptake into the brain rather than to local functional or structural alterations within the brain.

6. For follow-up studies, the residual amount of manganese from earlier applications and potential chronic effects in relation to it should be taken into account by adding appropriate control groups.

7. In most MEMRI studies reported so far, a certain degree of toxicity due to the applied dose of manganese is undeniable. Nevertheless, this technique may reveal new biological information that may otherwise be difficult to obtain or not at all achievable. Animal welfare should always be a prime concern, including a close monitoring of vital parameters, body weight or behavioral changes, and an adequate pain treatment.

References

1. Mendonca-Dias, M. H., Gaggelli, E., and Lauterbur, P. C. (1983) Paramagnetic contrast agents in nuclear magnetic resonance medical imaging. *Semin. Nucl. Med.* **13**, 364–76.
2. Koretsky, A. P., and Silva, A. C. (2004) Manganese-enhanced magnetic resonance imaging (MEMRI). *NMR Biomed.* **17**, 527–31.
3. Fabry, M. E., and Eisenstadt, M. (1975) Water exchange between red cells and plasma. Measurement by nuclear magnetic relaxation. *Biophys. J.* **15**, 1101–10.
4. Dwek, R. A., Radda, G. K., Richards, R. E., and Salmon, A. G. (1972) Probes for the conformational transitions of phosphorylase. Effect of ligands studied by proton-relaxation enhancement, and chemical reactivities. *Eur. J. Biochem.* **29**, 509–14.
5. Low, R. N. (1997) Contrast agents for MR imaging of the liver. *J. Magn. Reson. Imaging* **7**, 56–67.
6. Wang, C. (1998) Mangafodipir trisodium (MnDPDP)-enhanced magnetic resonance imaging of the liver and pancreas. *Acta Radiol. Suppl.* **415**, 1–31.
7. Diehl, S. J., Lehmann, K. J., Gaa, J., McGill, S., Hoffmann, V., and Georgi, M. (1999) MR imaging of pancreatic lesions. Comparison of manganese-DPDP

and gadolinium chelate. *Invest. Radiol.* **34**, 589–95.
8. Wendland, M. F. (2004) Applications of manganese-enhanced magnetic resonance imaging (MEMRI) to imaging of the heart. *NMR Biomed.* **17**, 581–94.
9. Aschner, J. L., and Aschner, M. (2005) Nutritional aspects of manganese homeostasis. *Mol. Aspects Med.* **26**, 353–62.
10. Prohaska, J. R. (1987) Functions of trace elements in brain metabolism. *Physiol. Rev.* **67**, 858–901.
11. Borgstahl, G. E., Parge, H. E., Hickey, M. J., Beyer, W. F., Jr., Hallewell, R. A., and Tainer, J. A. (1992) The structure of human mitochondrial manganese superoxide dismutase reveals a novel tetrameric interface of two 4-helix bundles. *Cell* **71**, 107–18.
12. Ingersoll, R. T., Montgomery, E. B., Jr., and Aposhian, H. V. (1999) Central nervous system toxicity of manganese. II: Cocaine or reserpine inhibit manganese concentration in the rat brain. *Neurotoxicology* **20**, 467–76.
13. Marklund, S. L. (1984) Extracellular superoxide dismutase in human tissues and human cell lines. *J. Clin. Invest.* **74**, 1398–403.
14. Wedler, F. C., Denman, R. B., and Roby, W. G. (1982) Glutamine synthetase from ovine brain is a manganese(II) enzyme. *Biochemistry* **21**, 6389–96.
15. Wedler, F. C., and Denman, R. B. (1984) Glutamine synthetase: The major Mn(II) enzyme in mammalian brain. *Curr. Top. Cell. Regul.* **24**, 153–69.
16. Takeda, A. (2003) Manganese action in brain function. *Brain Res. Brain Res. Rev.* **41**, 79–87.
17. Milne, D. B., Sims, R. L., and Ralston, N. V. (1990) Manganese content of the cellular components of blood. *Clin. Chem.* **36**, 450–2.
18. Arnaud, J., Bourlard, P., Denis, B., and Favier, A. E. (1996) Plasma and erythrocyte manganese concentrations. Influence of age and acute myocardial infarction. *Biol. Trace Elem. Res.* **53**, 129–36.
19. Davis, C. D., Zech, L., and Greger, J. L. (1993) Manganese metabolism in rats: An improved methodology for assessing gut endogenous losses. *Proc. Soc. Exp. Biol. Med.* **202**, 103–8.
20. Schmidt, P. P., Toft, K. G., Skotland, T., and Andersson, K. (2002) Stability and transmetallation of the magnetic resonance contrast agent MnDPDP measured by EPR. *J. Biol. Inorg. Chem.* **7**, 241–8.
21. Aisen, P., Aasa, R., and Redfield, A. G. (1969) The chromium, manganese, and cobalt complexes of transferrin. *J. Biol. Chem.* **244**, 4628–33.
22. Harris, W. R., and Chen, Y. (1994) Electron paramagnetic resonance and difference ultraviolet studies of Mn^{2+} binding to serum transferrin. *J. Inorg. Biochem.* **54**, 1–19.
23. Zheng, W., Kim, H., and Zhao, Q. (2000) Comparative toxicokinetics of manganese chloride and methylcyclopentadienyl manganese tricarbonyl (MMT) in Sprague-Dawley rats. *Toxicol. Sci.* **54**, 295–301.
24. Tjalve, H., Mejare, C., and Borg-Neczak, K. (1995) Uptake and transport of manganese in primary and secondary olfactory neurones in pike. *Pharmacol. Toxicol.* **77**, 23–31.
25. Tjalve, H., Henriksson, J., Tallkvist, J., Larsson, B. S., and Lindquist, N. G. (1996) Uptake of manganese and cadmium from the nasal mucosa into the central nervous system via olfactory pathways in rats. *Pharmacol. Toxicol.* **79**, 347–56.
26. Murphy, V. A., Wadhwani, K. C., Smith, Q. R., and Rapoport, S. I. (1991) Saturable transport of manganese(II) across the rat blood-brain barrier. *J. Neurochem.* **57**, 948–54.
27. Rabin, O., Hegedus, L., Bourre, J. M., and Smith, Q. R. (1993) Rapid brain uptake of manganese(II) across the blood-brain barrier. *J. Neurochem.* **61**, 509–17.
28. Crossgrove, J. S., Allen, D. D., Bukaveckas, B. L., Rhineheimer, S. S., and Yokel, R. A. (2003) Manganese distribution across the blood-brain barrier. I. Evidence for carrier-mediated influx of managanese citrate as well as manganese and manganese transferrin. *Neurotoxicology* **24**, 3–13.
29. Aschner, M., Guilarte, T. R., Schneider, J. S., and Zheng, W. (2007) Manganese: Recent advances in understanding its transport and neurotoxicity. *Toxicol. Appl. Pharmacol.* **221**, 131–47.
30. Drapeau, P., and Nachshen, D. A. (1984) Manganese fluxes and manganese-dependent neurotransmitter release in presynaptic nerve endings isolated from rat brain. *J. Physiol.* **348**, 493–510.
31. Narita, K., Kawasaki, F., and Kita, H. (1990) Mn and Mg influxes through Ca channels of motor nerve terminals are prevented by verapamil in frogs. *Brain Res.* **510**, 289–95.
32. Moos, T. (1996) Immunohistochemical localization of intraneuronal transferrin receptor immunoreactivity in the adult

mouse central nervous system. *J. Comp. Neurol.* **375**, 675–92.
33. Gunshin, H., Mackenzie, B., Berger, U. V., Gunshin, Y., Romero, M. F., Boron, W. F., Nussberger, S., Gollan, J. L., and Hediger, M. A. (1997) Cloning and characterization of a mammalian proton-coupled metal-ion transporter. *Nature* **388**, 482–8.
34. Aschner, M., Vrana, K. E., and Zheng, W. (1999) Manganese uptake and distribution in the central nervous system (CNS). *Neurotoxicology* **20**, 173–80.
35. Aschner, M., Gannon, M., and Kimelberg, H. K. (1992) Manganese uptake and efflux in cultured rat astrocytes. *J. Neurochem.* **58**, 730–5.
36. Gavin, C. E., Gunter, K. K., and Gunter, T. E. (1999) Manganese and calcium transport in mitochondria: Implications for manganese toxicity. *Neurotoxicology* **20**, 445–53.
37. Gavin, C. E., Gunter, K. K., and Gunter, T. E. (1990) Manganese and calcium efflux kinetics in brain mitochondria. Relevance to manganese toxicity. *Biochem. J.* **266**, 329–34.
38. Roels, H., Meiers, G., Delos, M., Ortega, I., Lauwerys, R., Buchet, J. P., and Lison, D. (1997) Influence of the route of administration and the chemical form ($MnCl_2$, MnO_2) on the absorption and cerebral distribution of manganese in rats. *Arch. Toxicol.* **71**, 223–30.
39. Sloot, W. N., and Gramsbergen, J. B. (1994) Axonal transport of manganese and its relevance to selective neurotoxicity in the rat basal ganglia. *Brain Res.* **657**, 124–32.
40. Takeda, A., Sotogaku, N., and Oku, N. (2002) Manganese influences the levels of neurotransmitters in synapses in rat brain. *Neuroscience* **114**, 669–74.
41. Serrano, F., Deshazer, M., Smith, K. D., Ananta, J. S., Wilson, L. J., and Pautler, R. G. (2008) Assessing transneuronal dysfunction utilizing manganese-enhanced MRI (MEMRI). *Magn. Reson. Med.* **60**, 169–75.
42. Yokel, R. A., Crossgrove, J. S., and Bukaveckas, B. L. (2003) Manganese distribution across the blood-brain barrier. II. Manganese efflux from the brain does not appear to be carrier mediated. *Neurotoxicology* **24**, 15–22.
43. Takeda, A., Sawashita, J., and Okada, S. (1995) Biological half-lives of zinc and manganese in rat brain. *Brain Res.* **695**, 53–8.
44. Crossgrove, J., and Zheng, W. (2004) Manganese toxicity upon overexposure. *NMR Biomed.* **17**, 544–53.
45. Chuang, K. H., Koretsky, A. P., and Sotak, C. H. (2009) Temporal changes in the T1 and T2 relaxation rates (DeltaR1 and DeltaR2) in the rat brain are consistent with the tissue-clearance rates of elemental manganese. *Magn. Reson. Med.* **61**, 1528–32.
46. Jiang, Y., and Zheng, W. (2005) Cardiovascular toxicities upon manganese exposure. *Cardiovasc. Toxicol.* **5**, 345–54.
47. Chen, J. Y., Tsao, G. C., Zhao, Q., and Zheng, W. (2001) Differential cytotoxicity of Mn(II) and Mn(III): Special reference to mitochondrial [Fe-S] containing enzymes. *Toxicol. Appl. Pharmacol.* **175**, 160–8.
48. Gavin, C. E., Gunter, K. K., and Gunter, T. E. (1992) Mn^{2+} sequestration by mitochondria and inhibition of oxidative phosphorylation. *Toxicol. Appl. Pharmacol.* **115**, 1–5.
49. Brouillet, E. P., Shinobu, L., McGarvey, U., Hochberg, F., and Beal, M. F. (1993) Manganese injection into the rat striatum produces excitotoxic lesions by impairing energy metabolism. *Exp. Neurol.* **120**, 89–94.
50. Hazell, A. S. (2002) Astrocytes and manganese neurotoxicity. *Neurochem. Int.* **41**, 271–7.
51. Zwingmann, C., Leibfritz, D., and Hazell, A. S. (2003) Energy metabolism in astrocytes and neurons treated with manganese: Relation among cell-specific energy failure, glucose metabolism, and intercellular trafficking using multinuclear NMR-spectroscopic analysis. *J. Cereb. Blood Flow Metab.* **23**, 756–71.
52. Erikson, K. M., Suber, R. L., and Aschner, M. (2002) Glutamate/aspartate transporter (GLAST), taurine transporter and metallothionein mRNA levels are differentially altered in astrocytes exposed to manganese chloride, manganese phosphate or manganese sulfate. *Neurotoxicology* **23**, 281–8.
53. Sidoryk-Wegrzynowicz, M., Lee, E., Albrecht, J., and Aschner, M. (2009) Manganese disrupts astrocyte glutamine transporter expression and function. *J. Neurochem.* **110**, 822–30.
54. Normandin, L., and Hazell, A. S. (2002) Manganese neurotoxicity: An update of pathophysiologic mechanisms. *Metab. Brain Dis.* **17**, 375–87.
55. Fitsanakis, V. A., Au, C., Erikson, K. M., and Aschner, M. (2006) The effects of manganese on glutamate, dopamine and gamma-aminobutyric acid regulation. *Neurochem. Int.* **48**, 426–33.
56. Lipe, G. W., Duhart, H., Newport, G. D., Slikker, W., Jr., and Ali, S. F. (1999)

Effect of manganese on the concentration of amino acids in different regions of the rat brain. *J. Environ. Sci. Health B.* **34**, 119–32.
57. Zwingmann, C., Leibfritz, D., and Hazell, A. S. (2007) NMR spectroscopic analysis of regional brain energy metabolism in manganese neurotoxicity. *Glia* **55**, 1610–7.
58. Bonilla, E., Arrieta, A., Castro, F., Davila, J. O., and Quiroz, I. (1994) Manganese toxicity: Free amino acids in the striatum and olfactory bulb of the mouse. *Invest. Clin.* **35**, 175–81.
59. Bonilla, E. (1978) Increased GABA content in caudate nucleus of rats after chronic manganese chloride administration. *J. Neurochem.* **31**, 551–2.
60. Gianutsos, G., and Murray, M. T. (1982) Alterations in brain dopamine and GABA following inorganic or organic manganese administration. *Neurotoxicology* **3**, 75–81.
61. Gwiazda, R. H., Lee, D., Sheridan, J., and Smith, D. R. (2002) Low cumulative manganese exposure affects striatal GABA but not dopamine. *Neurotoxicology* **23**, 69–76.
62. Reaney, S. H., Bench, G., and Smith, D. R. (2006) Brain accumulation and toxicity of Mn(II) and Mn(III) exposures. *Toxicol. Sci.* **93**, 114–24.
63. Seth, P. K., Hong, J. S., Kilts, C. D., and Bondy, S. C. (1981) Alteration of cerebral neurotransmitter receptor function by exposure of rats to manganese. *Toxicol. Lett.* **9**, 247–54.
64. Chandra, S. V., Malhotra, K. M., and Shukla, G. S. (1982) GABAergic neurochemistry in manganese exposed rats. *Acta Pharmacol. Toxicol. (Copenh).* **51**, 456–8.
65. Lai, J. C., Leung, T. K., and Lim, L. (1981) Brain regional distribution of glutamic acid decarboxylase, choline acetyltransferase, and acetylcholinesterase in the rat: Effects of chronic manganese chloride administration after two years. *J Neurochem.* **36**, 1443–8.
66. Struve, M. F., McManus, B. E., Wong, B. A., and Dorman, D. C. (2007) Basal ganglia neurotransmitter concentrations in rhesus monkeys following subchronic manganese sulfate inhalation. *Am. J. Ind. Med.* **50**, 772–8.
67. Gwiazda, R., Lucchini, R., and Smith, D. (2007) Adequacy and consistency of animal studies to evaluate the neurotoxicity of chronic low-level manganese exposure in humans. *J. Toxicol. Environ. Health A.* **70**, 594–605.
68. Calne, D. B., Chu, N. S., Huang, C. C., Lu, C. S., and Olanow, W. (1994) Manganism and idiopathic Parkinsonism: Similarities and differences. *Neurology* **44**, 1583–6.
69. Pal, P. K., Samii, A., and Calne, D. B. (1999) Manganese neurotoxicity: A review of clinical features, imaging and pathology. *Neurotoxicology* **20**, 227–38.
70. Perl, D. P., and Olanow, C. W. (2007) The neuropathology of manganese-induced Parkinsonism. *J. Neuropathol. Exp. Neurol.* **66**, 675–82.
71. Eriksson, H., Gillberg, P. G., Aquilonius, S. M., Hedstrom, K. G., and Heilbronn, E. (1992) Receptor alterations in manganese intoxicated monkeys. *Arch. Toxicol.* **66**, 359–64.
72. Nam, J., and Kim, K. (2008) Abnormal motor function and the expression of striatal dopamine D2 receptors in manganese-treated mice. *Biol. Pharm. Bull.* **31**, 1894–7.
73. Erikson, K. M., John, C. E., Jones, S. R., and Aschner, M. (2005) Manganese accumulation in striatum of mice exposed to toxic doses is dependent upon a functional dopamine transporter. *Environ. Toxicol. Pharmacol.* **20**, 390–394.
74. Burton, N. C., and Guilarte, T. R. (2009) Manganese neurotoxicity: Lessons learned from longitudinal studies in nonhuman primates. *Environ. Health Perspect.* **117**, 325–32.
75. Finkelstein, Y., Milatovic, D., and Aschner, M. (2007) Modulation of cholinergic systems by manganese. *Neurotoxicology* **28**, 1003–14.
76. Archibald, F. S., and Tyree, C. (1987) Manganese poisoning and the attack of trivalent manganese upon catecholamines. *Arch. Biochem. Biophys.* **256**, 638–50.
77. Gunter, T. E., Miller, L. M., Gavin, C. E., Eliseev, R., Salter, J., Buntinas, L., Alexandrov, A., Hammond, S., and Gunter, K. K. (2004) Determination of the oxidation states of manganese in brain, liver, and heart mitochondria. *J. Neurochem.* **88**, 266–80.
78. Rovetta, F., Catalani, S., Steimberg, N., Boniotti, J., Gilberti, M. E., Mariggio, M. A., and Mazzoleni, G. (2007) Organ-specific manganese toxicity: A comparative in vitro study on five cellular models exposed to $MnCl_2$. *Toxicol. In Vitro.* **21**, 284–92.
79. Komura, J., and Sakamoto, M. (1991) Short-term oral administration of several manganese compounds in mice: Physiological and behavioral alterations caused by different forms of manganese. *Bull. Environ. Contam. Toxicol.* **46**, 921–8.

80. Aschner, M., and Aschner, J. L. (1990) Manganese transport across the blood-brain barrier: Relationship to iron homeostasis. *Brain Res. Bull.* **24**, 857–60.
81. Davidsson, L., Cederblad, A., Lonnerdal, B., and Sandstrom, B. (1991) The effect of individual dietary components on manganese absorption in humans. *Am. J. Clin. Nutr.* **54**, 1065–70.
82. Keen, C. L., Bell, J. G., and Lonnerdal, B. (1986) The effect of age on manganese uptake and retention from milk and infant formulas in rats. *J. Nutr.* **116**, 395–402.
83. Zlotkin, S. H., Atkinson, S., and Lockitch, G. (1995) Trace elements in nutrition for premature infants. *Clin. Perinatol.* **22**, 223–40.
84. Dorner, K., Dziadzka, S., Hohn, A., Sievers, E., Oldigs, H. D., Schulz-Lell, G., and Schaub, J. (1989) Longitudinal manganese and copper balances in young infants and preterm infants fed on breast-milk and adapted cow's milk formulas. *Br. J. Nutr.* **61**, 559–72.
85. Britton, A. A., and Cotzias, G. C. (1966) Dependence of manganese turnover on intake. *Am. J. Physiol.* **211**, 203–6.
86. Dorman, D. C., Struve, M. F., James, R. A., McManus, B. E., Marshall, M. W., and Wong, B. A. (2001) Influence of dietary manganese on the pharmacokinetics of inhaled manganese sulfate in male CD rats. *Toxicol. Sci.* **60**, 242–51.
87. Finley, J. W., and Davis, C. D. (1999) Manganese deficiency and toxicity: Are high or low dietary amounts of manganese cause for concern? *Biofactors* **10**, 15–24.
88. Malecki, E. A., Radzanowski, G. M., Radzanowski, T. J., Gallaher, D. D., and Greger, J. L. (1996) Biliary manganese excretion in conscious rats is affected by acute and chronic manganese intake but not by dietary fat. *J. Nutr.* **126**, 489–98.
89. Wolf, G. L., and Baum, L. (1983) Cardiovascular toxicity and tissue proton T1 response to manganese injection in the dog and rabbit. *AJR Am. J. Roentgenol.* **141**, 193–7.
90. Lin, Y. J., and Koretsky, A. P. (1997) Manganese ion enhances T1-weighted MRI during brain activation: An approach to direct imaging of brain function. *Magn. Reson. Med.* **38**, 378–88.
91. Brurok, H., Schjott, J., Berg, K., Karlsson, J. O., and Jynge, P. (1997) Manganese and the heart: Acute cardiodepression and myocardial accumulation of manganese. *Acta Physiol. Scand.* **159**, 33–40.
92. Aoki, I., Tanaka, C., Takegami, T., Ebisu, T., Umeda, M., Fukunaga, M., Fukuda, K., Silva, A. C., Koretsky, A. P., and Naruse, S. (2002) Dynamic activity-induced manganese-dependent contrast magnetic resonance imaging (DAIM MRI). *Magn. Reson. Med.* **48**, 927–33.
93. Kuo, Y. T., Herlihy, A. H., So, P. W., Bhakoo, K. K., and Bell, J. D. (2005) In vivo measurements of T1 relaxation times in mouse brain associated with different modes of systemic administration of manganese chloride. *J. Magn. Reson. Imaging* **21**, 334–9.
94. Itoh, K., Sakata, M., Watanabe, M., Aikawa, Y., and Fujii, H. (2008) The entry of manganese ions into the brain is accelerated by the activation of N-methyl-D-aspartate receptors. *Neuroscience* **154**, 732–40.
95. Bock, N. A., Paiva, F. F., and Silva, A. C. (2008) Fractionated manganese-enhanced MRI. *NMR Biomed.* **21**, 473–8.
96. Sotogaku, N., Oku, N., and Takeda, A. (2000) Manganese concentration in mouse brain after intravenous injection. *J. Neurosci. Res.* **61**, 350–6.
97. Bock, N. A., Paiva, F. F., Nascimento, G. C., Newman, J. D., and Silva, A. C. (2008) Cerebrospinal fluid to brain transport of manganese in a non-human primate revealed by MRI. *Brain Res.* **1198**, 160–70.
98. Milatovic, D., Zaja-Milatovic, S., Gupta, R. C., Yu, Y., and Aschner, M. (2009) Oxidative damage and neurodegeneration in manganese-induced neurotoxicity. *Toxicol. Appl. Pharmacol.* **240**, 219–25
99. Canals, S., Beyerlein, M., Keller, A. L., Murayama, Y., and Logothetis, N. K. (2008) Magnetic resonance imaging of cortical connectivity in vivo. *Neuroimage* **40**, 458–72.
100. Kang, Y. S., and Gore, J. C. (1984) Studies of tissue NMR relaxation enhancement by manganese. Dose and time dependences. *Invest. Radiol.* **19**, 399–407.
101. Gallez, B., Demeure, R., Baudelet, C., Abdelouahab, N., Beghein, N., Jordan, B., Geurts, M., and Roels, H. A. (2001) Non invasive quantification of manganese deposits in the rat brain by local measurement of NMR proton T1 relaxation times. *Neurotoxicology* **22**, 387–92.
102. Nordhoy, W., Anthonsen, H. W., Bruvold, M., Jynge, P., Krane, J., and Brurok, H. (2003) Manganese ions as intracellular contrast agents: Proton relaxation and calcium interactions in rat myocardium. *NMR Biomed.* **16**, 82–95.

103. Nordhoy, W., Anthonsen, H. W., Bruvold, M., Brurok, H., Skarra, S., Krane, J., and Jynge, P. (2004) Intracellular manganese ions provide strong T1 relaxation in rat myocardium. *Magn. Reson. Med.* **52**, 506–14.
104. Caravan, P., Farrar, C. T., Frullano, L., and Uppal, R. (2009) Influence of molecular parameters and increasing magnetic field strength on relaxivity of gadolinium- and manganese-based T1 contrast agents. *Contrast Media Mol. Imaging* **4**, 89–100.
105. Sotak, C. H., Sharer, K., and Koretsky, A. P. (2008) Manganese cell labeling of murine hepatocytes using manganese(III)-transferrin. *Contrast Media Mol. Imaging* **3**, 95–105.
106. Spiller, M., Brown, R. D., 3rd, Koenig, S. H., and Wolf, G. L. (1988) Longitudinal proton relaxation rates in rabbit tissues after intravenous injection of free and chelated Mn^{2+}. *Magn. Reson. Med.* **8**, 293–313.
107. Takeda, A., Sawashita, J., and Okada, S. (1998) Manganese concentration in rat brain: Manganese transport from the peripheral tissues. *Neurosci. Lett.* **242**, 45–8.
108. Watanabe, T., Natt, O., Boretius, S., Frahm, J., and Michaelis, T. (2002) In vivo 3D MRI staining of mouse brain after subcutaneous application of $MnCl_2$. *Magn. Reson. Med.* **48**, 852–9.
109. Aoki, I., Wu, Y. J., Silva, A. C., Lynch, R. M., and Koretsky, A. P. (2004) In vivo detection of neuroarchitecture in the rodent brain using manganese-enhanced MRI. *Neuroimage* **22**, 1046–59.
110. Silva, A. C., Lee, J. H., Wu, C. W., Tucciarone, J., Pelled, G., Aoki, I., and Koretsky, A. P. (2008) Detection of cortical laminar architecture using manganese-enhanced MRI. *J. Neurosci. Methods* **167**, 246–57.
111. Saleem, K. S., Pauls, J. M., Augath, M., Trinath, T., Prause, B. A., Hashikawa, T., and Logothetis, N. K. (2002) Magnetic resonance imaging of neuronal connections in the macaque monkey. *Neuron* **34**, 685–700.
112. Murayama, Y., Weber, B., Saleem, K. S., Augath, M., and Logothetis, N. K. (2006) Tracing neural circuits in vivo with Mn-enhanced MRI. *Magn. Reson. Imaging* **24**, 349–58.
113. Simmons, J. M., Saad, Z. S., Lizak, M. J., Ortiz, M., Koretsky, A. P., and Richmond, B. J. (2008) Mapping prefrontal circuits in vivo with manganese-enhanced magnetic resonance imaging in monkeys. *J. Neurosci.* **28**, 7637–47.
114. Bock, N. A., Kocharyan, A., and Silva, A. C. (2009) Manganese-enhanced MRI visualizes V1 in the non-human primate visual cortex. *NMR Biomed.* **22**, 730–6.
115. Van der Linden, A., Verhoye, M., Van Meir, V., Tindemans, I., Eens, M., Absil, P., and Balthazart, J. (2002) In vivo manganese-enhanced magnetic resonance imaging reveals connections and functional properties of the songbird vocal control system. *Neuroscience* **112**, 467–74.
116. Van Meir, V., Pavlova, D., Verhoye, M., Pinxten, R., Balthazart, J., Eens, M., and Van der Linden, A. (2006) In vivo MR imaging of the seasonal volumetric and functional plasticity of song control nuclei in relation to song output in a female songbird. *Neuroimage* **31**, 981–92.
117. Tindemans, I., Verhoye, M., Balthazart, J., and Van Der Linden, A. (2003) In vivo dynamic ME-MRI reveals differential functional responses of RA- and area X-projecting neurons in the HVC of canaries exposed to conspecific song. *Eur. J. Neurosci.* **18**, 3352–60.
118. Watanabe, T., Schachtner, J., Krizan, M., Boretius, S., Frahm, J., and Michaelis, T. (2006) Manganese-enhanced 3D MRI of established and disrupted synaptic activity in the developing insect brain in vivo. *J. Neurosci. Methods* **158**, 50–5.
119. Null, B., Liu, C. W., Hedehus, M., Conolly, S., and Davis, R. W. (2008) High-resolution, in vivo magnetic resonance imaging of Drosophila at 18.8 Tesla. *PLoS One* **3**, e2817.
120. London, R. E., Toney, G., Gabel, S. A., and Funk, A. (1989) Magnetic resonance imaging studies of the brains of anesthetized rats treated with manganese chloride. *Brain Res. Bull.* **23**, 229–35.
121. Lee, J. H., Silva, A. C., Merkle, H., and Koretsky, A. P. (2005) Manganese-enhanced magnetic resonance imaging of mouse brain after systemic administration of $MnCl_2$: Dose-dependent and temporal evolution of T1 contrast. *Magn. Reson. Med.* **53**, 640–8.
122. Wadghiri, Y. Z., Blind, J. A., Duan, X., Moreno, C., Yu, X., Joyner, A. L., and Turnbull, D. H. (2004) Manganese-enhanced magnetic resonance imaging (MEMRI) of mouse brain development. *NMR Biomed.* **17**, 613–9.
123. Liu, C. H., D'Arceuil, H. E., and de Crespigny, A. J. (2004) Direct CSF injection of $MnCl_2$ for dynamic manganese-enhanced MRI. *Magn. Reson. Med.* **51**, 978–87.

124. Dastur, D. K., Manghani, D. K., and Raghavendran, K. V. (1971) Distribution and fate of 54Mn in the monkey: Studies of different parts of the central nervous system and other organs. *J. Clin. Invest.* **50**, 9–20.
125. Lee, D. Y., and Johnson, P. E. (1988) Factors affecting absorption and excretion of 54Mn in rats. *J. Nutr.* **118**, 1509–16.
126. Angenstein, F., Niessen, H. G., Goldschmidt, J., Lison, H., Altrock, W. D., Gundelfinger, E. D., and Scheich, H. (2007) Manganese-enhanced MRI reveals structural and functional changes in the cortex of Bassoon mutant mice. *Cereb. Cortex.* **17**, 28–36.
127. Boretius, S., Michaelis, T., Tammer, R., Ashery-Padan, R., Frahm, J., and Stoykova, A. (2009) In vivo MRI of altered brain anatomy and fiber connectivity in adult Pax6 deficient mice. *Cereb. Cortex.* **19**, 2838–47.
128. Gallez, B., Baudelet, C., and Geurts, M. (1998) Regional distribution of manganese found in the brain after injection of a single dose of manganese-based contrast agents. *Magn. Reson. Imaging* **16**, 1211–5.
129. Watanabe, T., Radulovic, J., Boretius, S., Frahm, J., and Michaelis, T. (2006) Mapping of the habenulo-interpeduncular pathway in living mice using manganese-enhanced 3D MRI. *Magn Reson Imaging* **24**, 209–15.
130. Alvestad, S., Goa, P. E., Qu, H., Risa, O., Brekken, C., Sonnewald, U., Haraldseth, O., Hammer, J., Ottersen, O. P., and Haberg, A. (2007) In vivo mapping of temporospatial changes in manganese enhancement in rat brain during epileptogenesis. *Neuroimage* **38**, 57–66.
131. Boretius, S., Gadjanski, I., Demmer, I., Bahr, M., Diem, R., Michaelis, T., and Frahm, J. (2008) MRI of optic neuritis in a rat model. *Neuroimage* **41**, 323–34.
132. Gadjanski, I., Boretius, S., Williams, S. K., Lingor, P., Knoferle, J., Sattler, M. B., Fairless, R., Hochmeister, S., Suhs, K. W., Michaelis, T., Frahm, J., Storch, M. K., Bahr, M., and Diem, R. (2009) Role of n-type voltage-dependent calcium channels in autoimmune optic neuritis. *Ann. Neurol.* **66**, 81–93.
133. Aoki, I., Naruse, S., and Tanaka, C. (2004) Manganese-enhanced magnetic resonance imaging (MEMRI) of brain activity and applications to early detection of brain ischemia. *NMR Biomed.* **17**, 569–80.
134. Berkowitz, B. A., Roberts, R., Goebel, D. J., and Luan, H. (2006) Noninvasive and simultaneous imaging of layer-specific retinal functional adaptation by manganese-enhanced MRI. *Invest. Ophthalmol. Vis. Sci.* **47**, 2668–74.
135. Fujioka, M., Taoka, T., Matsuo, Y., Mishima, K., Ogoshi, K., Kondo, Y., Tsuda, M., Fujiwara, M., Asano, T., Sakaki, T., Miyasaki, A., Park, D., and Siesjo, B. K. (2003) Magnetic resonance imaging shows delayed ischemic striatal neurodegeneration. *Ann. Neurol.* **54**, 732–47.
136. Yang, J., and Wu, E. X. (2008) Detection of cortical gray matter lesion in the late phase of mild hypoxic-ischemic injury by manganese-enhanced MRI. *Neuroimage* **39**, 669–79.
137. Haapanen, A., Ramadan, U. A., Autti, T., Joensuu, R., and Tyynela, J. (2007) In vivo MRI reveals the dynamics of pathological changes in the brains of cathepsin D-deficient mice and correlates changes in manganese-enhanced MRI with microglial activation. *Magn. Reson. Imaging* **25**, 1024–31.
138. Wideroe, M., Olsen, O., Pedersen, T. B., Goa, P. E., Kavelaars, A., Heijnen, C., Skranes, J., Brubakk, A. M., and Brekken, C. (2009) Manganese-enhanced magnetic resonance imaging of hypoxic-ischemic brain injury in the neonatal rat. *Neuroimage* **45**, 880–90.
139. Murphy, V. A., Rosenberg, J. M., Smith, Q. R., and Rapoport, S. I. (1991) Elevation of brain manganese in calcium-deficient rats. *Neurotoxicology* **12**, 255–63.
140. Pomier-Layrargues, G., Spahr, L., and Butterworth, R. F. (1995) Increased manganese concentrations in pallidum of cirrhotic patients. *Lancet* **345**, 735.
141. Yanik, M., Kocyigit, A., Tutkun, H., Vural, H., and Herken, H. (2004) Plasma manganese, selenium, zinc, copper, and iron concentrations in patients with schizophrenia. *Biol. Trace Elem. Res.* **98**, 109–17.
142. Finley, J. W., Johnson, P. E., and Johnson, L. K. (1994) Sex affects manganese absorption and retention by humans from a diet adequate in manganese. *Am. J. Clin. Nutr.* **60**, 949–55.
143. Barritt, G. J. (1999) Receptor-activated Ca^{2+} inflow in animal cells: A variety of pathways tailored to meet different intracellular Ca^{2+} signalling requirements. *Biochem. J.* **337**(Pt 2), 153–69.
144. Zamponi, G. W., and Snutch, T. P. (1998) Modulation of voltage-dependent calcium channels by G proteins. *Curr. Opin. Neurobiol.* **8**, 351–6.
145. Kita, H., Narita, K., and Van der Kloot, W. (1981) Tetanic stimulation increases the frequency of miniature end-plate potentials at

the frog neuromuscular junction in Mn^{2+}-, CO^{2+}-, and Ni^{2+}-saline solutions. *Brain Res.* **205**, 111–21.
146. Pautler, R. G., and Koretsky, A. P. (2002) Tracing odor-induced activation in the olfactory bulbs of mice using manganese-enhanced magnetic resonance imaging. *Neuroimage* **16**, 441–8.
147. Hsu, Y. H., Lee, W. T., and Chang, C. (2007) Multiparametric MRI evaluation of kainic acid-induced neuronal activation in rat hippocampus. *Brain* **130**, 3124–34.
148. Lu, H., Xi, Z. X., Gitajn, L., Rea, W., Yang, Y., and Stein, E. A. (2007) Cocaine-induced brain activation detected by dynamic manganese-enhanced magnetic resonance imaging (MEMRI). *Proc. Natl. Acad. Sci. U.S.A.* **104**, 2489–94.
149. Allen, T. J. (1990) The effects of manganese and changes in internal calcium on Na-Ca exchange fluxes in the intact squid giant axon. *Biochim. Biophys. Acta.* **1030**, 101–10.
150. Medina, D. C., Kirkland, D. M., Tavazoie, M. F., Springer, C. S., Jr., and Anderson, S. E. (2007) Na^+/Ca^{2+}-exchanger-mediated Mn^{2+}-enhanced (1)H_2O MRI in hypoxic, perfused rat myocardium. *Contrast Media Mol. Imaging* **2**, 248–57.
151. Hallam, T. J., and Rink, T. J. (1985) Agonists stimulate divalent cation channels in the plasma membrane of human platelets. *FEBS Lett.* **186**, 175–9.
152. Morita, H., Ogino, T., Seo, Y., Fujiki, N., Tanaka, K., Takamata, A., Nakamura, S., and Murakami, M. (2002) Detection of hypothalamic activation by manganese ion contrasted T(1)-weighted magnetic resonance imaging in rats. *Neurosci. Lett.* **326**, 101–4.
153. Brown, R. C., Egleton, R. D., and Davis, T. P. (2004) Mannitol opening of the blood-brain barrier: Regional variation in the permeability of sucrose, but not 86Rb+ or albumin. *Brain Res.* **1014**, 221–7.
154. Henning, E. C., Meng, X., Fisher, M., and Sotak, C. H. (2005) Visualization of cortical spreading depression using manganese-enhanced magnetic resonance imaging. *Magn. Reson. Med.* **53**, 851–7.
155. Morita, H., Ogino, T., Fujiki, N., Tanaka, K., Gotoh, T. M., Seo, Y., Takamata, A., Nakamura, S., and Murakami, M. (2004) Sequence of forebrain activation induced by intraventricular injection of hypertonic NaCl detected by Mn^{2+} contrasted T1-weighted MRI. *Auton. Neurosci.* **113**, 43–54.

156. Duong, T. Q., Silva, A. C., Lee, S. P., and Kim, S. G. (2000) Functional MRI of calcium-dependent synaptic activity: Cross correlation with CBF and BOLD measurements. *Magn. Reson. Med.* **43**, 383–92.
157. Weng, J. C., Chen, J. H., Yang, P. F., and Tseng, W. Y. (2007) Functional mapping of rat barrel activation following whisker stimulation using activity-induced manganese-dependent contrast. *Neuroimage* **36**, 1179–88.
158. Kuo, Y. T., Herlihy, A. H., So, P. W., and Bell, J. D. (2006) Manganese-enhanced magnetic resonance imaging (MEMRI) without compromise of the blood-brain barrier detects hypothalamic neuronal activity in vivo. *NMR Biomed.* **19**, 1028–34.
159. Parkinson, J. R., Chaudhri, O. B., Kuo, Y. T., Field, B. C., Herlihy, A. H., Dhillo, W. S., Ghatei, M. A., Bloom, S. R., and Bell, J. D. (2009) Differential patterns of neuronal activation in the brainstem and hypothalamus following peripheral injection of GLP-1, oxyntomodulin and lithium chloride in mice detected by manganese-enhanced magnetic resonance imaging (MEMRI). *Neuroimage* **44**, 1022–31.
160. Chuang, K. H., Lee, J. H., Silva, A. C., Belluscio, L., and Koretsky, A. P. (2009) Manganese enhanced MRI reveals functional circuitry in response to odorant stimuli. *Neuroimage* **44**, 363–72.
161. Bearer, E. L., Falzone, T. L., Zhang, X., Biris, O., Rasin, A., and Jacobs, R. E. (2007) Role of neuronal activity and kinesin on tract tracing by manganese-enhanced MRI (MEMRI). *Neuroimage* **37**(Suppl 1), S37–46.
162. Bissig, D., and Berkowitz, B. A. (2009) Manganese-enhanced MRI of layer-specific activity in the visual cortex from awake and free-moving rats. *Neuroimage* **44**, 627–35.
163. Watanabe, T., Frahm, J., and Michaelis, T. (2008) Manganese-enhanced MRI of the mouse auditory pathway. *Magn. Reson. Med.* **60**, 210–2.
164. Yu, X., Wadghiri, Y. Z., Sanes, D. H., and Turnbull, D. H. (2005) In vivo auditory brain mapping in mice with Mn-enhanced MRI. *Nat. Neurosci.* **8**, 961–8.
165. Yu, X., Sanes, D. H., Aristizabal, O., Wadghiri, Y. Z., and Turnbull, D. H. (2007) Large-scale reorganization of the tonotopic map in mouse auditory midbrain revealed by MRI. *Proc. Natl. Acad. Sci. U.S.A.* **104**, 12193–8.
166. Takeda, A., Kodama, Y., Ishiwatari, S., and Okada, S. (1998) Manganese transport in

the neural circuit of rat CNS. *Brain Res. Bull.* **45**, 149–52.
167. Smith, K. D., Kallhoff, V., Zheng, H., and Pautler, R. G. (2007) In vivo axonal transport rates decrease in a mouse model of Alzheimer's disease. *Neuroimage* **35**, 1401–8.
168. Pautler, R. G., Silva, A. C., and Koretsky, A. P. (1998) In vivo neuronal tract tracing using manganese-enhanced magnetic resonance imaging. *Magn. Reson. Med.* **40**, 740–8.
169. Watanabe, T., Michaelis, T., and Frahm, J. (2001) Mapping of retinal projections in the living rat using high-resolution 3D gradient-echo MRI with Mn^{2+}-induced contrast. *Magn. Reson. Med.* **46**, 424–9.
170. Thuen, M., Berry, M., Pedersen, T. B., Goa, P. E., Summerfield, M., Haraldseth, O., Sandvig, A., and Brekken, C. (2008) Manganese-enhanced MRI of the rat visual pathway: Acute neural toxicity, contrast enhancement, axon resolution, axonal transport, and clearance of Mn^{2+}. *J. Magn. Reson. Imaging* **28**, 855–65.
171. Lowe, A. S., Thompson, I. D., and Sibson, N. R. (2008) Quantitative manganese tract tracing: Dose-dependent and activity-independent terminal labelling in the mouse visual system. *NMR Biomed.* **21**, 859–67.
172. Takeda, A., Ishiwatari, S., and Okada, S. (1998) In vivo stimulation-induced release of manganese in rat amygdala. *Brain Res.* **811**, 147–51.
173. Chuang, K. H., and Koretsky, A. P. (2009) Accounting for nonspecific enhancement in neuronal tract tracing using manganese enhanced magnetic resonance imaging. *Magn. Reson. Imaging* **27**, 594–600.
174. Cross, D. J., Minoshima, S., Anzai, Y., Flexman, J. A., Keogh, B. P., Kim, Y., and Maravilla, K. R. (2004) Statistical mapping of functional olfactory connections of the rat brain in vivo. *Neuroimage* **23**, 1326–35.
175. Henriksson, J., Tallkvist, J., and Tjalve, H. (1999) Transport of manganese via the olfactory pathway in rats: Dosage dependency of the uptake and subcellular distribution of the metal in the olfactory epithelium and the brain. *Toxicol. Appl. Pharmacol.* **156**, 119–28.
176. Cross, D. J., Flexman, J. A., Anzai, Y., Morrow, T. J., Maravilla, K. R., and Minoshima, S. (2006) In vivo imaging of functional disruption, recovery and alteration in rat olfactory circuitry after lesion. *Neuroimage* **32**, 1265–72.
177. Natt, O., Watanabe, T., Boretius, S., Radulovic, J., Frahm, J., and Michaelis, T. (2002) High-resolution 3D MRI of mouse brain reveals small cerebral structures in vivo. *J. Neurosci. Methods* **120**, 203–9.
178. Thuen, M., Singstad, T. E., Pedersen, T. B., Haraldseth, O., Berry, M., Sandvig, A., and Brekken, C. (2005) Manganese-enhanced MRI of the optic visual pathway and optic nerve injury in adult rats. *J. Magn. Reson. Imaging* **22**, 492–500.
179. Ryu, S., Brown, S. L., Kolozsvary, A., Ewing, J. R., and Kim, J. H. (2002) Noninvasive detection of radiation-induced optic neuropathy by manganese-enhanced MRI. *Radiat. Res.* **157**, 500–5.
180. Chan, K. C., Fu, Q. L., Hui, E. S., So, K. F., and Wu, E. X. (2008) Evaluation of the retina and optic nerve in a rat model of chronic glaucoma using in vivo manganese-enhanced magnetic resonance imaging. *Neuroimage* **40**, 1166–74.
181. Lindsey, J. D., Scadeng, M., Dubowitz, D. J., Crowston, J. G., and Weinreb, R. N. (2007) Magnetic resonance imaging of the visual system in vivo: Transsynaptic illumination of V1 and V2 visual cortex. *Neuroimage* **34**, 1619–26.
182. Yu, D. Y., and Cringle, S. J. (2006) Oxygen distribution in the mouse retina. *Invest. Ophthalmol. Vis. Sci.* **47**, 1109–12.
183. Leergaard, T. B., Bjaalie, J. G., Devor, A., Wald, L. L., and Dale, A. M. (2003) In vivo tracing of major rat brain pathways using manganese-enhanced magnetic resonance imaging and three-dimensional digital atlasing. *Neuroimage* **20**, 1591–600.
184. Allegrini, P. R., and Wiessner, C. (2003) Three-dimensional MRI of cerebral projections in rat brain in vivo after intracortical injection of $MnCl_2$. *NMR Biomed.* **16**, 252–6.
185. Tucciarone, J., Chuang, K. H., Dodd, S. J., Silva, A., Pelled, G., and Koretsky, A. P. (2009) Layer specific tracing of corticocortical and thalamocortical connectivity in the rodent using manganese enhanced MRI. *Neuroimage* **44**, 923–31.
186. Nairismagi, J., Pitkanen, A., Narkilahti, S., Huttunen, J., Kauppinen, R. A., and Grohn, O. H. (2006) Manganese-enhanced magnetic resonance imaging of mossy fiber plasticity in vivo. *Neuroimage* **30**, 130–5.
187. van der Zijden, J. P., Wu, O., van der Toorn, A., Roeling, T. P., Bleys, R. L., and Dijkhuizen, R. M. (2007) Changes in neuronal connectivity after stroke in rats as studied by serial manganese-enhanced MRI. *Neuroimage* **34**, 1650–7.
188. Pautler, R. G., Mongeau, R., and Jacobs, R. E. (2003) In vivo trans-synaptic tract

tracing from the murine striatum and amygdala utilizing manganese enhanced MRI (MEMRI). *Magn. Reson. Med.* **50**, 33–9.

189. Bearer, E. L., Zhang, X., and Jacobs, R. E. (2007) Live imaging of neuronal connections by magnetic resonance: Robust transport in the hippocampal-septal memory circuit in a mouse model of Down syndrome. *Neuroimage* **37**, 230–42.

190. Walder, N., Petter-Puchner, A. H., Brejnikow, M., Redl, H., Essig, M., and Stieltjes, B. (2008) Manganese enhanced magnetic resonance imaging in a contusion model of spinal cord injury in rats: Correlation with motor function. *Invest. Radiol.* **43**, 277–83.

191. Stieltjes, B., Klussmann, S., Bock, M., Umathum, R., Mangalathu, J., Letellier, E., Rittgen, W., Edler, L., Krammer, P. H., Kauczor, H. U., Martin-Villalba, A., and Essig, M. (2006) Manganese-enhanced magnetic resonance imaging for in vivo assessment of damage and functional improvement following spinal cord injury in mice. *Magn. Reson. Med.* **55**, 1124–31.

192. Bilgen, M., Dancause, N., Al-Hafez, B., He, Y. Y., and Malone, T. M. (2005) Manganese-enhanced MRI of rat spinal cord injury. *Magn. Reson. Imaging* **23**, 829–32.

193. Bilgen, M., Peng, W., Al-Hafez, B., Dancause, N., He, Y. Y., and Cheney, P. D. (2006) Electrical stimulation of cortex improves corticospinal tract tracing in rat spinal cord using manganese-enhanced MRI. *J. Neurosci. Methods* **156**, 17–22.

194. Bilgen, M. (2006) Imaging corticospinal tract connectivity in injured rat spinal cord using manganese-enhanced MRI. *BMC Med. Imaging* **6**, 15.

195. Yu, X., Zou, J., Babb, J. S., Johnson, G., Sanes, D. H., and Turnbull, D. H. (2008) Statistical mapping of sound-evoked activity in the mouse auditory midbrain using Mn-enhanced MRI. *Neuroimage* **39**, 223–30.

196. Cross, D. J., Flexman, J. A., Anzai, Y., Maravilla, K. R., and Minoshima, S. (2008) Age-related decrease in axonal transport measured by MR imaging in vivo. *Neuroimage* **39**, 915–26.

197. Garrick, M. D., Dolan, K. G., Horbinski, C., Ghio, A. J., Higgins, D., Porubcin, M., Moore, E. G., Hainsworth, L. N., Umbreit, J. N., Conrad, M. E., Feng, L., Lis, A., Roth, J. A., Singleton, S., and Garrick, L. M. (2003) DMT1: A mammalian transporter for multiple metals. *Biometals* **16**, 41–54.

198. Crossgrove, J. S., and Yokel, R. A. (2004) Manganese distribution across the blood-brain barrier III. The divalent metal transporter-1 is not the major mechanism mediating brain manganese uptake. *Neurotoxicology* **25**, 451–60.

199. Au, C., Benedetto, A., and Aschner, M. (2008) Manganese transport in eukaryotes: The role of DMT1. *Neurotoxicology* **29**, 569–76.

200. Takeda, A., Ishiwatari, S., and Okada, S. (2000) Influence of transferrin on manganese uptake in rat brain. *J. Neurosci. Res.* **59**, 542–52.

201. Crossgrove, J. S., and Yokel, R. A. (2005) Manganese distribution across the blood-brain barrier. IV. Evidence for brain influx through store-operated calcium channels. *Neurotoxicology* **26**, 297–307.

202. Lockman, P. R., Roder, K. E., and Allen, D. D. (2001) Inhibition of the rat blood-brain barrier choline transporter by manganese chloride. *J. Neurochem.* **79**, 588–94.

203. Fitsanakis, V. A., Piccola, G., Marreilha dos Santos, A. P., Aschner, J. L., and Aschner, M. (2007) Putative proteins involved in manganese transport across the blood-brain barrier. *Hum. Exp. Toxicol.* **26**, 295–302.

204. Wang, X., Kim, S. U., van Breemen, C., and McLarnon, J. G. (2000) Activation of purinergic P2X receptors inhibits P2Y-mediated Ca^{2+} influx in human microglia. *Cell Calcium* **27**, 205–12.

205. Silva, A. C., Lee, J. H., Aoki, I., and Koretsky, A. P. (2004) Manganese-enhanced magnetic resonance imaging (MEMRI): Methodological and practical considerations. *NMR Biomed.* **17**, 532–43.

206. Deans, A. E., Wadghiri, Y. Z., Berrios-Otero, C. A., and Turnbull, D. H. (2008) Mn enhancement and respiratory gating for in utero MRI of the embryonic mouse central nervous system. *Magn. Reson. Med.* **59**, 1320–8.

Chapter 29

Spin Echo BOLD fMRI on Songbirds

Colline Poirier and Anne-Marie Van der Linden

Abstract

The advent of high-field MRI systems has allowed implementation of BOLD fMRI on small animals. Increased magnetic field improves the signal-to-noise ratio and thus allows improvement of spatial resolution. However, it also increases susceptibility artefacts in the commonly acquired gradient echo images. The problem is particularly challenging in songbirds due to the presence of numerous air cavities in the skull of birds. This problem can be solved by using spin echo BOLD fMRI. In this chapter, we describe how to use this technique in zebra finches, a small songbird of 15–25 g extensively studied in behavioural neurosciences of birdsong. The protocol implements auditory stimuli.

Key words: fMRI, BOLD, spin echo, songbird, auditory, birdsong.

1. Introduction

The neurobiology of birdsong, as a model for human speech, is an important area of research in behavioural neurosciences. Whereas electrophysiology and molecular approaches allow investigation of either different stimuli on few neurons, or one stimulus in large parts of the brain, BOLD fMRI allows the combination of both advantages, i.e. the comparison of the neural activation induced by different stimuli in the whole brain. Functional MRI in songbirds is challenging because of the small size of their brains and because their bones and especially their skull comprise numerous air cavities, inducing important susceptibility artefacts. Until now, gradient echo (GE) BOLD fMRI has been successfully developed in songbirds (for a review, see (1)). These studies focused on the primary and secondary auditory regions, regions which are free of susceptibility artefacts. However, because processes of interest may occur beyond these regions, whole brain BOLD fMRI is required. It can be achieved by using spin echo BOLD fMRI (2). In this chapter, we describe how to use this technique on zebra finches, a small songbird of 15–25 g extensively studied in behavioural neurosciences of birdsong. The main topic of fMRI studies on songbirds is song perception and song learning. The auditory nature of the stimuli combined with the weak sensitivity of SE BOLD fMRI (compared to GE BOLD fMRI) make the implementation of this technique very challenging.

2. Materials

2.1. Subject Preparation

1. 15–20 Zebra finches (*see* **Note 1**).
2. Isoflurane, (Isoflo, Abbott Laboratories Ltd, United Kingdom) to anaesthetize the bird (3% induction, 2% maintenance), mixed with oxygen (flow rate: 0.1 L/min) and nitrogen (flow rate: 0.2 L/min). Gas controller device (Sierra Instruments Inc., The Netherlands) (*see* **Note 2**).
3. Home-made mask and plastic tubes to connect the mask to the gas controller device.
4. Cloacal temperature probe, warm air flow device and feedback-controlled system (PC-sam software) (SA Instruments, Inc., Stony Brook, NY, USA) for maintaining the internal temperature of the bird constant. Pneumatic sensor (SA Instruments, Inc., Stony Brook, New York, USA) for measuring the respiration rate. (*see* **Note 3**).
5. A jacket for restraining the body of the bird.

2.2. Auditory Stimulation

1. Auditory stimuli (*see* **Note 4**).
2. Non-magnetic dynamic speakers (incorporated into the transmitting coil) and amplifier (MM1002a, Phonic Corporation, Taiwan) (*see* **Note 5**).
3. Presentation software (Neurobehavioral Systems Inc., USA) to deliver the stimuli and trigger the MRI image acquisition.

2.3. Data Acquisition

1. 7 T horizontal bore NMR microscope (Bruker BioSpin MRI GmbH, Germany) (*see* **Note 6**).
2. Homemade transmit coil (Helmholtz antenna, 45 mm) and receive coil (circular antenna, 15 mm).
3. ParaVision software (version 4, Bruker BioSpin MRI GmbH, Germany).

2.4. Data Processing

1. SPM software (http://www.fil.ion.ucl.ac.uk/spm/).
2. Zebra finch atlas (3).

3. Methods

3.1. Subject Preparation

1. Install the beak mask on the MRI bed and connect it to the gas controller device with plastic tubes. Open the gas bottles and switch on the gas controller device (**Fig. 1**).
2. Switch on the feedback controlled system and the warm air flow device.
3. Anaesthetize the bird by introducing its beak into the mask and fix its head with the circular antenna.
4. Position the pneumatic sensor under the chest and introduce the cloacal temperature probe. Close the jacket.
5. Position and fix the Helmholtz antenna which includes the dynamic speakers.
6. Connect the non-magnetic dynamic speakers to the amplifier.
7. Insert the entire setup into the bore of the magnet.

3.2. Data Acquisition

1. Acquire a set of one sagittal, one horizontal and one coronal gradient echo (GE) scout image (tri-pilot sequence) and sets of horizontal, coronal and sagittal multi-slice images (piloting RARE sequence) to determine the position of the brain in the magnet.
2. Decrease the noise of the gradients by increasing their ramp times to 600 μs.
3. Prepare the fMRI sequence: RARE T_2-weighted sequence – effective TE, 60 ms; TR, 2000 ms; RARE factor, 8; FOV, 16 mm; matrix size, 64 × 32; slice thickness, 0.75 mm; 15 continuous slices (*see* **Note 7**).

Fig. 1. Subject preparation. (**a**) Preparation of the home-made mask (1), the jacket (2), the cloacal temperature probe (3) and the pneumatic sensor (4) to measure the respiration rate of the bird. (**b**) Anaesthesia of the bird and fixation of its head with the circular antenna (5). (**c**) Positioning and fixation of the Helmholtz coil (6) which includes the headphones. The tube providing the warm air blown over the bird to maintain its temperature constant is visible on the *left* side of panel c (7).

4. Select the auditory protocol (auditory stimuli and timing of stimulus delivery) in the presentation software (*see* **Note 8**).
5. Run the auditory protocol and the fMRI sequence.
6. Zero-fill the data to 64 × 64 using trapezoid interpolation.
7. Take a first look at the results using the functional tool of ParaVision (option Processing/Functional Imaging...) (*see* **Note 9**).
8. Run an anatomical 3D RARE T_2-weighted sequence – effective TE, 60 ms; TR, 2000 ms; RARE factor, 8; FOV, 16 mm; matrix size, 256 × 128 × 64.
9. Zero-fill the data to 256 × 256 × 256 using trapezoid interpolation.

3.3. Data Processing

1. Convert the data into Analyze or Nifti format.
2. Adjust the voxel size in the heading of the files by multiplying the real voxel size by 10 (*see* **Note 10**).
3. Realign the fMRI data.
4. Co-register the anatomical 3D dataset to the zebra finch atlas (*see* **Note 11**). Apply the transformation matrix to the fMRI dataset.
5. Smooth the data with a 0.5-mm width Gaussian kernel.

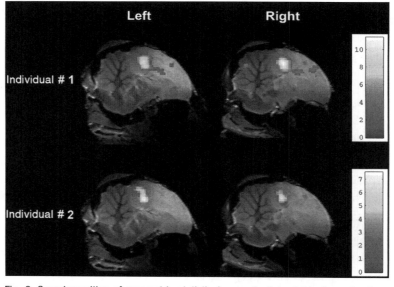

Fig. 2. Superimposition of parametric statistical maps (unilateral *t*-test) coming from two individuals on the zebra finch atlas. The maps correspond to the comparison 'ON' vs. 'OFF' (i.e. 'all stimuli' vs. 'rest periods') and illustrate the bilateral activation of the equivalent of the mammalian auditory cortex in the avian brain. Only voxels in which the *t*-test was found significant (uncorrected *p*-value < 0.001) are colour-coded (the correspondence between colours and *t*-values is indicated by the *colour bar*).

6. Carry out statistical voxel-based analyses using SPM5. Model the data as a box-car (no haemodynamic response function) (*see* **Note 12**). Estimate model parameters with the classical restricted maximum likelihood algorithm. Compute the mean effect of each auditory stimulus in each subject (fixed-effect analysis) and then compute classical statistics as wished for group analyses (mixed-effect analyses).

7. Project the parametric statistical map onto the zebra finch atlas to localize the functional activations (**Fig. 2**).

4. Notes

1. The choice of the species is of course dependent on the scientific question. However, other considerations like bird robustness to anaesthesia may also be taken into account. A safe minimal number of individuals is 15. This number takes into account the sensitivity of spin echo fMRI (*see* **Note 7**) and the natural inter-individual variability of biological phenomena measured in the experiment.

2. An important requirement for good fMRI results is the immobilization of the subject head. Fixing the head is usually not enough and is commonly associated with sedation or anaesthesia. When sedation is used, several elements, when correlated to the experimental paradigm (i.e. associated to auditory stimulation) can confound the fMRI results. These include stress-inducing noise, which is very high during the whole experiment but higher during stimulation periods than during rest periods (auditory stimuli being louder than gradient noise) and also eye movements induced by auditory stimulation and consequent changing visual stimulation (due to light remaining in the bore). Because anaesthesia drastically reduces these confounding factors, we prefer to use anaesthesia instead of sedation. The choice of the anaesthetic agent is important and depends on many parameters. These parameters include (a) the scientific question: if one wants to compare fMRI results with electro-physiology results, one may want to use urethane, an anaesthetic often used for electro-physiology recordings in birds. If one wants to investigate the role of noradrenalin in auditory discrimination, medetomidine, a potent α2-adrenoreceptor agonist, should be avoided. (b) The survival of the birds: if birds need to be reused in other experiments, or simply measured repeatedly during a long time, one will choose an anaesthetic maximizing the

survival rate of the birds. In this case, isoflurane is probably the best choice. In the case of gas being used, the anaesthetic needs to be mixed with nitrogen and oxygen.

We have previously compared the effects of three different anaesthetic agents (medetomidine, urethane and isoflurane) in zebra finches (4). The reader will find more information about the anaesthesia effect on fMRI results in that article. It is also important to keep the anaesthesia depth constant. It has been shown that anaesthetic level can profoundly change the BOLD response because of changes in neural activity (5) and in the BOLD transfer function (6). In the case of gas anaesthetic, it is important to use an accurate gas controller.

3. Functional MRI results are highly dependent on bird physiology. It is thus crucial to maintain it as stable as possible. With a cloacal temperature probe, a warm air flow device and a feedback-controlled system, we measure the internal temperature of the bird continuously and the temperature of warm air blown over the bird is adjusted to maintain the internal temperature of the bird at 40°C. We also record during the whole experiment the respiration rate of the bird thanks to a pneumatic sensor placed under its chest.

4. The bore is a confined space which can distort the auditory stimuli. It is thus important to first record the stimuli within the bore. In our system, auditory frequencies between 2500 and 5000 Hz have been found to be enhanced. To compensate this artificial enhancement, an equalizer function is applied to each stimulus using Wave-Lab software (Steinberg, Germany). The function consists of a Gaussian kernel with the following parameters: maximum amplitude, −20 dB, centred on 3750 Hz; width, 0.05 octaves (corresponding to the range 2500–5000 Hz).

5. The auditory device simply consists of headphones the magnets of which have been removed. Headphones are incorporated into the Helmholtz coil and connected to an amplifier.

6. The minimum magnetic field relevant to perform spin echo fMRI is 7 T (*see* **Note 7**).

7. Two types of sequences can be used to perform BOLD fMRI: spin echo and gradient echo sequences. Gradient echo sequences are more sensitive and thus more commonly used (for bird studies, *see* (1)). However, GE images contain susceptibility artefacts due to brain/air interface. These artefacts increase with magnetic field strength. At 7 T, less than 50% of the brain can be imaged. Susceptibility artefacts are drastically reduced with SE sequences

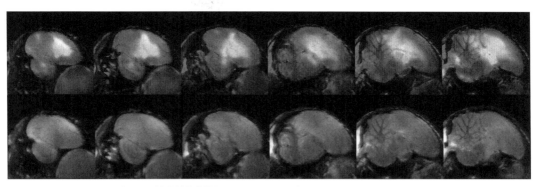

Fig. 3. Superimposition of GE and SE BOLD fMRI images (*in yellow/red*) on anatomical images (*in grey*). GE BOLD images (*top row*) cover around 50% of the brain, including the primary and secondary auditory regions, whereas SE BOLD images (*bottom row*) cover 99% of the brain.

(**Fig. 3**). SE BOLD fMRI is also more accurate than GE fMRI because the intra-vascular component of the BOLD signal is reduced and the extra-vascular component from large vessels is suppressed (7–9). The main limitation of SE fMRI consists in its weak sensitivity, requiring high magnetic fields (8, 10).

8. Due to the weak sensitivity of SE fMRI, we only used block designs alternating auditory stimulation periods (ON blocks) with resting periods (background noise of the scanner, OFF blocks), which are more sensitive than event-related designs. For the same reasons, it is important to use a block design that is as efficient as possible. To make a design efficient, we use the same rules as for human fMRI data (*see* documentation on http://www.fil.ion.ucl.ac.uk/spm/). We classically test three stimuli and use blocks of 16 s, which corresponds to the acquisition time of two images. Each stimulus type is presented 25 times, resulting in the acquisition of 50 images per stimulus and per subject.

9. We only use the ParaVision functional tool to check if we have been able to measure a differential BOLD response between the stimulation periods of the paradigm (ON blocks, all auditory stimuli together) and the baseline (OFF blocks). The reader should be aware that the experiment is not always successful. This is due to several factors including the low sensitivity of SE fMRI, the high level of the baseline (this problem only concerns auditory experiments) and potentially an unstable respiration rate. Repeating the experiment on the same subject another day sometimes solves the problem (if the respiration rate is more stable). In 25% of individuals, the experiment fails, despite several attempts.

10. SPM has been developed to process fMRI data acquired on human beings, that is, for voxels of around 2 mm. Numerous SPM settings are adapted to this approximate voxel size. If one does not want to change all these settings, the simplest way to proceed is to artificially increase the voxel size of bird fMRI data.

11. We are currently developing atlases for other songbird species as well.

12. The temporal resolution of SE fMRI does not allow measuring the shape of the haemodynamic response function. This shape has been calculated with GE fMRI in the auditory regions (11) but the coverage of GE fMRI does not allow testing it in other regions where this shape might be different. Until now, we thus have not convoluted the boxcar with a haemodynamic response function.

References

1. Van der Linden, A., Van Meir, V., Boumans, T., Poirier, C., and Balthazart, J. (2009) MRI in small brains displaying extensive plasticity. *Trends Neurosci.* **32**, 257–266.
2. Poirier, C., Boumans, T., Verhoye, M., Balthazart, J., and Van der Linden, A. (2009) Own-song recognition in the songbird auditory pathway: selectivity and lateralization. *J. Neurosci.* **29**, 2252–2258.
3. Poirier, C., Vellema, M., Verhoye, M., Van Meir, V., Wild, J. M., Balthazart, J., and Van Der Linden, A. (2008) A three-dimensional MRI atlas of the zebra finch brain in stereotaxic coordinates. *Neuroimage* **41**, 1–6.
4. Boumans, T., Theunissen, F. E., Poirier, C., and Van Der Linden, A. (2007) Neural representation of spectral and temporal features of song in the auditory forebrain of zebra finches as revealed by functional MRI. *Eur. J. Neurosci.* **26**, 2613–2626.
5. Richards, C. D. (2002) Anaesthetic modulation of synaptic transmission in the mammalian CNS. *Br. J. Anaesth.* **89**, 79–90.
6. Masamoto, K., Kim, T., Fukuda, M., Wang, P., and Kim, S.G. (2007) Relationship between neural, vascular, and BOLD signals in isoflurane-anesthetized rat somatosensory cortex. *Cereb. Cortex* **17**, 942–950.
7. Uludağ, K., Müller-Bierl, B., and Uğurbil, K. (2009) An integrative model for neuronal activity-induced signal changes for gradient and spin echo functional imaging. *Neuroimage* **48**, 150–165.
8. Duong, T. Q., Yacoub, E., Adriany, G., Hu, X., Ugurbil, K., and Kim, S. G. (2003) Microvascular BOLD contribution at 4 and 7 T in the human brain: gradient-echo and spin-echo fMRI with suppression of blood effects. *Magn. Reson. Med.* **49**, 1019–1027.
9. Yacoub, E., Duong, T. Q., Van De Moortele, P. F., Lindquist, M., Adriany, G., Kim, S. G., Uğurbil, K., and Hu, X. (2003) Spin-echo fMRI in humans using high spatial resolutions and high magnetic fields. *Magn. Reson. Med.* **49**, 655–664.
10. Lee, S. P., Silva, A. C., Ugurbil, K., and Kim, S. G. (1999) Diffusion-weighted spin-echo fMRI at 9.4 T: microvascular/tissue contribution to BOLD signal changes. *Magn. Reson. Med.* **42**, 919–928.
11. Van Meir, V., Boumans, T., De Groof, G., Van Audekerke, J., Smolders, A., Scheunders, P., Sijbers, J., Verhoye, M., Balthazart, J., and Van der Linden, A. (2005) Spatiotemporal properties of the BOLD response in the songbirds' auditory circuit during a variety of listening tasks. *Neuroimage* **25**, 1242–1255.

Section IX

Phenotyping

Chapter 30

MRI to Study Embryonic Development

Bianca Hogers

Abstract

Non-invasive imaging of embryonic development has been an ultimate goal for embryologists for many years. Due to advances in MRI hardware and software, the extremely high spatial resolution necessary to study embryos can now be obtained. Fixed embryos can be scanned to visualize the complex 3D morphology of the developing embryo in great detail, sometimes referred to as MR histology. As the sample remains intact, it is a suitable tool for the study of rare specimens, or for screening of huge numbers of transgenic embryos. In vivo MRI can be used for time course studies of either normal development or the progression of congenital malformations.

Key words: MR microscopy, embryo, development, heart, in vivo, gadolinium.

1. Introduction

Two main advantages led us to use magnetic resonance imaging for studying embryos. The first one is the high anatomical detail of fixed specimen as a diagnostic instrument for tracing the presence of congenital malformations (1–5). The second is to study non-invasively the development of single embryos in ovo (chicken, quail) (6, 7) as well as multiple pregnancies in utero (mouse) (8). The technology used does not interfere with normal development (6, 7) and can, therefore, reliably be applied. The main challenge for MRI of embryos is the small specimen size and, in case of mammals, the surrounding body of the mother. This chapter describes three applications of structural imaging of embryos with increasing complexity. With fixed embryos, resolutions of about 30 μm can be achieved, necessary to detect the smallest anatomical details (**Fig. 1**). However, low voxel

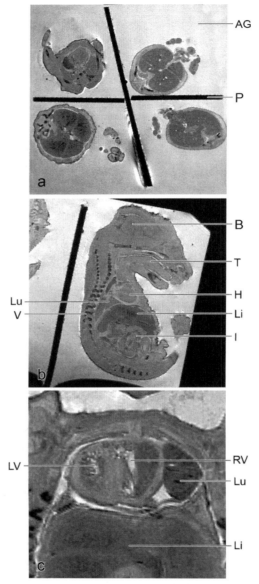

Fig. 1. 3D high-resolution whole body MRI of fixed mouse embryos (16.5 days of development). Multiple embryos in glass container, FOV $(25.6\ \text{mm})^3$ with an isotropic resolution of 50 μm. (a) Axial slices demonstrating the four embryos (transverse sections). The embryos are embedded in 1% agarose (AG) with gadolinium and separated by a crossed plastic partition (P). (b) Sagittal slice through one of the embryos. (c) Detail of the heart (transverse section) of the embryo in b. B, brain; H, heart; I, intestine; Li, liver; Lu, lung; LV, left ventricle; RV, right ventricle; T, tongue; V, vertebrae.

sizes yield low signal intensities making extended averaging and overnight scanning necessary. To compensate for the low tissue contrast due to lack of sufficient tissue differentiation, the contrast of the cardiovascular system can be enhanced by trapping

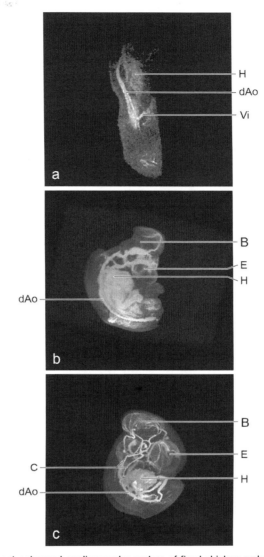

Fig. 2. Contrast-enhanced cardiovascular system of fixed chicken embryos, maximum intensity projections. (a) The vasculature of this young embryo of only 2.5 days of development consists of a looped heart (H) and a pair of dorsal aortae (dAo) leading to the vitelline (Vi) blood vessels. FOV 8 × 4 × 4 mm, isotropic resolution of 31 μm. (b) With 4 days of development the vasculature has developed considerably. Besides the blood vessels in the thorax and abdomen, those of the brain (B) and eyes (E) are demonstrated. FOV (10 mm)^3, isotropic resolution of 39 μm. (c) The final configuration of the major branches are developed at 5 days of development. FOV (14 mm)^3, isotropic resolution of 55 μm. C, carotid arteries.

gadolinium in the blood vessels resulting in 3D representations of the vascular system (**Fig. 2**). For in vivo MRI, the experimental time window is much shorter, firstly, because of the physiology of the animal under study, and secondly, because the embryonic size at least doubles overnight. Reduction of scan time is traded in

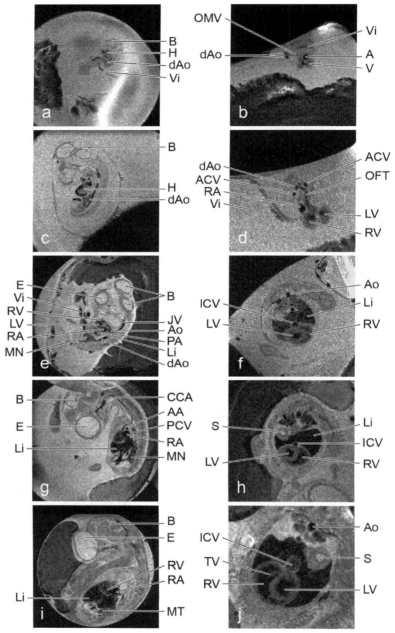

Fig. 3. In ovo time course imaging of quail development. *Left-hand* images are sagittal slices (0.5 mm) through the same quail embryo at successive embryonic days (ED) 3 (**a**), 5 (**c**), 7 (**e**), 9 (**g**) and 11 (**i**) with FOVs of (25.6 mm)^2 and a resolutions of 100×100 μm^2 while the *right-hand* images are the transverse sections (0.3 mm) through these hearts ED 3 (**b**), 5 (**d**), 7 (**f**), 9 (**h**), 11 (**j**) with FOVs of (22 mm)^2 and resolutions of 86×86 μm^2. A, atrium; AA, aortic arch; ACV, anterior cardinal vein; Ao, aorta; B, brain; CCA, common carotid artery; dAo, dorsal aorta; E, eye; H, heart; ICV, inferior caval vein; JV, jugular vein; Li, liver; LV, left ventricle; MN, mesonephros; MT, metanephros; OFT, outflow tract; OMV, omphalomesenteric vein; PA, pulmonary artery; PCV, posterior cardinal vein; RA, right atrium; RV, right ventricle; S, stomach; TV, tricuspid valve; V, ventricle; Vi, vitelline vessel.

exchange for resolution. Two-dimensional MRI sequences with a resolution of 100 μm result in adequate anatomical detail to perform prenatal diagnosis (**Fig. 3**). Finally, MRI of mouse embryos in utero provides additional complications. First, a mouse dam is involved that needs to be anesthetized and has to fit in an rf coil, which is unfavorable for the coil filling factor of the individual

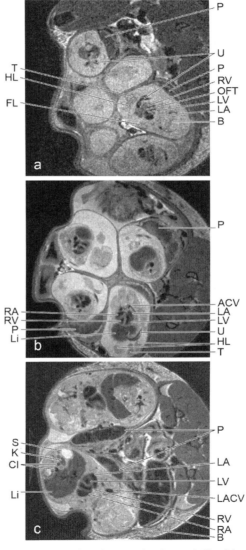

Fig. 4. In utero time course imaging of mouse development. Non-triggered axial slices (0.5 mm) through the abdomen of a pregnant dam at ED 12.5 (**a**), 14.5 (**b**) and 17.5 (**c**) with FOVs of $(25.6 \text{ mm})^2$ and resolutions of $100 \times 100 \text{ μm}^2$. ACV, anterior cardinal vein; B, brain; CI, common iliac arteries; FL, forelimb; HL, hind limb; K, kidney; LA, left atrium; LACV, left anterior cardinal vein; Li, liver; LV, left ventricle; OFT, outflow tract; P, placenta; RA, right atrium; RV, right ventricle; S, stomach; T, tail; U, umbilical vessels.

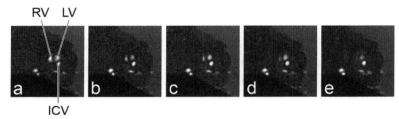

Fig. 5. In ovo cardiac imaging. Retrospectively gated (IntraAngio) movie frames of one transverse slice (0.3 mm) through an embryonic quail heart of 9 days of development with a FOV of (22 mm)^2 and a resolution of 172 μm. Frame 2/10 (**a**), 4/10 (**b**), 6/10 (**c**), 8/10 (**d**) and 10/10 (**e**). The different phases of the cardiac cycle are clearly visible, going from diastole (a) with completely filled right (RV) and left ventricles (LV) to systole (e) with almost empty ventricles. ICV, inferior caval vein.

embryos. A functional placenta is necessary for placental transfer of anesthesia, which is present from embryonic day 11.5 onwards. Before this moment, embryonic motion cannot be prevented and in utero MRI results in serious motion artifacts. For longitudinal studies, it is necessary to track the individual embryos over time and litters of 12 embryos are no exception. Although the localization in the abdomen is very flexible, the position of each embryo in the left and right uterine horns is fixed with respect to the next embryo, which can be used to identify individual embryos over time. Although embryonic ECG-triggering is not necessary for anatomical images (**Fig. 4**), retrospective gating (9) can be used to determine embryonic cardiac function in vivo. By using a saturation slice refocusing signal as navigator scan, cardiac images are co-registered with signal intensity dependent on the cardiac cycle through which the images can be reconstructed afterwards into a cardiac movie (**Fig. 5**).

2. Materials

2.1. Embryos In Vitro

1. Embryos of the desired stage of development
    a. Fertilized White Leghorn (*Gallus gallus domesticus*) or quail (*Coturnix coturnix japonica*) eggs incubated at 37°C and 60–70% relative humidity.
    b. Mouse or rat embryos excised from the uterus.
2. Rinsing and preparation fluid: Phosphate-buffered saline (PBS). Store at room temperature, use at 37°.
3. Glass containers, matching the size of the embryos.

2.1.1. Whole Body

1. Prepare 8% paraformaldehyde (500 mL): Place a conical flask with 400 mL purified water on top of a hot plate/stirrer

inside a fume hood. Set the heater on and use moderate stirring. Slowly add 40 g paraformaldehyde (*see* **Note 1**). Add 10 drops of 1 M NaOH after a few minutes. The solution will turn from cloudy to clear when ready. Switch off the heat but keep on stirring while cooling down. When cooled, fill up solution to 500 mL with purified water and filter when necessary. Store at 4°C.

2. Prepare 0.2 M phosphate buffer (500 mL): add 5.52 g $NaH_2PO_4 \cdot 1H_2O$ to 200 mL purified water (= 0.2 M base). Add 10.78 g $Na_2HPO_4 \cdot 2H_2O$ to 300 mL purified water (= 0.2 M acid). Add 0.2 M acid into the flask with 0.2 M base until pH is 7.4. Store at 4°C.

3. Prepare fresh 4% paraformaldehyde fixative (4%PFA) by adding equal amounts of 8%PFA to 0.2 M phosphate buffer.

4. Prepare fixative with gadolinium contrast agent: add 200 μL Dotarem (Guerbet, Paris, France) to 10 ml 4%PFA.

5. Embedding: 1% agarose LE Analytical Grade, dissolved in PBS, supplemented with 1:50 v/v Dotarem.

2.1.2. Cardiovascular System

1. PBS. Store at room temperature and heat before injection to 37°C.

2. Gadolinium-DOTA, 0.5 mM (Dotarem; Guerbet, Paris, France).

3. Gelatin (Merck, Darmstadt, Germany).

4. Horizontal pipette puller (Bachofer GmbH, Reutlingen, Germany).

5. Capillaries (2.0 mm O.D., 1.16 mm I.D.) (Clark Electromedical Instruments, Pangbourne, UK).

6. Disposable needles (21G) and disposable syringes (1 mL).

7. Dissecting microscope with a micromanipulator.

8. For young embryos: Petri dishes half-filled with 5% agarose.

9. Perfluoropolyether, e.g. Fomblin (FenS Chemicals, Goes, the Netherlands).

10. Prepare glass needles: First pull glass needles with tip diameters 4–20 μm from capillaries. Mount them on disposable injection needles (21G) with two-component glue (local hardware store) and connect each needle to a disposable 1 mL syringe.

2.2. Embryos In Ovo

1. Incubator installed in the scanner room.

2. Fertilized White Leghorn (*Gallus gallus domesticus*) or quail (*Coturnix coturnix japonica*) eggs incubated at 37°C and 60–70% relative humidity.

3. For embryos ≥ ED10: Ketanest-S, 5 mg/mL S-ketamine (Pfizer, Capelle, the Netherlands).

4. Disposable syringes, needles and tape.

2.3. Embryos In Utero

1. Pregnant mouse dam.
2. Respiration monitoring system (SA Instruments, Stony Brook, NY, USA).
3. Isoflurane inhalation equipment (UNO Roestvaststaal, Zevenaar, the Netherlands), including medical oxygen, medical air and isoflurane (PCH Pharmachemie, Haarlem, the Netherlands).

2.4. MRI

The MRI sequences provided here are based on a vertical 9.4 T micro-imaging system (Bruker Biospin, Rheinstetten, Germany) with an actively shielded Micro2.5 gradient (1 T/m, rise time < 110 μs), a BOS-II shim system, pre-emphasis and B_0 compensation. ParaVision 4.0 was used for image acquisition and analysis.

3. Methods

3.1. Embryos In Vitro

1. Remove embryo from egg or uterus and transfer it to a Petri dish filled with PBS (*see* **Note 2**).
2. Carefully remove all membranes.

3.1.1. Whole Body

1. Fix embryos overnight by immersion (*see* **Note 3**).
2. Prepare embedding solution: place a conical flask with 50 mL PBS on top of a hotplate/stirrer. Switch the heater on and use fast stirring. Add 1 mL Dotarem followed by 0.5 g agarose. Keep on heating until the solution is clear.
3. When multiple embryos are scanned simultaneously, take care to mark individual embryos, e.g. by removing (parts of) a right or left extremity. In addition, small plastic or glass rods can be inserted next to a particular embryo.
4. Transfer embryos to glass containers. To obtain the best sample-to-coil ratio the container diameter should be only a few millimeters larger than that of the embryo and use the smallest coil possible for the container. Take care that embryos are (remain) in a head-up position. For multiple embryos, the use of a plastic partition (*see* **Note 4**) will facilitate step 6 in preventing the embryos from falling while the agarose is poured in.
5. Gently pour the agarose solution in the containers (*see* **Note 5**). Immediately correct when embryos start floating.

6. After solidification, containers should be covered (lid or Parafilm) and stored at 4°C.

7. Place container inside rf coil.

8. MRI pilot scan to check proper position in center of coil:
   2D gradient echo with TR = 100 ms, TE = 4 ms, 1 average, rf excitation = sinc7H; 1 ms; FA = 15°, 3 orthogonal slices of 1 mm with FOV = (30 mm)^2, matrix 128 × 128, scan time 13 s (*see* **Note 6**).

9. Determine line width half-height (LWHH), using single-pulse spectroscopy with rf excitation = block pulse; 0.1 ms; FA = 10°, data points = 2,048, spectral width= 25 Hz, relaxation delay = 2,666 ms, no water suppression, scan time 2.67 s. Measure LWHH (*see* **Note 7**).

10. Adjust first and second-order shims.

11. Set frequency on resonance (*see* **Note 8**).

12. Determine LWHH again (*see* Step 2). Value should be improved.

13. High-resolution scan overnight: 3D gradient echo with TR = 14 ms, TE = 4.5 ms, rf excitation = sinc7H (*see* **Note 9**); 1 ms; FA = 15°, FOV = 14 × 14 × 7*, matrix 512 × 512 × 256*, BW = 50 kHz, 27 averages**, scan time 12 h 46 min** (*see* **Note 10**).

3.1.2. Cardiovascular System

1. As T_1 and T_2 are dependent on field strength, the optimal concentration of contrast agent has to be tested for your setup. Make a dilution series (add 2% gelatin to each tube, heat the solution until it turns clear and let it solidify at room temperature) and run the MRI protocol. The concentration with the highest signal intensity should be used for intravascular injection.

2. Place Petri dish with embryo under the dissection microscope and mount a needle on the micromanipulator.

3. Clear the blood from the vascular system by perfusing warm (37°C) PBS into the omphalomesenteric vein with a glass needle, followed by a 4% PFA (37°C) perfusion (*see* **Note 11**). Older embryos (≥ED7) can be injected directly into the cardiac ventricles.

4. Inject the warm contrast agent (*see* **Note 12**).

5. Transfer embryo to 4%PFA (4°C) to solidify the gelatin. Small embryos need 2 h, large embryos overnight.

6. Before imaging, transfer embryo to a suitable container matching the size of the embryo and fill container with Fomblin (*see* **Note 13**).

7. Place container inside rf coil.
8. MRI Pilot scan to check proper position in center of coil:
   2D gradient echo with TR = 100 ms, TE = 4 ms, NEX = 1, rf excitation = block pulse; 0.013 ms; FA = 15°, 3 orthogonal slices of 1 mm with FOV = 30 × 30 mm, matrix 128 × 128, scan time 13 s.
9. Determine line width half-height (LWHH), using spectroscopy: single pulse with TR = 2,666 ms, 1 average, rf excitation = block pulse; 0.1 ms; FA = 10°, no water suppression, scan time 2.67 s. Measure LWHH (*see* **Note 7**).
10. Adjust first- and second-order shims.
11. Set frequency on resonance (*see* **Note 8**).
12. Determine LWHH again (*see* Step 2). Value should be improved.
13. High resolution scan overnight.
    3D spin echo with TR = 200 ms, TE = 6 ms, rf excitation = block pulse; 0.013 ms; FOV = 10 × 10 × 10*, matrix 256 × 256 × 256, BW = 100 kHz, 4 averages, scan time 14 h 33 min 48 s (*see* **Note 10**).
14. Make a maximum intensity projection (MIP) of the dataset, resulting in a 3D representation of the cardiovascular system.

3.2. Embryos In Ovo

1. The eggs should retain the same (upward) orientation during incubation, as well as during all transfers or manipulations. Otherwise, the embryo will slowly float to the highest position introducing additional motion artifacts.
2. To prevent embryonic motion, cool down embryos (ED4–ED9) to room temperature for 30–45 min. Older embryos need anesthesia.
3. Anesthesia: Inject 150 μL Ketanest slowly into the air chamber (*see* **Note 14**). Seal the shell with a small piece of tape.
4. Mount the egg inside the rf coil (take care of Step 1).
5. MRI pilot scan to check fertilization of egg and proper position in the center of the coil:
   2D fast spin echo with TR = 4,000 ms, TE = 8 ms, turbo factor 8, 1 average, rf excitation = hermite; 1 ms; 3 orthogonal slices of 1 mm with FOV = (30 mm)^2, matrix 128 × 128, scan time 1 min 4 s.
6. MRI scan with longitudinal slices parallel to the body-axis:

2D fast spin echo with TR = 5,000 ms, TE = 9 ms, turbo factor 8, 1 average, rf excitation = hermite; 1 ms; x contiguous slices of 0.5 mm with FOV= $(25.6 \text{ mm})^2$, matrix 256 × 256, scan time 2 min 40 s.

7. MRI high resolution scan with slices transversally through heart:
2D fast spin echo with TR = 5,000 ms, TE = 9 ms, effective echo time = 36.5 ms, turbo factor 8, 8 averages, rf excitation = hermite; 1 ms; x contiguous slices of 0.3 mm with FOV = $(22 \text{ mm})^2$, matrix 256 × 256, scan time 21 min 20 s.

8. MRI for cardiac imaging in ovo.

9. Run automatic adjustments for IntraGate.

10. Run IntraGate pilot scan with TR = 9.22 ms, TE = 1.2 ms, 8 repetitions, rf excitation = hermite; 1 ms; FA = 45°, 3 orthogonal slices of 1 mm with FOV = $(30 \text{ mm})^2$, matrix 128 × 128, scan time 30 s.

11. Determine heart rate (1 slice low resolution).

12. Run IntraGate with TR = 7 ms, TE = 1.8 ms, 64 repetitions, rf excitation = hermite; 1 ms; FA = 60°, slice of 1 mm with FOV = 25.6 × 25.6 mm, matrix 128 × 128, aligned transversally to heart axis (short-axis view). Navigator: inslice 1 mm, rf pulse = gauss; 1 ms; FA = 10°, scan time 30 s. Expected heart rate 150, expected respiration rate 150, usable respiration signal 100%, 10 cardiac frames, 1 respiration frame.

13. Make cardiac movie via IntraGate (*see* **Note 15**). Use geometry values from Step 7 in this section (high-resolution fast spin echo slices) but change number of slices to one and position the slice at the upmost position of the heart (keep angles and slice offset identical). Number of repetitions = 128, expected heart rate = 50–100% of actual heart rate (determined in Step 11 above). Scan time 4 min 37 s.

14. Complete cardiac movies for whole heart. Clone scan, position the slice 0.3 mm lower etc.

15. Make cardiac movie via IntraAngio (*see* **Note 15**). Use the geometry values from Step 7 of this section (high-resolution fast spin echo slices) and change the matrix to 128 × 128. IntraAngio with TR = 7 ms, TE = 1.8 ms, 150 repetitions, rf excitation = sinc; 1 ms; FA = 40°. Navigator: in parallel saturation slice of 1 mm, rf pulse = hermite; 1 ms; FA = 3°. Expected heart rate = 50–100% of actual heart rate, 10 cardiac frames, 1 respiration frame. Scan time 1 min 41 s per slice.

3.3. Embryos In Utero

1. Anesthetize a pregnant dam with isoflurane (mixture of 0.3 L/min oxygen and 0.3 L/min air with 4% isoflurane for induction and 1.5% for maintenance) and mount the dam in an animal holder.
2. Place an air-pressure detection cushion under the abdomen/thorax and connect it to the monitoring equipment. Adjust depth of anesthesia continuously to maintain a stable respiration rate (50–60/min).
3. Tape the abdomen relatively tight to the animal holder to force the mouse to use the thorax for breathing (see **Note 16**).
4. MRI pilot scan to check proper position of embryos in the center of coil:
   2D fast spin echo with TR = 4,000 ms, TE = 8 ms, turbo factor 8, 1 average, rf excitation = hermite; 1 ms; 3 orthogonal slices of 1 mm with FOV = $(30 \text{ mm})^2$, matrix 128 × 128, scan time 1 min 4 s.
5. MRI scan with high-resolution slices transversally through mouse dam abdomen:
   2D fast spin echo with TR = 10,000 ms, TE = 5.7 ms, effective echo time = 23.16 ms, turbo factor 8, 4 averages, rf excitation = hermite; 1 ms; x contiguous slices of 0.5 mm with FOV = $(25.6 \text{ mm})^2$, matrix 256 × 256, scan time 21 min 20 s (see **Notes 17–19**).

4. Notes

1. Inspect regularly to avoid overheating and consequent spilling (it is a rapid fixer and toxic).
2. Do not remove the yolk sac vasculature in young avian embryos (≤ED4). Open the egg at the site of the air chamber. Pour the contents in a bowl filled with warm PBS. Take care to keep the yolk intact. Make a cut around the yolk sac blood vessels to remove the embryo from the yolk. Transfer embryo with yolk sac including intact blood vessels to another small bowl to remove excess yolk. Transfer clean embryo to a Petri dish half-filled with 5% agarose and covered with PBS. Carefully stretch the yolk sac membrane and pinch it, just outside the marginal sinus, into the agarose. Now the embryo is immobilized and ready for omphalomesenteric vein injection.
3. Perfusion fixation creates great risk of entrapment of air bubbles in smaller blood vessels.

4. Plasticizers (phthalates) in polyvinyl chloride create enormous chemical shift artifacts. Use glass containers or very rigid plastic. Partition walls can be easily created by two rectangular pieces of rigid plastic with a cut half-way through in the middle. Sliding the pieces at the side of the cuts in each other results in four separated chambers.

5. Let the agarose cool down a little (keep on shaking) before pouring it on the embryos but take care that the agar must not precipitate.

6. By using ParaVision Traffic Light function, adjustments like global shimming, setting to resonance frequency, rf power optimization and receiver gains are done automatically. When not available, these adjustments should be done manually.

7. Very convenient with the Bruker macro "calclinewidth."

8. Bruker systems, via "adjust basic frequency."

9. Using a shaped rf pulse, e.g. sinc with several lobes, prevents the formation of ripple artifacts.

10. FOV is dependent on size of embryo(s). Keep the resolution isotropic.
    The maximal available scan time determines the number of averages (NEX).

11. Keep the needle in place and change the syringe without introducing air in the needle.

12. A drop of dye (e.g., Nile blue sulfate) added to the contrast agent solution facilitates inspection of complete filling of the vasculature. A heating lamp above the glass needle prevents precocious solidification and thus clogging of the needle.

13. The size of the container should be chosen as small as possible to immobilize the embryo without disturbing its natural conformation. Take care not to entrap air bubbles while filling with Fomblin.

14. Make two small pinholes (with the needle of the syringe) in the air chamber. Inject the anesthesia into one hole, the second hole will reduce the internal pressure.

15. IntraGate uses either a slice-refocusing signal (in-slice) or a saturation slice-refocusing signal (in parallel or in oblique) as navigator scan. With a saturation slice navigator, multi-slice acquisition is possible but we found that acquisition of multiple single slices was superior to that of one multislice acquisition, especially for embryos. When no quantification of cardiac function is needed, this method can be used. Processing software requires that the first movie frame of all slices represent the same phase of the cardiac cycle, but

scanning of each slice is started randomly. Therefore, the phases of the first frame are also random and cannot be used for quantification. To tackle this problem, Bruker has developed the IntraAngio tool. This is a tool to obtain a multislice dataset via a slice-by-slice acquisition. IntraAngio employs one navigator slice at a fixed position resulting in identical movie frame starting points for each slice.

16. Do not tape directly on the skin, but first place a plastic sheet (5 × 4 cm) around the murine body. By restricting abdominal movement in this way, there is no need for triggering on maternal signals. In addition, no triggering on embryonic signals is necessary for anatomical images.

17. There is placental transfer of isoflurane from ED 11.5 onwards, preventing embryonic motion artifacts efficiently.

18. We found no harmful effects of repetitive scans (every 2 days) on embryonic development of normal embryos. However, transgenic mice can be more sensitive to the anesthesia load resulting in embryolethality.

19. For time course studies, individual embryos can be identified by their position (order) in the uterine horns (left and right). Count from the vagina upwards.

Acknowledgments

The author would like to thank Arno Nauerth for adapting the retrospective gating technique to embryonic applications, Kees Erkelens, Fons LeFeber and Linda van der Graaf for technical assistance, Jan Lens for preparing the figures and Robert Poelmann and Louise van der Weerd for their suggestions regarding the manuscript.

References

1. Johnson, G. A., Cofer, G. P., Fubara, B., Gewalt, S. L., Hedlund, L. W., and Maronpot, R. R. (2002) Magnetic resonance histology for morphologic phenotyping. *J. Magn. Reson. Imaging* **16**, 423–429.
2. Ichikawa, Y., Sumi, M., Ohwatari, N., Komori, T., Sumi, T., Shibata, H., Furuichi, T., Yamaguchi, A., and Nakamura, T. (2004) Evaluation of 9.4-T MR microimaging in assessing normal and defective fetal bone development: comparison of MR imaging and histological findings. *Bone* **34**, 619–628.
3. Yelbuz, T., M., Zhang, X., Choma, M. A., Stadt, H. A., Zdanowicz, D. V. M., Johnson, G. A., and Kirby, M. L. (2004) Approaching cardiac development in three dimensions by magnetic resonance microscopy. *Circulation* **108**, e154–155.
4. Schneider, J. E., Bamforth, S. D., Farthing, C. R., Clarke, K., Neubauer, S., and Bhattacharya, S. (2003) High-resolution imaging of normal anatomy, and neural and adrenal malformations in mouse embryos using magnetic resonance microscopy. *J. Anat.* **202**, 239–247.

5. Schneider, J. E., Bamforth, S.D., Farthing C. R., Clarke, K., Neubauer, S., and Bhattacharya, S. (2003) Rapid identification and 3D reconstruction of complex cardiac malformations in transgenic mouse embryos using fast gradient echo sequence magnetic resonance imaging. *J. Mol. Cell. Cardiol.* **35**, 217–222.
6. Bain, M. M., Fagan, A. J., Mullin, J. M., McNaught, I., Mclean, J., and Condon, B. (2007) Noninvasive monitoring of chick development in ovo using a 7 T MRI system from day 12 of incubation through to hatching. *J. Magn. Reson. Imaging.* **26**, 198–201.
7. Hogers, B., van der Weerd, L., Olofson, H., van der Graaf, L. M., DeRuiter, M. C., Gittenberger-de Groot, A. C., and Poelmann, R. E. (2009) Non-invasive tracking of avian development in vivo by MRI. *NMR Biomed.* **22**, 365–373.
8. Chapon, C., Franconi, F., Roux, J., Marescaux, L., Le Jeune, J. J., and Lemaire, L. (2002) In utero time-course assessment of mouse embryo development using high resolution magnetic resonance imaging. *Anat. Embryol.* **206**, 131–137.
9. Heijman, E., de Graaf, W., Niessen, P., Nauerth, A., van Eys, G., de Graaf, L., Nicolay, K., and Strijkers, G. J. (2007) Comparison between prospective and retrospective triggering for mouse cardiac MRI. *NMR Biomed.* **20**, 439–447.

Chapter 31

Mouse Phenotyping with MRI

X. Josette Chen and Brian J. Nieman

Abstract

The field of mouse phenotyping with magnetic resonance imaging (MRI) is rapidly growing, motivated by the need for improved tools for characterizing and evaluating mouse models of human disease. Image results can provide important comparisons of human conditions with mouse disease models, evaluations of treatment, development or disease progression, as well as direction for histological or other investigations. Effective mouse MRI studies require attention to many aspects of experiment design. In this chapter, we provide details and discussion of important practical considerations: hardware requirements, mouse handling for in vivo imaging, specimen preparation for ex vivo imaging, sequence and contrast agent selection, study size, and quantitative image analysis. We focus particularly on anatomical phenotyping, an important and accessible application that has shown a high potential for impact in many mouse models at our imaging center.

Key words: MRI, magnetic resonance imaging, mouse, phenotyping, anesthesia, mouse handling, monitoring, central nervous system, cardiac, whole-body perfusion, excised organs.

1. Introduction

The mouse was the first live animal to be imaged using magnetic resonance imaging (MRI) (1), but following this initial demonstration, it was not utilized extensively. Historically, rats and guinea pigs were preferred for small animal imaging experiments due to their larger size, which allows for more satisfactory visualization of anatomy at limited image resolutions. However, the completion of the draft human and mouse genome sequences in the last few years has renewed interest in mouse MRI as a tool to investigate models of human disease, for which the mouse is the mammal of choice.

Toward the end of understanding the genome, there is now a worldwide effort to study genotype and phenotype—the physical and biochemical manifestation of a given genotype—with particular focus on relevant models of human disease. Both random and targeted mutagenesis methods have been explored and there is currently a consortium aiming to create mutants through conditionally knocking out each of the 23,000 genes in the mouse. While some diseases are associated with a specific gene (e.g., cystic fibrosis, Huntington's disease, and sickle-cell anemia), it is probable that many functional disorders result from a combination of subtle gene mutations. Thus, a vast number of mutant mice require characterization and phenotyping to identify the most useful disease models. This will be a time-consuming step, one in which MRI can play an essential role.

Disease model phenotypes can be characterized by studying anatomy, physiology, behavior, or function. Many of these tests can be performed in vivo, but the final step is to perform histopathology to look for organ, biochemical, and cellular differences. Given the enormous numbers of mouse mutants produced, it is difficult to characterize every mouse exhaustively and the potential for overlooking interesting phenotypes is high. By adapting the techniques and tools of MRI, which is inherently noninvasive, we are able to take detailed images of the inside of a mouse without conventional "slicing and dicing" and then perform phenotyping analyses computationally.

This form of mouse phenotyping with MRI is rapidly growing. Early approaches used conventional MRI techniques in conjunction with other phenotyping tests (2–4). In recent years, many more methodologies have been developed to enhance mouse phenotyping. The application most comparable to diagnostic radiology uses anatomical scans to look for size or morphology differences in organs, structures, or pathologies (5–8). Beyond anatomy, functional assays are possible by measuring blood–tissue perfusion (9, 10), ejection fraction in the heart (11, 12), and BOLD (blood oxygen level-dependent) response in the brain (13, 14). Other quantitative measurements related to tissue microstructure can be made by studying physical properties of water molecules in tissue, such as diffusion (15), relaxation parameters T_1, T_2 and T_2^*, or magnetization transfer (16, 17). Cerebral blood volume and fraction can also be measured quantitatively (18). Metabolism can be assessed with magnetic resonance spectroscopy (19), and, more recently, this has been done with hyperpolarized ^{13}C (20).

Anatomical scans, with contrast intended to highlight anatomical features, will remain an important component of mouse imaging studies as they are broadly applicable and are readily adapted to large-scale studies. Findings on anatomical images can then lead to further investigation by guiding histology. In brain studies, for instance, it has been shown that anatomical

MRI is very sensitive, with 87% of behavioral mutants also showing a neuroanatomical phenotype (21). This represents a remarkable link between anatomy and function and suggests anatomical phenotyping is a powerful investigative tool. Even when qualitative radiological inspection cannot detect phenotypic differences, detailed computational analysis can reveal subtle differences by group comparisons (22).

The most vital part of image-based phenotyping starts with high-quality image data. Often, this will require images in several control and mutant animals, or in several animals at multiple time points. The group sizes required will depend on the subtlety of the phenotype in question, with larger groups required as phenotypes become more nuanced. As the number of mice studied increases, imaging throughput becomes essential and implementation of multiple-mouse MRI (23), in which several mice are imaged simultaneously in the same MRI scanner, is extremely beneficial. In this chapter, we will describe general methods to acquire quality images for mouse phenotyping and discuss subsequent analysis methods. We emphasize techniques that best allow for quantitative anatomical phenotyping in which differences between mice can be demonstrated and measured, but also mention other imaging methods providing alternative phenotyping possibilities.

2. Materials

2.1. MRI System

The most obvious difference between imaging mouse versus man is the size of the subject; the requisite increase in resolution for mouse images results in fewer protons per image voxel and lower signal-to-noise ratios in the mouse. To account for this, higher field magnets are preferred as they increase signal strength.

1. It is possible to begin mouse imaging using available clinical scanners and associated hardware and software (24). However, a dedicated *small-animal, high-field magnet* (>7 T) is much preferred. The two major vendors for small animal imaging are Bruker Biospin MRI GmbH (Ettlingen, Germany) and Agilent Technologies, Inc. (formerly Varian, Inc., Palo Alto, CA). The descriptions below about animal handling and imaging applications are largely independent of the magnet choice.

2. If only a clinical scanner is available, hardware limitations can be largely eliminated by using a *purpose-built insert gradient coil* with high slew rates and gradient strengths (25). In combination with the animal-specific hardware listed below, this will achieve significantly improved image resolution and

quality and may represent a good alternative to a dedicated animal system.

3. According to the organ or system being studied, an *application-specific radiofrequency coil* should be chosen. For example, in the case of the eye, a small surface coil is needed. If, on the contrary, the whole mouse is of interest then a birdcage volume coil encompassing the entire mouse is required. Small animal MRI system manufacturers, including Bruker and Varian, offer coils and major independent sellers include Doty Scientific, Inc. (Columbia, SC) and RAPID Biomedical GmbH (Rimpar, Germany). Multiple coil arrays for spatially-sensitive encoding, analogous to standard practice on human clinical systems, are likely to become more commonplace in animal systems as well.

4. The Mouse Imaging Centre has implemented a multiple-mouse MRI system on the Varian architecture, which is designed to allow their hardware to be scalable. The key benefit to such a system is the capability for more imaging-intensive, high-throughput studies. Along with a wide-bore, horizontal magnet, *multiple transmit/receive channels and multiple coils* are required (each channel accommodates one mouse).

2.2. Mouse Handling

1. All mouse research requires local ACC (Animal Care Committee), IACUC (Institutional Animal Care and Use Committee) or equivalent approval for mouse-handling procedures.

2. An *anesthetic* is needed to sedate the mouse during in vivo imaging experiments; gaseous isoflurane or injectable ketamine/xylazine mixtures are most common. For functional MRI studies, some investigators prefer α-chloralose as an alternative because it does not depress cardiovascular response as significantly as other agents.

3. If an inhalation anesthetic is used, then *oxygen* and associated *tubing* is necessary to deliver the anesthetic to the animal. A *scavenger* system is also required to properly and safely eliminate the anesthetic from the MRI room (there are passive and active models at Paragon Medical, Coral Springs, FL).

4. Mice under sedation cannot maintain their own body temperature, especially in air-conditioned magnet rooms; therefore a *heat source* (e.g., forced air or recirculating water bath), needs to be used. We specifically use model F7001 from Edemco Dryers, Colorado Springs, CO for multiple mice. A less powerful version is needed for single mice. In conjunction, a *proportional–integral–derivative*

controller (PID controller) is needed to create a feedback loop to maintain a constant temperature for the mouse (e.g., Omron Electronics, worldwide offices).

5. A *temperature probe* is used to monitor the temperature of the mouse. The internal temperature can be monitored with a rectal probe or the skin temperature can be monitored with a probe on the skin (we use a t-type thermocouple positioned at the abdomen from Omega Engineering Inc., Laval, Canada).

6. The monitoring of heart rate requires *ECG connectors* (pads, copper tape, or needle varieties are widely available. We use non-magnetic neonatal/pediatric ECG electrodes from ConMed Corp., Utica, NY). If using tape or pads, hair from the contact points (chest or limbs) must be removed with a *razor or depilatory cream*. Applying *ECG gel* helps improve electrical contact. Since the monitoring of the mouse will be taking place inside a magnetic field, non-magnetic electrical connections are preferred. Furthermore, to prevent interference from the gradients with physiological signals, either fiber-optic connections (26) or analog filters are beneficial.

7. The monitoring of the breathing rate requires *respiratory pillows or bellows* (we use pillows from Smiths Medical, various worldwide distribution sites).

8. While most of the monitoring equipment can be developed in house with basic electronics, *animal monitoring systems* can be purchased as a single unit from commercial vendors, such as the system made by SA Instruments, Inc. (Stony Brook, NY), Biopac Systems, Inc. (Goleta, CA) or Rapid Biomedical (Rimpar, Germany). These devices can also be interfaced with the MRI system for triggering or gating.

9. An *animal cradle* helps with reproducible mouse positioning and can have integrated physiological monitoring components. Sometimes *motion restraints* (e.g., Velcro or tape) are helpful to prevent motion artifacts (*see* **Fig. 1**). These can be developed in-house or are available from Dazai Research Instruments, Toronto, Canada.

10. Other miscellaneous items include *saline* and *syringes* as well as *gloves* and *laboratory coats or surgical gowns* for mouse handling.

2.3. Contrast Agents

The use of contrast agents is often beneficial to highlight particular anatomical features or to permit more rapid imaging.

1. The most common contrast agent is *gadolinium* (Gd), administered in a chelated form that keeps it in the extracellular space or bloodstream. Sometimes it is injected into

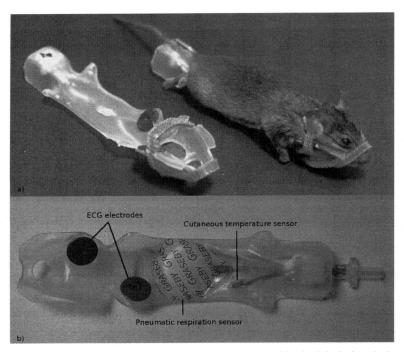

Fig. 1. The mouse sled used to position the mouse and provide physiological contacts (36). (**a**) The sled indicating the Velcro attachment to secure the mouse's head. (**b**) A close-up of the mouse sled with sensors labeled. Pictures provided courtesy of Jun Dazai.

the mouse before a scan to decrease T_1 relaxation times and permit faster imaging (e.g., 0.1 mM/kg gadopentetate dimeglumine (Berlex, Lachine, Quebec, Canada)) or to highlight pathology that breaks down the blood–brain barrier. If studying the vasculature or blood flow, various methods of time-of-flight angiography or perfusion measurements can be performed by saturating or inverting the signal in a volume of interest and then imaging signal effects during blood inflow (27, 28).

2. *Manganese* (Mn) is a useful contrast agent in the mouse that is not available to use in humans. It is most commonly used for imaging of the central nervous system, acting as a calcium analogue that is taken up across cellular calcium channels. The uptake of Mn by different parts of the brain depends on time and careful attention to the timing of injection can highlight specific parts of the brain (29). To generally decrease the T_1 time, we have injected 20 mg/kg $MnCl_2$ intraperitoneally 48 h prior to imaging (30).

3. In contrast to Gd and Mn agents, *iron oxide particles* create "negative contrast," a dark region in the MR image. These agents can be administered for functional vasculature measurements particularly in connection with solid tumors

(e.g., 0.5 mol Fe/L Resovist, Schering, Berlin, Germany). Additionally, they continue to be experimented with for cellular imaging applications, in which a cell population of interest is labeled either in vitro or in situ and then tracked for migration in normal or disease conditions (31–33).

2.4. Postmortem Imaging of Excised Organs

MRI can approach microscopic resolutions when motion is eliminated and scan time is not a limiting factor. Postmortem imaging of fixed tissue specimens therefore results in images that provide more detailed anatomy than can be achieved in vivo. Typically, an open-heart perfusion (*see* **Section 3.3.4**) is performed to fix the organs—and optionally to perfuse the mouse with contrast agent—and then the organs of interest are excised.

1. Injection anesthetic (e.g., Avertin, Sigma-Aldrich, worldwide offices)
2. Syringe and needle
3. Scalpel
4. Heparin (10 units/mL)
5. Saline
6. Contrast agent (optional, e.g., 2 mM ProHance, Bracco Diagnostics, Inc., worldwide offices)
7. Fixative (formalin)
8. Proton-free susceptibility matching fluid (e.g., Fluorinert, 3 M, Maplewood, MN)
9. NMR sample tube
10. Ultrasound biomicroscope (Vevo 770, VisualSonics, Toronto, Canada)
11. Gloves and goggles

2.5. Computation

After acquiring image data, a computing platform is required to visualize and analyze the data. The type of image analysis will depend heavily on the phenotypes of interest. We focus here on common needs for anatomical phenotyping.

1. A *workstation* with satisfactory memory, disk space, and speed is critical. It is most important to have ample RAM (random access memory) especially if the datasets are large. We typically work with images that are 0.5 GB in size, which require about 2–4 GB of RAM. It is also beneficial to have the fastest processors available (minimum 2 GHz) with ample diskspace. Having a good video card with minimum 512 MB facilitates smooth visualization of large 3D datasets. We have found a Linux-based workstation to be generally most cost effective, but it requires more computer savvy. Most commercial software licenses are available in both PC and Macintosh flavors.

2. To look at the data, *visualization software* is needed. There is freeware available on the internet, such as ImageJ (National Institutes of Health, http://rsbweb.nih.gov/ij/), OCCIviewer (University of Toronto, http://www.cardiacimaging.ca/technical_resources/occi_viewer/), and Display (McGill University, http://www.bic.mni.mcgill.ca/software/), though for more complicated 3D renderings, commercial software (e.g., Amira, http://www.amiravis.com/ or Analyze, http://www.analyzedirect.com/) is often better suited.

3. *Registration and segmentation software* are used to align and identify anatomical structures in images. Results from these image-processing steps are crucial for finding subtle phenotypes by computational methods. Popular open source software include SPM (Statistical Parametric Mapping, University College London), ITK (Insight Segmentation and Registration Toolkit), Automated Image Registration (University of California, Los Angeles), and Image Registration Toolkit (Imperial College, London); commercial versions are available as well (e.g., 3D-DOCTOR, Able Software Co., Lexington, MA and BrainVoyager, Maastricht, the Netherlands).

4. To conduct *statistical analysis*, software such as R and MATLAB (Mathworks, Natick, MA) are very powerful. There are a few pre-assembled MATLAB packages such as SurfStat (http://www.math.mcgill.ca/keith/surfstat/) and Partial Least Squares (http://www.rotman-baycrest.on.ca/index.php?section=84). Other packages are NIPY (Neuroimaging in Python, http://neuroimaging.scipy.org/site/index.html) and RMINC written in R (https://launchpad.net/rminc). We have found Python (http://www.python.org) and Perl (http://www.perl.org) to be good programming environments for establishing analysis pipelines for large multi-image datasets.

3. Methods

3.1. Animal Preparation for In Vivo Imaging

One of the advantages of MRI is the ability to image the mouse in vivo, which allows for longitudinal studies in individual subjects. In many experiments, this is preferred over imaging or histology of many different mice at set time points because it allows progress of disease—such as a spontaneous tumor—in the same animal to be followed over time (*see* **Note 1**). The overall goal in preparing an animal for live imaging is to place an anesthetized mouse in a consistent position (*see* **Note 2**) with various physiological monitoring devices attached. Rapid preparation time is required to accommodate several imaging sessions per day and dependable positioning facilitates image comparison after image acquisition.

1. To perform in vivo imaging requires maintaining and monitoring the physiology of the mouse during imaging sessions. A recent review discusses usage of different anesthetics for MR (34). In general, inhalation anesthetics, such as isoflurane and halothane, are employed as they are easy to control and are fairly gentle to the animal (*see* **Note 3**). Induction of the mouse is typically achieved with 3–4% isoflurane and then maintained at 1% isoflurane in 100% oxygen.

2. After induction, mice should be intraperitoneally injected with saline prior to imaging to prevent dehydration during long scanning sessions (roughly 0.2–0.4 mL/30 g/h).

3. Once under anesthesia, the mouse should be monitored via the temperature probe, ECG, and respiratory pillow. Supplemental monitors of physiological state may include exhaled CO_2, blood pressure, or blood oxygenation, although these are not routinely necessary for anatomical imaging.

4. In our experience, using forced air heating, we have found that C57/Bl6 mice are best maintained when the magnet bore temperature is at 27°C, which results in a skin temperature of 30°C (*see* **Note 4**). Maintaining good temperature control is not only important for the mouse, but in fact results in improved image quality (35). The type of heat source selected may be application dependent (*see* **Note 5**).

5. Beyond simple monitoring of the mouse, some experiments require gating on either or both the cardiac and the respiratory signals. The signal from the ECG and respiratory monitoring system are used for this purpose (*see* **Note 6**).

3.2. Considerations for Multiple Mice

Delays associated with animal handling are exacerbated when dealing with multiple mice simultaneously for a multiple-mouse MRI experiment. Since it is important to minimize the time each mouse is under anesthesia, every attempt should be made to streamline the process of preparing individual mice. This is so that the total duration of anesthesia for the mice is not unnecessarily extended because multiple mice are being prepared. In our laboratory, we have devised several shortcuts to reduce the preparation time for seven mice to less than half an hour (36). We direct interested readers to this reference for further details.

3.3. Data Acquisition

A powerful aspect of MRI is the flexibility to study various phenotypes in the mouse. Different organ systems can be studied with careful positioning of either a surface coil or a small volume coil. Once focused on the region of interest, both anatomy and function can be studied by choice of pulse sequence. A complete review of available pulse sequences is beyond the scope of this chapter. In general, MR sequences in routine use clinically provide visualization of anatomy, morphology, or pathology. Perfusion, diffusion, flow, and other functional measures require more sophisticated imaging sequences, although many of these are used routinely in research studies in both humans and mice (*see* **Note 7**). These are discussed in more detail in other chapters. In the next sections, we will discuss several imaging considerations and procedures for mouse phenotyping.

3.3.1. Anatomy

1. It is important in anatomical phenotyping to ensure good contrast between relevant structures. At higher field strengths, longitudinal relaxation times (T_1) increase and become less distinct, so for the most part, T_2-weighted images are preferred amongst the intrinsic image contrasts available. **Figure 2** shows several examples of T_2-weighted images.

2. Administration of Mn intraperitoneally 24 h prior to imaging of the brain is a popular way of obtaining neuroanatomical contrast in the mouse brain. Differential uptake of the Mn highlights different structures in T_1-weighted images because Mn is paramagnetic and locally shortens the T_1 relaxation time. Interestingly, Mn is transported along axons and can traverse synapses and has also been used for tract tracing after focal injections. **Figure 3** shows an example of a manganese-enhanced brain.

3. Another alternative for contrast within the central nervous system is to use diffusion-weighted imaging (15), which can differentiate diffusion rates of water parallel or perpendicular to the axis of a neuron. Recently, in vivo techniques have advanced so much so that fiber tractography, which follows fiber connectivity, can be performed in live mice (**Fig. 4a**) (37).

4. Another application of diffusion tensor imaging is to calculate the fractional anisotropy, which is the scalar value between 0 and 1 that describes the degree of anisotropy of white matter fibers. A value of 0 means that diffusion is isotropic and a value of 1 means that diffusion occurs only along one axis and is fully restricted along all other directions (**Fig. 4b**). Many other diffusion variants are also possible (38, 39).

    Although the majority of mouse studies to date have focused on the brain or the heart, most of the organs within the mouse have been imaged and phenotyped. A notable phenotyping endeavor is diabetes and obesity. In this

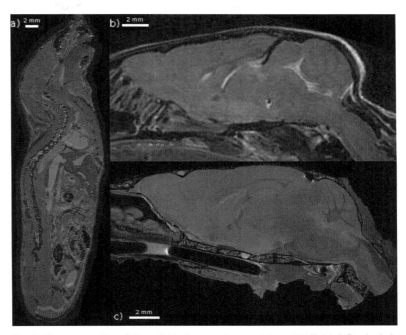

Fig. 2. Three examples of T_2-weighted mouse images taken at 7 T. (**a**) A fixed whole mouse prepared by perfusion fixation with ultrasound guidance (Ref. (101)) including 10 mM Magnevist. A spin echo sequence was used with TR/TE = 650/15 ms and $(100\ \mu m)^3$ voxels resulting in an imaging time of 13 h. (**b**) A brain imaged in vivo. A fast spin echo sequence was used with cylindrical acquisition, TR/TE = 900/12 ms, with TE_{eff} = 36 ms, echo train length 8, two averages, flip angle 40°, matrix size 400 × 240 × 240, resolution $(100\ \mu m)^3$ resulting in an imaging time of 2 h and 50 min (Ref. (119)). (**c**) A fixed brain prepared by perfusion fixation with 0.5 mM ProHance (Bracco Diagnostics, Inc.). After fixation, the skull is separated from the body. A fast spin echo sequence was used with TR/TE = 325/6 ms, with TE_{eff} = 30 ms, echo train length 6, four averages, matrix size 780 × 432 × 432, resolution $(32\ \mu m)^3$ resulting in an imaging time of 11.3 h. Figure provided courtesy of Christine Laliberté.

instance, anatomical images of the whole body need to be acquired.

5. In these applications, fat and water images can be separated based on their relative chemical shift via various methods (i.e., the Dixon technique). Volumetric measurements further permit the amount of fat to be quantified (40). Other sites of investigation have included the optic nerve (41), spinal cord (16, 42), liver (43, 44), kidneys (45), prostate (46), and pancreas (47). Initial selection of particular sequences for visualization in any of these organs or structures can be guided by existing literature studies.

3.3.2. Imaging in the Chest and Abdomen

1. In general, anatomy of interest distant from sites of physiological motion can be secured to prevent motion artifacts in images. For instance, we use specially cut Velcro to secure

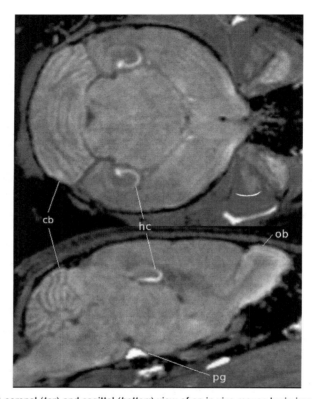

Fig. 3. A coronal (*top*) and sagittal (*bottom*) view of an in vivo mouse brain imaged with MnCl$_2$ injected as a contrast agent. The MnCl$_2$ was intraperitoneally injected 24 h prior to imaging. The T_1-weighted spin echo sequence had the following scan parameters: TR/TE = 90/3.8 ms, flip angle 55°, resolution = (115 μm)^3 and scan time = 2 h. With no manganese, the contrast is very flat but in the image it can be seen that the Mn^{2+} is taken up by the layers of the cerebellar folia (cerebellum, cb), the hippocampus (hc), the olfactory bulbs (ob) as well as in the pituitary gland (pg).

the head during brain imaging (*see* **Fig. 1**). Another simple method to minimize motion artifacts is to place the mouse in a supine position, such as for imaging of the kidneys (45) or spine.

Fig. 4. (continued) disorganization of fibers crossing the septal region. Fimbria: In mutants the enlarged hippocampal fimbria (fi) is shifted rostrally and exhibits a much thicker and less curved fiber structure as demonstrated by region-to-region fiber tracking (*bottom*). cing = cingulum; fx = fornix. Directional color code: *red* = left–right, *blue* = rostral–caudal, *green* = anterior–posterior. Reproduced with permission by Oxford University Press (Ref. (37)). (**b**) A coronal (*left*) and sagittal (*right*) view of the fractional anisotropy map in a fixed mouse brain. A 3D diffusion-weighted fast spin-echo sequence was used with an echo train length of 6 with a TR of 325 ms, first TE of 30 ms, and a TE of 6 ms for the remaining five echoes, ten averages, field of view 14 × 14 × 25 mm^3 and a matrix size of 120 × 120 × 214 yielding an image with (117 μm)^3 voxels. For the calculation, one $b = 0$ s/mm^2 image (with minimal diffusion weighting) and six high b-value images ($b = 1956$ s/mm^2) in six different directions [(1,1,0),(1,0,1),(0,1,1),(−1,1,0),(−1,0,1),(0,1,−1)] (Gx,Gy,Gz) were produced. Total imaging time was ∼16 h. *Yellows* indicate values closer to unity (fibers are very ordered in one direction) and *deep purples* indicate more isotropy. Figure provided courtesy of Dr Jacob Ellegood.

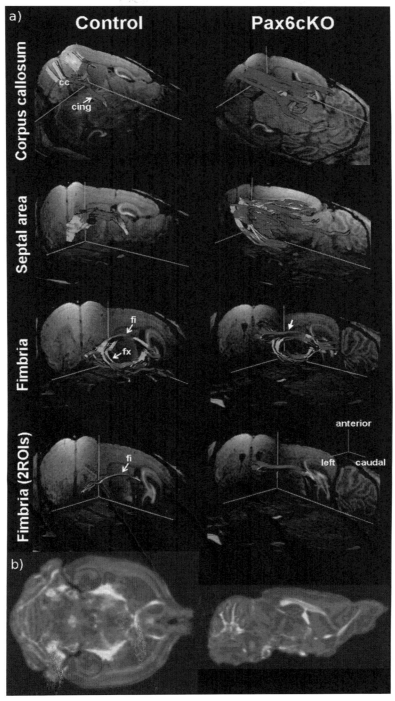

Fig. 4. Applications of diffusion-weighted imaging in mouse brain. For color figures, please refer to the on-line version, which is available through most institutional subscriptions. (**a**) 3D views of major fiber pathways of a control and Pax6cKO mouse in vivo. Corpus callosum: Although controls present with a pronounced connection of the 2 hemispheres via the corpus callosum (cc) and projections into the cingulate and motor cortex, mutants reveal a strong rostrocaudal fiber orientation and a nearly complete lack of interhemispheric projections. Septal area: Mutants are characterized by a vast

2. In most of the chest and abdomen, however, motion is pronounced and cannot be eliminated. Instead, imaging approaches aim to achieve a "snapshot" of the motion. This is particularly challenging as the murine heart rate is typically around 500 beats per minute and the anesthetized mouse breathes once every 1–2 s. In the past decade, several techniques have emerged to enable cardiac cine-MRI in which the ventricular size and shape can be assessed at multiple phases of the heart cycle and is discussed in much greater detail in the **Chapters 20, 21**, and **22**.

3. Briefly, there are typically two ways of taking images: prospectively and retrospectively. In prospective gating, the scanner is set to acquire data only during a specified phase of the heart cycle. This method is quite efficient and provides excellent cardiac images (48–50). In retrospective gating, data are continually acquired and the respiratory and ECG signals are recorded. Data are oversampled (e.g., data for more than one image are acquired) and then sorted according to cardiac phase and respiratory status retrospectively, retaining only data acquired during the quiescent respiratory period and at a specified cardiac phase (51, 52). A key benefit of the retrospective method is that multiple mice can be imaged in this fashion (53).

4. State-of-the-art techniques to date yield cardiac images on the order of 100×100 μm^2 with slice thicknesses at about 0.5 mm, with some groups achieving isotropic voxels (54, 55). Several disease models have been investigated, including models of myocardial infarction (56), transgenic models (57–59), and ischemia (60).

5. Another challenging area of research is imaging the lungs in the mouse. There are typically two ways of studying the lung: using hyperpolarized inert gases to look at the airspaces (61–63) or enhancing the water signal in the lung parenchyma (64, 65). The technical knowledge and skills for set up of a hyperpolarized gas system can be significant, but the reward is images of the airspaces exclusively (*see* **Chapters 10** and **24** for more details). Imaging the proton signal is also challenging as the microstructure of the parenchyma causes T_2^* to be on the order of milliseconds. The T_1 can be shortened by adding a gadolinium-based contrast agent (66) or having the mouse breathe in oxygen (65). This way signal averaging can be applied along with cardiac gating to yield images of the parenchyma (*see* **Fig. 5** from Watt et al.). The highest resolution images are 87×87 μm^2 in-plane (66) and a number of disease models in the mouse have been studied including lung cancer (67–69), lung inflammation (61, 70), and cystic fibrosis (64).

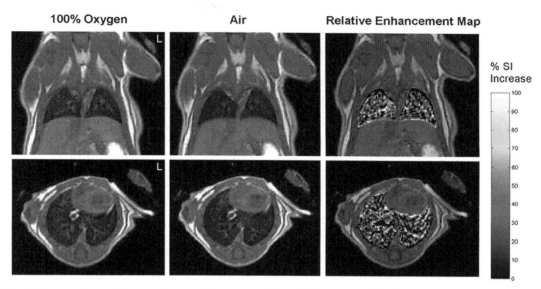

Fig. 5. Oxygen-enhanced images of a normal mouse breathing 100% oxygen and air. For color figures, please refer to the on-line version, which is available through most institutional subscriptions. Coronal (*top*) and axial (*bottom*) images were acquired using a three-dimensional (3D) fast spin-echo (FSE) cardiac-triggered, respiratory-gated sequence with dummy scans. The relative enhancement map shows the percent increase of signal intensity in the lungs observed with 100% oxygen inhalation relative to air for images at the same slice position. To highlight signal enhancement in the lung, the intensity changes in the rest of the body have been masked. All coronal images are shown at a position dorsal from the heart and away from major vessels. Reproduced with permission from Watt et al. (Ref. (65)).

3.3.3. Contrast Agent Administration

It is beneficial in some phenotyping applications to highlight anatomy or function by administration of contrast agents that shorten the T_1 or T_2 relaxation times, producing respectively a bright (positive contrast) or dark (negative contrast) signal. There are a number of classes of contrast agents that can be used with the aim of highlighting vasculature, elucidating tumor boundaries, assessing neuronal function, or even targeting or tracking cells. These are discussed more extensively in other chapters so we discuss them only briefly here.

1. Vascular imaging, although possible with simple pulse sequence selection (**Fig. 6a**), can benefit from administration of contrast agents, including chelated gadolinium (Gd) complexes and iron oxide agents. Functional information can be achieved by assessing blood inflow immediately following injection of contrast agent through a vein and then imaging dynamic changes in image contrast as the agent perfuses through the system (27, 28).

2. Alternatively, injection of agents that remain in the blood pool for extended periods can be used to acquire higher resolution anatomical images of both arteries and veins as well as vascular pathologies such as atherosclerosis (71, 72) (*see* **Fig. 6b–d**).

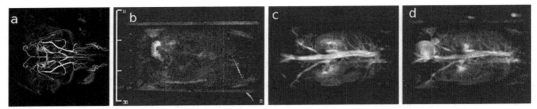

Fig. 6. Different strategies to image the vascular system. (a) A horizontal MIP image of the adult mouse brain achieved with an inversion recovery-prepared rapid gradient echo sequence, in which the inversion time was selected to null the tissue signal and highlight blood inflow. Sequence parameters: FOV = 25.6 × 25.6 × 13.6 mm, resolution of $(100\ \mu m)^3$, TR = 2 s, TE = 2.9 ms, TI = 620 ms, six averages, 64 acquisitions per TR each with 12° excitation spaced 10 ms apart, total acquisition time = 1 h 49 min. **Panels b–d**: 3D magnetic resonance angiography (MRA) before (l) and at 4 (m) and 55 min (n) after contrast injection. With kind permission from Springer Science+Business Media: MAGMA, Contrast enhancement in atherosclerosis development in a mouse model: in vivo results at 2 Tesla, 17, 2004, 191, L Chaabane et al., part of **Fig. 1** (Ref. (120)).

3. Assessment of functional perfusion or microvasculature parameters based on pre- and post-injection image evaluation is also being explored (73–77). Administration of vascular agents is most easily performed via intravenous tail vein injections, although alternative routes are necessary for rapid bolus injections.

4. Neuronal function may, in some cases, be explored via administration of Mn. An intrapertioneal injection of Mn followed by a period in which mice are exposed to stimulus can modify the uptake of Mn sufficiently to show functional differences between different stimulus conditions. This has been demonstrated most clearly in the auditory system (78, 79). Focal injections of Mn or exposure of the olfactory system to Mn can also provide measurements of axonal transport and regional activity (29, 80, 81) with sufficient sensitivity to show differences in transgenic mice (82).

Significant effort continues in the development of molecular imaging, which in MRI includes applications of cell tracking, localization of targeted contrast agents, or imaging of gene expression. Successful development in this area will produce new and important in vivo phenotyping opportunities. Several genetic agents, which produce or accumulate contrast via expression of a transgene, are being explored with the hope of elucidating spatial and temporal gene expression profiles in development or disease (83–89). Alternatively, agents designed to label particular cell populations by uptake or binding to specific surface markers are also in development. Current applications in molecular imaging include detection of transgene expression in tumor-bearing mice (88, 90), integrin expression in atherosclerotic plaques (91, 92), and finding specific peptides that will attach to amyloid plaques (93) or inflammatory lesions (94). The tracking of both injected (95–99) and endogenous (31–33) cell populations after labeling with iron oxide agents shows some promise, permitting the evaluation of cellular migration behaviors in vivo during homeostasis or disease. The potential impact of genetic and cellular imaging is remarkable, motivating needed developments of these MRI methods in molecular imaging.

3.3.4. Postmortem Imaging

In vivo imaging is an extremely useful aspect of MRI. However, some studies are best served by increasing scan time to achieve microscopic resolutions in ex vivo specimens. Since the scan times can approach tens of hours, it is necessary to fix tissue to avoid degradation.
1. To image individual organs like the brain, kidneys, and liver, the mouse is first fixed by conventional perfusion. In this procedure, mice are first anesthetized (e.g., by ketamine/xylazine) and then placed in a supine position.
2. A midline incision through the chest is followed by catheterization of the left ventricle. A cut is also made in the right atrium and then a saline/heparin flush is followed by perfusion of a fixative solution.
3. Including a contrast agent such as gadolinium in the perfusates also serves to reduce subsequent imaging time.
4. The organ of interest is then excised, immersed in a proton-free, susceptibility-matching fluid and then imaged.
5. Smaller, solenoidal RF coils can be used when imaging excised organs, which give better sensitivity.
6. In the case of embryos, they are removed from the mother and then immersion-fixed in formalin and 4 mM Multi-Hance (Bracco Diagnostics Inc, Princeton, NJ) for 3 days and then transferred to 1% agarose. **Figure 7** shows an example of a 15.5-dpc (days post coitum) embryo imaged in this fashion.

To image the whole mouse adult body, it is desirable to avoid damage to the thoracic cavity and two techniques have emerged to perfusion fix the mouse without breaching the chest cavity.
1. One method involves cannulating the right jugular vein and the left carotid artery and passing the perfusates through these points (100).
2. The other uses an ultrasound biomicroscope to guide the catheter into the left ventricle, allowing the mouse to be fixed in much the same fashion as the open-heart perfusion fixation, with the exception that the perfusates are drained through cuts in the femoral and jugular veins (101). **Figure 2a** shows an MR image of a mouse fixed in such a fashion.

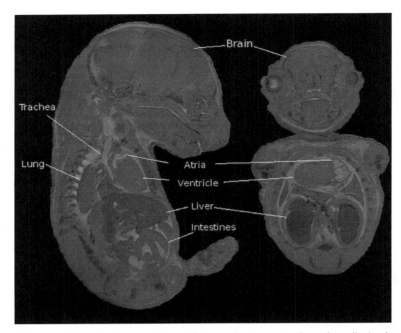

Fig. 7. Sagittal (*left*) and coronal views of a 15.5-dpc mouse embryo. A gradient echo scan was used with the following scan parameters: TR/TE =50/5.06 ms, flip angle 60°, FOV = 2.5 × 1.4 × 14 cm, resolution is $(32\ \mu m)^3$, six averages for a total scan time of about 14 h.

3.4. Anatomical Image Analysis

Figure 8 shows several examples of different phenotyping analyses for MRI that will be discussed below, as applied to neuroanatomical mutants (reproduced from Nieman et al. (21)). Briefly, the panels (a) and (b) show significant local volume changes; panels (c) and (d) show two illustrations in which a phenotype is immediately recognizable by eye; panels (e) and (f) are more subtle mutant phenotypes which require detailed statistical analyses for identification.

3.4.1. Basic Quantitative Measures

After acquiring image data for a study, the main challenge is quantitative data analysis. A human observer can make qualitative observations (e.g., "the left ventricular wall is thicker than the wildtype" or "the tumor looks bigger in one mouse compared to another") but the real power of MRI is that quantitative measures can be obtained from the images.

Fig. 8. (continued) (f) for the respective overlays. Data were acquired with in vivo magnetic resonance imaging (MRI) (panels a, d, and e), ex vivo specimen computed tomography (c), ex vivo specimen MRI (d), and in situ specimen MRI (f). All MRI images are single slices from three-dimensional isotropic datasets. Reproduced with permission from Nieman et al. (Ref. (21)).

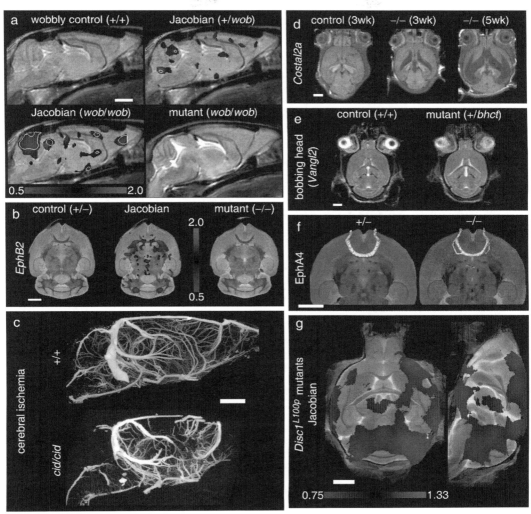

Fig. 8. Neuroimaging findings in mouse mutants. For color figures, please refer to the on-line version, which is available through most institutional subscriptions. In (**a**), an average control mouse image is shown with two Jacobian overlays indicating local volume changes in the heterozygous and homozygous wobbly mutants. Values greater or less than 1 in the Jacobian field represent growth or shrinkage, respectively. Jacobian overlays in panel (a) are shown in regions where $p < 0.05$ with white contour lines indicating $p = 0.002$ (a false discovery rate of 5% in the homozygous mutant). The average homozygous wobbly mutant image is also shown for reference and exhibits a small cerebellum. In (**b**), data from the EphB2 knockout study indicates an abnormality at the anterior commissure (the posterior portion is absent). Jacobian data in the middle image indicate regions where $p < 0.003$ (5% false discovery rate). Panel (**c**) shows two sample micro-CT datasets as maximum intensity projections. The cerebral ischemia mutant shows a striking defect in vascular perfusion as compared to the control. Panels (**d**) and (**e**) provide two examples of hydrocephalus. Three different individual images from the sonic hedgehog pathway mutation are shown (d) with massive expansion of the ventricles (shown as hypo-intense regions). In panel (e), control and mutant averages from the bobbing head study show more subtle ventricle expansion (shown as hyper-intense regions). In this case, statistical comparison of the mutant Jacobian values to a set of 20 control mice (four wildtype littermates supplemented with non-littermate wildtypes) shows expanded ventricles (regions indicate a 5% false discovery rate) and increased variability at the lateral ventricles (Jacobian variance ratio is shown in regions where $p < 0.05$). In the final panel (**f**), regions of abnormality representative of the disrupted-in-schizophrenia-1 mutants (Disc1) are shown as Jacobian data overlaid on an average image (regions indicate $p < 0.05$ by a Mann–Whitney U test). All white scale bars indicate 2 mm. Color scales are indicated in panels (a), (b), (e), and

1. Using freely available image processing packages, the previous examples can easily be quantified by measuring the wall thicknesses and segmenting tumor volumes. Most visualization packages have a manual segmentation feature and many have some form of semi-automatic segmentation.

2. If the subject numbers are few, manual segmentation is rapid, reliable, and widely utilized (7, 46, 51). An advantage is that it does not require any knowledge or development of more elaborate image processing methods. **Figure 9** shows an example of manual labeling of 62 substructures in the mouse brain (102).

3. Typically small, simple structures (e.g., the anterior commissure in the brain) can be segmented in tens of minutes as they only occupy a few slices within a dataset.

4. Segmenting more expansive structures like the corpus callosum or hippocampus can take more than an hour. If many mice are part of the study, manual segmentation on all subjects is prohibitive; hence using computer-aided image processing tools is prudent.

3.4.2. Registration

It is frequently impractical to perform manual segmentation of volumes for image analysis, and, in these cases, it is common to make use of registration algorithms to elucidate any disparities. The registration process produces transformed images that are aligned with one another such that common structures are placed on top of one another, in "register." The image transformations required to achieve such alignment encode phenotypic differences. The phenotypes can be extracted from registration-based analyses in several ways, based on structural volumes as described above or on more sophisticated voxel-by-voxel measurements of shape or size differences.

1. A number of registration packages are listed in **Section 2.5**, item 3. In general, the registration process begins by first globally aligning images through affine registration, which includes global translations, rotations, scales, and shears. These modifications are considered linear operations and are applied to the entire image (*see* **Note 8**).

2. With all images in this orientation, it is convenient to average the intensity images to produce a representative average image. This scaled average accounts for overall size and shear distortion differences. Although this average image space may be defined by a reference image, it is more satisfying to define an unbiased space incorporating all images under analysis as described in detail elsewhere (103).

3. The initial average after affine registration is generally fairly "blurred," due to imperfect alignment of anatomical features. An iterative nonlinear registration process can refine

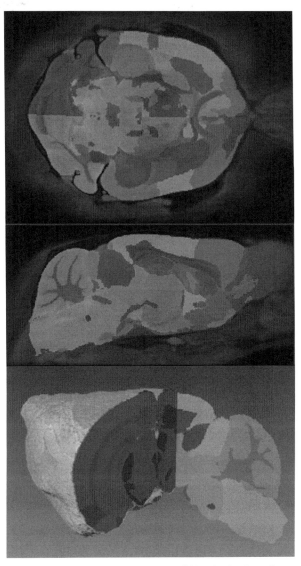

Fig. 9. Three views from a manual segmentation of 62 sub-structures from an average of 40 mouse brains (Ref. (102)). For color figures, please refer to the on-line version, which is available through most institutional subscriptions. The brains were from 20 male and 20 female 12-week-old C57Bl/6 J mice. Three brains were imaged in parallel with a fast spin echo sequence (TR/TE = 325/32 ms, four averages, $(32\,\mu m)^3$ resolution for a total scan time of 11.3 h. The *top* and *middle* panels show coronal and sagittal slices (respectively) of the atlas overlaid on top of the average. The *bottom* panel is a 3D volume rendering with a cutout showing the segmentation (a different colormap was used than the previous two panels).

the average and bring the component images into improved alignment. In this step, local regions of the image are shifted to match features with a target image (this could be a chosen representative image or the calculated average of the population (103)).

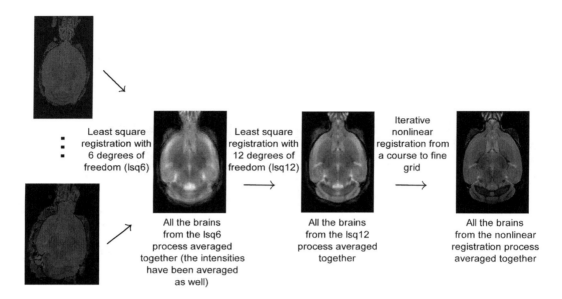

Fig. 10. The registration pipeline used at the Mouse Imaging Centre. This pipeline has been applied to fixed brains (Refs. (22, 102, 104)), live brains (Refs. (121, 122)), and embryos (unpublished).

4. The nonlinear registration is performed iteratively using a coarse-to-fine approach, with a gradual decrease of grid-point spacing (i.e., increasing registration resolution) until the desired level of refinement is achieved. **Figure 10** shows a sample registration pipeline.

5. The simplest way to see if a registration was successful is to visually verify the alignment of all images. If there are anomalies, then it could be that the registration procedure needs to be adjusted (*see* **Note 9**) or that artifacts in particular images are problematic. In some cases, however, it is actually very difficult to register two subjects together if the images are too dissimilar, for example, in **Fig. 8d**, the ventricles between the different genotypes are so different that the structures should be considered nonhomologous, a violation of assumptions inherent in the registration process. Nonetheless, it is often possible to use registration even in cases of mild nonhomology, as for instance in **Fig. 8b**—the posterior portion of the anterior commissure is absent in the mutant—in which registration was still largely successful.

3.4.3. Volumetrics

By combining manual segmentation with the registration transformations described, it is possible to determine the volumes of structures in many mouse images after segmentation of a single image, greatly improving study throughput (103, 104).

1. The typical procedure that we use is to manually label the average image and then to "back propagate" the segmented labels (which may include 40–60 substructures in the brain) along the transformation prescribed by the registration results.

2. For instance, after segmenting the whole brain in an average image and "back propagating" through the nonlinear and affine registrations to the original images, we measured the whole brain volume in C57/Bl6 mice and determined male brains were 2.5% larger than their female equivalents (22). This method, which uses "unscaled" comparisons of whole brains and cerebral substructures, gives absolute volume values and is preferred for performing survey studies, (i.e., screening genetically mutant mice).

3. An alternative basis for analysis uses volumetrics after affine registration—a "scaled" comparison—which may be more appropriate in some instances, particularly if specific hypotheses are to be tested (105). In this analysis, one compares a normalized or relative structure volume, which may also reveal subtle differences. The most appropriate comparison, based on "unscaled" or "scaled" image data, must be considered carefully.

4. This technique of finding substructural volumes in the brain is very difficult to accomplish histologically—although this remains the gold standard (*see* **Note 10**)—and a comparison between striatal volumes found with MRI and registration versus histology demonstrates the superiority of MRI for volume-based analyses of structure (104). A number of groups are using similar methods to study cerebral structures in various disease models (30, 106, 107).

5. One limitation of volumetric analysis is that it reduces complex shapes down to a single number. This line of inquiry will fail if multiple, opposing volumetric changes occur within a structure, minimizing or eliminating any cumulative volume change. Also, in the case that there is no volume change but rather a positional shift, volumetrics may also fail to detect the change. In this case, more sophisticated analyses are possible using the results of the registration.

6. The results of the registration provide not only a set of transformed images, but also a map encoding these image transformations with vectors or matrices. These maps may be analyzed computationally to look for group differences, without the need to reference segmented structure volumes. The vector representation of the image transforms is frequently referred to as the deformation field.

3.4.4. Analysis of the Deformation Field

Deformation-based analyses compare a metric computed from the deformation field on a per voxel basis, as opposed to reducing an entire image set to some defined number of segmented substructures. Below, we describe metrics to represent (1) positional shifts of anatomy; (2) anatomical variability; (3) volume changes in regions or structures, and (4) other methods to highlight local changes in shape.

1. The most straightforward deformation-field comparison of interest may be the vector difference from an individual to the group average from one genotype group to another.

2. A display of the magnitude of this difference provides a look at the regions of greatest positional shift between subjects and can be graphically displayed. In a set of genotypically identical mice, the set of all such fields can provide a sense of the inherent variability in the image data (22, 108).

3. This variability can be computed on a per voxel basis in a population by computing the arithmetic mean of all displacement vectors (**Fig. 11**) and represents the typical deviation (in millimeters displacement) from the population average.

4. Beyond a simple display of differences in the deformation fields themselves, local volume changes between images or

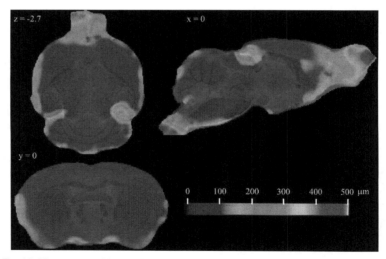

Fig. 11. The mean positional difference image created from the arithmetic mean of the deformation vectors used to create an average of nine excised 129S1/SvImj mouse brains. For color figures, please refer to the on-line version, which is available through most institutional subscriptions. Deformation magnitudes (measured in micrometers and shown in spectral color scale) are overlaid on top of the average image (*in grayscale*). *Purple* and *blue* regions (on the low end of the scale) indicate low variability, while *red* and *white* (on the high end of the scale) indicate high variability. Shown are three orthogonal slices with stereotaxic coordinates given in millimeters. With kind permission from Springer Science+Business Media: Cerebral Cortex, 15, 2004, 642, N Kovacevic et al., **Fig. 4** (Ref. (103)).

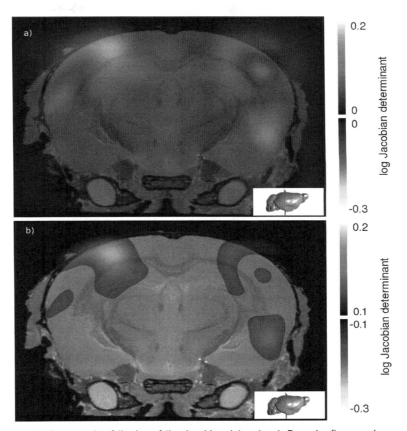

Fig. 12. An example of the log of the Jacobian determinant. For color figures, please refer to the on-line version, which is available through most institutional subscriptions. Using nonlinear image registration, an unbiased group average is created from 10 wild-type and 10 mutant mouse brain MRIs. The Jacobian determinant, calculated from the deformations needed to take each mouse's anatomy into the final group average space, is a measure that illustrates the expansion and contraction of tissue. In the colored areas, the log of the Jacobian determinant of a single wildtype brain is shown, overlaid on the anatomy of that subject resampled to the group average. The areas in *red* (positive scale) indicate areas where this particular brain had to grow to get to the group average, and the areas in *blue* (negative scale) where the brain had to shrink. The coronal level of the section is pictorially illustrated in the *bottom right corner*. (**a**) The full log of the Jacobian is shown and (**b**) the log of the Jacobian is thresholded to show more relevant areas of change. Figure provided courtesy of Matthijs van Eede.

populations can be determined by computing the determinant of the Jacobian matrix, formed by the matrix of deformation-field partial derivatives (*see* **Note 11**) (109). **Figure 12** shows an example image of the log of the Jacobian determinant, usually simply referred to as the Jacobian.

5. In the places where the log Jacobian is equal to 0, there is no local volume differences between groups; a value less than 0 indicates shrinkage and a value greater than 0 indicates

a growth. While valuable information can be gained from the Jacobian determinant, there is additional information encoded in the full Jacobian matrix, including anisotropic volume changes. Ongoing research therefore explores the use of alternative Jacobian-based metrics, such as strain matrices (110), to improve local detection.

6. Combined use of segmented structures and deformation-field measurements are another alternative for analysis. For instance, comparing the surface normal vector of a segmented structure with the local direction of the deformation field can determine local surface displacements. By calculating this vector at every point of a surface, a different metric for shape changes can be quantified (111).

3.4.5. Statistical Considerations

With the exception of gross anatomical phenotypes which are immediately and consistently obvious on simple inspection, it is necessary to isolate regions of significant differences in a statistically rigorous fashion. This is true independent of the metric—such as those in the previous section—selected as a basis for comparison. Cases that necessitate statistical comparison are among the more interesting to study by imaging, as the imaging results are instrumental in phenotyping these models. In studies where a measure of interest can be identified prospectively, such as an anatomical feature's volume, area, thickness, or length, it is beneficial to perform such measurements on each individual image and then compare the measured results. This analysis process has the benefit of condensing the most relevant image information into a single metric at the outset, simplifying statistical comparison to a single groupwise test. As in the comparison of segmented volumes, these measurements can be performed "manually" in each mouse or facilitated by the registration procedure to transfer segmentations and annotations from one image to all others. Thus, after defining the measures of interest in one image, the rest of the dataset can be generated computationally.

It is nevertheless common in imaging studies to be screening for possible phenotypes in a new mouse model, to seek secondary phenotypes in addition to more obvious ones, or to consider substructural features. In these cases, the unbiased nature of a voxel-by-voxel comparison is desirable. Voxel-by-voxel comparisons may be parametric, such as Student's t-tests or Hotelling's T-squared tests (112, 113), or nonparametric, such as U tests (114) or permutation tests (115), and are widely available in scientific or statistical software packages. It is necessary to note in voxel-by-voxel analyses, however, that the process of making many statistical comparisons increases the likelihood of false positives. In fact, given that a high-resolution (fixed, 32 μm isotropic) brain image contains ~18 million voxels, one is assured, using the customary $p < 0.05$ criterion, of finding incorrectly identifying 5% of them as significant (i.e., 90,000 false-positive voxels!).

For this reason, it is necessary in voxelwise statistical analyses to correct statistical results for multiple comparisons. While this is an area of ongoing research, particularly in neuroimaging applications, several methods have become fairly common. In our own studies, we have most frequently used the false discovery rate (FDR) (*see* **Note 12**), which controls the average proportion of false-positive voxels to true-positive voxels (*see* **Fig. 8a** and **b**). The benefit of the FDR is that it is easy to implement (i.e., also available in many existing software packages as described in **Section 2.5**, item 4), conceptually simple, and has been used quite widely. It is, however, quite conservative in cases where only a small number of voxels are affected and fails to take into account spatial relationships between significant voxels. The latter is particularly unsatisfying in deformation-based morphometry, where the registration procedure inherently constrains neighboring voxels to produce smooth deformations. An alternative method of adjusting statistical thresholds may be based upon treating the features of a statistical map collectively and considering the likelihood of such features occurring in a comparable random map. Such methods based on random field theory are also implemented in various software programs. In a similar spirit, authors have proposed various cluster-based analyses, in which groups of voxels with sufficiently similar responses are treated as a collective. This both increases the signal-to-noise ratio (after forming a score representative of all voxels in the cluster) and reduces the number of statistical comparisons, in part mitigating the multiple-testing problem. For voxelwise statistical tests in large datasets, it is important that one of these methods be employed to ensure reliable results.

Deformation-based analyses with statistical testing are particularly powerful for isolating subtle phenotypes. However, the ability to do so is a function of several factors. Underlying variability, be it biological or methodological (*see* **Note 13**) in nature, naturally masks small phenotypes and it is important to understand the limitations of a given protocol. With this aim, a variability map can be produced to indicate variability inherent in the population and/or method (particularly if specimen preparation is involved). For instance, **Fig. 11** shows a variability map of nine fixed, 129S1/SvImJ mouse brains reproduced from Kovacevic et al. (*see* **Note 14**). The variability in the inner structures of the brain is quite low (less than 120 μm or two voxels in a 60 μm image). If the phenotype involves detection of changes on this order, or the phenotype itself is highly variable (*see* **Note 15**), many more mice will be required. The process of estimating the variability and the number of mice required can be made more formal by specifically finding the "noise" intrinsic to a structure

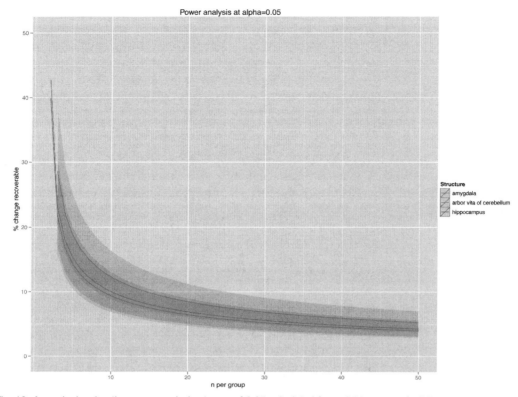

Fig. 13. A graph showing the power analysis at an α of 0.05 calculated from eight groups of wildtype mice (about 150 mice in total) for three structures in the brain: amygdala, arbor vita of the cerebellum, and the hippocampus. For color figures, please refer to the on-line version, which is available through most institutional subscriptions. The y-axis shows the expected change given the size of the groups on the x-axis. The *top limit* for each structure is the maximum variance, the *solid line* is the mean variance of the groups and the *bottom limit* is the minimum variance. As an example, if you expect that the volume difference in the arbor vita of your mutant population is about 20%, then you would need 9–10 mice per group to find a significant difference with $p < 0.05$. This graph quantitatively shows that more subtle differences require more animals per group. Figure provided courtesy of Dr Jason Lerch.

in question. As an example, Lerch and colleagues measured the mean and standard deviation of the cortical thickness in a group of 20 mice (116). From this, they created a power analysis graph that indicated 6–9 mice per group were sufficient to result in statistically significant results in their study. In **Fig. 13**, we have recreated a similar power analysis graph based on about eight groups of wildtype controls from different experiments; each group had between 7 and 20 mice. Within each group the variance in volumes was computed and the across-group comparisons were used to generate the group. Such information provides helpful guidelines for planning mouse-phenotyping studies.

3.5. Conclusions

In this chapter, we have described the techniques, tricks, and experiences we have gained over the last decade as a research group devoted to mouse imaging. During this period, the number of groups performing mouse MRI has continued to increase, each of these adding their own various implementations and techniques. We also listed some of the more obvious pitfalls that may be encountered by someone breaking into this exciting field of research. The live imaging is perhaps the most alluring aspect of mouse MRI, achieving isotropic resolutions of ~100 μm and permitting longitudinal evaluation. In some investigations, however, higher impact data may be acquired by preparing ex vivo samples, to achieve isotropic resolutions of less than 40 μm. While many measures of phenotype are possible using various quantitative and functional MRI methods, detection of anatomical changes remains a common and particularly powerful approach. Careful preparations and imaging methods permit computer-aided detection of anatomical phenotypes, making this method appropriate for high-throughput analyses of mutant mice. In this regard, MR-based mouse phenotyping stands to contribute a great deal to the understanding of the genetic basis of human diseases and the subsequent exploration of novel therapies.

4. Notes

1. The most significant benefit of noninvasive imaging is evaluation of temporal and spatial evolution of disease in individual mice. This increases statistical power because it eliminates heterogeneities that exist between mice. A good example is that of genetically engineered tumor models, in which time and location of tumor onset can vary from one mouse to the next. However, imaging protocols in compromised mice—young, old, or mutant—need to be judiciously chosen, as mice may be too fragile to survive long scan durations or frequently repeated sessions.

2. The biggest confounding factors in live imaging are motion artifacts. Because an image can take anywhere from a minute up to several hours, physiological motion or postural changes will compromise imaging results. In our lab, we have devised a custom-built, molded sled (**Fig. 1**) (36). The mouse lies prone on the sled and its head can be held in place for neuroimaging using Velcro straps. The sled also incorporates embedded ECG, respiratory, and temperature monitoring devices, greatly simplifying animal setup.

3. Injectable anesthetics (e.g., ketamine/xylazine, Avertin) can also be used, eliminating the need for gaseous delivery and scavenging hardware. However, these are only useful for short imaging sessions as the anesthetic wears off in 30–45 min. While it is possible, in principle, to deliver

anesthesia via a catheter in the tail vein during the imaging session, this adds a significant level of complication and is not recommended.

4. The amount of anesthesia and maintenance temperature will change with other strains of mice, which may metabolize anesthetic at different rates.

5. A water bath may be undesirable in some applications because it needs to be in close proximity to the mouse. This may necessitate a larger volume coil, which results in reduced SNR if the coil is used for both transmit and receive. If a small, localized surface coil is used then a water bath should have no impact on SNR if it can be positioned away from the receive coil (but still close to the mouse).

6. The ECG connections routinely show a respiratory signal superimposed with the cardiac signal. The two signals can, in principle, be separated electronically to provide separate cardiac and respiratory traces for monitoring and triggering. However, in practice, a "cleaner" respiratory signal is produced by a respiratory bellows or pillow system.

7. All vendors' platforms will perform standard image acquisitions and reconstructions automatically. However, for many sequences—particularly, specialized ones—it may be necessary or beneficial to implement site-specific image sequences and/or perform off-line reconstruction to ensure all acquisition and processing steps are understood and optimized.

8. It is in this step that it becomes obvious if the care in pre-planning of mouse positioning was sufficient. If different mice were not imaged in roughly similar positions, initial registration-based alignment will fail and some manual alignment will be necessary.

9. Troubleshooting a registration pipeline is methodical work. There are images produced at each step along the way and careful inspection of these images will reveal where in the pipeline a failure occurred. The parameters leading up to that step need to be tweaked. Examples of adjustments that can push along the pipeline include decreasing step sizes or increasing blurring of the input images.

10. Using MRI in biomedical research can be extremely powerful as samples can undergo subsequent histological verification. This is the gold standard in which a tissue is sectioned into a thin slice and studied with light microscopy, and remains a necessity to confirm mouse MRI results. A whole battery of histological stains further enhances the ability to highlight different microscopic structures. For example, if MR images indicate that plaques are seen in an Alzheimer's

mouse model, the iron and β-amyloid content can be verified with a Perl's Prussian Blue stain and then a thioflavin-S stain (117). The ability to do histology on mouse models frequently acts as a bridge between preclinical research and clinical diagnostic data.

11. The Jacobian matrix is the matrix of all first-order partial derivatives of a vector-valued function; in this case, the deformation field. Working with the Jacobian is a way of finding and displaying differences between deformation fields.

12. The false discovery threshold (q) is usually set at 0.05, although this seems largely guided by customary use of $p = 0.05$. Some authors have noted that a high q-value (i.e., 0.10) may be appropriate in some imaging experiments, although a strict convention is lacking.

13. If part of the experimental plan is to use registration as the tool for phenotyping, some thought needs to go into what signal-to-noise ratio (SNR) and resolution to use in the images. These are competing factors that must be traded-off with one another in determining how to spend imaging time. A recent study has determined that for the purposes of image registration, the best compromise is achieved by acquiring images with SNR ~20 and "spending" the rest of the time on resolution (118).

14. This particular example was done with mouse brains excised from the skull. Because the brain is very malleable, the variability shown is really more a map of specimen preparation—the regions of highest variability are at the exterior edges and the olfactory bulbs and brain stem. This again emphasizes how important sample preparation—live or fixed—is critical to computer-aided phenotyping. Our subsequent work after this study uses mouse brains in the skull.

15. In some cases, we have seen variable phenotypes especially with conditional knock-outs. This is when the at least two types of phenotype are seen with the same genotype. So sometimes even higher numbers are required to make sure all phenotypes are significantly represented.

Acknowledgments

The authors would like to thank the staff and students of the Mouse Imaging Centre, especially Jun Dazai, Dr Jacob Ellegood, Dr Jason Lerch, Christine Laliberte, Matthijs van Eede, and

Michael Wong for providing figures. The Mouse Imaging Centre (MICe) acknowledges funding from the Canada Foundation for Innovation and the Ontario Innovation Trust for providing facilities along with The Hospital for Sick Children. Operating funds from the Burroughs Wellcome Fund, the Canadian Institutes of Health Research, the National Cancer Institute of Canada—Terry Fox Program Projects, the National Institutes of Health, and the Ontario Research and Development Challenge Fund are gratefully acknowledged. The embryo work was funded by Genome Canada through the Ontario Genomics Institute (2008-OGI-TD-03).

References

1. Lauterbur P. C. (1974) Magnetic resonance zeugmatography. *Pure App Chem* **40**, 149–57.
2. Clapham J. C., Arch J. R. S., Chapman H., et al. (2000) Mice overexpressing human uncoupling protein-3 in skeletal muscle are hyperphagic and lean. *Nature* **406**, 415–8.
3. Jalanko A., Tenhunen K., McKinney C. E., et al. (1998) Mice with an aspartylglucosaminuria mutation similar to humans replicate the pathophysiology in patients. *Hum Mol Gen* **7**, 265–72.
4. Jones M. E., Thorburn A. W., Britt K. L., et al. (2000) Aromatase-deficient (ArKO) mice have a phenotype of increased adiposity. *PNAS* **97**, 12735–40.
5. Bock N. A., Zadeh G., Davidson L. M., et al. (2003) High-resolution longitudinal screening with magnetic resonance imaging in a murine brain cancer model. *Neoplasia* **5**, 546–54.
6. Fomchenko E. I. and Holland E. C. (2006) Mouse models of brain tumors and their applications in preclinical trials. *Clin Cancer Res* **12**, 5288–97.
7. McDaniel B., Sheng H., Warner D. S., Hedlund L. W., and Benveniste H. (2001) Tracking brain volume changes in C57BL/6 J and ApoE-deficient mice in a model of neurodegeneration: a 5-week longitudinal micro-MRI study. *NeuroImage* **14**, 1244–55.
8. Wei Q., Clarke L., Scheidenhelm D. K., et al. (2006) High-grade glioma formation results from postnatal pten loss or mutant epidermal growth factor receptor expression in a transgenic mouse glioma model. *Cancer Res* **66**, 7429–37.
9. Duhamel G., Callot V., Decherchi P., et al. (2009) Mouse lumbar and cervical spinal cord blood flow measurements by arterial spin labeling: sensitivity optimization and first application. *Magn Reson Med* **62**, 430–9.
10. Hillman G. G., Singh-Gupta V., Zhang H., et al. (2009) DCE-MRI of vascular changes induced by sunitinib in papillary renal cell carcinoma xenograft tumors. *Neoplasia* **11**, 910–20.
11. Berry C. J., Miller J. D., McGroary K., et al. (2009) Biventricular adaptation to volume overload in mice with aortic regurgitation. *J Cardiovas Magn Reson* **11**, 27.
12. Smits A. M., Van Laake L. W., Den Ouden K., et al. (2009) Human cardiomyocyte progenitor cell transplantation preserves long-term function of the infarcted mouse myocardium. *Cardiovas Res* **83**, 527–35.
13. Greve J. M., Williams S. P., Bernstein L. J., et al. (2008) Reactive hyperemia and BOLD MRI demonstrate that VEGF inhibition, age, and atherosclerosis adversely affect functional recovery in a murine model of peripheral artery disease. *J Magn Reson Imaging* **28**, 996–1004.
14. Schneider J. T. and Faber C. (2008) BOLD imaging in the mouse brain using a turboCRAZED sequence at high magnetic fields. *Magn Reson Med* **60**, 850–9.
15. Mori S. and Zhang J. (2006) Principles of diffusion tensor imaging and its applications to basic neuroscience research. *Neuron* **51**, 527–39.
16. McCreary C. R., Bjarnason T. A., Skihar V., Mitchell J. R., Yong V. W., and Dunn J. F. (2009) Multiexponential T_2 and magnetization transfer MRI of demyelination and remyelination in murine spinal cord. *NeuroImage* **45**, 1173–82.
17. Stanisz G. J., Odrobina E. E., Pun J., et al. (2005) T_1, T_2 relaxation and magnetization transfer in tissue at 3T. *Magn Reson Med* **54**, 507–12.

18. Luo F., Seifert T. R., Edalji R., et al. (2008) Non-invasive characterization of Beta-amyloid 1–40 vasoactivity by functional magnetic resonance imaging in mice. *Neuroscience* **155**, 263–9.
19. Heerschap A., Kan H. E., Nabuurs C. I. H. C., Renema W. K., Isbrandt D., and Wieringa B. (2007) In vivo magnetic resonance spectroscopy of transgenic mice with altered expression of guanidinoacetate methyltransferase and creatine kinase isoenzymes. *Sub-cell Biochem* **46**, 119–48.
20. Albers M. J., Bok R., Chen A. P., et al. (2008) Hyperpolarized ^{13}C lactate, pyruvate, and alanine: noninvasive biomarkers for prostate cancer detection and grading. *Cancer Res* **68**, 8607–15.
21. Nieman B. J., Lerch J. P., Bock N. A., Chen X. J., Sled J. G., and Henkelman R. M. (2007) Mouse behavioral mutants have neuroimaging abnormalities. *Hum. Brain Mapp* **28**, 567–75.
22. Spring S., Lerch J. P., and Henkelman R. M. (2007) Sexual dimorphism revealed in the structure of the mouse brain using three-dimensional magnetic resonance imaging. *NeuroImage* **35**, 1424–33.
23. Bock N. A., Nieman B. J., Bishop J. B., and Henkelman R. M. (2005) In vivo multiple-mouse MRI at 7 Tesla. *Magn Reson Med* **54**, 1311–6.
24. Brockmann M.-A., Giese A., Ulmer S., et al. (2006) Analysis of mouse brain using a clinical 1.5 T scanner and a standard small loop surface coil. *Brain Res* **1068**, 138–42.
25. Chronik B., Alejski A., and Rutt B. K. (2000) Design and fabrication of a three-axis multilayer gradient coil for magnetic resonance microscopy of mice. *Magma* **10**, 131–46.
26. Brau A. C. S., Wheeler C. T., Hedlund L. W., and Johnson G. A. (2002) Fiber-optic stethoscope: a cardiac monitoring and gating system for magnetic resonance microscopy. *Magn Reson Med* **47**, 314–21.
27. Herold V., Parczyk M., Mörchel P., et al. (2009) In vivo measurement of local aortic pulse-wave velocity in mice with MR microscopy at 17.6 Tesla. *Magn Reson Med* **61**, 1293–9.
28. Parzy E., Miraux S., Franconi J.-M., and Thiaudière E. (2009) In vivo quantification of blood velocity in mouse carotid and pulmonary arteries by ECG-triggered 3D time-resolved magnetic resonance angiography. *NMR Biomed* **22**, 532–7.
29. Pautler R. G., Silva A. C., and Koretsky A. P. (1998) In vivo neuronal tract tracing using manganese-enhanced magnetic resonance imaging. *Magn Reson Med* **40**, 740–8.
30. Bock N. A., Kovacevic N., Lipina T. V., Roder J. C., Ackerman S. L., and Henkelman R. M. (2006) In vivo magnetic resonance imaging and semiautomated image analysis extend the brain phenotype for cdf/cdf mice. *J Neurosci* **26**, 4455–9.
31. Shapiro E. M., Sharer K., Skrtic S., and Koretsky A. P. (2006) In vivo detection of single cells by MRI. *Magn Reson Med* **55**, 242–9.
32. Sumner J. P., Shapiro E. M., Maric D., Conroy R., and Koretsky A. P. (2009) In vivo labeling of adult neural progenitors for MRI with micron sized particles of iron oxide: quantification of labeled cell phenotype. *NeuroImage* **44**, 671–8.
33. Yang J., Liu J., Niu G., et al. (2009) In vivo MRI of endogenous stem/progenitor cell migration from subventricular zone in normal and injured developing brains. *NeuroImage* **48**, 319–28.
34. Berry C. J., Thedens D. R., Light-McGroary K., et al. (2009) Effects of deep sedation or general anesthesia on cardiac function in mice undergoing cardiovascular magnetic resonance. *J Cardiovas Magn Reson* **11**, 16.
35. Qiu H. H., Cofer G. P., Hedlund L. W., and Johnson G. A. (1997) Automated feedback control of body temperature for small animal studies with MR microscopy. *IEEE Trans Biomed Eng* **44**, 1107–13.
36. Dazai J., Bock N. A., Nieman B. J., Davidson L. M., Henkelman R. M., and Chen X. J. (2004) Multiple mouse biological loading and monitoring system for MRI. *Magn Reson Med* **52**, 709–15.
37. Boretius S., Michaelis T., Tammer R., Ashery-Padan R., Frahm J., and Stoykova A. (2009) In vivo MRI of altered brain anatomy and fiber connectivity in adult pax6 deficient mice. *Cerebral Cortex* **19**, 2838–47.
38. Cheung M. M., Hui E. S., Chan K. C., Helpern J. A., Qi L., and Wu E. X. (2009) Does diffusion kurtosis imaging lead to better neural tissue characterization? A rodent brain maturation study. *NeuroImage* **45**, 386–92.
39. Jensen J. H., Helpern J. A., Ramani A., Lu H., and Kaczynski K. (2005) Diffusional kurtosis imaging: the quantification of non-gaussian water diffusion by means of magnetic resonance imaging. *Magn Reson Med* **53**, 1432–40.

40. So P.-W., Yu W.-S., Kuo Y.-T., et al. (2007) Impact of resistant starch on body fat patterning and central appetite regulation. *PloS One* **2**, e1309.
41. Hegedus B., Hughes F. W., Garbow J. R., et al. (2009) Optic nerve dysfunction in a mouse model of neurofibromatosis-1 optic glioma. *J Neuropath Exp Neurol* **68**, 542–51.
42. Kim J. H., Budde M. D., Liang H.-F., et al. (2006) Detecting axon damage in spinal cord from a mouse model of multiple sclerosis. *Neurobiol Dis* **21**, 626–32.
43. Kalber T. L., Waterton J. C., Griffiths J. R., Ryan A. J., and Robinson S. P. (2008) Longitudinal in vivo susceptibility contrast MRI measurements of LS174T colorectal liver metastasis in nude mice. *J Magn Reson Imaging* **28**, 1451–8.
44. Sakata N., Hayes P., Tan A., et al. (2009) MRI assessment of ischemic liver after intraportal islet transplantation. *Transplantation* **87**, 825–30.
45. Sadick M., Schock D., Kraenzlin B., Gretz N., Schoenberg S. O., and Michaely H. J. (2009) Morphologic and dynamic renal imaging with assessment of glomerular filtration rate in a pcy-mouse model using a clinical 3.0 Tesla scanner. *Invest Radiol* **44**, 469–75.
46. Jennbacken K., Gustavsson H., Tesan T., et al. (2009) The prostatic environment suppresses growth of androgen-independent prostate cancer xenografts: an effect influenced by testosterone. *Prostate* **69**, 1164–75.
47. Olive K. P., Jacobetz M. A., Davidson C. J., et al. (2009) Inhibition of Hedgehog signaling enhances delivery of chemotherapy in a mouse model of pancreatic cancer. *Science (New York, N.Y.)* **324**, 1457–61.
48. Berr S. S., Roy R. J., French B. A., et al. (2005) Black blood gradient echo cine magnetic resonance imaging of the mouse heart. *Magn Reson Med* **53**, 1074–9.
49. Cassidy P. J., Schneider J. E., Grieve S. M., Lygate C., Neubauer S., and Clarke K. (2004) Assessment of motion gating strategies for mouse magnetic resonance at high magnetic fields. *J Magn Reson Imaging* **19**, 229–37.
50. Wiesmann F., Frydrychowicz A., Rautenberg J., et al. (2002) Analysis of right ventricular function in healthy mice and a murine model of heart failure by in vivo MRI. *Am J Physiol Heart Circ Phys* **283**, H1065–71.
51. Heijman E., Graaf W. D., Niessen P., and Nauerth A. (2007) Comparison between prospective and retrospective triggering for mouse cardiac MRI. *NMR Biomed* **20**, 439–47.
52. Hiba B., Richard N., Janier M., and Croisille P. (2006) Cardiac and respiratory double self-gated cine MRI in the mouse at 7 T. *Magn Reson Med* **55**, 506–13.
53. Bishop J., Feintuch A., Bock N. A., et al. (2006) Retrospective gating for mouse cardiac MRI. *Magn Reson Med* **55**, 472–7.
54. Bucholz E., Ghaghada K., Qi Y., Mukundan S., and Johnson G. A. (2008) Four-dimensional MR microscopy of the mouse heart using radial acquisition and liposomal gadolinium contrast agent. *Magn Reson Med* **60**, 111–8.
55. Feintuch A., Zhu Y., Bishop J., Davidson L., and Dazai J. (2007) 4D cardiac MRI in the mouse. *NMR Biomed* **5**, 360–5.
56. Ross A. J., Yang Z., Berr S. S., et al. (2002) Serial MRI evaluation of cardiac structure and function in mice after reperfused myocardial infarction. *Magn Reson Med* **47**, 1158–68.
57. Franco F., Thomas G. D., Giroir B., et al. (1999) Magnetic resonance imaging and invasive evaluation of development of heart failure in transgenic mice with myocardial expression of tumor necrosis factor-alpha. *Circulation* **99**, 448–54.
58. Nahrendorf M., Spindler M., Hu K., et al. (2005) Creatine kinase knockout mice show left ventricular hypertrophy and dilatation, but unaltered remodeling post-myocardial infarction. *Cardiovas Res* **65**, 419–27.
59. Schneider J. E., Stork L.-A., Bell J. T., et al. (2008) Cardiac structure and function during ageing in energetically compromised guanidinoacetate N-methyltransferase (GAMT)-knockout mice—a one year longitudinal MRI study. *J Cardiovas Magn Reson* **10**, 9.
60. Ten Hove M., Lygate C. A., Fischer A., et al. (2005) Reduced inotropic reserve and increased susceptibility to cardiac ischemia/reperfusion injury in phosphocreatine-deficient guanidino-acetate-N-methyltransferase-knockout mice. *Circulation* **111**, 2477–85.
61. Olsson L. E., Smailagic A., Onnervik P.-O., and Hockings P. D. (2009) (1)H and hyperpolarized (3)He MR imaging of mouse with LPS-induced inflammation. *J Magn Reson Imaging* **29**, 977–81.
62. Thomas A. C., Potts E. N., Chen B. T., Slipetz D. M., Foster W. M., and Driehuys B. (2009) A robust protocol for regional evaluation of methacholine challenge in

mouse models of allergic asthma using hyperpolarized ^{3}He MRI. *NMR Biomed* **22**, 502–15.

63. Wakayama T., Narazaki M., Kimura A., and Fujiwara H. (2008) Hyperpolarized ^{129}Xe phase-selective imaging of mouse lung at 9.4T using a continuous-flow hyperpolarizing system. *Magn Reson Med Sci* **7**, 65–72.

64. Sheth V. R., van Heeckeren R. C., Wilson A. G., and AM (2008) Monitoring infection and inflammation in murine models of cystic fibrosis with magnetic resonance imaging. *J Magn Reson Imaging* **28**, 527–32.

65. Watt K. N., Bishop J., Nieman B. J., Henkelman R. M., and Chen X. J. (2008) Oxygen-enhanced MR imaging of mice lungs. *Magn Reson Med* **59**, 1412–21.

66. Chen B. T., Yordanov A. T., and Johnson G. A. (2005) Ventilation-synchronous magnetic resonance microscopy of pulmonary structure and ventilation in mice. *Magn Reson Med* **53**, 69–75.

67. Carreno B. M., Garbow J. R., Kolar G. R., et al. (2009) Immunodeficient mouse strains display marked variability in growth of human melanoma lung metastases. *Clin Cancer Res* **15**, 3277–86.

68. Engelman J. A., Chen L., Tan X., Crosby K., and Guimaraes A.R. (2008) Effective use of PI3K and MEK inhibitors to treat mutant Kras G12D and PIK3CA H1047R murine lung cancers. *Nature Med* **14**, 1351–6.

69. Li L. Z., Zhou R., Xu H. N., et al. (2009) Quantitative magnetic resonance and optical imaging biomarkers of melanoma metastatic potential. *PNAS* **106**, 6608–13.

70. Blé F. -X., Cannet C., Zurbruegg S., et al. (2008) Allergen-induced lung inflammation in actively sensitized mice assessed with MR imaging. *Radiology* **248**, 834–43.

71. Alsaid H., Sabbah M., Bendahmane Z., et al. (2007) High-resolution contrast-enhanced MRI of atherosclerosis with digital cardiac and respiratory gating in mice. *Magn Reson Med* **58**, 1157–63.

72. Chaabane L., Pellet N., Bourdillon M. C., Desbleds-Mansard A., Sulaiman A., Hadour G., Thivolet-Béjui F., Roy P., Briguet A., Douek P., and Canet-Soulas E. (2004) Contrast enhancement in atherosclerosis development in a mouse model: in vivo results at 2 Tesla. *Magma* **17**, 188–95.

73. Cha S., Johnson G., Wadghiri Y. Z., et al. (2003) Dynamic, contrast-enhanced perfusion MRI in mouse gliomas: correlation with histopathology. *Magn Reson Med* **49**, 848–55.

74. Checkley D., Tessier J. J. L., Wedge S. R., et al. (2003) Dynamic contrast-enhanced MRI of vascular changes induced by the VEGF-signalling inhibitor ZD4190 in human tumour xenografts. *Magn Reson Imaging* **21**, 475–82.

75. Kiselev V. G., Strecker R., Ziyeh S., Speck O., and Hennig J. (2005) Vessel size imaging in humans. *Magn Reson Med* **53**, 553–63.

76. Neeman M. and Dafni H. (2003) Structural, functional, and molecular MR imaging of the microvasculature. *Ann Rev Biomed Eng* **5**, 29–56.

77. Zwick S., Strecker R., Kiselev V., et al. (2009) Assessment of vascular remodeling under antiangiogenic therapy using DCE-MRI and vessel size imaging. *J Magn Reson Imaging* **29**, 1125–33.

78. Yu X., Wadghiri Y. Z., Sanes D. H., and Turnbull D. H. (2005) In vivo auditory brain mapping in mice with Mn-enhanced MRI. *Nature Neurosci* **8**, 961–8.

79. Yu X., Sanes D. H., Aristizabal O., Wadghiri Y. Z., and Turnbull D. H. (2007) Large-scale reorganization of the tonotopic map in mouse auditory midbrain revealed by MRI. *PNAS* **104**, 12193–8.

80. Pautler R. G. and Koretsky A. P. (2002) Tracing odor-induced activation in the olfactory bulbs of mice using manganese-enhanced magnetic resonance imaging. *NeuroImage* **16**, 441–8.

81. Watanabe T., Radulovic J., Spiess J., et al. (2004) In vivo 3D MRI staining of the mouse hippocampal system using intracerebral injection of $MnCl_2$. *NeuroImage* **22**, 860–7.

82. Smith K. D. B., Kallhoff V., Zheng H., and Pautler R. G. (2007) In vivo axonal transport rates decrease in a mouse model of Alzheimer's disease. *NeuroImage* **35**, 1401–8.

83. Cohen B., Dafni H., Meir G., Harmelin A., and Neeman M. (2005) Ferritin as an endogenous MRI reporter for noninvasive imaging of gene expression in C6 glioma tumors. *Neoplasia* **7**, 109–17.

84. Deans A. E., Wadghiri Y. Z., Bernas L. M., Yu X., Rutt B. K., and Turnbull D. H. (2006) Cellular MRI contrast via coexpression of transferrin receptor and ferritin. *Magn Reson Med* **56**, 51–9.

85. Genove G., DeMarco U., Xu H., Goins W. F., and Ahrens E. T. (2005) A new transgene reporter for in vivo magnetic resonance imaging. *Nature Med* **11**, 450–4.

86. Gilad A. A., McMahon M. T., Walczak P., et al. (2007) Artificial reporter gene

providing MRI contrast based on proton exchange. *Nature Biotechnol* **25**, 217–9.
87. Gilad A. A., Winnard P. T., Van Zijl P. C. M., and Bulte J. W. M. (2007) Developing MR reporter genes: promises and pitfalls. *NMR Biomed* **20**, 275–90.
88. Weissleder R., Moore A., Mahmood U., et al. (2000) In vivo magnetic resonance imaging of transgene expression. *Nature Med* **6**, 351–5.
89. Zurkiya O., Chan A. W. S., and Hu X. (2008) MagA is sufficient for producing magnetic nanoparticles in mammalian cells, making it an MRI reporter. *Magn Reson Med* **59**, 1225–31.
90. Zhao M., Beauregard D. A., Loizou L., Davletov B., and Brindle K. M. (2001) Non-invasive detection of apoptosis using magnetic resonance imaging and a targeted contrast agent. *Nature Med* **7**, 1241–4.
91. Burtea C., Laurent S., Murariu O., et al. (2008) Molecular imaging of alpha v beta3 integrin expression in atherosclerotic plaques with a mimetic of RGD peptide grafted to Gd-DTPA. *Cardiovas Res* **78**, 148–57.
92. Winter P. M., Caruthers S. D., Yu X., et al. (2003) Improved molecular imaging contrast agent for detection of human thrombus. *Magn Reson Med* **50**, 411–6.
93. Larbanoix L., Burtea C., Laurent S., et al. (2008) Potential amyloid plaque-specific peptides for the diagnosis of Alzheimer's disease. *Neurobiol Aging* **31**, 1679–89.
94. Burtea C., Laurent S., Port M., et al. (2009) Magnetic resonance molecular imaging of vascular cell adhesion molecule-1 expression in inflammatory lesions using a peptide-vectorized paramagnetic imaging probe. *J Med Chem* **52**, 4725–42.
95. Anderson S. A., Shukaliak-Quandt J., Jordan E. K., et al. (2004) Magnetic resonance imaging of labeled T-cells in a mouse model of multiple sclerosis. *Ann Neurol* **55**, 654–9.
96. Bulte J. W. M. and Kraitchman D. L. (2004) Monitoring cell therapy using iron oxide MR contrast agents. *Curr Pharm Biotechnol* **5**, 567–84.
97. Heyn C., Ronald J. A., Mackenzie L. T., et al. (2006) In vivo magnetic resonance imaging of single cells in mouse brain with optical validation. *Magn Reson Med* **55**, 23–9.
98. Magnitsky S., Watson D. J., Walton R. M., et al. (2005) In vivo and ex vivo MRI detection of localized and disseminated neural stem cell grafts in the mouse brain. *NeuroImage* **26**, 744–54.
99. Shapiro E. M., Skrtic S., Sharer K., Hill J. M., Dunbar C. E., and Koretsky A. P. (2004) MRI detection of single particles for cellular imaging. *PNAS* **101**, 10901–6.
100. Johnson G. A., Cofer G. P., Gewalt S. L., and Hedlund L. W. (2002) Morphologic phenotyping with MR microscopy: the visible mouse. *Radiology* **222**, 789–93.
101. Zhou Y. Q., Davidson L., Henkelman R. M., et al. (2004) Ultrasound-guided left-ventricular catheterization: a novel method of whole mouse perfusion for microimaging. *Lab Invest* **84**, 385–9.
102. Dorr A., Lerch J. P., Spring S., Kabani N., and Henkelman R. (2008) High resolution three-dimensional brain atlas using an average magnetic resonance image of 40 adult C57Bl/6 J mice. *NeuroImage* **42**, 60–9.
103. Kovacević N., Henderson J. T., Chan E., et al. (2005) A three-dimensional MRI atlas of the mouse brain with estimates of the average and variability. *Cerebral Cortex* **15**, 639–45.
104. Lerch J. P., Carroll J., Spring S., et al. (2008) Automated deformation analysis in the YAC128 Huntington disease mouse model. *NeuroImage* **39**, 32–9.
105. Lerch J. P., Yiu A. P., Bohbot V. D., Henkelman R. M., Josseyln S. A., and Sled J. G. (2008) MRI can detect brain shape changes in mice caused by five days of learning. *International Society of Magnetic Resonance in Medicine 16th Scientific Meeting* 479.
106. Maheswaran S., Barjat H., Rueckert D., et al. (2009) Longitudinal regional brain volume changes quantified in normal aging and Alzheimer's APP x PS1 mice using MRI. *Brain Res* **1270**, 19–32.
107. Persson A.-S., Westman E., Wang F.-H., Khan F. H., Spenger C., and Lavebratt C. (2007) Kv1.1 null mice have enlarged hippocampus and ventral cortex. *BMC Neurosci* **8**, 10.
108. Thompson P. M., MacDonald D., Mega M. S., Holmes C. J., Evans A. C., and Toga A. W. (1997) Detection and mapping of abnormal brain structure with a probabilistic atlas of cortical surfaces. *J Comp Ass Tom* **21**, 567–81.
109. Chung M. K., Worsley K. J., Paus T., et al. (2001) A unified statistical approach to deformation-based morphometry. *NeuroImage* **14**, 595–606.
110. Lepore N., Brun C. A., Chiang M.-C., et al. (2006) Multivariate statistics of the Jacobian matrices in tensor based morphometry and their application to HIV/AIDS. *Med Image*

110. *Comput Comput Assist Interv Int Conf* **9**, 191–8.
111. Kim H., Besson P., Colliot O., Bernasconi A., and Bernasconi N. (2008) Surface-based vector analysis using heat equation interpolation: a new approach to quantify local hippocampal volume changes. *Med Image Comput Comput Assist Interv Int Conf* **11**, 1008–15.
112. Cao J. and Worsley K. J. (1999) The detection of local shape changes via the geometry of Hotelling's T-2 fields. *Ann Stat* **27**, 925–42.
113. Nieman B. J., Flenniken A. M., Adamson S. L., Henkelman R. M., and Sled J. G. (2006) Anatomical phenotyping in the brain and skull of a mutant mouse by magnetic resonance imaging and computed tomography. *Physiol Genomics* **24**, 154–62.
114. Clapcote S. J., Lipina T. V., Millar J. K., et al. (2007) Behavioral phenotypes of Disc1 missense mutations in mice. *Neuron* **54**, 387–402.
115. Nichols T. E. and Holmes A. P. (2002) Nonparametric permutation tests for functional neuroimaging: a primer with examples. *Hum Brain Mapp* **15**, 1–25.
116. Lerch J. P., Carroll J. B., Dorr A., et al. (2008) Cortical thickness measured from MRI in the YAC128 mouse model of Huntington's disease. *NeuroImage* **41**, 243–51.
117. Meadowcroft M. D., Connor J. R., Smith M. B., and Yang Q. X. (2009) MRI and histological analysis of beta-amyloid plaques in both human Alzheimer's disease and APP/PS1 transgenic mice. *J Magn Reson Imaging* **29**, 997–1007.
118. Kale S. C., Lerch J. P., Henkelman R. M., and Chen X. J. (2008) Optimization of the SNR-resolution tradeoff for registration of magnetic resonance images. *Hum Brain Mapp* **29**, 1147–58.
119. Nieman B. J., Bock N. A., Bishop J., Sled J. G., Chen X. J., and Henkelman R. M. (2005) Fast spin-echo for multiple mouse magnetic resonance phenotyping. *Magn Reson Med* **54**, 532–7.
120. Chaabane L., Pellet N., Bourdillon M. C., et al. (2004) Contrast enhancement in atherosclerosis development in a mouse model: in vivo results at 2 Tesla. *Magma* **17**, 188–95.
121. Lau J. C., Lerch J. P., Sled J. G., Henkelman R. M., Evans A. C., and Bedell B. J. (2008) Longitudinal neuroanatomical changes determined by deformation-based morphometry in a mouse model of Alzheimer's disease. *NeuroImage* **42**, 19–27.
122. Nieman B. J., Bock N. A., Bishop J., et al. (2005) Magnetic resonance imaging for detection and analysis of mouse phenotypes. *NMR Biomed* **18**, 447–68.

Chapter 32

Analysis of Freshly Fixed and Museum Invertebrate Specimens Using High-Resolution, High-Throughput MRI

Alexander Ziegler and Susanne Mueller

Abstract

Magnetic resonance imaging (MRI) is now considered a routine tool for comparative morphological imaging in small vertebrate model organisms. However, the application of high-resolution imaging protocols to visualize the anatomy of invertebrate organisms has not yet become a generally accepted tool among zoologists. Here, we describe MRI protocols that permit visualization of both the internal and the external anatomy of freshly fixed invertebrates and specimens from museum collections. The choice of protocols has been optimized to allow the assembly of the large numbers of datasets that are necessary for comparative morphological analyses. Although the primary focus of our work is on sea urchin internal anatomy, we also present results from a variety of other invertebrate taxa to demonstrate the principal feasibility of MRI studies to obtain anatomical information at high resolutions. Furthermore, we briefly describe procedures suitable for 3D modelling.

Key words: Magnetic resonance imaging, museum specimen, high-resolution, non-invasive, invertebrate, three-dimensional, visualization, reconstruction, comparative morphology, anatomy.

1. Introduction

The study of animal anatomy is a time- and specimen-consuming process when traditional invasive techniques such as dissection or histology are employed. However, a number of modern imaging techniques permit the study of both the internal anatomy and the external anatomy non-invasively (1, 2). Based on the acquired digital datasets, advanced three-dimensional (3D) visualization and reconstruction protocols can then be implemented, thereby greatly improving our understanding of complex structural relationships (3).

The use of magnetic resonance imaging (MRI) for the noninvasive study of small animals has long been in use, primarily for vertebrate model organisms such as the mouse, the rat, or the zebra finch. In addition, a recent study introduced a number of MRI protocols that were employed for the study of brain anatomy in endangered bird species (4). However, the advances in image resolution have also prompted invertebrate zoologists to employ MRI for the study of invertebrate anatomy (**Table 1**, (5–27)).

Since comparative studies can be complicated not only by technical but also by logistical aspects, we focus on the large-scale use of museum specimens for such comparative purposes. The

Table 1
List of invertebrate taxa scanned using MRI, with respective references

Higher taxon		Species	References	Scanning condition
Demospongiae	Haplosclerida	*Lubomirskia baicalensis*	(5)[a]	Ex vivo
Crustacea	Decapoda	*Callinectes sapidus*	(6)[a]	In vivo
		Cancer pagurus	(7)	In vivo
		Maja squinado	(7)	In vivo
		M. squinado	(8)	In vivo
		Procambarus clarkii	(9)	In vivo
		Cherax destructor	(10)[a]	In vivo
Arachnida	Araneae	*Eurypelma californicum*	(11)	In vivo
Insecta	Multiple taxa	Review article on all studies published prior to 2003	(12)	In vivo/ex vivo
	Lepidoptera	*Manduca sexta*	(13)[a]	In vivo
		M. sexta	(14)	In vivo
	Hymenoptera	*Apis mellifera carnica*	(15)[a]	Ex vivo
	Diptera	*Drosophila melanogaster*	(16)[a]	In vivo
	Coleoptera	*Ascioplaga mimeta*	(17)[a]	Ex vivo
Mollusca	Bivalvia	*Crassostrea gigas*	(18)	In vivo
		C. gigas	(19)[a]	In vivo
		Elliptio complanata	(20)[a]	In vivo
	Cephalopoda	*Sepia officinalis*	(21)[a]	Ex vivo
		Lolliguncula brevis	(22)	Ex vivo
Echinodermata	Echinoidea	*Psammechinus miliaris*	(23)[a]	In vivo/ex vivo
		P. miliaris	(24)[a]	Ex vivo
		Multiple echinoid species	(2, 25, 26)[a]	Ex vivo

[a]Studies with a predominantly morphological focus.

so-called wet collections of natural history institutions around the world house millions of invertebrate specimens that constitute an enormous resource that also includes endangered and recently extinct species.

Here, we give an account of MRI protocols that permit gathering data on invertebrate anatomy in a non-invasive manner, using freshly fixed or even century-old museum specimens. Since phylogenetic or ecological studies usually require the analysis of multiple organisms, we have optimized the protocols to allow for high-throughput scanning of large numbers of specimens. The resulting datasets permit interactive viewing as well as advanced automated or manual three-dimensional (3D) visualization of internal and external structures.

The primary focus of our work was the study of sea urchins (Echinodermata: Echinoidea). Therefore, this taxon of marine invertebrates serves here in an exemplary fashion. However, we have extended MRI studies to various other invertebrate groups in order to demonstrate that non-invasive MRI datasets can principally also be obtained from further taxa.

2. Materials

2.1. Invertebrate Specimens

1. Fresh specimens can be obtained from the field or through biological supply companies (e.g. Carolina Biological Supply Company, USA) (*see* **Note 1**).
2. Museum specimens are available from the so-called wet collections of natural history institutions worldwide (*see* **Note 2**).

2.2. Specimen Containers and Specimen Fixation Material

The choice of the specimen container depends on the inner diameter of the resonator as well as the general set-up of the MRI scanning system (e.g. horizontal or vertical bore magnet). Make sure that the caps of the specimen containers are tight as they might have to be placed horizontally.

1. Plastic tubes with a screw cap (e.g. 50 mL Falcon tubes, BD Biosciences, USA – **Fig. 1a, c** – try to obtain tubes that can stand on their own).
2. Custom-made containers with a screw cap (**Fig. 1a, b**). Materials that can be used for MRI include Plexiglas (acrylic glass), various kinds of plastics, and glassware (*see* **Note 3**).
3. Cap-sealed glassware NMR tubes (e.g. Sigma-Aldrich, USA). Conventional sizes have 3 mm, 5 mm, or 10 mm diameter, although NMR tubes with up to 30 mm diameter can be obtained.

Fig. 1. Specimen containers for invertebrates. (**a**) Linear ^{1}H-radiofrequency birdcage resonator (*left*) with custom-made specimen container of acrylic glass (*centre*) and 50 mL Falcon tube (*right*). (**b**) Custom-made specimen container inside resonator, only partly filled with distilled water to allow for better tuning. (**c**) 50 mL Falcon tube inside resonator. Conventional paper towels wrapped around the tube and fixed with adhesive tape can be used to fill the gap between specimen container and resonator.

4. A set of fine glass and/or plastic tubes of varying diameters such as one-way polyethylene Pasteur pipettes or glass stirrers (e.g. Carl Roth, Germany) for mechanical specimen fixation.

5. Rack for specimen preparation and transport of plastic tubes (e.g. Falcon tube rack, Formulatrix, USA).

2.3. Imaging Solutions

During our studies, distilled water was the predominant scanning solution. However, other solutions can be used for MRI as well.

1. Distilled or de-ionized water – use of distilled water reduces the risk of specimen contamination (*see* **Note 4**).

2. Formalin (aqueous solution of formaldehyde) – use a maximum concentration of 5% (v/v) for scanning. Available from chemical supply companies (e.g. Sigma-Aldrich, USA). Be aware of its toxicity!

3. Fomblin (a perfluoropolyether fluorocarbon) – proton-free oil (Solvay Solexis, Italy). Non-toxic (*see* **Note 5**).

4. Fluorinert (a fluorocarbon-based fluid) – proton-free liquid (e.g. FC-77 or FC-84 by 3M, USA). Its properties are similar to those of water, although its density is higher (1.6–1.9 g/cm^3) than that of water. Non-toxic (*see* **Note 5**).

2.4. Contrast Agents (Optional)

1. Nonspecific MRI contrast agents based on the element gadolinium. Commonly used products are Magnevist (Bayer HealthCare, Germany) (**Fig. 2**), Gadovist (Bayer Schering, Germany), or Magnegita (Insights Agents, Germany) (*see* **Note 6**).

2. Specific MRI contrast agent based on manganese. This paramagnetic metal is supplied, e.g. as MnCl$_2$ by chemical supply companies (e.g. Sigma-Aldrich, USA). (*see* **Note 7**).

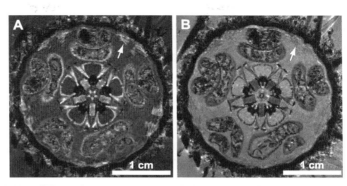

Fig. 2. Use of Magnevist as a contrast agent in invertebrate MRI. (**a**) Museum specimen of *Strongylocentrotus purpuratus* (CASIZ 5724) scanned with a 3D protocol (FLASH) at 81 μm isotropic resolution in distilled water. (**b**) Same specimen scanned with the same protocol after contrasting with 2 mM Magnevist. Differentiation of internal organs is improved, although some structures (e.g. tube feet ampullae (*arrows* in (**a**) and (**b**))) have become less visible.

2.5. Vacuum Pump

1. Conventional laboratory vacuum pump with tubing (e.g. Laboport, KNF, Germany).

2. Exsiccator with vacuum valve. The device needs to be large enough to fit the whole specimen container (e.g. Rotilabo, Carl Roth, Germany).

2.6. MRI Scanner Including Gradient System and Resonator

The imaging system described here has been used primarily for our high-throughput analyses in 2D and 3D. The choice of the imaging system and its subcomponents dictates parameters such as specimen size, maximum resolution, and scanning time.

1. 7 T PharmaScan 70/16 small animal imaging system (Bruker Biospin GmbH, Germany) with a ^{1}H-resonance frequency of 300 MHz. The system is equipped with a shielded gradient set with an inner diameter of 90 mm and a maximum gradient strength of 300 mT/m (*see* **Note 8**).

2. Linear ^{1}H-radiofrequency birdcage resonator (Bruker Biospin GmbH, Germany) with an outer diameter of 89 mm and an inner diameter of 38 mm.

3. Workstation operating the MRI scanner with pre-installed software (we used ParaVision 3.0.2 and ParaVision 4.0, Bruker Biospin GmbH, Germany).

2.7. Graphics Workstation for Image Processing

1. Desktop PC with CPU (e.g. 2 GHz), sufficient RAM (e.g. 2 GB), graphics card with OpenGL and texture mapping capabilities (e.g. GeForce series, NVIDIA, USA), sufficient hard drive capacity (e.g. 1 TB), and operating system (e.g. Mac OS or Windows). Additional components include two monitors, a mouse, a keyboard, a DVD burner, and a pen tablet for manual segmentation (e.g. Wacom, USA). The

set-up has to be able to smoothly handle image datasets of 100–500 MB size.

2. Image processing software with DICOM viewing capabilities (e.g. ImageJ, NIH, USA).

3. 3D visualization and modelling software (e.g. Amira, Visage Imaging GmbH, Germany).

3. Methods

The 2D and 3D MRI protocols described here will lead to initial results in most specimens. However, every invertebrate taxon will require its specific scanning parameters and the proposed protocols may have to be modified and optimized according to the needs of the study (*see* **Note 9**). Such adaptations should be done in close collaboration with the technical staff operating the MRI system. 3D model generation (**Fig. 3**) of internal and external structures usually requires the use of isotropic datasets (i.e. datasets which are composed of cubic voxels) that can be obtained using the 3D MRI protocols described below. The 2D MRI protocols provided here were used for high-throughput scanning, but resulted in non-isotropic virtual sections only partly suitable for 3D visualization.

3.1. Specimen Preparation

Note that the differences inherent in distinct fixation protocols and their effect on MRI scanning of invertebrates have not yet been established systematically.

1. Bring the specimen to room temperature if kept in a refrigerator.

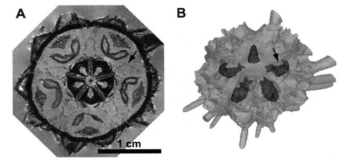

Fig. 3. Example of a surface-rendered 3D model based on a 3D MRI dataset. (**a**) Virtual horizontal section through a museum specimen of *Eucidaris metularia* (NHM 1969.5.1.15-40). (**b**) Surface-rendered 3D model based on the 3D MRI dataset of (a). Stewart's organs (*arrows* in (**a**) and (**b**)) can be seen through the semi-transparent endoskeleton. *See* Ref. (2) for colour figures as well as an interactive 3D PDF model.

2. (a) Specimens fixed and stored in formalin: this material can be scanned directly if the final formalin concentration is below 5% (v/v).

   (b) Formalin- or ethanol-fixed specimens that later were transferred into ethanol for long-term storage (i.e. most museum specimens): this material needs to be brought into distilled water in a gradual ethanol series (e.g. 75%, 40%, 20% (v/v) – each step for at least 2 h, and longer, if the specimen possesses a strong, encapsulating exoskeleton).

   (c) Specimens fixed in Bouin's solution that later were transferred into ethanol for long-term storage: proceed as in the previous step. (*see* **Note 10**).

3. Place the specimen inside the scanning container – if the previous step had not already been performed within the scanning container. The position of the specimen depends on the type of container used. In our set-up, the specimens were placed in the lower third to middle of a 50 mL Falcon tube. Use plastic or glass rods and tubing to carefully attach the specimen to one side or to the centre of the container (*see* **Note 11**). Make sure that hollow rods and tubes are completely immersed in and filled with the scanning solution in order to decrease the likelihood of boundary artefacts between air, scanning solution, and the specimen.

3.2. Contrasting of Specimens (Optional)

1. Manganese needs to be applied in vivo (*see* **Note 7**).
2. Inject the gadolinium-based contrast agent into the scanning solution with a final concentration of 2 mM. Let the sample rest for at least 2 h in order to achieve complete perfusion of the specimen – note that this parameter largely depends on the composition of the specimen and may therefore require more extensive trials.

3.3. Degassing of Specimens

1. Gently agitate the specimen container to remove air from the sample.
2. Evacuate for 30 min to 1 h in order to remove further air and to help fluid exchange.

3.4. Specimen Scanning

1. Place the specimen container inside the resonator (**Fig. 1b, c**) and place the resonator in the centre of the magnet.
2. Use a fast scout sequence to identify the position of the specimen inside the resonator (e.g. TriPilot_SE, **Table 2**).
3. Adjust the position of the scanning container accordingly and repeat the previous step, until the position of the specimen is satisfactory and well within the desired field of view (FOV) (*see* **Note 12**). Make sure that the position of the specimen is not inverted (supine) in the MRI control

Table 2
List of MRI protocols presented in this manuscript. These protocols have been employed on a 7 T Pharmascan 70/16 system (Bruker Biospin GmbH, Germany) using a linear birdcage resonator (Bruker Biospin GmbH, Germany) with an inner diameter of 38 mm

	TriPilot_SE	RARE_2D_low	RARE_2D_high	FLASH_3D_low	FLASH_3D_high	MDEFT_3D	RARE_3D
Software	ParaVision 4.0	ParaVision 4.0	ParaVision 4.0	ParaVision 4.0	ParaVision 4.0	ParaVision 3.0.2	ParaVision 3.0.2
Scan method	MSME	RARE	RARE	FLASH	FLASH	MDEFT	RARE
Signal weighting	T_1 (T_2)	PD (T_1)	PD (T_1)	T_1 (T_2^*)	T_1 (T_2^*)	T_1	T_2 (PD)
Repetition time (T_R) (ms)	200	1300	1300	30	30	21.16	2500
Echo time (T_E) (ms)	11	7.5	8.83	6	6.7	3.9	11.82
Average number (N_A)	1	1	4	1	12	6	1
Flip angle (α)	90°	90°	90°	35°	35°	19°	90°
RARE factor	–	4	4	–	–	–	4
Field of view (FOV) (cm)	5 × 5	3 × 3	3 × 3	3.12 × 3.12 × 3.12	3.12 × 3.12 × 3.12	3 × 3 × 3	3 × 3 × 3
Matrix size	128 × 128	256 × 256	384 × 384	64 × 64 × 64	384 × 384 × 384	256 × 256 × 256	256 × 256 × 256
Slice thickness (mm)	2	1	0.5	–	–	–	–
Slice number	3	10	20	–	–	–	–
Resolution ($X \times Y \times Z$) (μm)	391 × 391 × 2000	117 × 117 × 1000	78 × 78 × 500	488 × 488 × 488	81 × 81 × 81	117 × 117 × 117	117 × 117 × 117
Total scan time	25.6 s	1 min 2.4 s	6 min 14.4 s	2 min 2.9 s	14 h 44 min 44 s	5 h 4 min 16 s	11 h 22 min 40 s

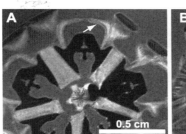
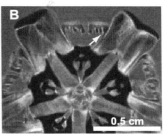

Fig. 4. High-throughput scanning of sea urchin specimens using a fast 2D protocol (RARE). (**a**) Museum specimen of *Echinus esculentus* (ZMK Mortensen collection) scanned in distilled water at 7 T with an in-plane resolution of 78 μm and a resolution of 500 μm in the Z plane. (**b**) Museum specimen of *Echinometra lucunter* (ZMB 5511) scanned using the same protocol. Note the absence (a) or presence (b) of frilled protractor muscles (*arrows*), a morphological feature for which over 50 sea urchin species have been checked by us during a single weekend using MRI.

software. Remember to initiate the auto adjustment and tuning protocols of the MRI system afterwards.

4. Low-resolution 2D overview imaging: use a relatively fast 2D protocol (RARE_2D_low, **Table 2**) in order to obtain a quick overview of the specimen's properties, particularly artefacts (*see* **Note 13**), and to optimize scan orientation (*see* **Note 14**).

5. High-resolution 2D imaging: use a 2D protocol with a higher matrix size and reduced slice thickness (RARE_2D_high, **Table 2**) (*see* **Note 15**). **Figure 4** shows an example derived from high-throughput, high-resolution 2D imaging.

6. Low-resolution 3D overview imaging: use a relatively fast 3D protocol (e.g. FLASH_3D_low, **Table 2**) in order to obtain a quick overview of the specimen's properties, particularly artefacts (*see* **Note 13**), and to optimize scan orientation (*see* **Note 14**).

7. High-resolution 3D imaging (**Fig. 5** demonstrates the effect of higher resolutions): use a 3D protocol with a higher matrix size (e.g. FLASH_3D_high, **Table 2**). Additional 3D protocols include MDEFT (e.g. MDEFT_3D, **Table 2**) or RARE (e.g. RARE_3D, **Table 2**). **Figure 6** depicts virtual sections derived from various MRI protocols (*see* **Notes 16–18**).

3.5. Image Processing

In order to generate images or 3D models of the acquired MRI datasets – such as in the images of various invertebrate taxa (**Fig. 7**) – the raw data need to be processed using a number of steps performed with graphics software. Note that organ designation and image interpretation can lead to mistaken

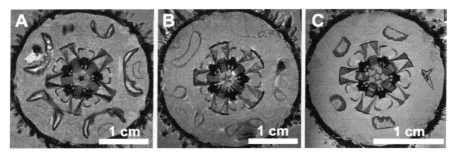

Fig. 5. Effect of higher resolutions demonstrated on freshly fixed specimens of *Psammechinus miliaris*. (a) Specimen scanned at 7 T using a 3D protocol (FLASH) with 117 μm isotropic resolution. (b) Specimen scanned at 7 T using a 3D protocol (FLASH) with 81 μm isotropic resolution. (c) Specimen scanned at 17.6 T using a 3D protocol (GEFI) with 44 μm isotropic resolution. Differentiation of sub-structures becomes easier with improved resolution.

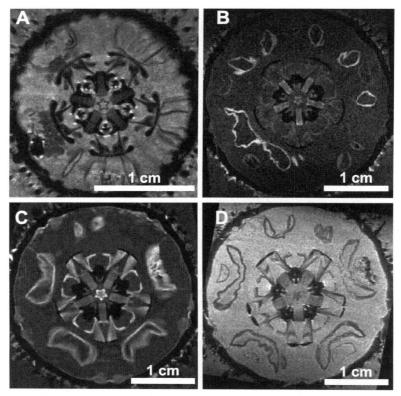

Fig. 6. Various examples of MRI protocols performed on freshly fixed specimens of *Psammechinus miliaris*. (a) Specimen scanned at 4.7 T using a 2D protocol (MSME) with 117 μm in-plane resolution. (b) Specimen scanned at 7 T using a 3D protocol (MDEFT) with 117 μm isotropic resolution. (c) Specimen scanned at 7 T using a 3D protocol (FLASH) with 117 μm isotropic resolution. (d) Specimen scanned at 7 T using a 3D protocol (RARE) with 117 μm isotropic resolution. All specimens were scanned in distilled water without the application of a contrast agent. Note the differences in liquid and soft tissue contrast.

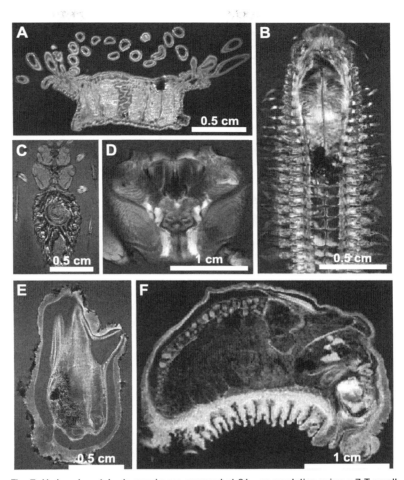

Fig. 7. Various invertebrate specimens scanned at 81 μm resolution using a 7 T small animal MRI system. (a) Sea anemone (Cnidaria: Actinaria). (b) Ragworm (Annelida: Nereididae). (c) Cockroach (Insecta: Blattodea). (d) Crab (Crustacea: Brachyura). (e) Sea squirt (Tunicata: Ascidiacea). (f) Sea slug (Gastropoda: Nudibranchia).

observations (27) and should therefore be checked using complementary techniques whenever possible (26).
1. Image processing: use a DICOM viewer (here: ImageJ) to open the MRI DICOM files (File:Import:Image Sequence). Reduce to 8-bit if provided in 16-bit format by the MRI system in order to reduce file size (Image:Type:8-bit). Crop the dataset (ROI rectangular selection:Image:Crop) and reduce slices (Plugins:Stacks:SliceRemover) to further reduce file size. If desired, use automated contrast enhancement (Process:Enhance Contrast). If you wish to generate an image stack with a standardized orientation (*see* **Note 14, Figs. 5 and 6**), use the image stack transformation tool (Plugins:TransformJ:TJ Rotate) – note that image stacks are

transformed using an interpolation algorithm which might result in a slight loss in resolution. When finished, save the image stack in TIFF format for subsequent 3D modelling (File:Save As:Image Sequence). At any given point in the above-mentioned process, you may use the interactive viewing tool (Plugins:3D:Volume Viewer) to analyze your dataset and to generate 2D images – note that this tool uses interpolation algorithms as well.

2. 3D reconstruction: use a 3D modelling software (here, Amira) to open the TIFF image sequence. 3D modelling of your dataset can be done using automated algorithms based on threshold greyscale values (surface-rendering: Isosurface tool, volume-rendering: Isovoxel tool) or using manual segmentation (surface-rendering: Segmentation Editor tool, **Fig. 3b**). *See* Ref. (3) for an in-depth description of the manual segmentation process of biological specimens.

3. Interactive 3D PDF generation: surface-rendered 3D models can be embedded and integrated into presentations as well as publications using the interactive 3D capabilities of PDF files (2, 28).

4. Notes

1. Specimens need to be fixed according to the procedures and protocols applicable to each taxon under study. Refer to the relevant literature for these fixation protocols. It is assumed here that whole specimens have been fixed either in ethanol, formalin, or Bouin's fluid. As specimen size needs to be selected according to the internal diameter of the available MRI resonator; specimens might have to be trimmed to fit the scanning container.

2. Wet collections constitute that part of museum collections where specimens are kept in alcohol for long-term storage, usually at around 70–80% (v/v) ethanol. Museum specimens can be shipped worldwide in accordance with the museum curators. However, keep in mind that air shipping regulations may require that the alcohol concentration is being lowered to a maximum of 24% (v/v). In addition, import and export regulations of the countries involved in the shipment of specimens need to be taken into careful consideration in order to avoid unexpected surprises. Specimen size needs to be selected according to the internal diameter of the available resonator (*see* **Note 1**). For comparison, **Figure 8** depicts virtual vertical MRI sections of a

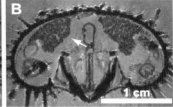

Fig. 8. Comparison of a freshly fixed and a museum specimen of *Psammechinus miliaris*. (**a**) Freshly fixed specimen scanned with a 3D protocol (FLASH) at 81 μm isotropic resolution and contrasted with Magnevist. (**b**) A 135-year-old museum specimen (ZMB 2011) scanned with the same protocol. This demonstrates that even century-old museum specimens can be integrated into comparative studies using MRI. The gonads (*arrows*) are more developed in (b) than in (a).

freshly fixed specimen and that of a museum specimen of the same species.

3. Most universities and research facilities will have a workshop that may be able to provide custom-made specimen containers for research purposes. Thread seal tape (a tape made out of polytetrafluoroethylene, also termed Teflon tape) is helpful for tight sealing of containers during scanning. The use of custom-made specimen containers filled with distilled water can result in problems during tuning of the MRI resonator due to a high load. If so, reduce the amount of water inside the specimen container (**Fig. 1b**), but make sure that the specimen is still fully immersed.

4. The handling of freshly fixed and museum specimens immersed in distilled water should be performed under clean or even near-sterile conditions to avoid contamination of the imaging solution. In our case, only a single specimen – of the hundreds of specimens analyzed – ever deteriorated; this specimen, fixed and then kept in ethanol for 3 years, had erroneously not been placed back into 75% (v/v) ethanol and showed heavy decay after 5 consecutive days of immersion in water.

5. Fomblin is a proton-free oil that is used to enhance contrast within the specimen by eliminating any signal from outside the specimen. Keep in mind that a certain amount of free hydrogen protons are necessary for a sufficiently strong MRI signal. Therefore, proton-free imaging solutions such as Fomblin or Fluorinert are more suitable for specimens with a higher content of signal-generating tissues. Note that Fomblin may prove to be unsuitable for valuable material since its oil-like properties may prevent its complete removal from the specimen surface after scanning!

6. Gadolinium (Gd) is a paramagnetic element that significantly decreases T_1 relaxation time. Because of the toxicity of the ion, it is provided in a chelated form such as gadopentetate dimeglumine or gado diethylenetriamine-pentaacetic acid (Gd-DTPA). It serves as an extracellular contrast agent and can be applied both in vivo and ex vivo. Final concentrations of 2 mM have been shown to lead to good results (**Fig. 2**, (2)). We recommend specimen scanning both before and after the application of the contrast agent whenever possible since some structures may become less visible after contrasting. Alternatively, you may also add the contrast agent directly to the fixative (e.g. 1:20 Magnevist/10% (v/v) formalin in buffered salt water solution). However, no systematic studies describing this approach in invertebrates have yet been published.

7. Manganese (Mn^{2+}) may serve as a Ca^{2+} analogue that enters neurons via activity-dependent calcium channels. The Mn^{2+}-rich areas can be visualized as hyper-intense regions in T_1-weighted MRI (29, 30). However, Mn^{2+} is an intracellular contrast agent and needs to be applied in vivo. Concentrations of 120 mM have been shown to lead to good results in crustaceans ((9, 10) – refer to these papers for examples of the application of the contrast agent in vivo).

8. Small animal MRI scanners are essential for scanning of smaller invertebrate specimens and access to these systems is improving. As of 2008, more than 1000 systems had been sold worldwide and sales are constantly increasing. Two companies (Bruker Biospin GmbH, Germany and Agilent Technologies, USA) are the current market leaders. Small animal MRI systems can be found primarily in neurological, neurobiological, or pharmaceutical research facilities. Most of these instruments are used during daytime, so that overnight and weekend scanning may constitute a resourceful application for fixed invertebrates.

9. Protocol choice is largely dependent on the desired contrast and resolution. The gradient echo 3D protocol advocated here (FLASH) has a major time advantage over spin echo 3D protocols such as RARE. However, FLASH protocols are more prone to artefacts (*see* **Note 13**). Since the signal intensity per voxel is mostly higher in non-isotropic datasets, RARE protocols seem to be much more suitable for high-throughput 2D screening of specimens (**Fig. 4**). *See* Ref. (5) for a description of differences between gradient and spin echo protocols in MRI of a crustacean.

10. Many specimens tend to float when exposed to the gradual re-watering process primarily due to the differing densities

of water and ethanol. If this happens, you need to wait long enough for the specimen to sink to the bottom of the container. Exposing the specimen to a vacuum may speed up the process of fluid interchange. Gentle agitation of the specimen container may additionally help to increase fluid interchange and also to release trapped air. However, make sure that the material is exposed to air for as short a duration as possible during the fluid interchange and the later handling process. Whenever possible, try to conduct specimen preparation inside the container in which the animal will be scanned in afterwards, e.g. in a 50 mL Falcon tube.

11. Mechanical specimen fixation is not always necessary, depending on the properties of the specimen. Heavier specimens (e.g. strongly calcified sea urchins) will sit at the bottom of the scanning container and might not be susceptible to the vibrations inherent to most small animal MRI scanners (vibrations inside the scanner are primarily caused by the gradient system – the stronger this system, the more vibrations will occur). However, the use of plastic or glass rods and tubing can become a problem when the outer parts of the animal need to be reconstructed three-dimensionally. Here, the structures used to fix the specimen inside the container might 'fuse' with the outer parts of the specimen if these show the same signal properties – as for example the proton-free glass rods do and the calcite sea urchin endoskeleton that is covered only by a thin epidermis which cannot be resolved by MRI. Even manual segmentation may then prove to be unsuitable to clearly separate these materials for 3D reconstruction. In some cases, it might even become necessary to embed the specimen inside the container in agarose in order to completely rule out movement artefacts. In that case, use agarose that solidifies at low temperatures (e.g. 45°C) in order to prevent tissue damage from exposure to heat. Before heating, the agarose can also be contrasted with gadolinium-based contrast agents.

12. Make sure that your specimen is completely contained within the FOV. Otherwise, aliasing (wrap-around) occurs when parts of the specimen are outside the FOV. In this case, that part of the specimen lying beyond the edge of the FOV is projected onto the other side of the image. To avoid this, increase the FOV (which will decrease resolution), use anti-aliasing, or apply saturation slices (which will result in longer scan times).

13. A number of artefacts inherent in MRI of animals can result in partially or even entirely unsuitable datasets (**Fig. 9**).

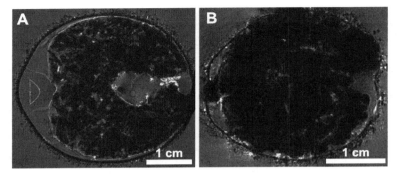

Fig. 9. Examples of artefacts due to ferro- or paramagnetic gut content. (a) Museum specimen of *Echinolampas depressa* (USNM E32955) with slight artefacts that are largely confined to the digestive tract. Some structures can still be discerned from others, and a 3D reconstruction of most structures may therefore still be possible. (b) Museum specimen of *Spatangus purpureus* (ZMB 3236) with heavy artefacts that prevent any differentiation of internal structures. A 3D reconstruction will be impossible in this case.

The major source of such artefacts seems to be ferro- or paramagnetic sediment or plant material inside the digestive tract (2). FLASH protocols are more prone to be susceptible to these artefacts than RARE protocols. However, the latter protocols take considerably longer (**Table 2** – note the differences in T_R). Other artefacts that can occur are boundary artefacts at tissue borders. The application of gadolinium-based contrast agents may in some cases lead to an improved tissue contrast in spite of artefacts (unpublished observation).

14. Try to identify a standard scan orientation that will be used for all subsequent scans – this may also involve angled scan geometries. This approach will greatly reduce the efforts during later image processing, e.g. manual segmentation and image alignment (**Figs. 5** and **6**). We use the 'coronal' orientation as our standard scan orientation.

15. Although much faster, 2D MRI protocols bear the risk of so-called partial volume effects caused by different resolutions along the X/Y and Z planes. An increase of the resolution in the Z plane will reduce such artefacts, although the possibility to do so may be technically limited. In addition, 2D image datasets are less suitable for 3D modelling and visualization.

16. In ParaVision 4.0, the output size of the DICOM image stack is set to the next higher matrix size as compared to the matrix size that was initially used (e.g. initial matrix size is $384 \times 384 \times 384$ – final output size in DICOM format is $512 \times 384 \times 384$). In this case you need to manually resize the DICOM images, e.g. in ImageJ (Image:Adjust:Size).

17. A typical technical problem that we encountered was overheating of the gradient system. If this occurs, the scan is interrupted and the data usually lost. Make sure that the protocols are suitable for your imaging and gradient system (matrix size is one of the major critical factors) and that the rooms which house the magnet as well as the gradient system controller are air-conditioned.

18. The achievable isotropic resolution increases with the strength of the magnetic field as well as the strength of the gradient system. A comparison of specimens scanned at different isotropic resolutions is shown in **Fig. 5**. Zero-filling, a standard method in MRI that has to be switched on or off prior to scanning, can be used to increase the resolution of the dataset by doubling the output matrix size using interpolation algorithms.

Acknowledgements

The authors would like to thank Cornelius Faber for scanning of specimens on a 17.6 T scanner (Würzburg, Germany) as well as Frank Angenstein for help with scanning of specimens on a 4.7 T scanner (Magdeburg, Germany). We would also like to thank the following curators for generous supply with museum specimens: Andrew Cabrinovic (Natural History Museum, London, UK), Danny Eibye-Jacobsen (Zoologisk Museum, Copenhagen, Denmark), Carsten Lüter (Naturkundemuseum, Berlin, Germany), Rich Mooi (California Academy of Sciences, San Francisco, USA), and David Pawson (National Museum of Natural History, Washington D.C., USA). We are grateful to the staff of the Biologische Anstalt Helgoland (Germany) for the supply with living sea urchin specimens.

References

1. Sutton, M.D. (2008) Tomographic techniques for the study of exceptionally preserved fossils. *Proc. Roy. Soc. B* **275**, 1587–1593.
2. Ziegler, A., Faber, C., Mueller, S., and Bartolomaeus, T. (2008) Systematic comparison and reconstruction of sea urchin (Echinoidea) internal anatomy: a novel approach using magnetic resonance imaging. *BMC Biol.* **6**, 33.
3. Ruthensteiner, B. (2008) Soft part 3D visualization by serial sectioning and computer reconstruction. *Zoosymposia* **1**, 63–100.
4. Corfield, J.R., Wild, J.M., Cowan, B.R., Parsons, S., and Kubke, M.F. (2008) MRI of post-mortem specimens of endangered species for comparative brain anatomy. *Nature Prot.* **3**, 597–605.
5. Müller, W.E.G., Kaluzhnaya, O.V., Belikov, S.I., Rothenberger, M., Schröder, H.C.,

Reiber, A., et al. (2006) Magnetic resonance imaging of the siliceous skeleton of the demosponge *Lubomirskia baicalensis*. *J. Struct. Biol.* **153**, 31–41.

6. Brouwer, M., Engel, D.W., Bonaventura, J., and Johnson, G.A. (1992) In vivo magnetic resonance imaging of the blue crab, *Callinectes sapidus*: effect of cadmium accumulation in tissues on protein relaxation properties. *J. Exp. Zool.* **263**, 32–40.
7. Fernández, M., Bock, C., and Pörtner, H.O. (2000) The cost of being a caring mother: the ignored factor in the reproduction of marine invertebrates. *Ecol. Lett.* **3**, 487–494.
8. Bock, C., Frederich, M., Wittig, R.M., and Pörtner, H.O. (2001) Simultaneous observations of haemolymph flow and ventilation in marine spider crabs at different temperatures: a flow weighted MRI study. *Magn. Reson. Imaging* **19**, 1113–1124.
9. Herberholz, J., Mims, C.J., Zhang, X., Hu, X., and Edwards, D.H. (2004) Anatomy of a live invertebrate revealed by manganese-enhanced magnetic resonance imaging. *J. Exp. Biol.* **207**, 4543–4550.
10. Brinkley, C.K., Kolodny, N.H., Kohler, S.J., Sandeman, D.C., and Beltz, B.S. (2005) Magnetic resonance imaging at 9.4 T as a tool for studying neural anatomy in non-vertebrates. *J. Neurosci. Meth.* **146**, 124–132.
11. Pohlmann, A., Möller, M., Decker, H., and Schreiber, W.G. (2007) MRI of tarantulas: morphological and perfusion imaging. *Magn. Reson. Imaging* **25**, 129–135.
12. Hart, A.G., Bowtell, R.W., Köckenberger, W., Wenseleers, T., and Ratnieks, F.L.W. (2003) Magnetic resonance imaging in entomology: a critical review. *J. Insect Sci.* **3**, 5.
13. Michaelis, T., Watanabe, T., Natt, O., Boretius, S., Frahm, J., et al. (2005) In vivo 3D MRI of insect brain: cerebral development during metamorphosis of *Manduca sexta*. *Neuroimage* **24**, 596–602.
14. Watanabe, T., Schachtner, J., Krizan, M., Boretius, S., Frahm, J., et al. (2006) Manganese-enhanced 3D MRI of established and disrupted synaptic activity in the developing insect brain in vivo. *J. Neurosci. Meth.* **158**, 50–55.
15. Haddad, D., Schaupp, F., Brandt, R., Manz, G., Menzel, R., and Haase, A (2004) NMR imaging of the honeybee brain. *J. Insect Sci.* **4**, 7.
16. Null, B., Liu, C.W., Hedehus, M., Conolly, S., and Davis, R.W. (2008) High-resolution, in vivo magnetic resonance imaging of *Drosophila* at 18.8 Tesla. *PLOS One* **3**, 7.
17. Hörnschemeyer, T., Goebbels, J., Weidemann, G., Faber, C., and Haase, A. (2006) The head morphology of *Ascioplaga mimeta* (Coleoptera: Archostemata) and the phylogeny of Archostemata. *Eur. J. Entomol.* **103**, 409–423.
18. Davenel, A., Quellec, S., and Pouvreau, S. (2006) Noninvasive characterization of gonad maturation and determination of the sex of Pacific oysters by MRI. *Magn. Reson. Imaging* **24**, 1103–1110.
19. Pouvreau, S., Rambeau, M., Cochard, J.C., and Robert, R. (2006) Investigation of marine bivalve morphology by in vivo MR imaging: first anatomical results of a promising technique. *Aquaculture* **259**, 415–423.
20. Holliman, F.M., Davis, D., Bogan, A.E., Kwak, T.J., Cope, W.G., and Levine, J.F. (2008) Magnetic resonance imaging of live freshwater mussels (Unionidae). *Inv. Biol.* **127**, 396–402.
21. Quast, M.J., Neumeister, H., Ezell, E.L., and Budelmann, B.U. (2001) MR microscopy of cobalt-labeled nerve cells and pathways in an invertebrate brain (*Sepia officinalis*, Cephalopoda). *Magn. Reson. Med.* **45**, 575–579.
22. Gozansky, E.K., Ezell, E.L., Budelmann, B.U., and Quast, M.J. (2003) Magnetic resonance histology: in situ single cell imaging of receptor cells in an invertebrate (*Lolliguncula brevis*, Cephalopoda) sense organ. *Magn. Reson. Imaging* **21**, 1019–1022.
23. Ziegler, A., and Angenstein, F. (2007) Analyse von Seeigeln (Echinoidea) mit Hilfe der bildgebenden Magnetresonanztomographie. *Mikrokosmos* **96**, 49–54.
24. Ziegler, A., Mueller, S., and Bartolomaeus, T. (2010) Sea urchin (Echinoidea) anatomy revealed by magnetic resonance imaging and 3D visualization. In: Harris, L., Boettger, S.A., Walker, C.W., and Lesser, M.P. (eds.), *Echinoderms*. Durham, CRC Press.
25. Ziegler, A. (2008) Non-invasive imaging and 3D visualization techniques for the study of sea urchin internal anatomy. PhD thesis, Freie Universität Berlin, Germany.
26. Ziegler, A., Faber, C., and Bartolomaeus, T. (2009) Comparative morphology of the axial complex and interdependence of internal organ systems in sea urchins (Echinodermata: Echinoidea). *Front. Zool.* **6**, 10.
27. Holland, N.D., and Ghiselin, M.T. (2009) Magnetic resonance imaging (MRI) has failed to distinguish between smaller gut regions and larger haemal sinuses in sea urchins (Echinodermata: Echinoidea). *BMC Biol.* **7**, 39.

28. Ruthensteiner, B., and Hess, M. (2008) Embedding 3D models of biological specimens in PDF publications. *Microsc. Res. Tech.* **71**, 778–786.
29. Pautler, R.G. (2006). Biological applications of manganese-enhanced magnetic resonance imaging. In: Prasad, P.V. (ed.), *Magnetic Resonance Imaging*. Humana Press.
30. Van der Linden, A., Van Meir, V., Boumans, T., Poirier, C., and Balthazart, J. (2009) MRI in small brains displaying extensive plasticity. *Trends Neurosci.* **32**, 257–266.

Section X

Metabolic and Targeted Imaging

Chapter 33

Applications of Hyperpolarized Agents in Solutions

Jan Henrik Ardenkjaer-Larsen, Haukur Jóhannesson,
J. Stefan Petersson, and Jan Wolber

Abstract

This chapter provides an overview of pulse sequences adapted to hyperpolarized MR imaging. Applications of hyperpolarized agents in aqueous solution are reviewed. Vascular (e.g., angiography, perfusion, and catheter tracking) as well as metabolic (e.g., oncology, cardiology, neurology, and pH mapping) applications are covered. Due to the rapid development of new applications for hyperpolarized agents, a review format has been used for this chapter instead of a strict protocol/procedure structure.

Key words: Hyperpolarized C-13, perfusion, angiography, metabolic imaging.

1. Introduction and General Considerations

Hyperpolarized agents in solution may enable a wide range of imaging applications. In this section, the relevant publications are summarized and the main clinical applications of hyperpolarized MR imaging are reviewed. These can be grouped into two types of applications:

- Angiography, perfusion, and interventional procedures
- Metabolic MR

The grouping is useful for defining the requirements for an optimal pulse sequence, as these requirements may depend on the specific application. A general limitation for all hyperpolarized imaging experiments is given by the in vivo relaxation times of the agent. The in vivo T_1 and T_2 relaxation times for the molecules reported in literature (1–4) have typically been in the range of 20–45 s and up to 5 s, respectively. Another important parameter that limits the available MR signal is the concentration of the hyperpolarized agent. The concentration of the hyperpolarized agents have been as high as 0.5 M (1–3, 5), and the dose in the range 0.1–0.3 mmol/kg. However, after the injection or infusion of the solution, the agent is rapidly diluted in the blood volume to a concentration of 10 mM or less.

When selecting an MR sampling scheme and during the implementation and adoption of pulse sequences for hyperpolarized imaging, care must be taken to ensure an efficient use of the available magnetization. The long T_2 relaxation time of ^{13}C-labeled molecules often allows the use of advanced multi-echo sequences. This is markedly different compared to hyperpolarized gas MR imaging, where a relatively short transverse in vivo relaxation time has led to the use of fast gradient echo sequences. However, it has been shown that even in hyperpolarized ^{3}He imaging, steady-state free precession sequences are feasible (6).

In order to set up a system for hyperpolarized imaging and/or spectroscopy, the following hardware modifications/options need to be considered:

- *Multi-nuclear spectroscopy (MNS) unit:* The unit needs to be equipped with a transmit and receive system that operates at the resonance frequency of the hyperpolarized nucleus in question. For ^{13}C and ^{15}N, the operating frequency compared to the one for proton is one quarter and one tenth, respectively, which has consequences for rf coil and gradient design. All the major MRI vendors are able to deliver this type of upgrade to their high-end clinical systems.

- *RF-coils:* The system needs to be equipped with a transmit and receive RF coil that is interfaced to the MNS-unit. Depending on the design of the experiment in question, the RF coil may either be a volume coil or a surface coil. Suitable MNS coils may need to be ordered directly from third-party coil vendors. A dual-tuned RF-coil is preferred as it may be tuned to both the resonance frequency of protons and the hyperpolarized nucleus. This type of coil makes it possible to obtain high-quality proton images that may be used to define area of interest before the injection of the hyperpolarized substance takes place.

- *High gradient performance:* As previously indicated, the gradient integral required to achieve a given field of view scales with γ^{-1}. Consequently, an MRI scanner system equipped with a gradient system with high amplitude and a high slew rate will make it possible to implement efficient pulse sequences. A preclinical animal system typically has gradient specifications that are a factor of ten (or more) higher than the ones found on clinical systems. As with all other demanding imaging applications (e.g., fMRI), the stability and the duty cycle of the gradient system will impact the outcome of an imaging experiment.

- *The operating field strength of the MR scanner:* The SNR obtained in an experiment using hyperpolarized nuclei is, as was described in **chapter 11**, independent of the scanner field strength. The separation of resonance peaks is a linear function of the external field. Consequently, the use of a high-field system when a metabolic experiment is performed will make it easier to separate the resonance peaks. It should also be noted that if a low magnetic field is used combined with high gradient amplitudes, artifacts due the concomitant gradients (also called the Maxwell gradients) could be introduced (7). In a clinical system operated at a main field of 1.5 T or 3 T and equipped with high amplitude gradients, one may detect the influence of the Maxwell terms as a phase change at large FOV. In a low-field system with high performance gradients, pixels may shift from their actual positions and slice profiles may be distorted (3).

- *Monitoring systems:* For ECG, respiration, temperature, and blood pressure, equipment which is compatible with the MR environment is available (e.g., the MR1025 from SA Instruments). For any monitored parameter, the normal range for the species (under the chosen anesthesia) should be decided in advance of the study so that an experiment can be aborted should the animal have abnormal physiological parameters. The general metabolic status of the animal will be affected if temperature, respiration, or heart rate drops below the normal physiological range. It is also important to carefully monitor the effect of the test item (hyperpolarized agent in the chosen formulation) on the animal's physiological parameters. The tolerance of the agent and formulation should be tested in advance of the study without involving hyperpolarization and isotopic enrichment. Finally, often respiratory and heart rates are used for gating of the MR imaging sequence.

  Anesthetized animals have poor temperature control. Especially smaller animals are prone to rapidly lose body heat and care should be taken to maintain normal physiological temperature during experiments by having an MR-compatible heating system. The temperature of the injectate should also be considered; this might be above body temperature immediately after dissolution due to the hyperpolarization process.

- *Injection system:* Due to the short half-life (relaxation time) of the hyperpolarization and the need for a minimal infusion line dead volume, no commercially available injection system has been found that allows automated infusion of the agent with control of injection rate and volume in a preclinical setting. Reference (8) describes a solution that might provide the required control, but the general method has been to inject the agent by hand. In order to ensure best possible reproducibility in data as well as the safety of the animal, it is recommended to practice the injection using a timer and dosing equipment identical to what will be used in-life before administering a hyperpolarized agent to an animal.

- *Safety (osmolarity, pH, temperature, dose, injection rate, etc.):* For formulations of hyperpolarized agents proper adjustment of pH to the physiological range (9–10), the temperature of the formulation, and the osmolarity should be carefully considered. The reason is obvious: the agent is injected in high concentration and volume and the tolerance of the formulation should be maximized. New formulations should be tested for osmolarity and administration of hypotonic solutions (osmolarity <290 mOsm/kg) should be avoided. Adjustment of osmolarity can be done with NaCl. It is encouraged to do a quality control of the polarization process for all these parameters (including polarization) before commencing any animal study.

A Good Practice Guide to the administration of substances was published 2001 by Diehl et al. (9). When considering injection rates and/or volumes in excess of that recommended refer to institutional guidelines and discuss with the veterinarian. Injection rates should be evaluated on the basis of cardiac output and injection volume compared to total blood volume. For a reference on physiological values for different species, see the paper by Davies and Morris from 1993 (10). The combination of the volume and rate should be inversely correlated so that larger volumes are given at lower injection rates. When administering multiple doses to an animal consider the total dose volume administered and refer to values given for repeated intravenous infusion (9). Remember to include in the volume, any saline dead volumes in the injection line.

	Mouse (rapid bolus i.v.)		Rat (rapid bolus i.v.)		Dog (rapid bolus i.v.)	
	Max vol. (mL/kg)	Max rate (mL/s)	Max vol. (mL/kg)	Max rate (mL/s)	Max vol. (mL/kg)	Max. rate
Ref. (9)	5	0.05	5	0.05	2.5	Dose given over 1 min

The physiological effect of a rapid bolus on heart and lung is buffered when it is administered in a peripheral vein or artery. The bolus is in this way more effectively diluted in blood and the peak concentration reaching key organs is reduced. For central venous administrations, it is strongly encouraged to confirm that the procedure is tolerated by administration of saline/osmotic control, before the study with the hyperpolarized agent is initiated.

For experiments of longer duration (>60 min) and experiments including surgical procedures, do ensure that the fluid balance of the animal is maintained according to standard practices.

1.1. Pulse Sequences for Angiography and Perfusion

In angiography and perfusion studies the imaging strategy is reduced to the visualization (spatial encoding) of the MR-active nuclei. The hyperpolarized agent should preferably not be metabolically active. First, this will facilitate quantitative analysis. Secondly, it means that the applied pulse sequence does not need to resolve the multiple resonance frequencies of the injected molecule and any metabolites. Indeed, it may even be desirable to suppress metabolite signals for vascular applications. Consequently, the requirements of a MR pulse sequence for these applications "vascular hyperpolarized imaging," are very similar to clinical routine angiographic or perfusion imaging, except for the differences imposed by the non-equilibrium magnetization. The intra-voxel spin dephasing caused by flow and acceleration needs to be suppressed and the sequence needs to be able to deal with inflow/outflow effects.

Most hyperpolarized MR vascular studies have been based on variations of fully balanced steady-state free precession (bSSFP) pulse sequences (11–14). The general outline of this sequence is shown in **Fig. 1**. When this sequence is used for conventional MR imaging, the flip angle is in the range 10–60° and a number of TR periods and pre-pulses are applied to establish steady state before starting the data collection, ensuring stable signal amplitude during the readout (15–17). In hyperpolarized imaging, the polarization (and thus the signal) irreversibly decays toward thermal equilibrium, which is essentially zero, and a steady state is never established (13). Instead, the collection of data takes place during a transient phase. In this situation, with an $\alpha/2$ pre-pulse (18) the optimal flip angle, α, is found to be 180° (19) resulting in a sequence that is similar to fast spin echo (FSE) (20) sequences.

The bSSFP sequence takes advantage of the long T_2-relaxation time of the hyperpolarized ^{13}C spins. Furthermore, the rewinding of the gradients in all directions during each TR interval minimizes loss of signal due to moving spins. The sequence is based on a single excitation; the 90° pre-pulse flips all spins within the imaged slice into the transverse plane and the following 180° pulses generate a series of spin echoes. By applying a centric phase-encoding scheme, the effect of signal amplitude decay due to imperfect refocusing pulses and finite T_2 relaxation is minimized. The signal is sampled as a full spin echo with its maximum in the center of the TR interval. This ensures that all effects from

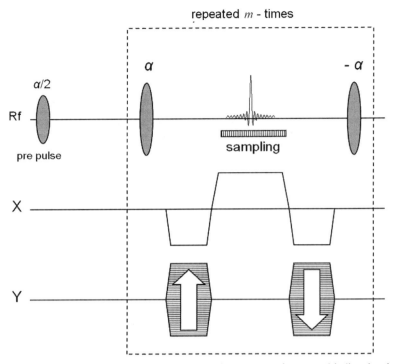

Fig. 1. The bSSFP sequence. In order to generate an $n \times m$ image matrix the signal during each sampling interval is sampled n times and the total number of echoes is m. The *box* indicated using *dashed line* shows the part of the sequence that is being repeated.

J-couplings are suppressed as in the case when the imaging agent hydroxyethyl-[1-^{13}C]propionate, generated through the PHIP (para-hydrogen induced polarization, cf. **Chapter 11**) method, is used.

It should be noted that the low gyromagnetic ratio of ^{13}C is a liming factor when reducing TE and TR times, as the gradient integral required to achieve a given field of view scales with γ^{-1}. However, implementation of bSSFP sequences on clinical MR scanners and the use of state-of-the-art gradient systems have been demonstrated successfully, and scan times of the order of 300 ms (1) with a matrix size of 128×64 (TR/TE of 6.6/3.3 ms) have been achieved.

1.2. Pulse Sequences Adapted for Metabolic Imaging and pH Mapping

A pulse sequence for metabolic imaging needs to be capable of resolving the resonance frequencies of the injected molecule as well as molecules generated by the metabolic processes. Therefore, in addition to the previous section, the *spectral* dimension has to be considered. The simplest approach is to apply a modified 2D or 3D chemical shift imaging (CSI) FID pulse sequence. These types of sequences require a number of excitation pulses

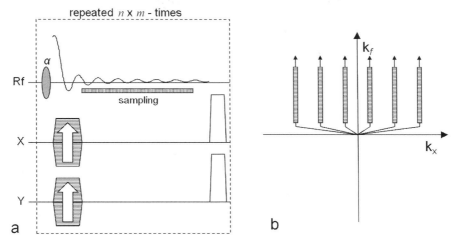

Fig. 2. The CSI pulse sequence. The left-hand side (**a**) outlines the sequence and the right-hand side (**b**) shows the corresponding k-space sampling scheme. In order to get an $n \times m$ image matrix the sequence is repeated $n \times m$ times. Both x and y gradients are used to phase encode the signal. The x gradient is changed n times and the y gradient m times. The FID signal is sampled for each repetition and the readout time determines the chemical shift frequency resolution. The position of the sampling window is indicated in (**a**) and its corresponding position in the k-space is indicated in (**b**). Note that during one repetition only data along one of the k-space track is collected.

that are equal to the number of voxels in the final raw data matrix (disregarding zero filling). Phase-encoding gradients are applied in two or three directions prior to the signal acquisition, which renders this sequence time consuming. However, the sequence is easy to adapt and consequently it has been used in most metabolic imaging with hyperpolarized molecules. It has also been used to demonstrate in vivo pH measurements (21). The sequence is outlined in **Fig. 2a** together with its k-space sampling scheme in **Fig. 2b**.

When proton CSI is performed the TR of the sequence may typically be several seconds. However, in the case of hyperpolarized CSI, the objective is to minimize TR, typically to less then 100 ms. The length of the sampling window is chosen to assure that the peaks from the injected substance and molecules created by metabolic processes are resolved and the tail of the FID may be truncated. This means that only a small part of the FID is sampled and the truncation introduces a line broadening. The flip angle of the excitation pulses are selected to match the total number of repetitions. Each sampling interval is followed by spoiler gradients in order to suppress the signal from magnetization remaining in the transverse plane. A sampling of the FID signal is used, and not a spin echo, as the application of a 180° pulse would alter the phase of not only the excited spins but also the spins still in the longitudinal direction.

In the k-space diagram in **Fig. 2**, only two dimensions are shown: the k_x-axis, which after the reconstruction will form the spatial x-dimension, and the k_f-axis, the evolution in the time dimension which after the reconstruction will form the frequency dimension (the chemical shift dimension). From the k-space diagram it follows that the center is not collected during this type of sequence and also that the spatial dimensions are sampled more coarsely than the time dimension, which results in low resolution spatial images.

In summary, the high resolution of the CSI sequence in the spectral domain makes it useful in situations where complicated spectra with several peaks are evaluated. The sequence puts emphasis on collecting spectral information with coarse localization, since the time window for sampling the signal is limited by the T_1 of the hyperpolarized nuclei and the dynamics of metabolism. A way of reducing the scan time is to not sample the corners of k-space. However, even a relatively low spatial resolution matrix of 16×16 will result in a total scan time of around 12 s (22).

In order to overcome the obvious inefficiencies of the basic CSI FID sequence in the context of hyperpolarized MR imaging, different multi-echo concepts have been developed (23). The CSI FID sequence may be extended to a multi-gradient echo sequence by adding alternating readout gradients. In the most extreme case, a large number of gradient echoes will result in a trajectory that covers a k-space volume sufficient to be reconstructed into a 3D/4D matrix with two/three spatial dimensions and one frequency dimension. The alternating gradient approach is similar to the way that the readout is performed in an echo planar imaging sequence. Consequently, when this is applied to a long echo train, in order to allow for spectroscopic information to be resolved and collected, the acronym EPSI, echo planar spectroscopic imaging, is used to describe this sequence (24). The total length of the echo train is only limited by the T_2^* of the imaged nucleus. In order to account for finite T_2^*, EPSI may be implemented as a multi-shot technique where a low flip angle excitation pulse is used and the sequence is repeated in order to obtain the desired spatial as well as spectroscopic resolution.

The echo train readout may be performed using the flyback technique (25) allowing for a read out k-space trajectory that is less sensitive to error in timing relative to the traditional alternating gradient readout used in most EPI implementations. This sequence has been applied for 2D and 3D hyperpolarized imaging (26). The data acquisition part of the sequence is outlined in **Fig. 3a** together with its k-trajectory in **Fig. 3b**. During the reconstruction, the tilted sample points are interpolated onto a rectilinear grid by the use of a sinc interpolation (27, 28). The equivalent of a sinc interpolation in the time dimension is a phase shift in the spectral dimension (29). Consequently, this interpolation may be performed by applying a 1D Fourier transform along the time dimension (the k_f-direction in **Fig. 3b**). The phase shift

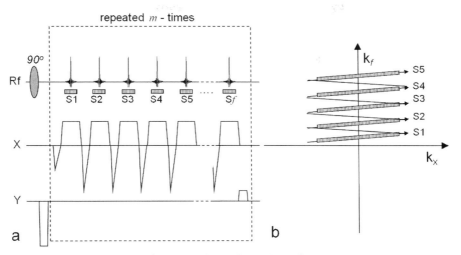

Fig. 3. The EPSI pulse sequence with a flyback gradient scheme. The left-hand side (**a**) outlines the sequence and the right-hand side (**b**) shows the corresponding k-space sampling scheme. In order to acquire a data matrix with the spatial dimensions n × m and f points in the frequency dimension, the readout scheme is repeated m times. During each repetition the phase-encoding gradient, y, applied before the readout is changed. The series of gradient echoes are then sampled during the intervals S1, S2,..., Sf. During each sampling window the signal is sampled n times. The corresponding k-space positions of the sampling windows (S1,..., Sf) during each repetition are indicated in (**b**).

procedure is then carried out on a sample point-to-sample point basis using (25):

$$p'_{sn} = p_{sn} e^{\left[\frac{i2\pi sn \Delta t B_s}{S}\right]} \qquad [1]$$

where p_{sn} is the nth sample within the spatial k-space sample s (the gradient echo no s), Δt is the distance between spatial k-space samples along the k_f-dimension, B_s is the spectral bandwidth, defined as the inverse of the time between the start of two consecutive sample intervals, and S is the total number of sampling intervals.

The EPI method has also been combined with spectral–spatial RF pulses in order to suppress the signal emanating from the outside of the volume of interest (30), multiband excitations in order to excite the metabolites with different effective flip angles (31), and ^{1}H decoupling techniques (32). A further extension of the k-space sampling is to replace the rapidly alternating readout gradient with a gradient scheme that will form a spiral trajectory in k-space (33, 34). This method has been applied to hyperpolarized imaging (35) and a temporal resolution of around 1 s has been demonstrated (36).

The EPSI sequence utilizes the available magnetization in a much more efficient way than the traditional CSI sequence with respect to spatial resolution. It also makes it possible to collect a 4D dataset (three spatial dimensions and one time dimension) (37) but it is still mainly a spectroscopic sequence and not an imaging sequence. In order to create metabolite maps, it depends

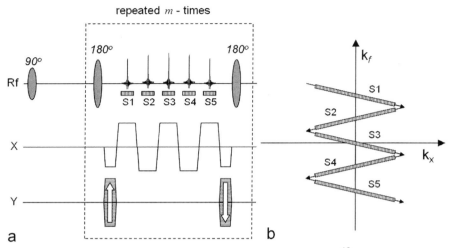

Fig. 4. The multi-echo bSSFP sequence used to generate maps of hyperpolarized [1-^{13}C]pyruvate and its metabolites. The left-hand side (**a**) outlines the sequence and the right-hand side (**b**) shows the corresponding k-space sampling scheme. In order to generate maps with the matrix dimension $n \times m$, the number of samples per echo is n and the number of phase-encoding interval is m. The number of echoes, S1, ..., S5, within a phase-encoding interval determines the number of resonance peaks that will be possible to resolve. The corresponding k-space positions of the sampling windows (S1, ..., Sf) during one repetition are indicated in (**b**). In this example, four metabolite maps may be calculated together with a B0 map.

on the resolution in the time/frequency dimension to separate information/signal from the injected substance and its metabolites.

When the spectrum is sparse and chemical shifts of the injected substance and its metabolites are known, a more efficient pulse sequence and post-processing procedure can be designed, which drastically reduces resolution in the spectral domain. This overcomes the trade-off between spatial and spectral sampling requirements by leaving more time for the spatial encoding. This idea has been applied in the case of hyperpolarized [1-^{13}C]pyruvate to create a multi-echo bSSFP pulse sequence (38). The basic outline of this sequence is shown in **Fig. 4a** together with a part of its k-space trajectory in **Fig. 4b**.

The main feature of this class of pulse sequences is that it acquires images with high resolution in the spatial domain. The gradient schema mimics the one used in the GREASE sequence (39) but the phase encoding is performed once in the beginning and rewound at the end of each interval between the 180° refocusing pulses. This results in the same phase encoding for all generated echoes. The generated echoes will be a superposition of the signals from all metabolites in the imaged volume. At the position of the central echo, which is a true spin echo, the signals from all metabolites will be in phase and this echo is also compensated for main field inhomogeneities. The k-space trajectory is designed to allow for separation of the signals arising from the different metabolites during post-processing. This requires prior knowledge of the phases of all the metabolites at the time of all echoes. The separation of the individual echoes is contingent on the specific chemical shift separation of the signals. As such, the suggested sampling scheme has many similarities with the Dixon technique (40, 41) suggested for separating water and fat in clinical proton imaging. An extension of the classical Dixon method has been suggested by Reeder et al. (42). In this approach, the minimum number of echoes required to reconstruct a certain number of MR signals equals the number of peaks that need to be resolved plus one. Furthermore, the reconstruction method is able to correct for phase errors caused by inhomogeneities in the main magnetic field. Alternatively, rather than making the B_0 map an output of the reconstruction, prior knowledge about the spatial variation of B_0 (from a ^{1}H MR field map of the same field of view) could be used to constrain the reconstruction. Different methods have been demonstrated for reconstructing the metabolic maps and detailed description may be found elsewhere (38, 43). Implementations of the multi-echo bSSFP pulse sequence combined with an interactive post-processing make it possible to reduce the scan time dramatically (38). In **Fig. 5**, a set of images generated using this approach is shown. The sequence was executed 25 s after an i.v. injection of hyperpolarized ^{13}C-pyruvate into a full size pig. The total scan time was 520 ms making it possible to collect all data within one heartbeat.

1.3. Summary

In this section, MR pulse sequences for hyperpolarized MR have been reviewed briefly. The main aspects to be aware of are as follows:

- Hyperpolarized MR is limited by the relaxation times of the non-equilibrium magnetization.
- Long T_2 times of hyperpolarized ^{13}C-labeled molecules in solutions allow for multi-echo sampling schemes.

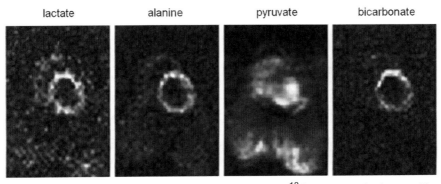

Fig. 5. Generated metabolite maps after the injection of hyperpolarized [1-^{13}C]pyruvate and using a multi-echo bSSFP sequence The matrix size is 64 × 32 and the images demonstrate a slice through the left ventricle of the pig. The lactate, alanine, and bicarbonate are all metabolites after the injected pyruvate. The total scan time was 520 ms.

- In applications such as metabolic imaging or pH mapping, a further trade-off is encountered due to the necessity to also sample the spectral domain.
- This trade-off can be partially overcome if prior knowledge about the chemical shifts of the MR signals of interest allows for sparse spectral sampling, thus retaining the spatial resolution of conventional MR imaging.

2. Angiography, Perfusion, and Interventional Procedures

In conventional vascular proton imaging, the contrast medium, CM, decreases the protons relaxation times in the vicinity of the paramagnetic center of the injected molecules. Depending on the applied pulse sequence, the passage of a CM bolus will result in an increase or a decrease of the MR signal and hence affect the contrast of the resulting MR image. However, the MR signal itself does not arise from the CM but the protons in the tissue. This is different compared to vascular imaging using hyperpolarized agents where the MR signal arises from the MR-active nuclei of the injected agent, e.g., ^{13}C. Due to the low concentration of ^{13}C in the body there is no background signal at all from tissues in the imaging volume.

In order to assure a strong signal in the image volume, a hyperpolarized agent for angiographic and perfusion imaging should preferably not be metabolically active. It should behave pharmacokinetically as an extra cellular fluid (ECF) MR contrast agent and mainly remain within the vascular bed at least during the first few re-circulations in the body. This requirement is fulfilled by hydroxyethyl-[1-^{13}C]propionate which may be polarized using the PHIP method. Consequently, the majority of vascular imaging results presented in the literature have been generated using this model substance. It is not intended for human clinical use but its toxicity profile allows it to be used for preclinical animal model work.

The results described in the following section are obtained with this molecule unless otherwise indicated. The molecular weight of this substance is ~130 g/mol and the in vivo ^{13}C T_1 and T_2 of the substance have been determined to 45 and 6 s, respectively (2). The preparation for a hyperpolarized angiography or perfusion imaging experiment performed in a pig model may include the following steps (2):

- The animal is premedicated with 10 mL ketamine (50 mg/mL Ketalar®, Pfizer) intramuscularly and then the anesthesia is induced 20 min later with 10–15 mL thiopental sodium (25 mg/mL Pentothal natrium®; Abbott) intravenously.

- The pig is intubated and connected to a volume-controlled respirator delivering a gas mixture of 70% N_2O and 30% O_2. Normocapnia is maintained with tidal volumes of 12–14 mL/kg at a respiratory rate of 20 breaths per minute.

- The anesthesia is maintained by intravenous infusion of a mixture (0.6–0.8 mL/min) of 10 mL midazolam (5 mg/mL Dormicum®; Pharma Hameln Gmbh) + 40 mL ketamine + 20 mL Vecuron (2 mg/mL Norcuron®; Organon) + 25 mL NaCl (9 mg/mL Fresenius Kabi).

- During the session, the respiratory status is monitored by measuring blood gases (pO_2 and pCO_2) and pH. Rectal temperature is monitored and kept close to 38°C by a warming lamp and a heating pad.

- Intravenous infusion is continuously given with Ringer acetate (Ringer-acetate®, Fresenius Kabi) 3 mL/min.

- In experiments involving catheterization, heparin 5000 IU (Heparin®, Lovens Kemiske Fabrik) is given for a time interval of 3 min before the start procedure.

- To confirm the position of catheters, one may inject 3 mL of iohexol 350 mg I/mL (Omnipaque®; Amersham Health AS) to obtain X-ray angiograms.

2.1. Angiography

The first angiographic images following an i.v. injection of a hyperpolarized substance in a rat were presented by Golman et al. (44, 45). This was followed by the first dynamic angiography also performed in a rat model by Svensson et al. (19). The small animal model allowed the imaging to be performed in a preclinical animal MR scanner where large gradient amplitude combined with high slew rate made it possible to suppress flow artifacts and to generate images with high in-plane resolution. Although the PHIP method has been used in generating hyperpolarized agents for many vascular applications, the first angiogram obtained using the endogenous substance [^{13}C]urea, was demonstrated using DNP (dynamic nuclear polarization, cf. **chapter 11** of this book) as a polarization method (46).

Dynamic time series obtained during intra-arterial injection in pig models has demonstrated the possibilities for coronary angiography (1). **Figure 7** shows images extracted from a dynamic acquisition, similar to the one described in (1), during which retrospective gating was applied. The images in **Fig. 6** show flow artifacts due to the high velocity when the injected substance leaves the tip of the catheter. The time series were generated using the bSSFP pulse sequence implemented on a clinical MRI scanner capable of 40 mT m^{-1} gradient amplitude. With maximum gradient amplitude and slew rate it was not possible to reach short enough TE to suppress the artifacts caused by moving ^{13}C spins.

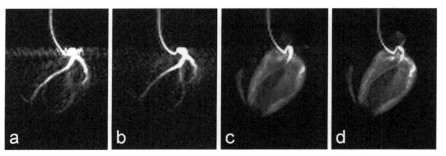

Fig. 6. Images obtained during an intra-arterial injection through a catheter placed in the left coronary arteries. Images (a) and (b) are from the early phase of the injection, while images (c) and (d) are obtained after the injection has ended and the hyperpolarized molecule has reached the myocardium.

2.2. Catheter Tracking

The possibility of visualizing a catheter in vivo, which contains flowing hyperpolarized ^{13}C signal agent, has been demonstrated by Magnusson et al. (47). In this pilot study, a commercial three-lumen catheter was modified in order for two of the lumens to form a loop where the hyperpolarized ^{13}C agent could flow without leaving the catheter. The third lumen was left open to enable injection of contrast agent into the endovascular space. Projection images of the catheter moving through the pig aorta to the renal arteries and to the aortic arch were acquired using a bSSFP pulse sequence. Two orthogonal projection images of the moving ^{13}C-filled catheter were acquired with a temporal resolution of 658 ms. In **Fig. 7**, two orthogonal ^{13}C images are shown. The catheter projections have been superimposed on proton road map images.

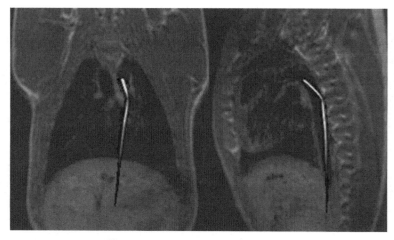

Fig. 7. Two orthogonal ^{13}C catheter-tracking images superimposed on proton road map images. The catheter was positioned inside the aorta.

Once the catheter is in the desired position, a bolus of hyperpolarized solution can be injected through the third lumen. In this way, the catheter-tracking technique may be combined with the previously described intra-arterial injection methods. It may also serve as a basis for obtaining a series of dynamic images to allow for perfusion mapping (1, 3).

2.3. Perfusion

Clinical proton MR imaging has been extended to allow for the extraction of perfusion information. Typically, this is accomplished by observing the signal change during the passage of a paramagnetic contrast media bolus. A series of images covering the volume of interest is collected. Retrospectively, software is used to calculate relevant perfusion parameters such as the tissue blood flow, F, the tissue blood volume, V_b, and the mean transit time, MTT (48, 49). A similar approach has been suggested for hyperpolarized imaging agents (50). However, when a series of images are collected during the passage of a bolus containing hyperpolarized spins, the signal will depend on the applied RF pulses, the inflow and outflow of spins, and the irreversible relaxation. The spin polarization decreases irreversibly due to T_1 relaxation. Therefore, the signal intensities as measured in a series of images will not directly be proportional to the concentration of the injected molecule. Consequently, it is not possible to correct for depolarization in the same manner in which decay is accounted for in imaging techniques based on radioactive tracers.

If the magnetization of the hyperpolarized compound is regarded as the tracer element, the concentration in a tissue voxel, $C_T(t)$, at a time, t, can, according to Johansson (51), be related via a convolution integral to the concentration in a supplying artery, $C_A(\tau)$, that has entered the volume during the infinitesimally short interval $d\tau$ at time τ:

$$C_T(t) = F \int_0^t C_A(\tau) R_M(t-\tau) d\tau \qquad [2]$$

where $R_M(t-\tau)$ is the residue function (52). It is defined as the fraction of tracer elements that entered the tissue volume at time $t = \tau$, and still remains in the tissue volume at time $t - \tau$. $C_A(t)$ is often referred to as the arterial input function (AIF). The residue function needs to include a polarization function P that takes into account both T_1 relaxation as well as polarization losses due to RF pulses. A detailed description of how to incorporate the effects of the depolarization during the image acquisitions and the influence of RF pulses in the case of perfusion assessment using a hyperpolarized tracer may be found in (51). If the concentration of the tracer in the tissue and in the arterial volume is known, the function $F \cdot R_M(t)$ may be determined by deconvolution and the tissue blood flow may be obtained. However, it should be noted that the depolarization rate, which is described by the evolution of P, needs to be known in order to determine CBV and MTT.

Hyperpolarized tracers may be used to study the tissue perfusion through the so-called washout or clearance (53, 54). A compact bolus of the molecule in question needs to be administrated directly into the organ. The administration may be performed using an injection through a catheter. During the outflow process, a series of images is then collected. If depolarization decays exponentially, the tissue concentration may be expressed according to (50):

$$C_T(t) = FC_0 e^{-\left(\frac{F}{\lambda'} + \frac{1}{T_D}\right)t} \qquad [3]$$

where C_0 is the concentration at $t = 0$, T_D is the depolarization time constant which needs to be known in order to determine the blood flow, and λ' is the equilibrium constant between the tissue concentration and that of the venous system.

The Kety–Schmidt method (55, 56) may be adopted for use with a hyperpolarized tracer. The advantage of this approach is that no assumptions need to be made about the shape of the AIF. Consequently, an i.v. injection may then be used. The tissue concentration may be calculated using the expression (50, 57–60):

$$C_T(t) = F\int_0^t C_A(\tau) e^{-\left(\frac{F}{\lambda'} + \frac{1}{T_D}\right)(t-\tau)} d\tau \qquad [4]$$

In this expression, the exponential factor $[F / \lambda' + 1/T_D]$ is constant and after sampling of C_T and C_A the parameter F, together with this constant, may be determined from Eq. [4] by a two-parameter fit. However, it should be noted that T_D needs to be known in order to estimate λ'. A series of images is generated during the passage of the hyperpolarized bolus and the perfusion map is the calculated on a pixel-to-pixel basis (3, 50).

Figure 8 shows an example of the results obtained in a cardiac pig model similar to the experiment described in (3). In this experiment, a bolus of 10 mL aqueous solution of 300 mM hydroxyethylpropionate with a polarization of 15–20% (PHIP) was injected at a rate of 1 mL/s. A bSSFP sequence was used to visualize the passage of the bolus through a slice placed as a two-chamber view. The imaging parameters were matrix size 64 × 64, slice thickness 10 mm, TR/TE 66.6 /33.3 ms resulting in a total scan time per image of 470 ms. In total, 31 images were collected with time interval of 2 s between the start of each image.

Johansson et al. (50, 61) have demonstrated a perfusion method that is unique for hyperpolarized tracers. In the bolus

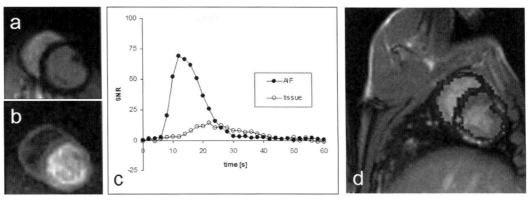

Fig. 8. Two images (**a** and **b**) from a series obtained during the passage of the hyperpolarized bolus for which a perfusion map may be calculated. The diagram (**c**) shows the AIF-signal measured in a pixel placed inside the left ventricle and the tissue signal measured in a pixel placed in the myocardium. In image (**d**) the perfusion map has been superimposed on a representative proton slice.

differentiation method, the tissue concentration is expressed according to:

$$C_T(t) = F \int_{t-\Delta t}^{t} C_A(\tau) R(t-\tau) d\tau \qquad [5]$$

This expression is almost identical to expression [**2**] but the idea is that if the time interval between two images, ΔT, is made short then P is essentially equal to R. Consequently, the residue function is no longer affected by organ-specific parameters. Instead, it is determined by the imaging sequence. For short ΔT, the given expression can be approximated by:

$$C_T(t) = F C_A(t - \Delta t/2) R(\Delta t/2) \Delta t \qquad [6]$$

This method has been applied to investigations of the blood flow in the renal cortex of rabbits (61). In **Fig. 9**, example images, obtained in a similar experimental setup using a rabbit model and evaluated using the bolus differentiation method, are shown. The flow map was calculated using a series of images generated using a bSSFP pulse sequence with centric phase encoding. The following sequence parameters were used: matrix size 64 × 64, slice thickness 10 mm, and TR/TE 8.9 /4.4 ms, resulting in a total scan time per image of 570 ms. It should be noted that this method (61) may be extended by using a single-shot sequence (e.g., EPI), that takes full advantage of all magnetization within the imaged slice.

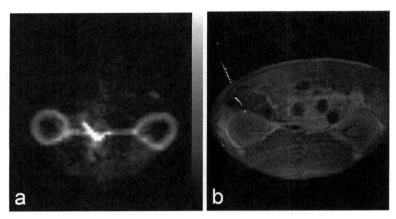

Fig. 9. Renal cortical blood flow map (**a**) and the corresponding proton image (**b**) obtained in a rabbit.

2.4. Summary

In this section, MR applications of hyperpolarized agents within the field of angiography, perfusion, and interventional applications, have been reviewed.

The main aspects to be aware of are the following:

- The character of hyperpolarized MR agents enables background-free angiographic and perfusion MR imaging.
- The catheter, during an interventional procedure, may be visualized using a flowing hyperpolarized MR agent.
- The MR signal is affected by the irreversible decay of the non-equilibrium polarization both due to locally varying T_1 and the influence of RF pulses during the acquisition of dynamic images. The models that are used to extract perfusion information need to be adapted accordingly.
- It is possible to quantify, e.g., cardiac or renal perfusion following intravenous injection of a hyperpolarized agent.

3. Metabolic Applications

3.1. Introduction

Most hyperpolarized MRI (metabolic MR) has been performed with [1-^{13}C]pyruvate (62–64). Several other agents have been investigated and will be referenced in this chapter. However, the biochemistry of pyruvate serves as a good introduction to the field of metabolic MR. Pyruvate is a substance that is at a central crossroad of energy metabolism (65), **Fig. 10**. Pyruvate is reduced to lactate by the coenzyme NADH, in the reaction catalyzed by the enzyme lactate dehydrogenase (LDH). The resulting NAD$^+$ is reestablished in the glycolytic pathway, which is an important

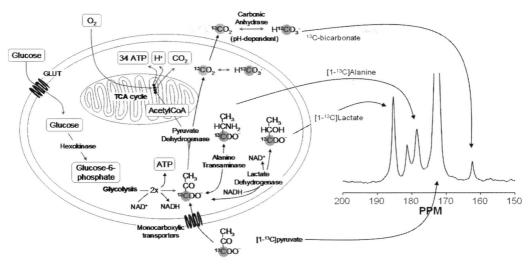

Fig. 10. Key metabolic pathways of pyruvate and illustration of isotopic labeling of metabolites. The labeled positions of the involved metabolites are highlighted by the gray circular shade. An in vivo 13C NMR spectrum of the metabolites is shown with molecular assignments. Chemical shifts (Ref. (99)) referenced to dioxane (external standard, 67.4 ppm) are CO_2 (124.5 ppm, not seen in spectrum), HCO_3 (160.9 ppm), pyruvate C-1 (170.6 ppm), lactate C-1 (183.2 ppm), alanine C-1 (176.5 ppm), and pyruvate hydrate C-1 (179.0 ppm).

source of ATP and also of biosynthetic intermediates. Oxidation of NADH by pyruvate allows the glycolytic pathway to produce ATP under anaerobic conditions, which is particularly important in tumors, which often have a poor and rapidly fluctuating blood supply and as a consequence are often hypoxic (66). Pyruvate can also be transaminated to form alanine in the reaction catalyzed by alanine aminotransferase (ALT). The source of the amino group for this reaction is glutamate. The reaction is particularly important in muscle, where, during fasting, amino groups resulting from protein degradation in the muscle are incorporated into alanine, which then passes via the circulation to the liver, where it is used to make glucose in the pathway of gluconeogenesis. The reactions catalyzed by LDH and ALT are thought to be near to equilibrium in most tissues and therefore these enzymes can catalyze rapid exchange of the ^{13}C label in [1-^{13}C]pyruvate with lactate and alanine respectively (67, 68). Pyruvate can also be oxidatively decarboxylated to produce acetyl CoA and $^{13}CO_2$ in the reaction catalyzed by pyruvate dehydrogenase (PDH). This irreversible reaction commits the carbon skeleton of pyruvate to oxidation in the tricarboxylic acid (TCA) cycle to form CO_2 and H_2O. The resulting reduced coenzymes, NADH and $FADH_2$, donate their electrons in the electron transport chain on the inner mitochondrial membrane to drive ATP synthesis. This is a fundamental energy process in all tissues, particularly those with a high energy demand, such as the heart. The NMR spectrum of these intermediates is shown in **Fig. 10** and the chemical shifts are given in the figure legend.

It has been shown that pyruvate (and lactate) can enter cells by three different routes (69):

- Transport on a specific H^+-monocarboxylate transporter (MCT)
- Through exchange for another anion
- Free diffusion of the undissociated acid

These processes are tissue specific and differ in their kinetic properties (70). Ninety percent of the pyruvate and lactate transport is facilitated by MCTs. Lactate and pyruvate transport across the plasma membrane of liver cells has been studied in the isolated perfused liver, the liver in situ, freshly isolated and cultured hepatocytes, and plasma membrane vesicles (70). The maximum rate of influx of lactate at 37°C was found to be 50 nmol/min per μL of intracellular volume. Studies have shown that pyruvate is transported more rapidly than lactate. The sarcolemmal pyruvate uptake in cardiomyocytes varies exponentially with the extracellular pyruvate concentration. Besides the H^+-linked carrier system, lactate appears to enter muscle cells quite rapidly by free diffusion of the undissociated acid at high concentrations (>10 mM). Brain capillary endothelial cells provide a barrier (the blood–brain barrier, BBB) for the exchange of solutes between the blood and cerebrospinal fluid. Lin et al. (71) found a wide distribution of several MCT isoforms in human tumor cell.

Lactate is not further metabolized in mammalian cells. Excess lactate is transported out of the cells and with the blood to the liver, where it is converted back to pyruvate and then into glucose by the gluconeogenic pathway. Thus, the liver restores glucose necessary for other organs to derive ATP from glycolysis. This constitutes the Cori cycle (65). Some cells, particularly cardiac, are highly permeable to lactate and pyruvate and since well oxygenated, can convert lactate back to pyruvate for further metabolism in the TCA cycle. Therefore, a contribution from inflow of $[1-^{13}C]$lactate signal from the blood cannot be excluded and may contribute to the lactate signal observed in a particular region.

The NMR signal involved in these metabolic processes can be modeled by first-order rate equations (72, 73). The pyruvate-to-lactate exchange is considered as an example.

$$\begin{aligned}\frac{dL_z}{dt} &= -R_L(L_z - L_\infty) + k_P P_z - k_L L_z \\ \frac{dP_z}{dt} &= -R_P(P_z - P_\infty) + k_L L_z - k_P P_z\end{aligned} \quad [7]$$

where L_z and P_z are the lactate and pyruvate magnetizations respectively. L_∞ and P_∞ can usually be taken as zero since the thermal equilibrium polarization is lower by many orders of magnitude over the initial pyruvate polarization. $R_{L,P}$ are the effective relaxation rates for the two molecules. The effective relaxation rate is the sum of the longitudinal relaxation rate, $T_{L,P}^{-1}$, and the depolarization due to the radio-frequency pulses

$$R_{L,P} = \frac{1}{T_{L,P}^1} - \frac{\ln\cos\alpha}{\Delta t} \qquad [8]$$

where α is the flip angle and Δt is the time between pulses. For example, a flip angle of 10° every second would contribute 0.015 s^{-1} (65 s) to the observed relaxation rate. If a longitudinal relaxation time, $T_{L,P}^{-1}$, of 40 s is assumed, the effective relaxation time would become 25 s. In (64) it is shown by simulation that the rate constants obtained from Eq. [7] are only moderately biased (correlated) by error in $R_{L,P}$. Therefore, it can be advantageous to assume values for $R_{L,P}$ when fitting Eq. [7] to experimental data.

Secondly, Eq. [7] has to consider the delivery of substrate (perfusion) to the tissue, since solving the rate equations with the initial condition $P_z(0) = P$ implies immediate delivery of substrate to the tissue and no washout (except for an exponential decay that can be lumped with $R_{L,P}$). A general and commonly used input function is the gamma-variate function

$$f(t) = \begin{cases} 0 & t \le t_0 \\ \dfrac{\beta^\alpha}{\Gamma(\alpha)}(t-t_0)^{\alpha-1} e^{-\beta(t-t_0)} & t > t_0 \end{cases} \qquad [9]$$

The gamma-variate function (normalized as written) is accepted to describe vascular dispersion of the bolus injection, but does not relate parameters to measures of tissue perfusion. Little work has been published to date on the appropriateness of standard perfusion models to hyperpolarization data. In the authors experience, the rate constants do not seem to be significantly biased by variations in tissue perfusion model.

The gamma-variate function can easily be incorporated into Eq. [7]

$$\frac{dL_z}{dt} = -R_L(L_z - L_\infty) + k_P P_z - k_L L_z$$
$$\frac{dP_z}{dt} = -R_P(P_z - P_\infty) + k_L L_z - k_P P_z + F(t-t_0)^{\alpha-1} e^{-\beta(t-t_0)}$$
$$\left(\tfrac{1}{t} - \beta\right) \qquad [10]$$

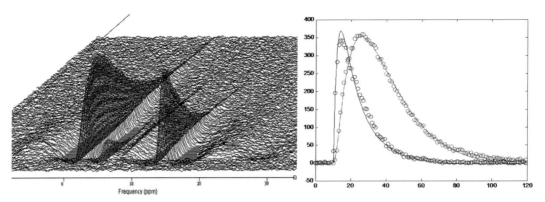

Fig. 11. Example of localized spectra from murine transgenic mammary tumor with 1 s temporal resolution (0–120 s) and 10° flip angle. The spectra are displayed and analyzed in jMRUI. The signal amplitudes of pyruvate and lactate were determined with AMARES quantification and prior knowledge of chemical shift difference (12.2 ppm), Gaussian line shape, identical phase, and time delay (first-order phase) of 1.73 ms. The amplitudes determined from the AMARES fitting are shown with a fit to Eq. [7]. The obtained fit parameters are $k_L = 0.022$ s^{-1}, $k_P = 0.136$ s^{-1}, scale = 204, $\alpha = 1.32$, $\beta = 0.039$ s^{-1}, $T_0 = 10.4$ s.

where F is a scale factor for the pyruvate bolus signal, and has been lumped with the β/Γ factor in Eq. [9]. Equation [10] can now be solved numerically and fitted to experimental data. The obtained rate constants will effectively be measures of cellular uptake by MCT and the enzyme exchange rate. Other ways of modeling the perfusion kinetics have been suggested (74, 75).

The NMR signals can be quantified in several ways depending on whether spectra or images are aquired. In the case of spectroscopic data with high temporal resolution, each spectrum can analyzed either by fitting the resonances to a model function (jMRUI (76), LCmodel (77)), simple integration, or just the signal amplitude. **Figure 11** shows an example of dynamic NMR acquisitions of mainly a murine transgenic mammary tumor (PymT). The signal is received with a surface coil to localize the signal to the tumor area and a volume coil is used for excitation. A flip angle of 10° is used with a temporal resolution of 1 s. As shown above the radio-frequency pulses significantly affect the signal decay, but do not bias the rate constants significantly. The NMR spectra in **Fig. 11** are displayed and quantified in jMRUI (AMARES) with prior knowledge of chemical shift difference (12.2 ppm), Gaussian line shape, fixed relative phase, and time delay of 1.73 ms (causing the phase oscillations in the spectra). The amplitudes obtained from the spectral fitting are then fitted to Eq. [7] to provide measures of exchange rates between pyruvate and lactate ($k_L = 0.022$ s^{-1}, $k_P = 0.136$ s^{-1}) and input function parameters ($\alpha = 1.32$, $\beta = 0.039$ s^{-1}).

3.2. Oncology

Cancer could be considered a metabolic disease as changes in metabolism clearly contribute to malignant progression (78).

Tumor cells have long been known to have high rates of aerobic glycolysis, the so-called "Warburg effect" (66), and our understanding of the biochemical basis for this increased flux has grown rapidly in the last few years. The most widely used technique clinically for imaging tumor metabolism and the response of tumors to treatment has been positron emission tomography (PET). The elevated aerobic glycolysis in tumors has been imaged indirectly using ^{18}F-fluorodeoxyglucose (FDG). FDG is a glucose analogue that is taken up by cells by the plasma membrane glucose transporter (GLUT), and then phosphorylated to fluorodeoxyglucose 6-phosphate by hexokinase, the first enzyme of the glycolytic pathway (**Fig. 10**). Since FDG is not significantly further metabolized the positron-emitting isotope is trapped in the cell, where it can be imaged using PET at a resolution of 3–5 mm. Magnetic resonance methods for detecting altered tumor metabolism include ^{1}H and ^{31}P MRS studies. MRS measurements of decreases in the levels of choline metabolites following treatment have been shown to be predictive of response in brain, prostate, and breast cancer in the clinic (79, 80). ^{1}H MRS measurements in animal models have demonstrated elevated tumor lactate levels and spectroscopic imaging experiments have demonstrated a decrease in tumor lactate concentration following drug treatment, consistent with a decrease in tumor glycolytic rate, increased lactate washout, or a decrease in tumor cellularity (81).

Intravenous injection of hyperpolarized [1-^{13}C]pyruvate has been shown, in experiments in mice and rats, to result in rapid labeling of tumor lactate (62–64). Experiments on isolated tumor cells have demonstrated that this labeling is due primarily to exchange of label between pyruvate and lactate in a near-equilibrium reaction catalyzed by LDH, rather than net chemical conversion of pyruvate to lactate (73). This is likely the case in vivo as well, since the reaction is thought to be near to equilibrium in vivo (82). The mechanism of the enzyme (ordered ternary complex, with binding of the coenzymes before lactate and pyruvate) means that label can exchange between lactate and pyruvate in the absence of dissociation of the coenzymes NAD$^+$ and NADH (83). These animal studies demonstrated that the majority of the lactate was labeled within the tumor, rather than being labeled in other tissues and then washed into the tumor via the circulation. In effect, the hyperpolarized [1-^{13}C]pyruvate labels by exchange the pools of tissue lactate, which are often associated with tumors and other pathologies that result in tissue hypoxia or ischemia. Passive exchange of the ^{13}C label into pre-existing metabolite pools probably also explains why hyperpolarized [1-^{13}C]pyruvate is effective at labeling alanine in tissues where it is abundant, such as muscle.

Experiments in mice have also demonstrated that the polarized pyruvate experiment can be used for treatment monitoring. In a murine lymphoma model, treatment with a cytotoxic drug (etoposide) resulted in a decrease in the flux of label from pyruvate to lactate in the tumor (64). This was shown to be due to a number of factors, including a decrease in tumor LDH and lactate concentrations, which probably resulted from a decrease in tumor cellularity. Experiments on isolated tumor cells also showed that there was a decrease in the NAD(H) pool as a result of poly-ADP ribose polymerase (PARP) activation. PARP, which is activated by DNA damage and ADP ribosylates nuclear proteins, uses as its substrate NAD$^+$, substantially depleting the dying cell of the coenzyme (84). In these experiments, the decrease in pyruvate–lactate label flux was demonstrated at 24 h after drug treatment, when there was already a ∼20% decrease in tumor volume. It will also be important to evaluate this method for detecting treatment response in comparison with other methods that are already being used clinically, such as FDG PET (85) and diffusion-weighted MRI (86).

Hyperpolarized pyruvate has also been used as a marker of tumor progression, with the levels of hyperpolarized lactate increasing with increasing tumor grade in the TRAMP model (87). Assessing tumor grade is normally limited by random biopsy sampling error, but this method could allow non-invasive determination of tumor grade and disease progression.

Another hyperpolarized substrate that could potentially be used for treatment monitoring is [5-^{13}C]glutamine. Glutamine utilization is closely correlated with tumor cell proliferation (88). The nitrogen in the molecule is used for biosynthetic purposes and the carbon skeleton is oxidized in the TCA cycle, where it makes a substantial contribution to cellular energy generation (89). Therefore, measurements of tumor [5-^{13}C]glutamine utilization might also be used to detect the effects of cytostatic drugs. The T_1 of the C-5 carbon is ∼16 s, shorter than both C-1 labeled glutamine and [1-^{13}C]pyruvate. However, the change in chemical shift when glutamine is converted to glutamate is sufficiently large (3.4 ppm) to make detection possible in vivo, which is not the case for C-1 labeling. In (90) it was demonstrated that in a human hepatoma cell line, HepG2, the cell uptake of glutamine and its net conversion to glutamate, in the irreversible reaction catalyzed by the intramitochondrial enzyme glutaminase, is sufficiently rapid to allow its detection with hyperpolarized [5-^{13}C]glutamine. Glutamine uptake by hepatomas is ∼30 times higher than in surrounding normal liver (88). Metabolic MR following i.v. injection of [5-^{13}C]glutamine might therefore be useful clinically for detection of hepatomas, but further preclinical evidence is required.

A recent study demonstrated that fructose (91) uptake and phosphorylation can be studied with metabolic MR. Hyperpolarized [2-^{13}C]fructose was injected into a transgenic model of prostate cancer (TRAMP) and demonstrated difference in uptake and metabolism in regions of tumor relative to surrounding tissue. The carbons of glucose have all short T_1, even when deuterated. Fructose is an isomeric mixture of five- and six-membered rings and has as its most stable isomer β-fructopyranose with a hemiketal in the C-2 position. The T_1 of this carbon was measured to be 13–15 s at 3 T. Fructose can enter glycolysis via hexokinase or fructokinase. Hexokinase phosphorylates fructose to fructose-6-phosphate, analogous to the first step of glycolysis, in which glucose is phosphorylated to glucose-6-phosphate (65). The metabolic flux to fructose-6-phosphate in the cell is related to the downstream glycolytic metabolic events as well as activity of the pentose phosphate pathway (65). Though the phosphorylated fructose chemical shift could not be separated from the β-fructofuranose isomer, the study in (91) demonstrated that [2-^{13}C]fructose metabolism was increased in tumor regions compared to the surrounding benign tissues. Further studies are required to fully understand the fructose metabolite signals and how they relate to malignancy.

Branched chain amino acid transferase (BCAT) activity was studied in (92), where hyperpolarized [1-^{13}C]ketoisocaproate was hyperpolarized and injected into tumor-bearing mice and rats. BCAT catalyzes the reaction ketoisocaproate to leucine and glutamate to α-ketoglutarate. BCAT is over-expressed in some cancers and linked to the c-myc and kras-2 oncogenes. The two tumor models in the study were both shown to have high [1-^{13}C]pyruvate to lactate conversion, but only one (murine lymphoma model) showed high [1-^{13}C]ketoisocaproate to leucine conversion. The conversion to leucine correlated with BCAT activity, which was biochemically assayed in tissue biopsies in the two tumor models. In the lymphoma model, the leucine signal was imageable with the CSI pulse sequence and showed similar signal-to-noise-ratio as the lactate image following pyruvate injection.

3.3. Cardiology

All cardiac metabolic MR to date has been performed with ^{13}C-labeled pyruvate as the substrate. Its metabolic fate in heart muscle has been described extensively, both under normal conditions (93) and in disease (94–96). Dynamic spectroscopy with surface coils has allowed the study of global or local changes to heart metabolism (97, 98) under various conditions and chemical shift imaging has allowed the imaging of changes to myocardial metabolism after local ischemia (22).

Hyperpolarized [1-^{13}C]pyruvate and its metabolites [1-^{13}C]lactate, [1-^{13}C]alanine, and ^{13}CO$_2$ have been measured in the heart both in vivo and in the isolated perfused rat heart (22, 99–101). $^{13}CO_2$ was visualized through the bicarbonate ($H^{13}CO_3^-$), which is in rapid exchange with $^{13}CO_2$ in the equilibrium catalyzed by carbonic anhydrase. The $^{13}CO_2$ produced arises exclusively from the first step of the TCA cycle catalyzed by PDH rather than from reactions in the TCA cycle when the pyruvate is C-1 labeled, as explained in **Section 3.1**.

The production of hyperpolarized $^{13}CO_2$ could be modulated by the co-administration of the medium-chain fatty acid octanoate, which is known to inhibit PDH (99). Octanoate suppressed $^{13}CO_2$ production and enhanced the lactate signal following the addition of hyperpolarized [1-^{13}C]pyruvate to the isolated perfused heart without changing TCA cycle flux, as determined by isotopomer analysis; PDH is therefore responsible for the $^{13}CO_2$ produced. The PDH enzyme complex determines the relative contribution of glucose and fatty acids to ATP production in the heart. It is inactivated by phosphorylation, catalyzed by the enzyme PDH kinase (PDK); PDK expression is induced by starvation, diabetes, and high levels of plasma lipids (102). Separate studies have shown that fasted animals also demonstrate a reduction in hyperpolarized $^{13}CO_2$ production compared to fed controls. A similar reduction was seen following the induction of type 1 diabetes and this decrease correlated with disease severity (101).

Following global ischemia in the isolated rat heart, subsequent injection of hyperpolarized [1-^{13}C]pyruvate resulted in the production of both [1-^{13}C]lactate and [1-^{13}C]alanine but not $H^{13}CO_3^-$ or $^{13}CO_2$ (103). The latter two metabolites appear following reperfusion suggesting that PDH flux is inhibited by ischemia initially and eventually recovers within 20 min. A study in pig has shown the feasibility of applying this technique in large animals (22). If the pyruvate is injected 2 h after a 15 min occlusion of the left circumflex artery (**Fig. 12**), CO_2 production is reduced in the region of ischemia. This effect seems to be more sensitive to ischemia than any of the other well-established parameters that characterize cardiac function (myocardial morphology, cardiac motion, perfusion, delayed enhancement). Following 15 min of coronary artery occlusion and 2 h of reperfusion, the cardiac muscle was "stunned": there was no change in the levels of cardiac enzymes in the plasma and gadolinium contrast medium perfusion was normal, with no delayed enhancement. However, there was a reduction in hyperpolarized ^{13}C-bicarbonate signal following injection of hyperpolarized [1-^{13}C]pyruvate. In contrast, following 45 min of coronary artery occlusion (**Fig. 13**), there were regional perfusion abnormalities demonstrated in several of the animals and delayed gadolinium enhancement in all of the animals as well as elevated levels of cardiac enzymes in the plasma. In these cases, there was complete loss of ^{13}C-bicarbonate

Applications of Hyperpolarized Agents in Solutions   681

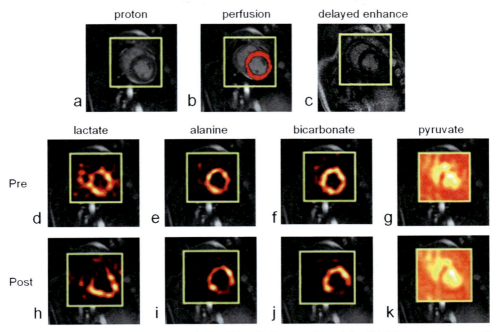

Fig. 12. The proton image (**a**), the semi-quantitative Gd-perfusion map (**b**) and the Gd delayed enhancement image (**c**) together with the lactate (**d**, **h**), the alanine (**e**, **i**), bicarbonate (**f**, **j**), and pyruvate (**g**, **k**) maps obtained in an animal after 15 min occlusion. The Gd-perfusion map, the lactate, the alanine map, the bicarbonate maps, and the pyruvate maps have been superimposed on the proton image.

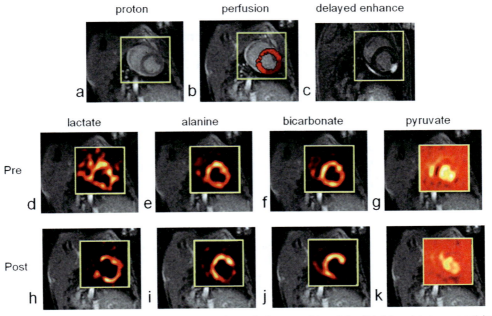

Fig. 13. The proton image (**a**), the semi-quantitative Gd-perfusion map (**b**) and the Gd delayed enhancement image (**c**) together with the lactate (**d**, **h**), the alanine (**e**, **i**), bicarbonate (**f**, **j**), and pyruvate (**g**, **k**) maps obtained in an animal after 45 min occlusion. The Gd-perfusion map, the lactate, the alanine map the bicarbonate maps, and the pyruvate maps have been superimposed on the proton image.

formation. These studies suggest that PDH activity can be measured indirectly in cardiac tissue of large animals and that this is a sensitive test for detection of cardiac ischemia.

Recently it has been demonstrated that C-2 labeling of pyruvate allows studying further intermediates of the TCA cycle (104). When the labeling position is shifted from C-1 to C-2, the label is no longer lost in the first step of the TCA cycle. The enzymatic conversion of pyruvate to lactate, acetylcarnitine, citrate, and glutamate was measured with 1 s temporal resolution. The appearance of ^{13}C-labeled glutamate was delayed compared with that of other metabolites, indicating that TCA cycle flux can be measured directly. The production of ^{13}C-labeled citrate and glutamate was decreased post-ischemia, as opposed to lactate, which was significantly elevated. These results showed that the control and fluxes of the Krebs cycle in heart disease can be studied using hyperpolarized [2-^{13}C]pyruvate.

3.4. Neurology

Imaging of the brain has been suggested as a potential application area for metabolic MR. Probably the aspiration originates from the fact that two-third of all MRI is in the neuro-area and that MRS has had some success in studying neurological disorders (105, 106) and brain tumors. To date, few studies have been published on brain metabolism with hyperpolarized agents. One reason is that the blood–brain barrier allows only a slow uptake of most agents. [1-^{13}C]Succinate (107) was polarized and injected into normal and 9 L tumor-bearing rats without any observation of other metabolites appearing on the timescale of T_1. Ex vivo brain biopsies did reveal conversion into glutamate and glutamine in the brain tumor tissue. [2-^{13}C]fructose (*see* **Section 3.3**) may be a potential candidate for brain metabolism studies as a fast uptake across the blood–brain barrier would be expected. A study with [1-^{13}C]pyruvate and the more lipophilic analogue ethyl-[1-^{13}C]pyruvate (108) has demonstrated unambiguously the uptake of both compounds in the normal rat brain and metabolic products. This opens up for the study of various neurological conditions with metabolic MR.

3.5. General Medicine

Alterations in liver metabolism were studied in (109). Pyruvate-to-lactate metabolism was studied in a rat liver model of ethanol metabolism. When the liver metabolizes ethanol, NAD$^+$ is oxidized to NADH thereby changing the ratio of LDH·NADH to LDH·NAD$^+$. Post the ethanol treatment an elevated lactate signal was observed relative to controls. A protocol was established where the animal was given hyperpolarized [1-^{13}C]pyruvate at baseline, then either saline or ethanol, and then, after a 45 min delay, a second [1-^{13}C]pyruvate injection. Spectra localized to a 15 mm axial slice across the liver with 5° flip angle and 3 s temporal resolution were acquired for data fitting to equations similar to Eq. [7]. A twofold increase in lactate signal was observed post-ethanol treatment. The study demonstrates that [1-^{13}C]pyruvate is a biomarker of changes in the NADH/NAD$^+$ ratio.

3.6. pH Mapping

It can be envisaged that hyperpolarized MR can be used to interrogate physiological parameters such as pH, pO$_2$, or redox potential. So far only pH has been demonstrated (110), and also for this type of imaging, spectral encoding is a likely requirement. pH can be estimated by measuring the relative concentration of bicarbonate and CO$_2$ in the tissue, since the equilibrium of this buffer is dependent on pH in the physiological range. Acid–base balance is tightly controlled in mammalian tissue, where the main extracellular pH buffer is bicarbonate (HCO$_3^-$). The reaction is catalyzed by carbonic anhydrase, which is essential in establishing the equilibrium between bicarbonate and carbon dioxide. The hyperpolarized ^{13}C label is rapidly exchanged between the two molecules even in the absence of the enzyme, but in the presence of carbonic anhydrase, equilibrium is achieved almost instantaneously. Many pathological states are associated with changes in tissue acid–base balance, including inflammation and ischemia (111, 112). For instance, most tumors have an acidic extracellular pH compared to normal tissue and this can be correlated with prognosis and response to treatment (113, 114). Despite the importance of pH and its relationship to disease, there is currently no clinical tool available to image the spatial distribution of pH in humans. A number of methods have demonstrated pH mapping in animals using MRS (115) and positron emission tomography (PET) (86). In principle, tissue pH could be determined from ^{13}C MRS measurements of endogenous H^{13}CO$_3^-$ and ^{13}CO$_2$, using the Henderson–Hasselbalch equation, if there was sufficient signal to noise:

$$\mathrm{pH} = \mathrm{p}K_\mathrm{a} + \log_{10}([\mathrm{H}^{13}\mathrm{CO}_3^-]/[^{13}\mathrm{CO}_2]) \qquad [11]$$

^{13}C-labeled bicarbonate has been hyperpolarized (110) and has been used to measure the extracellular pH. ^{13}C-labeled cesium bicarbonate was polarized to 16%; the cesium counter ion was chosen because of its high solubility and was removed using an ion-exchange column prior to injection. The ion-exchange does not affect polarization significantly and is similar to the step of removing the radical involved in the DNP process (*see* **Section 1** of this book). pH maps were generated of the tumor in a mouse EL-4 model. This method for measuring pH was compared with a well-established ^{31}P MRS pH probe (3-aminopropylphosphonate; 3-APP (115)), which measures extracellular pH. The comparison demonstrated that the tumor pH measured using hyperpolarized bicarbonate was predominately extracellular. pH images were obtained by calculating the relative concentrations of $H^{13}CO_3^-$ and $^{13}CO_2$ in each imaging voxel. This demonstrated that the extracellular pH in a lymphoma tumor was more acidic than the surrounding tissue. The method can be extended to imaging carbonic anhydrase activity: following $^{13}CO_2$ saturation, the relative drop in the $H^{13}CO_3^-$ signal on chemical shift imaging was used to image differing concentrations of the enzyme in a phantom (110). This was also shown in vivo by selectively saturating the hyperpolarized $^{13}CO_2$ signal following injection of hyperpolarized $H^{13}CO_3^-$, which resulted in a decrease in the hyperpolarized $H^{13}CO_3^-$ signal. The effect could be inhibited by administration of a carbonic anhydrase inhibitor, acetazolamide. The rate of decrease of $H^{13}CO_3^-$ was used to calculate the exchange flux; the flux catalyzed by the enzyme in the mouse lymphoma model was shown to be at least eightfold faster than in the non-enzyme catalyzed reaction.

3.7. Summary

In this section, metabolic MR applications of hyperpolarized agents have been reviewed.

- The central role of pyruvate in cellular energy metabolism makes it an ideal candidate for a variety of metabolic MR investigations. In all disease areas, applications of this compound have been demonstrated showing that the available time window for hyperpolarization is adequate to study alterations in cellular metabolism.

- Several other molecules have been investigated with promising results. The field of metabolic MR is in its infancy and will at no doubt expand further in the future.

- The method has the potential for clinical translation and [1-^{13}C]pyruvate is currently in clinical development.

References

1. Månsson, S., Johansson, E., Magnusson, P., Chai, C.M., Hansson, G., Petersson, J.S., Ståhlberg, F., and Golman, K. (2006) (13)C imaging – a new diagnostic platform. *Eur. Radiol.* **16**, 57–67.
2. Olsson, L.E., Chai, C.-M., Axelsson, O., Karlsson, M., Golman, K., and Petersson, J.S. (2006) MR coronary angiography in pigs with intra arterial injection of a hyperpolarized (13)C substance. *Magn. Reson. Med.* **55**, 731–737.
3. Golman, K. and Petersson, J.S. (2006) Metabolic imaging and other applications of hyperpolarized (13)C. *Acad. Radiol.* **13**, 932–942.
4. Ishii, M., Emami, K., Kadlecek, S., Petersson, J.S., Golman, K, et al. (2007) Hyperpolarized (13)C MRI of the pulmonary vasculature and parenchyma. *Magn. Reson. Med.* **57**, 459–463.
5. Golman, K., Olsson, L.E., Axelsson, O., Månsson, S., Karlsson, M., and Petersson, J.S. (2003) Molecular imaging using hyperpolarized (13)C. *Br. J. Radiol.* **76**, 118–127.
6. Wild, J.M., The, K., Woodhouse, N., Paley, M.N., de Zanche, N., and Kasuboski, L. (2006) Steady-state free precession with hyperpolarized ^{3}He: experiments and theory. *J. Magn. Reson.* **183**, 13–24.
7. Norris D.G. and Hutchison J.M.S. (1990) Concomitant magnetic field gradients and their effects on imaging at low magnetic field strengths. *Magn. Reson. Imaging* **8**, 33–37.

8. Comment, A., van den Brandt, B., Uffmann, K., et al. (2007) Design and performance of a DNP prepolarizer coupled to a rodent MRI scanner. *Conc. Magn. Reson.* **31**, 255–269.
9. Diehl, K.H., Hull, R., Morton, D., Pfister, R., Rabemampianina, Y., Smith, D., Vidal, J.M., and van de Vorstenbosch, C. (2001) European federation of pharmaceutical industries association and European centre for the validation of alternative methods. A good practice guide to the administration of substances and removal of blood, including routes and volumes. *J Appl Toxicol.* **21**, 15–23.
10. Davies, B. and Morris, T. (1993) Physiological parameters in laboratory animals and humans. *Pharm Res.* **10**, 1093–1095.
11. Oppelt, A., Graumann, R., Barfuss, H., et al. (1986) FISP—a new fast MRI sequence. *Electromedica.* **54**, 15–18.
12. Scheffler, K., Heid, O., and Hennig, J. (2001) Magnetization preparation during the steady state: fat-saturated 3D TrueFISP. *Magn. Reson. Med.* **45**, 1075–1080.
13. Scheffler, K. (2003) On the transient phase of balanced SSFP sequences. *Magn. Reson. Med.* **49**, 781–783.
14. Petersson, J.S. and Christoffersson, J.-O. (1997) A multi-dimensional partition analysis of SSFP image pulse sequences. *Magn. Reson. Imaging* **15**, 451–467.
15. Gyngell, M.L. (1989) The steady-state signals in short-repetition-time sequences. *J. Magn. Reson.* **81**, 474–483.
16. Patz, S. (1989) Steady-state free precession: an overview of basic concepts and applications. *Adv Magn Reson Imaging* **1**, 73–102.
17. Hennig, J. (1991) Echoes—how to generate, recognize, use or avoid them in MR-imaging sequences. Part II: echoes in imaging sequences. *Concepts Magn. Reson.* **3**, 179–192.
18. Deimling, M. and Heid, O. (1994) Magnetization prepared true FISP imaging. *Proc. Intl. Soc. Mag. Reson. Med.* 495.
19. Svensson, J., Månsson, S., Johansson, E., Petersson, J.S., and Olsson L.E. (2003) Hyperpolarized (13)C MR angiography using TrueFISP. *Magn. Reson. Med.* **50**, 256–262.
20. Hennig, J., Nauerth A., and Friedburg, H. (1986) RARE imaging: a fast imaging method for clinical MR. *Magn. Reson. Med.* **3**, 823–833.
21. Gallagher, F.A., Kettunen, M.I., Day, S.E., Hu, D. E., Ardenkjær-Larsen, J.H., et al. (2008) Magnetic resonance imaging of pH in vivo using hyperpolarized (13) C-labelled bicarbonate. *Nature* **453**, 940–944.
22. Golman, K., Petersson, J.S., Magnusson, P., Johansson, E., et al. (2008) Cardiac metabolism measured noninvasively by hyperpolarized (13)C MRI. *Magn. Reson. Med.* **59**, 1005–1013.
23. Yen, Y.-F., Kohler, S.J., Chen, A.P., Tropp, J., et al. (2009) Imaging considerations for in vivo (13)C metabolic mapping using hyperpolarized (13)C-pyruvate. *Magn. Reson. Med.* **62**, 1–10.
24. Ebel, A., Maudsley A.A., and Schuff, N. (2007) Correction of local B_0 shifts in 3D EPSI of the human brain at 4 T. *Magn. Reson.Imaging* **25**, 377–380.
25. Cunningham, C.H., Vigneron, D.B., Chen, A.P., et al. (2005) Design of flyback echo-planar readout gradients for magnetic resonance spectroscopic imaging. *Magn. Reson. Med.* **54**, 1286–1289.
26. Chen, A.P., Alberts, M.J., Chunningham, C.H., et al. (2007) Hyperpolarized C-13 spectroscopic imaging of the TRAMP mouse at 3T—initial experience. *Magn. Reson. Med.* **58**, 1099–1106.
27. Thévenaz, P., Blu, T., and Unser, M. (2000) Interpolaton revisited. *IEEE Trans. Med. Imaging* **19**, 739–758.
28. Yaroslavsky, L.P. (2001) Signal sinc-interpolation: a fast computer algorithm. *Bioimaging* **4**, 225–231.
29. Bracewell, R.N. (1986) *The Fourier Transform and Its Applications.* McGraw-Hill: New York.
30. Chen, P.A., Leung, K., Lam, W., Hurd, R.E., Vigneron, D.B., and Cunningham, C.H. (2009) Design of spectral-spatial outer volume suppression RF pulses for tissue specific metabolic characterization with hyperpolarized ^{13}C pyruvate. *J. Magn. Reson.* **200**, 344–348.
31. Larson, P.E.Z., Kerr, A.B., Chen, A.P., Lustig, M.S., Zierhut, M.L., Hu, S., Cunningham, C.H., Pauly, J.M., Kurhanewicz, J., and Vigneron, D.B. (2008) Multiband excitation pulses for hyperpolarized ^{13}C dynamic chemical-shift imaging. *J. Magn. Reson.* **194**, 121–127.
32. Chen, A.P., Tropp, J., Hurd, R.E., Van Criekinge, M., Carvajal, L.G., Xu, D., Kurhanewics, J., and Vigneron, D.B. (2009) In vivo hyperpolarized ^{13}C MR spectroscopic imaging with ^{1}H decoupling. *J. Magn. Reson.* **197**, 100–106.
33. Ljunggren, S. (1983) A simple graphical representation of Fourier-based imaging methods. *J. Magn. Reson.* **54**, 338–343.

34. Ahn, C.B., Kim, J.H., and Cho, Z.H. (1986) High-speed spiral-scan echo planar NMR imaging. *IEEE Trans. Med. Imaging* **5**, 2–7.
35. Levin, Y.S., Mayer, D., Yen, Y.-F., Hurd, R.E., and Spielman, D.M. (2007) Optimization of fast spiral chemical shift imaging using least squares reconstruction: application for hyperpolarized ^{13}C metabolic imaging. *Magn. Reson. Med.* **58**, 245–252.
36. Mayer, D., Yen, Y.-F., Tropp, J., Pfefferbaum, A., Hurd, R.E., and Spielman, D.M. (2009) Application of subsecond spiral chemical shift imaging to real-time multislice metabolic imaging of the rat in vivo after injection of hyperpolarized ^{13}C1-pyruvate. *Magn. Reson. Med.* **62**, 557–564.
37. Cunningham C.H., Chen, A.P., Lustig, M., Hargreaves, B.A., Lupo, J., Xu, D., Kurhanewics, J., Hurd, R.E., Pauly, J.M., Nelson, S.J., and Vigneron, D.B. (2008) Pulse sequence for dynamic volumetric imaging of hyperpolarized metabolic products. *J. Magn. Reson.* **193**, 139–146.
38. Leupold, J., Wieben, O., Månsson, S., Speck, O., Scheffler, K., Petersson, J.S., and Hennig J. (2006) Fast chemical shift mapping with multiecho balanced SSFP. *MAGMA* **19**, 267–273.
39. Vinitski, S., Mitchell, D.G., Szumowski, J., Burk, D.L., and Rifkin, M.D. (1990) Variable flip angle imaging and fat suppression in combined gradient and spin-echo (GREASE) techniques. *Magn. Reson. Imaging* **8**, 131–139.
40. Dixon, W.T. (1984) Simple proton spectroscopic imaging. *Radiology* **153**, 189–194.
41. Glover, G.H. (1991) Multipoint Dixon technique for water and fat proton and susceptibility imaging. *J. Magn. Reson. Imaging* **1**, 521–530.
42. Reeder, S.B., Wen, Z., Yu, H., Angel, A.R., et al. (2004) Multicoil Dixon chemical species separation with an iterative least-squares estimation method. *Magn. Reson. Med.* **51**, 35–45.
43. Leupold, J., Månsson, S., Petersson, J.S., Hennig, J., and Wieben, O. (2009) Fast multiecho balanced SSFP metabolite mapping (1)H and hyperpolarized (13)C compounds. *MAGMA* **22**, 251–256.
44. Golman, K., Axelsson, O., Jóhannesson, H., Månsson, S., Olofsson, C., and Petersson, J.S. (2001) Parahydrogen-induced polarization in imaging: subsecond (13)C angiography. *Magn. Reson. Med.* **46**, 1–5.
45. Golman, K., Ardenkjær-Larsen, J.H., Svensson, J., et al. (2002) (13)C-angiography. *Acad. Radiol.* **2**, 507–510.
46. Golman, K., Ardenkjær-Larsen, J.H., Petersson, J. S., Månsson, S., and Leunbach, I. (2003) Molecular imaging with endogenous substances. *Proc. Natl. Acad. Sci. U.S.A.* **100**, 10435–10439.
47. Magnusson, P., Johansson, E., Månsson, S., Petersson, J.S., et al. (2007) Passive catheter tracking during interventional MRI using hyperpolarized (13)C. *Magn. Reson. Med.* **57**, 1140–1147.
48. Østergaard, L., Weisskoff, R.M., Chesler, D.A., Gyldensted, C., and Rosen, B.R. (1996) High resolution measurement of cerebral blood flow using intravascular tracer bolus passages. Part I: Mathematical approach and statistical analysis. *Magn. Reson. Med.* **36**, 715–725.
49. Østergaard, L., Sorensen, A.G., Kwong, K.K., Weisskoff, R.M., Gyldensted, C., and Rosen, B.R. (1996) High resolution measurement of cerebral blood flow using intravascular tracer bolus passages. Part II: Experimental comparison and preliminary results. *Magn. Reson. Med.* **36**, 726–736.
50. Johansson, E. (2003) NMR imaging of flow and perfusion using hyperpolarized nuclei. PhD thesis, Lund University, Lund.
51. Johansson, E., Månsson, S., Wirenstam, R., Svensson, J., Petersson, J.S., Golman, K., and Ståhlberg, F. (2004) Cerebral perfusion assessment by bolus tracking using hyperpolarized (13)C. *Magn. Reson. Med.* **51**, 464–472.
52. Meier, P. and Zierler, K. (1954) On the theory of the indicator-dilution method for assessment of blood flow and volume. *J. Appl. Physiol.* **6**, 731–744.
53. Goodson, B.M., Song, Y.Q., Taylor, R.E., et al. (1997) In vivo NMR and MRI using injection delivery of laser-polarized xenon. *Proc. Natl. Acad. Sci. U.S.A.* **94**, 14725–14729.
54. Duhamel, G., Choquet, P., Grillon, E., et al. (2002) Global and regional cerebral blood flow measurements using NMR of injected hyperpolarized Xenon-129. *Acad. Radiol.* **9**, 498–500.
55. Kety, S.S. (1949) Measurements of regional circulation by the local clearance of radioactive sodium. *Am. Heart J.* **38**, 321–328.
56. Kety, S.S. (1951) The theory and applications of the exchange of inert gas at the lungs and tissues. *Pharmacol. Rev.* **3**, 1–41.
57. Peled, S., Jolesz, F.A., Tseng, C.H., Nascimben, L., Albert, M.S., and Walsworth R.L. (1996) Determinants of

tissue delivery for (129)Xe magnetic resonance in humans. *Magn. Reson. Med.* **36**, 340–344.
58. Martin, C.C., Williams, R.F., Gao, J.H., Nickerson, L.D.H., Xiong, J., and Fox, P.T. (1997) The pharmacokinetics of hyperpolarized xenon: implications for cerebral MRI. *J. Magn. Reson Imaging* **7**, 848–854.
59. Lavini, C., Payne, G.S., Leach, M.O., and Bifone, A. (2000) Intravenous delivery of hyperpolarized (129)Xe: a compartmental model. *NMR Biomed.* **13**, 238–244.
60. Killian, W., Seifert, F., and Rinneberg, H. (2002) Time resolved (129)Xe spectroscopy of human brain after inhaling hyperpolarized xenon gas. *Proc. Intl. Soc. Mag. Reson. Med.*
61. Johansson, E., Olsson, L.E., Månsson, S., Petersson, J.S., Golman, K., Ståhlberg, F., and Wirenstam, R. (2004) Perfusion assessment with bolus differentiation: a technique applicable to hyperpolarized tracers. *Magn. Reson. Med.* **52**, 1043–1051.
62. Golman, K., in't Zandt, R., and Thaning, M. (2006) Real-time metabolic imaging. *PNAS.* **103**, 11270–11275.
63. Golman, K., in't Zandt, R., Lerche, .M, Pehrson, R., and Ardenkjaer-Larsen, J.H. (2006) Metabolic imaging by hyperpolarized ^{13}C magnetic resonance imaging for in vivo tumor diagnosis. *Cancer Res.* **66**, 10855–10860.
64. Day, S.E., Kettunen, M.I., Gallagher, F.A., Hu, D.E., Lerche, M., Wolber, J., Golman, K., Ardenkjaer-Larsen, J.H., and Brindle, K.M. (2007) Detecting tumor response to treatment using hyperpolarized ^{13}C magnetic resonance imaging and spectroscopy. *Nat. Med.* **13**, 1382–1387.
65. Berg, J.M., Tymoczko, J.L., and Stryer, L. (2007) *Biochemistry*, sixth edition, W.H. Freeman and Company, New York, USA.
66. Gatenby, R.A. and Gillies, R.J. (2004). Why do cancers have high aerobic glycolysis? *Nat. Rev. Cancer* **4**, 891–899.
67. Yagi, G. and Hoberman, H.D. (1969) Rate of isotope exchange in enzyme-catalyzed reactions. *Biochemistry* **8**, 352–360.
68. Borgmann, U., Moon, T.W. and Laidler, K.J. (1974) Molecular kinetics of beef heart lactate dehydrogenase. *Biochemistry* **13**, 5152–5158.
69. Poole, R.C. and Halestrap, A.P. (1993) Transport of lactate and other monocarboxylates across mammalian plasma membranes. *Am. J. Physiol.* **264**, 761–782.
70. Jackson, V.N. and Halestrap, A.P. (1996) The kinetic, substrate and inhibitor specificity of the monocarboxylate (lactate) transporter of rat liver cells determined using the fluorescent intracellular pH indicator 2',7'-Bis(carboxyethyl)-5(6)-carboxyfluorescein. *J. Biol. Chem.* **271**, 861–868.
71. Lin, R.Y., Vera, J.C., Chaganti, R.S.K., and Golde, D.W. (1998) Human monocarboxylate transporter 2 (MCT2) is a high affinity pyruvate transporter. *J. Biol. Chem.* **273**, 28959–28965.
72. Brindle, K.M. (1988) NMR methods for measuring enzyme kinetics in vivo. *Prog. Nucl. Magn. Reson. Spectrosc.* **20**, 257–293.
73. Day, S.E., Kettunen, M.I., Gallagher, F.A., Hu, D.-E., Lerche, M., Wolber, J., Golman, K., Ardenkjaer-Larsen, J.H., and Brindle, K.M. (2007). Detecting tumor response to treatment using hyperpolarized ^{13}C magnetic resonance imaging and spectroscopy. *Nat. Med.* **13**, 1382–1387.
74. Zierhut, M.L., Yen, Y.F., Chen, A.P., Bok, R., Albers, M.J., Zhang, V., Tropp, J., Park, I., Vigneron, D.B., Kurhanewicz, J., Hurd, R.E., and Nelson, S.J. (2010) Kinetic modeling of hyperpolarized ^{13}C1-pyruvate metabolism in normal rats and TRAMP mice. *J. Magn. Reson.* **202**, 85–92.
75. Spielman, D.M., Mayer, D., Yen, Y.F., Tropp, J., Hurd, R.E., and Pfefferbaum, A. (2009) In vivo measurement of ethanol metabolism in the rat liver using magnetic resonance spectroscopy of hyperpolarized [1-^{13}C]pyruvate. *Magn. Reson. Med.* **62**, 307–313.
76. Naressi, A., Couturier, C., Devos, J.M., Janssen, M., Mangeat, C., de Beer, R., and Graveron-Demilly, D. (2001) Java-based graphical user interface for the MRUI quantitation package. *Magn. Reson. Mater. Biol., Phys., Med.* **12**, 141–152.
77. http://s-provencher.com/pages/lcmodel.shtml.
78. Shaw, R.J. (2006). Glucose metabolism and cancer. *Curr. Opin. Cell Biol.* **18**, 598–608.
79. Meisamy, S., Bolan, P.J., Baker, E.H., Bliss, R.L., Gulbahce, E., Everson, L.I., Nelson, M.T., Emory, T.H., Tuttle, T.M., Yee, D., et al. (2004). Neoadjuvant chemotherapy of locally advanced breast cancer: predicting response with *in vivo* H-1 MR spectroscopy – A pilot study. *Radiology* **233**, 424–431.
80. Kurhanewicz, J., Vigneron, D.B., and Nelson, S.J. (2000). Three-dimensional magnetic resonance spectroscopic imaging of brain and prostate cancer. *Neoplasia* **2**, 166–189.

81. Aboagye, E.O., Bhujwalla, Z.M., Shungu, D.C., and Glickson, J.D. (1998). Detection of tumour response to chemotherapy by ^{1}H nuclear magnetic resonance spectroscopy: effect of 5-fluorouracil on lactate levels in radiation-induced fibrosarcoma in tumours. *Cancer Res.* **58**, 1063–1067.
82. Veech, R.L., Lawson, J.W.R., Cornell, N.W., and Krebs, H.A. (1979) Cytosolic phosphorylation potential. *J. Biol. Chem.* **254**, 6538–6547.
83. Silverstein, E. and Boyer, P.D. (1964) Equilibrium reaction rates and the mechanism of bovine heart and rabbit muscle lactate dehydrogenase. *J. Biol. Chem.* **239**, 3901–3907.
84. Williams, S.N.O., Anthony, M.L., and Brindle, K.M. (1998) Induction of apoptosis in two mammalian cell lines results in increased levels of fructose-1,6-bisphosphate and CDP-choline as determined by ^{31}P MRS. *Magn. Reson. Med.* **40**, 411–420.
85. Witney, T.H., Kettunen, M.I., Day, S.E., Hu, D., Neves, A.A., Gallagher, F.A., Fulton, S.M., and Brindle, K.M. (2009) A comparison between radiolabeled fluorodeoxyglucose uptake and hyperpolarized ^{13}C-labeled pyruvate utilization as methods for detecting tumor response to treatment. *Neoplasia* **11**, 574–582.
86. Moffat, B.A., Chenevert, T.L., Lawrence, T.S., Meyer, C.R., Johnson, T.D., Dong, Q., Tsien, C., Mukherji, S., Quint, D.J., Gebarski, S.S., et al. (2005). Functional diffusion map: a noninvasive MRI biomarker for early stratification of clinical brain tumor response. *Proc. Natl. Acad. Sci. U.S.A.* **102**, 5524–5529.
87. Albers, M.J., Bok, R., Chen, A.P., Cunningham, C.H., et al. (2008) Hyperpolarized ^{13}C lactate, pyruvate, and alanine: noninvasive biomarkers for prostate cancer detection and grading. *Cancer Res.* **68**, 8607–8615.
88. Bode, B.P. and Souba, W.W. (1994) Modulation of cellular proliferation alters glutamine transport and metabolism in human hepatoma cells. *Ann. Surg.* **220**, 411–422.
89. Reitzer, L.J., Wice, B.M., and Kennell, D. (1979) Evidence that glutamine, not sugar, is the major energy source for cultured HeLa cells. *J. Biol. Chem.* **254**, 2669–2676.
90. Gallagher, F.A., Kettunen, M.I., Day, S.E., Lerche, M., and Brindle K.M. (2008) ^{13}C MR spectroscopy measurements of glutaminase activity in human hepatocellular carcinoma cells using hyperpolarized ^{13}C-labeled glutamine. *Magn. Reson. Med.* **60**, 253–257.
91. Keshari, K.R., Wilson, D.M., Chen, A.P., Bok, R., Larson, P.E.Z., Hu, S., Van Criekinge, M., Macdonald, J.M., Vigneron, D.B., and Kurhanewicz, J. (2009) Hyperpolarized [2-^{13}C]-fructose: a hemiketal DNP substrate for in vivo metabolic imaging. *J. Am. Chem. Soc.* Online.
92. Karlsson, M., Jensen, P.R., in't Zandt, R., Gisselsson, A., Hansson, G., Duus, J.Ø., Meier, S., and Lerche, M.H. (2009) Imaging of branched chain amino acid metabolism in tumors with hyperpolarized ^{13}C ketoisocaproate. *Inter. J. Cancer.* Online.
93. Panchal, A.R., Comte, B., Huang, H., Kerwin, T., Darvish, A., Des Rosiers, C., Brunengraber, H., and Stanley, W.C. (2000). Partitioning of pyruvate between oxidation and anaplerosis in swine hearts. *Am. J. Physiol. Heart Circ. Physiol.* **279**, 2390–2398.
94. Panchal, A.R., Comte, B., Huang, H., Dudar, B., Roth, B., Chandler, M., Des Rosiers, C., Brunengraber, H., and Stanley, W.C. (2001). Acute hibernation decreases myocardial pyruvate carboxylation and citrate release. *Am. J. Physiol. Heart Circ. Physiol.* **281**, 1613–1620.
95. Mallet, R.T. (2000) Pyruvate: metabolic protector of cardiac performance. *Proc. Soc. Exp. Biol. Med.* **223**, 136–148.
96. Mallet, R.T., Sun, J., Knott, E.M., Sharma, A.B., and Olivencia-Yurvati, A.H. (2005) Metabolic cardioprotection by pyruvate: recent progress. *Exp. Biol. Med.* **230**, 435–443.
97. Schroeder, M.A., Atherton, H.J., Cochlin, L.E., Clarke, K., Radda, G.K., and Tyler, D.J. (2009) The effect of hyperpolarized tracer concentration on myocardial uptake and metabolism. *Magn. Reson. Med.* **61**, 1007–1014.
98. Tyler, D.J., Schroeder, M.A., Cochlin, L.E., Clarke, K., and Radda, G.K (2008) Applications of hyperpolarized magnetic resonance in the study of cardiac metabolism. *Appl. Magn. Reson.* **34**, 523–531.
99. Merritt, M.E., Harrison, C., Storey, C., Jeffrey, F.M., Sherry, A.D., and Malloy, C.R. (2007) Hyperpolarized ^{13}C allows a direct measure of flux through a single enzyme-catalyzed step by NMR. *Proc. Natl. Acad. Sci. U.S.A.* **104**, 19773–19777.
100. Merritt, M.E., Harrison, C., Storey, C., Sherry, A.D., and Malloy, C.R. (2008)

Inhibition of carbohydrate oxidation during the first minute of reperfusion after brief ischemia: NMR detection of hyperpolarized $^{13}CO_2$ and $H^{13}CO_3$. *Magn. Reson. Med.* **60**, 1029–1036.

101. Schroeder, M.A., Cochlin, L.E., Heather, L.C., Clarke, K., Radda, G.K., and Tyler, D.J. (2008) In vivo assessment of pyruvate dehydrogenase flux in the heart using hyperpolarized carbon-13 magnetic resonance. *Proc. Natl. Acad. Sci. U.S.A.* **105**, 12051–12056.

102. Stanley, W.C., Recchia, F.A., and Lopaschuk, G.D. (2005) Myocardial substrate metabolism in the normal and failing heart. *Physiol. Rev.* **85**, 1093–1129.

103. Jager, P.L., Vaalburg, W., Pruim, J., de Vries, E.G.E., Langen, K.J., and Piers, D.A. (2001). Radiolabeled amino acids: basic aspects and clinical applications in oncology. *J. Nucl. Med.* **42**, 432–445.

104. Schroeder, M.A., Atherton, H.J., Ball, D.R., Cole, M.A., Heather, L.C., Griffin, J.L., Clarke, K., Radda, G.K., and Tyler, D.J. (2009) Real-time assessment of Krebs cycle metabolism using hyperpolarized ^{13}C magnetic resonance spectroscopy. *The FASEB J.* **23**.

105. Michaelis, T., Boretius, S., and Frahm, J. (2009) Localized proton MRS of animal brain in vivo: models of human disorders. *Progr. Nucl. Magn. Reson. Spectrosc.* **55**, 1–34.

106. Tran, T., Ross, B., and Lin, A. (2009) Magnetic resonance spectroscopy in neurological diagnosis. *Neurologic Clinics* **27**, 21–60.

107. Bhattacharya, P., Chekmenev, E.Y., Perman, W.H., Harris, K.C., Lin, A.P., Norton, V.A., Tan, C.T., Ross, B.D., and Weitekamp, D.P. (2007) Towards hyperpolarized ^{13}C-succinate imaging of brain cancer. *J. Magn. Reson.* **186**, 150–155.

108. Hurd, R.E, Yen,Y-F, Mayer, D., Chen, A., Wilson, D., Kohler, S., Bok, R., Vigneron, D., Kurhanewicz, J., Tropp, J., Spielman, D., and Pfefferbaum, A. (2010) Metabolic imaging in the anesthetized rat brain using hyperpolarized [1-^{13}C] pyruvate and [1-^{13}C] ethyl pyruvate. *Magn. Reson. Med.* **63**, 1137–1143.

109. Spielman, D.M., Mayer, D., Yen, Y.F., Tropp, J., Hurd, R.E., and Pfefferbaum, A. (2009) In vivo measurement of ethanol metabolism in the rat liver using magnetic resonance spectroscopy of hyperpolarized [1-^{13}C]pyruvate. *Magn. Reson. Med.* **62**, 307–313.

110. Gallagher, F.A., Kettunen, M.I., Day, S.E., Hu, D.E., Ardenkjaer-Larsen, J.H., Zandt, R., Jensen, P.R., Karlsson, M., Golman, K., Lerche, M.H., and Brindle, K.M. (2008) Magnetic resonance imaging of pH in vivo using hyperpolarized ^{13}C-labelled bicarbonate. *Nature* **453**, 940–944.

111. Adrogué, H.G., Gennari, F.J., Galla, J.H., and Madias, N.E. (2009) Assessing acid–base disorders. *Kidney Int.* **76**, 1239–1247.

112. Casey, J.R., Grinstein, S., and Orlowski, J. (2010) Sensors and regulators of intracellular pH. *Nat. Rev. Mol. Cell Biol.* **11**, 50–61.

113. Raghunand, N., He, X., van Sluis, R., Mahoney, B., Baggett, B., Taylor, C.W., Paine-Murrieta, G., Roe, D., Bhujwalla, Z.M., and Gillies, R.J. (1999) Enhancement of chemotherapy by manipulation of tumour pH. *Br. J. Cancer* **80**, 1005–1011.

114. Robey, I.F., Baggett, B.K., Kirkpatrick, N.D., Roe, D.J., Dosescu, J., Sloane, B.F., Hashim, A.I., Morse, D.L., Raghunand, N., Gatenby, R.A., and Gillies, R.J. (2009) Bicarbonate increases tumor pH and inhibits spontaneous metastases. *Cancer Res.* **69**, 2260–2268.

115. Gillies, R.J., Raghunand, N., Garcia-Martin, M.L., and Gatenby, R.A. (2004) pH imaging: a review of pH measurement methods and applications in cancers. *IEEE Eng. Med. Biol. Mag.* **23**, 57–64.

Chapter 34

Target-Specific Paramagnetic and Superparamagnetic Micelles for Molecular MR Imaging

Roel Straathof, Gustav J. Strijkers, and Klaas Nicolay

Abstract

Treatment of disease can only be effective when timely and accurate diagnosis of the pathology is achieved. More precise diagnosis can be accomplished if the underlying molecular processes involved in the pathology can be imaged in vivo. This is the field of molecular imaging, which aims to visualize cellular function and molecular processes in living organisms in a non-invasive way. With that aim, molecular markers are specifically targeted by imaging contrast agents. Molecular MRI needs powerful targeted contrast agents. For that purpose, target-specific gadolinium-containing paramagnetic and superparamagnetic, iron oxide-based micelles have been developed. Micelles are lipid-based nanoparticles which are biocompatible and carry a high payload of MR contrast-generating agent. The coupling of high-affinity ligands makes the micelles target-specific. Additionally, this lipid-based micelle platform allows for incorporation of contrast generating molecules for other imaging modalities, e.g., fluorescence or nuclear imaging. This permits applications for multiple imaging modalities, making micelles a highly versatile contrast agent.

Key words: Molecular MRI, contrast agent, micelles, targeting, phospholipids, gadolinium, iron oxide.

1. Introduction

The field of molecular imaging is a relatively recent extension to conventional imaging techniques (1). It seeks to gather information on molecular processes underlying pathologies. Ultimately, molecular imaging facilitates detecting and phenotyping of diseases in a much earlier development stage than possible with conventional imaging techniques.

Molecular MR imaging has the additional benefit of combining molecular insights with the broad range of other parameters that MR can provide, including anatomical, structural, and functional information. The above molecular information is elucidated by exogenous contrast agents (CA) targeted to disease-specific biomarkers using high-affinity ligands and thereby generating contrast as a disease characteristic.

MRI contrast agents have to meet the following requirements: they should bear a high payload of contrast-generating label, show biocompatibility, and have the ability to reach the targeted tissue. Lipid-based micelles are a successful example of such a contrast agent (2–4). By using lipids with a gadolinium chelate in their headgroup contrast enhancement is generated in T_1-weighted MR images. PEG-lipids are typically incorporated to increase circulation half lives in vivo and reduce interactions with plasma proteins. The relatively small diameter of micelles enables extravasation of the particles at inflammatory sites due to increased endothelial permeability.

Micelles can be prepared by two different approaches, namely, lipid film hydration (5, 6) or direct infusion of a lipid mixture in water (7). Both will be discussed in this chapter. These methods can also be employed for the encapsulation of a hydrophobic solid core, like iron oxide particles, resulting in particles coated with a monolayer of lipids.

Paramagnetic micelles are characterized by generating local signal enhancement in MR images, referred to as positive contrast. Superparamagnetic iron oxide-based micelles on the other hand, cause local signal attenuation on T_2-weighted scans, known as negative contrast. Changes in contrast are more readily observed for gadolinium-containing contrast agents, whereas the range of effect in space is larger for the superparamagnetic contrast agents. These factors have to be taken into account for a well-motivated choice between these two contrast-generating atoms.

The highly versatile micellar particles not only meet the aforementioned objectives but also can be equipped to allow their detection by different imaging modalities, e.g., PET, CT, or fluorescence microscopy (8, 9). Furthermore, the functional lipids can be coupled with different targeting ligands enabling visualizations of different disease-related molecular pathways.

As a specific example, collagen-targeted paramagnetic micelles were shown to be able to detect atherosclerotic plaques and, more importantly, discriminate stable atherosclerotic lesions from vulnerable ones (10). MR images showed clear and significant signal enhancement, while fluorescent microscopy proved a stronger accumulation of collagen-targeted micelles in collagen-rich atherosclerotic plaques. In **Fig. 1** MR and confocal laser scanning microscopy (CLSM) images of mouse carotid artery

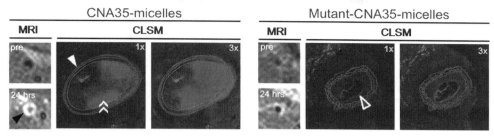

Fig. 1. T_1-weighted MR images and confocal laser scanning microscopy (CLSM) images taken from the carotid artery of mice injected with collagen-targeted paramagnetic micelles. These micelles were equipped with the high-affinity collagen-ligand CNA35 or a low-affinity ligand (mutant-CNA35), as a control. The MR images show a strong increase in signal intensity in the arterial vessel wall 24 h after CNA35 micelles injection (*black, solid arrow head*) compared to the pre-injection scan. CLSM fluorescence images show the carotid artery (24 h after injection) that can be identified by the elastic laminae (*white, solid arrowhead*). A strong accumulation of CNA35 micelles inside the vessel wall can be seen (*white, double, open arrow heads*). To stress the differences in accumulation between CNA35- and mutant-CNA35 micelles, fluorescence images are also shown with triple laser power, revealing nearly no accumulation for the latter group. The *white, single, open arrowhead* indicates the lumen. Adapted from (10).

atherosclerosis are shown, illustrating targeting of collagen with functionalized micelles.

Here, a comprehensive protocol is described for the preparation and functionalization of paramagnetic and superparamagnetic micelles. The proposed micelle formulations were proven to be successful in detecting several different disease processes in a non-invasive manner. This protocol should also enable the reader to choose other formulations in a well-motivated manner, depending on the specific study objectives. This protocol therefore lays the basis for the preparation of a class of highly versatile particles for molecular imaging.

2. Materials

2.1. Paramagnetic Micelles

1. Chloroform (this is a potentially toxic solvent and carries danger of serious damage to health by prolonged exposure through inhalation and if swallowed. Chloroform should be handled in a fume hood).

2. Methanol (this is highly flammable and toxic by inhalation, in contact with skin and if swallowed. Methanol should be handled in a fume hood).

3. HEPES-buffered saline (HBS) 1×: 2.38 g/L HEPES, 8.0 g/L NaCl in water, pH 6.7 (*see* **Notes 1 and 2**).

4. Gadolinium-DTPA-bis-stearyl-amide (Gd-DTPA-BSA) (Gateway Chemical Technology, St. Louis, MO, USA).

5. 1,2-Distearoyl-*sn*-glycero-3-phosphoethanolamine-*N*-[methoxy(polyethylene glycol)-2000] (ammonium salt)

(DSPE-PEG2000) (Lipoid GmbH, Ludwigshafen, Germany).

6. 1,2-Distearoyl-*sn*-glycero-3-phosphoethanolamine-*N*-[maleimide(polyethylene glycol)-2000] (ammonium salt) (DSPE-PEG(2000) maleimide) (Avanti Polar Lipids, Alabaster, AL, USA).

7. 1,2-Dipalmitoyl-*sn*-glycero-3-phosphoethanolamine-*N*-(lissamine rhodamine B sulfonyl) (ammonium salt) (Rhodamine-PE) (Avanti Polar Lipids, Alabaster, AL, USA).

8. Pear-shaped round-bottom flask, 250–500 mL (Schott AG, Mainz, Germany).

9. Parafilm® M (Pechiney Plastic Packaging, Menasha, WI, USA)

10. Rotary evaporation equipment Rotavapor R-200 with coupled water bath (Büchi AG, Flawil, Switzerland).

2.2. Superparamagnetic Micelles

1. Chloroform (this is a potentially toxic solvent and carries danger of serious damage to health by prolonged exposure through inhalation and if swallowed. Chloroform should be handled in a fume hood).

2. HEPES-buffered saline (HBS) 1×: 2.38 g/L HEPES, 8.0 g/L NaCl, pH 6.7 (*see* **Note 2**).

3. Iron oxide particles in hydrophobic solvent, e.g., magnetite in toluene as produced by thermal decomposition (11). Iron oxide particles can also be purchased as powder or dissolved in non-polar organic solvent (e.g., BioPAL, Worcester, MA, USA). Attention should be paid to the hydrophobicity of the coating present on the magnetite core.

4. 1,2-Distearoyl-*sn*-glycero-3-phosphoethanolamine-*N*-[methoxy(polyethylene glycol)-2000] (ammonium salt) (DSPE-PEG2000) (Lipoid GmbH, Ludwigshafen, Germany).

5. 1,2-Distearoyl-*sn*-glycero-3-phosphoethanolamine-*N*-[maleimide(polyethylene glycol)-2000] (ammonium salt) (DSPE-PEG(2000) maleimide) (Avanti Polar Lipids, Alabaster, AL, USA).

6. 1,2-Dipalmitoyl-*sn*-glycero-3-phosphoethanolamine-*N*-(lissamine rhodamine B sulfonyl) (ammonium salt) (Rhodamine-PE) (Avanti Polar Lipids, Alabaster, AL, USA).

7. 5 mL chloroform-resistant container and lid.

8. Glass syringe, Luer lock tip, 5 mL (Poulten & Graf, Wertheim, Germany).

9. Stainless steel needle, non-coring point, 6 inch, 21 gauges (Sigma Aldrich, Zwijndrecht, the Netherlands).
10. Syringe infusion pump (Pump 11 Pico Plus, Harvard Apparatus, Holliston, MA, USA).
11. Ultrasonicator (model 5510, Branson, Danbury, CT, USA).
12. Vivaspin 10,000 MWCO concentrator, 20 mL (Sartorius Stedim, Göttingen, Germany) (*see* **Note 3**).
13. Ultracentrifuge (Optima L-90 K, Beckman Coulter, Fullerton, CA, USA).

2.3. Coupling Ligands

1. Succinimidyl *S*-acetyl-thiol-acetate (SATA) (toxicity of SATA has not been determined; therefore it should be handled in a fume hood).
2. *N,N*-Dimethylformamide (DMF) (this compound is harmful by inhalation and contact with skin or eyes, may cause harm to the unborn child. DMF should be handled in a fume hood).
3. Ligand for targeting, e.g., antibody or peptide.
4. HEPES-buffered saline 1×, 2.38 g/L HEPES, 8.0 g/L NaCl in water, pH 7.4.
5. Vivaspin concentrator (Sartorius Stedim, Göttingen, Germany) (*see* **Note 3**).
6. Filter buffer: 2.38 g/L HEPES, 8.0 g/L NaCl, 0.37 g/L ethylenediamine tetraacetic acid (EDTA), pH 7.0.
7. De-acetylation buffer: add to 50 mL of water, 1.74 g hydroxyl amine (this is harmful when it comes into contact with the skin, use gloves), 5.95 g HEPES, 0.47 g ethylenediamine tetraacetic acid (EDTA), 1 g NaOH, pH 7.0. Prepare fresh.

2.4. Particle Characterization Methods

2.4.1. Particle Size Determination

1. Dynamic light scattering equipment (Malvern Zetasizer Nano S, Malvern Instruments, Malvern, UK).
2. Sterile filtered buffer, which has the same composition as the buffer containing the functionalized micelles.

2.4.2. Phosphate Determination

1. 70% perchloric acid (this compound is corrosive, will cause severe burns, and is very harmful by inhalation or ingestion or through skin contact. Perchloric acid should be handled in a fume hood).
2. Stock solution 50 mM sodiumbiphosphate ($NaH_2PO_4 \cdot 2H_2O$) (Fluka). This solution can be stored for several months at room temperature.

3. Working solution 0.5 mM sodiumbiphosphate ($NaH_2PO_4 \cdot 2H_2O$).
4. Hexa-ammonium molybdate ((NH_4)$_6MO_7O_{24} \cdot H_2O$) (Merck): 1.25% (w/v) solution. This should be prepared fresh.
5. L-ascorbic acid (Fluka): 5% (w/v) solution. Prepare fresh.
6. Glass test tubes (common model, 10 mL).
7. Glass marbles.
8. Block heater (BT5D-16, Grant, Hillsborough, NJ, USA).
9. Water bath.
10. Spectrophotometer (DU 800, Beckman Coulter, Fullerton, CA, USA).

2.4.3. Relaxivity Determination

1. Paramagnetic or superparamagnetic micelles in appropriate buffer (e.g., HBS buffer, pH 7.4) with a known concentration of contrast-generating atom (*see* **Note 4**).
2. Tabletop NMR spectrometer (Bruker Minispec MQ60 (1.41 T), Bruker, Ettlingen, Germany) and MRI scanner at which in vivo studies are performed.

2.5. In Vivo Use of Micelles

2.5.1. In Vivo Administration of Micelles in Mice

1. Approval for the use of mice for an in vivo experiment given by the local Animal Care and Use Committee according to local legislation.
2. Paramagnetic or superparamagnetic micelles in appropriate buffer (e.g., HBS buffer, pH 7.4) with a known concentration of contrast-generating atom (*see* **Note 4**).
3. Needles for intravenous injection, 27 G (0.4 mm) with Luer lock (Terumo Europe N.V., Leuven, Belgium) or intravenous injection catheter, 24 G (Abbocath®-T 24 G, Hospira Inc, Lake Forest, Illinois, USA).
4. Polyethylene catheter, inner diameter 0.4 mm (SIMS Portex, Hythe, UK).
5. Syringe, 1 mL (Terumo Europe N.V., Leuven, Belgium).
6. Warm water pad (T/Pump, Gaymar Industries, Orchard Park, NY, USA).
7. Anesthesia equipment (Sigma Delta Vaporizer, Penlon Ltd., Abingdon, UK).
8. Equipment for additional, local heat administration, e.g., gel pack, infrared light.

2.5.2. MRI Sequences for Micelle Detection

1. Experimental animal MRI scanner.
2. NMR/MRI acquisition software (e.g., ParaVision 5.0, Bruker Biospin GmbH, Ettlingen, Germany).

3. Methods

Anhydrous amphiphilic lipids need to be handled with care. Avoid contact with any water during the handling of the micelle constituents to prevent oxidation and hydration until the actual hydration step is carried out (**Fig. 2**). Lipid containers should therefore be thawed to room temperature before opening and stored at –25°C directly after use under a protective N_2 gas atmosphere.

The formation of micelles using amphiphilic molecules is primarily driven by the hydrophobic associative interactions of the non-polar hydrocarbon chains and the repulsive interactions between the hydrophilic headgroups (12). This leads to the self-assembly of well-defined structures in water in such a way that the hydrophilic headgroups face the water while the hydrophobic parts cluster together on the inside of the micelle (*see* **Fig. 2**). Whether micelles or other structures are formed depends on the chain length of the lipids used, the number of non-polar chains per molecule, and the size of the headgroup in relation to the cross-sectional area of the hydrocarbon chains (12, 13). For micelles, a relatively large headgroup is needed (*see* **Note 5**).

Because the self-assembly itself cannot be controlled, preparative steps should be carried out with precision to gain maximum control over this process. Close attention should be paid to those details that can reveal information on the self-assembly process. This will be discussed, especially in relation to the preparation of superparamagnetic micelles where unwanted products are more

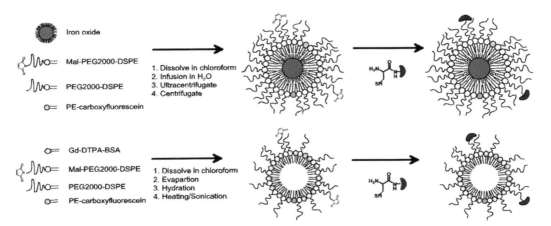

Fig. 2. Schematic presentations of preparation steps and coupling of the ligand (*dark grey* kidney shaped component with reactive thiol group). The *top row* depicts the formation of superparamagnetic iron oxide-based micelles, while the *bottom row* highlights the preparation of gadolinium-containing paramagnetic micelles. The products to the right show the configuration which is formed by the amphiphlic phospholipids through self-assembly.

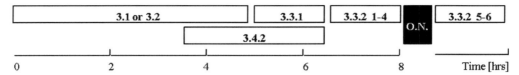

Fig. 3. Timeline for preparation and functionalization of paramagnetic or superparamagnetic micelles. If ligand conjugation is needed and thus DSPE-PEG2000-maleimide is used, **Sections 3.1 until 3.3.2.4** should be performed in direct consecutive order without delay. It is advised to start the phosphate determination (**3.4.2**) during the hydration step (**3.1.14**) or the ultracentrifugation steps (**3.2.14**). O.N., overnight.

readily formed in case of a poorly controlled preparation. This is because the hydrophobic character of the iron oxide core provokes aggregation of the iron oxide particles upon dispersion in water and this has to be prevented by using an excess of coating lipids, thereby maximizing the preparation of single-layered micelles containing one iron oxide particle.

The total process of preparation and functionalization typically takes one and a half days (*see* **Fig. 3**, section numbers in the text correspond with the numbers mentioned in **Fig. 3**). Some steps have to be carried out consecutively without delay, while other steps in the process can be paused or carried out overnight.

3.1. Preparation of Paramagnetic Micelles

1. Calculate the amount of lipids you need (*see* **Table 1**) (*see* **Note 6**).
2. Let the lipids warm to RT at least 15 min before opening the vials.
3. Thoroughly clean a 250 mL round-bottom flask. Rinse sequentially three times with demineralized water, two times with ethanol, and two times with acetone. Dry thoroughly with nitrogen.
4. Weigh the lipids and put them in the round-bottom flask.
5. To prevent oxidation and hydration of the stock lipids, carefully blow N_2 with minimal flow in the stock vials with lipids to replace air. Store lipids in freezer/dissicator.
6. Dissolve the lipids in 5–10 mL chloroform:methanol mixture (4:1 ratio). Stir flask until all lipids are completely dissolved.
7. Evaporate all chloroform and methanol using the rotary evaporation equipment. Set temperature of the coupled water bath to 30°C, pressure at 150 mbar (15 kPa; 2.17 psi) and the rotation speed between 100 and 140 rpm (*see* **Note 7**).
8. After all visible solvent has been evaporated, the film must be left on the rotary evaporation equipment for an extra 15 min at 0.0 mbar.

9. Put the flask with lipid film under N$_2$ flow for 1 h or preferably overnight to remove any residual chloroform and methanol.

10. Heat HBS buffer at 65°C for 10 min (*see* **Note 2**).

11. Heat the round-bottom flask with the dried lipid film in the water bath at 65°C for 3 min in a vertical position. Prevent any contact of the lipid film with water.

12. Hydrate the lipid film with a few milliliters of the heated HBS. The volume of the hydrating HBS buffer depends on the desired final concentration of lipids (*see* **Note 8**).

13. Seal the round-bottom flask with Parafilm® M and connect it onto the rotary evaporation equipment, thus preventing evaporation of the hydration buffer.

14. Heat the hydrated lipid film in the round-bottom flask at 65°C in the warm water bath, while slowly rotating between 60 and 100 rpm for 1 h (*see* **Note 9**). When the DSPE-PEG2000-maleimide will not be functionalized by ligand coupling, the paramagnetic micelles are ready and can be stored at 4°C without further preparation steps. It is important not to use the unfunctionalized micelles within 1 week after preparation to prevent the reactive maleimide group from undesired binding to other substances in vivo or even cause immunogenic responses.

15. Start the phosphate determination (*see* **Section 3.4.2**) to calculate DSPE-PEG2000-maleimide concentration in the micellar end product. Take into account that this procedure takes 3 h (*see* **Fig. 3**).

16. For functionalizing of the micelles, proceed without delay with coupling of the desired ligand (*see* **Section 3.3**), as the maleimide groups are subject to hydrolysis.

3.2. Preparation of Superparamagnetic Micelles

1. Calculate the amounts of lipids and iron oxide which are needed (*see* **Table 1**) (*see* **Notes 6 and 10**).

2. Let the container with lipids warm from the fridge at least 15 min before opening.

3. Clean glass syringe thoroughly. Rinse sequentially three times with demineralized water, two times with ethanol, and two times with acetone. Dry thoroughly with nitrogen.

4. Weigh the lipids and put them in a chloroform-resistant container with lid.

5. To prevent oxidation and hydration of the stock lipids, carefully blow N$_2$ with minimal flow in the stock vials with lipids to replace air. Store lipids in freezer/dissicator.

Table 1
Amounts of lipids needed and their ratios for preparation of paramagnetic and superparamagnetic micelles

A	Paramagnetic micelles		Superparamagnetic micelles	
	Needed	*Amounts*	*Needed*	*Amounts*
	In vivo	50–100 µmol Gd/kg BW	In vivo	10–50 mg Fe/kg BW
	In vitro	0.1–1 mM Gd	In vitro	10–500 µM Fe
	Starter example		*Starter example*	
	Gd	7.5 µmol	Fe	2 mg
	Total lipids	30 µmol	Total lipids	15 µmol
	Yield	80%	Yield	50%
	End product	6 µmol Gd	End product	1 mg Fe

B	Paramagnetic micelles			Superparamagnetic micelles		
		Molar ratio	MW		Molar ratio	MW
	Gd-DTPA-BSA	0.50	1,050	DSPE-PEG2000	0.89	2,806
	DSPE-PEG2000	0.39	2,806	DSPE-PEG2000-maleimide	0.10	2,941
	DSPE-PEG2000-maleimide	0.10	2,941	Rhodamine-PE	0.01	1,301
	Rhodamine-PE	0.01	1,301			

Table 1 (continued)

C		Units	Parameter		Units	Parameter
	Lipids per particle		A	Particle concentration	M	H
	Iron oxide core diameter	nm	B	Concentration iron oxide	g/mol	I
	Core + shell diameter	nm	$C = 4\pi(B/2)^2$	Molecular weight iron oxide	g/dm^3	J
	Surface particle	nm	D	Density iron oxide	dm^3/L	$K = H^*I/J$
	Lipid area	nm^2	$E = C/D$	Volume iron oxide core/L	nm^3	$L = 4/3\pi(a/2)^2$
	Number of lipids per particle		F	Volume iron oxide core	L^{-1}	$M = 10^{24*}K/L$
	Lipid excess factor		$G = E^* F$	Number of iron oxide cores/L		
	Number of lipids per particle			Number of lipids per L needed		$G^* M$

(A) Typical amounts of lipids needed for in vitro and in vivo experiments. The starter examples will give the amount of end product which is required for four mice with a body weight (BW) of 25 g. The amount of lipids needed for superparamagnetic micelles is based on the following parameters (*See* **Table 1C**): $A = 14$ nm; $B = 18$ nm; $D = 0.4$ nm^2; $F = 10$; $H = 80$ mM; $I = 231.25$ g/mol; $J = 5,170$ g/dm^3. (B) Molar ratios and molecular weight for different lipids used for paramagnetic and superparamagnetic micelles. These compositions have been used successfully but can be changed, e.g., to alter the type of fluorescent lipid or the different ratios. To determine the total amount of lipids needed for superparamagnetic micelles **Table 1C** is needed. (C) Calculation scheme to determine the number of lipids needed to coat the superparamagnetic iron oxide cores in 1 L of solvent. First the number of lipids needed for one particle is calculated (left-hand side) followed by calculating the number of iron oxide particles per liter (right-hand side). References A, B, H, I, and J should be provided by the user. It is advised not to change reference F.

6. Dissolve the lipids in 2 mL chloroform. Slowly stir container until all lipids are completely dissolved.

7. Add the calculated amount of iron oxide in the lipid-containing chloroform solution (*see* **Note 11**).

8. Sonicate the glass container for 5 min at room temperature.

9. Heat 50 mL of water on a hot plate at 80°C and use a magnetic stirrer.

10. Pull the chloroform solution into the clean glass syringe via the stainless steel needle, mount it onto the infusion pump, and place the tip of the needle inside the heated water (*see* **Note 12**).

11. Infuse the chloroform solution into the heated water at a rate of 0.5–2.0 mL/h (*see* **Note 13**).

12. Centrifuge the product twice at $4,000g$ for 10 min. Keep the supernatant and remove the pellet. The pellet contains iron oxide aggregates, while the required micellar iron oxide particles and empty micelles are in the supernatant.

13. Concentrate the second supernatant with Vivaspin concentrator and collect the micellar iron oxides (*see* **Note 3**).

14. Ultracentrifuge the concentrated sample three times at 65,000 rpm for 30 min. Remove the supernatant and keep the pellet. Replenish the supernatant with HBS 1×, pH 6.8 (*see* **Note 2**). Now the pellet contains the end product, while the supernatant predominantly contains empty micelles (*see* **Note 14**).

15. If DSPE-PEG2000-maleimide will be conjugated, start phosphate determination (*see* **Section 3.4.2**) to estimate DSPE-PEG2000-maleimide concentration in the micellar end product. Take into account that this procedure takes 3 h (*see* **Fig. 3**).

16. Centrifuge the final pellet at $4,000g$ for 5 min twice to remove any remaining aggregates and conserve the supernatant. When the DSPE-PEG2000-maleimide will not be functionalized by ligand coupling, the superparamagnetic micelles are ready and can be stored at 4°C without further preparation steps. It is important not to use the unfunctionalized micelles within 1 week after preparation to prevent the reactive maleimide group from undesired binding to other substances in vivo or even cause immunogenic responses.

17. For functionalizing the micellar iron oxide, proceed without delay with coupling of the desired ligand (*see* **Section 3.3**), as the maleimide groups are subject to hydrolysis.

3.3. Coupling Ligands

3.3.1. SATA Modification

1. Continue with the phosphate determination (*see* **Section 3.4.2**). When the phosphate determination is finished, you can proceed with the steps in **Section 3.3.2**.
2. Allow the SATA container to warm to room temperature before use.
3. Weigh 3 mg of SATA and dissolve in 130 µL of DMF (gives 0.1 M SATA solution).
4. Dissolve the ligand in the appropriate buffer and in the appropriate concentration (*see* **Note 15**).
5. Add SATA/DMF solution to the ligand solution (1:100 v/v and, for example, 20-fold excess of SATA) (*see* **Note 15**) and put the reaction mixture on a roller bench for 45 min.
6. Replace the buffer with filter buffer to remove unreacted SATA using Vivaspin concentrators (*see* **Note 3**), collect the SATA-modified ligand, and store the final solution at 4°C.

3.3.2. De-acetylation of Protected Thiol Groups

1. Add the freshly prepared de-acetylation buffer in 1:10 (v/v) ratio to the SATA-modified ligand solution.
2. Put this mixture on the roller bench for 60 min (*see* **Note 16**).
3. Calculate the amount of micelles that can be functionalized using the total amount of activated ligand (*see* **Note 17**) and mix these two volumes.
4. Put the reaction mixture under N_2 gas atmosphere and store at 4°C.
5. The next day, the unconjugated ligand should be washed away and the pH has to be increased to 7.4 using Vivapins concentrators (*see* **Notes 3 and 18**). The functionalized micelles should be collected from the Vivaspin slot.
6. Store the solution of end product at 4°C under a protective N_2 gas atmosphere.

3.4. Particle Characterization Methods

The characterization of the prepared micelles consists of two main elements: (1) quality control of the preparation and coupling steps and (2) determination of the efficacy of the contrast agent to lower relaxation times T_1 and T_2.

The quality control includes determination of particle size and total lipid concentration. Comparison of this final lipid content with the amount of starting lipid will give the overall yield. The relaxometric efficacy is defined as relaxivity r_i, $i \in \{1,2\}$ (in $mM^{-1}s^{-1}$), which defines the change in relaxation rate R_1 ($\equiv T_1^{-1}$) and R_2 ($\equiv T_2^{-1}$) as a function of the concentration of the contrast generating atom, i.e., gadolinium or iron. To determine the relaxivity, the relaxation times are measured as a function of the concentration of the contrast agent. This can be done at a desired temperature and magnetic field strength. For the concentration the result of the phosphate determination is required.

3.4.1. Particle Size Determination

1. Prepare three diluted samples with different concentrations of the functionalized micelles in sterile filtered buffer (*see* **Note 19**).
2. Leave the solutions to equilibrate before measuring, to make sure that only Brownian motion will be measured and not flow or convection (*see* **Note 20**).
3. Perform three measurements per dilution and calculate the diameter as the average of the three measurements. Examples of dynamic light scattering measurements of superparamagnetic micelles that illustrate the progress in the particle isolation process are shown in **Fig. 4**. An image of the micellar iron oxide preparation that was made with cryogenic transmission electron microscopy is shown in **Fig. 5**.

3.4.2. Phosphate Determination

1. Prepare micelle samples for calibration: add 0, 40, 60, 80, 100, 120, 160 µL from the 0.5 mM working solution. Prepare calibration samples in duplicate.
2. Prepare triple test samples. Add a sample volume which contains a phosphate amount between 10 and 80 nmol, preferably around 40 nmol.
3. Evaporate all samples in the block heater (180°C, 30 min) till dryness.
4. Let the tubes cool down to room temperature. Add 0.3 mL of perchloric acid to all tubes and place the marbles on top of the tubes.
5. Heat the tubes with marbles on top at 180°C in the block heater for least 60 min. Proceed until the solutions are clear. Let the tubes cool down afterward.
6. Prepare the water bath (100°C).
7. When the destruction (Step 5) is almost finished, prepare a fresh (5% (w/v)) solution of ascorbic acid.
8. Add to each sample 1.0 mL water, 0.5 mL 1.25% hexaammonium molybdate-solution, and 0.5 mL 5% (w/v) of ascorbic acid.
9. Vortex the tubes.
10. Place the tubes with marbles in the water bath at 100°C for 5 min.
11. Quickly cool down the tubes to room temperature, e.g., in a sink filled with cold tap water.

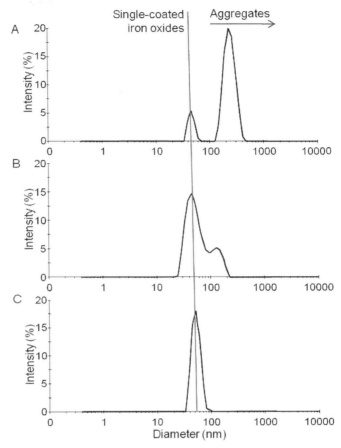

Fig. 4. Size distributions of superparamagnetic micelles, as measured by dynamic light scattering (DLS). The histograms show the shift in size distribution after consecutive centrifugation steps by removing aggregates: (a) Single-lipid monolayer-coated (*left peak*) and aggregated (*right peak*) iron oxide particles are observed directly after infusion in water. (b) After three ultracentrifugation rounds (65,000 rpm), less and smaller aggregates are found compared to (a). (c) A single, strong peak and no signs of aggregates are measured after the final centrifugation step (4,000g) compared to (b). These DLS intensity plots cannot be used to determine the relative amounts of single-coated iron oxide versus aggregates, because larger objects have a much larger scattering cross section (Signal intensity $\propto$ radius^6)

12. Determine the absorbance of each sample at 797 nm against the blank of the calibration curve, using a spectrophotometer. Calculate the amount of phosphate in the samples by means of the calibration curve. The calibration curve is linear up to absorbance readings of at least 1.0.

3.4.3. Determination of the Relaxivity

1. Prepare a dilution series of micelles within the range of 10 µM to 1 mM total amount of lipids. Also include a blank sample with only the solvent (*see* **Note 21**).

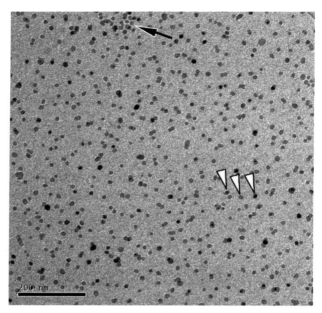

Fig. 5. Cryogenic transmission electron microscopy image showing mainly individual superparamagnetic micelles. Iron oxide cores are seen as hypo-intense spots (*white, solid arrow heads*). Occasional particle aggregates are identified (*black, solid arrow*). Total lipid concentration of this sample was 1 μM.

2. For each concentration, measure the T_1 and T_2 relaxation times. For the T_1 measurements an inversion recovery sequence (preferred) or a saturation recovery sequence can be used. T_2 relaxation times can be determined using the Carr–Purcell–Meiboom–Gill (CPMG) sequence. For accurate measurements field homogeneity, receiver gain, 90° and 180° pulses should be checked and calibrated for each sample.

3. Plot for each dilution series the relaxation rate R_1 ($\equiv T_1^{-1}$) or R_2 ($\equiv T_2^{-1}$) as a function of the known concentration of the contrast-generating atom, e.g., gadolinium (mM).

4. Fit a linear equation to the collected data points. The slope of the linear equation is the relaxivity r_1 or r_2 (mM^{-1}s^{-1}). An example of an r_2 estimation is shown in **Fig. 6**.

3.5. In Vivo Use of Micelles

3.5.1. Administration of Micelles

1. Prepare the micelle formulation with the appropriate amount of contrast-generating atoms, i.e., gadolinium or iron, for intravenous administration (**Table 1A**) via an injection catheter or a self-made injection system (*see* **Note 22**). The maximum volume which can be injected intravenously is approximately 10% of the circulating blood volume of the animal.

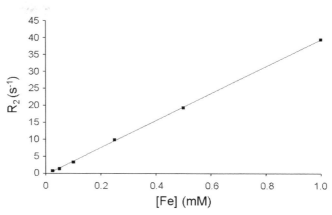

Fig. 6. Scatter plot of relaxation rate R_2 as function of iron concentration. Data points were fitted with a linear equation (*solid line*) to provide r_2 (*slope*).

2. Prepare the self-made injection system. This is composed of an injection needle, connected to a catheter and syringe. The needle can be obtained by removing the Luer lock from an injection needle (e.g., by burning it off using a cigarette lighter). The catheter has to be long enough to span the distance between the position of the animal inside the scanner and the syringe outside the scanner. The catheter and syringe will contain the prepared micelle formulation (*see* **Note 23**).

3. Put the mouse under anesthesia by inhalation anesthesia, e.g., isoflurane. Use a mixture of anesthetic and air and/or oxygen at an appropriate flow rate. Anesthetics are usually given as a 2–3% mixture with air and/or oxygen for anesthesia induction and 1–2% mixture for maintaining anesthesia. Flow rate of the mixture varies between 0.2 and 0.4 L/min for mice.

4. Maintain core body temperature of the mouse by positioning on a warm water pad. Apply additional heat to the tail of the mouse for vasodilation of the tail vein. Attention should be paid not to burn the skin or cause hyperthermia.

5. Insert the injection catheter or injection system in the tail vein and fixate the catheter or needle after confirming correct insertion.

6. After performing the pre-injection MRI scans, the micelle formulation can be injected. The injection volume can be administered within a period of 15–60 s depending on the volume. The additional venous volume load can cause the mouse to breathe more quickly just after injection, but the respiration rate will quickly return to normal.

3.5.2. MRI Sequences for Micelle Detection

3.5.2.1. Gadolinium-Containing Micelles

As gadolinium predominantly shortens T_1 relaxation time a T_1-weighted sequence is appropriate for the detection of gadolinium-containing micelles. To highlight differences in T_1 relaxation times as a tissue property, sequence timing parameters have to be optimized (*see* **Note 24**). The echo time (TE) has to be considerably shorter than the T_2 relaxation times of the tissue being imaged (TE << T_2) to avoid influence of gadolinium-induced T_2 contrast.

For a regular T_1-weighted spin-echo sequence the repetition time (TR) should be optimized to generate highest contrast between tissues without and with gadolinium-containing micelles. This generally results in a TR between the T_1 relaxation times of native tissue and the tissue containing the micelles. Typical parameters are TE = 5–15 ms and TR = 600–1,000 ms.

To generate even more contrast between regions with unaffected and shortened T_1 relaxation times, an inversion recovery spin-echo sequence can be used. In this case, the inversion time (TI) should be matched to the pre-contrast tissue T_1 relaxation time (TI ~ $T_1 \cdot \ln(2)$) to obtain highest contrast to noise after administration of the micelles. Typical parameter settings are TE = 5–15 ms, TR = 2,500–4,000 ms, and TI 800–1,200 ms.

Because the micelles are administered intravenously, undesired signals may arise from the lumen of blood vessels. Using a black-blood preparation sequence suppressing signals from the blood compartment can prevent this (*see* **Note 25**).

3.5.2.2. Iron Oxide-Containing Micelles

Iron oxide predominantly decreases the T_2 relaxation time. To visualize the differences in T_2 relaxation times, again a spin-echo sequence can be employed. To obtain T_2-weighting TE has to be in the range of the T_2 relaxation times (TE ~ T_2), whereas TR must be chosen appropriately longer than the T_1 relaxation times. Typical parameter settings are TE = 20–100 ms and TR = 2,000–4,000 ms.

Image contrast can also be generated based on the T_2^* relaxation time for which a gradient-echo sequence (e.g., a fast low-angle shot—FLASH) with long TE can be utilized. Typical parameters are excitation flip angle $\alpha = 15°$, TE = 5–20 ms, and TR 50–100 ms.

4. Notes

1. Unless stated otherwise, all solutions and buffers should be prepared in water that has a resistivity of at least 18.2 MΩ cm and total organic content of less than five parts per billion, e.g., as in MilliQ®. This standard is referred to as "water" in this chapter.

2. If the DSPE-PEG2000-maleimide lipid will not be conjugated, i.e., in case of preparation of unfunctionalized micelles, the pH of the HBS buffer should be set at 7.4.

3. Vivaspin concentrators are used in this protocol, but other methods for concentrating or separating products are possible as well, e.g., using size-exclusion chromatography, ultracentrifugation, or dialysis. For Vivaspin concentrators, this general protocol is used:
   a. Choose the molecular weight cutoff (MWCO) between the weights of the product(s) which are to be removed and which are to be kept. The MWCO should preferably be three times lower than the weight of the product which is to be kept, or three times higher than the material which is to be removed. The product that will be discarded is collected in the bottom part of the Vivaspin, while the top part with the filter will hold the desired product. Use centrifugation speeds recommended by the manufacturer.
   b. For concentrating the product, one spinning cycle is sufficient. If separation of the desired product is needed, three spinning cycles should be completed.
   c. For concentrating micellar iron oxides after infusion, choose a MWCO of 10,000 (*see* **Section 3.2**). For removing free SATA after SATA modification of the ligand, choose a MWCO at least three times lower than MW of the ligand (*see* **Section 3.3.1**). For removing free antibodies after coupling, choose a MWCO of 300,000 (*see* **Section 3.3.2**).

4. The concentration of the contrast-generating atom must be known to determine the relaxivity. Gadolinium concentration of paramagnetic micelles can be estimated using the phosphate concentration multiplied by the molar ratio of Gd-DTPA-BSA lipid or can be measured, e.g., by inductively coupled plasma mass spectrometry (ICP-MS). It is not possible to estimate the concentration of iron accurately and therefore it has to be measured, e.g., by ICP-AES, atomic absorption spectrometry (AAS), or calorimetric spectrometry using Prussian blue staining (14, 15).

5. To predict which type of structures will be formed with the desired amphiphilic molecules, the critical packing parameter (CPP) can be used. At low concentrations, for one-component systems this phenomenologically deducted formula in most cases correctly predicts the type of aggregate formed. The CPP is defined as

$$\mathrm{CPP} = \frac{v}{a_0 \cdot l_c},$$

where v is the volume of the hydrocarbon chain of the molecule, a_0 is the optimal surface area of the hydrophilic headgroup, and l_c is the critical chain length. Generally, micelles will be formed for a CPP < 1/3 (12, 13).

6. **Table 1A** shows the amounts of lipids needed for to produce sufficient amounts of micellar contrast agent material for typical in vitro or in vivo experiments. As an example, the starting amounts of lipids needed for the preparation of micelles for an in vivo experiment are shown. The phospholipids mentioned in **Table 1B** are proven to result in a micelle formulation that will work in molecular MR imaging. However, the ratios for each component, as well as the composition of the lipid mixture can be changed, e.g., in case a phospholipid labeled with a different fluorophore or an increased gadolinium content per paramagnetic micelle is preferred.

7. Evaporation is finished when the flask no longer has a cold surface. It is important that no inhomogeneities are visible after evaporation and that the film is evenly distributed over the bottom of the flask. The film will have a shiny appearance.

8. Assume a yield of total lipid content after evaporation and hydration between 80 and 90%.

9. Make sure the complete film is hydrated within these 60 min by changing the angle of the round-bottom flask on the rotary evaporation equipment. After hydration the turbidity of the solution must be checked. The final solution should be transparent.

10. **Table 1C** can be used to calculate the amount of lipids and iron oxide needed. Use an excess of lipids compared to the amount of iron oxide used. A tenfold excess is advised based on the surface of an iron oxide particle that has to be coated. Furthermore, assume a yield of 30–50% of iron content. For the overall calculation the following information is needed:
    a. Number of lipids needed to coat 1 iron oxide particle. Take into account the following:
        i. Diameter of a single iron oxide particle.
        ii. Additional layer(s) on iron oxide particle, e.g., an oleic acid coating, which further increases the diameter.
        iii. Assume iron oxide particle to be spherical.
        iv. Assume the cross-sectional area of one lipid to be 0.4 nm^2.
    b. Concentration of iron oxide cores in solution. Take into account the following:

i. iron oxide particle composition;
ii. particle density;
iii. volume of one iron oxide particle.

Together, these two numbers will provide the number of moles of total lipids needed to coat a desired number of iron oxide particles. This number of moles should be multiplied by the ratio of the different types of lipids that are to be used.

11. The total amount of lipids which will be added has to be dissolved in this volume. Therefore, the hydrophobic particle concentration should not be set too high. Iron concentration should be no higher than 30 mM to prevent aggregation or sedimentation.

12. If mounting of a filled syringe results in too much loss by dripping, one could start with mounting an empty syringe and then fill it. The continuous evaporation of chloroform in the syringe causes the iron oxide and lipid solution to be expelled out of the syringe. The tip of the needle should be placed as deep in the heated water as possible, but above the level of the magnetic stirrer, and close to wall of the beaker which holds the water.

13. A slow continuous stream of bubbles should be seen in the water. These bubbles contain the chloroform and the solvent of the iron oxide particles, e.g., toluene. The water should increasingly show a yellowish to brown discoloration during the infusion. Also, the water should not become turbid. If brown droplets attached to the tip of the needle are seen, which occasionally fall off and dissolve slowly, the temperature is not set high enough, or iron concentration in the syringe is too high and the organic solvent (e.g., hexane, toluene) does not evaporate quickly enough. Aggregates will form, which will result in a lower yield in the final product. Set the temperature at 85–90°C although this might cause loss of reactive maleimide groups for ligand coupling. If small brown particles are observed, aggregates are being formed and most likely the infusion rate is set too high.

14. Remove supernatant with glass pipette until the pellet starts moving, leaving about 1 mL of supernatant. The pellet is mobile and will flow like thick syrup when the ultracentrifuge tube is held in a skew position. When moved, the pellet will reveal a small black spot on the side of the ultracentrifuge tube. This spot represents new aggregates that result from the centrifugation step. Take care not to release these aggregates during resuspension of the pellet with HBS.

15. Calculate the concentration of the ligand which is needed:
    a. Determine the excess of SATA with respect to the ligand, e.g., 20:1. The higher the ratio, the higher the chance that the maleimide groups on the micelle react with the ligand. However, in this case, one also creates a higher chance of ligand–ligand coupling.
    b. Calculate the number of microliters of SATA/DMF solution needed for the total amount of ligand available. Take into account the determined excess of SATA from the previous step with respect to the ligand (MW of SATA: 231.25 and 0.1 M SATA solution).
    c. Bring the total amount of ligand in the appropriate buffer, which should have a volume one hundred times larger than the volume of SATA/DMF solution as calculated in the previous step.
16. When the reaction is complete, a sulfuric odor can be sensed. Make sure to continue without delay after this step, because free thiol groups are reactive and can form sulfur bridges.
17. Use the results from the phosphate determination to calculate the maleimide concentration in the micelle formulation. Only 1–10% of the total maleimide groups should react with the ligand.
18. In case of micellar iron oxides, purification is done after one and a half days of conjugation. During those 36 h, the reaction mixture can be stored at 4°C.
19. Sample concentration should cover a wide range, e.g., 1:10, 1:100, 1:1000 dilutions could be prepared. Make sure no bubbles are formed during preparation, i.e., do not use vortex mixer but gently pipette the solution up and down.
20. The temperature in the sample should be stabilized. Let the sample equilibrate for 3 min plus 1 min for every degree of difference in temperature between actual sample temperature and the measuring temperature.
21. Make sure the total lipid concentration remains above the critical micelle concentration (CMC) (16), below which micelles will disassemble.
22. When an injection catheter is used, the animal has to be taken out of the scanner after the initial pre-injection MRI scan. The repositioning of the animal after injection can have implications for the direct comparison of the pre- and post-injection MR images and does not allow measurement directly after injection. The use of a self-made injection system enables injection while the animal resides inside the

scanner, prevents movement of the animal, and allows for continuous scanning.

23. To prevent injection of residual air in the tubing, which can cause air embolisms, the assembled injection system has to be filled in the following order: 100 µL of saline solution for flushing; the prepared volume with the micelle formulation; and finally 50 µL of heparinized saline (5000 I.U. heparin in 100 mL of saline solution) to prevent blood coagulation inside the tubing after needle insertion. In between the three fluid compartments, a small volume of air can be kept to prevent mixing of the fluids. Also, this air bubble is very practical to observe movement of the fluids inside the tubing.

24. The sequence timing parameters TE, TR, TI, and α dictate the registered signal intensity from the measured object. For the spin-echo sequence the following equation holds:

$$S \propto \rho \, e^{-\frac{TE}{T_2}} \left(1 - e^{-\frac{TR}{T_1}}\right).$$

Here ρ is the proton density. If TE << T_2 the first exponential term can be neglected and the signal will mainly depend on T_1. If, on the other hand TR >> T_1, the second term within brackets reduces to 1, by which the signal becomes T_2 dependent.

For the inversion recovery spin-echo sequence, the following equation holds:

$$S \propto \rho \, e^{-\frac{TE}{T_2}} \left(1 - 2e^{-\frac{TI}{T_1}} + e^{-\frac{TR}{T_1}}\right).$$

Here TI and TR can be optimized as to induce highest contrast before and after injection of the contrast agent. For a gradient-echo sequence (with gradient and RF spoiling) the signal intensity is

$$S \propto \rho \, \frac{\left(1 - e^{-\frac{TR}{T_1}}\right) \sin(\alpha)}{1 - \cos(\alpha) \, e^{-\frac{TR}{T_1}}} e^{-\frac{TE}{T_2^*}}.$$

For low-excitation flip angle α, long TR, and long TE, the signal will mainly depend on the T_2^* relaxation time. For higher flip angle, short TR, and short TE, the gradient-echo sequence is T_1-weighted.

25. A black blood preparation sequence saturates the magnetization of the blood, thereby zeroing its signal. The saturation is accomplished by applying a slice-selective 90°

saturation pulse upstream of the imaging slices. The resulting transverse magnetization is spoiled by application of a crusher gradient. The saturated blood will flow into the imaging slices and therefore blood vessels will appear dark in the images. The thickness of the saturation slice may be chosen between a few millimeters to a few centimeters. The saturation slice can be placed on both sides of the imaging slices for more effective saturation of blood entering from both sides into the imaging slices.

Acknowledgments

The authors would like to thank Prof. Holger Grüll (Department of Biomedical Engineering, Eindhoven University of Technology, Eindhoven, The Netherlands) for kindly providing the magnetite nanoparticles used to prepare the superparamagnetic micelles and Dr. Erik Sanders (Biomedical NMR, Department of Biomedical Engineering, Eindhoven University of Technology, Eindhoven, The Netherlands) for the micelle characterization by cryogenic transmission electron microscopy. The presented work was supported by the Dutch BSIK program "Molecular Imaging of Ischemic Heart Disease" (grant number: BSIK03033); the Netherlands Heart Foundation program grant "Screening for rupture-prone atherosclerotic plaques with molecular MRI" (grant number: 2006T106); the EU-sponsored Network-of-Excellence Diagnostic Molecular Imaging (DiMI), aimed at the advancement of molecular imaging technologies for cardiovascular and neurological applications (grant number: LSHB-CT-2005-512146); and the EU-sponsored integrated project MEDITRANS, a multidisciplinary consortium dealing with image-guided, targeted delivery of novel nano-medicines for the treatment of chronic inflammation, cancer, and neurodegeneration (grant number: NMP4-CT-2006-026668).

References

1. Weissleder, R., and Mahmood, U. (2001) Molecular imaging. *Radiology* **219**, 316–333.
2. Mulder, W. J., Strijkers, G. J., van Tilborg, G. A., Griffioen, A. W., and Nicolay, K. (2006) Lipid-based nanoparticles for contrast-enhanced MRI and molecular imaging. *NMR Biomed* **19**, 142–164.
3. van Tilborg, G. A., Mulder, W. J., Deckers, N., Storm, G., Reutelingsperger, C. P., Strijkers, G. J., and Nicolay, K. (2006) Annexin A5-functionalized bimodal lipid-based contrast agents for the detection of apoptosis. *Bioconjug Chem* **17**, 741–749.
4. Mulder, W. J., Strijkers, G. J., Briley-Saboe, K. C., Frias, J. C., Aguinaldo, J. G., Vucic,

E., Amirbekian, V., Tang, C., Chin, P. T., Nicolay, K., and Fayad, Z. A. (2007) Molecular imaging of macrophages in atherosclerotic plaques using bimodal PEG-micelles. *Magn Reson Med* **58**, 1164–1170.
5. Nitin, N., LaConte, L. E., Zurkiya, O., Hu, X., and Bao, G. (2004) Functionalization and peptide-based delivery of magnetic nanoparticles as an intracellular MRI contrast agent. *J Biol Inorg Chem* **9**, 706–712.
6. Dubertret, B., Skourides, P., Norris, D. J., Noireaux, V., Brivanlou, A. H., and Libchaber, A. (2002) In vivo imaging of quantum dots encapsulated in phospholipid micelles. *Science* **298**, 1759–1762.
7. Koole, R., van Schooneveld, M. M., Hilhorst, J., Castermans, K., Cormode, D. P., Strijkers, G. J., de Mello Donega, C., Vanmaekelbergh, D., Griffioen, A. W., Nicolay, K., Fayad, Z. A., Meijerink, A., and Mulder, W. J. (2008) Paramagnetic lipid-coated silica nanoparticles with a fluorescent quantum dot core: a new contrast agent platform for multimodality imaging. *Bioconjug Chem* **19**, 2471–2479.
8. Duconge, F., Pons, T., Pestourie, C., Herin, L., Theze, B., Gombert, K., Mahler, B., Hinnen, F., Kuhnast, B., Dolle, F., Dubertret, B., and Tavitian, B. (2008) Fluorine-18-labeled phospholipid quantum dot micelles for in vivo multimodal imaging from whole body to cellular scales. *Bioconjug Chem* **19**, 1921–1926.
9. Torchilin, V. P. (2002) PEG-based micelles as carriers of contrast agents for different imaging modalities. *Adv Drug Deliv Rev* **54**, 235–252.
10. de Smet, M., van Bochove, G. S., Sanders, H., Arena, F., Mulder, W. J., Krams, R., Merkx, M., Strijkers, G. J., and Nicolay, K. (2008) Contrast-enhanced MRI of atherosclerosis with collagen targeted CNA35-micelles. *ISMRM 16th Scientific Meeting & Exhibition*, May 2008, Toronto, Canada.
11. Yu, W. W., Falkner, J. C., Yavuz, C. T., and Colvin, V. L. (2004) Synthesis of monodisperse iron oxide nanocrystals by thermal decomposition of iron carboxylate salts. *Chem Commun (Camb)*, **20**, 2306–2307.
12. Degiorgio, V. (1985) *Physics of Amphiphiles: Micelles, Vesicles, Microemulsions.* North-Holland Physics Publishing, Amsterdam.
13. Cevc, G. (1993) *Phospholipids Handbook.* Marcel Dekker Inc., New York.
14. van Tilborg, G. A., Mulder, W. J., Chin, P. T., Storm, G., Reutelingsperger, C. P., Nicolay, K., and Stijkers, G. J. (2006) Annexin A5-conjugated quantum dots with a paramagnetic lipidic coating for the multimodal detection of apoptotic cells. *Bioconjug Chem* **17**, 865–868.
15. van Tilborg, G. A., Mulder, W. J., Dekkers, N., Storm, G., Reutelingsperger, C. P., Stijkers, G. J., and Nicolay, K. (2006) Annexin A5-functionalized bimodal lipid-based contrast agents for the detection of apoptotis. *Bioconjug Chem* **17**, 741–749.
16. Ashok, B., Arleth, L., Hjelm, R. P., Rubinstein, I., and Onyuksel, H. (2004) In vitro characterization of PEGylated phospholipid micelles for improved drug solubilization: effects of PEG chain length and PC incorporation. *J Pharm Sci* **93**, 2476–2487.

Chapter 35

Tracking Transplanted Cells by MRI – Methods and Protocols

Michel Modo

Abstract

Cell tracking by magnetic resonance imaging (MRI) is an essential tool to understand the integration and migration of transplanted cells in vivo. At present, however, techniques to visualize cell transplants in patients are fairly limited and further development of cellular MRI is needed to advance the monitoring of grafted cells. The use of contrast agents to pre-label cells prior to transplantation is currently needed as transplanted cells integrate seamlessly into existing parenchyma and hence are indistinguishable from host cells. The development of appropriate contrast agents, as well as their in vitro incorporation into cells, is key to visualizing transplanted cells in vivo. We describe here procedures regarding how the in vitro incorporation of MR contrast agents can be tested, how they might affect cellular functions and how we can determine if sufficient contrast agent has been incorporated to allow detection. Before this technique can find its clinical application, in vitro and preclinical in vivo studies need to be conducted to determine the safety and specificity of this approach.

Key words: Cell transplant, cellular MRI, contrast agent, cell tracking, migration, stem cell, gadolinium, iron oxide, fluorine.

1. Introduction

Cell therapy is rapidly emerging as a novel paradigm to treat diseases characterized by cell loss or dysfunction (1). A major obstacle to monitor this exciting treatment opportunity, however, is how to identify, track and visualize the integration of grafted cells non-invasively in patients over many months (2). Post-graft survival and its contribution to functional improvements in many cases are thought to be crucial to the beneficial effects. Additionally, guiding the implantation of cells into deep-seated tissues, monitoring organs for potential neoplastic transformations of transplanted cells, as well as cellular rejection, are all important aspects of cell therapy that ideally is evaluated non-invasively by cellular MRI.

Incorporation of an MRI contrast agent is, nevertheless, required to afford a distinction between transplanted and host cells, as grafted cells integrate seamlessly into existing parenchyma. Several types of contrast agents based on gadolinium, manganese, iron oxide or fluorine have been described for cellular tracking (3). The most commonly used contrast media are ferumoxides, mostly due to their superparamagnetic relaxation properties that provide added sensitivity. Especially on T_2^*-weighted images, small quantities of iron cause a blooming effect (i.e. the affected area is about 50 times larger than the contrast particle) that permits the detection of a small numbers of cells (4). The use of micron-sized particles of iron oxide (MPIOs) even allows the visualization of single cells by MRI (5). In contrast, fluorine (^{19}F) agents have a comparatively poor signal-to-noise and hence a large quantity of cells is needed for detection. The advantage of ^{19}F agents is that they do not interfere with anatomical ^{1}H images typically used to assess pathology. The choice of an appropriate agent is largely dependent on the number of cells that need to be detected and on potential deleterious effects contrast agents might exert on cellular functions (**Fig. 1**) (3).

The development of bimodal agents, i.e. probes that can be detected by more than one imaging modality, provide a further development that especially for in vitro studies provides an efficient system to study the incorporation of a contrast agent and its location within the cells (6–8). It is important here to establish that a contrast agent is indeed inside the cells rather than loosely attached to the outer cell membrane as this could lead to the agent being dissociated from the cell and provide false positives upon implantation. It is also noteworthy that some contrast agents will have different relaxation characteristics depending on their cellular location (9). Although these agents provide additional benefits over currently available clinical probes, at present these agents are only used for experimental studies.

The clinical translation of cellular MRI has recently been described with a clinical trial assessing Endorem-labelled dendritic cell placement in patients with lymphomas (10) and the migration of stem cells in a patient with traumatic brain injury (11). Nevertheless, so far, cell tracking has not been robustly implemented in the clinical setting and still requires a further systematic pre-clinical validation to ensure the safety of this procedure. We present here the methodological framework in which the effects of MRI contrast agent incorporation in cells can be assessed prior to progressing to in vivo experiments.

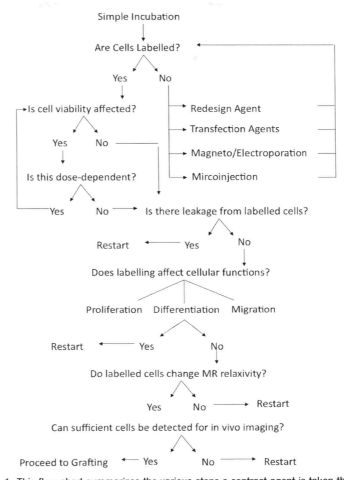

Fig. 1. This flow chart summarizes the various steps a contrast agent is taken through to ensure that it does not deleteriously affect cellular functions, while at the same time to ensure that it causes sufficient signal change to be detected by MRI.

2. Materials

2.1. Laboratory Consumables

1. 24 well plates (Corning, UK)
2. Anti-LAMP antibody (Abcam, UK)
3. Anti-dextran antibody (STEMCELL Technologies, USA)
4. Anti-Ki67 (Novocastra, UK),
5. FITC-conjugated antibody (Abcam, UK)
6. Alexa 488 (Molecular Probes, UK)
7. DAPI in Vectashield (Molecular Probes, UK)

8. Hoechst (Sigma, UK)
9. Sigma Fast DAB Tablets (Sigma, UK)
10. Potassium ferrocyanide stain (Sigma, UK)
11. DPX (Merck, UK)

2.2. Chemicals

1. Lipofectamine 2000 (Invitrogen, UK)
2. Trypan blue assay (Invitrogen, UK)
3. Fluorescein diacetate assay (Sigma, UK)
4. CyQUANT assay (Invitrogen, UK)
5. MTT assay (Sigma, UK)
6. [^{14}C]-Leucine (1 mL of 1.85 MBq; Amersham International, UK)
7. (5- and 6)-chloromethyl-2′,7′-dichlorodihydrofluorescein diacetate, acetyl ester (CM-H2DCFDA) solution (Invitrogen, UK)
8. Triton X-100 (Roche, UK)
9. Isopropanol (Sigma, UK)
10. HCl (Sigma, UK)
11. Nitric acid (Sigma, UK)
12. Indium (Sigma, UK)
13. Paraformaldehyde (Sigma, UK)

2.3. MRI Contrast Agents

1. Endorem/Feridex (Guerbet, France/Berlex, USA)
2. Gadophrin-2 (Schering, Germany)
3. Multihance (Bracco, Italy)
4. Primovist (Schering, Germany)
5. Resovist (Schering, Germany)
6. Teslascan (Amersham, USA)
7. CS-1000 (Celsense, USA)

2.4. Laboratory Hardware

1. Packard FilterMate (Packard Instruments, UK)
2. Packard Matrix 9600 β-counter (Packard Instruments, UK)
3. MRI scanner (Varian, UK) (**Note 1**)
4. DTX-880 Multimode detector (Becton Dickinson, UK)
5. Centrifuge 5804 R (Eppendorf, UK)
6. Oven incubator (Thermo Scientific, UK)
7. Inductively coupled mass spectroscopy (Varian, UK)
8. Hematocytometer (VWR, UK)

3. Methods

3.1. Cell Culturing and Labelling

3.1.1. Cell Culturing

To ensure that cell labelling with an MRI contrast agent is efficient and not affecting cellular functions, it is important to have cells growing in the lab. Most procedures for cell labelling take a few days and it is hence crucial to maintain viable cells during this time. For cell lines or frozen cells that are revived, it is important to proliferate or grow these for at least one passage in the laboratory to ensure that they are healthy and not infected prior to labelling. Once these baseline characteristics are established for cells, it is possible to proceed with the labelling (**Note 2**).

1. Grow cells according to the standard protocol as used in the lab.
2. Ensure that you grow the cells for at least one passage prior to labelling to ensure that they are healthy and not infected.
3. Cells are then labelled with the MRI contrast agent (*see* **Section 3.1.2**).

3.1.2. Labelling

The choice of contrast agent will depend on how many cells and which type of cell one intends to visualize after transplantation. Incorporation of clinically approved contrast agents that can be used for tracking grafted cells are Fe-based agents, Mn-based agents or Gd-based agents (12). Ideally, cellular MRI does not interfere with the general assessment of pathology. The use of alternative nuclei for MR imaging, such as ^{19}F, might therefore be exploited (13).

Pinocytosis. Some cells specifically take up particular contrast agents. For instance, hepatocytes easily take up cell-specific Gd- (e.g. Eovist) or Mn-contrast agents (e.g. Teslascan), whereas Kupffer cells take up Fe-based agents (e.g. Endorem/Feridex). Labelling a particular cell population can therefore be facilitated by choosing the appropriate agent designed to be taken up by this type of cell. Some contrast agents with a small molecular weight, such as Gd-based agents, might also get taken up in vitro into the cells through fluid phase pinocytosis. For this:

1. Culture cells according to standard protocol
2. Following overnight incubation of cultures, add the contrast agent at appropriate concentrations to fresh media and gently shake the culture to ensure good mixing of the contrast agent with the media. The concentration of contrast agent will depend on the contrast agent and type of cell. Clinically approved agents generally come in a prepared solution and it is recommended to start with three sets of concentrations 1:1, 1:10, 1:100. A further refinement of this dilution assay is needed to determine the best molar concentration

for cell labelling. Knowing the molar concentration will be important to assess the relaxivity characteristics of the contrast agent. Typically, iron oxide-based agents will be in the range of micromolars, whereas Gd-based agents will be in the range of millimolars.

3. Duration of incubation. Certain cells, such as Kupffer cells, rapidly incorporate contrast agents and incubation times of <2 h can be sufficient for cell labelling. However, duration of incubation also depends on the concentration of contrast agent in the media. The advantage of bimodal fluorescent agents is that during this process, it is possible to assess cellular uptake of the contrast agent dynamically under an inverted fluorescent microscope.

4. After sufficient contrast agent has been incorporated into the cells, wash the cells three times with PBS before adding media for further experimentation.

Transfection Agents. Although liver cells will easily take up various contrast agents, in some cases it might be desirable, for instance, to label hepatocytes with iron oxide particles that typically are not incorporated into these types of cells. The use of transfection agents can enable this process and ensure sufficient cellular uptake of particles to allow a reliable detection. For this:

1. Prepare transfection solution with 5 µL of Lipofectamine 2000 in 25 µL culture media for each well on a 24 well plate.

2. Mix contrast agent with transfection solution for 10 min on a shaker at room temperature. The contrast agent concentration will determine how much agent needs to be mixed with the transfection agent. A typical guidance is about 100 µg of ferumoxides to 5 µL of transfection agent (**Note 3**).

3. Incubate for 2–3 h in a 1:1 mixture of serum-free media and transfection agent-coated contrast agents. However, specific incubation times will depend on the contrast agent, the transfection agent and type of cells. These concentrations need to be refined experimentally.

4. Remove supernatant and wash cells three times with PBS before adding culture media for further experimentation.

3.2. Visualization of Contrast Agent Inside Cells by Microscopy

To ensure that the contrast particles were indeed incorporated into the cells it is necessary to visualize these using microscopy. Although relaxivity measurements (*see* **Section 3.4**) can indicate the presence of contrast media, it does not provide an indication if particles are indeed intracellular. Often contrast particles accumulate on the surface of the cells and even after a couple of washes these remain. It is therefore possible that it appears as if contrast agent incorporation occurred, although this is not the case. The only way to ensure that particles are inside the cells, rather than outside, is to perform microscopy. This can be through brightfield (e.g. Perl's stain to detect iron), fluorescence (anti-dextran antibody to recognize dextran used to chelate many contrast agents) or electron microscopy (EM). Brightfield and fluorescent microscopy are widely available in cell culture laboratory, whereas EM requires highly specialized equipment and access is more difficult. Nevertheless, EM can unequivocally identify metal particles inside the cells and even determine their location.

3.2.1. Detecting Iron Particles

Iron particles can be detected histologically by Perl's stain (14). For this:
1. Prepare Perl's solution with 1.0 g of potassium hexacyanoferrate (ferrocyanide), 25 mL of distilled water and 25 mL of 13% hydrochloric acid. This solution should be freshly prepared.
2. Wash cells/tissue with distilled water and add Perl's solution for 20–30 min.
4. Wash cells/tissue with distilled water and add neutral red to the tissue sections for 1–2 min. For cells, this step can be omitted as neutral red counterstains tissues in shades of red.
5. Rinse with tap water and dehydrate with graded alcohols (70%, 80%, 100%), clear and mount
6. Ferric iron will appear blue (**Note 4**).

3.2.2. Detecting Contrast Agents Based on Dextran

Many MRI contrast agents use dextran as a chelating agent and an immunocytochemical approach can therefore be used to specifically detect dextran-based contrast agent:
1. Label cells with contrast agent.
2. Permeabilize cells/tissue with a 0.1% Triton X solution for 5 min.
3. Rinse cells/tissue with PBS.
4. Add FITC-conjugated anti-dextran antibody at 1:1000 dilution in PBS to the cells/tissue.
5. Rinse with PBS.
6. Counterstain all cell nuclei with DAPI or Hoechst.
7. The contrast agent will appear in green, whereas cell nuclei in blue.

The green fluorescent signal should be clearly localized to particles within the cells. Nestin can serve as cytoplasmic dye in neural stem cells to determine if particles are inside or outside the cell. If there is a diffuse staining or an absence of staining, this could indicate that the dextran chelate is being downgraded within the cells or that there are no particles present within these cells, respectively.

3.2.3. The Use of Bimodal Agents

The use of bimodal contrast agents facilitates the visualization of cellular uptake (**Note 5**). Due to the fluorescent moieties in these agents, it is possible to directly visualize the contrast agent as it is taken up into the cells under an inverted fluorescent microscope. A bimodal agent currently undergoing clinical development consists of gadophrin-2 which can be used to label stem cells for cellular imaging after transplantation (8).

3.2.4. Intracellular Localization of Contrast Agent

It is also possible to easily determine the cellular compartments within which the agent is trapped within the cells (15). This is an important aspect in terms of the degradation of contrast particles within the cells, but is also relevant to the agent's relaxivity as this will depend on its access to bulk water or ability to freely tumble (for Gd- and Mn-agents). Iron oxide agent relaxivity is less affected by its intracellular localization.

1. Label cells with MRI contrast agent.
2. Wash three times with PBS prior to fixing cells with 4% paraformaldehyde.
3. Prepare primary antibody solution for cell organelles, such as lysosome-associated membrane protein 2 (LAMP2), by diluting the antibody 1:200 in PBS.
4. Add 250 μL of this solution to each well of a coverslip with contrast agent-labelled cells in a 24 well plate and leave overnight on a shaking platform (if a bimodal contrast agent is used, ensure that this will be in the dark to avoid photobleaching).
5. Remove primary antibody solution and replace with an appropriate secondary fluorescent antibody (e.g. Alexa 488 at 1:500) or develop using an appropriate brightfield method (e.g. DAB).
6. Use a microscope to determine if the contrast agent is inside or outside the cellular compartment.

3.3. Assaying the Effects of Contrast Agents on Cell Function

3.3.1. Cell Viability – Fluorescein Diacetate (FDA)

It is essential to measure the viability of the cells after cell labelling. Contrast agents contain metal particles that are rendered non-toxic through the use of chelating agents, such as dextran or albumin. Upon degradation of the protective coating, these metal particles can affect cell viability. In some cases, this will only be evident after long-term presence inside the cells and might only be detected after transplantation (16). Moreover, overloading of the cells with contrast agent can also lead to deterioration in cell viability. However, it is not sufficient to just measure cell viability straight after labelling, but ideally more protracted time points relevant to transplantation paradigms should also be investigated (1–7 days). A variety of cell viability assays are commercially available, such as trypan blue, marking all dead cells under brightfield or fluorescein diacetate (live/dead stain kit) to label all viable cells under fluorescence microscopy. This test requires the use of the FDA stock solution (5 mg/mL in DMSO) (17) and cells must be adherent to glass coverslips.

1. Following labelling with an MRI contrast agent, remove the culture medium and gently rinse twice with PBS.
2. Replace PBS with 200–300 μL medium containing FDA (2 μg/mL; final concentration) and incubate the cultures for 6 min at room temperature.
3. Remove medium and gently rinse twice with PBS.
4. Counterstain cell nuclei with DAPI or Hoechst.
5. Use a fluorescence microscope to check for green fluorescence in the cytoplasm of the viable cells only.
6. A quick semi-quantitative estimate of viable cells (green) could be carried out in, e.g. five random fields. DAPI-stained nuclei could be counted as this will help in estimating the approximate number of total cells in the counted field and determine the degree of toxicity of a particular contrast agent or protocol.

3.3.2. Cell Proliferation

Incorporation of contrast agents into cells will lead to their compartmentalization within the cells. As cells divide, the amount of contrast agent between cells will also decrease with time. However, contrast agents can also affect the cells' ability to proliferate by interfering with basic cell functions involved in mitosis. Either the counting of proliferating cells based on the number of cells in a culture dish or the use of antibodies, such as Ki67, labelling all dividing cells can be used for this. Moreover, various commercially available assays, such as CyQuant, are available that assess various aspects of proliferation and often are taken as a measure of proliferation. In some cases, the 3-(4,5-dimethylthiazol-2-yl)-2,5-diphenyltetrazolium bromide (MTT) assay is used to assess proliferation, but it is important to know that this test only indirectly reflects proliferation by measuring mitochondrial activity that is related to mitosis. Mitochondrial activity could be increased or decreased due to other reasons (such as the cells actively degrading iron particles) rather than due to proliferation.

3.3.3. Mitochondrial Activity

A commonly used assay to determine the overall metabolic activity of cells is based on mitochondrial dehydrogenases activity of the 3-(4,5-dimethylthiazol-2yl)-2,5-diphenyltetrazolium bromide (MTT) assay (18):

1. Label Cells with MRI contrast agent.
2. Prepare the MTT assay solution with 5 mg/mL in PBS (pH adjusted to 7.2 and filtered through 0.2 μm filter). This solution can be stored in the dark at 4°C for up to 2 weeks.

3. Dilute MTT solution in culture medium (1:10) and incubate with cells for 4 h.

4. Remove media with MTT and place 20 μL of 0.25% trypsin per well in a 96 well plate and place on a shaker for 5 min at high speed.

5. Add 100 μL of isopropanol with 0.04 N HCl and place on shaker for 15 min. This will dissolve the formazan.

6. Measure absorbance at 595–655 nm to quantify mitochondrial activity.

It is important to include appropriate control conditions (such as no MTT and no contrast agent) and express fluorescence absorbance in relation to these controls.

3.3.4. Protein Synthesis – [^{14}C]-Leucine Incorporation Assay

This assay is used as an indirect functional assay that reflects the overall synthetic activity of the cells (14). [^{14}C]-leucine is a radioactive amino acid that gets incorporated in proteins. Following incubation of the cell cultures, the cells are harvested onto a glass membrane and the membrane is dried followed by the radioactivity being counted using a Packard Matrix 9600 β-counter (Packard Instruments, Berkshire, UK). The counts are presented as counts per minute (cpm). If the MRI contrast agent used has cytotoxic effects, it is expected to result in lower counts (19).

This assay requires the use of a 96-well cell culture plate. At the time of replacing the culture medium with medium containing the contrast agent:

1. Add [^{14}C]-leucine solution to the medium to give a final dilution of 0.2 μCi/well.

2. Label plate with appropriate radioactivity warning signs (**Note 6**).

3. Incubate the plate for the required period of labelling.

4. Post incubation, the cells are harvested and their membranes are analysed (*see* Step 5) or the plate could be sealed with parafilm to be stored at −20°C for later analysis.

5. At the time of harvesting cells, the plate temperature should approximate room temperature, i.e. frozen plates must be completely defrosted.

6. Cells are harvested onto a glass membrane using the cell harvester.

7. The membrane must be dried in an oven (50–60°C) for 2–3 h. Ensure that the membrane is completely dry; otherwise, contamination of the counter will occur.

8. Count the radioactivity of the membrane using the β-counter for 6 min and calculate counts per minute (cpm).

High counts mean a high level of protein synthesis and this will indirectly reflect the level of cellular synthetic activity.

3.3.5. Reactive Oxygen Species

Reactive oxygen species (ROS) should be measured in response to cell labelling to determine if the procedure produces any stressors to the cells. ROS should be measured straight after labelling and at least 24 h post-labelling to determine if these cells are undergoing a continued stress or if it is only a transient phenomenon associated with the labelling procedure rather than the presence of the contrast agent. It is possible that some of the contrast agents are degraded inside the cells. This can lead to the production of ROS and result in cell death.

1. Label cells with MRI contrast agent.
2. Prepare the (5- and 6)-chloromethyl-2′,7′-dichlorodihydrofluorescein diacetate, acetyl ester (CM-H2DCFDA) solution by diluting 50 µg in 8.6 mL of PBS.
3. Add 1 mL of this solution to each well of a 24 well plate and leave on cultures for 60 min.
4. Wash three times with PBS prior to fixing cells with 4% paraformaldehyde.
5. Quantify green fluorescence in FITC channel. If ROS are present these will emit a green fluorescent light.

If no ROS are present, no green fluorescence will be emitted. It is important to include a control condition to gain an estimate of the natural background.

3.3.6. Cellular Differentiation

In the case of undifferentiated cells, such as stem cells, it is an essential component of the in vitro testing regime that any potentially deleterious effects on differentiation are investigated. Although mostly, this functional aspect is omitted from most studies, the overall consensus is that the differentiation of most cell types is not affected by contrast agents. Nevertheless, a few reports indicate that it is possible for contrast agents to significantly affect differentiation and it is therefore important to establish in each type of cell if this might be the case. For instance, the use of iron oxide particles has been linked to an inhibition of chondrogenesis in mesenchymal stem cells (20). However, the appropriate conditions for these experiments are highly dependent on the type of cell. It is therefore best to first establish the standard differentiation properties of cells and then to investigate if contrast agent labelling significantly affects this process.

3.4. MR Relaxometry

Magnetic resonance images are very dependent on the sequences used to acquire the images. Sequences can be designed to highlight fluids (such as on T_2-weighted images) or to be fairly insensitive to fluids (such as on T_1-weighted images). The effect of contrast agents on the signal in images will depend on the type of sequences that are being used to scan a sample, but are also dependent on the strength of the magnet. As field strength increases, the signal-to-noise and spatial resolution increases. However, contrast agents do not necessarily follow the same principle. It is therefore important to bear these factors in mind if clinical translation is envisaged. Most pre-clinical and cell-labelling experiments are conducted on high field strength magnets (>4.7 T), whereas most clinical studies are conducted at either 1.5 T or 3 T possibly resulting in too little signal to detect transplanted cells.

To determine if sufficient contrast agent has been incorporated into the cell to effect a signal change on MRI, relaxometry needs to be conducted to quantify the relaxation signal on an MR image. For this:

1. Label cells with contrast agent.

2. Cells are placed into a vial (e.g. Eppendorf tube) with media or PBS.

3. Comparisons should include cells with no contrast agent, media/PBS and distilled water to determine the specific change that the contrast agent induces inside the cells.

4. Insert Eppendorf into the coil of the scanner using either custom-made holders for the Eppendorfs or embed Eppendorfs into agarose gel for scanning. It is also possible to add cells directly into agarose gels and to insert these into the coils for scanning. In some cases, agarose gels are preferred as the signal of these often resembles that of tissue. Ideally, several comparisons can be run in one scanning session (a standard control should be included for all scans). For measuring the relaxivity of a particular contrast agent, a 1 M solution can be used to express the molar relaxivity of the compound.

5. Scanning parameters will depend on the scanner hardware. On a Varian 7 T system, the following can be used to measure T_1 (fast spin echo sequence: TRs = 100, 500, 1000, 2500, 5000, 10,000 ms; TE = 10 ms; FOV = 50 mm × 50 mm; matrix 128 × 128, two averages) or T_2 (multi-slice multi-echo sequence: TR = 2000 ms; TE = 15 ms; number of echoes = 8, FOV = 50 mm × 50 mm; matrix 256 × 256, two averages). However, it should be noted that different methods are available to measure relaxivity and the specific choice will depend on its application. T_1 and T_2 relaxivity should be measured for all conditions. It is recommended that scanning sequences for relaxometry should be set up by an experienced MR physicist.

6. Create a T_1 or T_2 map in VnmrJ (Varian software). Use these maps to measure relaxivity. Alternatively, measure signal values for each image and plot these prior to calculating relaxivity (**Fig. 2**).

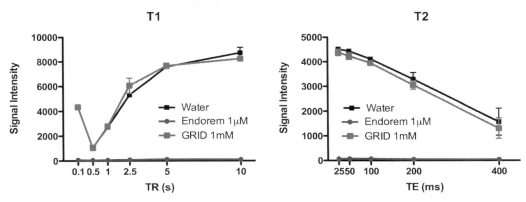

Fig. 2. A comparison of T_1 and T_2 relaxivity measurements of a gadolinium-based (GRID), an iron oxide-based contrast agent (Endorem) and distilled water. GRID causes a signal increase with a very short TR on T_1, but does not significantly affect T_2. In contrast, Endorem results in a dramatic signal loss in both T_1 and T_2. This strong signal decrease in T_2 can easily be exploited to see a small number of cells, whereas the signal increase on T_1 is much more subtle and makes it challenging to achieve a sufficient detection of transplanted cells.

7. On maps, measure the signal for each condition. The inverse of the mean value will be the relaxivity in milliseconds.

8. These values can be plotted to determine what concentration of agent will yield the strongest decrease in signal for T_2 agents or the highest increase for T_1 agents (**Note 7**). It will further help to determine what parameters might be best to visualize the labelled cells in vivo.

Calculating the relaxivity of contrast agents within cells will be essential to determine how many cells can be detected. This will also provide the basis for deciding which contrast agent is more effective for identifying transplanted cells in vivo by MRI. Once the best concentration of contrast agent has been determined, it is recommended to establish a curve that determines how many cells can be detected using this approach. Typically, a signal change of >10% is the target for these experiments.

3.5. Quantifying Cellular Uptake of MRI Contrast Agents

Inductively coupled plasma mass spectrometry (ICP-MS) can be used to quantify the uptake of various contrast agents into cells. For this:

1. Label cells with contrast agent of interest.

2. Use sufficient quantities to yield >60,000 cells in 250 µL of media. However, it is essential to count the number of cells per condition using a hematocytometer. An exact number of cells is needed to calculate how much contrast agent was taken up per cell.

3. Ideally, include control conditions with only cells, media plus contrast agent and only media to define the background or noise in the measurements.

4. Digest the sample overnight with an equal volume of nitric acid.

5. From this digested sample, add 0.25 mL to a mixture of 0.05 mL indium (serving as internal control), 0.3 mL of nitric acid and 9.4 mL of distilled water.

6. A sample of this mixture is then sprayed into the ICP-MS (this needs to be done by an experienced researcher).

7. From these results, calculate the amount of contrast agent per cell by dividing the total amount of particles (expressed in moles or milligrams) in the total sample by the number of cells to yield a concentration of mol/cell or mg/cell.

Knowing the amount of mol/cell will provide the basis to calculate after how many cell divisions it will no longer be possible to detect cells by MRI and it will also help to determine if the amount of contrast agent per cell needs to be increased to ensure a more reliable detection. This measure will also be essential to determine how effective a particular agent is in changing relaxivity. If a large quantity of intracellular contrast agent is needed to affect relaxivity, it is preferable to chose a contrast agent that is more effective and would require less cellular uptake.

4. Notes

1. Various MRI vendors for animal imaging are available. Relaxivity measurements can also be performed on lower field strength magnets. It is important to bear in mind though that in vitro measurements of cellular relaxivity are best conducted at the same field strength than in vivo measurements, as signal changes due to contrast agents are field dependent. Typically higher fields (>7 T) are preferred.

2. It is important to consider what type of cells one is labelling. Phagocytic cells (e.g. macrophages) happily take up even large MRI particles, whereas non-phagocytic cells (e.g. neural stem cells) are more difficult to label. This will also determine if additional strategies (e.g. transfection agents, electroporation) are needed to incorporate agent efficiently into the cells.

3. Uptake of contrast agents and the influence of transfection agents is highly cell-type dependent and concentrations need to be determined empirically. It is also important to note that different transfection agents have very distinct properties that might be more desirable for one type of cell, but not for another. It is important to initially dedicate sufficient

time to these issues to ensure a robust labelling protocol can be set up.

4. Perl's stain detects all iron and therefore will also pick up other cells that contain iron, such as macrophages or blood cells. Ideally, this method is therefore only used in vitro or in tissue that does not have cells that naturally contain large quantities of iron. As liver tissue is generally used as a positive control for Perl's stain due to its naturally high content of iron, it is not recommended to use this method to detect iron-based contrast agents inside liver tissue, but in tissues, such as the brain, that are mostly devoid of iron, it can be an effective method to detect iron particles inside cells.

5. Bimodality might also refer to other combinations of imaging modalities and care should therefore be taken that the appropriate imaging modalities can be used to visualize the agent of interest.

6. While using radioactivity, it is important to consider the safety of the experimenter and other people in the laboratory. It is worth noting that any work involving radioactivity requires adherence to institutional guidelines and needs to be discussed with a radiation protection officer.

7. T_1 agents will also affect T_2 values and in some cases either, or a combination of both, can be used in vivo.

Acknowledgements

MM is currently supported by a RCUK fellowship. Funding for the MR imaging of cell transplants by NIBIB Quantum Grant (1 P20 EB007076-01), EU FP VII (201842 – ENCITE) and MRC (G0802552) is gratefully acknowledged.

References

1. Modo. M. (2008). Brain repair how stem cells are changing neurology. *Bull. Soc. Sci. Med. Grand Duche Luxemb.* **2**, 217–257.
2. Modo, M. (2008). Noninvasive imaging of transplanted cells. *Curr. Opin. Organ Transplant* **13**, 654–658.
3. Modo, M., Hoehn, M., Bulte, J.W. (2005) Cellular MR imaging. *Mol. Imaging* **4**, 143–164.
4. Bulte, J.W., Kraitchman, D.L. (2004) Monitoring cell therapy using iron oxide MR contrast agents. *Curr. Pharm. Biotechnol.* **5**, 567–584.
5. Shapiro, E.M., Sharer, K., Skrtic, S., Koretsky, A.P. (2006) In vivo detection of single cells by MRI. *Magn. Reson. Med.* **55**, 242–249.
6. Modo, M., Cash, D., Mellodew, K., Williams, S.C., Fraser, S.E., Meade, T.J.,

Price, J., Hodges, H. (2002) Tracking transplanted stem cell migration using bifunctional, contrast agent-enhanced, magnetic resonance imaging. *Neuroimage* **17**, 803–811.

7. Mulder, W.J., Koole, R., Brandwijk, R.J., Storm, G., Chin, P.T., Strijkers, G.J., de Mello Donega, C., Nicolay, K., Griffioen, A.W. (2006) Quantum dots with a paramagnetic coating as a bimodal molecular imaging probe. *Nano Lett.* **6**, 1–6.

8. Daldrup-Link, H.E., Rudelius, M., Metz, S., Piontek, G., Pichler, B., Settles, M., Heinzmann, U., Schlegel, J., Oostendorp, R.A., Rummeny, E.J. (2004) Cell tracking with gadophrin-2: a bifunctional contrast agent for MR imaging, optical imaging, and fluorescence microscopy. *Eur. J. Nucl. Med. Mol. Imaging* **31**, 1312–1321.

9. Terreno, E., Geninatti Crich, S., Belfiore, S., Biancone, L., Cabella, C., Esposito, G., Manazza, A.D., Aime, S. (2006) Effect of the intracellular localization of a Gd-based imaging probe on the relaxation enhancement of water protons. *Magn. Reson. Med.* **55**, 491–497.

10. de Vries, I.J., Lesterhuis, W.J., Barentsz, J.O., Verdijk, P., van Krieken, J.H., Boerman, O.C., Oyen, W.J., Bonenkamp, J.J., Boezeman, J.B., Adema, G.J., Bulte, J.W., Scheenen, T.W., Punt, C.J., Heerschap, A., Figdor, C.G. (2005) Magnetic resonance tracking of dendritic cells in melanoma patients for monitoring of cellular therapy. *Nat. Biotechnol.* **23**, 1407–1413.

11. Zhu, J., Zhou, L., XingWu, F. (2006) Tracking neural stem cells in patients with brain trauma. *N. Engl. J. Med.* **355**, 2376–2378.

12. Karabulut, N., Elmas, N. (2006) Contrast agents used in MR imaging of the liver. *Diagn. Interv. Radiol.* **12**, 22–30.

13. Ahrens, E.T., Flores, R., Xu, H., Morel, P.A. (2005) In vivo imaging platform for tracking immunotherapeutic cells. *Nat. Biotechnol.* **23**, 983–987.

14. Perls, M. (1867) Nachweis von Eisenoxyd in gewissen Pigmenten. *Virchows Archiven der Pathologie, Anatomie und Physiologie* **39**, 42–48.

15. Brekke, C., Morgan, S.C., Lowe, A.S., Meade, T.J., Price, J., Williams, S.C., Modo, M. (2007) The in vitro effects of a bimodal contrast agent on cellular functions and relaxometry. *NMR Biomed.* **20**(2), 77–89.

16. Modo, M., Beech, J.S., Meade, T.J., Williams, S.C., Price, J. (2009) A chronic 1 year assessment of MRI contrast agent-labelled neural stem cell transplants in stroke. *Neuroimage* **47**(Suppl. 2), T133–142.

17. Friend, J.R., Wu, F.J., Hansen, L.K., Remmel, R.P., Hu, W.S. (1999) Formation and characterization of hepatocyte spheroids. In: Morgan, J.R., Yarmush. M.L. (eds.), *Methods in Molecular Medicine: Tissue Engineering Methods and Protocols* (pp. 248–249). Totowa, N.J: Humana Press Inc.

18. Mitry, R.R., Hughes, R.D., Bansal, S., Lehec, S.C., Wendon, J.A., Dhawan, A. (2005) Effects of serum from patients with acute liver failure due to paracetamol overdose on human hepatocytes in vitro. *Transplant. Proc.* **37**, 2391–2394.

19. Puppi, J., Mitry, R.R., Modo, M., Dhawan, A., Raja, K., Hughes, R.D. (in press) Use of clinically-approved iron oxide MRI contrast agent to label human hepatocytes. *Cell Transplant.*

20. Kostura, L., Kraitchman, D.L., Mackay, A.M., Pittenger, M.F., Bulte, J.W. (2004) Feridex labeling of mesenchymal stem cells inhibits chondrogenesis but not adipogenesis or osteogenesis. *NMR Biomed.* **17**, 513–517.

Chapter 36

MRI of CEST-Based Reporter Gene

Guanshu Liu and Assaf A. Gilad

Abstract

In recent years, several reporter genes have been designed for non-invasive magnetic resonance imaging (MRI). Here, we offer a brief summary of recent advances in MRI reporter gene technology, as well as elaborated protocols for cloning, expression, and imaging of reporter genes based on a chemical exchange saturation transfer (CEST) method. These protocols emphasize new developments in CEST-MRI data acquisition and processing.

Key words: MRI, CEST, MRI reporter gene, in vivo MR imaging, molecular imaging.

1. Introduction

One of the major challenges of the post-genome era is to study the expression of genes in a physiological context. This will involve not just the histological analysis and extraction of DNA, RNA, and protein from tissue specimens, but rather, imaging gene expression patterns in the whole organism. The development of recombinant DNA cloning techniques at the beginning of the 1970s enabled the transfer of genetic material from one organism to another. This subsequently led to the developing of reporter genes. A reporter gene is a gene whose product can be readily detected and is either fused to the gene of interest or replaces it. The main applications for these reporters include:

Monitoring gene expression levels, investigating dynamic molecular interactions between proteins, studying cellular interactions, tracking cells in normal/abnormal development or in cell transplantation therapy, and monitoring gene replacement therapy. The majority of reporter genes were developed for optical imaging. These genes can encode to proteins that emit light upon absorption of photons at a specific frequency (fluorescence) or at the cleavage of a substrate (bioluminescence). In both cases, the released photon is detected in a frequency-dependent manner using a charge-coupled device (CCD) to generate an image. However, optical imaging modalities are limited by the short depth of light penetration through tissue (few millimeters) and thus are useful only for imaging superficial tissues. Throughout the years, reporter genes have been also developed for other imaging modalities, including nuclear imaging (1).

The quest to develop reporter genes for magnetic resonance (MR) began with over-expressing creatine kinase (2) and arginine kinase (3). The transgene expression was detected via metabolic changes with MR spectroscopy. Later, the over-expression of proteins that are naturally involved in iron metabolism (4, 5) and storage (6, 7) was used for MR imaging. The rationale was that iron is a (super) paramagnetic metal, which can provide MRI contrast. A different strategy was to express a transgenic enzyme, β-galactosidase, which converts certain compounds to (super) paramagnetic contrast agents (8, 9). With the exception of ferritin (6, 7), all other MRI reporter genes rely on the administration of a substrate that may not be accessible to all tissues, especially to the central nervous system (CNS).

An alternative reporter gene was based on previous observations that poly-L-lysine has a uniquely high chemical exchange rate, suitable for CEST MRI (10, 11). An artificial gene called LRP (lysine-rich protein), which has a high percentage of lysine residues, was designed de novo. This gene provided contrast based on CEST MRI that was sufficient to distinguish 9L rat glioma cells over-expressing the transgene from control 9L cells (12). A broader overview on MR reporter genes can be found in the reviews (13, 14).

Chemical exchange saturation transfer (CEST) is a new type of MRI contrast that has been recently developed to highlight the exchangeable protons by reducing the bulk water signal through chemical exchange (15). CEST is implemented by applying a saturation radiofrequency (RF) pulse, or pulses, at the exchangeable proton resonance frequency for a long duration to saturate the proton's magnetization through proton chemical exchange. Since these protons constantly exchange with bulk water protons, they can be detected as a reduction in the water proton MR signal (16). In addition to a vast number of applications (a comprehensive summary can be found in recent papers (16, 17) *see* also **Chapter 9**), CEST imaging shows great promise for the imaging of reporter genes (12). There are three major advantages for using a CEST reporter gene. First, biocompatible peptides produced through the genetic manipulation can be used, rather than metal-based exogenous contrast agents, as MRI-visible reporters, which can reduce the risk of disturbing the microenvironment of cells. Second, CEST contrast is switchable, which does not interfere with other MRI contrasts, such as T_1 and T_2, when the saturation pulse is switched off. The most desirable advantage, however, is the potential to create different "colors," by applying a saturation pulse at different frequencies, for simultaneously imaging more than one target, as optical imaging often does (18). With the aid of this multicolor approach, it is possible to highlight multiple cells or the expression of multiple genes in a CEST MRI image.

As a new MRI technique, CEST imaging is still experiencing rapid development and several major technical hurdles must be overcome before it can quickly expand to a wider range of applications. One such hurdle is the main magnetic field B_0 inhomogeneity. Magnetic field inhomogeneity could eliminate the real contrast or create large artificial contrast due to the frequency misregistration across the region of interest (19, 20). Post-acquisition processing to correct B_0 inhomogeneity, therefore, is crucial in CEST in addition to a standard shimming procedure. *Water saturation shift reference* (WASSR) is a new B_0 mapping technique (20) that can accurately determine the absolute water frequency of each pixel, based on sweeping a low power, short saturation pulse at a variety of frequencies near the water resonance. This approach uses the same pulse sequence as the CEST imaging to avoid an additional image registration procedure. In order to obtain a "pure" direct water saturation spectrum with minimal interference from CEST and conventional magnetization transfer (MT), which is essential to determine the correct center of the water spectra, the WASSR method employs short and weak saturation pulses. The procedure for acquiring and processing WASSR data will be introduced in this protocol.

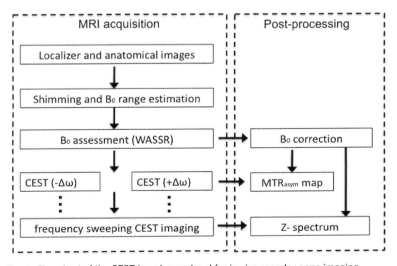

Fig. 1. Flowchart of the CEST imaging protocol for in vivo reporter gene imaging.

The current paradigm for quantitative analysis of CEST is to use the z-spectrum so that comprehensive information about the saturation behaviors of a target at all frequencies can be obtained through one study (11). However, such an approach requires a fairly long acquisition time, and therefore, may not be practical for certain applications. Instead, acquisition time can be greatly shortened by using an alternative approach that acquires only several CEST-weighted images around the desired offsets (21). In this protocol, both CEST imaging and z-spectrum analysis will be discussed, from which readers can choose according to their specific purposes. **Figure 1** provides an overview of the entire protocol, with a listing of the necessary steps to acquire CEST data and produce CEST contrast maps.

2. Materials

2.1. General Requirements

1. A high-field (4.7 T and above), small animal narrow bore or clinical MR scanner (3 T and above) with a relatively homogeneous main magnetic field, fast and reliable gradient coils, and a high signal-to-noise RF coil.
2. All the devices for animal anesthesia, motion restraint, physiological monitoring, and body temperature and respiration maintenance are required, for example, an animal holder specifically designed for mouse body imaging with nosepiece adaptor for anesthesia (1–2% isoflurane mixed with air), a physiological (respiration) monitoring and gating system (e.g., SA instruments, Inc., Stony Brook, NY), and an anesthesia induction chamber for use with isoflurane.
3. For CEST imaging, data post-processing is a crucial step to extract CEST information from the raw images; an image processing software is therefore required and can be chosen from a variety of commercial software packages, including MATLAB (Mathworks, Natick, MA), IDL, etc.

2.2. Gene Cloning, Vector Construction, Cell Culture Conditions, and Transfection

The main advantage of the artificial reporter genes based on CEST is the large variability in the amino acid sequence, since different sequences can provide different CEST contrast.
1. Lysine-rich protein (LRP) (12) was a prototype reporter gene and a higher level of expression might be achieved due to codon optimization (minimizing repeats) and amino acid sequence optimization (18). The gene can be generated by cloning, in tandem, of DNA oligomers encoding to lysine (AAA or AAG). Two complimentary synthetic oligonucleotides (84 base pairs long), encoding the artificial desired sequence are required, In addition, a start (ATG)

codon is required to initiate translation, preferably in the context of a Kozak sequence. Cloning the gene in the frame with an antigen tag (such as HA, V5, myc, etc.) could be useful for subsequent detection with immunohistochemistry.

Alternatively, genes can be synthesized and purchased from biotech companies (e.g., Blue Heron Biotechnology, Bothell, WA).

2. The expression vectors, either plasmids or viral vectors, should be selected for the specific experiment. The vector should have a promoter that allows the expression of the protein in the right type of cells. For example the cytomegalovirus (CMV) promoter can give expression in most of the mammalian cells but the glial fibrillary acidic protein (GFAP) promoter will drive expression of transgene only in astrocytes in the central nervous system.

3. For cloning, the following restriction endonucleases are required: BamHI and BglII (New England Biolabs, MA, USA).

4. Lipofectamine 2000 from Invitrogen is suggested in this chapter; however, different transfection reagents from different manufacturers can be used. Additional reagents and material required for the transfection are Opti-MEM® (reduced serum media, Invitrogen), phosphate-buffered saline (PBS) medium (with serum), and 10-cm culture dishes and 8-well glass chamber slides (Lab-Tek II, Nalge Nunc, USA). For selection of stably expressing cells, the appropriate selection antibiotics should be used, based on the antibiotic resistance gene carried by the expression vector (e.g., G418, Hygromycin B, Puromycin).

5. For validation by immunofluorescence the following reagents are needed: a specific antibody that recognizes the fused epitope (e.g., anti-V5 anti-HA, etc.), PBS, cold acetone, Tris-buffered saline Tween-20 (TBST); blocking solution (10% w/v serum in TBST), secondary antibody conjugated to fluorescent dye (e.g., FITC or TRITC) or enzyme (e.g., horseradish peroxidase (HRP) or alkaline phosphatase (AP)).

For protein extraction, the following reagents are required: M-PER® (mammalian protein extraction reagent; Pierce Protein Research Products), dialysis tube (cutoff = 3.5 kDa), protease inhibitor cocktail (e.g., Halt^{TM}; Pierce Protein Research Products), and Bradford assay (Pierce Protein Research Products).

2.3. MRI

The following pulse sequences should be available.
1. A fast 3D localizer sequence and sequences for anatomic imaging, such as multi-slice spin echo (SE) and multi-slice gradient echo (GRE).

2. A localized spectroscopy sequence, such as PRESS (point-resolved spectroscopy).

3. A fast spin echo sequence, such as RARE (rapid acquisition with relaxation enhancement), or a fast gradient echo sequence, such as EPI (echo planar imaging), including a magnetization transfer (MT) module, with which a saturation pulse can be manipulated with respect to the desired pulse shape, power, duration, and offset. The MRI scanner should be equipped with an isoflurane inhalation setup and animal cradle.

2.4. General Lab Equipment

Humidified incubator set at $37°C$, benchtop centrifuges, upright and inverted microscopes.

2.5. B_0 Correcting Algorithm

The analytic expression for the steady state saturation lineshape for the longitudinal magnetization $\left(M_Z^{SS}\right)$ of a single proton, is given by (22):

$$M_Z^{SS} = \frac{M_Z^0}{1 + \left(\frac{\omega_1}{\Delta\omega}\right)^2 \left(\frac{T_1}{T_2}\right)},$$

where M_z^0 is the initial longitudinal magnetization, ω_1 is the strength of the applied RF pulse, which resonates at $\Delta\omega$ (frequency with respect to the proton center frequency). This formula is implemented into a MATLAB algorithm (*see* **Section 3.5**).

3. Methods

3.1. Cloning and Transfection

3.1.1. Cloning of CEST Reporter Gene

Two complimentary synthetic oligonucleotides (84 base pairs long), encoding the artificial desired sequence, should be designed, so that, after annealing, they will retain corresponding endonuclease restriction site overhangs (e.g., to BglII at the 5′-end and BamHI at the 3′-end). This double-strand DNA is referred to as the monomer. The monomer should be cloned into the expression vector in the context of the desired promoter that will allow expression in the target cells. After cloning, the new vector should be digested with BglII and BamHI, and the released insert should be ligated into the BglII sites of the parental vector (*See* **Note 1**). The ligation will result in a dimer (i.e., a sequence encoding to polypeptide with double length). Next, the new vector should be digested with BglII and BamHI and the released insert (dimer) should be ligated into the BglII sites of the parental vector to form a tetramer. This process should be repeated until the DNA sequence is sufficiently long (*see* **Note 2**).

3.1.2. Cell Culture and Transfection

The cell line of choice can be transfected with the new construct. In general, any transfection or viral infection protocol is adequate; here, we briefly describe a general protocol for transfection using Lipofectamine 2000 from Invitrogen (*See* **Note 3**).

A day before transfection, the cells should be plated in 10-cm tissue culture dish in a dilution such that the cells will be at 80% confluence on the transfection day.

1. Lipofectamine 2000 (60 μl) should be diluted in 1 mL opti-MEM, mixed gently, and allowed to stand for 5 min. Then, 24 μg DNA should be added to 1 mL optiMEM. The diluted DNA and lipofectamine solution should be combined and incubated for 20 min at room temperature.

    Next, the cells should be washed with optiMEM or PBS and the medium should be replaced with 8 mL optiMEM.

2. The DNA transfection mixture (2 mL) should then be added gently to each well and incubated at 37°C for 5–6 h.

3. Then, regular medium (with serum) should be used.

4. In order to generate stable clones, the media should be supplemented with the appropriate selection of antibiotics (e.g., G418, hygromycin B, puromycin).

3.1.3. Validation by Immunofluorescence

In order to validate the protein expression, the cells could be stained using a specific antibody that recognizes the fused epitope (e.g., anti-V5 anti-HA, etc.). The protocol below can be modified and optimized depending on the antibodies and cell (*see* **Note 4**).

The cells can be grown on 8-well glass chamber slides (Lab-Tek II, Nalge Nunc, USA) overnight. The volumes may need to be adjusted for different plates.

1. The cell culture medium can then be removed, the cells washed with PBS, and fixed with cold acetone for 10 min at −20°C.

2. After air-drying for 15 min, the cells should be washed twice, for 5 min each time, with TBST (0.2 mL/wash).

3. Next, 0.1 mL of blocking solution (PBS containing serum) should be added and the mixture should be incubated for 60 min at room temperature to reduce non-specific binding of antibody.

4. The blocking solution should be removed and 0.1 mL of PBS should be added (containing serum), with the appropriate first antibody (1:100 to 1:400 dilution of antibody). This should be incubated for 60 min at room temperature or overnight at 4°C.

5. The cells should be washed three times, for 5 min each time, with PBS and then incubated in 0.1 mL of PBS containing the appropriate secondary antibody conjugated to

fluorescent dye or enzyme for 60 min at room temperature in the dark.

6. Then, wash cells three times, for 5 min each time, with PBS again, and continue counter-staining.

7. Finally, the cells should be mounted with mounting media, covered with a cover-slip, and the results should be analyzed with microscopy.

3.1.4. Protein Extraction

As CEST reporter genes are extremely sensitive, the changes in pH and the exchange rate are dramatically reduced when the cells are fixed. The best way to measure the CEST contrast in vitro is to extract the proteins. The cells should be washed twice and collected in 10 mM of ice-cold PBS (pH = 7.1, without Mg^{2+}/Ca^{2+}). After centrifugation at $2,500g$ for 10 min at 4°C, the pellet should be suspended in 1 mL of M-PER® (mammalian protein extraction reagent Pierce Protein Research Products) and shaken gently for 10 min at 4°C. Cell debris should be removed by centrifugation at $14,000g$ for 15 min at 4°C. The supernatant should be transferred to a dialysis tube (cutoff = 3.5 kDa) and dialyzed twice against PBS. A protease inhibitor cocktail should be added (e.g., Pierce Protein Research Products; Halt^{TM}) and the protein extraction should be stored at –80°C. Protein concentrations can be determined using the Bradford assay (Pierce Protein Research Products).

3.1.5. Transplantation

Animal procedures should be conducted in accordance with the guidelines for the care and use of research animals. The cells should be collected and suspended in a low volume of PBS or saline. The cells should be inculcated into the target tissue while the animal is kept anesthetized by isoflurane inhalation (1–2%).

3.2. General MRI Protocol

3.2.1. Localizer

1. In vivo MR imaging should be performed several days after cell transplantation. The animals should be restrained in an animal cradle, centered at both the center of the RF coil and the center of the magnet, and kept anesthetized using 1–2% isoflurane gas throughout the imaging procedure.

2. After the animal is positioned appropriately so that the region of interest is set as close to the magnetic field and transmitter RF coil isocenter as possible, which often provides favorable image quality and minimal B_1 and B_0 inhomogeneities, tuning and matching settings must be manually adjusted using the tuning and matching function associated with scanner interface (i.e., "wobble" function in Bruker ParaVision software) so that the scanner is in a resonant condition.

3. Next, a localizer sequence (triplot RARE (rapid acquisition with relaxation enhancement), MSME (multi-slice

multi-echo), or FLASH (fast low-angle shot) on Bruker small animal scanners) should be obtained, with large FOVs to acquire three images along the *XY*, *YZ*, and *XZ* planes.

4. The optimal MR acquisition parameters should be determined, including, B_0 homogeneity, resonance frequency offset, transmit gain for 90° and 180° flip angles, and receiving gain. This can be achieved by the automatic global parameter toolbox provided by the manufacturer.

5. T_2w (T_2 weighted) anatomical images should then be acquired, using a multiple-slice fast SE sequence (i.e., RARE), and the subsequent region of interest for single-slice CEST imaging should be defined. The following typical parameters can be used: acquisition bandwidth = 50 kHz; 15 slices; 0.7 mm slice thickness; TE as short as possible ($\sim$ 4–5 ms); TR= 1,000 ms; RARE factor (echo number) = 16; FOV=20 mm × 20 mm; and 128 × 64 matrix size.

6. The desired slice for the CEST experiment should be chosen, based on the anatomic indicators, such as needle track, distance to the center, or T_1/T_2 contrast. Due to the fact that CEST imaging employs a long saturation pulse, a single-slice approach is typically preferred in a CEST acquisition to save time. A single slice RARE sequence without turning on the saturation pulse is then tested to ensure that the defined single-slice geometry sufficiently covers the region of interest.

3.3. Determining B_0 Inhomogeneity

3.3.1. Shimming

Because CEST relies on manipulation of the frequency chemical shift, a good shimming is, therefore, crucial to accurately measure CEST. This can be achieved by either an automatic shimming procedure or a manual shimming procedure. Regardless, slice shimming with a geometry approximating that of CEST imaging should be used, and the B_0 field should be shimmed to maximize the FID signal for the first and second order for optimal magnetic field homogeneity across the slice(*see* **Note 5**).

3.3.2. Estimating B_0 Inhomogeneity

Using a geometry approximating the CEST imaging except for a 50% increase in slice thickness, a point-resolved spectroscopy (PRESS) sequence without suppression should be performed to determine the range of water broadening at the shimming condition. Typical acquisition parameters used are TR/TE = 1,000/16.7 ms; spectrum acquisition size = 8,192; sweep width = 10,080 Hz (25 ppm at 9.4 μT); and NA = 1. The spectrum can be processed in-line using the the Topspin spectrum processing package for the full spectrum width of water peak, which is the approximate frequency range of B_0 inhomogeneity.

3.3.3. WASSR Acquisition

A series of magnetization transfer (MT)-weighted RARE sequences should be employed to acquire WASSR images at the saturation frequency range that has been determined in a previous procedure (e.g., from −1 ppm to +1 ppm with respect to water). The typical imaging parameters used are acquisition bandwidth = 50 kHz; single slice; 0.7 mm slice thickness; TE = 5 ms; TR = 1,500 ms; RARE factor = 8; FOV = 20 mm × 20 mm; and matrix size set to 128 frequency-encoding steps and 64 phase-encoding steps. The saturation pulse prior to each phase-encoding step is realized by a 200–500 ms continuous wave (CW) pulse at a B_1 strength of 0.5 μT (21.3 Hz, see **Note 6**).

A 0.1 ppm increment in saturation frequency between each scan is used under typical shimming conditions. The total acquisition times for generating an absolute B_0 map varies from 3 to 10 min. The error of B_0 estimation is under the hertz level for each pixel at a reasonable SNR (i.e., SNR/pixel > 15), using the parameters listed above.

3.4. CEST Imaging and z-Spectra Acquisition

3.4.1. CEST Imaging

The same magnetization transfer (MT)-weighted RARE sequences that were used to acquire the WASSR scans can be utilized to acquire CEST images at particular irradiation frequencies. The typical imaging parameters used are acquisition bandwidth = 50 kHz; single slice; 0.7 mm slice thickness; TE = 5 ms, TR = 5,000 ms; RARE factor = 8; FOV = 20 mm × 20 mm; 128 × 64 matrix size; 3,000 ms continuous wave (CW) pulse; B_1 strength=3.6 μT (153 Hz); and NA=2. This leads to an acquisition time of approximately 80 s.

3.4.2. Z-Spectrum Acquisition

The same method used to obtain CEST images is used to collect a series of CEST-weighted imaging at a range of saturation offsets, which is the so-called "z-spectra" approach. The typical saturation offset can be swept from −5 ppm to +5 ppm in steps of 0.2 ppm. The total scan time required to generate a z-spectrum is 67 min using the imaging parameters suggested here (see **Note 7**).

3.5. Image Postprocessing and Analysis

3.5.1. Generation of Saturation-Based B_0 Maps (see **Note 8**)

1. Raw WASSR image data should be processed using a custom-written MATLAB script for the B_0 correction algorithm (see **Section 2.5**).

2. In a pixel-wise manner, the vector of MR signal magnitude, $S_{\exp}(x, y)$, corresponding to the saturation offset vector, $\Delta\omega(x, y)$, as shown in **Fig. 2a**, is fit to the equation

$$S_{\exp}(x,y) = \left\{ \eta^2 + \left[\frac{M_0(x,y)}{1 + \left(\frac{\omega_1(x,y)}{\Delta\omega(x,y) - \delta\omega_0(x,y)}\right)^2 \frac{T_1(x,y)}{T_2(x,y)}} \right]^2 \right\}^{1/2}$$

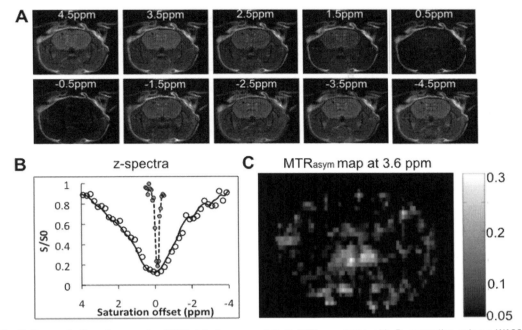

Fig. 2. Demonstration of processing CEST data from raw data to MTR$_{asym}$ maps with B_0 correction using a WASSR approach. (**a**) Representative CEST-weighted axial MR images of a mouse brain injected with iPS cells; the respective saturation offsets are marked at the top of each image; (**b**) the z-spectra display of WASSR data (*dashed line with solid circles*), raw CEST data (*hollow circles*), and B_0 shift-corrected CEST data (*solid line without markers*) of a manually selected ROI; and (**c**) representative CEST contrast map at 3.6 ppm, processed by pixel-wise B_0 correction and interpolation.

to estimate $\delta\omega_0(x, y)$. The fitting is performed using the non-linear fitting function (*lsqcurvefit*) in MATLAB, with M_0, $\delta\omega_0$, and $\left(\frac{T_1}{T_2}\omega_1^2\right)$ as the free parameters, and the experimental noise (η) estimated from the mean signal of air within the image.

3. The absolute water resonant frequency B_0 map is constructed by assigning $\delta\omega_0(x, y)$ to corresponding pixels, providing the B_0 shift information and basis for B_0 correction in the latter part.

3.5.2. Processing Raw CEST Data

1. The saturation offsets for CEST images should be corrected pixel-wise using the B_0 map determined previously, using the algorithm $\Delta\omega(x, y)_{\text{corrected}} = \Delta\omega(x, y) - \Delta\omega_0(x, y)$.

2. The raw CEST-weighted images should be interpolated for signal intensities, $S_{x,y}(\Delta\omega_{\text{interp}})$, at the desired offsets, $\Delta\omega(x, y)_{\text{desired}}$, using the cubic-spline fitting (*spline*) function in MATLAB.

3. The CEST z-spectrum can be plotted by the relative signal intensities, $S^{\Delta\omega}/S_0$, as a function of saturation frequency

offset with respect to water. Typically, ROI masks are manually drawn and the mean intensities of the selected ROI are used to plot the z-spectra (**Fig. 2b**).

4. The CEST contrast is quantified by calculating the asymmetry in the magnetization transfer ratio (MTR_{asym}), as defined by $MTR_{asym} = (S^{-\Delta\omega} - S^{\Delta\omega})/S^{-\Delta\omega}$. The MTR_{asym} for each pixel should be calculated and used to construct the parametric map, MTR_{asym} map (**Fig. 2c**). (*see* **Note 9**).

4. Notes

1. The pair of restriction enzymes, BamHI and BglII, have compatible ends, while BamHI cuts after the first G of the sequence, GGATCC; BglII cuts after the first A in the sequence, AGATCA. Therefore, after ligation, the newly formed sequence is GGATCA if BamHI is located in the 3′ end, or AGATCA if BglII is in the 3′ end. In both cases, the new sequence cannot be cleaved any further, which is required for elongation of the gene. It is critical to make sure that at least one of these enzymes do not cut anywhere in the expression vector. If both enzymes cut, there are other pairs of restriction enzymes that can be used.

2. A reporter that encodes to a longer protein will give a higher CEST signal since the signal is proportional to the number of the exchangeable NH protons. However, since the gene is constructed from repetitive elements, cloning and amplifying the plasmid in bacteria becomes more and more complicated with each cloning cycle. One way to address this problem is to grow the plasmid *Escherichia coli* strands, which are designed for cloning direct repeats, such as Stbl3 (Invitrogen). An alternative way is to order a full-length synthetic gene from biotech companies (e.g., Blue Heron Biotechnology, Bothell, WA).

3. Varying the concentrations of DNA and transfection reagent could optimize the transfection efficiency. It is important not to disturb the transfection complexes (DNA and reagent) by mixing or pipetting.

4. If the protein is rich with lysine residues, the isoelectric point (pI) will be too high to perform immuno-blotting; therefore, immunofluorescence is the method of choice. In addition, immunofluorescence can provide information about the special localization of the protein in the cells.

5. For manual shimming, the procedure must be performed in an iterative manner since the shimming coils are coupled to each other and affect all three dimensions.

6. For converting tesla to hertz for ^{1}H proton, equation $f = \gamma B_0$ should be used in which γ is the gyromagnetic ratio and is 42.58 MHz/T for ^{1}H protons.

7. When using a clinical imager for the consecutive WASSR-CEST acquisitions, the automated pre-scan for shimming and frequency adjustment must be turned off between the scans. This is not an issue for high-resolution spectrometers, where offsets and shims are not adjusted automatically before each scan.

8. In order to reduce the processing time, non-signal voxels in the MRI images should be removed using a signal-to-noise ratio threshold of SNR $\geq$ 15.

9. Additional data processing procedures, such as noise, filtering, and threshold, may be used to further improve the image quality.

References

1. Serganova, I., Ponomarev, V., and Blasberg, R. (2007) Human reporter genes: potential use in clinical studies. *Nucl Med Biol* 34, 791–807.

2. Koretsky, A. P., Brosnan, M. J., Chen, L. H., Chen, J. D., and Van Dyke, T. (1990) NMR detection of creatine kinase expressed in liver of transgenic mice: determination of free ADP levels. *Proc Natl Acad Sci USA* 87, 3112–3116.

3. Walter, G., Barton, E. R., and Sweeney, H. L. (2000) Noninvasive measurement of gene expression in skeletal muscle. *Proc Natl Acad Sci USA* 97, 5151–5155.

4. Alfke, H., Stoppler, H., Nocken, F., Heverhagen, J. T., Kleb, B., Czubayko, F., and Klose, K. J. (2003) In vitro MR imaging of regulated gene expression. *Radiology* 228, 488–492.

5. Weissleder, R., Moore, A., Mahmood, U., Bhorade, R., Benveniste, H., Chiocca, E. A., and Basilion, J. P. (2000) In vivo magnetic resonance imaging of transgene expression. *Nat Med* 6, 351–355.

6. Cohen, B., Ziv, K., Plaks, V., Israely, T., Kalchenko, V., Harmelin, A., Benjamin, L. E., and Neeman, M. (2007) MRI detection of transcriptional regulation of gene expression in transgenic mice. *Nat Med* 13, 498–503.

7. Genove, G., Demarco, U., Xu, H., Goins, W. F., and Ahrens, E. T. (2005) A new transgene reporter for in vivo magnetic resonance imaging. *Nat Med* 4, 450–454.

8. Kodibagkar, V. D., Yu, J., Liu, L., Hetherington, H. P., and Mason, R. P. (2006) Imaging beta-galactosidase activity using (19)F chemical shift imaging of LacZ gene-reporter molecule 2-fluoro-4-nitrophenol-beta-d-galactopyranoside. *Magn Reson Imaging* 24, 959–962.

9. Louie, A. Y., Huber, M. M., Ahrens, E. T., Rothbacher, U., Moats, R., Jacobs, R. E., Fraser, S. E., and Meade, T. J. (2000) In vivo visualization of gene expression using magnetic resonance imaging. *Nat Biotechnol* 18, 321–325.

10. Goffeney, N., Bulte, J. W., Duyn, J., Bryant, L. H., Jr., and van Zijl, P. C. (2001) Sensitive NMR detection of cationic-polymer-based gene delivery systems using saturation transfer via proton exchange. *J Am Chem Soc* 123, 8628–8629.

11. McMahon, M., Gilad, A., Zhou, J., Sun, P., Bulte, J., and van Zijl, P. (2006) Quantifying exchange rates in chemical exchange saturation transfer agents using the saturation time and saturation power dependencies of the magnetization transfer effect on the magnetic resonance imaging signal (QUEST and QUESP): pH calibration for poly-L-lysine and a starburst dendrimer. *Magn Reson Med* 55.

12. Gilad, A., McMahon, M., Walczak, P., Winnard, P., Raman, V., van Laarhoven, H., Skoglund, C., Bulte, J., and van Zijl, P. (2007) Artificial reporter gene providing

MRI contrast based on proton exchange. *Nat Biotechnol* 25, 217–219.

13. Gilad, A. A., Winnard, P. T., Jr., van Zijl, P. C., and Bulte, J. W. (2007) Developing MR reporter genes: promises and pitfalls. *NMR Biomed* 20, 275–290.

14. Gilad, A. A., Ziv, K., McMahon, M. T., van Zijl, P. C., Neeman, M., and Bulte, J. W. (2008) MRI Reporter Genes. *J Nucl Med* 12, 1905–1908.

15. Ward, K. M., Aletras, A. H., and Balaban, R. S. (2000) A new class of contrast agents for MRI based on proton chemical exchange dependent saturation transfer (CEST). *J Magn Reson* 143, 79–87.

16. Zhou, J., and van Zijl, P. C. (2006) Chemical exchange saturation transfer imaging and spectroscopy. *Prog Nucl Magn Reson Spectrosc* 48, 109–136.

17. Sherry, A. D., and Woods, M. (2008) Chemical exchange saturation transfer contrast agents for magnetic resonance imaging. *Annu Rev Biomed Eng* 10, 391–411.

18. McMahon, M. T., Gilad, A. A., DeLiso, M. A., Berman, S. M., Bulte, J. W., and van Zijl, P. C. (2008) New "multicolor" polypeptide diamagnetic chemical exchange saturation transfer (DIACEST) contrast agents for MRI. *Magn Reson Med* 60, 803–812.

19. Sun, P., Farrar, C., and Sorensen, A. (2007) Correction for artifacts induced by B_0 and B_1 field inhomogeneities in pH-sensitive chemical exchange saturation transfer (CEST) imaging. *Magn Reson Med* 58.

20. Kim, M., Gillen, J., Landman, B. A., Zhou, J., and van Zijl, P. C. M. (2009) WAter Saturation Shift Referencing (WASSR) for chemical exchange saturation transfer experiments. *Mag Res Med* 61, 1441–1450.

21. Zhou, J., Blakeley, J., Hua, J., Kim, M., Laterra, J., Pomper, M., and van Zijl, P. (2008) Practical data acquisition method for human brain tumor amide proton transfer (APT) imaging. *Magn Reson Med* 60.

22. Morrison, C., and Henkelman, R. M. (1995) A model for magnetization transfer in tissues. *Magn Reson Med* 33, 475–482.

INDEX

A

Adiabatic half passage (AHP) 247
Affective disorders 310–312
Alzheimer's disease 119, 227, 293, 550
Amide proton transfer (APT) 174
Amygdala 310, 555–557, 622
Amyloid plaque 293–307, 610
Amyloid precursor protein (APP) 294–295, 305
Anesthesia
    avertin................................... 601, 623
    halothane 110, 325, 603
    induction chamber.......... 97–99, 106, 129, 145, 319, 736
    isoflurane 97, 99–100, 110, 119, 121, 129, 132, 135, 144–145, 230, 295–296, 299, 306, 314, 316, 319, 325–326, 340, 346, 359, 389–392, 396, 401, 410, 412–413, 418, 440–441, 452, 471, 478, 480, 491–492, 513–514, 516, 520, 552, 554, 570, 574, 586, 590, 598, 603, 707, 736, 738, 740
    ketamine..97
    pentobarbital ... 299, 325, 425, 432, 471, 491–492, 501
    xylazine 97, 132, 359, 464–465, 467, 471, 598, 611, 623
Angiogenesis 367, 369, 478, 512
Animal monitoring equipment 143, 295, 441, 599
Anisotropy 148–149, 211, 221, 245, 364, 368, 375, 377, 604, 606
Apolipoprotein E (ApoE) 407, 411–412
Apparent diffusion coefficient (ADC).... 140–141, 146, 262–264, 460, 494–499, 504–505
Arterial input function (AIF) 106, 123–124, 260, 496, 498, 505, 669–671
Arterial spin labeling (ASL)
    continuous arterial spin labeling (CASL) 261
    labeling efficiency 123, 133–134
    labeling plane 126
    pulsed arterial spin labeling (PASL) 118, 123, 126–128, 133–134
    transit time 123, 133–134
Arterial stenosis 259
Astrocyte 265, 322, 327, 532–533, 536, 550
Atherosclerosis 166, 407, 511, 513, 523, 526, 609–610, 693
Atlas 64, 280, 322, 358, 570, 572–573, 576, 615
Attention deficit/hyperactivity disorder (ADHD) ... 310
Auditory 165, 569–570, 572–575, 610
Avertin, *see* Anesthesia, avertin

Avidin 297, 303
Axon 356, 365, 551, 555–556, 604
Azan stain 419

B

Bandwidth 19–20, 25–26, 28, 31, 48–49, 55, 65, 79, 181, 340, 343, 345, 347–349, 375, 401, 414–415, 425, 442, 469, 480, 482, 497, 502–503, 543, 563, 741–742
B_0 field .. 8–9, 12, 16, 17, 23, 222, 345, 348, 401, 417, 741
B_1 field homogeneity 8–9, 16, 25, 27, 82, 85, 87, 133, 340, 343–344, 346, 349
Bimodal contrast agent 724
Biomarker 213, 294, 542, 683, 692
Biosecurity 91, 107
Biotin....................................... 297, 303
Bird 570–571, 573–574, 576, 634
Birdcage coils 295–296, 362, 373, 400–401
Black blood sequence 708, 713
Bloch equations 8, 125, 142, 173–174, 343
Blood–brain barrier (BBB) 54, 58, 63, 160, 264, 266–268, 270, 533–534, 538, 541, 545, 552, 558, 600, 674, 682
Blood-cerebrospinal fluid-barrier (BCB) 533, 545, 549
Blood flow 4, 105, 118–119, 122–123, 125–127, 130, 132–133, 135, 157–159, 163, 165–167, 230, 232, 256–258, 260–262, 268, 300, 313, 365–367, 401, 415, 422, 439, 449, 451–452, 490, 497–498, 502, 512–513, 550, 669–672
Blood oxygenation 103, 119, 268, 550, 603
Blood oxygenation level dependent (BOLD)
    extravascular component 159
    intravascular component 160
Blood velocity 4, 133, 159, 415
Blood volume 63, 118, 156, 158–159, 162–163, 166, 228, 260, 478, 482–485, 490, 497–498, 502, 505, 596, 656, 658, 669, 706
BOLD, *see* Blood oxygenation level dependent (BOLD)
Bolus 63, 106, 118, 134, 260, 371, 425–427, 429–430, 432, 493–494, 502, 533, 541, 552, 610, 658, 666, 669–671, 675–676
Bolus tracking 118
Bonferroni correction 282
Brain atlas 64, 280, 322
Brain tumor 478–479, 682
B-value 366, 376–377, 494–495, 497, 504–505, 606

C

^{13}C 6, 191, 206, 208–213, 215–216, 218–219, 221–223, 327, 596, 656, 659–660, 664–668, 672–674, 677–680, 682–684
Calcium (Ca^{2+}) 258, 261, 263, 327, 416, 532–535, 539, 549–551, 554, 559, 600, 646, 740
Cancer 477–486, 489–506, 608, 676–677, 679
Carotid arteries 257, 263, 444, 449, 451, 479, 551, 581–582, 611, 692–693
Carr-Purcell-Meiboom-Gill pulse train 486, 706
Cellular MRI 718, 721
^{13}C enriched substance 667, 672
Cerebral blood flow (CBF) 118–120, 122–124, 130–135, 260–262, 268–269, 365, 367–368, 370, 550, 552
Cerebral blood volume (CBV) 118, 260, 596
Cerebral ischemia 256, 267, 451, 456, 613
Cerebral metabolic rate (CMR) 268
Cerebral metabolism 269, 338, 682
Cerebral perfusion 117–135
Cerebrospinal fluid (CSF) 50, 228, 267–268, 285, 288, 356, 376, 533, 546, 674
Cerebrovascular reactivity 325
C-fos 551
Chemical exchange saturation transfer (CEST) 171–184, 733–745
Chemical shift 4, 13–15, 39–40, 87, 174–175, 191, 211–212, 221, 339–340, 343–345, 348–349, 444, 461, 591, 605, 660–662, 664–666, 673–674, 676, 678–679, 684, 741
Chemical shift imaging (CSI) 39–40, 87, 461, 660–663, 679, 684
Chemical shift selective (CHESS) 339, 345, 349, 503
Choline (Cho) 177, 183, 311, 321–322, 327, 378, 534, 536, 677
Chronic obstructive pulmonary disease (COPD) 462
Cine 388, 390–391, 395, 397–398, 401, 414, 426, 429, 431, 461, 466–467, 608
Circadian Rhythm 319, 324, 328
Circle of Willis 257, 259, 451
CMR, *see* Cerebral metabolic rate (CMR)
Congo red 297–298, 302–303, 305
Continuous ASL (CASL) 118, 122–124, 126–128, 133–134, 261
Contrast agent
 biodistribution 178
 extracellular agent 58, 599, 646
 intracellular agent 646, 730
 intravascular agent 55, 231
 low-molecular-weight contrast agent 533
 macromolecular contrast agent (MMCA) 51
 multimodal contrast agents 724
 passive targeting 408
 targeted contrast agent 610
Corpus callosum 120, 310, 607, 614
Cortical thickness 622
CRAZED 242–244, 246–248, 250–251
Creatine (Cr) 15, 65, 310, 321–322, 327, 338, 378, 444, 451, 734
Creatine kinase system 734
Cryogenic Probe 358

D

DAPI stain 419, 725
Deformation-based morphometry (DBM) 285–286, 621
Dental gyrus 546, 548
Deoxyhemoglobin 104, 155–163, 166, 231, 233, 268, 512
DIACEST 175, 177–178, 181
Diamagnetism 155
Diaminobenzidine (DAB) 298, 302, 304, 307, 720, 724
Diffusion coefficient 140–142, 191–192, 262, 460, 494, 497–498
Diffusion tensor imaging (DTI) 139–151, 167, 359–360, 364–368, 375–377, 604
Diffusion-weighted imaging (DWI) 260–264, 364, 490, 493–495, 497, 504–505, 604, 607
Dynamic contrast enhancement (DCE) . . 105–106, 369, 371, 490, 494–495, 497, 500
Dynamic Nuclear Polarization (DNP) 172, 190, 207–213, 667
Dynamic susceptibility contrast (DSC) . . . 118, 260, 490, 501

E

Echo planar imaging (EPI) 36, 147, 229, 249, 358, 374, 494, 497, 662, 738
Eddy currents 41, 110, 147, 338, 375
Edema 258, 261–264, 266, 361, 463, 470
Eigenvalue 141–142, 148, 150–151, 376
Ejection Fraction (EF) 397, 399–400, 596
Electrocardiograph (ECG) 103–104, 112, 143, 390–392, 401–402, 410–413, 425–426, 428–429, 441, 452–453, 463, 584, 599, 603, 608, 623–624, 657
Electron Paramagnetic Resonance (EPR) 208–209, 223
Embryo 150, 579–588, 590–592, 611–612, 616, 626
Emphysema 460–463
End-diastolic volume (EDV) 399
End-systolic volume (ESV) 399
Europium (Eu) 177
Excised organs 601, 611
Experimental autoimmune encephalomyelitis (EAE) 355, 371–372

F

^{19}F6 . . . . 74, 172, 191, 212, 228–229, 233, 235, 418, 718, 721
Fast low angle single shot (FLASH) 34–35, 56, 85–86, 130, 178–179, 235–236, 301, 414, 425–426, 431, 444, 545, 637, 640–642, 645–646, 648, 708, 741
FASTMAP 320, 331, 378, 442, 444
Fast spin echo (FSE) 39, 178, 180, 299, 374, 493–494, 497, 516, 521–522, 527, 588–590, 605–606, 609, 615, 659, 728, 738
Femoral artery 450–451, 514, 516–518, 524–527
Ferromagnetic materials 54, 66, 155
Ferumoxytol 718, 722
Fick's law 139
Fiducial markers 91, 94, 96, 109

Field inhomogeneities ... 53–54, 56, 90, 363, 389, 413, 484, 664
Field strength 3, 14–15, 19–20, 27, 36–37, 51, 53, 69–72, 86–87, 119, 123–125, 128–129, 133, 143, 161, 163, 166–167, 178, 190, 197–198, 200, 206, 208–210, 216–217, 219, 220, 231, 241–242, 315, 318, 348, 375, 389, 400, 408–409, 417, 465, 485, 490, 503, 543, 557, 574, 587, 604, 657, 704, 727–728, 730
Filling factor 82, 583
First pass perfusion MRI 502
Fixation ... 95, 105, 129, 143, 146, 299–302, 306, 377, 419, 571, 590, 605, 611, 635, 636, 638, 644, 647
Flip angle (FA) 9–13, 25, 27, 35, 38–39, 47, 52, 56, 60–61, 64, 82, 86, 131, 180, 235–236, 244, 247, 280, 302, 344–347, 349, 373, 401, 418, 425, 442–444, 465, 469, 545, 605–606, 612, 640, 659, 661–663, 675–676, 682, 708, 713, 741
Flow sensitive alternating inversion recovery (FAIR) 122–124, 128, 130, 132, 134–135, 260, 368
Fluid-attenuated Inversion Recovery (FLAIR) 316, 318, 327
Fluorine MRI (^{19}F MRI) 228, 418
Fluorinert 297, 301, 601, 636, 645
Fourier transform 4, 17–18, 25–26, 28, 32, 41–42, 78–79, 148, 245, 338–339, 341, 446, 503, 662
Fractional anisotropy (FA) 148, 364, 375–376, 604, 606
Free induction decay (FID) 9, 11–13, 18–19, 34–36, 39, 48, 52, 78, 346, 497, 660–662, 741
Frequency encoding 27–31, 33, 181, 415, 742
Full width half maximum (FWHM) 84, 505
Functional MRI (fMRI) 37, 100, 118–119, 132, 158, 160, 163–164, 227–228, 232, 237, 268–270, 356, 374, 511–529, 531–559, 569–576, 598, 623, 656

G

GABA 15, 310, 312, 321–322, 327, 536
Gadolinium (Gd) chelates 51, 58, 66, 599
Gadopentetate dimeglumine 501, 600, 646
β-Galactosidase (βgal) 734
Gamma-variate function 675
Gd-DOTA-4AmP 425–426, 429
Generalized autocalibrating partially parallel acquisition (GRAPPA) 42, 497, 504
Ghosting artifacts 249, 374
Glioma 480, 734
Glucose 256, 258, 327–328, 538–539, 673–674, 677, 679–680
Glutamate (Glu) 213, 258, 263, 310, 312, 321–322, 327, 338, 378, 532, 536, 550–552, 673, 678–679, 682
Glutamine (Gln) ... 213, 310, 321–322, 327, 338, 378, 532, 678, 682
Glycolysis 228, 263, 674, 677, 679
Gradient coil 23, 72–74, 120, 176, 243, 400, 597, 736
Gradient echo (GE, GRE) 28–29, 34–38, 41, 54, 56, 60–61, 85, 132, 147, 149, 160–163, 167, 231–232, 234, 261, 280, 301–302, 320, 360, 363–364, 378, 388–390, 426, 442–444, 446, 454, 462, 467, 469–470, 479–480, 482–484, 486, 493–494, 527, 543, 569, 571–574, 587–588, 610, 612, 646, 656, 662–663, 708, 713, 737–738
Gradient and spin echo (GRASE) 261, 646
Gray matter (GM) 285, 356, 361, 365, 370, 373, 377, 532, 550
Gyromagnetic ratio 5–6, 65, 141, 183, 190–192, 206–207, 212, 229, 341, 343, 660, 745

H

^{1}H 5–6, 52, 65, 74, 171–172, 179, 183, 191, 206, 208, 212, 218, 221–223, 228–229, 235–236, 238, 243, 310–311, 321, 327, 343, 372, 378, 440, 442, 636–637, 663, 665, 677, 718, 745
^{3}He 6, 189, 191–194, 206, 460–465, 467–468, 470, 656
Hematocrit (Hc) 156
Hemodynamic response 268–269
Hemoglobin 153, 156–157, 166–167, 228, 230–231, 237, 268, 528
HEPA 107, 112
High angular resolution diffusion imaging, (HARDI) 147
Hippocampus 302–303, 310–311, 338, 534, 547, 549, 557, 606, 614, 622
Huntington's disease 596
HYPER-CEST 175
Hyperpolarization 175, 190, 207, 209–210, 212–213, 220–223, 657, 675, 684
Hypoxia/ischemia 174, 231, 256, 258, 261–262, 267, 366, 369, 450–451, 512, 550, 554, 608, 613, 677, 679–680, 682–683

I

Image processing 177, 278–280, 283–284, 287, 290, 304, 321, 323, 486, 516, 522, 602, 614, 637–638, 641–644, 648, 736
Image reconstruction 41, 426, 435, 466
Image registration 278, 602, 619, 625, 735
Induction chamber 97–99, 106, 129, 145, 319, 736
Infarct size 397, 399–400, 403
Inflammation 256, 361, 372, 378, 408, 462–464, 467–470, 683, 714
Intermolecular multiple-quantum coherences (iMQC) 241–251
Intravascular agent 119
In-utero MRI 584
Inversion recovery (IR) 49, 57, 61–62, 123–124, 135, 229, 235, 238, 260, 361, 368, 388, 503, 610, 706, 708, 713
Invertebrate 633–649
Iron 58, 63–64, 70, 111, 155–156, 172, 261, 267, 298, 303–305, 363, 408, 478–479, 533, 550–551, 600, 609, 611, 625, 692, 694, 697–699, 701–712, 718, 722–723, 725, 727, 729, 731, 734
Ischemia 174, 256, 258, 261–262, 267, 366, 369, 450–451, 512, 550, 554, 608, 613, 677, 679–680, 682–683

Isoflurane 97, 99–100, 110, 119, 121, 129, 132, 135, 144–145, 230, 295–296, 299, 306, 314, 316, 319, 325–326, 340, 346, 359, 389–392, 396, 401, 410, 412–413, 418, 440–441, 452, 471, 478, 480, 491–492, 513–514, 516, 520, 552, 554, 570, 574, 586, 590, 592, 598, 603, 707, 736, 738, 740

K

Ketamine....... 97, 132, 325, 359, 424, 427, 464–465, 467, 471, 552, 586, 598, 611, 623, 666–667
Kidney 58, 142, 160, 164–165, 444–445, 545, 583, 605–606, 611, 697
K-space............. 30–33, 37–38, 41–42, 84–85, 245, 341, 390, 415, 471, 504, 661–664

L

Labeling of cells 363, 721–722, 724, 727–728
Lactate (Lac) 213, 228, 321, 378, 665, 672–674, 676–683
Larmor frequency 5, 8–9, 16–17, 24, 72, 75, 77, 82, 174, 179, 208–209, 218, 221, 343
Larmor precession 13, 82
LCModel.................... 315, 323, 347, 360, 676
Left coronary artery (LCA) 388
Legislation 696
Line broadening.......................... 79, 444, 661
Line width 223, 321, 331, 587–588
Lipid suppression 349
LIPOCEST 175–176, 178
Liposomes 175, 177–179, 181, 183
Localization by adiabatic selective refocusing (LASER) 344
Look-locker method............... 123–124, 128, 132
LPS.......................... 463–464, 467, 469–470
Lung 62, 104, 189–191, 459–471, 608–609, 658

M

Magnetic field strength 3, 14–15, 19, 53, 167, 197, 200, 208–210, 216–217, 219, 231, 241–242, 315, 389, 400–401, 408, 417, 465, 574, 704
Magnetic labeling 118, 122
Magnetic moment 3, 5–8, 46, 51, 154–155, 208
Magnetic resonance
  angular momentum 4–6, 190, 193
  free induction decay (FID) 9, 78, 346
  gyromagnetic ratio 5–6, 74–75, 141, 183, 190–192, 206–207, 212, 229, 341, 343, 660, 745
  laboratory frame 8, 78
  Larmor frequency............ 5, 8–9, 16–17, 24, 72, 75, 77, 82, 174, 179, 208–209, 218, 221, 343
  Larmor precession............................ 13, 82
  longitudinal relaxation 10–11, 49, 51–54, 60, 118, 123, 191, 209, 211, 250, 604, 675
  magnetic moment 3, 5–8, 46, 51, 154–155, 208
  nuclear spin 51, 189–191, 193, 207–209, 212–213, 215–216, 222–223, 341, 343
  quantum mechanical description................ 5, 9
  rotating frame........................... 85, 222
  safety 71, 108, 433–434, 436
  shielding constant 14
  signal contrast 482

signal detection 79
signal-to-noise ratio (SNR) 20, 45, 108, 128, 146–147, 181, 206, 261, 304, 325, 338, 345, 400, 408, 417, 454, 463, 485, 502, 597, 621, 625, 679, 745
spin echo 4, 11–13, 33–34, 38–40, 54, 56–57, 59–60, 85–86, 140, 146–147, 160–163, 166, 178, 245–247, 261, 280, 289, 299, 317–318, 346, 349, 360–362, 374, 415–416, 479, 482–484, 486, 493, 497, 516, 521–522, 527–528, 543, 569–576, 588–590, 605–606, 609, 615, 646, 659, 661, 664, 708, 713, 728, 737
spin relaxation 11, 53, 249, 336
spin–spin coupling............................. 7
static magnetic fields 108, 408
T_1 .... 10–12, 35–36, 49–54, 56–63, 85–86, 119, 121, 123–125, 129, 133, 135, 160, 172, 175, 181, 191, 198, 211–212, 218, 220–223, 228–229, 231–236, 238, 243, 260–261, 263, 266–267, 270, 280, 323, 325–326, 337, 342, 345, 347, 349, 356, 359–364, 367, 401, 416–417, 426, 431, 434, 461, 463, 469, 478, 480, 482, 490, 497–498, 531–532, 541–545, 547, 549–550, 552, 556–558, 587, 596–600, 604, 606, 608–609, 646, 655, 662, 666, 669, 672, 678–679, 692–693, 703, 706, 708, 713, 727–729, 731, 735, 738, 741–743
T_2 .... 11–12, 18, 35–39, 50–55, 59–63, 130–131, 133, 146, 161, 172, 181, 218, 229–232, 242, 247, 249, 260–261, 263–264, 266–270, 279–280, 282, 299, 301–302, 323, 325–326, 337, 345, 349, 356, 359–364, 370, 417, 478, 482, 484–486, 490, 494, 497–498, 501, 505, 527, 531, 542–544, 571–572, 587, 596, 604–605, 609, 655–656, 659, 665–666, 692, 703, 706, 708, 713, 718, 727–729, 735, 738, 741–743
T_2^* 12, 18, 53–54, 56, 60–61, 146, 229, 231–232, 301–302, 349, 478, 484–486, 490, 494, 501, 527, 596, 608
Magnetic resonance angiography (MRA) 117, 259–260, 425–426, 430–431, 433, 439, 449, 452–453, 455, 610
Magnetic resonance imaging
  contrast 48, 105, 541–542, 636, 692, 718, 720–721, 723–727, 729, 734–735
  diffusion 139–151
  echo time (TE) 11, 13, 38, 54, 56, 60, 62, 161, 231–233, 245–246, 249, 259, 344, 349, 360, 375, 401, 414, 425, 460, 462, 464, 469, 480, 483, 493, 496–497, 505, 527, 543, 589–590, 708
  endogenous contrast 56, 171
  exogenous contrast 66, 167, 267, 692, 734
  magnetization transfer 55, 57, 63, 122, 172, 178, 180, 183, 337, 356, 372–373, 443, 486, 558, 596, 735, 738, 742, 744
relaxivity 339, 460, 696
T_1-weighted............................ 51, 57, 60, 260, 263, 266–267, 270, 280, 290, 361, 416, 426, 431, 434, 469, 478, 480, 482, 490, 497, 498, 500, 532, 543, 547, 556–558, 604, 606, 646, 692–693, 708, 713, 727
T_2^*-weighted 301–302, 490, 494, 501, 505

T₂-weighted 38–39, 63, 119, 130–131, 264, 266–267, 270, 279–280, 289–290, 299, 356, 360–361, 370, 482, 485, 497–498, 505, 571–572, 604–605, 692, 727
Magnetic resonance spectroscopic imaging (MRSI) 338–341, 344–350
Magnetic resonance spectroscopy (MRS)
  chemical shift imaging (CSI) ... 39–40, 87, 660, 684
  fluorine 234, 718
  localization 310, 320, 327
  metabolite concentration 4, 346
  phosphorous spectroscopy 65
  point-resolved spectroscopy (PRESS) 178, 180, 318, 738, 741
  proton spectroscopy 14
  stimulated echo acquisition mode (STEAM) 342–344, 346–349
Magnetic susceptibility
  diamagnetic 155, 166, 228, 268
  ferromagnetic materials 54, 66, 155
  paramagnetic 51, 54, 58, 63, 155–156, 158, 160, 166, 172, 175, 200, 208–212, 228, 260, 263, 267–268, 423, 461, 463, 478, 531, 604, 636, 646, 648, 666, 691–714
Magnetization
  fast spin echo (FSE) 39, 178, 299, 374, 493, 497, 516, 521–522, 527, 588–590, 605–606, 609, 659, 728, 738
  field of view (FOV) 28, 49, 65, 87, 110, 302, 341, 395, 401, 413–415, 425, 465, 480, 493–494, 497, 519, 527
  Fourier transform 57, 61–62, 123, 361, 610, 708
  frequency encoding ... 27–31, 33–34, 181, 415, 742
  gradients 262
  image reconstruction 41, 435, 466
  inversion recovery 49, 57, 61, 123, 229, 235, 238, 260, 361–362, 368, 388, 503, 610, 706, 708, 713
  inversion time (TI) 57
  k-space 85
  parallel imaging 41–42, 147, 340, 359, 435, 493–494, 497, 502–503
  pulse sequence 45, 47, 51, 53, 56, 59–64, 146, 217–219, 222–223, 242–243, 262, 339, 408, 411, 486, 512, 529, 604, 609, 655–656, 659–661, 663–668, 671, 679, 735, 737
  rapid acquisition using radiofrequency echoes (RARE) 39, 130, 178–181, 279, 282, 299, 444, 533, 571–572, 640–642, 646, 648, 738, 740–742
  repetition time (TR) 35, 38, 40, 50–52, 54, 56, 62, 128, 244, 246, 249–250, 259, 299, 302, 341, 360, 389, 395, 400–401, 414–415, 425, 442, 452, 465, 480, 493, 497, 527, 543–544, 640, 708
  saturation recovery pulse sequence 229, 362, 706
  slice-selective excitation 342
  transverse component 9, 16, 35, 47–48, 61
  volume element (voxel) 338
Magnetization transfer (MT) 55, 57, 63, 122, 172, 178, 180, 183, 337, 356, 372–373, 443, 486, 558, 596, 735, 738, 742, 744

Magnetization transfer ratio (MTR) 57, 63, 183, 373, 744
Magnevist 243, 371, 425, 479, 501, 605, 636–637, 645–646
Manganese 263, 327, 531–559, 600, 604, 606, 636, 639, 646, 718
Manganese-enhanced MRI (MEMRI) ... 263–264, 270, 372, 532, 536–537, 539–541, 543–559
Mannitol 538, 552
Mean transit time 133, 498, 502, 669
Metastases 479–480, 482–483
Methacholine 461–462, 464–465, 467–468
Micelles 691–714
Micron-sized particles of iron oxide (MPIO) 718
Misregistration 283, 289, 443, 454, 735
Mitochondria 141, 258, 532, 534, 536, 539, 554, 673, 678, 725–726
MnCl₂ 263, 535, 538–542, 545–548, 551–554, 556, 600, 606, 636
Molecular MRI 714
Morphometry
  deformation-based 285–286, 618, 621
  shape-based 285–287
  volume-based 617
Motion artifacts 94, 97, 101, 129, 314, 359, 462, 500, 584, 588, 592, 599, 605–606, 623
Mouse embryo 150, 580, 583, 612
Mouse phenotyping 595–626
Multiple sclerosis (MS) 164, 355, 365
Museum specimen 637–638, 641, 645, 648
Myo-inositol (mI) 15, 310, 321, 327, 338, 372, 378

N

N-acetyl aspartate (NAA) 15, 310–312, 321–322, 327, 338, 378
Nanoparticles 54, 59, 61, 63–64, 66, 177, 261, 267, 714
Neurodegeneration 293–307, 356, 555, 714
Neuronal activation 158, 163, 166, 268, 327
Neuronal plasticity 268, 270
Neurovascular coupling 268
Nissl staining 547
Nitric oxide 258, 523, 529
Normal-appearing white matter (NAWM) 363
Nuclear Overhauser effect (NOE) 7

O

Olfactory bulb 300, 533, 542, 546–547, 549, 553, 555, 606, 625
Overhauser effect 7, 207
Oxygenation 103–104, 119, 227–228, 230–233, 236, 268, 425, 428, 550, 603
Oxygen consumption (CMRO₂) 158, 461
Oxyhemoglobin 104, 156, 228, 237

P

³¹P 6, 52, 65, 191, 212, 440, 444, 450–451, 677, 683
PARACEST 175, 177–178

Para-Hydrogen Induced Polarization (PHIP)......191, 207, 213, 215, 220–222, 660, 666–667, 670, 697, 709
Parallel imaging............41–42, 147, 340, 359, 435, 493–494, 497, 502–503
Paramagnetism......................155, 157, 166, 543
Parametric maps.........183, 287, 466, 468, 602, 744
Parkinson's disease...........................534, 550
PCO_2..667
Perfusion.........4, 58, 63, 117–135, 144, 191, 220, 256–257, 259–261, 263, 265–266, 268–270, 296, 299–300, 306, 337, 356, 360, 365, 367–370, 377, 461, 463, 490, 495–496, 500–502, 504–505, 512, 534, 539, 587, 590, 596, 601, 604–605, 610–611, 613, 639, 655, 659, 666, 669–672, 675–676, 680–681
Peripheral artery disease (PAD)...............511–529
Peripheral nerve stimulation (PNS)................265
Perl's staining/Prussian blue staining....303, 305, 709
Pharmacological MRI (phMRI).....118, 311, 374, 462
Phase contrast......................................388
Phased array...41, 77, 84, 96, 358–359, 428–429, 491, 493, 497
Phase encoding..........24, 29–40, 52, 181, 231, 328, 340–341, 348, 390, 395, 401, 441, 443, 453–454, 659, 661, 663–664, 671, 742
Phenotyping............277, 287, 579–592, 595–626, 633–649, 691
Phospho-creatine (PCr).......321, 338, 444, 450–451
Phospholipids.................................697, 710
Physiological monitoring....90, 94–95, 102–105, 112, 231, 515, 599, 603, 736
Pixel-by-pixel analysis.............................121
PO_2.................229–230, 234–237, 461, 667, 683
Point-resolved spectroscopy (PRESS)....178, 180, 318, 320–321, 342, 344, 346–347, 349, 372, 738, 741
Poly-L-lysine................................177, 734
Presenilin....................................294, 305
Principal component analysis (PCA)................377
Proliferation...................265, 372, 378, 678, 725
Proton density............46–47, 56, 59–60, 266, 323, 328, 362, 459, 713
Proton Echo Planar Spectroscopic Imaging (PEPSI)......................................342
Proton resonance frequency (PRF).................734
Pseudo CASL (pCASL).......................123, 134
Pulsed ASL (PASL).......................118, 122–125
Purkinje cells......................................547
Pyruvate.......212, 221, 223, 532, 664–665, 672–684

Q

Q-factor..84
Quadrature detection......................17, 78, 80

R

R_2^*.............228–229, 231–232, 237–238, 478, 483–486, 543
Radiofrequency
 birdcage coil................................362
 quadrature detection......................78, 80
 solenoid coil........................16, 83, 219
 surface coil...................................423
 volume coils....................95, 100, 124

Radiofrequency excitation
 absorption....................................444
 emission.......................................123
 hard pulse...13, 24–25, 47, 49, 54–55, 62, 65, 118, 126, 128, 134, 147, 174, 217, 219–220, 232, 238, 242–244, 247, 250–259, 262, 342–348, 442, 521, 589, 591, 663, 669, 672, 734, 738
Radiofrequency field homogeneity, see B_1 field homogeneity
Rapid acquisition using radiofrequency echoes (RARE), see Magnetization
Relaxivity.........51, 58–59, 65, 236, 483, 696, 703, 705–709, 722, 724, 728–730
Reporter gene..............................733–745
$\Delta R_2^*/\Delta R_2$...484
Ringer's solution............................144, 146

S

Saturation...................25, 27, 48–49, 57, 63–65, 123, 134, 172–178, 180–183, 206, 227–232, 235, 237, 260, 361–362, 369, 372–373, 378, 426, 443–444, 447, 493–494, 503, 546, 555, 584, 589, 591, 647, 706, 713–714, 734–736, 738, 741–743
Schizophrenia.........................310–311, 550, 613
Sedation................324, 424, 427, 436, 573, 598
Segmentation..........280, 283–285, 288–289, 304, 397, 402, 418–419, 440, 446, 602, 614–616, 637, 644, 647–648
Sensitivity encoding (SENSE).......................42
Shimming...................72, 90, 93, 130, 146, 163, 180, 183–184, 320–321, 330–331, 339, 345–346, 349, 360, 375, 378, 389, 391, 442, 444, 503, 521, 591, 735, 741–742, 744–745
Signal to noise ratio (SNR)..........20, 45, 108, 128, 146–147, 181, 206, 261, 304, 325, 328, 338, 345, 400, 408, 417, 454, 463, 485, 502, 597, 621, 625, 679, 745
Solenoid coils..............................16, 83, 219
Songbird...576
Spectroscopic imaging (SI)...........64–65, 337–350, 662, 677
Spinal cord (SC).............149, 355–378, 557, 605
Spin echo (SE)............4, 11–13, 33–34, 38–40, 54, 56–57, 59–60, 85–86, 140, 146–147, 160–163, 166, 178, 245–247, 261, 280, 289, 299, 317–318, 346, 349, 360–362, 374, 415–416, 479, 482–484, 486, 493, 497, 516, 521–522, 527–528, 543, 569–576, 588–590, 605–606, 609, 615, 646, 659, 661, 664, 708, 713, 728, 737
Static dephasing regime (SDR)..........161–163, 166
Statistical parametric map (SPM)..............439, 449
Stenosis...259
Stent.......................................422–426, 429–436
Stimulated echo acquisition mode (STEAM)....................342–344, 346–349
Stroke............117, 119, 227, 255–271, 399, 453, 490
Superior colliculus........................120, 553–556
Superparamagnetic iron oxide (SPIO)....58, 261, 267, 692, 697, 701

Surface coil, *see* Radiofrequency, surface coil
Susceptibility artefact 569, 574
Susceptibility matching 601, 611

T

T1, *see* Magnetic resonance, spin relaxation, longitudinal relaxation
T1-weighted, *see* Magnetic resonance imaging, contrast, T_1-weighted
T_2^*, *see* Magnetic resonance, spin relaxation, T_2^*
T_2^* weighted, *see* Magnetic resonance imaging, contrast, T_2^* weighted
T2, *see* Magnetic resonance, spin relaxation, transverse relaxation
T2-weighted, *see* Magnetic resonance imaging, contrast, T_2 weighted
TE, *see* Magnetic resonance imaging, echo time
Throughput 90, 93, 105–106, 108, 181, 490, 597–598, 616, 623, 633–649
Time-of-flight (TOF) 259, 367, 452, 529, 600
Transfection agent 722, 730
Transferrin 59, 221, 520, 533–534, 539, 551
Transit time 123, 133–134, 498, 502, 669
Tumor 112, 119, 167, 234, 265, 477–486, 489–506, 602, 609–610, 612, 614, 623, 674, 676–679, 682–684
Tumor oxygenation 227
Tune & match 151, 393, 413

U

Ultra-short echo time (UTE) 62, 462–464, 469–470
Ultrasmall superparamagnetic iron oxide (USPIO) 58, 261, 267, 477–486

V

Van Gieson stain 419
Vascular Endothelial Growth Factor-A (VEGF-A) 478, 480–484, 512
Vascular volume 478
Ventilation 230, 423–424, 428, 436, 452, 460–463, 465–468
Ventilation-perfusion imaging 461, 463
Ventilation-perfusion (V/A) ratio 463
Very small iron-oxide particle (VSOP) 58
Voltage gated calcium channels (VGCC) 327, 534, 549–551, 559
Volume element (Voxel) 31–33, 37, 39–40, 66, 125–127, 140, 161, 178, 180, 229, 231, 233–234, 236, 238, 242, 249, 262, 278–289, 318, 320–321, 323, 328, 330, 338–342, 345–349, 372, 378, 442, 446, 454, 480, 482, 494, 497, 505, 557, 572–573, 579, 597, 614, 618, 620, 646, 659, 669, 684
Volumetry 277–290, 323, 355

W

WAG/Rij rats 491–492, 498–499
Water suppression 320–321, 339, 344–345, 347–349, 372, 587–588
White matter (WM) 50, 120, 142, 285, 288, 356, 362–364, 369–370, 376–377, 532, 547–548, 604

X

^{129}Xe 6, 189, 191–193, 206
Xylazine, *see* Anesthesia, xylazine

Z

Zebra finch 46, 570, 572–574, 634
Z-spectrum 183, 736, 742–743